Organometallic Compounds

Literature Survey 1937-1959

Edited by

Michael Dub

Monsanto Chemical Company
Research and Engineering Division

Volume III

Organic Compounds of Arsenic, Antimony and Bismuth

Springer-Verlag Berlin Heidelberg GmbH

1962

ISBN 978-3-662-23217-0   ISBN 978-3-662-25225-3 (eBook)
DOI 10.1007/978-3-662-25225-3

TABLE OF CONTENTS

I. INTRODUCTION

SCOPE   This non-critical compilation of literature data on organic drivatives of arsenic, antimony, and bismuth, reported after 1936, was prepared to provide an easy reference to the methods of their synthesis and to their physical and chemical properties. Biological properties of the organic derivatives of these three elements were not considered. The presence volume is based upon Chemical Abstracts (CA), Volume 31 (1937) through Volume 53(1959), and upon "Current Chemical Papers," published by the Chemical Society (London), issues for January through June 1960. For references published prior to 1937 the reader is referred to "Die Chemie der metall-organischen Verbindugen," by E. Krause and A. von Grosse, Verlag von Gebrüder Borntraeger, Berlin, 1937.

It should be pointed out that only compounds containing at least one carbon-arsenic, carbon-antimony, and carbon-bismuth bond, respectively, are included in the survey.

HIGHLIGHTS   The discovery of the chemotherapeutic efficacy of monosodium p-arsanilate, 3,3'-diamino-4,4'-dihydroxyarsenobenzene, and 3-amino-4,4'-dihydroxy-3'-(sulfinomethylamino)arsenobenzene sodium salt (Atoxyl, Salvarsan, and Neosalvarson, respectively) provided inpetus to extensive research of organoarsenicals, which slakened only after the discovery of antibiotics. After 1936, a great variety of substituted arenearsonic and diarylarsinic acids was prepared and reduced to various types of trivalent arsenic derivatives. Further modifications of the Bart method were made, and aryldiazonium fluoroborates, chlorozincates, and chloroferrates, which are more stable than aryldiazonium chlorides, were introduced as the arylat-ing agents in the preparation of aromatic derivatives of arsenic. Organo-lithium and -sodium compounds were used in the preparation of aliphatic and aromatic arsenic derivatives. To improve the solubility of arsenobenzene derivatives, various solu-bilizing substituents, such as hydroxyalkyl, hydroxyalkoxy, sulfo, and sulfino groups, were introduced into the molecules. Moreover, the existence of arsonous acids, $RAs(OH)_2$, was established; the compounds were previously considered as hydrates of the arsenoso derivatives. The synthesis of alkylidenetriorganoarsenic and homologs with a double bond between the arsenic and a carbon atom, and of penta-phenylarsenic are very significant contributions to the arsenic chemistry of recent years. Moreover, a series of heterocyclic arsenic compounds with the five-, six-, seven-, eight-, nine-, and ten-membered ring systems were synthesized. Also, spirocyclic compounds with the arsenic links were prepared. Significant progress was made in the studies of the molecular structure of arseno compounds. Earlier investigations of arsenomethane (Steinkopf et al., Ber. 59, 1468(1926)) led to the conclusion that its molecule has a structure of a five-membered ring of tri-valent arsenic. More recent studies of aromatic arseno compounds (Kraft et al., Doklady Akad. Nauk S.S.S.R. 65, 509-12(1949); ibid., 131, 1074-6, 1342-4(1960)) indicate that colorless, crystalline arsenoarene derivatives exict as six-mem-bered arsenic rings, while the colored, amorphous arseno compounds are chain poly-mers containing from about give to about fifty RAs< units. New classes of organo-arsenic derivatives comprise compounds with the As-N, As=N, As-Mg, and As-B bonds. Tertiary arsines were used as ligands in the valence studies of transition metals.

In general the synthetic methods for the preparation of organoarsenicals are also used in the synthesis of organoantimony compounds. Aryldiazonium and diarylchloronium, -bromonium, and -iodonium chlorides complexed with antimony trichloride were introduced as the arylating agents for antimony compounds. The existence of primary and secondary alkylstibines, $RSbH_2$ and R_2SbH, was reported but remains questionable. Such compounds were apparently prepared at very low temperatures, but they were not isolated. Phenylstibine. ($PhSbH_2$) and diphenylstibine (Ph_2SbH) are the only representatives of these two classes that were isolated and identified. In the class of tetraorganostibonium compounds, derivatives with four aryl substituents were isolated. Pentaorganoantimony compounds, a new class, includes pentaalkyl, pentaalkenyl, and pentaaryl derivatives. Although many organic derivatives of antimony have been prepared, mainly in search for new chemotherapeuticals, the achievements have remained far behind those made in the chemistry of organoarsenicals.

Attempts to prepare primary and secondary bismuthines, $RBiH_2$ and R_2BiH, were unsuccessful. Although some evidence for the existence of tetramethyldibismuthine was reported in 1935 (F. A. Paneth and H. Loleit, J. Chem. Soc. 1935, 366) no other compounds of this type were described to date. Bismuth analogs of the arseno compounds are unknown. Pentaphenylbismuth was isolated but it was found to be less stable than the antimony analog. Several tetraphenylbismuth salts were prepared, but only the perchlorate, nitrate, and tetraphenylborate were found to be stable at higher temperatures. Diarylbismuth-alkali metal derivatives were prepared in liquid ammonia, but not isolated. They were used as intermediates in the preparation of asymmetrical tertiary bismuthines. Complex salts of aryldiazonium chlorides and diaryliodonium halides with bismuth tri- and pentachloride have been introduced as the arylating agents in the preparation of organobismuth compounds. In general, little progress has been made in the chemistry or organic derivatives of bismuth.

ARRANGEMENT OF THE DATA SECTION   All compounds in this volume, separated in three groups (As, Sb, and Bi), are classified according to their chemical structure. Methods for their preparation are summarized at the opening of the various classes. Compounds for which extensive preparation and/or physical and/or chemical data were reported are listed individually, followed by others, for which limited information was published, arranged in tables. The individually listed compounds as well as those compiled in the tables within each class are arranged in the order of increasing number of carbon, hydrogen, and other element atoms per molecule. To find a compound in the compilation, the user should locate the proper class via the Detailed Table of Contents and then scan pertinent pages in the Data Section for the searched compound, starting at the individual listing.

In the interest of brevity, numerical values of some properties are not always reported, but in such instances specific references are made to the original literature and to CA.

NOMENCLATURE   An attempt was made to name all compounds listed individually according to the rules compiled in Chemical Abstracts 39, 5867-5979(1945).

SYMBOLS AND ABBREVIATIONS   Practically all compounds compiled in the tables are represented by their formulae written in the linear form. In addition to the group formulae which are commonly used, C_5H_5N stands for pyridine and C_4H_8O for tetrahydrofuran.

Throughout this volume, the following abbreviations are used besides those used by CA:
    b.   boiling point or boils
    (dec.) following temperature data   melts (or boils) with decomposition
    m.   melting point or melts
    obtd.   obtained
    r. t.   room temperature
    rxn(s).   reaction(s)

FORMULA INDEX   A formula index to all three volumes is being prepared and will be issued shortly upon completion.

IV

A. A R S E N I C

1. COMPOUNDS OF TRIVALENT ARSENIC

PRIMARY ARSINES

The arsines are prepared by:

1. Reacting alkali metal dihydroarsenides, $MAsH_2$, or alkaline earth dihydro-arsenides, $M(AsH_2)_2$, with alkyl halides (Method I).
2. Reducing dihaloörganoarsines, $RAsX_2$, with zinc and aqueous or methanolic hydrochloric acid (Method II).
3. Reducing dihaloörganoarsines, $RAsX_2$, or organoarsenic tetrahalides, $RAsX_4$, with $LiAlH_4$ or $LiBH_4$ in ether at very low temperatures (Method III).
4. Reducing arsonic acids, $RAsO(OH)_2$, with amalgamated zinc and hydrochloric acid in an inert atmosphere (Method IV).
5. Electrolytic reduction of arsonic acids, $RAsO(OH)_2$, in hydrochloric acid. (Method V). The reaction proceeds via the arsenoso and arseno derivatives.
6. Passing arsine, AsH_3, through alkyl halides (Method VI).
7. Reacting arsine, AsH_3, with an organometallic compound, $RR'R'COMgI$, obtained from a ketone and a Grignard compound (Method VII).
8. Reacting an "α-ynecarboxylic" acid, $RC:CCO_2H$, with arsenic chloride, $AsCl_3$, under reflux and treating the product with N KOH at 0° (Method VIII). The reaction occurs by addition of $AsCl_3$ to the triple bond, elimination of CO_2, and reduction of the $AsCl_2$ group.

TABLE I
PRIMARY ARSINES

Primary Arsine $RAsH_2$	Prepd. from	Method	Yield %	Props. & Remarks	Ref.
C_1 CF_3AsH_2	$RAsI_2 + LiAlH_4$	III	49	b. -12.5°/753 mm.,	333-4
	$RAsI_2$	II	98	Decomp. at 330°[1]	333-4
MeAsH$_2$	$Ca(AsH_2)_2 + MeCl$	I	80		9
	$KAsH_2 + MeCl$	I		m. -143°, b. 2°[2],[3],[4],[5]	607

1. Rxn. with aq. NaOH → CHF_3, F ion, and a bright yellow solid, chiefly As (333).
2. Rxn. with CHBr in C_5H_5N → $(MeAsC)_2 \cdot 3HBr$ (9).
3. Rxn. with $CH_2:CHCN$ → $MeAs(CH_2CH_2CN)_2$ (243).
4. Rxn. with K in liq. NH_3 → MeAsHK, which with NH_4Br yields $MeAsH_2$ (607).
5. Rxn. with B_2H_6 at -78° → $MeAsH_2 \cdot BH_3$, which dissoc. at r. t.; on heating gradually decomp. into $MeAsB_{1.01}H_{2.9}$ and $MeAsB_{1.03}H_{0.79}$ (1169).

TABLE I (cont'd.)
PRIMARY ARSINES

Primary Arsine RAsH$_2$	Prepd. from	Method	Yield %	Props.& Remarks	Ref.
C$_2$ CF$_3$CF$_2$AsH$_2$	RAsCl$_2$	II		b. 26-27°; n$_D^{20}$ 1.331	135
C$_3$ ClCH:C(Me)As-H$_2$ (?)	MeC!CCO$_2$H + AsCl$_3$	VIII	?	Colorless needles, m. 91.5-92°	857
C$_5$ n-C$_5$H$_{11}$AsH$_2$	RAsO(OH)$_2$	V	30	b. 125-127°/730 mm., very sensitive to O	44
EtMe$_2$CAsH$_2$	AsH$_3$ + EtMe$_2$COMgI	VII			962
C$_6$ 4-ClC$_6$H$_4$AsH$_2$	RAsO(OH)$_2$	IV	38	b. 92°/14 mm.[6,7]	242
PhAsH$_2$	RAsCl$_4$ + LiBH$_4$ or LiAlH$_4$	III			1289
	RAsCl$_2$ + LiBH$_4$ or LiAlH$_4$	III	70	Colorless oil, m. -47°[8,9,10]	1293
4-H$_2$NC$_6$H$_4$AsH$_2$	RAsO(OH)$_2$	IV	77	b. 84-92°/0.33 mm.[11,12]	242
3,4-H$_2$N(HO)C$_6$-H$_4$AsH$_2$	RAsO(OH)$_2$	V		Forms a stable HCl salt[13]	1259
4-MeC$_6$H$_4$AsH$_4$				Magnetic props.[14]	1029

6. Rxn. with atm. O → 4-ClC$_6$H$_4$AsO(OH)$_2$ (242).
7. Rxn. with CH$_2$:CHCN → 4-ClC$_6$H$_4$As(CH$_2$CH$_2$CN)$_2$ (242).
8. Thermodynamic data (896).
9. Magnetic props. (1029).
10. Rxns. with: SOCl$_2$ → PhAsCl$_2$ and (PhAsS)$_4$ (16, 17).
    PhNSO → PhAsS, PhAsO, and PhNH$_2$ (16); PhAsSO and PhNH$_2$ (17).
    SO$_2$Cl$_2$ → (PhAs)n (17).
    CH$_2$CH$_2$O, followed by SOCl$_2$ → PhAs(CH$_2$CH$_2$OH)$_2$ (?) and PhAsCl$_2$ (122).
    RMgBr → PhAs(MgBr)$_2$ (68).
    CH$_2$:CHCN → PhAs(CH$_2$CH$_2$CN)$_2$ (242, 804).
    Na followed by RCl → PhRAsH and PhR$_2$As (808).
    I in CCl$_4$ → PhAsI$_2$ (thermodynamic data) (869, 1293).
    O in Et$_2$O → PhAsO (1293).
11. Rxn. with atm. O → (4-H$_2$NC$_6$H$_4$As:)n (242).
12. Rxn. with CH$_2$:CHCN → 4-H$_2$NC$_6$H$_4$As(CH$_2$CH$_2$CN)$_2$ (242).
13. Rxn. with RAsO(OH)$_2$ or RAsO in HCl → the corresponding arseno derivs. (1259).
14. Rxn. with PhNSO → 4-MeC$_6$H$_4$AsS and 4-MeC$_6$H$_4$AsO (16).

TABLE I (cont'd.)
PRIMARY ARSINES

Primary Arsine RAsH$_2$	Prepd. from	Method	Yield %	Props. & Remarks	Ref.
C$_7$ 4,3-HO(NaOSO-CH$_2$NH)C$_6$H$_3$As-H$_2$	3,4-H$_2$N(HO)C$_6$H$_3$AsH$_2$			Prepd. by treating with HO-CH$_2$OSONa in 5% HCl	363, 370
3-MeOC$_6$H$_4$AsH$_2$	RAsO(OH)$_2$	IV	70	b. 107-108°/16 mm.; b. 60-61°/1 mm.[15]	811
4,3-HO(MeNH)-C$_6$H$_3$AsH$_2$	RAsO(OH)$_2$	IV		Colorless crystals,[16,17]	362
C$_8$ 4,3-HO(NaOSO-CH$_2$NMe)C$_6$H$_3$-AsH$_2$				[18]	362
C$_9$ 4,3-HO[HOCH$_2$-CH(OH)CH$_2$NH]-C$_6$H$_3$AsH$_2$	RAsH$_2$	IV		Colorless cryst. powder[19]	362
C$_{11}$ PhCH(OH)(CH$_2$)$_4$-AsH$_2$	PhCH(OH)(CH$_2$)$_4$Cl + AsH$_3$	VI			961

14. Rxn. with PhNSO → 4-MeC$_6$H$_4$AsS and 4-MeC$_6$H$_4$AsO (16).
15. Rxn. with atm. O → rapid oxidation (811).
16. Rxn. with RAsO → 4,3-HO(MeNH)C$_6$H$_3$As:AsR (362).
17. Rxn. with HOCH$_2$OSONa → 4,3-HO(NaOSOCH$_2$NMe)C$_6$H$_3$AsH$_2$ (362).
18. Rxn. with RAsO → 4,3-HO(NaOSOCH$_2$NMe)C$_6$H$_3$As:AsR (362).
19. Rxn. with RAsO → 4,3-HO[HOCH$_2$CH(OH)CH$_2$NH]C$_6$H$_3$As:AsR (362).

SECONDARY ARSINES

The arsines are prepared by:

1. Reducing arsinic acids, R$_2$AsOOH, with amalgamated zinc in hydrochloric acid or in aqueous sodium hydroxide (Method Ia) or treating a mixture of arsinic acid and zinc in methanol with mercuric chloride in an inert atmosphere, followed by hydrochloric acid (Method Ib).
2. Treating primary arsines, RAsH$_2$, with one mole-equivalent of metallic sodium or calcium in liquid ammonia and reacting the metal derivatives, RAsHNa or (RAsH)$_2$Ca, with an alkyl halide (Method II).

3. Reducing halodiorganoarsines, R_2AsX, with zinc or platinized zinc in hydrochloric acid in an inert atmosphere (Method IIIa) or in methanol under reflux (Method IIIb).
4. Reducing halodiorganoarsines, R_2AsX, or diorganoarsenic trihalides, R_2AsX_3, with $LiAlH_4$ or $LiBH_4$ in ethyl or butyl ether at a low temperature (Method IV).

TABLE II
SECONDARY ARSINES

Secondary Arsine	Prepd. from	Method	Yield %	Props., Reactions and Remarks	Ref.
C_2 $(CF_3)_2AsH$	$(CF_3)_2AsI$	IV	16	b. 19°, disproportionates at 200° to $(CF_3)_3As$, HCF_3, and As	333, 334
	$(CF_3)_2AsI$	IIIa		Rxn. with aq. NaOH → HCF_3 and F ion	333
Me_2AsH	Me_2AsCl	IIIa		b. 36°	9
	$(MeAsH)_2Ca$	II	44	m. -136.1°, b. (calcd.) 37.1°; rxn.[1]	1169
C_4 $\overline{CH_2(CH_2)_3}AsH$	$\overline{CH_2(CH_2)_3}AsCl$	IV	55	Colorless, water-clear liquid, m. -29 to -28°, very reactive with O to form a white oxidation prod.	1297
C_5 $\overline{CH_2(CH_2)_4}AsH$	$\overline{CH_2(CH_2)_4}AsCl$[2]	IV	77	Colorless, skin-irritating liq. with phosphine-like odor. m. -11°, very sensitive to O[3]	1297

1. Rxns. with: AcBr in Et_2O + $BaCO_3$ → Me_2AsBr, Me_2AsAc, $(Me_2As)_2$, and probably $[Me_2Ac_3As_2]Br(9)$.
   $AcNH_2.BF_3$ → Me_2AsAc (9).
   $RMgBr$ → $Me_2AsMgBr$ (9).
   $CH_2:CO$ → Me_2AsAc (9).
   Cl_3COCl → Me_2AsCl and $Me_2AsCOCCl_3$ (9).
   CS_2 at 20° → $(Me_2As)_2CS$ (9).
   B_2H_6 at -78.5° → $Me_2AsH.BH_3$ which decomp. at 50° to form a white solid consisting of $(Me_2AsBH_2)_3$ and $(Me_2AsBH_2)_4$ (1169).
2. The redn. with $LiBH_4$ in Et_2O is carried out at -50° and with $LiAlH_4$ at 60° (1297).
3. Rxns. with : O_2 → $\overline{CH_2(CH_2)_4}AsO_2H$ (1297).
   I_2 → $\overline{CH_2(CH_2)_4}AsI$ (1297).

TABLE II (cont'd.)
SECONDARY ARSINES

Secondary Arsine	Prepd. from	Method	Yield %	Props., Reactions and Remarks	Ref.
C_7 MePhAsH	MePhAsCl	IIIb	61	Colorless liq., b. 108-111°/ 37 mm.	251
	MePhAsO$_2$H	Ia	22	b. 71-72°/15 mm.; rxn. with CH$_2$:CHCN → MePhAsCH$_2$CH$_2$CN	811
C_8 EtPhAsH	PhAsHNa	II	31	b. 86.5-87.5°/14 mm.	808
Me(3-MeOC$_6$H$_4$)-AsH	3-MeOC$_6$H$_4$AsHNa	II	85	b. 117-119°/15 mm.; rxn. with CH$_2$:CHCN → Me(3-MeOC$_6$H$_4$)AsCH$_2$CH$_2$CN	811
C_9 PrPhAsH	PhAsHNa	II	51	b. 96-97°/12 mm.	808
C_{10} BuPhAsH	PhAsHNa	II	68	b. 117°/15.5 mm.	808
C_{12} Ph$_2$AsH	Ph$_2$AsO$_2$H	Ia	82	b. 161-162°/20 mm.	242 1289
	Ph$_2$AsCl$_3$	IV			
	Ph$_2$AsCl$_3$	IV		Colorless liq., m. -17°; rxns.[4]	1293
Ph(2-H$_2$NC$_6$H$_4$)-C$_6$H$_4$)AsH	Ph(2-O$_2$NC$_6$H$_4$)AsO$_2$H	Ib	43	b. 140°/0.6 mm., very sensitive to O; rxn.[5]	243
	Ph(2-H$_2$NC$_6$H$_4$)AsO$_2$H	Ia[6]		b. 218-220°/35 mm.; rxn.[7]	94

4. Rxns. with: CH$_2$CH$_2$O → (Ph$_2$AsCH$_2$-)$_2$ and Ph$_2$AsCH$_2$CH$_2$OH (122).
CH$_2$:CRCN → Ph$_2$AsCH$_2$CHRCN and (Ph$_2$As)$_2$ (242).
O in Et$_2$O → Ph$_2$AsO$_2$H (1293).
Br in Et$_2$O → Ph$_2$AsBr$_3$ (1293).
5. Rxn. with CH$_2$:CHCN → Ph(2-H$_2$NC$_6$H$_4$)AsCH$_2$CH$_2$CH (243).
6. Redn. carried out in aq. NaOH (94).
7. Rxn. with atm. O → [Ph(2-H$_2$NC$_6$H$_4$)As]$_2$O (94).

TERTIARY ARSINES

Some of the methods for the preparation of tertiary arsines reported below were known before 1937 and were reported by E. Krause and A. von Grosse in "Die Chemie der metall-organischen Verbindungen", Verlag von Gebrüder Borntraeger, Berlin, 1937, pp. 455-8 and 497-504.

SYMMETRICAL ALIPHATIC AND ALICYCLIC ARSINES

Symmetrical trialkyl-, trialkenyl-, triaralkenyl-, and triaralkynylarsines are prepared by:

1. Reacting arsenic halides, AsX_3, with organometallic compounds such as alkyl- and alkenyllithium or Grignard compounds.
2. Metathesis of arsenic halides, AsX_3, or alkylhaloarsines, $R_{2-n}AsX_{1+n}$, with trialkylaluminum or organomercury halides in suitable solvents.
3. Reacting alkyl halides with arsenic at elevated temperatures.
4. Treating arsenic halides, AsX_3, and alkyl halides, RX, mixed in a ratio of 1:3 with metallic sodium.
5. Reacting arsenic halides, AsX_3, with acetylene in the presence of mercuric chloride.
6. Heating arsenic trioxide, As_2O_3, with an alkali carboxylate in the corresponding carboxylic anhydride and pyrolyzing the reaction product.
7. Cleavage of tetraalkyldiarsines, $(R_2As)_2$, with alkyl halides.

TRICYANOARSINE C_3AsN_3 $As(CN)_3$
Prepn.: By reacting $AsCl_3$ with Me_3SiCN (1:3) in xylene under N; 64% yield (87a). The compound is apparently also formed by electrolysis of AgCN in C_5H_5N with an As anode (1108).
Props.: white powder (87a).

TRIS(TRIFLUOROMETHYL)ARSINE C_3F_9As $(CF_3)_3As$
Prepn.: By reacting ClF_3 with As at 220-240°; 70% yield, along with $(CF_3)_2AsI$ and CF_3AsI_2 (74, 117). At 220°, 78% yield (333, 1266).
By cleaving $[(CF_3)_2As]_2$ with ClF_3 at 75° in a sealed tube (254).
By pyrolysis of $(CF_3)_2AsNHMe$ (253).
Props.: B. 33.3° (74, 117, 334). Stable in O (1266). Stable to H_2O at r. t. but hydrolyzes in alk. buffer soln. (333). Forms equimol. azeotropic mixt. with MeI, b. 25° (333). Decomp. at 350-410° in a silica vessel → C_2F_6, C_3H_8-C_4H_{10}, and SiF_4-CO mixt.; the rate of pyrolysis was detd. (23). Dissocn. energies (869a). Parameters detd. by electron-diffraction method (112). Decomp. in UV light → As and C_2F_6 (333).
Rxns. with: Aq. NaOH → complete hydrolysis (117, 333, 521).
I_2 at 100° → ClF_3, $(CF_3)_2AsI$, CF_3AsI_2, and AsI_3 (117, 333, 1266).
Cl_2 (1:1) at r. t. in a sealed tube → $(CF_3)_3AsCl_2$ and $(CF_3)_2AsCl_3$ (333).
Cl_2 at 125° → $CClF_3$ and $AsCl_3$ (333).
Cl_2 (1:0.65) in the vapor phase → a mixt. $(CF_3)_{2-n}AsCl_{1+n}$, $CClF_3$, and $AsCl_3$ (333).
Br_2 in a sealed tube → $(CF_3)_2AsBr$ and CF_3AsBr_2 (333).
AsI_3 at 230-240° → CF_3I, $(CF_3)_2AsI$, and CF_3AsI_2 (333, 12).
CoF_3 in N at 100° → CF_4, AsF_3, and $(CF_3)_2AsF$ (333).
NH_3 at -64° → $(CF_3)_2AsNH_2$ and CHF_3 (253).
NH_3 at 20° → $[(CF_3)_2As]_2NH$, $(CF_3)_2AsNH_2$, and CHF_3 (253).
$MeNH_2$ at 20° → CHF_3 (253).
$EtNH_2$ (excess) in a Carius tube → CHF_3, two solids, and a volatile fraction, probably CF_3AsNEt (253).

H_2 at 200-240° → As and CHF_3, but in UV light C_2F_6 is also formed (333).
MeI in UV light or at 235° → C_2F_6, CF_3I, and $(CF_3)_2AsMe$ along with CF_3AsMe_2 (333, 334, 335).

TRIMETHYLARSINE C_3H_9As Me_3As

Prepn.: By reacting $AsCl_3$ dissolved in xylene with MeMgI in isopentyl ether in a
N atm., 13.2% yield (1042) or with MeMgBr in Et_2O (1314).
By treating As_2O_2 with KOAc in Ac_2O and heating the prod. at 300°; isolated
as $Me_3As(HgCl_2)_2$ (1214).
By biological methylation of As_2O_3 in bread cultures by Scopulariopsis bre-
vicaulis (194) or Aspergillus niger (195).
Props.: B. 50-52°(1042). m. -87.3°(extrapol.); b. 50.4° (extrapol); d. 1.150 (747).
Mol. structure and bond length (1158). Intramol. potential (1131). Thermodynamic
data (747, 748, 1081). Parachor (717). Force const. (1138). Hybridization and
dipole moment (461). Raman spectrum (359, 994, 1080). Microwave spectrum (735).
Relation between polarity and reactivity with O, CO_2, and H_2O (1098). Vibrational
energy level splitting and optical isomerism. (1285). Dissocn. energies (1089).
Decomp. at 410-450° → CH_4, C_2H_6, C_2H_4, C_3H_6, and H_2; rate of decompn. (23, see
also 749).
Rxns. with: CF_3I (1:1) → Me_2AsCF_3 (520).
H_2SO_4 → Me_3AsO (991, 993).
CH_2CH_2O in H_2O → $Me_3As(OH)CH_2CH_2OH$ (118).
PCl_5 → Me_3AsPCl_3 and Me_3AsCl_2 (561).
$SbCl_5$ → $Me_3AsSbCl_3$ and Me_3AsCl_2 (561).
RX (X = Cl or I) → Me_3RAsX (1042, 1314).
HNO_3 in Et_2O → $Me_3As(OH)ONO_2$ (1053).
Br_2 → Me_3AsBr_2 (1314).
Use: Catalyst for polymerization of vinyl monomers (247).
Complex compds.: $Me_3AsSbCl_3$ (561).
$MeAsPCl_3$ (561).
Me_3AsBH_3, m. 72° (dec.), unstable above 40° (540).
$Me_3AsTlMe_3$, vapor pressure (236).
$Me_3AsInMe_3$, m. 28.2-28.8°, b. 155° (extrapol.); dissocn. data (336).
$Me_3AsGaMe_3$, dissocn. data (235).
Me_3AsSiH_3I (22).
$Me_3As.2HgCl_2$, m. 250-258° (1214).
$Me_3As.AuI$, m. 176-178° (dec.), crystal structure (800).
$Me_3AsAuCl$; m. 165-175° (dec.), unstable in air, crystal structure (800).
$Me_3AsAuSCN$, m. 127-128° (dec.) (800).
[Cu(Me_3As)I]$_4$, white crystals, decomp.>300°, unstable in air (797).
[Pd(Me_3As)$_2Cl_2$], cis-trans equilibrium (205). Prepn. (801).
[Pd(Me_3As)$_2Br_2$], m. 229° (801).
[Pd(Me_3As)$_2$(SCN)$_2$], m. 124°, structure and configuration (801).
[Pd(Me_3As)$_2$(NO$_2$)$_2$], lemon-colored crystals, m. 234°, structure and configuration (801).
[Pd(Me_3As)]NO$_2$Cl]$_2$, m. 186-188°, -NO$_2$ bridged structure (801).
[Pd(Me_3As)Br$_2$]$_2$, diffraction data (405); crystal structure (72, 801, 1282); Br-bridged
structure (801).
[Pd(Me_3As)Cl$_2$]$_2$, m. 252-260° (dec.), prepd. by heating [Pd(Me_3As)$_2Cl_2$] in vacuo
(786); diffraction data (405); clathrates with H and N (257); crystal structure
(72, 801); Cl-bridged structure (220, 796, 801).

[Pd(Me$_3$As)BrCl]$_2$, decomp. 248°; crystal structure; Br-bridged mol.(801).
[Pd(Me$_3$As)Br(SCN)]$_2$, m. 189-190°; SCN-bridged mol.(801).
[Pd(Me$_3$As)Cl(SCN)]$_2$, m. 192-193°; SCN-bridged mol.(801).

TRIS(TRIFLUOROVINYL)ARSINE C$_6$F$_9$As (CF$_2$:CF)$_3$As
Prepn.: By reacting AsCl$_3$ with CF$_2$:CFMgBr (619) or with CF$_2$:CFMgI in Et$_2$O at
-15°, warming the rxn. mixt. to 20°, and decompg. at -10° with 17% HCl; 40%
yield (1165).
Props.: B. 57.5°/83mm., stable to bases (619); b. 58°/95mm., b. 50°/ 70mm., b.
110-111°/746mm., n$_D^{18}$ 1.3938, d.$_8$ 1.8400 (1165).
Rxn. with: Liq. HCl at -110° → CF$_2$:CFAsCl$_2$ (1165).

TRIS(PENTAFLUOROETHYL)ARSINE C$_6$AsF$_{15}$ (C$_2$F$_5$)$_3$As
Props.: Decomp. 280° → fluorocarbons, rate of decompn. (23), Dissocn. energies (869a).

TRIS(2-CHLOROVINYL)ARSINE C$_6$H$_6$AsCl$_3$ (ClCH:CH)$_3$As
Prepn.: By reacting C$_2$H$_2$ with a mixt. of HgCl$_2$, 25% HCl, Me$_2$CO, and AsCl$_3$ at
50-55° (615). Rxn. mechanism (1033).
    By-prod. in the rxn. of PhAsCl$_2$ with C$_2$H$_2$ in the presence of AlCl$_3$ (45).
Rxn. mechanism (1033).
Props.: B. 107-108°/4mm (45); b. 128-129° (615). Dipole moment (855). Light
absorption data (856). Analysis (1031).
Use: Catalyst in the prepn. of CH$_2$:CHCl from C$_2$H$_2$ and HCl (115).

TRIVINYLARSINE C$_6$H$_9$As (CH$_2$:CH)$_3$As
Prepn.: By reacting CH$_2$:CHMgBr with AsCl$_3$ in C$_4$H$_8$O under reflux; 62% yield (773,
774),
By reacting CH$_2$:CHLi with AlCl$_3$ in Et$_2$O; 51% yield (1121a).
Props.: B. 45-46°/41mm. (773); b. 129.8°, readily oxidized in air but thermally
stable in the absence of O (773,774).
Rxns. with: HCl in Et$_2$O at -20° → (CH$_2$:CH)$_3$As.HCl (773).
I$_2$ in CCl$_4$ → (CH$_2$:CH)$_3$AsI$_2$ (773).
AsBr$_3$ → (CH$_2$:CH)$_2$AsBr and CH$_2$:CHAsBr$_2$ (773, 774).
AsCl$_3$ → (CH$_2$:CH)$_2$AsCl and CH$_2$:CHAsCl$_2$ (773, 774).
Derivs.: Methiodide, m. 154-156° (decompn.) (773).
Ethiodide, m. 184-185°(773).
Hydrochloride, hydroscopic crystals (773).
Complex compds.: cis{Pt[(CH$_2$:CH)$_3$As]$_2$}Cl$_2$, yellow solid, m. 90° (773, 774).

TRI(1-CHLOROETHYL)ARSINE C$_6$H$_{12}$AsCl$_3$ (MeCHCl)$_3$As
Prepn.: By reacting AsCl$_3$ with MeCHN$_2$ in C$_6$H$_6$ at 3-5° (1320, 1321).
Props.: B. 81-82°/2mm.; d$_4^{20}$ 1.445; n$_D^{25}$ 1.5307 (1320, 1321); b. 94-95°/4mm. (1321).

TRIETHYLARSINE C$_6$H$_{15}$As Et$_3$As
Prepn.: By reacting AsF$_3$ with Et$_3$Al in n-C$_6$H$_{14}$ (622).
Props.: B. 140-142°/756mm. (622). Dipole moment (716). Heat of combustion (725b).
Parachor (717). Catalytic effect on the NaCl elimination in esterification of PhCH$_2$
Cl with NaOAc (1323a).

Rxns. with: $MeCHBrCO_2Et \rightarrow MeCH(CO_2Et)Et_3AsBr$ (81).
$Br_2 \rightarrow Et_3AsBr_2$ (516).
$AsBr_3$ (1:2) $\rightarrow EtAsBr_2$ (773).
Use: Catalyst for polymerization of vinyl monomers (247).
Complex compds.: $[Cd(Et_3As)_2I_2]$, m. 79-81°, prepd. by reacting Et_3As with CdI_2 in a cold aq. soln., was transformed into the I-bridged $[Cd_2(Et_3As)_2I_2]_2$, m. 80-81°, when kept in vacuo over H_2SO_4 (352).
$[Cd_2(Et_3As)_2Br_2]Br_2$, m. 175-178°, Br-bridged structure (352).
$[Et_3As.CuI]_4$, prepd. by reacting Et_3As with Cu_2I_2 in aq. KI, decomp. 190-240° (797); crystal structure (797, 1281).
$[Et_3As.CuBr]_4$, m. 234-235° (dec.) (797).
$[Et_3As.AuCl]$, m. 94-95°, crystallographic data (800).
$[Et_3As.AuI]$, m. 77°, crystallographic data (800).
$Et_3As:BH_3$, m. 40° (decompn.) (540).
$[Hg_2(Et_3As)_2Br_2]Br_2$, fine needles, m. 162-163° (352).
$[Hg_2(Et_3As)_2I_2]I_2$, white prisms, m. 87-88° (352).
$[Hg_2(Et_3As)_3I_3]I$, m. 58-70° (352).
$[Hg_3(Et_3As)_2I_4]I_2$, yellow crystals, m. 114-115° (352).
$[Hg_4(Et_3As)_2Cl_6]Cl_2$, m. 138° (352).
$[Pd(Et_3As)_2Cl_2]$, cis-trans equilibrium study (205, 210).
$[Pd(Et_3As)Cl_2]$, a Cl-bridged, non-ionic complex, m. 212°, prepd. by heating the preceding complex in vacuo (796).
$[Pd(Et_3As)(C_5H_{11}N)Cl_2]$, yellow-orange crystals, m. 68-70° (213).
$[Pt(Et_3As)_2Cl_2]$, prepd. by reacting Et_3As with K_2PtCl_4; occurs in cis- and trans-form, m. 154-155° and 120-121°, resp.; dipole moment (204).
  cis-Isomer, m. 142-142.5°; dipole moment (600). Cond. (601).
$[Pt(Et_3As)Cl_2]_2$, a Cl-bridged, orange salt, m. 208-209° (decomp.), prepd. by reacting powdered $PtCl_2$ with molten cis- and/or trans-$[Pt(Et_3As)_2Cl_2]$ (207).
cis-$[Pt(Et_3As)_2Br_2]$, m. 113-114°, prepd. by treating cis-$Pt(Et_3As)_2Cl_2$ + KBr (1052)
trans-$[Pt(Et_3As)_2Br_2]$, yellow, m. 120-121°; dipole moment (600). M. 107-108°, prepd. by reacting Et_3As with K_4PtBr_6 (1052).
trans-$[Pt(Et_3As)_2I_2]$, orange-yellow, m. 93-94°, dipole moment (600).
cis-$[Pt(Et_3As)_2(NO_2)_2]$, white, m. 169-170° (transition into the trans-form) (600).
trans-$[Pt(Et_3As)_2(NO_2)_2]$, white, m. 199-200°, dipole moment (600).
cis-$[Pt(Et_3As)_2I_2]$, m. 87-88°, prepd. from cis-$[Pt(Et_3As)_2Cl_2]$ + KI (1052).
trans-$[Pt(Et_3As)_2I_2]$, orange, m. 80-81°, prepd. by treating K_2PtI_4 with Et_3As (1052
cis-$[Pt(Et_3As)_2Cl_4]$, m. 125° (925).
trans-$[Pt(Et_3As)_2Cl_4]$, m. 155° (929).
trans-$[Pt(Et_3As)_2Br_4]$, m. 173° (929).
trans-$[Pt(Et_3As)_2I_4]$, m. 144° (929).
$[Pt(Et_3As)_2HCl]$, prepd. by reducing trans-$[Pt(Et_3As)_2Cl_2]$ (214).
$[Pt(Et_3As)_2(CH_2NH_2)_2](PtCl_4)$, rose-colored salt, prepd. by dissolving cis-$[Pt(Et_3As)_2Cl_2]$ in $(CH_2NH_2)_2$ and treating with K_2PtCl_4 (1052).
cis- and trans- $\{Pt(Et_3As)_2[SC(NH_2)_2]_2\}Cl_2$ (1052).
cis- and trans- $\{Pt(Et_3As)_2[SC(NH_2)_2]_2\}Br_2$ (1052).
cis- and trans- $\{Pt(Et_3As)_2[SC(NH_2)_2]_2\}I_2$ (1052).
$[Et_3As.AgI]_4$, colorless crystals, m. 182-185° (800). Diffraction data (405). Crystal structure (800, 1281).
$(Et_3As)_2(SnBr_4)$, m. 126-132° (12).
$(Et_3As)_2(SnCl_4)$, m. 155-159° (12).

TRIALLYLARSINE C$_9$H$_{15}$As (CH$_2$:CHCH$_2$)$_3$As
Prepn.: by reacting CH$_2$:CHCH$_2$MgBr with AsCl$_3$ at -15° to r. t.; 40% yield along with a by-prod., probably [(CH$_2$:CHCH$_2$)$_2$As-]$_2$ (1262).
Rxn. with allyl bromide in MeCN → (CH$_2$:CHCH$_2$)$_4$AsBr (?) (1262).

TRIPROPYLARSINE C$_9$H$_{21}$As Pr$_3$As
Props.: dipole moment (716). Parachor (717).
Complex compds.: [Cd(Pr$_3$As)$_2$I$_2$], m. 27-29°, on heating with CdI$_2$ in alc. forms the I-bridged [Cd(Pr$_3$As)I]$_2$I$_2$, fine needles, m. 114-116.5° (352).
[Cu(Pr$_3$As)I]$_4$, white crystals, m. 205-212° (decomp.) (797).
[Au(Pr$_3$As)X], (X = Cl,I, or SCN), are monomolec. compds. (798).
[Hg(Pr$_3$As)Br]$_2$Br$_2$, m. 91-92° (352).
[Hg(Pr$_3$As)I]$_2$I$_2$, m. 107-108° (352).
[Hg$_2$(Pr$_3$As)$_3$I$_3$]I, pale yellow leaflets, m. 84-85.5° (352).
[Hg$_3$(Pr$_3$As)$_2$Cl$_4$]Cl$_2$, m. 105° (352).
[CdHg(Pr$_3$As)(Pr$_3$P)I$_2$]I$_2$, white crystals, m. 121-123°; I-bridged structure (802).
[(Pr$_3$As)BrHgBr$_2$PdBr(Pr$_3$As)], orange crystals, m. 89-90° (802).
[Pd(Pr$_3$As)$_2$Cl$_2$], cis-trans equilibrium (205, 210).
[Pd(Pr$_3$As)Cl]$_2$Cl$_2$, dark orange, m. 159-160°, with Cl-bridged structure, prepd. by reacting Pr$_3$As with Na$_2$PdCl$_4$ in Me$_2$CO (212) or by heating [Pd(Pr$_3$As)$_2$Cl$_2$] in vacuo, m. 160° (796).
[Pd(Pr$_3$As)$_2$(SCN)$_2$][PdCl$_2$], m. 151° (796).
[Pd(Pr$_3$As)$_2$(SCN)$_2$], m. 90-91.5° (796).
[Pd(Pr$_3$As)$_2$(SCN)$_2$][Pd(SCN)$_2$], m. 142° (796).
[Pt(Pr$_3$As)Cl]$_2$Cl$_2$, red orange, m. 156.5-157.5°, prepd. from [Pt(C$_2$H$_4$Cl$_2$]$_2$ and Pr$_3$-As in Me$_2$CO(202), Prepd. by reacting PtCl$_2$ powder with a melt of cis- and/or trans-[Pt(Pr$_3$As)Cl$_2$] (207). Prepd. from Pr$_3$As and K$_2$PtCl$_4$, occurs in the cis- and trans-form (203). Reacts with toluidine to form trans-[Pt(Pr$_3$As)(C$_7$H$_7$NH$_2$)Cl$_2$ yellow, m. 94°; trans-[Pt(Pr$_3$As)(Pr$_3$N)Cl$_2$], m. 104-105°, prepd. in an analogous manner; with 2,2'-bipyridyl an unstable [Pt(Pr$_3$As)(C$_5$H$_4$N)$_2$Cl][Pt(Pr$_3$As)Cl$_3$] was obtd., which decomp. to form [Pt(Pr$_3$As)$_2$Cl$_2$] (202).
trans-[Pt(Pr$_3$As)$_2$Cl$_2$], yellow, m. 51.5-52.5°, dipole moment (600).
trans-[Pt(Pr$_3$As)(C$_7$H$_7$NH$_2$)Cl$_2$], assocn. const. (206, 211). IR spectrum (209, 211).
trans-[Pt(Pr$_3$As)(C$_5$H$_{11}$N)Cl$_2$], mustard yellow prisms, m. 96.5-98°, prepd. by reacting trans-[Pt(Pr$_3$As)Cl]$_2$Cl$_2$ with piperidine (208). IR spectrum (209). Uv spectrum (215).
trans-[Pt(Pr$_3$As)(NH$_3$)Cl]$_2$Cl$_2$, mustard yellow plates, m. 91-92° (208).
[(Pr$_3$As)·AgI]$_4$, m. 219-221° (800). Isomorphous with the CuI homolog (798).

TRIS(2-METHYLALLYL)ARSINE C$_{12}$H$_{21}$As (CH$_2$:CMeCH$_2$)$_3$As
Prepn.: by reacting CH$_2$:CMeCH$_2$MgBr with AsCl$_3$ (616).
Props.: B. 114°/15mm. (616).
Complex compd. with HgCl$_2$, needles, m. 96-97° (616).

TRIBUTYLARSINE C$_{12}$H$_{27}$As Bu$_3$As
Prepn.: by reacting BuMgBr with AsCl$_3$ in Et$_2$O; 50% yield (1115).
Props.: B. 113-115°/10mm. (1115). Dipole moment (716). Parachor (717).
Complex compds.: [Cu(Bu$_3$As)I]$_4$, colorless needles, m. 61°; the arsine is displaced from the complex by one mol. of bipyridyl (797).
[Hg(Bu$_3$As)Br]$_2$Br$_2$, m. 86-87° (352).
[Hg(Bu$_3$As)I]$_2$I$_2$, yellow crystals, m. 55-56° (352).

[$Hg_2(Bu_3As)_3I_3$]I, m. 74-75° (352).
[$Hg_3(Bu_3As)_2Br_4$]Br$_2$, m. 62-64° (352).
[$Hg_3(Bu_3As)_2I_4$]I$_2$, bright yellow crystals, m. 63-65° (352).
[$Pd(Bu_3As)_2Cl_2$], orange crystals, m. 52-53° (455). Cis-trans equilibrium (205, 210).
[$Pd(Bu_3As)_2(NO_2)_2$], m. 97-98° (455).
[$Pd(Bu_3As)_2(SCN)_2$], m. 74-75° (455, 796). Prepd. by treating ($Pd(Bu_3As)_2Cl_2$] with KCNS; reacts with ($NH_4)_2PdCl_4$ to form [$Pd(Bu_3As)(SCN)$]$_2Cl_2$, which with KCNS gives [$Pd(Bu_3As)_2(SCN)_2$] and [$Pd(Bu_3As)(SCN)$]$_2$, m. 111° (796).
[$Pd(Bu_3As)(NH_3)Cl_2$] orange crystals, m. 73-74° (dec.) (455).
[$Pd(Bu_3As)Cl$]$_2Cl_2$, red crystals, m. 128-129° (455); m. 128° (220). Cl-bridged structure (220, 455). Prepd. by heating [$Pd(Bu_3As)_2Cl_2$] in vacuo; reacts with $NaNO_2$ to form[$Pd(Bu_3As)(NO_2)$]$_2(NO_2)_2$, m. 96°, and with NH_4CNS to form [$Pd(Bu_3-As)(SCN)$]$_2Cl_2$, m. 101° (796).
[$Pt(Bu_3As)Cl$]$_2Cl_2$, orange crystals, m. 126-127°, Cl-bridged structure, prepd. by treating powdered $PtCl_2$ with molten cis- and/or trans-[$Pt(Bu_3As)_2Cl_2$] (207).
trans-[$Pt(Bu_3As)_2Cl_2$], yellow crystals, m. 53.5-54.5°, dipole moment (600).

TRIS(TRIMETHYLSILYLMETHYL)ARSINE $C_{12}H_{33}AsSi_3$ ($Me_3SiCH_2)_3As$
Prepn.: by reacting Me_3SiCH_2MgCl dissolved in C_4H_8O with $AsCl_3$ dissolved in C_6H_6; 80% yield (1121).
Props.: M. 67-68.5° (1121).
Rxns. with: RI → R($Me_3SiCH_2)_3As$ I (1121).
Halogens → ($Me_3SiCH_2)_3As$ dihalides (1121).
Complex compd. with $HgCl_2$, decomp. 176-176.8° (1121).

TRICYCLOPENTADIENYLARSINE $C_{15}H_{15}As$ ($C_5H_5)_3As$
Prepn.: by reacting C_5H_5Na with $AsCl_3$ in C_4H_8O at a low temp. (282a).
Props.: Black substance (382a).

TRIS(2- ETHYLHEXYL)ARSINE $C_{24}H_{51}As$ ($EtBuCHCH_2)_3As$
Use: Curing agent for polyepoxides (1132).

TRIOCTYLARSINE $C_{24}H_{51}As$ (n-$C_8H_{17})_3As$
Use: Catalyst for polymerization of acrolein (1133).

TRIS(3,3,5-TRIMETHYLCYCLOHEXYL)ARSINE $C_{27}H_{51}As$ ($3,3,5$-$Me_3C_6H_8)_3As$
Use: Lead scavenger for gasoline (1328).

TRI(PHENYLETHYNYL)ARSINE $C_{24}H_{15}As$ (PhC:C)$_3As$
Prepn.: by reacting $AsCl_3$ with PhC:CMgBr or PhC:CNa in C_4H_8O-C_6H_6 (517).
Props.: Crystals, m. 127° (517).
Rxn. with aq. alk. → evolution of PhC:CH (517).

TRISTYRYLARSINE $C_{24}H_{21}As$ (PhCH:CH)$_3As$
Prepn.: by adding a mixt. of PhCH:CHBr and $AsCl_3$ to Na wire in C_6H_6 contg. AcOEt (265).
Props.: Needles, m. 82° (265).
Rxn. with $AsCl_3$ at 180° → 1-chloroarsindole (265).

TRISDODECYLARSINE $C_{36}H_{75}As$ $(n-C_{12}H_{25})_3As$
Prepn.: by reacting $C_{12}H_{25}MgBr$ with $AsBr_3$; 56% yield (837).
Props.: B. 200°/0.08mm.; b. 200°/0.08mm.; d_{20}^{35} 0.900; n_D^{25} 1.4740(837).

TRISTETRADECYLARSINE $C_{42}H_{87}As$ $(n-C_{14}H_{29})_3As$
Prepn.: by reacting $n-C_{14}H_{29}MgBr$ with $AsBr_3$ (837).
Props.: d_{20}^{25} 0.908; n_D^{25} 1.4740 (837).

TRIS(2,4-DI-tert'-BUTYL-6-METHYLCYCLOHEXYL)ARSINE $C_{45}H_{87}As$ [2,4,6-Me(Me$_3$C)$_2$-C$_6$H$_8$]$_3$As
Use: Lead scavenger · for gasoline (1328).

SYMMETRICAL AROMATIC ARSINES

Symmetrical aromatic tertiary arsines are prepared by:

1. Reacting diazonium double salts with arsenic trichloride (3:1) or with aryldihalo- or diarylhaloarsines in acetone in the presence of metals such as Fe or Zn. (Method I).
2. Reacting arsenic trihalides, or esters of arsenous acid, or aryldihalo- or diarylhaloarsines with organometallic compounds, such as aromatic Grignard derivatives or aryllithium compounds (Method II).
3. Reducing symmetrical tertiary arsine oxides catalytically or chemically with H_3PO_2 or $SnCl_2$ + HCl (Method III).
4. Treating an arsenic trihalide with three equivalents of an aryl halide and metallic sodium (Method IV).
5. Disproportionation of diarylhalo- or aryldihaloarsines or of arsinous acids with alkali (Method V).
6. Heating aromatic amines with arsenic trichloride at 100° (Method VI).

TRI(4-BROMOPHENYL)ARSINE $C_{18}H_{12}AsBr_3$ $(4-BrC_6H_4)_3As$
Prepn.: By reacting $4-BrC_6H_4MgBr$ in Et_2O with $4-BrC_6H_4AsCl_2$, 68% yield (99); with $4-BrC_6H_4AsBr_2$, 3% yield (1261).
        By reacting $4-BrC_6H_4MgBr$ with $AsCl_3$, 21% yield (1261).
Props.: M. 132-134°, b. 284-290°/7mm. (99); m. 93-95° (1261).
Rxns. with: $KMnO_4$ in Me_2CO → $(4-BrC_6H_4)_3AsO$ (99).
MeI → Me$(4-BrC_6H_4)_3$AsI (101).
H_2O_2 in Me_2CO → $(4-BrC_6H_4)_3AsO$ (1261).

TRI(4-CHLOROPHENYL)ARSINE $C_{18}H_{12}AsCl_3$ $(4-ClC_6H_4)_3As$
Prepn.: by treating a mixt. of $(4-ClC_6H_4N_2)_2ZnCl_4$ and $AsCl_3$ with Zn in Me_2CO at 5° and completing the rxn. at r. t. or under reflux (507).
Props.: Needles, m. 75° (507).
Complex compds.: with $HgCl_2$, white needles, m. 235° (507).
{Co[$(ClC_6H_4)_3As](NO)(CO)_2$}, red crystals, m. 135° (781, 783).
{[$(ClC_6H_4)_3As]_4Pt(O)$}, colorless, m. 178° (784).
{Rh[$(ClC_6H_4)_3As]_2(CO)Cl_3$}, decomp. 230-240°; structure (1253).
{Ph[$(ClC_6H_4)_3As]_2(CO)Cl$}, yellow prisms, non-electrolyte, diamagnetic, decomp. 230-245° (1252).

TRI(3-NITROPHENYL)ARSINE $C_{18}H_{12}AsN_3O_6$ $(3-O_2NC_6H_4)As$
Prepn.: by reducing $(3-O_2NC_6H_4)_3AsO$ with H_3PO_3 contg. HI (96).
Props.: M. 206-207° (96).
Rxn. with $KMnO_4$ → $(3-O_2NC_6H_4)_3AsO$ (96).

TRIPHENYLARSINE $C_{18}H_{15}As$ Ph_3As
Prepn.: By reacting $(PhN_2)_2 \cdot ZnCl_4$ with $AlCl_3$ (3:1) in Me_2CO in the presence of
Fe powder; 43% yield. If Fe was added to a suspension of $(PhN_2)_2 \cdot ZnCl_2$ and $AsCl_3$
in Me_2CO, 18% yield was obtd. along with Ph_3AsO (1066). $(PhN_2)_2 \cdot ZnCl_4$ may be
used in a mixt. with PhN_2Cl, and Ph_2AsCl or $PhAsCl_2$ may be used instead of $AsCl_3$,
while Zn dust may be employed instead of Fe powder. The rxn. is carried out at
5° and completed at r. t. or under reflux (507; see also 1272).
        By reacting PhBr and Na-Hg with PhAsO, $(Ph_2As)_2O$, or $(Ph_2As)_2$
(96).
        By treating a mixt. of $AsCl_3$ and PhCl (1:3) with Na in C_6H_6 under re-
flux; 80% yield (1197) and 88-91% yield (1136).
        By refluxing PhLi or PhMgBr with As powder in ether or xylene; 3.6%
yield (1192).
            By-prod. in the rxn. of $PhAsCl_2$ with C_2H_2 in the presence of $AlCl_3$ (45).
            By-prod. in the rxn. of $PhNHNH_2$ with H_3AsO_4 in the presence of Cu_2O (56).
            By-prod. in the rxn. of $PhNHNH_2$ with $PhAsO_3H_2$ in H_2O at 70-75° (129).
            Formation by neutron irradiation of $AsCl_3$ in C_6H_6 (907).
Props.: B. 205° (45). Vapor pressure (386). Magnetic data (975, 1029).
Dipole moment (1147b). Crystal structure (593, 1286). Dielectric relaxation
time(1035). UV spectrum (598, 809). Near UV spectrum (1050).
IR spectrum (711, 821, 1050). NMR spectrum (61).
Use: Promoter in hydrocyanation of diene (21). Wear inhibitor for Fe carbonyl
(63). Catalyst for polymerization of acrolein (1133). Catalyst for tetrameri-
zation of C_2H_2 to cycloöctatetraene (1323b). Curing agent component for poly-
epoxides (1132).
Rxns. with: $AlCl_3$ in the presence or absence of PhBr, followed by KI → Ph_4-
AsI (201).
$AlCl_3$ at 250° → Ph_2AsCl, equilibrium study (353, 384).
$AlCl_3 + H_2O_2$ → As(V) deriv. (1030).
H_2O_2 in alk. soln. → no rxn. (1030, 1032).
H_2O_2 in Me_2CO → Ph_3AsO (1136).
SeO_2 → Ph_3AsO and Ph_3AsSe (842).
RX (X=halogen) → Ph_3RAsX (98, 710, 805).
RX (X=halogen) $AlCl_3$, followed by KI → Ph_3RAsI (755).
AcCl + $AlCl_3$ → Ph_2AsCl, $PhAsCl_2$, Ph_3As, and PhAc (788).
RMgX, followed by aq. HX (X=halogen) → Ph_3RAsX (1136).
MeOCONClNa → MeOCON:$AsPh_3$ + NaCl (191).
Chloramine-T → $4-MeC_6H_4SO_2N$:$AsPh_3$ (799).
BuLi in EtOH → $3-LiC_6H_4AsPh_2$ (473).
Li in C_4H_8O, followed by Me_3SiCl and hydrolysis → Ph_2AsO_2H, $(Ph_2As)_2O$, and
    Me_3SiPh (1309).
Ph_2IBF_4 → Ph_4AsBF_4 (779, 905).
$[C_8H_8 \cdot Fe(CO)_3]$ → $[Fe(C_8H_8)(Ph_3As)(CO)_3]$ (814).
$[Fe(NO)_2(CO)_2]$ → $[Fe(Ph_3As)(NO)_2(CO)]$ (783).

$(4\text{-MeC}_6\text{H}_4\text{NC})_2\text{Pd} \rightarrow (\text{Ph}_3\text{As})_4\text{Pd}$, m. 80-100° (782).

Thermal neutrons in vacuo $\rightarrow$ Ph$_2$As and PhAs radicals (771, 772).

<u>Complex compds.</u>: Ph$_3$As:AlCl$_3$, very reactive but stable in an inert atm. up to 250°, is oxidized by atm. O to Ph$_3$AsO (755).

[Cr(Ph$_3$As)(CO)$_5$], m. 135-135.5°, prepd. by treating Cr(CO)$_6$ with Ph$_3$As in (CH$_2$-OMe)$_2$ under reflux (833a).

[Cu(Ph$_3$As)[P(OPh)$_3$]Br], m. 116-118°, prepd. by heating (PhO)$_3$P·CuBr with Ph$_3$-As(23a).

[Fe(CO)$_2$(Ph$_3$As)$_2$Br$_2$] (542).

[Fe(CO)$_2$(Ph$_3$As)$_2$Cl$_2$] (542).

[Fe(CO)$_3$(Ph$_3$As)I$_2$] (542).

[Fe(Ph$_3$As)(NO)(CO)], red crystals, decomp. 100-110° (783).

[Fe(Ph$_3$As)(C$_8$H$_8$)(CO)$_2$], red oil (814).

[Hg(Ph$_3$As)$_2$Br$_2$], m. 182-212° (352).

[Hg(Ph$_3$As)$_2$Cl$_2$], m.244-245° (96).

[Hg(Ph$_3$As)$_2$I$_2$], m. 197° (352).

[Hg(Ph$_3$As)Br]$_2$Br$_2$, m. 219° (352).

[Hg(Ph$_3$As)Cl]$_2$Cl$_2$, m. 251-253° (352).

Hg[Co(CO)$_3$(Ph$_3$As)]$_2$, yellow fine crystals, stable in air and in non-oxidizing conc. acids; IR spectrum (544).

[Mn(Ph$_3$As)$_2$(CO)$_3$Cl], yellow; Mn(Ph$_3$As)$_2$(CO)$_3$Br , orange; Mn(Ph$_3$As)$_2$(CO)$_3$I , dark orange; IR spectra (3).

[Mo(Ph$_3$As)(CO)$_3$], yellow crystals, sol. in CHCl$_3$, CH$_2$Cl$_2$, and Me$_2$CO; instable in solns., darkens at 150°, decomp. 170° (2).

[Ni(Ph$_3$As)(CO)$_3$], m. 105° (1058, 1323); m. 96-105° (1111).

[Os(Ph$_3$As)$_3$Cl], colorless, diamagnetic crystals, stable in air (1260).

[Os(PhAs)$_3$Br], colorless, diamagnetic crystals, stable in air (1260).

(Ph$_3$As)$_4$Pd(0), m. 80-100° (782).

(Ph$_3$As)$_4$Pt(0), colorless, m. 183° (784).

cis-[Pt(Ph$_3$As)$_2$Cl$_2$], almost colorless crystals, dipole moment (600).

trans-[Pt(Ph$_3$As)$_2$Cl$_2$], yellow crystals, dipole moment (600).

trans(?)- Pt(Ph$_3$As)$_2$Cl$_4$, m. 290° (929).

Pt(Ph$_3$As)Cl$_4$, M. 284° (1192).

[Re(Ph$_3$As)$_2$(CO)$_3$Cl], white; Re(Ph$_3$As)$_2$(CO)$_3$I , pale yellow; IR spectra (3).

[Rh(Ph$_3$As)$_2$(CO)Cl], yellow prisms, decomp. 242-245° (1242); decomp. 230-240° (1254); (See also 541, 545).

[Rh(Ph$_3$As)$_2$(CO)Cl$_3$], decomp. 250-255°, other props. and structure (1253).

[Rh(Ph$_3$As)$_2$(CO)ClI$_2$], decomp. 200-205°, other props. and structure (1253).

[Rh(Ph$_3$As)$_2$(CO)I$_3$], decomp. 195-210°, other props. and structure (1253).

[Ru(Ph$_3$As)$_2$(CO)$_2$I$_2$], non-ionic complex (541, 543).

TRI-o-TOLYLARSINE C$_{21}$H$_{21}$As (2-MeC$_6$H$_4$)$_3$As

<u>Prepn.</u>: From AsCl$_3$ and 2-MeC$_6$H$_4$MgBr; 39% (1198).

<u>Props.</u>: M. 108-110° (1198).

<u>Rxns. with:</u> 2-MeC$_6$H$_4$Br + AlCl$_3$ $\rightarrow$ (2-MeC$_6$H$_4$)$_3$As(OH)OAsBr(C$_6$H$_4$Me-2)$_3$ (755).

Chloramine-T $\rightarrow$ (2-MeC$_6$H$_4$)$_3$As:NSO$_2$C$_6$H$_4$Me-4 (799).

Br$_2$ in AcOH $\rightarrow$ (2-MeC$_6$H$_4$)$_3$AsBr$_2$ (1198).

TRI-p-TOLYLARSINE $C_{21}H_{21}As$ $(4-MeC_6H_4)_3As$

Prepn.: by treating $AsCl_3$ with $4-MeC_6H_4MgBr$ in C_6H_6; 46% yield (1198).
Props.: M. 147-148° (1198). Magnetic data (1029).
Rxns. with: $PhBr + AlCl_3 \rightarrow [Ph(4-MeC_6H_4)_3As][AlBrCl_3]$ (755).
$4-MeC_6H_4Br + AlCl_3$, followed by $KI \rightarrow (4-MeC_6H_4)_4AsI$ (755).
Br_2, followed by $NaOH \rightarrow (4-MeC_6H_4)_3As(OH)_2$ (1198).
Complex compds.: $\{Co[(4-MeC_6H_4)_3As](No)(CO)_2\}$, red crystals, m. ~125° (781, 783);
IR spectrum (733a).
$\{Fe[4-MeC_6H_4)_3As](NO)_2(CO)\}$, red crystals, decomp. ~130° (783).
$\{Rh[(4-MeC_6H_4)_3As]_2(CO)Cl\}$, deep yellow prisms, decomp. 235-242° (1252).
$\{Rh[(4-MeC_6H_4)_3As]_2(CO)Cl_3\}$, decomp. 250-260°, other props. and structure (1253).

TRI-3-ANISYLARSINE $C_{21}H_{21}AsO_3$ $(3-MeOC_6H_4)_3As$

Prepn.: By reacting $AsCl_3$ with $3-MeOC_6H_4MgI$ in $Et_2O-C_6H_6$ (96).
By treating $(3-NaOC_6H_4)_3As$ with Me_2SO_4 (96).
Props.: M. 112-113° (96).
Rxn. with MeI $\rightarrow$ Me$(3-MeOC_6H_4)_3AsI$ (101).
Complex compd. with $HgCl_2$, m. 177-178° (96).

TRI-4-ANISYLARSINE $C_{21}H_{21}AsO_3$ $(4-MeOC_6H_4)_3As$

Prepn.: by reacting $AsCl_3$ with $4-MeOC_6H_4MgI$ in $Et_2O-C_6H_6$ (96).
Props.: M. 157-159° (96). Magnetic data (1029).
Rxn. with $KMnO_4 \rightarrow (4-MeOC_6H_4)_3As(OH)_2$ (96).
Complex compd.: $\{Rh[(4-MeOC_6H_4)_3As_2]_2(CO)Cl\}$, yellow prisms, decomp. 235-240° (1252).

TABLE III
SYMMETRICAL AROMATIC ARSINES

Triarylarsine	Prepd. from	Method	Yield %	Props.& Remarks	Ref.
C_{18} $(2,4,6-Br_3C_6-H_2)_3As$	$2,4,6-Br_3C_6H_2N_2Cl.-FeCl_3 + AsCl_3$	I		m. 81-83°	1066
$[4,3-Br(O_2N)-C_6H_3]_3As$	$[4,3-Br(O_2N)C_6H_3]_3-AsO + H_3PO_2$ in AcOH + HI	III	87	m. 189-191°	99
$(3-HOC_6H_4)_3As$	$(3-HOC_6H_4)_3AsO + H_3-PO_2 + HI$	III		m. 187-188°; rxn. with $Me_2-SO_4 \rightarrow (3-MeOC_6H_4)_3As$	96
$(3-HO_3SC_6H_4)_3-As$ (?)				Affinity for Ag in aq. soln.	7
$(3-H_2NC_6H_4)_3As$	$(3-O_2NC_6H_4)_3AsO$ $SnCl_2 + HCl$ in ACOH			m. 178-179°; rxn. with Ac_2O $(3-AcNHC_6H_4)_3As$	96
$(4-H_2NC_6H_4)_3As$	$(4-HCl.H_2NC_6H_4)_2As-Cl$ or $(4-H_2NC_6H_4)_2-AsOH + H$	V		Crystalline powder, m. 174-175°; rxn. with I $\rightarrow$ $(4-H_2NC_6H_4)_3AsO$	64

TABLE III (cont'd.)
SYMMETRICAL AROMATIC ARSINES

Triarylarsine	Prepd. from	Method	Yield %	Props. & Remarks	Ref.
C_{21} $(PhCH_2)_2As$	$PhCH_2MgCl + (BuO)_3-$ As in Et_2O at $-70°$			m. 102-104° (formed as by-prod.)	727
$(4-H_2NCONHC_6-H_4)_3As$	$4-H_2NCONHC_6H_4AsO_3-$ $H_2 + SO_2$ in aq. HI, followed by alkali			Decomp. 180°	456
C_{24} $(3-AcNHC_6H_4)_3-$ As	$(3-H_2NC_6H_4)_3As +$ Ac_2O			Rxn. with $KMnO_4$ in H_2O-Me_2- CO → $(3-AcNHC_6H_4)_3As(OH)_2$	96
$(4-EtOC_6H_4)_3As$				Magnetic props.	1029
$[3,4-H_2N(EtO)-C_6H_3]_3As$	$[3,4-H_2N(EtO)C_6H_3]_3-$ AsO + H_2/Ni in Et-OH at 80-100/1000 p.s.i.	III	52	m. 198-199° tripicrate, m. 168-169° (dec.), 3HCl, m. 196-198° (dec.)	1261
C_{27} $(4-MeEtNC_6H_4)_3-$ As	$MeEtNPh + AsCl_3$ at 100°	VI		m. 206°	1256
$[2,4-Me(Me_2N)-C_6H_3]_3As$	$3-MeC_6H_4NMe_2 + AsCl_3$ at 100°	VI		m. 98°	1256
C_{36} $(2-PhC_6H_4)_3As$	$2-PhC_6H_4Cl + AsCl_3$ + Na in C_6H_6	IV		m. 190°; rxn. with halogens → the arsine dihalides	1318
$(4-Me_2NC_{10}H_6)_3-$ As	$1-C_{10}H_7NMe_2 + AsCl_3$ at 100°	VI		m. 148°	1256

SYMMETRICAL HETEROCYCLYLARSINES

Symmetrical tertiary heterocyclylarsines are prepared by:

1. Treating mixtures of arsenic trichloride and a halogen-substituted heterocyclic compound (1:3) with metallic sodium.
2. Reacting arsenic trichloride or dihaloheterocyclyl- or halodiheterocyclyl-arsine, $RAsX_2$ or R_2AsX, with an organometallic compound, such as Grignard derivatives of heterocyclic compounds, heterocyclylmercury halides, and heterocyclyllithium compounds.

TRI(5-BROMO-2-FURYL)ARSINE $C_{12}H_6AsBr_3O_3$ $(\overline{O}CBr:CHCH:\overset{\frown}{C})_3As$
Prepn.: by reacting $\overline{O}CBr:CHCH:\overset{\frown}{C}HgCl$ with $AsCl_3$ in C_6H_6 under reflux (345, 753).
Props.: M. 105° (345); m. 106° (753).
Rxn. with $AsCl_3 \rightarrow$ a resin (345).

TRI(5-CHLORO-2-FURYL)ARSINE $C_{12}H_6AsCl_3O_3$ $(\overline{O}CCl:CHCH:\overset{\frown}{C})_3As$
Prepn.: by reacting $\overline{O}CCl:CHCH:\overset{\frown}{C}HgCl$ with $AsCl_3$ in C_6H_6 under reflux (345, 753).
Props.: M. 70°(345); m. 63° (753).
Rxns. with: $AsCl_3 \rightarrow$ $(\overline{O}CCl:CHCH:\overset{\frown}{C})_2AsCl$ and $\overline{O}CCl:CHCH:\overset{\frown}{C}AsCl$ (345).
$HgCl_2 \rightarrow$ $\overline{O}CCl:CHCH:\overset{\frown}{C}HgCl$ (753).

TRI-2-FURYLARSINE $C_{12}H_9AsO_3$ $(\overline{O}CH:CHCH:\overset{\frown}{C})_3As$
Prepn.: By reacting $\overline{O}CH:CHCH:\overset{\frown}{C}MgBr$ with $AsCl_3$ (345).
By reacting $\overline{O}CH:CHCH:\overset{\frown}{C}X$ (X=Cl or Br) with $AsCl_3$ (3:1) and Na (345).
By refluxing $\overline{O}CH:CHCH:\overset{\frown}{C}HgCl$ with $AsCl_3$ (3:1) in C_6H_6; 59% yield (866).
Props.: M. 31-32° (sealed tube), b. 158-159°/12mm. (345). M. 33.5°, b.
153°/3mm., darkens on prolonged exposure to sunlight (866).
Rxns. with: $AsCl_3$ at 160° → $\overline{O}CH:CHCH:\overset{\frown}{C}AsCl_2$ and $(\overline{O}CH:CHCH:\overset{\frown}{C})_2AsCl$ (345, 866).
$Cl_2 \rightarrow$ $(\overline{O}CH:CHCH:\overset{\frown}{C})_3AsCl_2$ (346); $AsCl_3$ and $\overline{O}CHClCHClCHClC Cl_2$ (753).
$H_2O_2 \rightarrow$ $(\overline{O}CH:CHCH:\overset{\frown}{C})_3AsO$ (346, 347).
Oxidizing agents in alk. soln. → $(\overline{O}CH:CHCH:\overset{\frown}{C})_2AsO_2H$ (346).

TRI-3-FURYLARSINE $C_{12}H_9AsO_3$ $(\overset{\frown}{C}H:CHOCH:\overset{\frown}{C})_3As$
Prepn.: by refluxing $\overset{\frown}{C}H:CHOCH:\overset{\frown}{C}HgCl$ with $AsCl_3$ (3:1) for 3 hrs., 64% yield (67).
Props.: Oil (67).
Complex with $HgCl_2$, m. 152-153° (67).

TRI-2-THIENYLARSINE $C_{12}H_9AsS_3$ $(\overline{S}CH:CHCH:\overset{\frown}{C})_3As$
Prepn.: By reacting 2-thienyl-MgBr with $AsCl_3$ in Et_2O and heating the rxn. mixt.
at 100°, 65% yield (349).
By reacting a mixt. of dichloro-2-thienylarsine and chlorodi-2-
thienylarsine with 2-thienylmercury chloride in toluene under reflux (494).
By-prod. in the rxn. of 2-thienylmercury chloride with $AsCl_3$ in
MePh under reflux (494).
Props.: M. 25-26°, b. 195-198°/4mm. (349); b. 193-198°/0.55mm. (494).
Rxns. with: $AsCl_3$ at 150° → $(\overline{S}CH:CHCH:\overset{\frown}{C})_2AsCl$ and $\overline{S}CH:CHCH:\overset{\frown}{C}AsCl_2$ (349).
H_2O_2 at 100° → $(\overline{S}CH:CHCH:\overset{\frown}{C})_3AsO$ (349).

TRI-2-PYRIDYLARSINE $C_{15}H_{12}AsN_3$ $(C_5H_4N)_3As$
Prepn.: By reacting 2-pyridylmagnesium bromide with $AsCl_3$ in C_6H_6 (273, 806,
1022).
By reacting 2-pyridyllithium with $AsCl_3$ in Et_2O at -20°; 46% yield
(479).
Props.: M. 85° (273, 1022); m. 76-78°(479); m. 85-85.5° (806).
Dihydrate, m. 80-83° (479).
Derivatives: Trihydrochloride, m. 152°; Dipicrate, m. 152-153°; Dimethiodide
monohydrate, m. 188-197°; Dimethiodide, m. 213-215°; Dimethopicrate, m. 180-
182°; Trimethiodide, m. 201.5-204.5° (806).

ASYMMETRICAL TERTIARY ARSINES

Asymmetrical tertiary arsines are prepared by:

1. Reacting haloörganoarsines, $RAsX_2$, R_2AsX, or $RR'AsX$, and alkyl halides, $R''X$, with mercury in a sealed tube at room temperature or with zinc powder in ether (Method I).
2. Treating haloörganoarsines, R_2AsX, and alkyl halides, $R'X$, with metallic sodium in a suitable solvent (Method II).
3. Reacting haloörganoarsines with organolithium or organosodium compounds (Method III).
4. Reacting haloörganoarsines, R_2AsX, $RR'AsX$, or $RAsX_2$ with Grignard compounds, $R''MgX$, or treating diorgano-arsinomagnesium halides R_2AsMgX, or organoarsylenebismagnesium halides, $RAs(MgX)_2$, with alkyl or aryl halides, $R''X$ (Method IV).
5. Displacement of one or two organic groups from symmetrical or asymmetrical tertiary arsines, R_3As or $R_2R'As$, by means of alkyl halides, $R''X$, at an elevated temperature (240°) or in ultraviolet light (Method V).
6. Treating mixtures of halodiorganoarsines, R_2AsX or $RR'AsX$, and alkyl halides, $R''X$, with aqueous sodium hydroxide and reducing the product with sulfur dioxide in an acid medium containing the iodide ions (Method VI).
7. Reacting haloörganoarsines, R_2AsX or $RAsX_2$, with Grignard compounds. The haloörganoarsines, R_2AsX and $RAsX_2$, may be prepared by treating tertiary arsines with arsenic trichloride at an elevated temperature (Method VII).
8. Cleaving tetraörganodiarsines $(R_2As)_2$ with alkyl halides (Method VIII).
9. Condensation of haloörganoarsines, $RAsX_2$ or $RR'AsX$, with acetylene in the presence of aluminum chloride to yield the corresponding tertiary arsines containing 2-chlorovinyl groups (Method IX).
10. Decomposition of a mixture of aryldiazonium double salts, $RN_2X.MX_n$, and haloörganoarsines, $R'AsX_2$, R'_2AsX,   or $R'R''AsX$, with a metal powder, such as zinc or iron in acetone at a low temperature (Method X).
11. Pyrolysis of arsonous acid anhydrides with carboxylic acids, $R_2AsOO-CR$ (Method XI).
12. Reduction of asymmetric tertiary arsine oxides or sulfides, $RR'R''AsO$, or $RR'R''AsS$, with hypophosphorous acid in an acid medium
      in the presence of the iodide ions or with sulfur dioxide in a chloroform-hydrochloric acid mixture or in an aqueous solution containing iodine (Method XII).
13. Treating bis(diorganoarsenic) oxides with carboxylic anhydrides, $(RCO)_2O$ and potassium carbonate at 100° (decarboxylation) (Method XII).
14. Reacting secondary arsines, R_2AsH, with carbon disulfide at 20° under nitrogen (Method XIV).
15. Reacting primary or secondary arsines, $RAsH_2$ or $RR'AsH$, with α-olefins (in the presence of sodium alkoxides) (Method XV).
16. Treating primary arsines with sodium in liquid ammonia and reacting the sodium derivatives with alkyl halides (Method XVI).

17. Reacting secondary arsines, R_2AsH, with alkylene oxides, $RCHCH_2O$, at 130° in a sealed tube in CO_2 (Method XVII).
18. Reacting secondary arsines, R_2AsH, with acyl halides in ether in the presence of calcium carbonate (Method XVIII).
19. Biological methylation of arsinic acids, R_2AsO_2H, by Penicillium brevicaula in bread cultures (Method XIX).
20. Introducing a substituent into one or two organic groups of a symmetrical tertiary arsine (Method XX).
21. Reacting secondary arsines with ketene to form acyldialkylarsines (Method XXI).

ARSINES $R_2R'As$

BIS(TRIFLUOROMETHYL)METHYLARSINE $Me(CF_3)_2As$ $C_3H_3AsF_6$
Prepn.: By reacting $MeAsI_2$ with CF_3I and Hg in a sealed tube at r. t.; 14% yield (254).
By reacting $[(CF_3)_2As]_2$ with MeI in a sealed tube at 75-105° (254).
By reacting $(CF_3)_3As$ with MeI at 235° or in UV light; up to 50% yield (333, 334, 335, 521).
By reacting $(CF_3)_2AsI$ with MeMg I; 63% yield (333).
By reacting Me_2AsCF_3 with CF_3I at 240°, 82% yield with 38% conversion (521).
Props.: B. 52° (333); b. 53° (335).
Rxns. with: Aq. NaOH → partial hydrolysis (333, 551).
MeI (1:1) at 240° → Me_2AsCF_3, CF_3I, CHF_3, and $MeAs(CF_3)_2$ (521).
CF_3I (1:2) at 240° → no rxn. (521).

(TRIFLUOROMETHYL)DIMETHYLARSINE $C_3H_6AsF_3$ Me_2AsCF_3
Prepn.: By reacting Me_2AsI and CF_3I with Hg in a sealed tube, 40% yield; with Me_2AsCl instead of Me_2AsI trace amts. of the prod. were formed (254).
By treating $(Me_2As)_2$ with CF_3I in a sealed tube, 98% yield (254).
By heating $Me_2AsOOCCF_3$ at 205°, 23% yield (256).
By reacting Me_3As with CF_3I (1:1) at 50-55° in a sealed tube (520).
By reacting CF_3AsI_2 with MeMgI (335).
By-prod. in the rxn. of $(CF_3)_3As$ with MeI at 235-240° or in UV light (333, 521).
Props.: B. 58°(254, 335, 520); $\log_{10}P = 7.3737-1487/T$ (520).
Rxns. with: Alk. → hydrolysis (335, 521).
AgI → no rxn. (520).
CF_3I (1:1) at 240° → $MeAs(CF_3)_2$, MeI, CHF_3, and Me_2AsCF_3 (521).

DIMETHYLACETYLARSINE C_4H_9AsO Me_2AsAc
Prepn.: By reacting Me_2AsH with $AcNH_2.BF_3$; 10-15% yield (9).
By adding dropwise AcCl to $Me_2AsMgBr$ in Et_2O; 20% yield (9).
By passing $CH_2:CO$ into Me_2AsH (9).
By-prod. in the rxn. of Me_2AsH with AcBr and $BaCO_3$ (9).
Props.: B. 39.5°/14mm (9).

3-DIMETHYLARSINO-1-PROPANETHIOL $C_5H_{13}AsS$ $Me_2As(CH_2)_3SH$
Prepn.: by reacting $Cl(CH_2)_3AsMe_2$ with NaHS in EtOH under reflux in a coal
gas atm. (739).
Props.: B. 54-58°/3mm., n_D^{20} 1.5296 (739).
Complex metal mercaptides:
$Co[S(CH_2)_3AsMe_2]_2$ (?) (740).
$Ni[S(CH_2)_3AsMe_2]_2$, green prisms, m. 85-86° (740).
$Pd[S(CH_2)_3AsMe_2]_2$, orange crystals, m. 180° (739).
$\{Pd[S(CH_2)_3AsMe_2]Cl\}_2$, yellow crystals, decomp. 250°, S,S'-bridged structure (741).
$\{Pd[S(CH_2)_3AsMe_2]Br\}_2$, orange-yellow prisms, turn red at 200°, S,S'-bridged
structure (741).
$\{Pd[S(CH_2)_3AsMe_2]NO_2\}_2$, pale yellow needles, S,S'-bridged structure (741).
$\{Pd[S(CH_2)_3AsMe_2]OOC-\}_2$, orange plates, m. 180°, S,S'-bridged structure (741).
$Pt[S(CH_2)_3AsMe_2]_2$, yellow crystals, m. 187° (739).
$\{Pt[S(CH_2)_3AsMe_2]Cl\}_2$, yellow crystals, m. 310°, S,S'-bridged structure (742).

β,β'-(METHYLARSYLENE)DIPROPIONITRILE $C_7H_{11}AsN_2$ $MeAs(CH_2CH_2CN)_2$
Prepn.: by reacting $MeAsH_2$ with $CH_2:CHCN$ at a low temp. in the presence of
MeONa (243).
Props.: B. 148-149°/0.1mm. (243).
Rxn. with $Cl_2 \rightarrow ClAs(CH_2CH_2CN)_2$ (243).
Complex compd.: $\{Pd[MeAs(CH_2CH_2CN)_2]Cl_2\}$, yellow brown, m. 129° (243).

2-BROMOPHENYLDIMETHYLARSINE $C_8H_{10}AsBr$ $2-BrC_6H_4AsMe_2$
Prepn.: By reacting Me_2AsI with $2-BrC_6H_4MgI$; 52% yield (523).
By reacting $2-BrC_6H_4AsCl_2$ with MeMgI; 75% yield (612).
Props.: B. 85-86°, is readily oxidized by air (612).
Rxns. with: BuLi, followed by $Me_2AsI \rightarrow o-C_6H_4(AsMe_2)_2$ (612).
BuLi, followed by $Et_2PCl \rightarrow 2-Et_2PC_6H_4AsMe_2$ and $PhAsMe_2$ (621).
Complex compd.: $Pd[2-BrC_6H_4(AsMe_2)]Br_2$, m. 183-185° (523).

4-BROMOPHENYLDIMETHYLARSINE $C_8H_{10}AsBr$ $4-B4C_6H_4AsMe_2$
Prepn.: by reacting $4-BrC_6H_4AsBr_2$ with MeMgI; 83% yield (99) or $4-BrC_6H_4-$
$AsCl_2$ with MeMgI, 91% yield (480).
Props.: B. 120-125°/11mm. (99); 1. 130-131°/17mm., n_D^{2} 1.6105; d_{20} 1.6082
(480).
Rxns. with: $MeI \rightarrow Me_3(4-BrC_6H_4)AsI$ (100).
Mg, followed by CO_2 or by $Me_2AsI \rightarrow 4-HO_2CC_6H_4AsMe$ and $4-C_6H_4(AsMe_2)_2$, resp. (480).
BuLi, followed by $CO_2 \rightarrow 4-HO_2CC_6H_4AsMe_2$.

DIMETHYLPHENYLARSINE $C_8H_{11}As$ $PhAsMe_2$
Prepn.: By reacting $PhAsCl_2$ with MeMgI, 90% yield (757).
By-prod. formed in the rxn. of Me_2AsI with a Grignard compd. from
$2-BrC_6H_4I$ and 3 equivs. of Mg (523).
Props.: UV spectrum (111). Basic strength (272).
Rxns. with: $PbCl_4 \rightarrow Me_2PhAs(OH)Cl$ (144).
$PhCH_2Cl$, followed by picric acid $\rightarrow Me_2Ph(PhCH_2)AsCl \rightarrow Me_2PhCH_2)AsOC_6H_2(NO_2)_3$
(193).
$Br(CH_2)_xCN \rightarrow Me_2Ph[NC(CH_2)_x]AsBr$ (242).

2-PhC$_6$H$_4$CH$_2$Br → delisquescent glass, characterized as Me$_2$Ph(2-PhC$_6$H$_4$CH$_2$)As-
picrate. If the quaternization is carried out at 200°/14mm. in CO$_2$ → (2-
PhC$_6$H$_4$CH$_2$)$_2$AsPh (244).
4-PhC$_6$H$_4$CH$_2$Br → Me$_2$Ph(4-PhC$_6$H$_4$CH$_2$)AsBr (244).
o-C$_6$H$_4$(CH$_2$Br)$_2$ → o-xylylenebis(dimethylphenylarsonium) deriv. and 2-phenyliso-
arsindoline (757).

Complex compds. (in the following complexes Ars stands for PhAsMe$_2$):
[CdArsCl$_2$], m. 220° (143; see also 187, 188).
[CdArs Br$_2$], m. 186° (143; see also 187, 188).
[CdArsI$_2$], m. 108° (143; see also 187, 188).
[CuArsI], pale yellow crystals, m. 127°, soly. data (142).
[CuArs$_2$I], white prisms, m. 94°, soly. data (142).
[CuArs$_2$Cl], white prisms, m. 127°, soly. data (142).
[CuArsBr], white prisms, m. 106°, soly. data (142).
Complex compds. with halides of Zn, Co, and Ni (187, 188).
[AuArsCl], white needles, m. 121° (319).
[AuArsBr], white needles, m. 120.5° (319).
[AuArsI], white needles, m. 130.5° (319).
[IrArs$_3$X$_3$] (X = Cl, Br, and I) are formed in a low yield (314).
H[IrArs$_2$X$_4$] (X = Cl, Br, and I) were pptd. as the ammonium and pyridinium
salts (314).
[HgArsCl$_2$], m. 193-194° (193); m. 201° (13).
[HgArsBr$_2$], m. 171°; [HgArsI$_2$], yellow, m. 144° (13).
[HgArs$_2$Br$_2$], white, m. 115°; [HgArs$_2$I$_2$], yellow, m. 104° (13; see also 187,
188).
[OsArs$_3$Br$_3$], red, m. 163°; on heating with HBr → red crystals, m. 173° (321).
[OsArs$_3$Cl$_3$], pinkish orange, m. 185° (321).
[OsArs$_4$Br$_2$], brownish yellow, m. 131° (321).
[OsArs$_4$Cl$_2$], yellow, m. 133-135° (321).
[OsArs$_3$I$_2$]$_2$, purple black, m. 175° (321).
[PdArs$_2$Cl$_2$]$_2$, orange, m. 170-173° (757).
[RhArs$_4$Br$_2$], pale yellow, m. 68-69° (310).
[RhArs$_4$I$_2$], yellow microcryst. powder, m. 79-80° (310).
[RhArs$_4$Cl$_2$], almost colorless oil (310).
[RhArs$_6$] [PhArs$_2$I$_4$], m. 210° (310).
[RhArs$_6$] [RhArsI$_5$] (310).
[RhArs$_6$] [RhI$_6$] (310).
[RhArs$_3$Cl$_2$]$_2$, yellow powder, m. 88° (310).

DIMETHYL-α-PICOLYLARSINE C$_8$H$_{12}$AsN $\overline{N:CHCH:CHCH:C}$CH$_2$AsMe$_3$
Prepn.: by reacting Me$_2$AsI with α-picolyllithium in dry Et$_2$O under N (494a).
Props.: Colorless oil, b. 86-90° 2mm., can be stored in an inert atm. in dark-
ness; on exposure to air is rapidly oxidized (494a).
Salt: picrate, elongated needles, m. 106° (494a).
Rxns. with: O → picoline and cacodylic acid (494a).
Cu$_2$Cl$_2$ in satd. KCl soln. cont. HCl → the dichlorocuprate complex (494a).
CuSO$_4$.5H$_2$O in alc., followed by NaClO$_4$ → bis(α-picolyldimethylarsine oxide)-
copper (I) perchlorate, pale cream, almost white crystals (494a).

Complex compds.:

$$\left[\begin{array}{c} \text{Me} \quad \text{Me} \qquad \text{Me} \quad \text{Me} \\ \diagdown \text{As} \diagup \qquad \diagdown \text{As} \diagup \\ H_2C \qquad M \qquad CH_2 \\ \diagdown N \diagup \qquad \diagdown N \diagup \end{array} \right] MCl_2$$, where M is Cu and Ag, both are diamagnetic complexes (494a).

(In the following formulae Ars stands for dimethyl$^\alpha$-picolylarsine).
[PdArs$_2$](ClO$_4$)$_2$, pale yellow crystals (494a).
[PtArs$_2$](ClO$_4$)$_2$, pale cream-colored crystals (494a).
[RuArs$_2$Cl$_2$]Cl.3H$_2$O, bright green plates; cond. and soly. data (494a).
[RuArs$_2$Cl$_2$]ClO$_4$ (494a).
[RuArsCl$_2$], bright orange-red, diamagnetic crystals; soly. data (494a).

4-(DIMETHYLARSINO)BENZOIC ACID   C$_9$H$_{11}$AsO$_2$ 4-HO$_2$CC$_6$H$_4$AsMe$_2$
Prepn.:   by reacting 4-BrC$_6$H$_4$AsMe$_2$ with Mg or BuLi and treating the corresp.
metal deriv. with solid CO$_2$, 80 and 63% yield, resp. (480).
Props.:   M. 143-144° (480).

DIMETHYL-o-TOLYLARSINE   C$_9$H$_{13}$As   2-MeC$_6$H$_4$AsMe$_2$
Complex compds.:   [Cd(2-MeC$_6$H$_4$AsMe$_2$)]I$_2$, m. 187° (143).
[Fe(2-MeC$_6$H$_4$AsMe$_2$)$_4$Cl$_3$]$_2$ (928).

DIMETHYL-p-TOLYLARSINE   C$_9$H$_{13}$As   4-MeC$_6$H$_4$AsMe$_2$
Complex compds.:   (Ars in the following formulae stands for 4-MeC$_6$H$_4$AsMe$_2$).
[CdArs]I$_2$, m. 126° (143).
[IrArs$_3$Cl]$_2$Cl$_2$, m. 97° (311).
[IrArs$_3$Br]$_2$Br$_2$, m. 96° (311).
[IrArs$_3$I]$_3$I$_2$, m. 86-88° (311).
[(FeCl$_3$)$_2$.Ars$_3$]$_2$, m. 100-101° (928).
[RhArs$_4$Br$_2$], pale yellow, m. 68-70° (310).
[RhArs$_4$I$_2$], orange yellow crystalline powder, m. 80-82° (310).
[RhArs$_3$Cl]$_2$Cl$_2$, m. 87-88° (310).
[RhArs$_3$]Cl$_3$, yellow crystals, m. 86-88° (308).
[RhArs$_3$]Br$_3$, red, m. 109° (308).
[RhArs$_3$]I$_3$, purplish red, m. 85-86° (308).
[RhArs$_6$]RhI$_6$, bright red, m. 200° (308).
[RhArs$_3$(SnCl$_2$)Cl$_3$], yellow needles, m. 111° (309).
[RhArs$_6$][RhArsI$_5$] (310).

DIMETHYL(2-METHYLMERCAPTOPHENYL)ARSINE   C$_9$H$_{13}$AsS   2-MeSC$_6$H$_4$AsMe$_2$
Prepn.:   by reacting 2-MeSC$_6$H$_4$AsCl$_2$ with MeMgI (743).
Props.:   B. 122-124° 3.5mm., n^{25} 1.6278 (743).
Complex compds.:   (Ars in the following formulae stands for 2-MeSC$_6$H$_4$AsMe$_2$).
The arsine forms two types of coördination compds. with monovalent copper:
(1) [CuArs][CuX$_2$], where X = Cl, Br, or I, and (2) [CuArs$_2$X], where X = Cl, Br,
I, or ClO$_4$. Both types are white crystalline or powdery substances. [CuArs-
SCN], an insol. complex was isolated. Mol. cond. of the complexes in PhNO$_2$ was
detd. and IR spectra obtd. Mol. wt. detn. precludes a dimeric structure for
the [CuArs$_2$X] compds. (223).

Bivalent Co forms two types of the complex compds.: [CoArs₂(ClO₄)₂] and
[CoArs₂][CoX₄], where X = Cl, Br, or I. [CoArs₂(ClO₄)₂]·4H₂O, brown crystals,
lose H₂O over P₂O₅; [CoArs₂][CoCl₄]·H₂O, green crystals; [CoArs₂][CoBr₄],
bright green crystals; [CoArs₂][CoI₄], purple crystals. All four complexes
prepd. by reacting alc. solns. of the arsine with the corresponding Co salts
in H₂O to form the hydrate and in alc. to form the anhyd. compds. [CoArs(SCN)₂]
0.5EtOH, greenish black crystals. [CoArs₂I₂]·2H₂O, blue crystals (224).
Gold salts form sol. complexes of the following two types: [AuArsX], where X =
Cl, Br, or I, and [AuArs₃ClO₄]. While for the former type monomeric structure
with bicovalent Au is suggested, nothing definite is known about the structure
of the latter type (223).
Complex salts with Ni: [NiArs₂Cl₂], blue crystals; [NiArs₂Br₂], green crystals;
[NiArs₂I₂], brown crystals (743).
Complexes with Pd: [PdArsCl₂], pale yellow; [PdArsBr₂], yellowish-orange; [Pd-
ArsI₂], deep red; [PdArs(SCN)₂], yellowish orange; [PdArs₂Cl₂], deep orange;
[PdArs₂Br₂], orange brown; [PdArs₂I₂], reddish-brown; [PdArs₂(ClO₄)₂], yellow
orange (743).
Complexes with Pt salts: [PtArsI]I·2H₂O, pale yellow, prepd. by adding the
arsine to aq. K₂PtCl₄ + KI. [PtArsCl]Cl·H₂O, pale yellow, prepd. by reflux-
ing [PtArs₂][PtCl₄] with the arsine in Me₂CO-H₂O. [PtArs₂Br]Br·2H₂O, pale
yellow. [PtArs₂Cl]ClO₄, cream-colored crystals, prepn. by treating [PtArs₂-
Cl]Cl in alc. with 60% HClO₄. [PtArs₂Br]ClO₄, cream-colored crystals, prepd.
by analogous method. [PtArs₂I]ClO₄, buff. [PtArs₂][PtCl₄], pale pink, prepd.
from the arsine and aq. K₂PtCl₄ in Me₂CO. [PtArs₂][PtBr₄], pale brown. [Pt-
Ars₂][PtCl₆], orange, prepd. from K₂PtCl₆. [PtArs₂][PtBr₆], brownish orange
(225).
Silver salts form insol. complexes [AgArs₂X], where X = Cl, Br, or I, of unknown structure (223).

BIS(2-CHLOROVINYL)PHENYLARSINE C₁₀H₉AsCl₂ (ClCH:CH)₂AsPh

Prepn.: By reacting PhAsCl₂ with C₂H₂ in the presence of AlCl₃ (45, 263).
          By reacting (ClCH:CH)₂AsCl with PhMgI in Et₂O (674).
Props.: B. 175-177°/4mm. (45); b. 170-178° 5mm. (263, 674).
Rxns. with: RI → RPh(ClCH:CH)₂AsI (263).
Alc.KOH → PhAsCl₂ (674).
Complex compds.: with HgCl₂, m. 157-158°; with AgNO₃, m. 205-212° (263).

(PHENYLARSYLENE)DIACETIC ACID C₁₀H₁₁AsO₄ PhAs(CH₂CO₂H)₂

Prepn.: by reacting PhAs(CH₂CO₂H)Cl with ClCH₂CO₂H in aq. NaOH and reducing the
rxn. mixt. with SO₂ in an acid medium contg. HI; 82% yield (119).
Props.: M. 126-127° (119).
Rxn. with H₂O₂ in Me₂CO → (HO₂CCH₂)₂AsPhO (119).

DIMETHYL-β-STYRYLARSINE C₁₀H₁₃As PhCH:CHAsMe₂

Prepn.: By treating a mixt. of PhCH:CHBr and Me₂AsBr with Na in C₆H₆ (265).
          By reacting Me₂AsBr with PhCH:CHMgBr (265).
Props.: Yellow oil, b. 125-135°/5mm. (265).
Rxn. with Cl₂ in CCl₄ → Me₂(PhCH:CH)AsCl₂ (265).
Complex with HgCl₂, m. 131° (265).

2-(DIMETHYLARSINO)-N,N-DIMETHYLANILINE $C_{10}H_{16}As$ $2\text{-}Me_2NC_6H_4AsMe_2$
Prepn.: by reacting Me_2AsI with $2\text{-}Me_2NC_6H_4MgBr$ in Et_2O; 68% yield (810).
Props.: B. 110-114°/11mm. (810).
Rxns. with: $Br(CH_2)_xBr \rightarrow [2\text{-}Me_2NC_6H_4AsMe_2(CH_2)x/_2\text{-}]_2Br_2$ (813).
$2\text{-}MeOCH_2C_6H_4CH_2Cl \rightarrow [2\text{-}Me_2NC_6H_4AsMe_2CH_2C_6H_4CH_2OMe]Cl$ (813).
$o\text{-}C_6H_4(CH_2Br_2)_2 \rightarrow [o\text{-}C_6H_4((CH_2)Me_2AsC_6H_4NMe_2\text{-}o)_2]Br_2$ (813).
$o\text{-}C_6H_4(CH_2Br)_2$ in MeOH $\rightarrow [2\text{-}Me_2NC_6H_4AsMe_2CH_2C_6H_4CH_2OMe\text{-}o]Br$ (813).
Derivs.: Monomethiodide, m. 221-222°; monomethopicrate, m. 178-181° (810).
Complex compds.: (Ars stands for $2\text{-}Me_2NC_6H_4AsMe_2$).
$[PdArsI]_2Br_2$, deep orange crystals, m. 229-230°, I-bridged structure (810).
$[PdArsI]_2I_2$, deep red crystals, m. 221-222°, I-bridged structure (810).
$[PdArs_2(NO_3)_2]$, faintly greenish yellow crystals, m. 180° (810).
$\{PdArs_2[OC_6H_2(NO_2)_3]_2\}$, m. 188-190° (810).
$[PdArs_2Br]Br$, ruby-red prisms, m. 170-172° (810).
$[PdArs_2Br]OC_6H_2(NO_2)_3$, m. 174-175° (810).
$[PdArs_2Br_2]$, orange yellow tablets, m. 170-172° (810).
$[PdArs_2I_2]$, deep red crystals, m. 206-207° (810).
$[PdArs_2Cl_2]$, pale yellow plates turning black at 215-220°(810).
$[PdArs_2Cl]Cl$, orange red prisms, m. 187-188° (810).
$[PdArs_2Cl]OC_6H_2(NO_2)_3$, m. 171-172° (810).
$[PdArsBr_2]$, yellowish orange needles darkening at~220°, effervesc. at 240° (810).
$[PdArs_2SO_4].3H_2O$, pale-buff powder, m. 178-180° (810).

4-(DIMETHYLARSINO)-N,N-DIMETHYLANILINE $C_{10}H_{16}AsN$ $4\text{-}Me_2NC_6H_4AsMe_2$
Prepn.: by reacting Me_2AsI with $4\text{-}Me_2NC_6H_4Li$, 73% yield (480).
Props.: B. 136-137°/15mm.; picrate, m. 159-160° (480).
Complex compds. with: $AgI(2:1)$, m. 146-148°, instable in solns. (187, 188).
$AgI(1:1)$, stable in solns. (187, 188).
$CuI(2:1)$ and with halides of Zn, Cd, Hg, Co, and Ni (187, 188).

BIS(3-DIMETHYLARSINOPROPYL)METHYLARSINE $C_{11}H_{27}As_3$ $MeAs(CH_2CH_2CH_2AsMe_2)_2$
Complex compds.: $\{Co[MeAs(CH_2CH_2CH_2AsMe_2)_2](CO)_3\}$, $\{Mo[MeAs(CH_2CH_2CH_2AsMe_2)_2]\text{-}(CO)_3\}$, and $\{W[MeAs(CH_2CH_2CH_2AsMe_2)_2](CO)_3\}$ (969a).

DIMETHYL-1-NAPHTHYLARSINE $C_{12}H_{13}As$ $Me_2[1\text{-}C_{10}H_7]As$
Rxns. with: Cl_2 in $CCl_4 \rightarrow$ the arsine dichloride (1155).
Br_2 in $Cl_4 \rightarrow$ the arsine dibromide (1156).
BrCN in petr. ether $\rightarrow Me(1\text{-}C_{10}H_7)AsCN$ (1156).
Concd. HCl or HBr $\rightarrow$ cleavage of $C_{10}H_8$ (1156).

β,β'-(PHENYLARSYLENE)DIPROPIONITRILE $C_{12}H_{13}AsN_2$ $PhAs(CH_2CH_2CN)_2$
Prepn.: by reacting $PhAsH_2$ with $CH_2{:}CHCN$ under reflux in a N atm., 79% yield (242, 804).
Props.: Needles, m. 59-60°, b. 200-210°/0.1mm. (242).
Rxns. with: HCl in EtOH-$CHCl_3$, followed by alc. $NH_3 \rightarrow PhAs[CH_2CH_2C({:}NH)NH_2]_2$ (242, 804).
KOH in aq. alc. $\rightarrow PhAs(CH_2CH_2CO_2H)_2$ (242).
$PhMgBr$ in $C_6H_6 \rightarrow PhAs(CH_2CH_2Bz)_2$ (243).
Complex compds.: (Ars in the following formulae stands for $PhAs(CH_2CH_2CN)_2$):
$[PdArs_2Cl_2]$, fine brownish yellow crystals, m. 189-195° (242).
$[HgArsCl_2]$, needles, m. 139-140° (242).

(p-AMINOPHENYL)BIS(2-CYANOETHYL)ARSINE $C_{12}H_{14}AsN_3$ p-$H_2NC_6H_4As(CH_2CH_2CN)_2$
Prepn.: by adding dropwise CH_2:CHCN to p-$H_2NC_6H_4AsH_2$ at 55°, and heating the
mixt at 100°; 96% yield (242).
Props.: Needles, m. 77.5-78.5° (242).
Rxns.with: Ac_2O-AcOH → 4-$AcNHC_6H_4As(CH_2CH_2CN)_2$ (242).
$ClCH_2COCl$ → 4-$ClCH_2CONHC_6H_4As(CH_2CH_2CN)_2$ (242).
NaOCN → 4-$H_2NCONHC_6H_4As(CH_2CH_2CN)_2$ (242).
$NaNO_2$, followed by β-naphthol → 2-$HOC_{10}H_6$-1-N:N-$C_6H_4As(CH_2CH_2CN)_2$ (242).
HCl in EtOH-$CHCl_3$, followed by NH_3 → 4-$H_2NC_6H_4As[CH_2CH_2C(:NH)NH_2]_2$ (242).

p-CHLOROPHENYLBIS(2-GUANYLETHYL)ARSINE $C_{12}H_{18}AsClN_4$ p-$ClC_6H_4As[CH_2CH_2C(:NH)$-
$NH_2]_2$
Prepn.: by reacting p-$ClC_6H_4As(CH_2CH_2CN)_2$ with dry HCl in EtOH-$CHCl_3$ and
treating the rxn. prod. with alc. NH_3; isolated as a salt (242).
Salts: Dinitrate monoalcoholate, plates, m. 103-105° (242).
Dipicrate monohydrate, yellow crystals, m. 202° (242).

3-DIETHYLARSINOPHENOL METHYLCARBAMATE $C_{12}H_{18}AsNO_2$ 3-$MeNHCO_2C_6H_4AsEt_2$
Prepn.: by reacting 3-$HOC_6H_4AsEt_2$ with MeNCO at room temp. (481).
Props.: gummy solid (481).

PHENYLBIS(2-GUANYLETHYL)ARSINE $C_{12}H_{19}AsN_4$ PhAs$[CH_2CH_2C(:NH)NH_2]_2$
Prepn.: by treating PhAs$(CH_2CH_2CN)_2$ with dry HCl in EtOH-$CHCl_3$ at r. t. and
reacting the prod. with alc. NH_3; isolated as dinitrate (242, 804).
Salts: Dinitrate monohydrate, needles, m. 156-157° (242).
Dipicrate, orange yellow crystals, m. 189-190° (242).
Disalicylate monohydrate, m. 98.5-103.5°; bis(2,5-dinitrobenzoate), very pale
yellow, m. 90-105° (242).
Rxn. with 1% aq. KOH → PhAs$(CH_2CH_2CO_2H)_2$ (242).

p-AMINOPHENYLBIS(2-GUANYLETHYL)ARSINE $C_{12}H_{20}AsN_5$ p-$H_2NC_6H_4As[CH_2C(:NH)NH_2]_2$
Prepn.: by reacting p-$H_2NC_6H_4As(CH_2CH_2CN)_2$ with dry HCl in EtOH-$CHCl_3$ at r. t.
and treating the rxn. mixt. with alc. NH_3, isolated as tripicrate (242).
Salt: Tripicrate, yellow crystals, m. 138° (dec.) (242).
Rxn. with Ac_2O-AcOH → the p-AcNH derivs. (242).

(o-DIMETHYLARSINOPHENYL)DIETHYLPHOSPHINE $C_{12}H_{20}AsP$ 2-$Me_2AsC_6H_4PEt_2$
Prepn.: by reacting 2-$BrC_6H_4AsMe_2$ with BuLi in petr. ether under reflux and
treating the rxn. prod. with Et_2PCl (612).
Props.: Crude, b. 152-170°/22mm; b. 105-106°/0.6mm. (612).
Derivs.: Monomethiodide. H_2O, m. 162-163° (612).
Rxn. with Br$(CH_2)_x$Br under N → $[Me_2AsC_6H_4P(Et)_2(CH_2)x]Br_2$ (612).
Complex compds.: (AsP in the following formulae stands for $Me_2AsC_6H_4PEt_2$):
$[Pd(AsP)_2][PdBr_4]$, reddish brown ppt. (612).
$[Pd(AsP)Br_2]$, yellow, m. 308° (dec.) (612).
$[Pd(AsP)_2]Br_2$, yellow crystals, m. 266-267°, dipicrate, m. 236° (dec.) (612).

TRIFLUOROMETHYLDIPHENYLARSINE $C_{13}H_{10}AsF_3$ Ph_2AsCF_3
Prepn.: by reacting Ph_2AsI with CF_3I and Hg in a sealed tube at 20°; 76% yield
(255).

Props.: B. 86-88°/0.001mm.; IR spectrum; at 302° decomp. → CHF_3, C_6H_6, CO_2, and a solid prod. (255).
Rxns. with: Alk. soln. → CHF_3 (255).
Br_2 → $Ph_2(CF_3)AsBr_2$ (255).
MeI → CHF_3, CF_3I, and other decompn. prods. (255).

DI-(4-BROMOPHENYL)METHYLARSINE $C_{13}H_{11}AsBr_2$ (4 BrC_6H_4)$_2$AsMe
Prepn.: by reacting Me(4-BrC_6H_4)AsI with BrC_6H_4MgBr, 79% yield (99).
Props.: B. 230-240°/14mm.; m. 71-73° (99).
Rxns. with: $KMnO_4$ in acetone → Me(BrC_6H_4)$_2$AsO (99).
HNO_3-H_2SO_4 → the 3-nitro deriv. (99).
MeI → Me_2(4-BrC_6H_4)$_2$AsI (100).

METHYLDIPHENYLARSINE $C_{13}H_{13}As$ MePh$_2$As
Prepn.: By reacting Ph_2AsBr with MeMgI (96).
By heating (Ph_2As)$_2$O with K_2CO_3 or KOAc and Ac_2O at 100-125° (1215).
Props.: B. 186-188°/21mm. (96); b. 168-172°/10mm. (1215).
Rxns. with: $KMnO_4$ in Me_2CO → MePh$_2$AsO (96).
$PbCl_4$ → MePh$_2$AsCl$_2$ (144).
Br in CCl_4 → MePh$_2$AsBr$_2$ or MePh$_2$AsBr$_4$ (146).
I in $CHCl_3$ → MePh$_2$AsI$_2$ (146).
Complex compds. (Ars in the following formulae stands for MePh$_2$As):
[CdArsCl$_2$], m. 292°; [CdArsBr$_2$], m. 257°; [CdArsI$_2$], m. 100° (143).
[CuArs$_3$][CuCl$_3$], forms brown and blue crystals, m. 245°; [CuArs$_3$][CuBr$_3$], dark green to black crystals, m. 202°, soly. data (142).
[CuArsCl]$_4$, microcrystalline powder, m. 109° (939).
[CuArsBr]$_4$, microcrystalline powder, m. 170° (939).
[CuArsI]$_4$, white powder, m. 200° (939).
[CuArs$_3$Cl], rhombic crystals, m. 117° (939).
[CuArs$_3$Br], m. 134° (939).
[CuArs$_3$I], m. 128° (939).
[CuArsCl], white prisms, m. 116°, soly. data (142).
[CuArsBr], white prisms, m. 133°, soly. data (142).
[CuArsNO$_3$], white prisms, m. 107°, soly. data (142).
[CuArs$_4$][CuCl$_2$], m. 127° (939).
[CuArs$_4$][CuBr$_2$], white ppt. (939).
[CuArs$_4$]I, white ppt., m. 117° (939).
[CuArs$_4$]ClO$_4$, white, flaky crystals, m. 187° (939); ionization equil. (139).
[CuArs$_4$]NO$_3$, white crystals, m. 115° (939).
[AuArsCl], white needles, m. 121° (319).
[AuArsBr], white needles, m. 118.5° (319).
[AuArsI], white needles, m. 128° (319).
[AuArsCN] (?), white, unstable substance; [AuArsCN]x (?), yellow, insol. polymer (319).
[IrArs$_3$Cl$_2$]$_2$, m. 115-116°; [IrArs$_3$Br$_2$]$_2$, m. 245°; [IrArs$_3$I$_2$]$_2$, m. 80-82° (311).
[IrArs$_3$Cl$_3$] (?), yellow powder, m. 95-96°; [IrArs$_3$Br$_3$], brown-yellow, m. 105°, [IrArs$_3$I$_3$], yellow, m. 244° (314).
H[IrArs$_2$Cl$_4$], m. 144°; H[IrArs$_2$Br$_4$], reddish pink, m. 144° (314).
NH_4[IrArs$_2$Cl$_4$], red-pink; NH_4[IrArs$_2$Br$_4$], m. 144° (314).

$C_5H_4N \cdot H[IrArs_2Cl_4]$, pink-yellow, m. 198°; $C_5H_4N \cdot H[IrArs_2Br_4]$, pink-yellow (314).

$[Fe_2Ars_3Cl_6]$, microcryst. yellow plates, m. 93°; $[Fe_2Ars_4Cl_6]$, brown crystals, m. 213° (928).

$[HgArsCl_2]$, colorless needles, m. 186°; $[HgArsBr_2]$, m. 142; $[HgArsI_2]$, cream-colored, m. 116°; $[HgArs_2Cl_2]$, m. 131°; $[HgArs_2Br_2]$, m. 100.5°; $[HgArs_2I_2]$, m. 83° (13).

$[OsArs_3Br_3]$, red, m. 205°, on boiling with $H_3PO_2 \rightarrow [OsArs_4Br_2]$, brownish yellow, m. 100° (321).

$[OsArs_3Cl_3]$, pink, m. 120°, on boiling with $H_3PO_2 \rightarrow [OsArs_4Cl_2]$, yellow, m. 113° (321).

$[OsArs_3I_2]_2$, dark purple, prepd. by treating $[OsArs_3X_3]$ with HI (321).

$[PdArs_2Br_2]$, orange ppt., m. 178°; $[PdArs_3Br]Br$, black crystals with a reddish streak, m. 109° (930).

$[PtArs_2Br_2]$, m. 201°; $[PtArs_3Br]Br$, orange crystals, m. 151° (930).

$[PtArs_2Cl_2]$ was isolated in two forms, one almost white cryst. substance, m. 223°, practically insol. in org. solvents, and the other is light yellow, more sol., m. 214° after softening at 207°; on standing the latter is transformed into the former, which is apparently the cis-isomer (929).

$[PtArs_2Br_2]$, bright yellow crystals, m. 201°, almost certainly the cis-form (929).

trans-(?)-$[PtArs_2I_2]$, pinkish-orange cryst. powder, m. 229° (929).

cis-trans-$[PtArs_2Cl_4]$, m. 205-215° (929).

trans(?)-$[PtArs_2Br_4]$, orange crystals, m. 227°; mixt. of cis and trans-form, m. 214° (929).

trans(?)-$[PtArs_2I_4]$, purplish-black crystals, m. 193° (929).

cis-$[PtArs_2Br_2Cl_2]$, orange crystals, m. 203° (929).

trans-$[PtArs_2Cl_2I_2]$, brown crystals, m. 163° (929).

$[PtArs_2Br_2I_2]$, m. 185° (929).

$[PtArs(CO)Cl_2]$; $[PtArs(CO)Br_2]$; $[PtArs(CO)I_2]$ (582).

$[RhArs_3Cl_3]$ occurs in two forms (?), orange yellow, insol. crystals, m. 176-178° (309) and lemon-yellow microcryst. powder, m. 122-124°, extremely sol. in org. solvents (308).

$[RhArs_3Br_3]$ (?), orange-red microcrystalline powder, sol. in org. solvents, m. 116° (308).

$[RhArs_3I_3]$, purplish red (308).

$[RhArs_6][RhCl_4]$ (?), orange leaflets and rhombs, m. 171°; $[RhArs_6][RhBr_4]$ (?), reddish brown leaflets m. 180°; $[RhArs_6][RhI_4]$ (?), red leaflets, m. 168° (307).

$[RhArs_6][RhCl_6]$ (?), orange cryst. mass, m. 176-178° (308). $[RhArs_6][RhBr_6]$ (?), bright red, microcryst. powder, m. 191°; both sparingly sol. in org. solvents (307).

$[RhArs_6][RhI_6]$, purplish red, m. 200° (308).

$[RhCl_2(Ars)_3SnCl_2]_2$ occurs in two modifications: yellow amorphous ppt., m. 149°, and yellow microcryst. powder, m. 129° (309).

$[RhCl_3(Ars)_3]_2SnCl_2$, yellow, m. 169°, on recrystn. from EtOH gave $[RhArs_3Cl_3]$ (309).

$[Ru_2Ars_3X_3]$, where X = Cl, Br, or I, deeply colored solids, which are reduced by H_3PO_2 to $[Ru_2Ars_4X_2]$ (316).

p-CARBAMIDOPHENYLBIS(2-CYANOETHYL)ARSINE $C_{13}H_{15}AsN_4O$ p-$H_2NCONHC_6H_4As(CH_2CH_2CN)_2$
Prepn.: by mixing a soln. of p-$H_2NC_6H_4As(CH_2CH_2CN)_2$ in HCl with NaOCN in H_2O; 100% yield (242).

Props.: Needles, m. 142-147° (242).
Rxn. with HCl in EtOH-CHCl$_3$, followed by NH$_3$ → p-H$_2$NCONHC$_6$H$_4$As[CH$_2$CH$_2$C(:NH)-NH$_2$]$_2$ (242).

3,3'-(3-METHOXYPHENYLARSYLENE)DIPROPIONIC ACID C$_{13}$H$_{17}$AsO$_5$ 3-MeOC$_6$H$_4$As(CH$_2$-CH$_2$CO$_2$H)$_2$

Prepn.: by hydrolyzing 3-MeOC$_6$H$_4$As(CH$_2$CH$_2$CN)$_2$ with KOH in aq. EtOH under reflux in N; 95% yield (811).
Props.: M. 54° (811).
Rxns. with: P$_2$O$_5$ → 2,3,5,6-tetrahydro-8-methoxy-1H,7H-benz[ij]arsinolizine-1,7-dione (811).
P$_2$O$_5$-H$_3$PO$_4$ → the preceding benz[ij]arsinolizenedione and probably 1-(2-carboxy-ethyl)-1,2,3,4-tetrahydro-7-methoxy-4-oxoarsinoline (811).

p-CARBAMIDOPHENYLBIS(2-GUANYLETHYL)ARSINE C$_{13}$H$_{21}$AsN$_6$O   p-H$_2$NCONHC$_6$H$_4$As[CH$_2$-CH$_2$C(:NH)NH$_2$]$_2$

Prepn.: by reacting p-H$_2$NCONHC$_6$H$_4$As(CH$_2$CH$_2$CN)$_2$ with dry HCl in EtOH-HCCl$_3$ and treating the rxn. prod. withNH$_3$; isolated as dipicrate (242).
Salts.: Dipicrate monohydrate, orange crystals, m. 118-120° (242).

2-CHLOROVINYLDIPHENYLARSINE C$_{14}$H$_{12}$AsCl Ph$_2$AsCH:CHCl

Prepn.: By treating a mixture of (PhN$_2$)$_2$ZnCl$_4$ and ClCH:CHAsCl$_2$ (1:1) in acetone with Zn dust at 0° and completing the rxn. at r. t. or under reflux (507).
By reacting Ph(ClCH:CH)AsCl or ClCH:CHAsCl$_2$ with PhMgBr (260, 262, 263)
By-prod. in the rxn. of PhAsCl$_2$ with C$_2$H$_2$ in the presence of AlCl$_3$ (45, 263).
By-prod. in the rxn. of ClCH:CHAsCl$_2$ with C$_6$H$_6$ in the presence or absence of AlCl$_3$ (260, 262).
By reacting ClCH:CHAsCl$_2$ with PhMgBr in Et$_2$O; 57% yield (674).
Props.: Oil, b. 161°/2mm.; b. 169-170°/3mm (507); b. 194-195°/4mm (44); b. 190-195°/5mm (263); b. 195-198°/3mm (260, 262); b. 176-178°/3mm (674).
Complex compd. with HgCl$_2$, m. 238° (263).
Rxns. with: Aq. NaOH → (Ph$_2$As)$_2$O and C$_2$H$_2$ (507).
Alc. KOH, followed by HCl → Ph$_2$AsCl (674).

2-BIPHENYLYLDIMETHYLARSINE C$_{14}$H$_{15}$As 2-PhC$_6$H$_4$AsMe$_2$

Prepn.: By reacting Me$_2$AsI with a Grignard compd. obtd. from o-C$_6$H$_4$I$_2$ or o-Br-C$_6$H$_4$I and Mg in C$_6$H$_6$ (522).
By reacting Me$_2$AsI with 2-PhC$_6$H$_4$MgI in C$_6$H$_6$ under reflux, 46% yield (522).
Props.: B. 114°/0.5 mm (522).
Complex compd.: [Pd(2-PhC$_6$H$_4$AsMe$_2$)$_2$Br$_2$], m. 223-224° (523).

p-CHLOROACETAMIDOBIS(2-CYANOETHYL)ARSINE C$_{14}$H$_{15}$AsClN$_3$O   p-ClCH$_2$CONHC$_6$H$_4$As-(CH$_2$CH$_2$CN)$_2$

Prepn.: by adding dropwise ClCH$_2$COCl to an aq. suspension of p-H$_2$NC$_6$H$_4$As(CH$_2$-CH$_2$CN)$_2$ contg. Na$_2$CO$_3$; 90% yield (242).
Props.: m. 96° (242).

p-ACETAMIDOPHENYLBIS(2-CYANOETHYL)ARSINE $C_{14}H_{16}AsN_3O$ $p\text{-}AcNHC_6H_4As(CH_2CH_2CN)_2$

Prepn.: by refluxing $p\text{-}H_2NC_6H_4As(CH_2CH_2CN)_2$ in $Ac_2O\text{-}AcOH$ (242).
Props.: M. 117-118.5° (242).
Rxn. with HCl in $EtOH\text{-}CHCl_3$, followed by $NH_3 \rightarrow p\text{-}AcNHC_6H_4As[CH_2CH_2C(:NH)NH_2]_2$ (242).

p-ACETAMIDOPHENYLBIS(2-GUANYLETHYL)ARSINE $C_{14}H_{22}AsN_5O$ $p\text{-}AcNHC_6H_4As[CH_2CH_2C\text{-}$
$(:NH)NH_2]_2$

Prepn.: 1) By reacting $p\text{-}AcNHC_6H_4As(CH_2CH_2CN)_2$ with dry HCl in $EtOH\text{-}CHCl_3$ and treating the rxn. prod. with NH_3; isolated as dinitrate (242).
         2) By treating $p\text{-}H_2NC_6H_4As[CH_2CH_2C(:NH)NH_2.HCl]_2$ with $AcOH\text{-}Ac_2O$ (242).
Props.: Dinitrate monohydrate, m. 136°; bis-3,5-dinitrobenzoate dihydrate, pale yellow microcryst. powder, m. 115-155° (242).

DIBUTYLPHENYLARSINE $C_{14}H_{23}As$ Bu_2PhAs
Prepn.: by reacting $PhAsCl_2$ with BuMgBr in Et_2O (455).
Props.: Colorless liquid, b. 158-161°/21mm (455).
Complex compds.:
$[Pd(Bu_2PhAs)_2Cl_2]$, non-ionic, orange crystals, m. 47° (455).
$[Pd(Bu_2PhAs)Cl]_2Cl_2$, bright red crystals, m. 166°, Cl-bridged structure (455).

(2-DIETHYLARSINOPHENYL)DIETHYLPHOSPHINE $C_{14}H_{24}AsP$ $2\text{-}Et_2AsC_6H_4PEt_2$
Prepn.: by reacting $2\text{-}BrC_6H_4AsEt_2$ with BuLi in light petr., followed by Et_2PCl, 63% yield (612).
Props.: B. 136°/1mm (612).
Rxn. with $Br(CH_2)_xBr \rightarrow [Et_2\overline{AsC_6H_4PEt_2(CH_2)_x}]Br_2$ (612).
Complex compds.: (ArsP in the following formulae stands for $2\text{-}Et_2AsC_6H_4PEt_2$):
$[Cu(ArsP)_2Cl]$, m. 211-212° (238).
$[Cu(ArsP)_2][CuCl_2$, m. 194-196° (238).
$[Cu_3(ArsP)_2Cl_3]$ (?), m. 163-169° (open tube), m. 183.5-184.5° (evac. tube) (238).
$[Cu(ArsP)_2Br]\cdot0.5$ EtOAc, loses the solvent of crystn. at 35°/0.5 mm.; at 100°/0.2 mm. the complex is converted into $[Cu(ArsP)_2][CuBr_2]$, m. 222-240° (238).
$[Cu(ArsP)_2I]$, m. 205.5-207.5°, stable at elevated temps. (238).
$[Cu(ArsP)_2\text{-bromocamphosulfonate}]$, m. 155.5-157.0°, was converted to optically inactive iodide by treating with KI in aq. Me_2CO (238).
$[Au(ArsP)_2Cl]$, oil, prepd. by treating (ArsP) with $HAuCl_4.4H_2O$ in H_2O, was converted into $[Au(ArsP)_2I]$, m. 212-213° by treating with KI. The chloride crystillizes from EtOAc in the form of $[Au(ArsP)_2Cl]\cdot0.5EtOAc$, m. 224-230°, which on heating loses EtOAc and in EtOH yields readily the picrate, m. 118-120° (238).
$H[Au(ArsP)_2SO_4]\cdot H_2O$, prepd. by treating the iodide with Ag_2SO_4, m. 214-216° (slow heating), m. 200° (instantaneous heating); decomp. on standing in dild. H_2SO_4. Treatment with $Ba(OH)_2$ yields $[Au(ArsP)_2OH]$, m. 222-224° (238).
$[Au(ArsP)_2\text{-camphosulfonate}]$, m. 225-226°, prepd. from the AuI complex and Ag (+)-camphosulfonate, gave an optically inactive picrate (238).
$[Au(ArsP)_2\text{-bromocamphosulfonate}]$, m. 165-165.5°, $[M]_D + 306°$ (238).
$[Au(ArsP)_2\text{-bitartrate}]$, m. 188-190.5°, $[M]_D + 29°$, gave an inactive picrate (238).
$[Au(ArsP)_2\text{-tartrate}]\cdot H_2O$, m. 185-187° (slow heating), m. 140° (rapid heating), $[M]_D + 89°$ (238).
$[Au(ArsP)_2\text{-menthoxyacetate}]$, m. 171.5-173.5° (238).

[Au(ArsP)$_2$-(+)-camphonatronate], m. 173-182° (238).
[Au(ArsP)$_2$-(-)-menthylphthalamate], m. 86.5-88.5 (238).
[Au(ArsP)$_2$-(-)-N-(PhMeCH)-phthalamate], m. 108-110°, [M]$_D$-128° (1.039 soln.) (238)
[Pt(ArsP)$_2$][PtCl$_4$], prepd. from the arsinophosphine and Na$_2$PtCl$_6$, was converted int
[Pt(ArsP)$_2$Cl$_2$] when heated at 200°/760mm. for 18 hrs. (238).
[Ag(ArsP)$_2$][AgI$_2$], m. 253-255°, on treating with the arsinophosphine gave [Ag(ArsP)
I], m. 158-162°, which at 100°/0.4mm was reverted into Ag(ArsP)$_2$[AgCl$_2$] (238).

(2-CYANOETHYL)DIPHENYLARSINE C$_{15}$H$_{14}$AsN Ph$_2$AsCH$_2$CH$_2$CN
Prepn.: by refluxing Ph$_2$AsH with CH$_2$:CHCN; 83% yield along with a small amt.
of (Ph$_2$As)$_2$ (242).
Props.: Needles, m. 37-39° (242).
Rxns. with: HCl in EtOH-HCCl$_3$, followed by NH$_3$ → Ph$_2$AsCH$_2$CH$_2$C(:NH)NH$_2$ (242).
KOH in aq. alc. → Ph$_2$AsCH$_2$CH$_2$CO$_2$H (242)
PhMgBr in C$_6$H$_6$ → Ph$_2$AsCH$_2$CH$_2$Bz (243).

3-(DIPHENYLARSINO)PROPIONIC ACID C$_{15}$H$_{15}$AsO$_2$ Ph$_2$AsCH$_2$CH$_2$CO$_2$H
Prepn.: by hydrolyzing Ph$_2$AsCH$_2$CH$_2$CN with KOH in 50% aq. alc. (242).
Props.: m. 104-106° (242).
Rxns. with: SOCl$_2$ → Ph$_2$AsCH$_2$CH$_2$COCl (243).
HgO → Ph$_2$As(O)CH$_2$CH$_2$CO$_2$H (243).
Br$_2$ followed by H$_2$S → Ph$_2$As(S)CH$_2$CH$_2$CO$_2$H (243).

DIMETHYL(2-PHENYLBENZYL)ARSINE C$_{15}$H$_{17}$As Me$_2$(2-PhC$_6$H$_4$CH$_2$)As
Prepn.: by reacting Me$_2$AsI with 2-PhC$_6$H$_4$MgCl in Et$_2$O-C$_6$H$_6$ under reflux, 60% yield
(244).
Props.: B. 124-126°/0.4mm., on exposure to air becomes turbid (244).
Rxn. with Cl$_2$ in CCl$_4$, followed by AlCl$_3$ → 5,6-dihydro-5-methylarsanthridine (244)
Complex compd.: {Pd[Me(2-PhC$_6$H$_4$CH$_2$)As]Cl$_2$}, pale orange, m. 150° (244).

(2-GUANYLETHYL)DIPHENYLARSINE C$_{15}$H$_{17}$AsN$_2$ Ph$_2$AsCH$_2$CH$_2$C(:NH)NH$_2$
Prepn.: by reacting Ph$_2$AsCH$_2$CH$_2$CN with dry HCl in EtOH-HCCl$_3$ and treating the
rxn. prod. with alc. NH$_3$ (242).
Salts: Nitrate hemihydrate, needles, m. 162-164° (dec.); picrate, yellow needles,
m. 193-195° (242).

(2-CYANOPROPYL)DIPHENYLARSINE C$_{16}$H$_{16}$AsN Ph$_2$AsCH$_2$CHMeCN
Prepn.: by refluxing Ph$_2$AsH with CH$_2$:CMeCN (242).
Props.: Oil(242).
Rxn. with KOH in aq. alc. → Ph$_2$AsCH$_2$CHMeCO$_2$H (242).
Complex compd.: [Pd(Ph$_2$AsCH$_2$CHMeCN)$_2$Cl$_2$], yellowish brown crystals, m. 172-174°
(dec.) (242).

DIMETHYL(1,2-DIPHENYLETHYL)ARSINE C$_{16}$H$_{19}$As Me$_2$(PhCH$_2$CHPh)As
Prepn.: by treating Me$_2$(PhCH$_2$)$_2$AsBr with PhLi in Et$_2$O and hydrolyzing the prod.
(1313).
Rxn. with conc. HCl → cleavage of stilbene and bibenzyl (1313).

4-BROMOPHENYLDIPHENYLARSINE C$_{18}$H$_{14}$AsBr 4-BrC$_6$H$_4$AsPh$_2$
Prepn.: By reacting (4-BrC$_6$H$_4$)PhAsBr with PhMgBr (96).

By reacting Ph(4-BrC$_6$H$_4$)AsCl with PhN$_2$Cl·ZnCl$_2$ in Me$_2$CO at 0°, followed by Fe powder; 81% yield (908).
Props.: M. 64-65° (96); HgCl$_2$ adduct, m. 172-173° (908).
Rxnx. with: KMnO$_4$ in Me$_2$CO → (4-BrC$_6$H$_4$)Ph$_2$As(OH)$_2$ (96).
Mg, followed by MeSi(OMe)$_3$ → Ph$_2$AsC$_6$H$_4$SiMe(OMe)$_2$ (1120).

2-(DIPHENYLARSINO)BENZOIC ACID C$_{19}$H$_{15}$AsO$_2$ 2-(Ph$_2$As)C$_6$H$_4$CO$_2$H
Prepn.: by reducing Ph$_2$(2-HO$_2$CC$_6$H$_4$)AsO with SO$_2$ in HCl contg. KI (613).
Props.: M. 135-138°, is readily oxidized to the arsine oxide (613).
Complex Compd.: [Pd(Ph$_2$AsC$_6$H$_4$CO$_2$H)$_2$Br$_2$], m. 225° (dec.) (613).

DIPHENYL-p-TOLYLARSINE C$_{19}$H$_{17}$As Ph$_2$(4-MeC$_6$H$_4$)As
Prepn.: by reacting Ph(4-MeC$_6$H$_4$)AsCl with (PhN$_2$Cl)$_2$·ZnCl$_2$ in Me$_2$CO at 0° in the presence of Fe powder (908).
Props.: M. 48°, HgHl$_2$ adduct, m. 182-183° (908).
Rxn . with: PhBr + AlCl$_3$ → [Ph$_3$(4-MeC$_6$H$_4$)$_2$As][AlBrCl$_3$] (755).

β-(DIPHENYLARSINO)PROPIOPHENONE C$_{20}$H$_{19}$As Ph$_2$AsCH$_2$CH$_2$Bz
Prepn.: by reacting Ph$_2$AsCH$_2$CH$_2$CN with PhMgBr in C$_6$H$_6$ (243).
Props.: Colorless needles, m. 83-84° (243).
Derivs.: 2,4-Dinitrophenylhydrazone, orange-red crystals, m. 124-126° (243).
Methopicrate, yellow needles, m. 114-115° (243).
Rxn. with: Isatin → 3-(diphenylarsinomethyl)cinchophen oxide (243).

2-(4-DIPROPYLARSINOPHENYL)QUINOLINE C$_{21}$H$_{24}$AsN —AsPr$_2$

Prepn.: by treating 4-BrC$_6$H$_4$AsPr$_2$ with BuLi and reacting the rxn. mixt. with quinoline at 0° in Et$_2$O; 60% yield (480).
Props.: B. 218-219°/0.08mm. (480).

β,β'-[4-(2-HYDROXY-1-NAPHTHYLAZO)PHENYLARSYLENE]DIPROPIONITRILE C$_{22}$H$_{19}$AsN$_4$O
2-HOC$_{10}$H$_6$N$_2$C$_6$H$_4$As(CH$_2$CH$_2$CN)$_2$
Prepn.: by adding diazotized 4-H$_2$NC$_6$H$_4$As(CH$_2$CH$_2$CN)$_2$ in HCl to alk. soln. of -naphthol (242).
Props.: M. 86-90° (242).

8-METHYL-2-(4-DIPROPYLARSINOPHENYL)QUINOLINE C$_{22}$H$_{26}$AsN —AsPr$_2$
Me
Prepn.: by treating 4-BrC$_6$H$_4$AsPr$_2$ with BuLi and reacting the rxn. mixt. with 8-methylquinoline at 0° in Et$_2$O; 79% yield (480).
Props.: M. 84-85° (480).

6-METHOXY-2-(4-DIPROPYLARSINOPHENYL)QUINOLINE C$_{22}$H$_{26}$AsNO —AsPr$_2$
MeO
Prepn.: by treating 4-BrC$_6$H$_4$AsPr$_2$ with BuLi and reacting the rxn. mixt. with 6-methoxyquinoline at 0° in Et$_2$O; 76% yield (480).
Props.: M. 95-96° (480).

β,β'-(PHENYLARSYLENE)DIPROPIOPHENONE C$_{24}$H$_{23}$AsO$_z$ PhAs(CH$_2$CH$_2$Bz)$_2$
Prepn.: by reacting PhAs(CH$_2$CH$_2$CN)$_2$ with PhMgBr in C$_6$H$_6$; 25% yield (243).
Props.: Pale yellow needles, m. 83-84° (243).

PHENYLBIS(2-PHENYLBENZYL)ARSINE $C_{32}H_{27}As$ $Ph(2-PhC_6H_4CH_2)_2As$

Prepn.: By reacting $PhAs(MgBr)_2$ with 2-PHC_6H_4Br in C_6H_4Br in $C_6H_6-Et_2O$, 94% yield (244).

By reacting $PhAsMe_2$ with 2-$PhC_6H_4CH_2Br$ in CC_2 at 200°/14mm. (244).

Props.: M. 70-77°, b. 225°/0.005mm. (dec.) (244).

Rxn. with Cl_2 in CCl_4, followed by $AlCl_3$ in CS_2 → 5,6-dihydro-5-phenylarsanthridine (244).

Complex compds.: $\{Pd[Ph(2-PhC_6H_4CH_2)_2As]_2Cl_2\}\cdot 2.5H_2O$, orange, m. 188-190° (244). $HgCl_2\cdot Ph(2-PhC_6H_4CH_2)_2As$, decomp. 172-173° (244).

Other asymmetrical arsines $R_2R'As$ are compiled in Table IV.

TABLE IV

ASYMMETRICAL TERTIARY ARSINES $R_2R'As$

$R_2R'As$	Prepd. from	Method *	Yield %	Props., Rxns.	Ref.
C_4					
$Me_2(Cl_3CCO)As$	$Me_2AsH + Cl_3CCOCl$	XVIII		m. 32-35°, b. 68°/14mm., sublimes at 30° in vac.	9
$Me_2(ClCH:CH)As$	$ClCH:CHAsCl_2 + MeMgI$	IV	51	b. 42°/13mm., d_{20} 1.3932; rxn.[1]	674
Me_2EtAs	$Me_2AsMgBr + EtBr$	IV		b. 86°;	9
				cis-, trans- $Pt(Me_2EtAs)_2Cl_2$	210
$Me_2(CF_3CF_2CF_2)As$	$Me_2AsI + CF_3CF_2CF_2I + Hg$	I	16	b. 93°, IR spectrum, rxn.[2]	254
C_5					
$(ClCH:CH)_2AsMe$	$Me(ClCH:CH)AsCl + C_2H_2 + AlCl_3$	IX		b. 140-145°/10mm.	261
$Me_2(ClCH_2CH_2CH_2)As$	$MeAsCl_2 + C_2H_2 + AlCl_3$ $Cl(CH_2)_3AsCl_2 + MeMgI$	IX IV		by-prod. rxn. 3,4 b. 59-62°/10mm.	261 55 739
$(Me_2As)_2CS$	$Me_2AsH + CS_2$	XIV			9
Me_2PrAs	$MePrAsO_2H$	XIX		b. 105-107°/12mm.	193
C_7					
$Me_2(\overline{CH:CHCH:CHN:C})-As$	$Me_2AsI + 2\text{-pyridyl-Li}$	III	81	b. 90-91°/14mm.	480
$Me_2(\overline{CH:CHCH:NCH:C})-As$	$Me_2AsI + 3\text{-pyridyl-Li}$	III	58	b. 105-107°/16mm.	480
$(ClCH:CH)_2(Me_2CH)As$	$(ClCH:CH)_2AsCl + Me_2CHMgBr$	IV	80	b. 115°/10mm., rxn.[5]	674
$(ClCH:CH)_2PrAs$	$(ClCH:CH)_2AsCl + PrMgI$	IV	89	b. 112°/4mm., d_{20} 1.4174, n_D^{20} 1.5522, rxn.[6]	674

1. Rxn. with alc. KOH → $(Me_2As)_2O + C_2H_2$ (674).
2. Rxn. with 10% NaOH → $CF_3CF_2CF_2H$ (254).
3. Rxn. with Mg + $MeAsI_2$ → $MeAs(CH_2CH_2CH_2AsMe_2)_2$ (55).
4. Rxn. with NaHS → $HSCH_2CH_2CH_2AsMe_2$ (739).
5. Rxn. with alc. KOH → $Me_2CHAsCl_2 + C_2H_2$ (674).
6. Rxn. with alc. KOH → $PrAsCl_2 + C_2H_2$ (674).
* See page 18.

TABLE IV (cont'd.)

ASYMMETRICAL TERTIARY ARSINES R₂R'As

R₂R'As	Prepd. from	Method	Yield %	Props. Rxns.	Ref.
C₈					
$(CF_3)_2PhAs$	$PhAsI_2 + CF_3I + Hg$	I	10	b. 160°, IR spectrum, rxns.7,8	255
$Me_2(3\text{-}O_2NC_6H_4)As$	$3\text{-}O_2NC_6H_4AsCl_2 + MeMgI$	IV		with RI → $RMe_2(3\text{-}O_2NC_6H_4AsI$	998
$Me_2(2\text{-}HOC_6H_4)As$	$Me_2AsI + 2\text{-}HOC_6H_4Li$	III	47	b. 85-87°/0.3mm.	481
$Me_2(3\text{-}HOC_6H_4)As$	$Me_2AsI + 3\text{-}HOC_6H_4Li$	III	80	b. 87-90°/0.3mm., rxn.9	481
$Me_2(4\text{-}HOC_6H_4)As$	$Me_2AsI + 4\text{-}HOC_6H_4Li$	III	28	rxn. 10	481
$(ClCH:CH)_2BuAs$	$(ClCH:CH)_2AsCl + BuMgI$	IV	89.3	b. 105-106°/8mm., d_{20} 1.3660, n^{20} 1.5448, rxn. 11	674
$(CH_2:CH)_2BuAs$	$(CH_2:CH)_3As + AsCl_3 +$ $BuMgBr$	VII		b. 56°/12mm.	773,774
C₉					
$Me_2(4\text{-}CF_3C_6H_4)As$				Complex compds. with the halides of Zn, Cd, Hg, Co, and Ni m. 133°	187,188
$Me_2(2\text{-}HO_2CC_6H_4)As$	$2\text{-}HO_2CC_6H_4AsCl_2 + MeMgI$	IV		b. 97-100°/9mm., rxn. 12	54
$Me_2(PhCH_2)As$	$Me_2AsI + PhCH_2MgCl$	IV		b. 123-124°/15mm.	605
$Me_2(3\text{-}MeOC_6H_4)As$	$3\text{-}MeOC_6H_4AsH_2 + Na +$ MeI	XVI			811
$(ClCH:CH)_2(Me_2CHC\text{-}H_2CH_2)As$	$(ClCH:CH)_2AsCl + iso\text{-}$ $C_5H_{11}MgI$	IV	87.5	b. 125-126°/3mm., d_{20} 1.2859, $n^{2}D$ 1.5390, rxn. 13	674
C₁₀					
$Et_2(2\text{-}BrC_6H_4)As$	$2\text{-}BrC_6H_4AsCl_2 + EtMgI$	IV		b. 109-110°/1mm., rxn. 14	612
Et_2PhAs	$PhAsH_2 + Na + EtCl$	XVI		b. 100.5-101°, rxn.15	808
$Et_2(3\text{-}HOC_6H_4)As$	$Et_2AsI + 3\text{-}HOC_6H_4Li$	III	34	b. 112°/0.15mm., rxn.16	481
$Bu_2(CH_2:CH)As$	$(CH_2:CH)_3As + AsCl_3$ $BuMgCl$	VII	51	b. 56-84°/12-10mm.	773,774
$Me_2(n\text{-}C_8H_{17})As$	$Me_2AsI + n\text{-}C_8H_{17}MgI$	IV	50	b. 102-104°/11mm.	605
C₁₁					
$(ClCH:CH)_2(4\text{-}MeC_6H_4)As$	$(ClCH:CH)_2AsCl + 4\text{-}Me\text{-}$ C_6H_4MgI	IV	78.3	n^{20}_D 1.5990, rxn.17	674
$Me_2(3\text{-}C_9H_6N)As$	$Me_2AsI + 3\text{-}quinolyl\text{-}Li$	III	74	b. 125-127°/0.5mm.	480
C₁₂					
$(ClCH:CH)_2(PhCH:\!\text{-}CH)As$	$(ClCH:CH)_2AsCl + PhCH:\text{-}$ CHMgI	IV	44	b.125-135°/3mm., d_{20} 1.4183	674

34

(NCCH₂)(4-ClC₆-H₄)As	4-ClC₆H₄AsH₂ + CH₂:-CHCN	XV	92	m. 71-72°, rxn. 19,20	242
(CH₂:CHCH₂)₂(4-Br-C₆H₄)As	4-BrC₆H₄AsCl₂ + CH₂:-CHCH₂MgBr	IV	73	b. 105-106°/0.06mm.	480
(HO₂CCH₂)₂(4-ClC₆H₄)As	(NCCH₂)₂(4-ClC₆H₄)-As + KOH in aq. alc.			m. 67-69°	242
(CH₂:CHCH₂)₂PhAs	PhAsI₂ + CH₂:CHCH₂MgBr	IV	84	b. 131°/10mm.,[21]	616
(HO₂CCH₂)₂PhAs	(NCCH₂)₂PhAs + KOH in aq. alc.			Needles, m. 83-84.5°	242
Pr₂(4-BrC₆H₄)As	4-BrC₆H₄AsCl₂ + PrMgBr	IV	94	b. 106-107°/0.05mm., rxn.[22]	480
C₁₃					
[3,4-Br(O₂N)C₆H₃]₂-MeAs	[3,4-Br(O₂N)C₆H₃]₂MeAsO + H₃PO₂ in AcOH + HI	XII	86	m. 82-84°, rxn.[23]	99
(NCCH₂CH₂)₂(3-MeO-C₆H₄)As	3-MeOC₆H₄AsH₂ + CH₂:C-HCN	XV	100	b. 206-212°/0.03., rxn.[24]	811
(cyclo-C₆H₁₁)₂MeAs				b. 107°/20mm.	1053
C₁₄					
Ph₂EtAs	(Ph₂As)₂O + EtCO₂K + (EtCO)₂O	XIII	53.6	b. 162-165°/10mm.	1215
Me₂(2-PhOC₆H₄)As	2-PhOC₆H₄AsCl₂ + MeMgI	IV		b. 205-207°/30mm., rxn.[23]	271

7. Rxn. with alk. solns. → CHF₃ (255).
8. Rxn. with Br → CF₃PhAsBr (255).
9. Rxn. with MeNCO, followed by MeI → Me₃(3-MeNHCO₂C₆H₄)AsI (481).
10. Rxn. with Me₂NCOCl, followed by Me I → Me₃(4-Me₂NCO₂C₆H₄)AsI (481).
11. Rxn. with alc. KOH → BuAsCl₂ + C₂H₄ (674).
12. Rxn. with PhCH₂Br → Me₂(PhCH₂)₂AsBr (1313).
13. Rxn. with alc. KOH → iso-C₅H₁₁AsCl₂ + C₂H₂ (674).
14. Rxn. with BuLi, followed by R₂AsI → 2-R₂AsC₆H₄AsEt₂ (238).
15. Rxn. with (CH₂Br)₂ → Et₂Ph(BrCH₂CH₂)AsBr and (Et₂PhAsCH₂-)₂Br₂ (609).
16. Rxn. with MeNCO → Et₂(3-MeNHCO₂C₆H₄)As (481).
17. Rxn. with alc. KOH → 4-MeC₆H₄AsCl₂ + C₂H₂ (674).
18. Rxn. with alc. KOH → PhCH:CHAsCl₂ + C₂H₂ (674).
19. Rxn. with HCl in EtOH-HCCl₃, followed by NH₃ → 4-ClC₆H₄As[CH₂CH₂C(:NH)NH₂]₂ (242).
20. Rxn. with KOH in aq. alc. → 4-ClC₆H₄As(CH₂CH₂CO₂H)₂ (242).
21. Complex with HgCl₂, needles m. 120° (616).
22. Rxn. with BuLi, followed by quinoline → 2-[4-(dipropylarsino)phenyl]quinoline (480).
23. Rxn. with MeI → the methiodide (100).
24. Rxn. with KOH in aq. EtOH → hydrolysis of the CN group (811).

TABLE IV (cont'd.)

ASYMMETRICAL TERTIARY ARSINES $R_2R'As$

$R_2R'As$	Prepd. from	Method	Yield %	Props. Rxns.	Ref.
$Me_2(4\text{-}PhOC_6H_4)As$	$Me_2AsI + 4\text{-}PhOC_6H_4MgBr$	IV		b. 189-190°/10mm., d^{17} 1.2690, n^{17}_D 1.6122, $[R_L]_D$ 75.09, rxns. 25, 26	271
$Ph_2(HOCH_2CH_2)As$	$Ph_2AsH + \overline{CH_2CH_2O}$	XVII		b. 240-247° (dec.), rxn.[25,27]	122
$(CH_2{:}CMeCH_2)_2PhAs$	$PhAsI_2 + CH_2{:}CMeCH_2MgX$	IV		b. 153°/13mm., d^{25} 1.120	616
$(EtO_2CCH_2)_2PhAs$	$PhAs(MgBr)_2 + BrCH_2CO_2Et$	IV	5	[Pd $PhAs(CH_2CO_2Et)_2]_2Br_2$), orange crystals, m. 110-111°	68
$[HO_2C(CH_2)_3]_2PhAs$	$[HO_2C(CH_2)_3]_2PhAsO + SO_2$	XII		m. 71-74°, di-Ag salt	242
$Me_2(n\text{-}C_{12}H_{25})As$	$Me_2AsI + n\text{-}C_{12}H_{25}MgCl$	IV	43	b. 149-150°/10mm., d^{18} 1.467	605
C_{15} $Ph_2(ClOCCH_2CH_2)As$	$Ph_2AsCH_2CH_2CO_2H + SOCl_2$ in $CHCl_3$			m. 143-144°	243
Ph_2PrAs	$(Ph_2As)_2O + PrCO_2K + (PrCO)_2O$	XIII	50	b. 178-181°/12mm.	1215
$(4\text{-}MeC_6H_4)_2MeAs$				$[Fe_2\{(4\text{-}MeC_6H_4)_2MeAs\}_3Cl_3]_2$, m. 148°	928
C_{16} $(\overline{CH{:}CHCH{:}CHN{:}C})_2\text{-}PhAs$	$PhAsCl_2 + 2\text{-pyridyl-}MgBr$	IV	10	m. 88°, derivs.[28]	806
$Ph_2(HO_2CCHMeCH_2)As$	$Ph_2AsCH_2CHMeCN + KOH$ in 50% aq. alc.			m. 82°, rxn.[29]	242
C_{17} $Ph_2(\overline{CH{:}CHCH{:}CHN{:}C})\text{-}As$	$Ph_2AsCl + 2\text{-pyridyl-}MgBr$	IV	9.4	m. 62°, derivs.[30]	806
C_{18} $(2,4,6\text{-}Br_3C_6H_2)_2\text{-}PhAs$	$PhAsCl_2 + (2,4,6\text{-}Br_3C_6H_2N_2)_2ZnCl_4$ in $Me_2CO + Fe$	X	30	m. 185°	1066
$Ph_2(4\text{-}O_2NC_6H_4)As$	$Ph(4\text{-}O_2NC_6H_4)AsCl + (PhN_2)_2ZnCl_4$ in $Me_2CO + Fe$	X	51	m. 109°, adduct[31]	908
$Ph_2(2\text{-}BrC_6H_4)As$	$2\text{-}BrC_6H_4AsCl_2 + PhMgBr$	IV		m. 101.5-103.5°, rxn.[32]	238

Compound	Preparation	Method	Yield	Properties	Ref.
$Ph_2(4\text{-}ClC_6H_4)As$	$Ph(4\text{-}ClC_6H_4)AsCl + (Ph\text{-}N_2)_2ZnCl_4$ in Me_2CO + Fe	X	62	b. 225-230°/6mm., adduct 33	908
$(4\text{-}H_2NC_6H_4)_2PhAs$	$PhAsCl_2 + 4\text{-}H_2NC_6H_4Li$ at -60°	III	91	m. 69°, rxn.[34]	472
$Ph_2(cyclo\text{-}C_6H_{11})As$	$Ph_2AsBr + cyclo\text{-}C_6H_{11}\text{-}MgI$	IV	96	b. 200-203°/4mm., rxns.[25,35]	96
$Me_2(n\text{-}C_{16}H_{33})As$	$Me_2AsI + n\text{-}C_{16}H_{33}MgCl$	IV	61	b. 200-202°/11mm.	605
C19					
$Ph_2(2\text{-}MeC_6H_4)As$	$Ph_2AsCl + 2\text{-}MeC_6H_4MgBr$	IV	72	m. 63-64°, rxn.[35]	613
$Ph_2(3\text{-}MeC_6H_4)As$	$Ph_2AsCl + 3\text{-}MeC_6H_4MgBr$	IV		b. 170-173°	473
	Ph_3As + BuLi, followed by Me_2SO_4	XX		HgCl₂ adduct, m. 201-202°, rxn. 35	473
C20					
$Ph_2(PhC:C)As$	$Ph_2AsCl + PhC:CNa$	III	73	m. 43°, b. 168°/3mm., stable to light and moisture; rxn.[36]	518
	$Ph_2AsCl + PhC:CMgBr$	IV		m. 92°, HgCl₂-adduct	518
$Ph_2(2\text{-}MeO_2CC_6H_4)As$	$Ph(2\text{-}MeO_2CC_6H_4)AsCl + ZnCl_2$ in Me_2CO + Fe	X		m. 231-232° (dec.)	908
$(4\text{-}MeC_6H_4)_2(4\text{-}ClC_6H_4)As$	$(4\text{-}MeC_6H_4)_2AsCl + 4\text{-}ClC_6H_4MgI$	IV		m. 125-126.5°, rxn.[25]	805
$(4\text{-}MeOC_6H_4)_2PhAs$	$PhAsCl_2 + (4\text{-}MeOC_6H_4\text{-}N_2)_2ZnCl_4$ in Me_2CO + Fe	X	40	m. 76-79°	1066
C21					
$Ph_2[4\text{-}(MeO)_2MeSi\text{-}C_6H_4]As$	$Ph_2AsC_6H_4Br + Mg +$	XX	59.7	b. 180-200°/0.45mm., n_D^{25} 1.243	1120
C22					
$(1\text{-}C_{10}H_7)_2(CH:C)As$	$(1\text{-}C_{10}H_7)_2AsCl + HC:C\text{-}MgBr$	IV		m. 123°, stable to moisture, rxn. 37	518

25. Rxn. with alkyl halides → the corresp. arsonium halides (96, 122, 242, 271, 805).
26. Rxn. with I → $Me_2(4\text{-}PhOC_6H_4)AsI_2$ (271).
27. Rxn. with $SOCl_2$ in $CHCl_3$ → Ph_2AsCl (122).
28. Dihydrochloride, m. 146-147°; dipicrate, m. 142-143°; dimethiodide, m. 193-195°; dimethopicrate, m. 190-191° (806).
29. Rxn. with HgO → $Ph_2(HO_2CCHMeCH_2)AsO$ (243).
30. Methiodide, m. 160-162°; dimethopicrate·H_2O, m. 153° (806).
31. HgCl₂-adduct, m. 145-146° (908).
32. Rxn. with BuLi in petr. ether, followed by Me_2AsI → $2\text{-}MeAsC_6H_4AsPh_2$ (238).

TABLE IV (cont'd.)
ASYMMETRICAL TERTIARY ARSINES $R_2R'As$

$R_2R'As$	Prepd. from	Method	Yield %	Props. Rxns.	Ref.
$Ph_2(1\text{-}C_{10}H_7)As$	$Ph_2AsBr + 1\text{-}C_{10}H_7MgBr$	IV		m. 110-111°, rxns. 25,35	96
$Ph_2(2\text{-}C_{10}H_7)As$	$Ph_2AsBr + 2\text{-}C_{10}H_7MgBr$	IV		m. 90-91°	96
C_{26}					
$(1\text{-}C_{10}H_7)_2PhAs$	$PhAsCl_2 + 1\text{-}C_{10}H_7MgBr$	IV		m. 205-206, rxn.38	96
$[PhCH_2O(CH_2)_3]_2\text{-}PhAs$	$PhAsCl_2 + PhCH_2O(CH_2)_3\text{-}MgCl$	IV		PdCl₂ complex, golden-yellow plates, m. 69°, rxn.25	242
C_{28}					
$(1\text{-}C_{10}H_7)_2(PhC\text{:}C)As$	$(1\text{-}C_{10}H_7)_2AsCl + PhC\text{:}C\text{-}Na$	III		m. 142°, stable to moisture	518
	$(1\text{-}C_{10}H_7)_2AsCl + PhC\text{:}C\text{-}MgBr$	IV		rxn.36	518
C_{34}					
$(AcNHC_6H_4SO_2NHC_6H_4)_2PhAs$	$(4\text{-}H_2NC_6H_4)_2PhAs + 4\text{-}AcNHC_6H_4SO_2Cl$ in C_5H_5N	XX	96	m. 184°	472

33. HgCl₂-adduct, m. 178° (908).
34. Rxn. with RSO_2Cl in C_5H_5N → $(4\text{-}RSO_2NHC_6H_4)_2AsPh$ (472).
35. Rxn. with $KMnO_4$ in Me_2CO → the corresponding arsine oxide (96, 473, 613).
36. Rxn. with aq. alk. → $PhC\text{:}CH$ (518).
37. Rxn. with aq. alk. → $CH\text{:}CH$ (518).
38. Rxn. with $KMnO_4$ in Me_2CO → $(1\text{-}C_{10}H_7)_2PhAs(OH)_2$ (96).

TABLE V
ASYMMETRICAL TERTIARY ARSINES RR'R"As

RR'R"As	Prepd. from	Method	Yield %	Props., Rxns., Remarks	Ref.
C_6 MeEtPrAs	EtPrAsI + MeMgI	IV	37	b. 129-130° MeEtPrAs·2HgCl₂, m. 171-172°, de- comp. 188-189°; MeEtPrAs·HgCl₂, m. 137-138°; rxns.[1],[2]	193
	EtPrAsO₂H + Penicillium brevicaule	XIX			193
C_7 MeEt(EtO₂CCH₂)As	MeEtAsI + BrCH₂CO₂Et + Zn	I	40	Colorless liquid of bad odor, b. 83-85°/11mm.	630
C_8 (CF₃)MePhAs	MePhAsI + CF₃I + Hg	I	67	b. 186°, IR spectrum, rxns.[3],[4]	255
C_9 Me(ClCH:CH)PhAs	Me(ClCH:CH)AsCl + Ph-MgBr	IV		Light yellow oil of unpleasant odor	261
	Ph(ClCH:CH)AsCl + Me-MgI	IV			263
MeEtPhAs EtPr(EtO₂CCH₂)As	EtPrAsI + EtO₂CCH₂MgBr	IV	40.1	Rxn. with PbCl₄ → MeEtPhAsCl₂; b. 122-124°, d16 1.2359, powerful sternutator; addn. compd.[5]; rxn.[6]	144 630
C_{10} Me(NCCH₂CH₂)PhAs	MePhAsH + CH₂:CHCN	XV	85	Rxn.[7]; b. 117°/0.5mm., b. 172-174°/14mm., darkens on exposure to air	811 810
Me(HO₂CCH₂CH₂)PhAs	MePhAsCH₂CH₂CN + alc. KOH		99	b. 138°/0.3mm., m. 32°, rxn.[8]	811
C_{11} Me(NCCH₂CH₂)(3-Me-OC₆H₄)As	Me(3-MeOC₆H₄)AsH + CH₂:CHCN at 100°	XV	90	Odorless oil, b. 148-152°, rxn.[7]	811

1. Rxn. with PhCH₂Cl, followed by picric acid → MeEtPr(PhCH₂)AsCl → MeEtPr(PhCH₂)AsOC₆H₂(NO₂)₃ (193).
2. Rxn. with picric acid → the picrate (193).
3. Rxn. with alk. solns. → hydrolysis and CHF₃ formation (255).
4. Rxn. with Br → (CF₃)PhAsBr (?) (255).

TABLE V (cont'd.)

ASYMMETRICAL TERTIARY ARSINES RR'R"As

RR'R"As	Prepd. from	Method	Yield %	Props., Rxns., Remarks	Ref.
Me(HO$_2$CCH$_2$CH$_2$)(3-MeOC$_6$H$_4$)As	Me(3-MeOC$_6$H$_4$)AsCH$_2$CH$_2$-CN + alc. KOH		97	b. 166-173°/0.02mm., b. 138-146°/0.003mm., rxn.[9]	811
Me(EtO$_2$CCH$_2$)PhAs	MePhAsI + EtO$_2$CCH$_2$Mg-Br	IV	50	Colorless liq. of bad odor, b. 146-148°/10mm.	630
C$_{13}$					
Et(EtO$_2$CCH$_2$CH$_2$)PhAs	EtPhAsMgI + EtO$_2$CCH$_2$-CH$_2$I	IV	44		
EtBu(4-MeC$_6$H$_4$)As	Et(4-MeC$_6$H$_4$)AsI + Bu-MgBr	IV		b. 152-154°/17mm., d^{20} 1.0991, n$^{20}_D$ 1.5390, dipole moment	716
C$_{14}$					
MePh(2-HO$_2$CC$_6$H$_4$)As	MePh(2HO$_2$CC$_6$H$_4$)AsS + SO$_2$	XII		m. 149-151, salts[10]	623
MePh(4-HO$_2$CC$_6$H$_4$)As	MePh(4-HO$_2$CC$_6$H$_4$)AsO + SO$_2$	XII			623
MePh(3-MeC$_6$H$_4$)As	MePhAsI + 3-MeC$_6$H$_4$Br + Mg	IV		b. 165-166°/9mm., d^0 1.2671, d$^{11}_1$ 1.2563; n^{18} 1.6199; salts,[11] rxn.[12]	623
MePh(2-MeC$_6$H$_4$)As	MePhAsI + 2-MeC$_6$H$_4$Br + Mg	IV		b. 162-163°,D d^{18} 1.2595, d^0 1.2765, n$^{18}_D$ 1.6210, rxn.[12]	623
MePh(4-MeC$_6$H$_4$)As	MePhAsI + 4-MeC$_6$H$_4$Br + Mg	IV		b. 166-167mm., rxn.[12]	623
C$_{15}$					
(NCCH$_2$CH$_2$)Ph(2-H$_2$-NC$_6$H$_4$)As	Ph(2-H$_2$NC$_6$H$_4$)AsH + C-H$_2$:CHCN	XV	46	b. 197-200°/0.35mm., turns red on exposure to air	243
EtPh(4-HO$_2$CC$_6$H$_4$)As	EtPh(4-HO$_2$CC$_6$H$_4$)AsO + SO$_2$	IV		m. 124-125°; salts[13]	624
EtPh(4-MeC$_6$H$_4$)As	EtPhAsI + 4-MeC$_6$H$_4$Br + Mg	IV		b. 173.5-174.5°, d$^{20}_0$ 1.2188, n$^{19}_D$ 1.611, rxn.[12]	624
C$_{18}$					
Ph(4-BrC$_6$H$_4$)(4-Cl-C$_6$H$_4$)As	Ph(4-BrC$_6$H$_4$)AsCl + 4-ClC$_6$H$_4$MgI	IV	85	b. 64-65°/0.5mm., rxns.[14,15]	805

	Reaction	Method	Yield (%)	Properties	Ref.
C_{19}					
$Ph(4\text{-}BrC_6H_4)(4\text{-}MeC_6H_4)As$	$Ph(4\text{-}BrC_6H_4)AsCl + (4\text{-}MeC_6H_4N_2)_2ZnCl_4 + Fe$	X	52	Adduct with $HgCl_2$, m. 196-197°	908
$Ph(4\text{-}ClC_6H_4)(MeC_6H_4)As$	$Ph(4\text{-}ClC_6H_4)AsCl + (4\text{-}MeC_6H_4)_2 \cdot ZnCl_2 + Fe$	X	38	Viscous liq. $HgCl_2$ adduct, m. 188-189°	908
	$Ph(4\text{-}MeC_6H_4)AsCl + 4\text{-}ClC_6H_4MgI$	IV			
$Ph(4\text{-}O_2NC_6H_4)(4\text{-}MeC_6H_4)As$	$Ph(4\text{-}MeC_6H_4)AsCl + (4\text{-}O_2NC_6H_4N_2)FeCl_4$	X	42	$HgCl_2$ adduct, m. 193-194°	908
$Et(1\text{-}C_{10}H_7)(4\text{-}MeC_6H_4)As$	$Et(4\text{-}MeC_6H_4AsI + 1\text{-}C_{10}H_7MgBr$	IV	62.5	b. 235-236°/10mm., d_0^{16} 1.2454, n_D^{16} 1.6623, rxns.17,18	627
$Et(2\text{-}C_{10}H_7)(4\text{-}MeC_6H_4)As$	$Et(4\text{-}MeC_6H_4)AsI + 2\text{-}C_{10}H_7MgBr$	IV	55.9	b. 214-215°/8mm., n_D^{20} 1.6572, rxns.19,20	627
C_{20}					
$Ph(4\text{-}BrC_6H_4)(2\text{-}MeO_2CC_6H_4)As$	$Ph(2\text{-}MeO_2CC_6H_4)AsCl + (4\text{-}BrC_6H_4N_2)_2 \cdot ZnCl_2 + Fe$	X	79		908
C_{21}					
$Ph(4\text{-}MeC_6H_4)(2\text{-}MeO_2CC_6H_4)As$	$Ph(2\text{-}MeO_2CC_6H_4)AsCl + (4\text{-}MeC_6H_4N_2)_2 \cdot ZnCl_4 + Fe$	X	52	$HgCl_2$ adduct, m. 192°	908

5. Addn. compd. with Cu_2Br_2, m. 106-106.5° (630).
6. Rxn. with 10% HCl → non-crystallizable syrup (630).
7. Rxn. with alc. KOH → hydrolysis of the CN group (810, 811).
8. Rxn. with H_2SO_4 at 100° → $Me(3\text{-}HO_3SC_6H_4)AsCH_2CH_2CO_2H$ (?), m. 158-159° (811).
9. Rxn. with P_2O_5 in xylene → 1,2,3,4-tetrahydro-7-methoxy-1-methyl-4-oxoarsinoline (811).
10. Salts: NH_4, sol. in H_2O; strychnine salt, m. 183-185°; quinine salt, m. 210-211° (623).
11. Salts with: strychnine, m. 138-139°, $[\alpha]_D$18-19,8°; quinine, m. 169-170° $[\alpha]_D$17-83,01° (623).
12. Rxn. with $KMnO_4$ → the arsine oxide (623).
13. Salts with: strychnine, m. 204-205°; quinine, m. 182-183° (624).
14. Rxn. with Br_2 in AcOH, followed by NaOH → the arsine oxide (805).
15. Rxn. with RBr in the presence of $AlCl_3$ → $RPh(4\text{-}BrC_6H_4)(4\text{-}ClC_6H_4)AsBr$ (805).
16. Rxn. with RBr → $RPh(4\text{-}ClC_6H_4)(4\text{-}MeC_6H_4)AsBr$ (805).
17. Rxn. with $KMnO_4$, followed by H_2S → $Et(1\text{-}C_{10}H_7)(4\text{-}HO_2CC_6H_4)AsS$ (627).
18. Rxn. with $PhCH_2Br$ → $Et(1\text{-}C_{10}H_7)(4\text{-}MeC_6H_4)(PhCH_2)AsBr$ (627).
19. Rxn. with $KMnO_4$, followed by H_2S → $Et(2\text{-}C_{10}H_7)(4HO_2CC_6H_4)AsS$ (627).
20. Rxn. with $PhCH_2Br$ → $Et(2\text{-}C_{10}H_7)(4\text{-}MeC_6H_4)(PhCH_2)AsBr$ (627).

TABLE V (cont'd.)

ASYMMETRICAL TERTIARY ARSINES RR'R"As

RR'R"As	Prepd. from	Method	Yield %	Props., Rxns., Remarks	Ref.
Ph(4-EtC$_6$H$_4$)(4-MeC$_6$H$_4$)As	Ph(4-MeC$_6$H$_4$)AsCl + 4-EtC$_6$H$_4$MgBr	IV		m. 73°, b. 196-198°/0.2mm., rxns. 21, 22	805
C$_{24}$ Ph(4-PhC$_6$H$_4$)(4-ClC$_6$H$_4$)As	Ph(4-PhC$_6$H$_4$)AsCl + 4-ClC$_6$H$_4$MgI	IV	90	b. 279-290°/0.3mm., rxns. 14,23	805

21. Rxn. with Br in AcOH, followed by NaOH → [Ph(4-EtC$_6$H$_4$)(4-MeC$_6$H$_4$)As(OH)]$_2$O (805).
22. Rxn. with RBr → RPh(4-EtC$_6$H$_4$)(4-MeC$_6$H$_4$)AsBr (805).
23. Rxn. with RBr → RPh(4-PhC$_6$H$_4$)(4-ClC$_6$H$_4$)AsBr (805).

DI- AND TRITERTIARY ARSINES

The arsines are prepared by:

1. Reacting alkylene- or arylene-bis(dihaloarsines) or -bis(haloörganoarsines), ($X_2AsYAsX_2$ or $XRAsYAsRX$, where Y is an aliphatic or aromatic divalent radical) with Grignard compounds in an inert atmosphere.
2. Reacting (diorganoarsino)aryl halides, R_2AsYX (where Y is an arylene radical and X a halogen), with magnesium, and treating the resulting Grignard compounds with halodiorganoarsines, R_2AsX.
3. Reacting ethynylenebis(magnesium halide), $(\dot{c}CMgX)_2$, with halodiorgano-arsines, R_2AsX, in a suitable solvent.
4. Reacting arylenebis(magnesium halides) with halodiorganoarsines.
5. Treating (diorganoarsino)aryl halides with alkyllithium compounds and reacting the resulting (diorganoarsino)aryllithium derivatives with halodiorganoarsines.
6. Reacting arylenedilithium with halodiorganoarsines.
7. Reacting sodium acetylide with halodiorganoarsines in liquid ammonia.
8. Heating secondary arsines, R_2AsH, with alkylene oxides, $R\dot{C}HCH\dot{O}$.
9. Reducing alkylenebis(diorganoarsine) oxides with sulfur dioxide.
10. Treating (diorganoarsino)alkyl halides with magnesium and reacting the resulting Grignard compounds, $R_2As(CH_2)_nMgX$, with dihaloörganoarsines, $RAsX_2$, to form tertiary arsines containing three arsine groups.
11. Heating at 125° bis(diorganoarsenic) oxides, $(R_2As)_2O$, with alkali metal carboxylates (RCO_2M) and the corresponding carboxylic anhydrides, $(RCO)_2O$; tertiary arsines with two arsine groups are formed as by-products.

o-PHENYLENEBIS(DIMETHYLARSINE) $C_{10}H_{16}As_2$ o-$C_6H_4(AsMe_2)_2$
Prepn.: By reacting o-$C_6H_4(AsCl_2)_2$ with MeMgI (1:6) in Et_2O under N (199) or under A, 87.6% yield (327).
By treating 2-$BrC_6H_4AsMe_2$ with BuLi in petr. ether under reflux and reacting the resulting Li deriv. with Me_2AsI (612).
Props.: Pale yellow liquid, m. -12°, b. 153-158°/20mm., n_D^{20} 1.6204, can be stored under A (327); b. 150-156°/15mm. (612); b. 156°/20mm., sensitive to O (199).
Rxns. with: $Br(CH_2)_nBr \rightarrow [Me_2As\overline{C_6H_4AsMe_2(CH_2)_n}]Br_2$ (489, 807).
o-$C_6H_4(CH_2Br)_2 \rightarrow [Me_2AsC_6H_4As(Me)_2CH_2C_6H_4CH_2]Br_2$ (610).
Cold $HNO_3 \rightarrow$ the di(hydroxy nitrate) anhydride (610).
Complex compds. (DA in the following formulae stands for o-$C_6H_4(AsMe_2)_2$.
[Sb(DA)Cl_3], cond. and mol. wt. (1179b).
[As(DA)Cl_3], cond. and mol. wt. (1179b).
[Bi(DA)Cl_3], cond. and mol. wt. (1179b).
[Bi(DA)_3Cl_3](ClO_4)_3, cond. and mol. wt. (1179b).
[Cr(DA)Cl_3·H_2O], blue powder, decomp. 240-245° (943).
[Cr(DA)Br_3·H_2O] (943).
[Cr(DA)Cl_2(EtOH)(H_2O)]Cl, magnetic moment and bonding (151).
[Cr(DA)_2Cl_2]ClO_4, pale green ppt. (943).
[Cr(DA)_2Br_2]ClO_4, (943).

[Cr(DA)$_2$I$_2$]ClO$_4$, pale green ppt. (943).
[Cr(DA)$_2$Cl$_2$][Cr(DA)Cl$_4$], green powder (943).
[Cr(DA)$_2$Br$_2$][Cr(DA)Br$_4$] (943).
[Cr(DA)$_2$I$_2$][Cr(DA)I$_4$] (943).
[Cr(DA)(CO)$_4$], pale yellow, m. 170° (vacuo), diamagnetic, monomeric in C$_6$H$_6$, poor
  electrolyte (919, 921).
[Cr(DA)$_2$(CO)$_2$], yellow crystals, m. 220° (dec.), diamagnetic, monomeric in C$_6$H$_6$,
  poor electrolyte (919); m. 226° (dec.) (921).
[Co(DA)$_3$](ClO$_4$)$_2$, magnetic moment and bonding (151).
[Co(DA)$_3$](ClO$_4$)$_3$, magnetic moment and bonding (151).
[Co(DA)$_2$Cl$_2$](ClO$_4$), greenish blue ppt. (935), Magnetic moment and bonding (151).
[Co(DA)$_2$]Cl$_2$, greenish yellow powder (933, 935).
[Co(DA)$_2$]Br$_2$, brownish yellow ppt. (935).
[Co(DA)$_2$]I$_2$, brown ppt. (933, 935).
[Co(DA)$_2$Cl$_2$]Cl, bright green needles (935).
[Co(DA)$_2$Br$_2$]Br, dark green, rhombic crystals (935).
[Co(DA)$_2$I$_2$]I, black crystals (933, 935).
[Co(DA)$_2$Br$_2$]ClO$_4$, light green ppt. (935).
[Co(DA)$_2$](SCN)$_2$, gray ppt. (935).
[Co(DA)$_2$(SCN)$_2$]ClO$_4$, orange ppt. (935).
[Co(DA)$_2$[CoCl$_4$], green crystals, (935).
[Co(DA)$_2$(SCN)$_2$]$_2$[Co(SCN)$_4$], greenish black crystals (935).
[Cu(DA)$_2$]Br, ionization equil. in C$_6$H$_6$ (139); m. 182° (618).
[Cu(DA)$_2$]I, m. 189° (618).
[Cu(DA)$_2$]ClO$_4$, m. 300° (dec.) (618).
[Cu(DA)$_2$][CuCl$_2$], m. 266° (618).
[Cu(DA)$_2$][CuBr$_2$], m. 232° (618); ionization equil. in C$_6$H$_6$ (139).
[Cu(DA)$_2$][CuI$_2$], m. 174° (618); ionization equil. in C$_6$H$_6$ and PhNO$_2$ (140).
[Au(DA)$_2$]ClO$_4$, white ppt. prepd. by treating HAuCl$_3$ with (DA) in aq. alc. and
  reducing with H$_3$PO$_2$, explodes >300° (509, 512, 937).
[Au(DA)$_2$]Br, yellow crystals, m. 226° (509, 512, 937).
[Au(DA)$_2$]I (509, 512).
[Au(DA)$_2$]OC$_6$H$_2$(NO$_2$)$_3$ (509, 512).
[Au(DA)$_2$][AuI$_2$] (509, 512).
[Au(DA)$_2$][CuI$_2$] (509, 512).
[Au(DA)$_2$I$_2$]I, dark red compd., m. 246°, 6-coördinated Au(III) (509, 512, 937).
[Au(DA)$_2$I$_2$]ClO$_4$, dark red (509, 512).
[Au(DA)$_2$I](ClO$_4$)$_2$, red compd. with 5-coördinated Au(III) (509, 512).
[Au(DA)$_2$NO$_3$](NO$_3$)$_2$, colorless (509, 512).
[Au(DA)$_2$](ClO$_4$)$_3$, white (509, 512).
[Fe(DA)$_2$Br$_2$] (931); magnetic moment and bonding (151).
[Fe(DA)$_2$I$_2$] (931).
[Fe(DA)$_2$(SCN)$_2$] (931).
[Fe(DA)$_2$Br$_2$]Br (931).
[Fe(DA)$_2$Cl$_2$]ClO$_4$, decomp. 210° (931); magnetic moment and bonding (151).
[Fe(DA)$_2$Cl$_2$][FeCl$_4$], m. 219° (931, 942).
[Fe(DA)$_2$Cl$_2$][FeCl$_4$]$_2$, black ppt. (942).
[Fe(DA)$_2$Br$_2$][FeBr$_4$], m. 209° (931).
[Fe(DA)(CO)$_3$], golden yellow crystals, m. 131° (in vacuo), diamagnetic, monomeric
  in C$_6$H$_6$, poor electrolyte (919, 921).

[Fe(DA)$_2$(CO)], yellow crystals, m. 150° (dec.), diamagnetic, monomeric in C$_6$H$_6$, poor electrolyte (919, 920).

[Fe(DA)(CO)$_2$I], brown yellow, diamagnetic crystalline powder (919, 920).

[Fe(DA)(CO)$_2$I$_2$], brown, shiny microcrystals, m. 186°, diamagnetic, monomeric in C$_6$H$_6$, poor electrolyte (919, 920).

[Fe(DA)$_2$I$_2$], brown yellow ppt. (920).

[Fe(DA)(CO)$_2$Br], orange yellow crystals, m. 225° (dec.), diamagnetic, monomeric in C$_6$H$_6$, poor electrolyte (919, 920).

[Mn(DA)$_2$Cl$_2$] (?), white crystals (944).

[Mn(DA)$_2$Br$_2$], white, rhombic crystals, decomp. 196° (944).

[Mn(DA)$_2$I$_2$], pale yellow crystals, decomp. 194° (944).

[Mn(DA)$_2$(H$_2$O)Cl$_2$], purplish red crystals, m. 202° (944).

[Mn(DA)$_2$(H$_2$O)Br$_2$]ClO$_4$, green crystals (944).

[Mo(DA)(CO)$_4$], white crystals, m. 158° (in vacuo), diamagnetic, monomeric in C$_6$H$_6$, poor electrolyte (919, 921).

[Mo(DA)$_2$(CO)$_2$], yellow crystals, m. 180° (dec.), diamagnetic, monomeric in C$_6$H$_6$, poor electrolyte (919); m. 231° (in vacuo) (921).

[Mo(DA)(CO)$_3$I$_2$], golden yellow crystals, decomp. ~200° (922).

[Mo(DA)(CO)$_3$Br$_2$], orange ppt., decomp. ~200° (922).

[Mo(DA)Br$_4$], orange-brown crystals, m. 210° (in vacuo) (922).

[Mo(DA)$_2$(CO)$_2$I]I, pale yellow crystals, m. 195° (in vacuo), electrolyte (922).

[Mo(DA)$_2$(CO)$_2$Br]Br, pale yellow crystals, m. 200° (in vacuo), electrolyte (922).

[Mo(DA)(CO)$_2$Br$_4$], yellow ppt., m. 210° (in vacuo) (922).

[Ni(DA)$_2$](SCN)$_2$, reddish brown (934).

[Ni(DA)$_2$]Br$_2$, brown (934); [Ni(DA)$_2$]Cl$_2$, red ppt. (933, 934).

[Ni(DA)$_2$]I$_2$, dark brown (934); decomp. ~140° (940).

[Ni(DA)$_2$](ClO$_4$)$_2$. buff-pink (934).

[Ni(DA)$_2$I$_2$], red ppt. (940).

[Ni(DA)I$_4$], black crystals. (940).

[Ni(DA)$_2$Cl]ClO$_4$ (510).

[Ni(DA)$_2$Br]ClO$_4$ (510).

[Ni(DA)$_2$I]ClO$_4$ (510).

[Ni(DA)(CO)$_2$], colorless crystals, m. 125° (940). Structure (941). Dipole moment (213a).

[Ni(DA)$_2$I$_2$]I, black ppt. (934), [Ni(DA)$_2$Cl$_2$]Cl, greenish yellow crystals (933, 934). Magnetic moment and bonding (151).

[Ni(DA)$_3$](ClO$_4$)$_2$, brown ppt. (934).

[Ni(DA)$_2$Br$_2$]Br, orange, (934).

[Ni(DA)$_2$(SCN)$_2$]SCN, orange brown (934).

[Ni(DA)$_2$Cl$_2$]ClO$_4$, yellow (934).

[Ni(DA)$_2$Cl$_2$](ClO$_4$)$_2$, quadrivalent Ni (938).

[Ni(DA)$_2$Br$_2$](ClO$_4$)$_2$, deep green (938).

[Ni(DA)$_2$Cl$_2$]$_2$[PtCl$_6$], yellow ppt. (934).

[Ni(DA)Br$_2$] (?), pink powder (940).

[Ni(DA)Br$_3$] (?), brown ppt. (940).

[Os(DA)$_2$Cl$_2$], pale yellow ppt.; stable up to 360° (946).

[Os(DA)$_2$Br$_2$], fawn-brown crystals, stable to above 360° (946).

[Os(DA)$_2$I$_2$], cinnamon-brown ppt., stable to above 360° (946).

[Os(DA)$_2$(SCN)$_2$], light brown powder, stable to above 360° (946).

[Os(DA)$_2$Cl$_2$]Cl, red, monoclinic crystals, decomp. 255° (946).

[Os(DA)₂Br₂]Br, purple-blue crystals, decomp. 304° (946).

$[Os(DA)_2Br_2]Br$, purple-blue crystals, decomp. 304° (946).
$[Os(DA)_2I_2]I$, blue-green ppt., decomp. 288° (946).
$[Os(DA)_2Cl_2]ClO_4$, red powdery crystals, decomp. 293° (explosively) (946).
$[Os(DA)_2Br_2]ClO_4$, purple-blue crystals, decomp. 302° (946).
$[Os(DA)_2I_2]Br$, blue-green ppt., decomp. 288° (946).
$[Os(DA)_2I_2]ClO_4$, decomp. 273° (946).
$[Os(DA)_2(SCN)_2]ClO_4$, decomp. 265° (explos.) (946).
$[Os(DA)_2Cl_2](ClO_4)_2$, blue-black crystals, decomp. 160° (946).
$[Os(DA)_2Br_2](ClO_4)_2$, black shiny crystals, decomp. 194° (946).
$[Os(DA)_2I_2](ClO_4)_2$, dark brown crystals, decomp. 210° (946).
$[Pd(DA)_2]Cl_2$, deep yellow crystals, ionic compd., forms pale yellow almost colorless hydrate (455).
$[Pd(DA)_2Cl]Cl$ (511).
$[Pd(DA)_2Cl_2]$, greenish yellow crystals, non-ionic (455).
$[Pd(DA)_2][PdCl_4]$, deep red crystals (455).
$[Pd(DA)_2Cl]ClO_4$ (510, 511).
$[Pd(DA)_2Br]Br$ (511).
$[Pd(DA)_2]Br_2$, m. 182° (523).
$[Pd(DA)_2Br]ClO_4$ (510, 511).
$[Pd(DA)_2I]I$ (511).
$[Pd(DA)_2I]ClO_4$ (510, 511).
$[Pd(DA)_2(SCN)](SCN)$ (511).
$[Pd(DA)_2NO_2]NO_2$ (511).
$[Re(DA)_2Cl_2]$, yellow-brown (256a),
$[Re(DA)_2Br_2]$, green (256a).
$[Re(DA)_2I_2]$, green (256a).
$[Re(DA)_2Cl_2]ClO_4$, golden-yellow (256a).
$[Re(DA)_2Br_2]ClO_4$, orange (256a).
$[Re(DA)_2I_2]ClO_4$, red (256a).
$[Re(DA)_2Cl_4]ClO_4$, purple substance, prepd. by oxidation of $[Re(DA)_2Cl_2]Cl$; diamagnetic uni-valent electrolyte; redn. with Ti(III) yields the trivalent Re deriv.
  (376b).
$[Re(DA)_2Br_4]Br_3$, green substance (376b).
$[Rh(DA)_2Cl_2]Cl$ (932),
$[Rh(DA)_2Cl_2]Cl$ (932).
$[Rh(DA)_2Br_2]Br$ (932).
$[Rh(DA)_2I_2]I$ (932).
cis-$[Ru(DA)_2Cl_2]$ was reduced by $LiAlH_4$ to trans-$[Ru(DA)_2HCl]$ (215a).
cis-$[Ru(DA)_2Br_2]$ was reduced by $LiAlH_4$ to trans-$[Ru(DA)_2HBr]$ (215a).
cis-$[Ru(DA)_2I_2]$ was reduced by $LiAlH_4$ to trans-$[Ru(DA)_2HI]$ (215a).
trans(?)-$[Ru(DA)_2Cl_2]$, pale yellow crystals, stable to above 360° (945).
trans(?)-$[Ru(DA)_2Br_2]$, orange pink, acicular crystals, stable to above 360° (945).
trans(?)-$[Ru(DA)_2I_2]$, occurs in two modifications; one is orange and the other
  purple. The latter isomerizes at 370° to the orange form (945).
trans(?)-$[Ru(DA)_2(SCN)_2]$, silvery gray crystals (945).
$[Ru(DA)_2Cl_2]Cl$, green ppt., decomp. ~200° (945).
$[Ru(DA)_2Cl_2]ClO_4$, m. 265° (945).
$[Ru(DA)_2Br_2]Br$, green crystals, m. 230° (dec). (945).
$[Ru(DA)_2I_2]ClO_4$, sage-green crystals (945).
$[Ru(DA)_2Cl_2][Ru(DA)Cl_4]$, bluish green crystals, decomp. 233° (945),

[Ru(DA)$_2$Br$_2$][Ru(DA)Br$_4$], bright green crystals, decomp. 258° (945).
[Tc(DA)$_2$Cl$_2$]Cl, orange substance; magnetic, cond., and spectrum data (376c).
[Tc(DA)$_2$Br$_2$]Br, red substance; magnetic, cond., and spectrum data (376c).
[Tc(DA)$_2$I$_2$]I, deep red to black; magnetic, cond., and spectrum data (376c).
[Tc(DA)$_2$I$_2$], brown substance; magnetic, cond., and spectrum data (376a).
[Sn(DA)Cl$_4$], m. 214-224° (12).
[Ti(DA)Br$_4$], golden cryst.plates; cond. and mol. wt. (1179c).
[Ti(DA)Cl$_4$], golden cryst. plates; cond. and mol. wt. (1179c).
[Ti(DA)(H$_2$O)Br$_3$], red powder; cond. and mol. wt. (1179c).
[Ti(DA)(H$_2$O)Cl$_3$], red powder; cond. and mol. wt. (1179c).
[W(DA)(CO)$_4$], pale yellow, m. 168° (919); m. 186° (in vacuo) (921).
[W(DA)$_2$(CO)$_2$], yellow ppt., m. 200°(dec.) (919); m. 237° (in vacuo) (921).

p-PHENYLENEBIS(DIMETHYLARSINE) C$_{10}$H$_{16}$As$_2$ p-C$_6$H$_4$(AsMe$_2$)$_2$
Prepn.: by reacting 4-Me$_2$AsC$_6$H$_4$MgBr with Me$_2$AsI, 91% yield (480).
Props.: B. 146-148°/16 mm. (480); b. 99-101°/0.5 mm. (610).

ETHYNYLENEBIS(DIETHYLARSINE) C$_{10}$H$_{20}$As$_2$ Et$_2$AsC:CAsEt$_2$
Prepn.: by reacting Et$_2$AsBr with (:CMgBr)$_2$ in CHCl$_3$ (516).
·Props.: B. 130°/8 mm., sensitive to air (516).
Rxns. with: H$_2$O → C$_2$H$_2$ evolution (516).
MeI → [Et$_2$AsC:CAsEt$_2$Me]I (516).
BzO$_2$H → cleavage of the As-C bond (516).

METHYLBIS[3-(DIMETHYLARSINO)PROPYL]ARSINE C$_{11}$H$_{27}$As$_3$ MeAs(CH$_2$CH$_2$CH$_2$AsMe$_2$)$_2$
Prepn.: by treating ClCH$_2$CH$_2$CH$_2$AsMe$_2$ with Mg and reacting the resulting Grignard
compound with MeAsI$_2$ (55).
Props.:. Colorless oil, b. 196-198°/16 mm., readily oxidized by atm. O to a solid
(55).
Rxns. with transition metal salts in alc. → highly colored complex compds. (55).
Complex compds. (TArs in the following formulae stands for MeAs(CH$_2$CH$_2$CH$_2$AsMe$_2$)$_2$):
[Co(TArs)I$_2$], brownish black, paramagnetic solid; oxidation by air gives reddish
   black [Co(TArs)I$_3$] which is diamagnetic (55).
[Co(TArs)$_2$(ClO$_4$)$_2$], unstable, green, paramagnetic solid (55).
[Cu(TArs)I], diamagnetic, monomeric non-electrolyte (in PhNO$_2$) (55).
[Fe(TArs)(SCN)$_3$], deep green, paramagnetic solid (55).
[Ni(TArs)I$_2$], diamagnetic, monomeric non-electrolyte (in PhNO$_2$) (55).
[Ni(TArs)Br$_2$], diamagnetic, monomeric non-electrolyte (in PhNO$_2$), is converted to
   a purple powder, probably [Ni(TArs)$_2$(ClO$_4$)$_2$] by HClO$_4$ in aq. alc.; with Br in
   CHCl$_3$ yields [Ni(TArs)Br$_3$] (55).

As,As-DIETHYL-As',As'-DIMETHYL-o-PHENYLENEDIARSINE C$_{12}$H$_{20}$As$_2$ 2-Me$_2$AsC$_6$H$_4$AsEt$_2$
Prepn.: by treating 2-BrC$_6$H$_4$AsEt$_2$ with BuLi and reacting the resulting Li deriv.
with Me$_2$AsI in petr. ether (238).
Props.: B. 97.5-98.5°/0.15 mm. (238).
Complex compds. (DArs in the following formulae stands for 2-Me$_2$AsC$_6$H$_4$AsEt$_2$):
[Au(DArs)$_2$]I, m. 157-157.5°, prepd. by reacting the diarsine with HAuCl$_3$·4H$_2$O and
   treating the rxn. mixt. with KI (238).
[Au(DArs)$_2$] (+)-camphosulfonate, m. 184.5-186°, [M]$_D$ + 54° (238).
[Au(DArs)$_2$] (+)-bromocamphosulfonate, m. 175-177.5°, gave an inactive picrate (238).

o-PHENYLENEBIS(DIETHYLARSINE) $C_{14}H_{24}As_2$ o-$C_6H_4(AsEt_2)_2$
Prepn.: by treating 2-$BrC_6H_4AsEt_2$ with BuLi in petr. ether and reacting the rxn.
prod. with Et_2AsI (238).
Props.: B. 116.5-117.0°/0.3 mm. (238).
Complex compds.: {Au[o-$C_6H_4(AsEt_2)_2]_2Br$}, 181-182°, prepd. by reacting the diar-
sine with $NaAuCl_4$ and NaH_2PO_2 and treating the prod. with NaBr (238).
{Au[o-$C_6H_4(AsEt_2)_2]_2I$}, m. 166-167°, prepd. in the same manner as the preceding
  compd., except that NaI was used instead of NaBr (238).

3,4-BIS(DIETHYLARSINO)TOLUENE $C_{15}H_{26}As_2$ 3,4-$(Et_2As)_2C_6H_3Me$
Prepn.: by reacting 3,4-$IMg(BrMg)C_6H_3Me$ with Et_2AsI, 23% yield (513a).
Props.: B. 100-110°/0.2 mm. (513a).
Complex compd.: {Cu[$MeC_6H_3(AsEt_2)_2]_2$]CuI_2, m. 229-231° (513a).

ETHYLENEBIS(METHYLPHENYLARSINE) $C_{16}H_{20}As_2$ (-$CH_2AsMePh)_2$
Prepn.: by reacting (-$CH_2AsPhCl)_2$ with MeMgI in Et_2O-C_6H_6 under N, 74% yield (609)
Props.: B. 163-164°/0.2 mm. (609).
Rxns. with: H_2O_2 → the dioxide (609).
(-$CH_2Br)_2$ → [$MePhAsCH_2CH_2AsMePhCH_2CH_2$.]$Br_2$ (609).
o-$C_6H_4(CH_2Br)_2$ → 1,4-diphenyl-1,4-diarsa-6,7-benzocycloöct-2-ene dimethobromide
(609).

2,2'-BIPHENYLENEBIS(DIMETHYLARSINE) $C_{16}H_{20}As_2$ (2-$C_6H_4AsMe_2)_2$
Prepn.: By reacting (2-$C_6H_4Li)_2$ with Me_2AsI in C_6H_6 under reflux, 63% yield (524).
By-prod. in the rxn. of 2-BrC_6H_4I with 3 equivs. of Mg, followed by 4 equivs. of
Me_2AsI (523).
Props.: M. 46-46.5°, b. 110°/0.05 mm. in N (523); b. 146-152°/0.7 mm. (524).
Rxns. with alkylene dihalides → the corresponding cyclic diarsonium derivs. (524).
Complex with $PdBr_2$, [Pd(2-$C_6H_4AsMe_2)_2Br_2$], decomp. 270° (523).

ETHYLENEBIS(DIBUTYLARSINE) $C_{18}H_{40}As_2$ $Bu_2AsCH_2CH_2AsBu_2$
Prepn.: by reacting(-$CH_2AsBuCl)_2$ with BuMgBr under H in C_6H_6-Et_2O (455).
Props.: Colorless liq., b. 161-162°/0.04 mm. (455).
Complex with $PdCl_2$, [Pd(-$CH_2AsBu_2)_2Cl_2$]; yellow crystals, m. 221° (455).

As,As-DIMETHYL-As',As'-DIPHENYL-o-PHENYLENEDIARSINE $C_{20}H_{20}As_2$ 2-$Me_2AsC_6H_4AsPh_2$
Prepn.: by treating 2-$BrC_6H_4AsPh_2$ with BuLi and reacting the prod. with Me_2AsI
(238).
Props.: M. 103-104°, b. 194-200°/1 mm. (238).
Complex compds.: [Cu($Me_2AsC_6H_4AsPh_2$)][CuI_2], m. 214-215.5°, prepd. along with
[Cu($Me_2AsC_6H_4AsPh_2)_2I$], m. 179.5-180.5, from the diarsine and Cu_2I_2 in H_2O in
the presence of KI. The latter was converted into the former by refluxing in
EtOH (238).
[Cu($Me_2AsC_6H_4AsPh_2)_2$]-(+)-bromocamphosulfonate, oil (238).

o-TERPHENYLENE-2,2"-BIS(DIMETHYLARSINE) $C_{22}H_{24}As_2$ o-$C_6H_4(C_6H_4AsMe_2)_2$
Prepn.: by-prod. in the rxn. of 2-BrC_6H_4I with 3 equivs. of Mg followed by excess
of Me_2AsI (523).
Props.: M. 92.5-93.5° (523).
Complex compds.: {Pd[2-$C_6H_4(C_6H_4$-2-$AsMe_2)_2]Br_2$}, orange, decomp. 350° (523).
{Au_2[2-$C_6H_4(C_6H_4$-2-$AsMe_2)]Cl_2$}, m. 180° (523).

ETHYLENEBIS(BUTYLPHENYLARSINE) $C_{22}H_{32}As_2$ $(-CH_2AsBuPh)_2$
Prepn.: by reacting $(-CH_2AsPhCl)_2$ with BuMgBr in C_6H_6 (199).
Props.: Colorless liq., b. 184-186°/0.06 mm. (199).
Rxns. with: MeI → dimethiodide (199).
Br_2, followed by Na_2S or H_2S → $[-CH_2AsBuPh(S)]_2$ (199).
Complex with $PdCl_2$, $[Ph(-CH_2AsBuPh)_2Cl_2]$, non-ionic compd., occurs in two modifications, the α-form, pale yellow crystals, m. 172-174°, and the β-form, m. 185-186° (455)

o-PHENYLENEBIS(DIBUTYLARSINE) $C_{22}H_{40}As_2$ $o-C_6H_4(AsBu_2)_2$
Prepn.: by reacting $o-C_6H_4(AsCl_2)_2$ with BuMgI under N (199).
Props.: B. 245-247°/20 mm. (199).
Complex compds.: $\{Pd[o-C_6H_4(AsBu_2)_2]_2\}Cl_2$, deep yellow crystals; tetrahydrate, colorless crystals (455).
$\{Pd[o-C_6H_4(AsBu_2)_2]Cl_2\}$, yellow crystals, m. 273-275°, non-ionic (455).
$\{Pd[o-C_6H_4(AsBu_2)_2]_2\}(PdCl_4)$, reddish brown ppt. (455).

METHYLENEBIS(DIPHENYLARSINE) $C_{25}H_{22}As_2$ $Ph_2AsCH_2AsPh_2$
Prepn.: By reacting $CH_2(AsCl_2)_2$ with PhMgBr in C_6H_6 under reflux, 71.4% yield (12
By heating $(Ph_2As)_2O$ with KOAc and Ac_2O at 125°; 13% yield along with Ph_2AsMe (12
By reducing $Ph_2AsCH_2As(O)Ph_2$ with SO_2 in $CHCl_3$ (245).
Props.: White needles, m. 96-97° (245); m. 97-97.5° (1215).
Rxns. with: Concd. HNO_3 → Ph_2AsO_2H (1215).
MeI → $CH_2(AsMePh_2)_2I_2$ (1215).

ETHYNYLENEBIS(DIPHENYLARSINE) $C_{26}H_{20}As_2$ $Ph_2AsC:CAsPh_2$
Prepn.: by reacting Ph_2AsCl with NaC:CH in liq. NH_3 or with $(:CMgBr)_2$ in $CHCl_3$-Et_2O (514, 516).
Props.: Needles, m. 105° (516); m. 100.5° (514).
Rxn. with aq. KOH → evolution of C_2H_2 (514, 516).

ETHYLENEBIS(DIPHENYLARSINE) $C_{26}H_{24}As_2$ $Ph_2AsCH_2CH_2AsPh_2$
Prepn.: By reacting $(-CH_2AsPhCl)_2$ with PhMgBr in C_6H_6 (199).
By reacting Ph_2AsH with $\overline{CH_2CH_2O}$ at 130° in a sealed tube under N (122).
Props.: Colorless crystals, m. 99-102° (199); m. 100-103°, stable to H_2O, sol. in org. solvents (122).
Complex with $PdCl_2$: $[Pd(Ph_2AsCH_2CH_2AsPh_2)Cl_2]$, yellow crystals, decomp. at highe temps. without melting (455).

ETHYNYLBIS(DICYCLOHEXYLARSINE) $C_{26}H_{44}As_2$ $[:CAs(C_6H_{11})_2]_2$
Prepn.: by reacting $(C_6H_{11})_2AsCl$ with $(:CMgBr)_2$ in $CHCl_3$ (516).
Props.: Rhombic crystals, m. 72° (516).
Rxns. with: Aq. KOH → evolution of C_2H_2 (516).
BzO_2H → cleavage of the As-C bonds (516).
MeI → $[(C_6H_{11})_2AsC:CAsMe(C_6H_{11})_2]I$ (516).

ETHYNYLENEBIS(DI-1-NAPHTHYLARSINE) $C_{42}H_{28}As_2$ $(1-C_{10}H_7)_2AsC:CAs(C_{10}H_7-1)_2$
Prepn.: by reacting $(1-C_{10}H_7)_2AsCl$ with NaC:CH in liq. NH_3 or with $(:CMgBr)_2$ in $CHCl_3$ (516)
Props.: Needles, m. 232° (516).
Rxns. with: $AgNO_3$ → cleavage of the As-C bond (516).
BzO_2H → $[CAs(O)(1-C_{10}H_7)_2]_2$ (516).

CYANO- AND THIOCYANATOARSINES

Cyano- and thiocyanatoarsines, $RAs(CN)_2$, R_2AsCN, $RR'AsCN$, R_2AsCNS, and $RR'AsCNS$, are prepared by:

1. Reacting organohaloarsines, $RAsX_2$, R_2AsX or $RR'AsX$, with silver, sodium, or potassium cyanide or thiocyanate in the presence or absence of a suitable solvent (Method I).
2. Treating bis(diarylarsenic) oxides with anhydrous HCN (Method II).
3. Reacting tertiary arsines, R_2AsR', with cyanogen bromide, BrCN, and subjecting the reaction mixture to dry distillation (Method III).

The dicyanoarsines are compiled in Table VI and cyano- and thiocyanotoarsines in Table VII.

TABLE VI
DICYANOARSINES

$RAs(CN)_2$	Prepd. from	Method	Yield %	Props. and Rxns.	Ref.
$PrAs(CN)_2$	$PrAsCl_2 + AgCN$	I	72	m. 82-86°	44
$OCH:CHCH:CAs(CN)_2$	$OCH:CHCH:CAsCl_2 + AgCN$	I	61	Colorless needles, m. 105°; rxn.1	346, 347
$BuAs(CN)_2$	$BuAsCl_2 + AgCN$	I		m. 61-63°	44
$n-C_5H_{11}As(CN)_2$	$n-C_5H_{11}AsCl_2 + Ag-$	I		m. 69-69.5°	44
$n-C_6H_{13}As(CN)_2$	$n-C_6H_{13}AsCl_2 + Ag-$	I		m. 67.8-69.8°	44

1. Rxns. with O → $\overline{OCH:CHCH:CH}AsO$ (347).

TABLE VII
CYANO- AND THIOCYANATOARSINES

RR'AsCN or RR'AsCNS	Prepd. from	Method	Yield %	Props., Rxns., Remarks	Ref.
C₃					
(CF₃)₂AsCN	R₂AsI + AgCN	I		b. 89.5°, n_D^{20} IR spectrum, rxns. D₁,₂	333, 334, 1266
(CF₃)₂AsCNS	R₂AsI + AgCNS	I		b. 117°, n_D^{20} 1.445, IR spectrum	333
Me₂AsCN	R₂AsCl + AgCN	I	80	b. 116-118°/757 mm., rxn.2	1266
	R₂AsI + AgCN	I	76	b. 159-160°/730 mm.	44
				b. 162°/770 mm.	256
C₄					
Me(ClCH:CH)AsCN	RR'AsCl + KCN	I		Light yellow oil	261
C₅					
(ClCH:CH)₂AsCN	R₂AsCl + KCN	I		b. 120°/12 mm., skin irritant	734
Et₂AsCN	R₂AsCl + AgCN	I	81	b. 80-81/13 mm.	44
C₆					
EtPrAsCN	RR'AsCl + AgCN	I	72	b. 110-113°/27 mm. Vapor pressure and volatility	44, 1057
EtPrAsCNS	RR'AsCl + KCNS	I	52	b. 102-110°/0.65 mm.	44
C₇					
BuEtAsCN	RR'AsCl + AgCN	I	65	b. 112-112.5°/65 mm.	44
C₈					
Me(3-O₂NC₆H₄)AsCN	RR'AsCl + AgCN	I	66	Light yellow solid, m. 79.5-80.5°	251
MePhAsCN	RR'AsCl + NaCN	I	63	Colorless liq., b. 147-148°/20 mm.	251
MePhAsCNS	RR'AsCl + NaCNS	I	49	Pale yellow liq., b. 176-179°/18 mm.	251

1. Rxn. with H₂O at 72° and 104° → CHF₃, HCN, and H₃AsO₃ (333).
2. Rxn. with aq. alkali → CHF₃ (333).

TABLE VII (cont'd.)

CYANO- AND THIOCYANATOARSINES

RR'AsCN or RR'AsCNS.	Prepd. from	Method	Yield %	Props., Rxns., Remarks	Ref.
C_9					
(OCH:CHCH:C)$_2$AsCN	R$_2$AsCl + AgCN	I	70	Colorless, hydrolyzable crystals, m. 32°, b. 166-167/20 mm. b. 142-143°/2.3 mm.	346, 347
(SCH:CHCH:C)$_2$AsCN	R$_2$AsCl + NaCN	I		Colorless plates, m. 51-55°, b. 175-200°/0.3 mm.	866 494
Ph(ClCH:CH)AsCN	RR'AsCl + KCN	I			263
C_{11}					
Ph(SCH:CHCH:C)AsCN	RR'AsCl + NaCN	I		Almost colorless oil, m. 49-51°, b. 168-174°/0.5 mm.	494
C_{12}					
Me(1-C$_{10}$H$_7$)AsCN	Me$_2$(1-C$_{10}$H$_7$)As + BrCN	III		m. 85°, b. 190-191°/8 mm., rxn.3	1156
Me(1-C$_{10}$H$_7$)AsCNS	RR'AsCl + NaCNS in Me$_2$CO	I		m. 50-50.5°, b. 211-213°/8 mm., rxn.3	1156
C_{13}					
(2-ClC$_6$H$_4$)$_2$AsCN	R$_2$AsCl + NaCN	I		m. 85-87°	494
(4-ClC$_6$H$_4$)$_2$AsCN	R$_2$AsCl + NaCN	I		Colorless prisms, m. 58-60°, b. 198-200°/0.5 mm.	494
(3-O$_2$NC$_6$H$_4$)$_2$AsCN	R$_2$AsCl + KCN in MeOH	I		m. 151-152°	865
(3-O$_2$NC$_6$H$_4$)$_2$AsCNS	R$_2$AsCl+ NaCNS in MeOH	I		m. 103-105°	865
Ph(2-ClC$_6$H$_4$)AsCN	RR'AsCl + NaCN in H$_2$O	I		Pale yellow, m. 40-42°	494
Ph(4-ClC$_6$H$_4$)AsCN	RR'AsCl + NaCN in H$_2$O	I		m. 102°	494
Ph$_2$AsCN	R$_2$AsCl+ KCN or NaCN	I	85	b. 184-186°	56
				Dipole moment	1147b
	(Ph$_2$As)$_2$O + HCN	III		Props.4, rxn.5	56
Ph(2-H$_2$NC$_6$H$_4$)AsCN	RR'AsCl + NaCN	I		Hydroscopic solid, m. 53°	494

$(3\text{-}H_2NC_6H_4)_2AsCN$	$R_2AsCl + NaCN$	I	Colorless plates, m. 114°	494
C_{14} $Ph(3\text{-}OCHC_6H_4)AsCN$	$RR'AsCl + NaCN$	I	Viscous, golden-yellow oil	494
$(2\text{-}ClC_6H_4)(2\text{-}MeC_6H_4)As\text{-}CN$	$RR'AsCl + NaCN$	I	Colorless prisms, m. 69°	494
$(3\text{-}ClC_6H_4)(2\text{-}MeC_6H_4)As\text{-}CN$	$RR'AsCl + NaCN$	I	Colorless prisms, m. 64°	494
$(4\text{-}ClC_6H_4)(2\text{-}MeC_6H_4)As\text{-}CN$	$RR'AsCl + NaCN$	I	Colorless prisms, m. 64°	494
$(2\text{-}ClC_6H_4)(4\text{-}MeC_6H_4)\text{-}AsCN$	$RR'AsCl + NaCN$	I	Brown oil	494
$(3\text{-}ClC_6H_4)(4\text{-}MeC_6H_4)\text{-}AsCN$	$RR'AsCl + NaCN$	I	Brown oil	494
$(4\text{-}ClC_6H_4)(4\text{-}MeC_6H_4)\text{-}AsCN$	$RR'AsCl + NaCN$	I	Brown oil	494
$(2\text{-}O_2NC_6H_4)(2\text{-}MeC_6H_4)\text{-}AsCN$	$RR'AsCl + NaCN$	I	Yellow prisms, m. 105-107°	494
$(3\text{-}O_2NC_6H_4)(2\text{-}MeC_6H_4)\text{-}AsCN$	$RR'AsCl + NaCN$	I	Colorless prisms, m. 99-101°	494
$(4\text{-}O_2NC_6H_4)(2\text{-}MeC_6H_4)\text{-}AsCN$	$RR'AsCl + NaCN$	I	Colorless prisms, m. 103-105°	494
$(4\text{-}O_2NC_6H_4)(4\text{-}MeC_6H_4)\text{-}AsCN$	$RR'AsCl + NaCN$	I	Yellow needles, m. 88-90°	494
$Ph(2\text{-}MeC_6H_4)AsCN$	$RR'AsCl + NaCN$	I	Reddish, viscous oil	494
$Ph(3\text{-}MeC_6H_4)AsCN$	$RR'AsCl + NaCN$	I	Brownish, yellow liq.	494
$Ph(4\text{-}MeC_6H_4)AsCN$	$RR'AsCl + NaCN$	I	Golden yellow liq.	494
C_{15} $Ph(3\text{-}AcC_6H_4)AsCN$	$RR'AsCl + NaCN$	I	Colorless prisms, m. 57-59°	494
$Ph(3\text{-}MeO_2CC_6H_4)AsCN$	$RR'AsCl + NaCN$	I	Pale brown oil	494
$Ph(4\text{-}MeO_2CC_6H_4)AsCN$	$RR'AsCl + NaCN$	I	Pale brown oil	494
$(2\text{-}MeC_6H_4)_2AsCN$	$R_2AsCl + NaCN$	I	Colorless prisms, m. 74°	494
$(3\text{-}MeC_6H_4)_2AsCN$	$R_2AsCl + NaCN$	I	Brownish yellow, m. 43-47°	494
$(4\text{-}MeC_6H_4)AsCN$	$R_2AsCl + NaCN$	I	Colorless plates, m. 62°	494
$(2\text{-}MeC_6H_4)(3\text{-}MeC_6H_4)\text{-}AsCN$	$RR'AsCl + NaCN$	I	Pale brown oil	494

Pure with conc. KCl or HBr ↑ liberation of C_6H_6 (1156).

53

TABLE VII (cont'd.)
CYANO- AND THIOCYANATOARSINES

RR'AsCN or RR'AsCNS	Prepd. from	Method	Yield %	Props., Rxns., Remarks	Ref.
(2-MeC$_6$H$_4$)(4-MeC$_6$H$_4$)AsCN	RR'AsCl + NaCN	I		Pale brown	494
(3-MeC$_6$H$_4$)(4-MeC$_6$H$_4$)AsCN	RR'AsCl + NaCN	I		Reddish brown oil	494
Ph(2,4-Me$_2$C$_6$H$_3$)AsCN	RR'AsCl + NaCN	I		Viscous brown oil	494
C$_{16}$					
(2-MeC$_6$H$_4$)(2,4-Me$_2$C$_6$H$_3$)-AsCN	RR'AsCl + NaCN	I		Brown oil	494
(4-MeC$_6$H$_4$)(2,4-Me$_2$C$_6$H$_3$)-AsCN	RR'AsCl + NaCN	I		Brown oil	494
C$_{17}$					
Ph(1-C$_{10}$H$_7$)AsCN	RR'AsCl + NaCN	I		Yellow plates, m. 98-99°	494
Ph(2-C$_{10}$H$_7$)AsCN	RR'AsCl + NaCN	I		m. 59-61°	494
C$_{18}$					
(4-MeC$_6$H$_4$)(1-C$_{10}$H$_7$)AsCN	RR'AsCl + NaCN	I		Colorless prisms, m. 83-84°	494
(4-MeC$_6$H$_4$)(2-C$_{10}$H$_7$)AsCN	RR'AsCl + NaCN	I		Colorless prisms, m. 99°	494
C$_{21}$					
(1-C$_{10}$H$_7$)$_2$AsCN	R$_2$AsCl + NaCN	I		Colorless plates, m. 191°	494
(2-C$_{10}$H$_7$)$_2$AsCN	R$_2$AsCl + NaCN	I		Colorless prisms, m. 89-91°	494

4. Dipole moment (854, 1148). Light absorption data (853, 856). Identification (249).
5. Rxn. with AcCl + AlCl$_3$ → cleavage of the C-As bonds with formation of PhAc and AcCO$_2$H (788).

DIORGANOHALOARSINES

The compounds of the general structure R_2AsX or $RR'AsX$ are prepared or formed by:

1. Reducing arsinic acids, R_2AsO_2H, with hypophosphorous acid in a hydrohalic acid, or with sulfur dioxide in an acid medium, such as hydrohalic acids, in the presence of catalytic amounts of an alkali metal iodide, or with hydriodic acid in acetic acid (Method I).
2. Reacting tertiary arsines, R_3As, with arsenic halides to form mixtures of the R_2AsX and $RAsX_2$ type compounds (Method II).
3. Heating arsenic powder with alkyl halides at elevated temperatures to form mixtures of the R_2AsX and $RAsX_2$ type compounds. While trifluoroiodomethane reacts with arsenic at 200°, methyl halides and homologs require considerably higher temperatures and the presence of copper (Method III).
4. Reacting sodium arsenite with alkyl halides in 80% ethanol and treating the reaction product with sulfur dioxide and sodium iodide, hydrolyzing the $RAsI_2$ compound with aqueous sodium hydroxide, then treating the arsonous acid with an alkyl iodide and reducing once more with sulfur dioxide (Method IV).
5. Condensation of organodihaloarsines, $RAsX_2$, with acetylene in the presence of aluminum chloride at 10-15° to form mixtures of organohalo(2-halovinyl)- and organodi(2-halovinyl)arsines, $RR'AsX$ and $RAsR'_2$ (Method V).
6. Decomposition of tertiary arsines containing a 2-halo-1-alkenyl group with alcoholic potassium hydroxide, followed by a treatment with hydrohalic acids (Method VI).
7. Cleaving tertiary arsines with halogens in a sealed tube or in suitable solvents or with a cobalt trihalide under nitrogen (Method VII).
8. Thermal decomposition of tertiary arsine dihalides in vacuo. Asymmetrical trialkylarsine dihalides lose the shorter or shortest alkyl group and alkylarylarsine dihalides one alkyl group. The arsine dihalides can be prepared by treating tertiary arsines with halogens in halogenated hydrocarbons (Method VIII).
9. Reacting arsenic trihalides with diazoalkanes to form mixtures of (haloalkyl)dihalo-, bis(haloalkyl)halo-, and tris(haloalkyl)arsines (Method IX).
10. Cleaving tetraalkyl- or tetraaryldiarsines with alkylhalides or halogens in methanol (Method X).
11. Treating bis(dialkyl- or diarylarsenic) oxides with hydrohalic acids, or with thionyl chloride, or with mercury(II) chloride, or with iron(III) chloride-hydrochloric acid (Method XI).
12. Heating arsenic trioxide with potassium carboxylates at 300-400° and passing the distillate into hydrated iron(III) chloride in concentrated hydrochloric acid (Method XII).
13. Metathesis of arsenic trihalides or organodihaloarsines (in the preparation of the $RR'AsX$ type compounds) with organometallic compounds, such as $Bu_2Sn(CH:CH_2)_2$, R_3Al, R_4Pb, and $RHgX$, to form mixtures of monohaloarsines, dihaloarsines, and tertiary arsines (Method XIII).

14. Treating arsonic acids with acetyl chloride to form methylenebis(organo-haloarsines) (Method XIV).
15. Reacting arsenoso compounds, RAsO, with alkylene oxides, $R\overline{CHCH_2O}$, reducing the product with sulfur dioxide in aqueous solution containing catalytic amounts of potassium iodide, and treating the reaction mixture with thionyl chloride (Method XV).
16. Reacting alkyl halides with organodihaloarsines, $RAsX_2$, or with arsenoso compounds, RAsO, in alcoholic sodium hydroxide under reflux and, when arsenoso compounds are employed, neutralizing the reaction mixture and treating with sulfur dioxide (Method XVI).
17. Treating diaryl(2-hydroxyethyl)arsines with thionyl chloride (Method XVII).
18. Decomposing aryldiazonium tetrahaloarsenites with copper(II) chloride in the presence of phenylhydrazine or by adding an aryldiazonium salt to arsenic trichloride in concentrate hydrochloric acid containing iron(III) chloride or copper(II) chloride (Method XVIII).
19. Bubbling air through a mixture of phenylhydrazine and arsenic trichloride in hydrochloric acid containing copper(II) chloride (Method XIX).
20. Decomposing a mixture of arsenic trichloride or organodihaloarsine with a double salt of aryldiazonium halide-metal halide, e. g., $(PhN_2Cl)_2 \cdot ZnCl_2$, in acetone in the presence of iron or zinc powder (Method XX).
21. Reacting arsenic acid, H_3AsO_4, with arylhydrazines in aqueous solution in the presence of copper compounds and treating the product with fuming hydrochloric acid (Method XXI).
22. Heating organodihaloarsines, $RAsX_2$, with arsenoso compounds, RAsO, in a CO_2 atmosphere in the presence or absence of ZnO, Fe, $FeCl_3$, or $ZnCl_2$ (Method XXII).
23. Reacting organodihaloarsines, $RAsX_2$, with benzene in the presence or absence of aluminum chloride to form the RPhAsX type arsines (Method XXIII).
24. Neutron irradiation of arsenic trichloride in benzene leads to the formation of mixtures containing chlorodiphenylarsine (Method XXIV).

Dialkyl- and diaryliodoarsines can be converted into the corresponding chloro-, bromo-, or fluoroarsines by treating the iodoarsines with silver chloride, bromide, and fluoride, respectively. The chloroarsines are converted into the iodoarsines by treating the former with sodium iodide in acetone or with aqueous sodium bicarbonate and triturating the precipitate with hydriodic acid.

BROMOBIS(TRIFLUOROMETHYL)ARSINE C_2AsBrF_6 $(CF_3)_2AsBr$
Prepn.: by reacting $(CF_3)_3As$ with Br_2 in a sealed tube; 17% yield (333, 334).
Props.: B. $59.5°/745$ mm., n_D^{20} 1.398 (333, 334).
Rxn. with aq. NaOH → CHF_3 (333).

CHLOROBIS(TRIFLUOROMETHYL)ARSINE C_2AsClF_6 $(CF_3)_2AsCl$
Prepn.: By heating $(CF_3)_3AsCl_2$ at 125°, a mixt. of $(CF_3)_2AsCl$ and CF_3AsCl_2 is formed. (333).
By reacting $(CF_3)_3AsI$ with AgCl (334, 1266).
Props.: B. 46° (calcd.), n_D^{18} 1.351, stable to H_2O at r. t. (333, 334, 1266).
Rxns. with: Dry NH_3 at -64° → $[(CF_3)_2As]_2NH$, while the rxn. in the gase phase gives also $(CF_3)_2AsNH_2$ (253).
Excess NH_3 at 20° → CHF_3 and $As(NH_2)_3$ (253).

MeNH$_2$ in the gase phase at 20° → (CF$_3$)$_2$AsNHMe (253).
EtNH$_2$ in the gase phase → (CF$_3$)$_2$AsNHEt (253).
Me$_2$NH in the gase phase → (CF$_3$)$_2$AsNMe$_2$ (253).
Aq. NaOH → CHF$_3$ (333).

FLUOROBIS(TRIFLUOROMETHYL)ARSINE C$_2$AsF$_7$ (CF$_3$)$_2$AsF
<u>Prepn.</u>: By heating (CF$_3$)$_3$As with CoF$_3$ at 100° in a N atm. (333).
  By reacting (CF$_3$)$_2$AsI with AgF at 25° (333, 334, 1266).
<u>Props.</u>: B. 25° (calcd.) (333, 334, 1266).
<u>Rxn.</u> with aq. NaOH → CHF$_3$ (333).

BIS(TRIFLUOROMETHYL)IODOARSINE C$_2$AsF$_6$I (CF$_3$)$_2$AsI
<u>Prepn.</u>: By treating (CF$_3$)$_3$As with I$_2$ at 100° (117, 333, 1266).
  By treating [(CF$_3$)$_2$As]$_2$ with CF$_3$I or CH$_3$I (254).
  By-prod. in the rxn. of As with CF$_3$I at 220° (74, 117, 333, 1266).
  By-prod. in the rxn. of (CF$_3$)$_3$As with AsI$_3$ (333).
<u>Props.</u>: B. 14°/54 mm. (74); b. 92°, n$_D^{25}$ 1.425 (117); stable to H$_2$O at r. t.
(333); b. 92° (334).
<u>Rxns. with:</u> Aq. NaOH → HCF$_3$ (117, 333).
Hg in a sealed tube → [(CF$_3$)$_2$As]$_2$ (117).
AgF at 20° → (CF$_3$)$_2$AsF (333, 1266).
LiAlH$_4$ → (CF$_3$)$_2$AsH (333).
MeMgI → (CF$_3$)$_2$AsMe (333).
HgO → [(CF$_3$)$_2$As]$_2$O (333, 1266) and [CF$_3$)$_2$AsO]$_2$Hg (333).
Ag$_2$O → (CF$_3$)$_2$AsOAg (333).
H$_2$O in a sealed tube → (CF$_3$)$_2$AsO$_2$H and I$_2$ (333).
AgX (X = F, Cl, Br, or CN) → (CF$_3$)$_2$AsX (334, 1266).
AgSCN at 100° → (CF$_3$)$_2$AsSCN (1266).

BROMODIMETHYLARSINE C$_2$H$_6$AsBr Me$_2$AsBr
<u>Prepn.</u>: By reacting Me$_2$AsI with AgBr at 60°; 81% yield (256).
  By reducing Me$_2$AsOONa with H$_3$PO$_2$ in 60% HBr at 40°; 68% yield (660).
  By heating MeBr with As powder at 350° in the presence of Cu; a mixt.
of MeAsBr$_2$ and Me$_2$AsBr is formed (774a).
<u>Props.</u>: B. 129° (255); b. 128-129 (660). Raman spectrum (659). Electron dif
fraction data (1146).
<u>Rxns. with:</u> PhCH:CHBr + Na → PhCH:CHAsMe$_2$ (265).
PhCH:CHMgBr → PhCH:CHAsMe$_2$ (265).
(RO)$_3$P → Me$_2$AsPO(OR)$_2$ (634).
Me$_2$NH → Me$_2$AsNMe$_2$ (851).

CHLORODIMETHYLARSINE C$_2$H$_6$AsCl Me$_2$AsCl
<u>Prepn.</u>: By reacting Me$_2$AsI with AgCl at 60°, 90% yield (256).
  By treating (Me$_2$As)$_2$O with HCl-FeCl$_3$ or HgCl$_2$ (447).
  By reducing Me$_2$AsOONa with H$_3$PO$_2$ in 37% HCl at 60°, 70% yield (660).
  By treating Me$_2$AsCH:CHCl with alc. KOH, followed by HCl, 52.8% yield
(674).
  By heating MeCl with As powder at 350° in the presence of Cu (774a).
  By heating a mixt. of AcOK with As$_2$O$_3$ at 300-400° and passing the dis-
tillate into hydrated FeCl$_3$ in concd. HCl, 28% yield (1307).

Props.: B. 107° (256); b. 106.5° (660). Raman spectrum (659). Electron diffraction data (1146).
Counteracts the effect of antiknock agents in gasoline (952).
Rxns. with: AgCN → Me_2AsCN (44).
Et_3P → Me_2AsPEt_3Cl (237).
NH_3 at -46° → solids insol. in C_6H_6 and Et_2O (253).
Me_2NH → Me_2AsNMe_2 (253, 256).
CF_3CO_2Ag → $Me_2AsOOCCF_3$ (256).
NaOH → $(Me_2As)_2O$ (602).
KSC(S)OR → $Me_2AsSC(S)OR$ (987, 988).
$HSP(S)(OR)_2$ → $Me_2AsP(S)(OR)_2$ (988).

DIMETHYLIODOARSINE C_2H_6AsI Me_2AsI
Prepn.: By reducing Me_2AsOOH with SO_2 in HCl or H_2SO_4 in the presence of KI (254, 660).

By reacting MeI with Na_3AsO_3 in 80% EtOH and treating stepwise the prod. with SO_2 and MeI in HCl soln., 46% yield (480).

By treating $(Me_2As-)_2$ with I in MeOH (868).

By heating MeI with As powder at 280° in the presence of Cu; the prod. contains Me_2AsI + $MeAsI_2$ (774a).
Props.: B. 154-155° (480, 660). Raman spectrum (659). Thermodynamic data (868). Electron diffraction data (1146).
Rxns. with: PR_3 → Me_2AsPR_3I (237).
RLi → $RAsMe_2$ (238, 480, 481, 524, 612).
RMgX → $RAsMe_2$ (244, 271, 522, 523, 605).
CF_3I + Hg → Me_2AsCF_3 (254).
$CF_3CF_2CF_2I$ → $Me_2AsCF_2CF_2$ (254).
AgX (X = F, Cl, Br, and CN) → Me_2AsX (256).
Grignard reagent from o-BrC_6H_4I and 3 equivs. Mg → $PhAsMe_2$, 2-$BrC_6H_4AsMe_2$, 2-$C_6H_4(AsMe_2)_2$, 2-$PhC_6H_4AsMe_2$, $(2-C_6H_4Me_2)_2$, and 2-$C_6H_4(C_6H_4AsMe_2)_2$ (523).
α-Picolinyllithium → dimethyl-α-picolinylarsine (494a).

CHLORO(2-CHLOROVINYL)METHYLARSINE $C_3H_5AsCl_2$ $Me(ClCH:CH)AsCl$
Prepn.: by reacting $MeAsCl_2$ with CH⋮CH in the presence of $AlCl_3$ at 10-15° (261).
Props.: B. 112-115°/10 mm. (261).
Rxns. with: PhMgBr → $Me(ClCH:CH)AsPh$ (261).
CH⋮CH in the presence of $AlCl_3$ → $MeAs(CH:CHCl)_2$ (261).
H_2O_2 → $Me(ClCH:CH)AsOOH$ (261).
H_2S → $[Me(ClCH:CH)As]_2S$ (261).
KCN → $Me(ClCH:CH)AsCN$ (261).
EtONa → $[Me(ClCH:CH)As]_2O$ (261).

CHLORO(2-CHLOROETHYL)METHYLARSINE $C_3H_7AsCl_2$ $Me(ClCH_2CH_2)AsCl$
Prepn.: by treating $[Me(HOCH_2CH_2)As]_2O$ with $SOCl_2$ (715).
Props.: B. 60-61°/1 mm. (715).
Rxns. with: H_2O → $[Me(HOCH_2CH_2)As]_2O$ (715).
NaOH → $MeAs(ONa)_2$ (715).

METHYLENEBIS(CHLOROMETHYLARSINE) $C_3H_8As_2Cl_2$ $CH_2(AsClMe)_2$
Prepn.: by treating $MeAsO_3H_2$ with AcCl (1024).

CHLOROBIS(2-CHLOROVINYL)ARSINE $C_4H_4AsCl_3$ $(ClCH:CH)_2AsCl$
Prepn.: By-prod. in the rxn. of $AsCl_3$ with C_2H_2 in 25-30% HCl in the presence of Cu_2Cl_2 (19).
By-prod. in the rxn. of $PhAsCl_2$ with C_2H_2 in the presence of $AlCl_3$ (45).
Rxn. mechanism (1033).
Props.: B. 35-107°/.4 mm. (45). Dipole moment (855). Light absorption data (856).
Analysis (1031).
Use: Catalyst for the prepn. of $CH_2:CHCl$ from C_2H_2 + HCl (115).
Rxn. with: $AsCl_3$ in the presence of colza oil catalyst $\rightarrow$ $ClCH:CHAsCl_2$ (19).
RMgX $\rightarrow$ $(ClCH:CH)_2AsR$ (674).
KCN $\rightarrow$ $(ClCH:CH)_2AsCN$ (734).
NaOH $\rightarrow$ $(ClCH:CH)_2AsOH$ (1031).
Iodo-Cu(I) reagent $\rightarrow$ ppt. (trace analysis) (300, 301).

BROMODIVINYLARSINE C_4H_6AsBr $(CH_2:CH)_2AsBr$
Prepn.: by adding $(CH_2CH)_3As$ to $AsBr_3$ (2:1) and completing the exothermic rxn. under reflux at 60°/15 mm; a mixt of $(CH_2:CH)_2AsBr$ and $CH_2:CHAsBr_2$ is formed. If the ratio $(CH_2:CH)_3As:AsBr_3$ = 2:1, the title compd. is the main prod.; 81% yield of the title prod. was obtd. by reacting $AsBr_3$ with $Bu_2Sn(CH:CH_2)_2$ (1:2) on a steam bath (773, 774).
Props.: B. 59°/19 mm.; b. 60° 20 mm. (773).

CHLORO(2-CHLOROETHYL)ETHYLARSINE $C_4H_9AsCl_2$ $Et(ClCH_2CH_2)AsCl$
Prepn.: by reacting EtAsO with $\overline{CH_2CH_2O}$ in H_2O, reducing the prod. with SO_2 in H_2O contg. KI, and treating the oxide with $SOCl_2$ (715).

CHLORODIETHYLARSINE $C_4H_{10}AsCl$ Et_2AsCl
Prepn.: By reducing Et_2AsO_2H dissolved in concd. HCl with SO_2 in the presence of KI, 60% yield (44).
By treating $EtAsCl_2$ with Et_4Pb (2:1) at 120° under N, 78% yield (661, 662, 663).
By heating EtCl with As powder in the presence of Cu (774a).
Props.: b. 156°/736 mm. (44); b. 74-78°/74 mm.; d_8^{28} 1.215 (661, 663).
Rxns. with: AgCN $\rightarrow$ Et_2AsCN (44).
$(RO)_3P$ $\rightarrow$ $Et_2AsPO(OR)_2$ (634).
ROH in $Et_2O-C_5H_5N$ $\rightarrow$ Et_2AsOR (650).
Air $\rightarrow$ white solid (661).

DIETHYLIODOARSINE $C_4H_{10}AsI$ Et_2AsI
Rxns. with: EtONa $\rightarrow$ Et_2AsOEt (648).
MeONa in Et_2O $\rightarrow$ $Et_2MeAs(OMe)I$ (648).
RMgX $\rightarrow$ Et_2AsR (513a).

METHYLPHENYLCHLOROARSINE C_7H_8AsCl $MePhAsCl$
Prepn.: by reducing $MePhAsO_2H$ with SO_2 in HCl contg. KI; 61% yield (251).
Props.: Colorless liquid, b. 127°/23 mm. (251).

Rxns. with: NaCN → MePhAsCN (251).
NaCNS → MePhAsCNS (251).
NaOMe → MePhAsOMe (251).
NaOAc → MePhAsOAc (251).
Zn-Hg in MeOH → MeAsH (251).

IODOMETHYLPHENYLARSINE C_7H_8AsI MePhAsI
Prepn.: by reacting PhAsO with MeI in alc. NaOH under reflux, neutralizing the rxn. mixt., and reducing with SO_2 in HCl (623).
Props.: B. 130-131°/7 mm. (623).
Rxns. with: RMgX → MePhAsR (623, 630).
NaOH → (MePhAs)$_2$O (626).
NaCH(CO$_2$Et)$_2$ → [CH(CO$_2$Et)$_2$]$_2$ and (MePhAs)$_2$ or (MePhAs)$_2$O (628).

CHLORODI-2-FURYLARSINE $C_8H_6AsClO_2$ $(2-C_4H_3O)_2AsCl$
Prepn.: by reacting $(2-C_4H_3O)_3As$ with $AsCl_3$ under reflux, a mixt. of $(2-C_4H_3O)_2$-AsCl with $(2-C_4H_3O)AsCl_2$ is formed (345, 866).
Props.: B. 152-145°/18 mm.; n$_D^{20}$ 1.6153; d$_4^{20}$ 1.598 (345); b. 122-127°/1 mm., relatively stable if stored cold in the dark; $HgCl_2$ catalyzes disproportionation to $(C_4H_3O)_3As$ and $AsCl_3$ (866).
Rxns. with: Aq. alk. → [$(2-C_4H_3O)_2As]_2O$ (346, 347).
AgCN → $(2-C_4H_3O)_2AsCN$ (346, 347, 866).
Cl_2, followed by hydrolysis → di-2-furylarsinic acid and tri-2-furylarsine dichloride (753).

CHLORODI-2-THIENYLARSINE $C_8H_6AsClS_2$ $(2-C_4H_3S)_2AsCl$
Prepn.: By adding portion-wise $AsCl_3$ to $(2-C_4H_3S)_3S$ at 150°, a mixt. of $(2-C_4$-$H_3S)AsCl_2$ with $(2-C_4H_3S)_2AsCl$ is formed (349).
By refluxing 2-thienylmercury chloride with $AsCl_3$ in MePh, a mixt. of $(2-C_4H_3S)AsCl_2$, $(2-C_4H_3S)_2AsCl_2$, and $(2-C_4H_3S)_3As$ is formed (494).
Props.: B. 185-195°/15 mm. (349); b. 195-210°/13 mm. (494).

CHLOROCHLOROVINYLPHENYLARSINE $C_8H_7AsCl_2$ Ph(ClCH:CH)AsCl
Prepn.: By heating $ClCH:CHAsCl_2$ with C_6H_6 in the presence or absence of $AlCl_3$ (260, 262).
By reacting $PhAsCl_2$ with C_2H_2 in the presence of $AlCl_3$ (45, 263).
Props.: B. 138-142°/3 mm. (260, 262); b. 165°/4 mm. (45); b. 135-145°/4 mm. (263).
Rxns. with: RMgX → Ph(ClCH:CH)AsR (260, 262, 263).
$AlCl_3$ in CS_2 → 1-chloroarsindole (260, 262, 263).
H_2O_2 → Ph(ClCH:CH)AsO$_2$H (263).
Alc. EtONa → Ph(ClCH:CH)AsOEt (263).
Alc. KCN → Ph(ClCH:CH)AsCN (263).
H_2S → [Ph(ClCH:CH)As]$_2$S (263).

ETHYLPHENYLIODOARSINE $C_8H_{10}AsI$ EtPhAsI
Prepn.: by reacting PhAsO with EtI in alc. NaOH, neutralizing the rxn. prod. with HCl, and reducing with SO_2 in HCl (624).
Props.: B. 139-140°/8 mm., d^{20} 1.8406 (624).
Rxns. with: RMgI → EtPhAsR (624).

NaOH → $(EtPhAs)_2O$ (262).
NaOP(OEt)$_2$ → EtPhAsPO(OEt)$_2$ (629).

ETHYLIODO-p-TOLYLARSINE $C_9H_{12}AsI$ Et(4-MeC$_6$H$_4$)AsI
Prepn.: by adding EtI to a mixt.of 4-MeC$_6$H$_4$AsO and 10% NaOH in 96% EtOH, neutrali-
zing the rxn. prod. with HCl,and reducing with SO$_2$; 74.3% yield (627).
Props.: B. 157.5-158°/10 mm., d^{18} 1.7332 (627).
Rxns. with: RMgX → Et(4-MeC$_6$H$_4$)AsR (627, 716).
NaOH → [Et(4-MeC$_6$H$_4$)As]$_2$O (627).

CHLORODI-3-PYRIDYLARSINE DIHYDROCHLORIDE $C_{10}H_8AsClN_2 \cdot 2HCl$ (C$_5$H$_4$N)$_2$AsCl·2HCl
Prepn.: by heating at 200° 3-(dichloroarsino)pyridine hydrochloride with 3 mole
equivs. of 3-arsenosopyridine; isolated as dihydrochloride (589).
Rxns. with: Alkali (equiv. amt.) → 40% (C$_5$H$_4$N)$_2$AsCl.HCl and 50% (C$_5$H$_4$N)$_2$AsO$_2$H
(590).
Alkali (100% excess) → 85% (C$_5$H$_4$N)$_2$AsO$_2$H and 30% C$_5$H$_4$NAsO$_3$H$_2$ (590).
KCN or AgCN → (C$_5$H$_4$N)$_2$AsCl.HCl, (C$_5$H$_4$N)$_2$AsO$_2$H, C$_5$H$_5$NAsO$_3$H$_2$,and HCN (590).
Hg(CN)$_2$ → (C$_5$H$_4$N)$_2$AsCl.HgCl$_2$.2H$_2$O, m. 160-161° (591).

CHLOROMETHYL-1-NAPHTHYLARSINE $C_{11}H_{10}AsCl$ Me(1-C$_{10}$H$_7$)AsCl
Prepn.: by dry distn. of Me$_2$(1-C$_{10}$H$_7$)AsCl$_2$ (1155).
Props.: Yellow, m. 62.2-62.7°, b. 165.5-166°/6 mm. (1155).
Rxns. with: KOH in aq. MeOH → [Me(1-C$_{10}$H$_7$)As]$_2$O (1155).
NaI in Me$_2$CO → Me(1-C$_{10}$H$_7$)AsI (1156).
NaCNS in Me$_2$CO → Me(1-C$_{10}$H$_7$)AsCNS (1156).
NaOH in 50% EtOH, followed by 50% H$_2$O$_2$ → Me(1-C$_{10}$H$_7$)AsOOH (1156).
Concd. HCl or HBr → cleavage of C$_{10}$H$_8$ (1156).

CHLORO(4-NITROPHENYL)PHENYLARSINE $C_{12}H_9AsClNO_2$ Ph(4-O$_2$NC$_6$H$_4$)AsCl
Prepn.: by treating [Ph(4-O$_2$NC$_6$H$_4$)As]$_2$O with HCl (908).
Props.: B. 170-190°/6 mm. (908).
Rxns. with: H$_2$O$_2$ → Ph(4-O$_2$NC$_6$H$_4$)AsO$_2$H (908).
RN$_2$Cl.ZnCl$_2$ + Fe → Ph(4-O$_2$NC$_6$H$_4$)RAs (908).

CHLORODIPHENYLARSINE, Clark I, $C_{12}H_{10}AsCl$ Ph$_2$AsCl
Prepn.: By condensing PhAsCl$_2$ with PhAsO at 290-300° in a continuous manner; 82%
yield along with small amts. of Ph$_3$As and As$_2$O$_3$ (57).
        By heating to 120° a mixt. of PhAsCl$_2$ and PhAsO (1:3) in the presence of
ZnO; 93.6% yield. Fe, FeCl$_3$, and ZnCl$_2$ may be used instead of ZnO (59).
        By treating Ph$_2$AsCH$_2$CH$_2$OH with SOCl$_2$ in CHCl$_3$ (122).
        By treating Ph$_3$As with AsCl$_3$ at 250°; equilibrium study (353).
        By treating Ph$_2$AsCH:CHCl with alc. KOH followed by HCl; 62.8% yield (674).
        By oxidation of PhNHNH$_2$ with H$_3$AsO$_4$ in the presence of Cu$_2$O or other
catalysts such as Cu(OH)$_2$, Cu, CuSO$_4$, or Cu bronze, followed by a treatment with
aq. HCl; 69-73% crude yield, along with small amts. of PhAsCl$_2$ and Ph$_3$As (56).
        By diazotization of a mixt. of PhNH$_2$.HCl and AsCl$_3$ in EtOH, followed by a
treatment of the PhN$_2$AsCl$_4$ complex with CuCl$_2$ in the presence of PhNHNH$_2$; 25% yield
(136).
        By bubbling air through a mixt. of PhNHNH$_2$ and AsCl$_3$ in aq. HCl in the

presence of CuCl$_2$; 28% yield, along with some PhAsCl$_2$ (136).

By adding PhN$_2$Cl to AsCl$_3$ in concd. HCl and treating the complex with FeCl$_3$·and CuCl$_2$; 40% yield along with 8% PhAsCl$_2$ (136).

By reducing (Ph$_2$As)$_2$O or Ph$_2$AsO$_2$H with SO$_2$ in HCl (136).

By pyrolysis of Ph$_3$AsCl$_2$ at 230° in vacuo; 36% yield (516).

By reacting PhHgCl with AsCl$_3$ (2:1) at 110°; 75% yield (585, 896).

By-prod. in the rxn. of (PhN$_2$)$_2$ZnCl$_4$ with AsCl$_3$ in Me$_2$CO in the presence of Zn (607).

By-prod. from AsCl$_3$ irradiated with neutrons in C$_6$H$_6$ (907, 1139).

By treating Ph$_2$AsO$_2$H with AcCl (1024).

Props.: B. 170-172° (122); m. 38-39°, b. 172-175°/11 mm. (136); b. 177-178°/3 mm., m. 37.8-38.3° (674); m. 44-45° (896); soly. data (252), Disproportionation (353). Decompn. kinetics (384, 973). Identification (249). Light absorption data (853, 856).

Dipole moment (854, 1147b). Magnetic data (975).

Decomp. at 215-280° → PhAsCl$_2$ and Ph$_3$As; at 320° Ph$_3$As and AsCl$_3$ are formed (973).

Rxns. with: KCN or NaCN → Ph$_2$AsCn (56).

AlCl$_3$ at 200°, followed by KI → Ph$_4$AsI (201).

CH$_2$N$_2$ → Ph$_2$AsCH$_2$As(O)Ph$_2$ (245).

Iodo-Cu(I) reagent (from NaI + CuSO$_4$) → ppt. (300, 301).

RMgX → Ph$_2$AsR (473, 613, 806).

NaC⋮CH → Ph$_2$AsC⋮CAsPh$_2$ (514).

(⋮CMgBr)$_2$ → Ph$_2$AsC⋮CAsPh$_2$ (514, 516).

RC⋮CNa or RC⋮CMgBr → Ph$_2$AsC⋮CR (518).

I$_2$ + 2H$_2$O → Ph$_2$AsO$_2$H (anal. method) (530).

NaOR → Ph$_2$AsOR (625).

H$_2$O$_2$ → Ph$_2$AsO$_2$H (674, 1030, 1032).

AlCl$_3$→ Ph$_2$AsCl.AlCl$_3$, a sirupy liquid, decompg. at 200° into Ph$_4$AsCl, 2AlCl$_3$, As, and Cl (755).

RCOCl + AlCl$_3$ → cleavage of the Ph groups and formation of PhCOR (788, 789).

KSC(S)OR → Ph$_2$AsSC(S)OR (988).

HSP(S)(OR)$_2$ + H$_2$CO$_3$ → Ph$_2$AsSP(S)(OR)$_2$ (988).

AgF → Ph$_2$AsF (1100).

RCOSK → Ph$_2$AsSC(O)R (1249).

LiBH$_4$ or LiAlH$_4$ → Ph$_2$AsH (1293).

(2-ACETAMIDOPHENYL)IODOPHENYLARSINE C$_{14}$H$_{13}$AsINO Ph(2-AcNHC$_6$H$_4$)AsI

Prepn.: By treating [Ph(2-AcNHC$_6$H$_4$)As]$_2$O with HI (93).

By reacting [Ph(2-H$_2$NC$_6$H$_4$)As]$_2$ with Ac$_2$O and treating the rxn. prod. with I$_2$ (94).

Props.: M. 147-148° (93, 94).

(3-ACETAMIDOPHENYL)IODOPHENYLARSINE C$_{14}$H$_{13}$AsINO Ph(3-AcNHC$_6$H$_4$)AsI

Prepn.: By acetylation of [Ph(3-H$_2$NC$_6$H$_4$)As]$_2$O, followed by a treatment with HI (93).

By acetylation of [Ph(3-H$_2$NC$_6$H$_4$)As]$_2$, followed by a treatment with I$_2$ (94).

Props.: M. 146-147° (93).

ETHYLENEBIS(CHLOROPHENYLARSINE) $C_{14}H_{14}As_2Cl_2$ $[CH_2As(Ph)Cl]_2$

Prepn.: by reducing $[CH_2As(Ph)O_2H]_2$ with SO_2 in concd. HCl contg. KI (199).

Props.: Colorless solid, m. 91-93°, is not appreciably hydrolyzed by damp air, can be recrystallized from EtOH (199).

Rxns. with: Hot HCl → $PhAsCl_2$ (199).
Hot NaOH → slow hydrolysis (199).
RMgX → $(CH_2AsPhR)_2$ (199, 609).
$(CH_2Br)_2$ in NaOH → $\overline{As(Ph)CH_2CH_2As(Ph)O}$ (609).

Other diorganohaloarsines are compiled in Table VIII starting on the following page.

TABLE VIII

DIORGANOHALOARSINES

R₂AsX	Prepd. from	Method	Yield %	Props., Rxns., Remarks	Ref.
C₂					
(ClCH₂)₂AsCl	AsCl₃ + CH₂N₂	IX		b. 86-88°/16 mm., d^{20}_{15} 1.8485; rxn.1	121
Me₂AsF	Me₂AsI + AgF at 20°		15	b. 75°	256
C₄					
(CH₂:CH)₂AsI	(CH₂:CH)₃AsI₂	VII		b. 83°/17 mm.	773
(MeCHCl)₂AsCl	AsCl₃ + MeCHN₂ in C₆H₆ at 3-5°			b. 51.5°/1 mm., d^{20}_{20} 1.5870	1320
Et₂AsBr	Et₃AsBr₂ at 100°		92	b. 61-61.5°/3 mm. Rxn. with (:CMgBr)₂ → (:CAsEt₂)₂	1321, 516
C₅					
EtPrAsCl	EtPrAsO₃H₂ + SO₂	I	21	b. 176°/729 mm., rxns.2,3 Rxns.4,5	44
EtPrAsI					
Et(Me₂CH)AsI	EtAsI₂ + Me₂CHI	XVI		b. 87-88°, d^{15} 1.7955, n^{20}_D 1.830, lacrimator, rxn. 6	626
C₆					
(NCCH₂CH₂)₂AsCl	MeAs(CH₂CH₂CN)₂	VII		b. 190-195°/0.5 mm., rxn.1	243
EtBuAsCl	EtBuAsO₂H + SO₂	I		b. 89-92°/19 mm., rxn.2	44
EtBuAsI	EtAsI₂ + BuBr	XVI	35	b. 107-108°/9 mm., b. 112-113°/12 mm., d^{25} 1.6818, rxn. 7,8	629
Et(Me₂CHCH₂)AsI	EtAsI₂ + Me₂CHCH₂Br	XVI	69.8	b. 103-104°/13 mm., d^{20}_{0} 1.7629	255
C₇					
(CF₃)PhAsBr	PhAs(CF₃)₂ + Br₂	VII		Light yellow liq., b. 174-176°/0.85 mm., rxn.2	255
Me(3-O₂NC₆H₄)AsBr	RR'AsO₂H + SO₂	I	54	HCl salt, m. 178-180°, rxn.s	251
Me[3,4-H₂N(HO)C₆H₃]AsCl	RR'AsO₂H + SO₂	I		Light yellow liq., b. 62°/0.8 mm., rxn.2	93
Me(3-O₂NC₆H₄)AsCl	RR'AsO₂H + SO₂	I	52		251

Compound	Reaction	Method	Yield	Properties	Ref.
Me[3,4-H$_2$N(HO)C$_6$H$_3$]AsI	RR'AsCl + NaHCO$_3$ followed by HI	XVI		HI salt, m. 136-137°	93
Et(Me$_2$CHCH$_2$CH$_2$)AsI	Me$_2$CHCH$_2$CH$_2$Br + EtAsI$_2$	XVI	79.5	b. 118-119°/10 mm., d$_Q^{18}$ 1.7867, rxn.8	629
C$_8$					
(HO$_2$CCH$_2$)PhAsCl				Rxn. with ClCH$_2$CO$_2$H in aq. NaOH → Ph(HO$_2$CCH$_2$)$_2$AsO	119
(Me$_2$CHCH$_2$)$_2$AsCl	AsCl$_3$ + (Me$_2$CHCH$_2$)$_3$Al	XIII		b. 76-77°/8 mm., n$_D^{20.5}$ 1.4862, d$_{20}$ 1.1265.	1330
C$_9$					
(CH$_2$:CHCH$_2$)PhAsBr	PhAsO + CH$_2$:CHCH$_2$Br	XVI	50	b. 138-139°/10 mm., d$_0^{15}$ 1.4736	629
C$_{10}$					
Ph(ŚCH:CHCH:C̈)AsCl	PhAsCl$_2$ + (ŚCH:CHCH:C̈)-HgCl	XIII		b. 150-156°/0.7 mm.	494
	(ŚCH:CHCH:C̈)AsCl$_2$ + Ph-HgCl	XIII			494
[HO$_2$C(CH$_2$)$_3$]PhAsCl	RR'AsO$_2$H + SO$_2$	I	88	Needles, m. 45°, rxn.10	242
BuPhAsBr	PhAs(OH)$_2$ + BuBr	XVI	75	b. 147-148°/10 mm., d$_0^{17}$ 1.3692, rxn.8	629
[-CH$_2$(Bu)AsCl]$_2$	[-CH$_2$(Bu)AsO$_2$H]$_2$ + SO$_2$	I		Pale yellow liq., b. 160-165°/0.05 mm., rxn.4	455

1. Rxn. with oxidizing agents (aq. H$_2$O$_2$) → the corresponding arsinic acid, R$_2$AsO$_2$H (121, 243, 1066, 1100).
2. Rxn. with AgCN → the corresponding cyanoarsine, R$_2$AsCN or RR'AsCN (44, 251).
3. Rxn. with KCNS → the corresponding thiocyanatoarsine, R$_2$AsCNS (44).
4. Rxn. with RMgX → the corresponding tertiary arsine (193, 455, 630, 805).
5. Rxn. with NaCH(CO$_2$Et)$_2$ → [CH(CO$_2$Et)$_2$]$_2$ and (EtPrAs)$_2$ or (EtPrAs)$_2$O (628).
6. Rxn. with aq. alkali or ammonia → the corresponding bis(dialkylarsenic) oxide (95, 626, 1100).
7. Rxn. (EtO)$_3$P → [EtBuAsP(OEt)$_3$]I (629).
8. Rxn. with NaOP(OR)$_2$ → the R'R"AsPO(OR)$_2$ type compounds (629, 631).
9. Rxn. with NaHCO$_3$ followed by HI → Me[3,4-H$_2$N(HO)C$_6$H$_3$]AsI (93).
10. Rxn. with Br(CH$_2$)$_3$CN → Ph[HO$_2$C(CH$_2$)$_3$]$_2$AsO (242).

TABLE VIII (cont'd.)
DIORGANOHALOARSINES

R_2AsX	Prepd. from	Method	Yield %	Props., Rxns., Remarks	Ref.
C_{11}					
$Me(1-C_{10}H_7)AsBr$	$[Me(1-C_{10}H_7)As]_2O + HBr$	XI			1156
	$Me_2(1-C_{10}H_7)As + Br_2$ and pyrolysis	VIII		Rxn. 11	1156
$Me(1-C_{10}H_7)AsF$	$Me(1-C_{10}H_7)AsO + HF$			b. 139.5-140°/6 mm., d. 1.4402, rxn. 11	1156
$Me(1-C_{10}H_7)AsI$	$Me(1-C_{10}H_7)AsCl + NaI$ in Me_2CO			m. 102-102.4°, rxn. 11	1156
C_{12}					
$[4,3-HO(O_2N)C_6H_3]_2AsBr$	$R_2AsO_2H + SO_2$	I		Yellow crystals, m. 131-132°	95
$(4-BrC_6H_4)_2AsCl$	$H_3AsO_4 + 4-BrC_6H_4NHNH_2$	XXI		m. 57-58°, rxn. 12	1336
	$R_2AsO_2H + SO_2$	I			1336
$(4-FC_6H_4)_2AsCl$	$H_3AsO_4 + 4-FC_6H_4NHNH_2$	XXI		m. 39°, b. 160-160°/12 mm., rxn. 12	1336
$(3-O_2NC_6H_4)_2AsCl$	$R_2AsO_2H + SO_2$	I	91	m. 112-113°, rxns. 2,3	865
$[4,3-HO(O_2N)C_6H_3]_2AsCl$	$R_2AsO_2H + SO_2$	I		Yellow crystals, m. 142-143°, rxn. 13	95
$(4-ClC_6H_4)_2AsCl$	$4-ClC_6H_4AsCl_2 + 4-ClC_6H_4AsO$	XXII		m. 45-48°, b. 175-179°/0.6 mm.	494
$(2-ClC_6H_4)_2AsCl$	$H_3AsO_4 + 4-ClC_6H_4NHNH_2$	XXI		m. 51°, b. 225°/15 mm.	1336
	$2-ClC_6H_4AsCl_2 + 2-ClC_6H_4AsO$	XXII		m. 73-75°	494
$[4,3-HO(O_2N)C_6H_3]_2AsI$	$[4,3-HO(O_2N)C_6H_3]_2AsCl + NaI$ in Me_2CO	I	65	m. 126-128°	99
$Ph(4-BrC_6H_4)AsCl$	$RR'AsO_2H + SO_2$	I	65	b. 46-47°, b. 184-186°/0.3 mm.,	805
	$[Ph(4-BrC_6H_4)As]_2O + HCl$	XI		b. 178-190°/6 mm., rxns. 4,14	908
	$PhAsCl_2 + 4-BrC_6H_4N_2Cl. ZnCl_2$	XX		b. 215-222°/5 mm., rxn. 12	1066
$Ph(4-BrC_6H_4)AsBr$	$RR'AsO_2H + SO_2$	I		m. 45-46°	96
$Ph(2-ClC_6H_4)AsCl$	$RR'AsO_2H + SO_2$	I		m. 30-35°	494
$Ph(4-ClC_6H_4)AsCl$	$[Ph(4-ClC_6H_4)As]_2O + HCl$	XI		b. 180-190°/6 mm., rxn. 12,14	908

Compound	Reaction	Method	Yield (%)	Properties	Ref.
Ph(2-O₂NC₆H₄)AsI	[Ph(2-O₂NC₆H₄)As]₂O + HI	XI	54	m. 113–114°	93
Ph₂AsF	Ph₂AsCl + AgF			b. 157–157.5°, rxns.[1,6]	1100
Ph₂AsI				Rxns. with CF₃I + Hg → Ph₂AsCF₃	255
Ph(4-H₂NO₂SC₆H₄)AsBr	RR'AsO₂H + SO₂	I		m. 100–101°, rxn.[6]	957
Ph(2-H₂NC₆H₄)AsCl	RR'AsO₂H + SO₂	I		Rxn.[16]	93
Ph(3-H₂NC₆H₄)AsCl	RR'AsO₂H + SO₂	I		HCl salt, m. 173–175°, rxn.[15]	93
Ph(4-H₂NO₂SC₆H₄)AsCl	RR'AsO₂H + SO₂	I		m. 106–107°	957
Ph(4-H₂NO₂SC₆H₄)AsI	RR'AsO₂H + HI	I		m. 121–122°	957
(4-H₂NC₆H₄)₂AsCl	R₂AsO₂H + SO₂	I	93	2HCl salt, colorless crystals decompg. in aq. soln. at 90° to (4-H₂NC₆H₄)₃As; rxn.[16]	64
[3,4-H₂N(HO)C₆H₃]₂AsCl	R₂AsO₂H + SO₂	I		Pearly flakes, m. 214–215°, rxn.[6]	95
(cyclo-C₆H₁₁)₂AsCl	R₃AsCl₂ at 210°	VIII	59	Rxn. with (:CMgBr)₂ → [:CAs(C₅H₁₁)₂]₂	516
C₁₃ Ph(2-HO₂CC₆H₄)AsCl	RR'AsO₂H + SO₂	I		Colorless prisms, m. 134–136°, rxn.[17]	494
Ph(4-HO₂CC₆H₄)AsCl	RR'AsO₂H + SO₂	I		Colorless prisms, m. 115–117°, rxn.[17]	494
(2-O₂NC₆H₄)(2-MeC₆H₄)AsCl	RR'AsO₂H + SO₂	I		Yellow prisms, m. 93–96°	494
(4-O₂NC₆H₄)(4-M₃C₆H₄)AsCl	RR'AsO₂H + SO₂	I		Yellow needles, m. 94–96°	494
Ph(4-MeC₆H₄)AsCl	RR'AsO₂H + SO₂	I	76	Pale yellow liq., b. 175–180°/0.05 mm.	805
	[Ph(4-MeC₆H₄)As]₂O + HCl	XI		b. 190–225°/18 mm., rxn.[4,12]	908
Ph(2-MeOC₆H₄)AsI	RR'AsO₂H + HI	I	37	m. 68–69°	93

11. Rxn. with conc. HCl or HBr → cleavage of C₁₀H₈ (1156).
12. Rxn. with H₂O₂ → the corresponding arsinic acid (908, 1066, 1336).
13. Rxn. with NaI in Me₂CO → the corresponding iodoarsine (99, 908, 1066).
14. Rxn. with RN₂Cl:ZnCl₂ + Fe → the corresponding tertiary arsine (908).
15. Rxn. with H₂O followed by H₃PO₂ → [Ph(3-H₂NC₆H₄)As:]n (94).
16. Rxn. with NH₄OH → (4-H₂NC₆H₄)₂AsOH (64).
17. Rxn. with MeOH + HCl → Ph(MeO₂CC₆H₄)AsCl (494).

TABLE VII (cont'd.)
DIORGANOHALOARSINES

R₂AsX	Prepd. from	Method	Yield %	Props., Rxns., Remarks	Ref.
CH₂(AsClPh)₂	PhAsO₃H₂ + AcCl	XIV			1024
C₁₄					
Ph(3-ClCH₂COC₆H₄)AsCl	Ph(3-AcC₆H₄)AsCl + Cl₂ in CCl₄ at 100°			Viscous yellow oil	494
Ph(3-AcC₆H₄)AsCl	RR'AsO₂H + SO₂	I		Colorless prisms, m. 71-72°	494
Ph(3-MeO₂CC₆H₄)AsCl	Ph(3-HO₂CC₆H₄)AsCl + MeOH + HCl			Brown oil	494
Ph(4-MeO₂CC₆H₄)AsCl	Ph(4-HO₂CC₆H₄)AsCl + MeOH + HCl			Brown oil	494
Ph(2-AcNHC₆H₄)AsCl	RR'AsO₂H + SO₂	I		Colorless needles, m. 162-164°	494
(4-MeOC₆H₄)₂AsBr	R₂AsO₂H + SO₂	I		m. 63-64°, rxn.16	95
Ph(PhCH₂CH₂)AsCl	RR'AsO₂H + SO₂	I		b. 137-140°/0.3 mm., rxn.18	614
Et(2-PhC₆H₄)AsCl	RR'AsO₂H + SO₂	I	70	b. 136-137°/0.2 mm, rxn.19	245
Ph(4-EtOC₆H₄)AsCl	PhAsCl₂ + (4-EtOC₆H₄N₂-Cl)₂.ZnCl₂	XX		Rxn.	1066
(-CH₂AsPhI)₂	[-CH₂As(Ph)O₂H]₂ + SO₂	I		Yellow crystals, m. 82°	199
C₁₆					
(4-AcC₆H₄)₂AsCl	R₂AsO₂H + SO₂	I		m. 124-125°, b. 210°/11-12 mm. (dec.)	1188
C₁₈					
Ph(4-PhC₆H₄)AsCl	RR'AsO₂H + SO₂	I	87	m. 87-88°, b. 211-212°/0.4 mm.	805
C₂₀					
(1-C₁₀H₇)₂AsCl	RAsCl₂ + RAsO	XXII		m. 163-165°	494
(2-C₁₀H₇)₂AsCl	2-C₁₀H₇AsCl₂ + 2-C₁₀H₇AsO	XXII		Colorless plates, m. 88°	494
(4-HO₂CC₆H₄)[2-(4-MeOC₆H₄)-C₆H₄]AsCl	RR'AsO₂H + SO₂	I		m. 200-204°, rxn.20	179

18. Rxn. with AlCl₃ at 100° → 1-phenylarsindoline and a prod., b. 90-95/0.5 mm. (614).
19. Rxn. with piperidine N-pentamethylenedithiourethane → 2-biphenylylethylarsino-N-pentamethyl-enedithiourethane, m. 124-125° (245).
20. Pyrolysis at 250° 14 mm. → 4-(3-methoxy-5-dibenzarsenolyl)benzoic acid (179).

ORGANODIHALOARSINES

Alkyl-, alkenyl-, aryl-, and heterocyclyl-dihaloarsines are prepared by:

1. Treating arsenoso compounds, RAsO, or arsonous acids, RAs(OH)$_2$, with hydrohalic acids at elevated temperatures, or with hydrogen halides dissolved
   in alcohol-ether mixtures at 0°, or with halogens in chloroform (Method I).
2. Reducing arsonic acids, RAsO$_3$H$_2$, with sulfur dioxide in hydrohalic acids,
   preferably in the presence of catalytic amounts of potassium iodide, or
   with phosphorus trichloride in carbon tetrachloride or acetic acid, or
   with a mixture of phosphorus trichloride with phosphorus pentachloride,
   or with 50% hydriodic acid, or with tin dichloride in diluted hydrochloric
   acid (Method II).
3. Treating arsonic acids, RAsO$_3$H$_2$, with acyl halides (Method III).
4. Reducing arsenic acid, H$_3$AsO$_4$, with arylhydrazines, RNHNH$_2$, in aqueous solution in the presence of copper bronze and treating the product with fumic
   hydrochloric acid (Method IV).
5. Reacting arsenic trichloride, AsCl$_3$, or arsenic trioxide, As$_2$O$_3$, in an
   acid medium with aryldiazonium salts in the presence of copper or copper
   salts, and reducing the resulting arylarsenic tetrahalide with sulfur
   dioxide in a hydrohalic acid containing catalytic amounts of iodine or
   of an iodine salt (Method V).
6. Heating a mixture of arsenic triacetate and arsenic trichloride or arsenic
   trichloride and acetyl chloride with aluminum chloride at 150-175° (Method
   VI).
7. Heating arsenic trioxide, As$_2$O$_3$, with alkyl halides in aqueous sodium hydroxide and reducing the reaction product (pentavalent arsenic derivative)
   with sulfur dioxide in concentrated hydrochloric acid (Method VII).
8. Treating arsenic trihalides with one mole-equivalent of a Grignard compound (Method VIII).
9. Treating tertiary arsines, such as (CF$_2$:CF)$_3$As, with hydrogen halides
   (Method IX).
10. Treating arsenic trichloride with olefins, e.g. C$_2$F$_4$, in the presence
    of aluminum chloride, or with acetylene or acetylene derivatives (to form
    the corresponding 2-chloroalkenyl derivatives) in the presence of copper(I)
    chloride or mercury(II) chloride, with or without promoters such as SbCl$_3$,
    SbCl$_5$, CdCl$_2$, ZnCl$_4$, or FeCl$_3$ (Method X).
11. Reacting halodiorganoarsines, R$_2$AsX, with arsenic trichloride in a ratio
    of 1:1 at 150° in the presence of 2% of colza oil (Method XI).
12. Heating tertiary arsines, R$_3$As, with arsenic trihalides in a ratio of
    1:2 (Method XII).
13. Metathesis of arsenic trihalides with organometallic compounds, such as
    Bu$_2$Sn(CH:CH$_2$)$_2$, R$_4$Pb, RHgCl, and R$_3$Al (Method XIII).
14. Decomposing alkylbis(α-alkenyl)arsines with alcoholic potassium hydroxide
    (evolution of acetylenic compounds) and heating the reaction mixture in a
    stream of a hydrogen halide (Method XIV).
15. Treating primary arsines, RAsH$_2$, with iodine in carbon tetrachloride (Method
    XV).
16. Treating arylbis(2-hydroxyethyl)arsines with thionyl chloride (substitution
    of the OH group by Cl, followed by elimination of the ethyl group) (Method
    XVI).

17. Cleavage of the arseno derivatives, RAs:AsR, with dry halogens or sulfuryl chloride, or by oxidation of the arseno compounds with hydrogen peroxide in hydrochloric acid (Method XVII).
18. Cleavage of tertiary arsines, R_3As, with halogens; mixtures of haloorgano-arsines, $RAsX_2$ and R_2AsX, are formed (Method XVIII).
19. Pyrolysis of tertiary arsine dihalides, R_3AsX_2, (elimination of alkyl halides and disproportionation of R_2AsX) (Method XIX).
20. Heating above 200° arsenic powder with alkyl halides in the presence or absence of copper (Method XX).
21. Reacting arsenic trichloride with diazoalkanes to form mixtures of (halo-alkyl)arsenic halides and tris(haloalkyl)arsines, $(RCHX)_{3-n}AsX_n$ (Method XXI).
22. Treating primary arsines with thionyl chloride (Method XXII).
23. Reacting diazonium double salts, $RN_2.MXn+_1$ (where M is a metal such as Zn or Fe, X a halogen, and n the valency number of M), with an arsenic trihalide in acetone in the presence of a metal (Method XXIII).
24. Neutron-irradiation of arsenic trichloride in benzene (Method XXIV).
25. Heating tertiary aromatic amines with arsenic trichloride at 100° to form N,N-disubstituted 4-aminoaryldichloroarsines (Method XXV).

TRIFLUOROMETHYLDIIODOARSINE $CAsF_3I_2$ CF_3AsI_2
Prepn.: By-prod. in rxn. of CF_3I with As at 220° (74, 117, 333, 1266).
  By-prod. in the rxn. of $(CF_3)_3As$ with I_2 at 100°, 5% yield (333).
  By-prod. in the rxn. of $(CF_3)_3As$ with AsI_3 at 230-240° (333).
Props.: Yellow eye-irritating liq. of bad odor, b. 100°/48.5 mm., n_D^{20} 1.6888 (1266); b. 183°/760 mm. (dec.), stable to H_2O at r. t. (333); b. 154° (334).
Rxns. with: $LiAlH_4 \rightarrow CF_3AsH_2$ (333).
$Zn + HCl \rightarrow CF_3AsH_2$ (333).
Aq. NaOH $\rightarrow CHF_3$ (333).
$H_2O_2 \rightarrow CF_3AsO_3H_2$ (334, 335).
$RMgX \rightarrow R_2AsCF_3$ (335).

METHYLENEBIS(DICHLOROARSINE) $CH_2As_2Cl_4$ $CH_2(AsCl_2)_2$
Prepn.: By treating a mixt. of $As(OAc)_3$ and $AsCl_3$ with $AlCl_3$ and heating the prod. with AdCl at 150-175°; 35% yield (499).
  By heating a mixt. of AcCl, As_2O_3, and $AlCl_3$ 2 hrs. at 170-5° and treat-ing the product, after cooling, with $SOCl_2$ (1024).
  By heating $CH_2(AsO)_2$ with conc. HCl at 60-70° (1213).
Props.: Flaky leaflets, m. 73°, b. 80°/4 mm. (499). Long, hard needles, m. 73°, b. 138-140°/12 mm., b. 255°/760 mm. (1024); m. 72-72.5° (1213).
Rxns. with: $NaOH \rightarrow CH_2(AsO)_2$ (1024).
$H_2S \rightarrow CH_2(AsS)_2$ (1213).
$RMgBr \rightarrow CH_2(AsR_2)_2$ (1215).

DICHLOROMETHYLARSINE CH_3AsCl_2 $MeAsCl_2$
Prepn.: By reducing $MeAsO_3H_2$ with SO_2 in hydrochloric acid contg. KI, 50% yield (44), 70% yield (660, 744).
  By treating $MeAsO_3H_2$ with AcCl (1024).
  By heating at 350° MeCl with powdered As in the presence of Cu (774a).

Props.: B. 129.5-130°/740 mm. (44); b. 130.7-131.5° (660); b. 132-133° (744).
Vapor pressure and volatility (1057). Identification (249, 530). Raman spectrum
(659).
Ihibits the effect of antiknock compds. in gasoline (952).
Rxns. with: Ergosterol in $CHCl_3$ → golden-yellow color (55).
RN_2-sulfate in the presence of Cu_2Cl_2, followed by hydrolysis → $MeRAsO_2H$ (251).
CH:CH in the presence of $AlCl_3$ → $Me(ClCH:CH)AsCl$ and $(ClCH:CH)_2AsMe$ (261).
$2-BrCH_2C_6H_4CH_2CH_2Br$ and Na → $\overline{CH_2C_6H_4CH_2CH_2}AsMe$ (557).
NH_4F → $MeAsF_2$ (744).
RCOSK → $MeAs[SC(O)R]_2$ (1249).
Alc., ROH, in $n-C_6H_{14}$ in the presence of Et_3N → $MeAs(OR)_2$ (1283).
Iodo-Cu(I) reagent (from NaI + $CuSO_4$) → ppt. (trace detn.) (301).

DIIODOMETHYLARSINE CH_3AsI_2 $MeAsI_2$
Prepn.: By reducing $MeAsO(CH)_2$ with SO_2 in HCl in the presence of KI (254).
By heating MeI with As powder at 280° in the presence of Cu; the prod.
contains $MeAsI_2$ + Me_2AsI (774a).
Props.: Raman spectrum (659).
Rxns. with: RMgI → $MeAsR_2$ (55).
CF_3I + Hg → $MeAs(CF_3)_2$ (254).
ROH in Et_2O contg. pyridine → $MeAs(OR)_2$ (646).
Me_2NH → $MeAs(NMe_2)_2$ (851).

DICHLORO(PENTAFLUOROETHYL)ARSINE $C_2AsCl_2F_5$ $C_2F_5AsCl_2$
Prepn.: By reacting $AsCl_3$ with C_2F_3 in the presence of $AlCl_3$ at r. t. then at 70-90°/9-14 atm.; 50-52% yield (135).
By reducing $C_2F_5AsO(OH)_2$ by SO_2 in HCl contg. HI (135).
Props.: B. 86-87°; d_{20} 1.9221; n_D^{20} 1.4063 (135).
Rxns. with: KOH → C_2F_5H (135).
H_2O_2 → $C_2F_5AsO(OH)_2$ (135).
NH_4F → $C_5F_5AsF_2$ (135).
Zn in MeOH-HCl → $C_2F_5AsH_2$ (135).

DICHLORO(2-CHLOROVINYL)ARSINE, LEWISITE, $C_2H_2AsCl_3$ $ClCH:CHAsCl_2$
Prepn.: By passing C_2H_2 at 30-40° through a mixt. of $AsCl_3$ and Cu_2Cl_2 dissolved
in 25-30% HCl; 85% yield (18).
By treating $AsCl_3$ with C_2H_2 in the presence of $HgCl_2$ in aq. HCl contg.
small amts. of promoters, such as $SbCl_3$, $SbCl_5$, $CdCl_2$, $ZnCl_2$, $SnCl_4$, or $FeCl_3$
(65); reaction mechanism (1033).
By passing C_2H_2 into a stirred mixt. of $AsCl_3$ in CCl_4 and a concd. soln.
of Cu_2Cl_2 in aq. $HOCH_2CH_2NH_2 \cdot HCl$ (539).
By treating $(ClCH:CH)_2AsCl$ with $AsCl_3$ (2:3) at 150° in the presence of
colza oil (19).
By-prod. in the rxn. of $PhAsCl_2$ with C_2H_2 in the presence of $AlCl_3$ (45).
By passing C_2H_2 into a mixt. of $AsCl_3$ with a soln. of $HgCl_2$ in Et_2OH,
18% HCl, or satd. NaCl soln.; 80-85% yield; investigation of various factors
affecting the rxn. (615).
Props.: The substance occurs in two isomeric forms. The product prepd. in the
presence of $AlCl_3$ contains more of the trans isomer, while that prepd. in the
presence of Cu_2Cl_2 or $HgCl_2$ contains more of the cis form (538). The cis isomer,

m. -46.6°, b. 51-52°/5 mm., is sol. in cold NaOH without evolution of gas with removal of 2Cl atoms by hydrolysis; addn. of HCl regenerates the trans isomer (538). Electron-diffraction study indicates that the prod. b. 150.2°/760 mm. is the cis isomer, but the possibility of the 1-chlorovinyl structure of this form is not excluded (298). The trans isomer, b. 190°/760 mm., as determined by the electron-diffraction method (298), is apparently more visicant; it dissolves in cold NaOH soln. with evolution of C_2H_2 and elimination of 3Cl atoms (538). Vapor pressure and volatility (1057).
Cis-isomer freezing point - 44.7°; thermodynamic data (1287).
Trans-isomer freezing point -1.2°; thermodynamic data (1287).
Dipole moments of the cis and trans isomers(759, 855).
Refr. index, density, and mol. refr. data for both isomers (915).
Identification (249, 530, 1326). Analytical procedure (784a, 1031).
Light absorption data (853, 856).
Rxns. with: 1,2- and 1,3-Dithiols → 2-(2-chlorovinyl)-1,3-dithio-2-arsa-cyclo-pentanes and -cyclohexanes, resp. (1167).
Keratin → the cyclic ClCH:CHAsS-R-S type compds. (1168).
NaOR → ClCH:CHAs(OR)$_2$ (1275).
KCN in MeOH → ClCH:CHAs(OMe)$_2$ (125).
NH_4F → FCH:CHAsF$_2$ (744).
PCl$_5$(1:1) → PCl$_3$, AsCl$_3$, and (CHCl$_2$)$_2$ (80).
C_6H_6 in the presence or absence of AlCl$_3$ → Ph(ClCH:CH)AsCl and Ph$_2$AsCH:CHCl (260, 262).
RMgX → R(ClCH:CH)AsCl and R$_2$(ClCH:CH)·As (260, 262, 674),
Cl$_2$-H$_2$O or H$_2$O$_2$ → ClCH:CHAsO$_3$H$_2$, mainly trans-isomer (538).
H$_2$O → ClCH:CHAsO (538, 1275).
[CH$_2$NHC(S)SH]$_2$ → ClCH:CHAsSC(S)NHCH$_2$CH$_2$NHC(S)S (548).
HSCH$_2$CH(SH)CH$_2$SO$_3$Na → ClCH:CHAsSCH$_2$CH(CH$_2$SO$_3$H)S (997).
NaOH the cis form → hydrolysis of two Cl atoms, without evolution of C_2H_2, while the trans form → hydrolysis of 3Cl atoms + C_2H_2 evolution (538).
Iodo-Cu(I) reagent (from NaI + CuSO$_4$) → ppt. (trace detn.) (300, 301).
Br$_2$ (Methyl red) titration (925).
Use: Catalyst for the prepn. of CH$_2$:CHCl form C_2H_2 + HCl (115).
Adduct with dioxane, long colorless needles, m. 58-59° (1275).

DICHLOROETHYLARSINE C$_2$H$_5$AsCl$_2$ EtAsCl$_2$
Prepn.: By adding slowly Et$_4$Pb to AsCl$_3$ (1:2) at 100° under N; 95-97% yield (661, 662, 663).
By reducing EtAsO$_3$H$_2$ with SO$_2$ in HCl contg. KI (44, 744).
By heating EtCl with As powder in the presence of Cu (774a).
Props.: B. 82.5°/75 mm.; d$_{20}^{22}$ 1.6570 (663); b. 152-155°/740 mm.(44); d. 1.6735-1.6799 (661).
Vapor pressure and volatility (1057).
Identification (249, 530, 1326).
Light absorption data (853, 856). Dipole moment (855, 1147b).
Rxns. with: RSH or RSNa → EtAs(SR)$_2$ (645, 650).
RONa → EtAs(OR)$_2$ (44, 650).
K$_2$CO$_3$ → (EtAsO)$_3$ (661).
Et$_4$Pb → Et$_2$AsCl (661, 663).

NH$_4$F → EtAsF$_2$ (744).
Br$_2$ (methyl red), titration (925).
PhC$_6$H$_4$N:NSCNHNHC$_6$H$_4$Ph, spectral analysis (1199).
EtNH$_2$ in CHCl$_3$ → EtAs:NEt (44).

DICHLORO-2-FURYLARSINE C$_4$H$_3$AsCl$_2$O $\overline{OCH:CHCH:C}$AsCl$_2$
Prepn.: by reacting (2-C$_4$H$_3$O)$_3$As with AsCl$_3$ at 160°; (2-C$_4$H$_3$O)$_2$AsCl is also formed (345, 866).
Props.: B. 102-103°; n$_D^{20}$ 1.6134; d$_4^{20}$ 1.817 (345); b. 80-93°/0.7 mm., relatively stable if stored cold in the dark; HgCl$_2$ catalyzes disproportionation to (C$_4$H$_3$O)$_3$As and AsCl$_3$ (866).
Rxns. with: Aq. alk. → (2-C$_4$H$_3$O)AsO (346, 347).
AgCN → (2-C$_4$H$_3$O)As(CN)$_2$ (346).
H$_2$S in EtOH → (2-C$_4$H$_3$O)AsS (347).
Cl$_2$ → AsCl$_3$ (753).

3-(DICHLOROARSINO)PYRIDINE C$_5$H$_4$AsCl$_2$N C$_5$H$_4$NAsCl$_2$
Prepn.: By treating a mixt. of 3-aminopyridine, AsCl$_3$, and Cu$_2$Cl$_2$ in concd. HCl with aq. NaNO$_2$ and reducing pyridyl-arsine tetrachloride with SO$_2$ in the presence of KI; isolated as monohydrate (85, 86, 589).
        By oxidizing 3,3'-arsenopyridine with 3% H$_2$O$_2$ in concd. HCl; 56% yield (85).
Props.: M. 226-227°, sinters at 217° (85).
Complex compds.: with heavy metal salts (87).
Rxns. with: H$_3$PO$_2$ + HI → 3,3'-arsenopyridine (85).
NH$_4$OH → 3-arsenosopyridine (85, 589).

DICHLORO-3-NITROPHENYLARSINE C$_6$H$_4$AsCl$_2$NO$_2$ 3-O$_2$NC$_6$H$_4$AsCl$_2$
Prepn.: by reducing 3-O$_2$NC$_6$H$_4$AsO$_3$H$_2$ with SO$_2$ in HCl at r. t. and crystallizing the prod. at low temp. (584); 90% (865).
Props.: M. 46° (584); m. 55° (865).
Rxns. with: 2-MeC$_6$H$_4$N$_2$Cl → 3,3'-azoxydi(benzenearsonic acid) (613).
RMgI → R$_2$(3-O$_2$NC$_6$H$_4$)As (998).
H$_2$S in EtOH → 3-O$_2$NC$_6$H$_4$AsS (552).

DICHLORO-4-NITROPHENYLARSINE C$_6$H$_4$AsCl$_2$NO$_2$ O$_2$NC$_6$H$_4$AsCl$_2$
Prepn.: by reducing 4-O$_2$NC$_6$H$_4$AsO(OH)$_2$ by SO$_2$ in HCl; isolated as O$_2$NC$_6$H$_4$As[SCSN-(CH$_2$)$_5$]$_2$ (93); redn. in HCl contg. KI gave 75% yield of the dichloroarsine (865, 1188).
Props.: Light yellow solid, m. 54-55° (865). Needles, m. 53-54° (1188); b. 189°/0.4 mm. (757).
Rxns. with: RCOCl + AlCl$_3$ → cleavage of the C-As bond with formation of 4-O$_2$-NC$_6$H$_4$COR (788).
KSC(S)OEt → 4-O$_2$NC$_6$H$_4$As[SC(S)OEt]$_2$ (988).
K$_2$S in Me$_2$CO → ppt. (989).
H$_2$S in EtOH → 4-O$_2$NC$_6$H$_4$AsS (552).
RN$_2$Cl, followed by hydrolysis → R(4-O$_2$NC$_6$H$_4$)AsO$_2$H (1188, 1228).

DICHLORO(4-CHLOROPHENYLARSINE C$_6$H$_4$AsCl$_3$ ClC$_6$H$_4$AsCl$_2$
Prepn.: By heating 4-ClC$_6$H$_4$HgCl with AsCl$_3$ (1:1) at 110° (896).

By-prod. in the rxn. of H_3AsO_4 with $4\text{-}ClC_6H_4NHNH_2$ in aq. soln. in the presence of Cu bronze, followed by treating the prod. with fuming HCl (1336).
Props.: B. 142-145° /15 mm. (896); b. 160-170°/20 mm., b. 277° (1336).
Rxns. with: $RCOCl + AlCl_3 \rightarrow$ cleavage of the C-As bond and formation of $4\text{-}ClC_6\text{-}H_4COR$ (788).
Cl in $H_2O \rightarrow 4\text{-}ClC_6H_4AsO(OH)_2$ (896).
$KSC(S)OEt \rightarrow 4\text{-}ClC_6H_4As[SC(S)OEt]_2$ (988).
Aq. $NH_4OH \rightarrow 4\text{-}ClC_6H_4AsO$ (1336).

DICHLOROPHENYLARSINE $C_6H_5AsCl_2$ $PhAsCl_2$
Prepn.: By reducing $PhAsO_3H_2$ with SO_2 in HCl in a continuous process, 95-98% yield (57), and in a batch operation in the presence of I_2 (58, 136).
By bubbling air through a stirred mixt. of $AsCl_3$, $PhNHNH_2$, $CuCl_2$, and $FeCl_3$ in HCl; 65.5% yield (136).
By diazotizing a mixt. of $PhNH_2 \cdot HCl$ and $AsCl_3$, dissolving the resulting $PhN_2Cl.AsCl_3$ in concd. HCl, and treating the soln. first with $PhNHNH_2$ and $FeCl_3$ and then with $CuCl_2$ (Ph_2AsCl is also formed) (136).
By treating $PhAs(CH:CHCl)_2$ with alc. KOH and then with HCl; 79.4% yield (674).
By adding PhN_2SO_3Na to a stirred soln. of As_2O_3 in HCl contg. $CuSO_4$ at 65°; 71% yield (58).
By treating PhAs:AsPh with SO_2Cl_2 (17).
By reacting $PhAsH_2$ with $\overline{CH_2CH_2O}$ and treating the prod. with $SOCl_2$ (122).
By adding dropwise PhMgBr to $AlCl_3$ (1:1) in Et_2O under N (728).
By-prod. in the rxn. of $PhAsO_3H_2$ with AcCl (1024).
By-prod. in the rxn. of $PhAsH_2$ with $SOCl_2$ (16, 17).
By-prod. in the rxn. of $PhNHNH_2$ with H_3AsO_4 in the presence of Cu_2O, followed by a treatment with HCl (56).
By-prod. in the rxn. of $(PhN_2)_2ZnCl_4$ with $AsCl_3$ in Me_2CO in the presence of Zn (507).
By-prod. in the redn. of $Ph(PhCH_2CH_2)AsO_2H$ with SO_2 in HCl contg. KI (614).
By-prod. in the rxn. of PhHgCl with $AsCl_3$ at 110° (896).
By neutron irradiation of $AsCl_3$ in C_6H_6 (907, 1139).
Props.: B. 127°/4 mm. (45); b. 115-117°/10 mm. (58); b. 119-121°/15 mm. (122); b. 116°/6 mm. (135); b. 60°/0.2 mm. (614); b. 100-102°/2 mm (674). b. 252°/760 mm.; d_{15} 1.54 (728). Vapor pressure and volatility (1057). Disproportionation (353, 384). Dipole moment (1148).
Rxn. with: H_2O (partial hydrolysis) $\rightarrow$ PhAsO and $PhAsCl_2$ (57, 58, 1275).
C_2H_2 at 0-5° in the presence of $AlCl_3 \rightarrow$ ClCH:CHAsCl$_2$, (ClCH:CH)$_2$AsCl, (ClCH:-CH)$_3$As, Ph(ClCH:CH)AsCl, (ClCH:CH)$_2$AsPh, ClCH:CHAsPh$_2$, and Ph$_3$As (45, 263).
PhAsO at 290-300° $\rightarrow$ Ph$_2$AsCl, Ph$_3$As, and As$_2$O$_3$ (57, 58).
PhAsO (1:3) at 120° in the presence of ZnO $\rightarrow$ Ph$_2$AsCl (59).
$PhNH_2.HCl$ followed by $NaNO_2$ and then by $NaHCO_3 \rightarrow Ph_2AsO_2H$ (136).
$Me_2CHCH(CO_2H)NHCOCH_2Cl \rightarrow Me_2CHCH(CO_2H)NHCOCH_2As(Ph)O_2H$ (75).
$1,3,4\text{-}MeC_6H_3(SH)_2 \rightarrow$ 5-methyl-2-phenyl-1,3-dithia-2-arsindan (180).
$AlCl_3$ at 200-280°, followed by KI $\rightarrow Ph_4AsI$ (201).
$Br(CH_2)_xCN$ in 40% alk. $\rightarrow Ph[HO_2C(CH_2)_x]AsO_2H$ (242).
Alkyl or aralkyl halides (RX) in alk. soln. $\rightarrow PhRAsO_2H$ (242, 244, 614).

RMgX → PhAsR$_2$ (242, 455, 757, 806).
Me$_2$SO$_4$(1:1) → MePhAsO$_2$H (251).
PhCH:CHBr + NaOH, followed by SO$_2$ in HCl → 1-phenylarsindole (266).
PhC:CH at 140-150° → 1-phenyl-3-chloroarsindole (266).
(CH$_2$COCl)$_2$ + Na in C$_6$H$_6$ → (CH$_2$CO)$_2$AsPh (267).
Iodo-Cu(I) reagent (from NaI + CuSO$_4$) → ppt. (trace detn.) (300, 301).
RLi → PhAsR$_2$ (472).
2-BrCH$_2$C$_6$H$_4$CH$_2$CH$_2$Br + Na → tetrahydro-2-phenylisoarsinoline (557).
NaOR → PhAs(OR)$_2$ (625).
Glycols → PhAsO(CH$_2$)$_x$O (632).
(EtO)$_3$Sb → PhAs(OEt)$_2^x$ + EtOSbCl$_2$ (643).
o-C$_6$H$_4$(OH)$_2$ at 250° in a sealed tube → o-OC$_6$H$_4$OAsCl (644).
Radioactive AsCl$_3$ (907).
NaOH → PhAsO (728).
NaOH, followed by RN$_2$Cl → PhRAsO$_2$H (1187).
RN$_2$X, followed by Cu$_2$Br$_2$ in 95% EtOH → PhRAsO$_2$H (957).
Cl$_2$ → PhAsCl$_4$ (728).
KSC(S)OR → PhAs[SC(S)OR]$_2$ (988).
H$_2$O$_2$ → PhAsO$_3$H$_2$ (778, 1030, 1032).
NH$_4$F → PhAsF$_2$ (744).
AlCl$_3$, followed by pyrolysis → Ph$_4$AsCl (744).
Na in Et$_2$O → [PhAs:]$_2$ (756).
RCHCH$_2$O → PhAs(OCHRCH$_2$Cl)$_2$ or PhAsCl(OCHRCH$_2$Cl) (787).
RCOCl + AlCl$_3$ → cleavage of the C-As bond and RCOPh formation (788, 789).
(RN$_2$Cl)$_2$ZnCl$_2$, followed by KOH → (PhRAs)$_2$O (908).
Br$_2$ (titration) (925).
KSP(S)(OR)$_2$ → PhAs[SP(S)(OR)$_2$]$_2$ (988).
HSCH$_2$CH(SH)CH$_2$SO$_3$Na → SAs(Ph)SCH$_2$CHCH$_2$SO$_3$H (997).
H$_2$S in EtOH → PhAsS (552).
RN$_2$Cl, followed by hydrolysis → PhRAsO$_2$H (1188).
RCOSK → PhAs[SC(O)R]$_2$ (1249).
LiAlH$_4$ or LiBH$_4$ → PhAsH$_2$ (1293).

DIIODOPHENYLARSINE C$_6$H$_5$AsI$_2$ PhAsI$_2$
Prepn.: by treating PhAsH$_2$ in CCl$_4$ (869).
Props.: n$_D$ 1.84 (106). Thermodynamic data (869).. Volatility (1104).
Rxns. with: CF$_3$I + Hg → PhAs(CF$_3$)$_2$ (255).
RMgX → PhAsR$_2$ (616).
RN$_2$Cl.ZnCl$_2$, followed by KOH → (PhRAs)$_2$O (908).

2-AMINO-4-DICHLOROARSINOPHENOL C$_6$H$_6$AsCl$_2$NO 2,4-H$_2$N(Cl$_2$As)C$_6$H$_3$OH
Prepn.: by reducing 3,4-H$_2$N(OH)C$_6$H$_3$AsO$_3$H$_2$ with SO$_2$ in HI (872).
Props.: Neutralization curve (1232).
Hydrochloride, decomp. 146-148° (1259).
Rxns. with: COCl$_2$ → {2,5-HO[(HO)$_2$As]C$_6$H$_3$NH}$_2$CO (233).
Zn → [4,3-HO(H$_2$N)C$_6$H$_3$As:]$_2$ (752).
HSCH$_2$CH(SH)CH$_2$SO$_3$Na → 3,4-H$_2$N(HO)C$_6$H$_3$AsSCH$_2$CH(S)CH$_2$SO$_3$H (997).

4-DICHLOROARSINOBENZOYL CHLORIDE C$_7$H$_4$AsCl$_3$O 4-ClOCC$_6$H$_4$AsCl$_2$
Rxns. with: RNH$_2$ in the presence of alkali → RNHCOC$_6$H$_4$AsO (288, 766, 1164).

RNH$_2$ in C$_6$H$_6$ followed by 95% EtOH → RNHCOC$_6$H$_4$AsO (828).
RNH$_2$ in C$_6$H$_6$ followed by abs. MeOH → RNHCOC$_6$H$_4$AsCl$_2$ (828).
NaHNCO$_2$Et in Et$_2$O → EtO$_2$CN(COC$_6$H$_4$AsO$_3$H$_2$)$_2$ (1164).
Styrene-divinylbenzene polymer → ion-exchange resin (1203).

DICHLORO-p-TOLYLARSINE C$_7$H$_7$AsCl$_2$ 4-MeC$_6$H$_4$AsCl$_2$
Prepn.: By treating 4-MeC$_6$H$_4$As(CH:CHCl)$_2$ with alc. KOH, followed by HCl; 92.7%
yield (674).
      By reducing 4-MeC$_6$H$_4$AsO$_3$H$_2$ with SO$_2$ in concd. HCl contg. KI (1188).
Props.: Plates, m. 31° (1188); m. 42-43° (674).
Rxns. with: Na in Et$_2$O → (4-MeC$_6$H$_4$As:)$_2$ (756).
o-C$_6$H$_4$(CH$_2$Br)$_2$ + Na → 2-p-tolylisoarsindoline (757).
RN$_2$Cl, followed by hydrolysis → (MeC$_6$H$_4$)RAsO$_2$H(1188).

4-(DICHLOROARSINO)BENZOIC ACID C$_7$H$_5$AsCl$_2$O$_2$ 4-HO$_2$CC$_6$H$_4$AsCl$_2$
Prepn.: by reducing 4-HO$_2$CC$_6$H$_4$AsO$_3$H$_2$ with SO$_2$ in HCl (179).
Props.: M. 156-162° (179).
Rxns. with: RN$_2$Cl → R(4-HO$_2$CC$_6$H$_4$)AsCl$_3$ (179).
1,3,4-MeC$_6$H$_3$(SH)$_2$ → 2-(4-carboxyphenyl)-5-methyl-1,3-dithia-2-arsindan (179).

N-(4-AMINO-2-DICHLOROARSINOPHENYL)MELAMINE C$_9$H$_{10}$AsCl$_2$N$_7$ 5,2-H$_2$N[N:C(NH$_2$)N:C-
                                                        (NH$_2$)N:CNH]C$_6$H$_3$AsCl$_2$
Prepn.: by reducing 4,2-H$_2$N(Cl)C$_6$H$_3$AsO(OH)$_2$ with SO$_2$ in dild. HCl contg. HCl
and reacting under reflux the arsenoso deriv. with triamino-s-triazine (419).

N-(2-AMINO-4-DICHLOROARSINOPHENYL)MELAMINE C$_9$H$_{10}$AsCl$_2$N$_7$ 3,4-H$_2$N[N:C(NH$_2$)N:C-
                                                        (NH$_2$)N:CNH]C$_6$H$_3$AsCl$_2$
Prepn.: By reacting 3,4-O$_2$N(H$_2$N)C$_6$H$_3$AsO$_3$H$_2$ with 2,4-diamino-6-chlorotriazine in
aq. alk. and reducing the prod. with SnCl$_2$ in dild. HCl (420, 421).
      By condensing 3,4-(H$_2$N)$_2$C$_6$H$_3$AsO$_3$H$_2$ with 2,4-diamino-6-chloro-s-triazine
in aq. alk. and reducing the prod. with SnCl$_2$ in dild. HCl (419).

2,4-DIAMINO-6-[4-AMINO-3-(DICHLOROARSINO)ANILINO]-s-TRIAZINE C$_9$H$_{10}$AsCl$_2$N$_7$
                                    2,5-H$_2$N[N:C(NH$_2$)N:C(NH$_2$)N:CNH]C$_6$H$_3$AsCl$_2$
Prepn.: by reducing N-(4,6-diamino-s-triazin-2-yl)-6-nitro-m-arsanilic acid with
SnCl$_2$ in cond. HCl (419).

2-(4-AMINO-6-HYDRAZINO-s-TRIAZIN-2-YlAMINO)-4-(DICHLOROARSINO)PHENOL C$_9$H$_{10}$As-
                        Cl$_2$N$_7$O 4,3-HO[N:C(NH$_2$)N:C(NHNH$_2$)N:CNH]C$_6$H$_3$AsCl$_2$
Prepn.: by reducing the corresponding arsonic acid in concd. HCl contg. HI at 60°
(418).
Props.: White cryst. ppt. (418).
Rxns. with: Alk.→ the corresponding arsenoso deriv. (418).

N-(2,6-DIAMINO-4-DICHLOROARSINOPHENYL)MELAMINE C$_9$H$_{11}$AsCl$_2$N$_8$ 3,5,4-(H$_2$N)$_2$[N:C-
                                                (NH$_2$)N:C(NH$_2$)N:CNH]C$_6$H$_2$AsCl$_2$
Prepn.: by treating N-(4,6-dichloro-s-triazin-2-yl)-3,5-dinitroarsanilic acid
with aq. ammonia and reducing the prod. with SnCl$_2$ in HCl cont. HI (419).

2-(4-DICHLOROARSINOBENZAMIDO)THIAZOLE C$_{10}$H$_7$AsCl$_2$N$_2$OS 4-(SCH:CHN:CNHCO)C$_6$H$_4$AsCl
Prepn.: by reacting 4-(Cl$_2$As)C$_6$H$_4$COCl with 2-aminothiazole in C$_6$H$_6$ and treating
the prod. with abs. MeOH (828).

2-(4-DICHLOROARSINOBENZAMIDO)THIAZOLINE $C_{10}H_9AsCl_2N_2OS$
4-($\overline{SCH:CHNH\dot{C}HCNHCO})C_6H_4AsCl_2$
Prepn.: by reacting 4-(Cl$_2$As)C$_6$H$_4$COCl with 2-aminothiazoline in C$_6$H$_6$ and treating the prod. with abs. MeOH (828).

β-(DICHLOROARSINO)CAMPHOR $C_{10}H_{15}AsCl_2O$ $\overline{C(:O)CH_2CHCMe_2\dot{C}CH_2AsCl_2}$
$\quad\quad\quad\quad\quad\quad\quad\quad\quad\quad\quad\quad\quad\quad\quad\quad\quad CH_2CH_2\rfloor$
Prepn.: by heating β-camphorsulfinic acid or bis-10-camphorylmercury with AsCl$_3$ at 100° and 190-200°, respectively (751).
Rxns. with: conc. NaOH → the arsonous acid (751).

5-DICHLOROARSINO-2-THIENYL PHENYL KETONE $C_{11}H_7AsCl_2O_2$ $\overline{SC(COPh):CHCH:CAsCl_2}$
Prepn.: by reacting AsCl$_3$ with 5-chloromercury-2-thienyl phenyl ketone (1280).
Props.: Greenish yellow, m. 113° (1280).
Rxns. with: NaOH soln. → the arsenoso deriv. (1280).
3% H$_2$O$_2$ → the arsonic acid (1280).

N-(2-DICHLOROARSINO-5-DIMETHYLAMINOPHENYL)MELAMINE $C_{11}H_{14}AsCl_2N_7$
5,2-Me$_2$N-[$\dot{N}$:C(NH$_2$)N:$\dot{C}$(NH$_2$)N:$\dot{C}$NH]C$_6$H$_3$AsCl$_2$
Prepn.: by reacting under reflux 2,4-Cl(Me$_2$N)C$_6$H$_3$AsO(OH)$_2$ with triamino-s-triazine (1:1) and reducing the arsonic acid with SO$_2$ in dild. HCl contg. HI (419).

3-DICHLOROARSINO-8-NITRODIBENZOFURAN $CL_2H_6AsCl_2NO_2$

Prepn.: by boiling 8-nitro-3-dibenzofuranarsonic acid with PCl$_3$ in AcOH; 80% yield (1145).
Props.: M. 152° (1145).
Rxns. with: HgCl$_2$ at 350°→ cleavage of the C-As bond (1145).
H$_2$O → the arsenoso deriv. (1145).

3-DICHLOROARSINODIBENZOFURAN $C_{12}H_7AsCl_2O$ $\overline{C_6H_4OC_6H_3}$-3-AsCl$_2$
Prepn.: by refluxing 3-dibenzofuranarsonic acid with PCl$_3$ in AcOH; 85% yield (1145).
Props.: White solid, m. 130° (1145).
Rxn. with: H$_2$O→ the arsenoso deriv. (1145).

1-DICHLOROARSINO-7-NITRO-9-FLUORENONE $CL_3H_6AsCl_2NO_3$

Prepn.: by reducing 7-nitro-9-oxo-1-fluorenearsonic acid with PCl$_3$ in HOAc under reflux; 60% yield (450); 90% yield (1049).
Props.: Yellow needles, m. 215-217° (450, 1049).

3-DICHLOROARSINO-7-NITRO-9-FLUORENONE $C_{13}H_6AsCl_2NO_3$

Prepn.: by reducing the corresponding arsonic acid with PCl$_3$ in AcOH near boiling; 85% yield (1049).
Props.: m. 231-232° (1049).
Rxns. with: Fe(OH)$_2$ in alk. soln. at 90°→ 7-amino-9-oxo-4-fluorenearsonic acid (1049).

2-[4,6-BIS (DIMETHYLAMINO)-s-TRIAZIN-2-YLAMINO]-5-DICHLOROARSINO-1,3-PHENYLENE-DIAMINE $C_{13}H_{19}AsCl_2N_8$ 3,5,4-(H$_2$N)$_2$[$\dot{N}$:C(NMe$_2$)N:C(NMe$_2$)N:$\dot{C}$NH]C$_6$H$_2$AsCl$_2$
Prepn.: by reacting 3,5,4-(O$_2$N)[$\dot{N}$:CCIN:CCIN:$\dot{C}$NH]C$_6$H$_2$AsO$_3$H$_2$ with aq. Me$_2$NH and reducing the prod. with SnCl$_2$ in HCl contg. HI (419).

77

TABLE IX

ORGANODIHALOARSINES

RAsX₂	Prepd. from	Method	Yield %	Props., Rxns., Remarks	Ref.
C₁					
CF₃AsBr₂	(CF₃)₃As + Br₂	XVIII	12	b. 118°/745 mm., n_D^{20} 1.528	333
CF₃AsCl₂	(CF₃)₃AsCl₂ at 125°	XIX		b. 71° (extrapolated); stable to H₂O at r. t., rxns.[1,2]	333 334, 266
	CF₃AsI₂ + AgCl				
ClCH₂AsCl₂	AsCl₃ + CH₂N₂	XXI		b. 57-58°/16 mm., d_{15}^{17} 1.9806; rxn.[3]	121
CH₂(AsBr₂)₂	CH₂(AsO)₂ + 48% HBr	I	80	m. 87.5°	213
MeAsBr₂	MeAsO₃H₂ + SO₂ in 60% HBr	II		m. -70°, b.180.5-181.5°	660
	MeBr + As + Cu	XX			774a
MeAsF₂	MeAsCl₂ + NH₄F at 80°		95	Raman spectrum; m. -29.7°, b. 76.5°, fumes in air and etches glass at r.t., rxn.[4]; Microwave spectrum	659 744 926
C₂					
CF₂:CFAsCl₂	(CF₂:CF)₃As + HCl at -110°	IX	92.5	b. 115°, n_D^{20} 1.4820, d_{20} 1.9800	165
CF₃CF₂AsF₂	CF₃CF₂AsCl₂ + NH₄F on steam bath			b. 47-48°, d_{20} 2.0754, n_D^{20} 1.3472	135
ClCH:CHAsF₂	ClCH:CHAsCl₂ + NH₄F at 100°	XIII		b. 43.5°/14.5 mm.	744 773-4
CH₂:CHAsBr₂	AsBr₃ + Bu₂Sn(CH:CH₂)₂ (2:1)	XII	39.4	b. 73-76°/14 mm.	773
	(CH₂:CH)₃As + AsBr₃				
HO₂CCH₂AsCl₂	RAsO₃H₂ + PCl₃	II		Crystals, turn brown at 90° m. 112° (dec.)	825
	(:AsCH₂CO₂H)₂ + dry Cl in Et₂O	XVII			825
ClCH₂CH₂AsCl₂	HOCH₂CH₂AsO₃H₂ + SO₂	II		b. 79-80°/12 mm.	218
	ClCH₂CH₂AsO₃H₂ + SO₂	II	60	b. 99.8°-100°/18 mm.	44
(CH₂AsI₂)₂	PhAsCH₂CH₂As(Ph)CH₂CH₂ + HI	IX		Yellow crystals, m. 136°, rxn.[5]	609
EtAsBr₂	Et₃As + AsBr₃	XII	51	b. 200-201°	773
	EtAsO₃H₂ + SO₂	II	44	m. 87-88°	44

Compound	Preparation	Method	Yield	Vapor pressure and volatility	Ref.
EtAsF₂	EtAsCl₂ + NH₄F at 80°			m. -38.7°; b. 94°; b. 74°/400 mm.; thermodynamic data; fumes in air, etches glass at r. t., rxn.[4]	1057, 744
EtAsI₂				Rxn. with RI + NaOH → R-EtAsI	629
				Rxn. with ROH → EtAs(OR)₂	631, 646
C₃					
ClCH:CHCH₂AsCl₂	AsO₃ + ClCH:CHCH₂Cl + NaOH	VII		b. 104-105°, rxn.[6]	221
CH₂:CHCH₂AsCl₂	RAsO₃H₂ + SO₂	II	52	b. 42°/4.5 mm. (dec.)	44
ClCH₂CH₂CH₂AsCl₂	HO(CH₂)₃AsO₃H₂ + SO₂ in HCl, followed by SOCl₂	II		Rxn. with MeMgI → Cl(CH₂)₃AsMe₂; b. 132-136°/25 mm.	55, 739
PrAsCl₂	PrAs(CH:CHCl)₂ + alc. KOH	XIV	74.1	b. 90-93°/46-52 mm., d_{20}^{20} 1.5123, n_D^{20} 1.5358	674
	PrAsO₃H₂ + SO₂	II	47	b. 98.3-99.5°/70 mm., rxn.[7,8]	44
Me₂CHAsCl₂	(ClCH:CH)₂AsCHMe₂ + alc. KOH	XIV	58	Vapor pressure, volatility	1057
				b. 64-66°/12-13 mm., d_{20} 1.5722, n_D^{20} 1.5289	674
MeOCH₂CH₂AsCl₂	RAsO₃H₂ + SO₂	II	40	b. 94-95°/6 mm.	44
C₄					
SCH:CHCH:CAsCl₂	SCH:CHCH:CHgCl + AsCl₃ at 100°	XIII	43	b. 106-110°/4 mm., b. 125-130°/15 mm.	349, 494
	(SCH:CHCH:C)₃As + AsCl₃ at 150°	XII			349
ClCH₂O(CH₂)AsCl₂	RAsO₃H₂ + SO₂	II	36	b. 136-137°/5 mm.[9]	44
SO₂(CH₂CH₂AsCl₂)₂	SO₂(CH₂CH₂AsO₃H₂)₂ + SO₂	II	55	b. 79.5-80.5°	44
EtOCH₂CH₂AsCl₂	RAsO₃H₂ + SO₂	II	20	b. 95-97°/10 mm.	44

1. Rxn. with NH₃ at 20° → CHF₃ (253).
2. Rxn. with aq. NaOH → CHF₃ (333).
3. Rxn. with H₂O₂ → the corresponding arsonic acid, (RAsO₃H₂) (121, 674, 1128, 1336).
4. Rxn. with H₂O → the arsenoso deriv., RAsO (744).
5. Rxn. with piperidinium 1-piperidinecarbodithioate → [CH₂As(SC(S)N(CH₂)₄CH₂)₂]₂ (690).
6. Rxn. with PCl₅ → AsCl₃, PCl₃, and Cl₂CHCHClCH₃ (80).
7. Rxn. with AgCN → PrAs(CN)₂ (44).
8. Rxn. with AgOAc → PrAs(OAc)₂ (44).
9. The compd. was also prepared by adding HO(CH₂)₃As[O(CH₂)₃AsCl₂]₂ to a 40% HCHO soln. and satg. the mixt. with HCl (44).

TABLE IX (cont'd.)
ORGANODIHALOARSINES

$RAsX_2$	Prepd. from	Method	Yield %	Props., Rxns., Remarks	Ref.
$BuAsCl_2$	$BuAsO_3H_2 + SO_2$ in HCl + KI	II	45	b. 120°/60 mm., rxn. 10,11	44
$Me_2CHCH_2AsCl_2$	$BuAs(CH:CHCl)_2 + KOH$	XIV	89.9	b. 58-60°/5 mm., n^{20} 1.5238,	674
	$(Me_2CHCH_2)_3Al + AsCl_3$	XII		b. 57-58°/8 mm., n^{20}0.5, 1.5108, d_{20} 1.4126	1330
C_5					
$CH:CHN:CHCH:CAsCl_2$	$RAsO_3H_2 + SO_2$ in HCl	II		m. 201-202°	874
$CH:CHCONHCH:CAsCl_2$	$RAsO_3H_2 + H_3PO_2$	II		HCl salt hemihydrate, sinters at 92-110°, m. 154°,112	85
$C:NC(NH_2):CHCH:CAsCl_2$	$C:NC(NO_2):CHCH:CAsO_3H_2 +$ SO_2 in HCl	XVII			590
$CH_2:CH(CH_2)_3AsCl_2$	$RAsO_3H_2 + SO_2$	II	32	b. 102.5-105.5°/3.5 mm. (dec.)	44
$n-C_5H_{11}AsBr_2$	$RAsO_3H_2 + SO_2$	II	35	b. 125.5-127°/18 mm.	44
$n-C_5H_{11}AsCl_2$	$RAsO_3H_2 + SO_2$	II	40	b. 115°/60 mm.	44
$Me_2CHCH_2CH_2AsCl_2$	$RAsO_3H_2 + SO_2$	II	38	b. 120°/60 mm.	44
	$Me_2CHCH_2CH_2As(CH:CHCl)_2 +$ alc. KOH	XIV	74.6	b. 91-92°/8 mm., d_{20} 1.3863 n^{20} 1.5170, rxn.13	674
$EtMeCHCH_2AsCl_2$	$RAsO_3H_2 + SO_2$	II	38	b. 101-105°/21 mm., d^{25}_{25} 1.4302	44
C_6					
$2,4-(O_2N)_2C_6H_3AsCl_2$	$RAsO_3H_2 + SO_2$	II	75	Light yellow, m. 69-70.5°	865
$2,5-Cl_2C_6H_3AsCl_2$	$RAsO_3H_2 + SO_2$	II		m. 56-57°, b. 177-178°/20 mm.	120
$2,4-Cl_2C_6H_3AsCl_2$	$RAsO_3H_2 + SO_2$	II		b. 167-168°/12 mm.	120
$3,4-Cl_2C_6H_3AsCl_2$	$RAsO_3H_2 + SO_2$	II		b. 175-176°/12 mm.	120
$2,5-(Cl_2As)_2C_6H_3NO_2$	$2,5-(OAs)_2C_6H_3NO_2 + Cl_2$	I		Pale yellow powder, m. 73°	79
$2,1,4-ClC_6H_3(AsCl_2)_2$	$2,1,4-ClC_6H_3(AsO_3H_2)_2 + SO_2$	II	48	Yellow, m. 72.6-74.3°	865
$2-BrC_6H_4AsCl_2$	$RAsO_3H_2 + SO_2$	II	95	Pale yellow crystals, m. 110°/0.5 mm., rxn.14	612
$4-BrC_6H_4AsCl_2$	$H_3AsO_4 + 4-BrC_6H_4NHNH_2$	IV	80	b. 138-165°/0.4-0.8 mm., rxn.13	1336
	$RAsO_3H_2 + PCl_3$	II		b. 90-91°/3 mm., rxn.15	480

Compound	Preparation	Type	Yield (%)	Properties	Ref.
4-BrC₆H₄AsBr₂	RAsO₃H₂ + SO₂	II	86	b. 18C-185°/.2 mm.	99, 1261
CH:C(CO₂Me)OCBr:CAsBr₂	AsBr₃ + CH:C(CO₂Me)OCBr:CHgCl	XIII	50	m. 35-36°, rxn.15	67
2-O₂NC₆H₄AsCl₂	RAsO₃H₂ + SO₂	II	88	m. 95-96° Light yellow, m. 49°	865 1188
3,4-HO(O₂N)C₆H₃AsCl₂	RAsO₃H₂ + SO₂	II		Soly. data	1188
4,3-HO(O₂N)C₆H₃AsCl₂	RAsO₃H₂ + SO₂			Rxn. with RN₂Cl → R[4,3-HO(O₂N)C₆H₃]AsO₂H m. 44-45°	865
2-ClC₆H₄AsCl₂	RAsO₃H₂ + SO₂ Chlorination of 2-H₂NC₆H₄AsCl₂		86	via diazotization and decompn. in the presence of Cu₂Cl₂	865
4-Cl₂SC₆H₄AsO₃HNa	4-HO₃SC₆H₄AsO₃HNa + PCl₃ + PCl₅			m. 84-85°, rxn.16	956
4-ClSC₆H₄AsCl₂	(4-SC₆H₄AsCl₂)₂ + dry Cl₂	II	62	Rxn. with C₂H₂ → 4-ClCH₂-SC₆H₄AsCl₂	960
o-C₆H₄(AsCl₂)₂	C₆H₄As(Cl)OAs + SOCl₂			Crystallizes with 2 moles of dioxane, m. 76-86°, on exposure ot air evolves HCl, rxn.17	199
	o-C₆H₄(AsO₃H₂)₂ + SO₂	II	64.7		434 327 283
2-HO₃SC₆H₄AsBr₂	RAsO₃H₂ + SO₂	II		m. 96°/from (CS₂)	
		II		Crystals; rxn. with NaOH 2-HO₃SC₆H₄AsO	283
4-HO₃SC₆H₄AsBr₂	RAsO₃H₂ + SO₂	II		Rxn. with NaOH → 4-HO₃-SC₆H₄AsO	955
2-HOC₆H₄AsCl₂	RAsO₃H₂ + SO₂			Rxn.12	283
3-HOC₆H₄AsCl₂	RAsO₃H₂ + SO₂			Rxn.12	283
3,4-(HO)₂C₆H₃AsCl₂					286
3,4-H₂N(Cl)C₆H₃AsCl₂	RAsO₃H₂ + SO₂	II	20	Rxn. with COCl₂ → CO[NH-(Cl)C₆H₃As(OH)₂]₂	233
PhAsF₂	PhAsCl₂ + NH₄F			m. 42°, b. 110°/48 mm.	744
4-HO₃SC₆H₄AsI₂	RAsO₃H₂ + 50% HI	II		Rxn.18	955

10. Rxn. with AgCN → BuAs(CN)₂ (44).
11. Complex salts with heavy metals (87).
12. Rxn. with NH₄OH → the corresponding arsenoso deriv. (85, 283, 418, 506, 956, 958, 959).
13. Rxn. with HNO₃ → Me₂CHCH₂CH₂AsO₃H₂ (674).
14. Rxn. with RMgX (1:2) → 2-BrC₆H₄AsR₂ (238, 612).
15. Rxn. with RMgX (1:2) → 4-BrC₆H₄AsR₂ (99, 480, 1261).
16. Rxn. with Cl₂ in CHCl₃, followed by hydrolysis → 4-ClO₂SC₆H₄AsO₃H₂ (956).
17. Rxn. with RMgI → o-C₆H₄(AsR)₂ (199, 327).
18. Rxn. with CH₂(CH₂)₄NH·HSC(S)N(CH₂)₄CH₂ → CH₂(CH₂)₄CH₂ → CH₂(CH₂)₄NH·NO₃SC₆H₄As[SC(S)N(CH₂)₄CH₂]₂ (955).

TABLE IX (cont'd.)
ORGANODIHALOARSINES

RAsX₂	Prepd. from	Method	Yield %	Props., Rxns, Remarks	Ref.
4-H₂NO₂SC₆H₄AsBr₂	RAsO₃H₂ + SO₂	II		m. 191-192°, rxn.12	956
3-H₂NC₆H₄AsCl₂	RAsO₃H₂ + SO₂	II		HCl salt, white cryst. solid rxn.12	283
2,3-H₂N(HO)C₆H₃AsCl₂	RAsO₃H₂ + SO₂	II	77	HCl salt, needles, m. 136-137°	286
5,2-H₂N(HO)C₆H₃AsCl₂	RAsO₃H₂ + SO₂	II		HCl salt, m. 183-183.4°, rxn.19	289
2,5-H₂N(HO)C₆H₃AsCl₂	RAsO₃H₂ + SO₂	II		HCl salt, m. 128-128.2°	289
4,3-H₂N(HO)C₆H₃AsCl₂	RAsO₃H₂ + SO₂	II	55	HCl salt, needles, m. 144-145°	286
4-H₂NO₂SC₆H₄AsCl₂	RAsO₃H₂ + SO₂ or PCl₃	II		m. 176-177°	956
	4-H₂NO₂SC₆H₄AsO + 10% HCl	I		Rxn.12	956
4-H₂NO₂SC₆H₄AsI₂	4-H₂NO₂SC₆H₄AsO + 50% HI	II		m. 192-193°	956
3,4-(H₂N)₂C₆H₃AsCl₂	RAsO₃H₂ + SO₂	II	49	2HCl salt, diamond-shaped plates	286
n-C₆H₁₃AsCl₂	RAsO₃H₂ + SO₂	II	10	b. 125-127°/28 mm., rxn.20	44
C₇					
4-NCHNC₆H₄AsBr₂	RAsO + aq. HBr	I		HCl salt, white crystals	51
4,3,5-Me(O₂N)₂C₆H₂AsCl₂	RAsO₃H₂ + SO₂	II	86	m. 126-127.5°	865
2-HO₂CC₆H₄AsCl₂	RAsO₃H₂ + SO₂	II		Crystals, m. 156°, rxn. with RMgX → 2-HO₂CC₆H₄AsR₂	54
CH:C(CO₂Et)OCBr:CAsBr₂	CH:C(CO₂Et)OCBr:CHgCl + As-Cl₃	XIII	39	m. 52-53°	67
2,3-H₂N(HO₂C)C₆H₃AsCl₂	RAsO₃H₂ + SO₂	II	99	HCl·H₂O salt, needles, m. 124-124.5°, rxn.21	290
3,4-H₂N(HO₂C)C₆H₃AsCl₂	RAsO₃H₂ + SO₂	II	90	HCl salt, rectangular prisms m. 195°, rxn.21	290
3,2-H₂N(HO₂C)C₆H₃AsCl₂	RAsO₃H₂ + SO₂	II	93	HCl salt, plates, m. 195°, rxn.21	290
2,6-H₂N(HO₂C)C₆H₃AsCl₂	RAsO₃H₂ + SO₂	II	47	HCl salt, needles, sinter 149-150°, rxn.21	290
2,4-H₂N(HO₂C)C₆H₃AsCl₂	RAsO₃H₂ + SO₂	II	96	HCl salt, needles, m. 220°, rxn.21	290
3,5-H₂N(HO₂C)C₆H₃AsCl₂	RAsO₃H₂ + SO₂	II	87	HCl·H₂O salt, needles, m. 221-222°, rxn.21	290
4,3-H₂N(HO₂C)C₆H₃AsCl₂	RAsO₃H₂ + SO₂	II	60	HCl salt, m. 120-121°, rxn21	290

Compound	Preparation	Method	Yield (%)	Properties	Ref.
4,2-H₂N(HO₂C)C₆H₃AsCl₂	$RAsO_3H_2 + SO_2$	II	80	HCl·H₂O salt, rectangular prisms, m. 219°, rxn.[21]	290
5,2-H₂N(HO₂C)C₆H₃AsCl₂	$RAsO_3H_2 + SO_2$	II	98	HCl salt, plates, m. 191.5°, rxn.[21]	290
2,5-H₂N(HO₂C)C₆H₃AsCl₂	$RAsO_3H_2 + SO_2$	II	90	HCl·H₂O, needles, m. 215°, rxn.[21]	290
4-ClCH₂C₆H₄AsCl₂	4-HOCH₂C₆H₄AsCl₂ + PCl₃			m. 29-30°, b. 157-160°/7 mm., rxn.[3]	1128
4-H₂NC(:NH)C₆H₄AsBr₂	RAs(OH)₂ + HBr	I		HBr salt, m. 219° (dec.)	738
2-MeC₆H₄AsCl₂	$RAsO_3H_2 + SO_2$	II	89	Plates, m. 48-50°, rxns[22,23]	1187
4-H₂NC(:NH)C₆H₄AsCl₂	RAs(OH)₂ + HCl	I		HCl salt, m. 208° (dec.)	738
3,4,5-H₂N(HO)(H₂NCO)C₆H₂-AsCl₂	3,4,5-H₂N(HO)(MeO₂C)C₆H₂As-O₃H₂ + NH₃, followed by SO₂			m. 177-178°	288
4-HOCH₂C₆H₄AsCl₂	$RAsO_3H_2 + SO_2$	II	30	Greenish oil, rxn.[24]	1128
3-MeOC₆H₄AsCl₂	$RAsO_3H_2 + SO_2$	II	70	b. 157°/15 mm., rxn.[25]	613
4-MeOC₆H₄AsCl₂	RAsO + HCl	I		m. 49-50°, rxn.[26]	93
				m. 47-48°, b. 138.5-139°/0.5 mm.	757
2-MeSC₆H₄AsCl₂	$RAsO_3H_2 + SO_2$	II		Colorless crystals, m. 66-67°, rxn.[27]	743
n-C₇H₁₅AsCl₂	$RAsO_3H_2 + SO_2$	II	6	b. 131-131.5°/4 mm., d25 1.296	44
C₈					
4-NCCH₂C₆H₄AsCl₂	$RAsO_3H_2 + SO_2$	II		Colorless needles, m. 56-57°, rxn.[21]	1128
				b. 215-218°/15 mm.	865
3-ClCH₂COC₆H₄AsCl₂	$RAsO_3H_2 + SO_2$	II	94	b. 170-180°/0.8 mm., $n_D^{21.5}$ 1.6302	189
				b. 210-211°/11-12 mm., m. 55-56°	1118

19. Rxn. with RSH → 5,2-H₂N(HO)C₆H₃As(SR)₂ (356).
20. Rxn. with AgCN → n-C₆H₁₃As(CN)₂ (44).
21. Rxn. with aq. alkali → the corresponding arsenoso deriv. (288, 290, 417, 419-421, 424, 426, 429, 1128).
22. Rxn. with PbCl₄ → 2-MeC₆H₄AsCl₄ (144).
23. Rxn. with RN₂Cl, followed by hydrolysis → R(2-MeC₆H₄)AsO₂H (1188).
24. Rxn. with PCl₃ → 4-ClCH₂C₆H₄AsCl₂ (1128).
25. Rxn. with RN₂Cl, followed by hydrolysis → R(3-MeOC₆H₄)AsO₂H (619).
26. Rxn. with o-C₆H₄(CH₂Br)₂ + Na → 2-(4-methoxyphenyl)isoarsindoline (757).
27. Rxn. with RMgX → 2-MeSC₆H₄AsR₂ (743).

TABLE IX (cont'd.)
ORGANODIHALOARSINES

RAsX$_2$	Prepd. from	Method	Yield %	Props., Rxn., Remarks	Ref.
$PhCH{:}CHAsCl_2$	$PhCH{:}CHAs(CH{:}CHCl)_2$ + alc. KOH	XIV	41.8	b. 108-110°/5 mm.	674
$2\text{-}CH_2{:}CHC_6H_4AsCl_2$	$RAsO_3H_2 + SO_2$	II		m. 55°, rxn. with AlCl$_3$ in CS$_2$ → 1-chloroarsindole	264
$4,3,5\text{-}EtO(O_2N)_2C_6H_2AsCl_2$	$RAsO_3H_2 + SO_2$	II	74	m. 81-82.3°	865
$4\text{-}AcC_6H_4AsCl_2$	$RAsO_3H_2 + SO_2$	II		m. 100°, b. 202°/13-15 mm., rxn.[28]	1118
$4\text{-}HO_2CCH_2C_6H_4AsCl_2$	$4\text{-}H_2NCOCH_2C_6H_4AsCl_2$ + 20% HCl at 80°	II		m. 107-109°	1128
$4\text{-}HO_2CCHOHC_6H_4AsCl_2$	$RAsO_3H_2 + SO_2$	II	60	m. 167°	393
$4\text{-}H_2NCOCH_2C_6H_4AsCl_2$	$RAsO_3H_2 + SO_2$	II		m. 143-145°, rxn. with 20% HCl → $4\text{-}HO_2CCH_2C_6H_4AsCl_2$	1128
$4\text{-}ClCH_2CH_2SC_6H_4AsCl_2$	$RAsO_3H_2$	II	68	b. 186-193°/0.25 mm.	865
	$4\text{-}ClSC_6H_4AsCl_2 + C_2H_4$	II	96	b. 148-152°/0.04 mm.	960
		II	62		960
$4\text{-}H_2NCH_2CONHC_6H_4AsCl_2$	$RAsO_3H_2 + SO_2$	II		Stable to hydrolysis	288
$4\text{-}HOCH_2CH_2OC_6H_4AsCl_2$	$RAsO_3H_2 + SO_2$	II		Stable to hydrolysis	284
$4\text{-}Me_2NO_2SC_6H_4AsI_2$	$RAsO_3H_2 + HI$	II		m. 132.5-134°, rxn.[12]	958
$3,4\text{-}H_2N(HOCH_2CH_2O)C_6H_3AsCl_2$	$RAsO_3H_2 + SO_2$	II	54	HCl salt, white cryst. solid, m. 174°	502
$Me(CH_2)_5CCl{:}CHAsCl_2$	$Me(CH_2)_5C{:}CCu + AsCl_3$	X		Rxn. with H$_2$O$_2$ → RAsO$_3$H$_2$	445
$O(CH_2CH_2OCH_2CH_2AsCl_2)_2$	$O(CH_2CH_2OCH_2CH_2AsO_3H_2)_2 + SO_2$	II	5	b. 185-190°/0.1 mm.	44
C$_9$ $3,5,4\text{-}(H_2N)_2(N{:}CClN{:}CClN{:}C\text{-}NH)C_6H_2AsCl_2$	$4,3,5\text{-}H_2N(O_2N)_2C_6H_2AsO_3H_2$ + (ClCN)$_3$ in H$_2$O at 0°, followed by SnCl$_2$				422
$N{:}C(NH_2)N{:}C(NH_2)N{:}C(NH_2)N{:}CSC_6H_4AsCl_2$	$RAsO_3H_2 + SO_2$	II		White crystals, rxn.[21]	426
$4,3\text{-}HC[N{:}C(NH_2)N{:}C(NH_2)N{:}CNH]C_6H_3AsCl_2$	$RAsO_3H_2 + SO_2$	II		HCl salt, cryst. solid, rxn.[21]	417, 419, 424, 429
$4\text{-}MeO_2CCH_2C_6H_4AsCl_2$	By-prod. in the prepn. of $4\text{-}NCCH_2C_6H_4AsO_3H_2$ by the Bart-Scheller method			m. 89-90°, rxn.[3]	1128

Compound	Reaction	Method	Yield (%)	Properties	References
4-EtOC(:NH)C6H4AsCl2	4-NCC6H4AsO + HCl gas in Et2O-EtOH		53	HCl salt, m. 141°, rxn.21	288
EtMeNC6H4AsBr2	EtMePhN + AlCl3	XXV		HBr salt, m. 143°	1256
EtMeNC6H4AsCl2	RAsO3H2 + SO2	II		HCl salt, m. 99°	1256
4-[N:C(NHNH2)N:C(NHNH2)NC-NHNH]C6H4AsCl2				White cryst. ppt., rxn.12	418
4-EtMeNC6H4AsI2				HI salt, light yellow, un-stable	1256
C10					
2-C10H7AsCl2	RAsO3H2 + SO2 / RAsO + HCl	I		m. 67-68°, complex[29]	136, 39
4-[N:C(NH2)N:CHCH:CNH]C6H4-AsCl2	RAsO3H2 + SO2	II		White crystals, m. 250°	506
4-[N:CHCH:C(NH2)N:CNH]C6H4-AsCl2	RAsO + HCl	I			39
4,3-[N:C(NH2)N:CHCH:CNH]C6-H3AsCl2	RAsO3H2 + SO2	II	27	HCl salt, m. 163-164°	393
3,4-H2N[HO2C(CH2)3]C6H3AsCl2	RAsO2H2 + SnCl	II			420, 421
2,5,4-H2N(MeO)[N:C(NH2)N:C-(NH2)N:CNH]C6H2AsCl2	RAsO3H2 + SO2	II	74	m. 167.5-169.5°	1276, 419
4-[O(CH2CH2)2NO2S]C6H4AsI2	RAsO3H2 + SO2	II			1276
4,5,2-MeNH(H2N)[N:C(NH2)N:C(NH2)N:CNH]C6H2AsCl2					
C11					
4-[CH2(CH2)4NO2S]C6H4AsI2	RAsO3H2 + SO2	II	50	m. 127-129°	1276
C12					
2-(4-BrC6H4O)C6H4AsCl2	RAsO3H2 + SO2	III		Prisms, m. 76-77° Rxns.30,31	858
HNC6H4C6H3AsCl2	RAsO3H2 + SO2	III			1127
2-(4-O2NC6H4)C6H4AsCl2	RAsO3H2 + PCl3	III	76.5	m. 81-82°	449
2,4-Ph(O2N)C6H3AsCl2	RAsO3H2 + PCl3	III	70.9	m. 105-106°	449
(:NC6H4AsCl2)2	(:NRAsO3H2)2 + SO2	III	78	Yellow crystals, m. 129-131°	611
O(:NC6H4AsCl2)2	O(:NRAsO3H2)2 + SO2	II	81	Yellow, m. 119-120°	865

28. Rxn. with Cl2 in EtOH, followed by hydrolysis → the corresponding arsonic acid (231, 1118).
29. 2-C10H7AsCl2.C10H7N2Cl, complex salt, m. 88° (dec.) (131).
30. Rxn. with hot conc. HCl in EtOH → cleavage of the C-As bond (1127).
31. Rxn. with NH4OH in Me2CO → the arsonous acid (1127).

TABLE IX (cont'd.)

ORGANODIHALOARSINES

RAsX₂	Prepd. from	Method	Yield %	Props., Rxns., Remarks	Ref.
(-SC₆H₄AsCl₂)₂	(4-H₂O₃AsC₆H₄S-)₂ + SO₂	II	84	m. 125-126.5°, Cl in CHCl₃ cleaves the S-S bond	960
2-PhC₆H₄AsCl₂	RAsO₃H₂ + SO₂	II	88	White needles, m. 47.5-48.5°, rxns. 36;37	245
2-PhOC₆H₄AsCl₂	RAsO₃H₂ + SO₂	II		Orange-red oil; dehydrohalogenation32; rxn.33	271
4-PhOC₆H₄AsCl₂	RAsO₃H₂ + SO₂	II		Pale yellow cryst. mass, m. 66° Rxn. 12	271
4-(4-HO₃SC₆H₄)C₆H₄AsI₂	RAsO₃H₂ + 50% HI	II		Rxn. 12	959
2-(PhClAs)C₆H₄AsCl₂	2-(PhAsO₂H)C₆H₄AsO₃H₂ + SO₂	II		Dehydrohalogenation34	200
4-(H₂NO₂SC₆H₄)C₆H₄AsI₂	RAsO₃H.NH₄ + 50% HI	II		Rxns.12	959
4-(PhNHO₂S)C₆H₄AsI₂	RAsO₃H₂ + HI	II		m. 125-126°; rxn.12	958
4-(4-H₂NO₂SC₆H₄NHO₂S)C₆H₄-AsI₂	RAsO₃H₂ + HI	II		m. 195-197°, rxn.12	958
4-Me₂NC₁₀H₆AsBr₂				HBr salt, soly. data	1256
4-Me₂NC₁₀H₆AsCl₂				HCl salt, m. 110-112°	1256
4-Me₂NC₁₀H₆AsI₂				HI salt, pale yellow, m. 119-120°	1256
C₁₃					
3-Dichloroarsino-9-fluorenone	RAsO₃H₂ + PCl₃	II	80	m. 161-163°	1049
4-(HO₂CC₆H₄NHO₂S)C₆H₄AsI₂	RAsO₃H₂ + HI	II	75	m. 234-236°; rxn.31	958
2-(3-MeC₆H₄O)C₆H₄AsCl₂	RAsO₃H₂ + SO₂	II		Oil	858
2-(2-MeC₆H₄O)C₆H₄AsCl₂	RAsO₃H₂ + SO₂	II		Needles, m. 73-74°	858
2-(4-MeC₆H₄O)C₆H₄AsCl₂	RAsO₃H₂ + SO₂	II		Needles, m. 73°	858
2-(4-MeOC₆H₄)C₆H₄AsCl₂	RAsO₃H₂ + SO₂ in HCCl₃-HCl	II		m. 63-67°; rxn.35	179
	RAsO₃H₂ + SO₂ in HCl	II		m. 63-64°	858
4-(MePhNO₂S)C₆H₄AsI₂	RAsO₃H₂ + HI	II	85	m. 98-99°	1276
C₁₄					
2-(4-ClCH₂COC₆H₄)C₆H₄AsCl₂	RAsO₃H₂ + SO₂	II		Pale yellow, m. 105-106°	189
2-(2,4-Me₂C₆H₃O)C₆H₄AsCl₂	RAsO₃H₂ + SO₂	II		Prisms, m. 52.5-54°	858
2-(2,5-Me₂C₆H₃O)C₆H₄AsCl₂	RAsO₃H₂ + SO₂	II		Prisms, m. 70-71.5°	858
2-(2,3-Me₂C₆H₃O)C₆H₄AsCl₂	RAsO₃H₂ + SO₂	II		Prisms, m. 71.5-73°	858

C_{18}					
4-(4-PhC$_6$H$_4$NHC$_2$S)C$_6$H$_4$AsI$_2$	RAsO$_3$H$_2$ + H	II	75	m. 146-147°	1276
C_{19}					
4-[Ph(PhCH$_2$)NO$_2$S]C$_6$H$_4$AsI$_2$	RAsO$_3$H$_2$ + HI	II	80	m. 154-156°	1276

32. On heating at 200° 10 mm. → 10-chlorophenoxarsine (271).
33. Rxn. with RMgI → 2-PhOC$_6$H$_4$AsMe$_2$ (271).
34. On heating under reflux at 12 mm. HCl is eliminated and 5,10-dichloro-5,10-dihydroarsanthrene along with tri-o-phenylenediarsine is formed (200).
35. On heating at 200° → 5-chloro-3-methoxydibenzarsenole (179).
36. Rxn. with piperidine N-pentamethylenedithiocarbamate → 2-biphenylarsylenebis(N-pentamethylene-dithiocarbamate, m. 216° (245).
37. Rxn. with EtBr in et. NaOH → Et(2-PhC$_6$H$_4$)AsO$_2$H (245).

ARSENOSO AND ARSONOUS ACIDS DERIVATIVES

Organoarsenic compounds formed by alkaline hydrolysis organodihaloarsines or of the reaction products obtained by reduction of arsonic acids with sulfur dioxide in the presence of the iodide ions in an acid medium occur in the form of arsonous acids, $RAs(OH)_2$, or of arsenoso compounds, $(RAsO)n$. Until about 1940 all organoarsenic compounds prepared as mentioned above were considered to be the arsenoso compounds. E. Krause and A. von Grosse described, in their monumental treatise "Die Chemie der metall-organischer Verbindungen," Verlag von Gebrüder Borntraeger, Berlin, 1937, pp. 470 and 511, the products of alkaline hydrolysis of organodihaloarsines as the arsenoso compounds, and no mention was made of the existence of the arsonous acids.

In many instances the arsenoso compounds reported in the literature were isolated in the crude state and were not properly identified. The general empirical formulae of the compounds, $RAsO$, correspond to the anhyrides of arsonous acids, but they exist apparently as dimeric, trimeric, or tetrameric aggregates in which each arsenic atoms forms two single bonds with oxygen atoms:

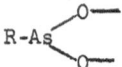

Identification of these two types of compounds is difficult because some arsenoso compounds were reported to crystallize with various amount of water of crystallization, and it is not certain whether they are hydrates of the arsenoso compounds or arsonous acids and their semianhydrides, respectively.

In 1940 Scott et al. (U. S. 2,221,817) reported the preparation of 2-amino-4-arsenosophenol of a relatively high purity (99%) with properties quite different from those previously reported for the compounds. The data were reinvestigated by Banks et al. (J. Am. Chem. Soc. 69, 5-11(1947), and they found that the compound reported by Scott et al. was 3-amino-4-hydroxybenzenearsonous acid. The latter was also converted into the corresponding arsenoso derivative. Arsonous acids can be dehydrated step-wise to the corresponding semianhydrides, $(RAsOH)_2O$ and then to the arsenoso derivatives, $(RAs{<}^{O-})n$.

Moreover, arsonous acids were converted to the corresponding mono- and dithioarsonous acids in reactions with hydrogen sulfide at pH 6 and ammonium sulfide in diluted hydrochloric acid, respectively. Most arsonous acids reported to date contain a substituted aromatic group as the organic moiety. Numerous arsonic acids were converted into the corresponding arsenoso derivative and arsonous acids. Arsonic acids carrying a free carboxylic group in the aromatic ring show a greater tendency toward the formation of arsonous acids than do the corresponding carbamylarsonic acids, although several carbamylarenearsonic acids were converted into the corresponding arsenoso and arsonous acid derivatives.

ARSENOSO COMPOUNDS

Organoarsenic oxides of the general formula $(RasO)n$, exist probably in the form of dimeric, trimeric, or tetrameric aggregates in which each arsenic atom forms two single bonds with two oxygen atoms.

The arsenoso compounds are prepared or formed by:

1. Reducing arsonic acids, $RAsO_3H_2$, with:
   (a) Sulfur dioxide in aqueous or methanolic hydrochloric acid or in sulfuric acid, in the presence of catalytic amounts of the iodide ions, and neutralizing the reaction product with aqueous alkali or ammonia.
   (b) Phosphorus trichloride or a mixture of the trichloride with pentachloride in an organic solvent, such as acetic acid, ethyl acetate, chloroform, or ether, and neutralizing the reaction product with aqueous alkali or ammonia.
   (c) Phosphorus oxychloride at 110-115° without any solvent and treating the reaction product with aqueous alkali.
   (d) Aqueous sodium bisulfite and precipitating the arsenoso compound with an acid.
   (e) Phenylhydrazine in methanol.
2. Hydrolysis of organodihaloarsines, $RAsX_2$, with aqueous alkali or ammonia, or with amines in benzene and recrystallizing the product from 95% ethanol.
3. Oxidation of primary arsines, $RAsH_2$, with oxygen in ether.
4. Dehydration of arsonous acids by azeotropic distillation with ethanol-benzene mixture, or by drying in vacuo at a temperature above 100° or over P_2O_5.
5. Transesterification of arsonous acid esters with arsenosophenyl derivatives containing an acidic group on the aromatic ring.
6. Reacting tertiary aromatic amines with arsenic trichloride at 100° and neutralizing the reaction product with aqueous sodium carbonate solution to yield a mixture of the corresponding arsenosoaniline, $RR'NC_6H_4AsO$, and tertiary aminoarylarsine, $(RR'NC_6H_4)_3As$.
7. Reacting arsenic trichloride with acetylenic compounds at 120-130° and treating the reaction mixture with potassium hydroxide solution to form arsenosohaloalkenes of the general structure: $R(Cl)C:C(R')AsO$.
8. Heating arsenic trioxide with acyl chlorides, $RCOCl$, at 210° in the presence of aluminum chloride, distilling the reaction product in vacuo, and treating it with aqueous ammonia.
9. Reacting arylazocarboxylic acids salts, RN_2CO_2K, with arsenic trichloride in acetone and neutralizing the reaction product.
10. Treating arseno compounds, $(RAs:)x$, with sulfur dioxide, sulfuric or an arsonic acid.
11. Heating arsenic trioxide with potassium carbonate, acetic anhydride, and acetic acid at 135-138° diarsenosomethane is formed.

DIARSENOSOMETHANE $CH_2As_2O_2$ $CH_2(AsO)_2$
Prepn.: By treating $CH_2(AsCl_2)_2$ with 2N NaOH and pptg. the prod. with H_2SO_4 or NH_4Cl (1024).

By heating As_2O_3 with K_2CO_3, Ac_2O, and AcOH at 135-138°, 64% yield (1213).
Props.: Occurs in two forms, one is sol. and the other insol. in C_6H_6 (apparently a dimer); both modifications m. 218-220° (dec.) (1213); m. 265° (dec.) (1024). On heating to 300° → $(Me_2As)_2O$, which was isolated as a complex with $2HgCl_2$ (1214).
Rxns. with: H_2O_2 → $CH_2(AsO_3H_2)_2$ (1024, 1213).

$Na_2S_2O_4 \rightarrow$ yellow ppt. $(:AsCH_2As:)x$ (?) (1024).
$NaH_2PO_2 \rightarrow (:AsCH_2As:)x$ (1213).
$Na_2Fe(CN)_5NO.H_2O \rightarrow$ black brown coloration (1024).
Concd. HCl at 60-70° $\rightarrow CH_2(AsCl_2)_2$ (1213).
48% HBr at 100° $\rightarrow CH_2(AsBr_2)_2$ (1213).

ARSENOSOMETHANE CH_3AsO MeAsO

Prepn.: by treating $(MeAs:)x$ with H_2SO_4 at r. t. (993).
Props.: Raman spectrum (994).
Rxns. with: $[3,4-H_2N(HO)C_6H_3As:]x$ in 3% HCl $\rightarrow (MeAs:)x$ and $3,4-H_2N(HO)C_6H_3AsO$ (693).
$\overline{CH_2CH_2O}$ in $H_2O \rightarrow Me(HOCH_2CH_2)AsO_2H$ (715).
H_2SO_4 at an elevated temp. $\rightarrow As_2O_3$ (993).
$RR'NC(S)SNa \rightarrow MeAs[SC(S)NRR']_2$ (1248).
$RCOSH \rightarrow MeAs[SC(O)R]_2$ (1249).
Complex compds.: $MeAsO.H_2SO_4$, white crystals, prepd. by treating $(MeAs:)x$ with the minimum amt. of H_2SO_4 required for oxidation (993).

2-ARSENOSOVINYL CHLORIDE C_2H_2AsClO ClCH:CHAsO

Prepn.: By shaking cis- or trans-$ClCH:CHAsCl_2$ with a large volume of H_2O (538).
        By shaking trans-$ClCH:CHAsCl_2$ dissolved in C_6H_6 with H_2O to form a white polymeric solid; the rxn. proceeds apparently through the $ClCH:CHAs(OH)_2$ and monomeric $ClCH:CHAsO$ steps (1275).
Props.: Occurs in two stereoisomeric forms; the cis-form, prepd. from the cis-arsine, m. 131°, and the trans-form, prepd. from the trans-arsine, m. 143° (538).

ARSENOSOETHANE C_2H_5AsO EtAsO

Prepn.: by adding $EtAsCl_2$ under CO_2 to a suspension of K_2CO_3 in C_6H_6 contg. a small amt. of H_2O and heating the mixt. under reflux; 80% yield (661).
Props.: Colorless liquid, b. 158°/10 mm., exists as $(EtAsO)_3$ (661).
Rxns. with: $\overline{CH_2CH_2O} \rightarrow Et(HOCH_2CH_2)AsOOH$ (715).

ARSENOSOBENZENE C_6H_5AsO PhAsO

Prepn.: By partial hydrolysis of $PhAsCl_2$ with H_2O at 50-70° or aq. NaOH; isolated as an oil mixt. contg. PhAsO and $PhAsCl_2$ (57, 58).
        By reducing $PhAsO_3H_2$ with SO_2 in aq. HCl contg. I_2 at 70°; isolated as a mixt. of PhAsO with $PhAsCl_2$ (58).
        By reducing $PhAsO_3H_2$ with $NaHSO_3$ in the presence of Cl^-, I^-, or SO_4^{--} (58).
        By reducing $PhAsO_3H_2$ with SO_2 in H_2O at 80°; 95% yield (651).
        By hydrolyzing $PhAsCl_2$ with hot NaOH and neutralizing the rxn. prod.
        By reducing $PhAsO_3H_2$ with $POCl_3$ at 110-115°, then treating the rxn. mixt. with ice-water and alkali, and neutralizing the soln. (981).
        By reacting PhN_2CO_2K with $AsCl_3$ in Me_2CO and treating the product with HCl, followed by H_2O, then heating with KOH at 90-100°, and satg. the soluble fraction with NH_4Cl; 42% yield (1062).
        By reducing $PhAsO_3H_2$ with $NaHSO_3$ in alk. soln. at 75-80°; 86% yield (549).
        By-prod. in the rxn. of $PhN_2Cl.AsCl_3$ with $PhNHNH_2.HCl$ in the presence of $CuCl_2$ and $FeCl_3$; 15.5% yield (136).
        By oxidation of $PhAsH_2$ with O in Et_2O (1293).

Props.: M. 140-143° (16); m. 127-130° (1062); m. 145° (1293).
UV spectrum (598).
Rxns. with: PhAsCl$_2$ at 290-300° → Ph$_2$AsCl + Ph$_3$As + As$_2$O$_3$ (57, 58).
PhAsCl$_2$ (3:1) at 120° in the presence of ZnO → Ph$_2$AsCl (59).
RN$_2$Cl → RPhAsOOH (93, 243, 494, 957).
PhBr + NaHg → Ph$_3$As (96).
SO$_2$ in HCl → PhAsCl$_2$ (136).
RSH → PhAs(SR)$_2$ (650).
RI in alc. NaOH followed by SO$_2$ in HCl → RPhAsI (623, 624).
RBr in aq.-alc. NaOH → RPhAsBr (629).
Hg(OAc)$_2$ in AcOH at r. t. → 4-AcHgC$_6$H$_4$AsO (549).
PhC$_6$H$_4$N:NCSNHNHC$_6$H$_4$Ph, spectral analysis (1199).
Me$_2$NC(S)SNa → PhAs[SC(S)NMe$_2$]$_2$ (1248).
RCOSH → PhAs[SC(O)R]$_2$ (1249).

4-ARSENOSOPHENOL C$_6$H$_5$AsO$_2$ 4-HOC$_6$H$_4$AsO
Rxns. with: Diazonium salts → the 3-diazo deriv. (428).
HOCH$_2$CHSHCH$_2$SH → 4-HOC$_6$H$_4$AsSCH$_2$CH(CH$_2$OH)S (442).
RN$_2$Cl in alk. soln. → PhRAsOOH (805).

4-ARSENOSOANILINE C$_6$H$_6$AsNO 4-H$_2$NC$_6$H$_4$AsO
Rxns. with: RSO$_2$Cl → RSO$_2$NHC$_6$H$_4$AsO (274, 388).
NaNO$_2$, followed by coumarin + CuCl$_2$ → 3-(p-arsenosophenyl)cumarin (?) (400, 404).
HOCH$_2$CHSHCH$_2$SH → 4-H$_2$NC$_6$H$_4$AsSCH$_2$CH(CH$_2$OH)S (441, 442).
NaNO$_2$, followed by CuCN → 4-NCC$_6$H$_4$AsO (738).
RNHAc + PCl$_3$, followed by NaOH and HCl → RN:C(Me)NHC$_6$H$_4$As(OH)$_2$ (978).
Me$_2$NC(S)SNa → 4-H$_2$NC$_6$H$_4$As[SC(S)NMe$_2$]$_2$ (1248).

2-AMINO-4-ARSENOSOPHENOL C$_6$H$_6$AsNO$_2$ 3,4-H$_2$N(HO)C$_6$H$_3$AsO
Prepn.: By azeotropic dehydration of 3,4-H$_2$N(HO)C$_6$H$_3$As(OH)$_2$ (43, 1233).
By dehydration of 3,4-HCl.H$_2$N(HO)C$_6$H$_3$As(OH)$_2$ over P$_2$O$_5$ at 70° in vacuo;
isolated in the form of HCl salt (43).
By reacting [3,4-H$_2$N(HO)C$_6$H$_3$As:]$_2$ with 3,4-H$_2$N(HO)C$_6$H$_3$AsO$_3$H$_2$ in 10% HCl
in the presence of HI or I (693).
By treating [3,4-H$_2$N(HO)C$_6$H$_3$As:]$_2$ with SO$_2$ (694).
By reducing 3,4-H$_2$N(HO)C$_6$H$_3$AsO$_3$H$_2$ with SO$_2$ in HCl contg. KI at 15° and
neutralizing the rxn. mixt. with NH$_4$OH to pH 5-7 (379, 684, 1112).
The compd. is purified by dissolving in EtOH, pptg. impurities with
Et$_2$O and the prod. with C$_6$H$_6$ (1113).
Props.: Bright yellow, probably amorphous solid, sol. in abs. and 95% EtOH, turns
white in H$_2$O (43). Monohydrate forms white rectangular prisms (1112). Soly. data
(1233). Decomp. 133° (1259).
Salts: Hydrochloride is entered separately below.
Hydrobromide, greenish yellow; hydriodide, orange yellow; sulfate, yellow green;
nitrate, yellow; tartrate, light yellow; lactate, light yellow; sulfamate, light
yellow; ascorbate, light yellow; citrate, light yellow (43). Hydrohalides were
prepd. from the arsenoso compd. and hydrohalic acids in the absence of EtOH
(1114). Hydrochloride crystallizes from EtOH as hemialcoholate (1233).
Rxns. with: 0.5 equiv. of H$_2$O → [3,4-H$_2$N(HO)C$_6$H$_3$AsOH]$_2$O (43).

Concd. HCl at 120° → red coloration (186).
(ClCN)$_3$ → 4-arsenoso-2-(4,6-dichloro-s-triazin-2-ylamino)phenol (417, 419, 421, 424).
NaNO$_2$ in HCl → 3,4-H$_2$N(HO)C$_6$H$_3$AsO$_3$H$_2$ (379).
HSCH$_2$CHSHCH$_2$OH → 3,4-H$_2$N(HO)C$_6$H$_3$AsSCH$_2$CH(CH$_2$OH)S (441).
HOCH$_2$SO$_2$Na → 4,3-HO(NaOSOCH$_2$NH)C$_6$H$_3$AsO (684).
RCOCl → 4,3-HO(RCONH)C$_6$H$_3$AsO (1134).
HS-RCO$_2$R' → 3,4-H$_2$N(HO)C$_6$H$_3$As(S-RCO$_2$R')$_2$ (1206).
Dextrose + NaHSO$_3$ → the N-dextrose Na sulfonate deriv. (1205).

2-AMINO-4-ARSENOSOPHENOL HYDROCHLORIDE, MARPHARSEN C$_6$H$_6$AsNO$_2$.HCl
4,3-HO(HCl.H$_2$N)C$_6$H$_3$AsO

Prepn.: By dehydration of 4,3-HO(HCl.H$_2$N(C$_6$H$_3$As(OH)$_2$ over P$_2$O$_5$ at 70° in vacuo (43
By dissolving 3,4-H$_2$N(HO)C$_6$H$_3$AsO in dil. HCl and freezing out the salt (1231).
By mixing 4,3-HO(HCl.H$_2$N)C$_6$H$_3$AsCl$_2$ with 3,4-H$_2$N(HO)C$_6$H$_3$AsO.H$_2$O in H$_2$O at pH 3.2-3.4, cooling the mixt. to -30°, evaptg. in vacuo, and drying over CaCl$_2$ to the monohydrate state, which on further drying in vacuo over P$_2$O$_5$ loses 1/2 mol. of H$_2$O and at 70° 1 mm. forms anhydrous salt (1239).
By electrolytic redn. of 3,4-H$_2$N(HO)C$_6$H$_3$AsO$_3$H$_2$ in 3 N HCl (1259).
Props.: Greenish yellow solid, readily sol. in H$_2$O and EtOH; on standing over hydrated CaCl$_2$ in N → [4,3-HO(HCl.H$_2$N)C$_6$H$_3$AsOH]$_2$O (43).
Decomp. 147° (1259). Stability (49, 50).
Rxns. with: H$_2$O → cleavage of the As-C bond (49, 50).
H$_2$O + O → aminophenol, iminoquinone, isophenoxazone, and arsenoisophenoxazone (49, 50).
Electrolytic redn. → the arseno deriv. (1249).

4-CYANAMIDOPHENYLARSINE OXIDE C$_7$H$_5$AsN$_2$O 4-NCNHC$_6$H$_4$AsO
Prepn.: by reducing the corresponding arsonic acid with SO$_2$ in HCl in the presence of HI (47, 51).
Props.: Monohydrate forms white solid, which loses H$_2$O at 100° in vacuo (51).
Salts: Hydrochloride monohydrate and anhydrous (51).
Rxns. with: Mercaptans → the corresponding thio esters (51).
Concd. HCl → NCNHC$_6$H$_4$AsCl$_2$.HCl (51).
Concd. HBr → NCNHC$_6$H$_4$AsBr$_2$.HBr (51).

4-ARSENOSO-2-MERCAPTOBENZIMIDAZOLE C$_7$H$_5$AsN$_2$OS

Prepn.: by reducing the corresponding arsonic acid with SO$_2$ in the presence or absence of HI or with PhNHNH$_2$ (535, 581).

5-ARSENOSO-2-MERCAPTOBENZIMIDAZOLE C$_7$H$_5$AsN$_2$OS

Prepn.: by reducing the corresponding arsonic acid dissolved in glycerol or aq. glycerol by SO$_2$ in the presence of NaI (371) or with PhNHNH$_2$ (535, 581).

p-ARSENOSOTOLUENE C_7H_7AsO p-MeC$_6$H$_4$AsO
<u>Prepn.:</u> By reducing 4-MeC$_6$H$_4$AsO$_3$H$_2$ with POCl$_3$ at 110-115°, decomp. the rxn. with ice-water, treating with alk., and neutralizing (981).
        By reacting 4-MeC$_6$H$_4$N$_2$CO$_2$K with AsCl$_3$ in Me$_2$CO, treating the rxn. prod. with HCl, H$_2$O, KOH, and satg. the sol. prod. with NH$_4$Cl; 32% yield (1062).
<u>Props.:</u> M. 182-185° (16); m. 167-169° (1062).
<u>Rxns. with:</u> RI in alc. NaOH, followed by SO$_2$ → 4-MeC$_6$H$_4$RAsI (627).
Hg(OAc)$_2$ in AcOH → 4,2-Me(AcOHg)C$_6$H$_3$AsO → 4-MeC$_6$H$_4$HgOAc (549, 552).

7-ARSENOSO-4-HYDROXYQUINAZOLINE $C_8H_5AsN_2O_2$

<u>Prepn.:</u> by reducing 4-hydroxy-7-quinazolinearsonic acid with SO$_2$ in dild. HCl contg. KI and neutralizing the rxn. prod. with NH$_4$OH, 69.1% yield (1319).
<u>Props.:</u> m. 295° (1319).

4-ARSENOSO-3-HYDROXYACETANILIDE $C_8H_8AsNO_3$ 2,4-HO(AcNH)C$_6$H$_3$AsO
<u>Prepn.:</u> By reducing 2,4-HO(AcNH)C$_6$H$_3$AsO$_3$H$_2$ with SO$_2$ in HCl and neutralization of the rxn. prod. (1003).
        By transesterification of 2,4-HO(AcNH)C$_6$H$_3$As(SCH$_2$CO$_2$H)$_2$ with any arsenoso- or arsonosophenyl derivative, containing an acidic group attached to the arom. ring, in a neutral aq. medium; isolated as hemi-hydrate (1004).
<u>Rxns. with:</u> RSH → ·2,4-HO(AcNH)C$_6$H$_3$As(SR)$_2$ (355, 356).
HOCH$_2$CHSHCH$_2$SH → 2,4-HO(AcNH)C$_6$H$_3$AsSCH$_2$CH(CH$_2$OH)S (441).

4-ARSENOSO-2-(4,6-DICHLORO-s-TRIAZIN-2-YLAMINO)PHENOL C$_9$H$_5$AsCl$_2$N$_4$O 4-($\overline{N:CClN:}$ CClN:CNH)C$_6$H$_4$AsO
<u>Prepn.:</u> by reacting 3,4-H$_2$N(HO)C$_6$H$_3$AsO with (ClCN)$_3$ (1:1) in aq. soln. at 0° under exclusion of O (419).
<u>Rxns. with:</u> Bases (NH$_3$ or amines) → the 4,6-diamino derivs. (419).

4-ARSENOSO-2-(4,6-DICHLORO-s-TRIAZIN-2-YLAMINO)PHENOL C$_9$H$_5$AsCl$_2$N$_4$O$_2$ 4,3-HO($\overline{N:}$ CClN:CClN:CNH)C$_6$H$_3$AsO

<u>Prepn.:</u> by reacting 3,4-H$_2$N(HO)C$_6$H$_3$AsO with 2,4,6-trichlorotriazine in H$_2$O under exclusion of O (417, 420, 421, 424).

4-ARSENOSO-N-(CARBAMYLMETHYL)BENZAMIDE C$_9$H$_9$AsN$_2$O$_3$ 4-(NH$_2$COCH$_2$NHCO)C$_6$H$_4$AsO
<u>Prepn.:</u> By reducing 4-[(carbamylmethyl)carbamyl]benzenearsonic acid with SO$_2$, 65% yield (1164).
        By condensation of 4-ClOCC$_6$H$_4$AsCl$_2$ with H$_2$NCH$_2$CONH$_2$.HCl in 10% Na$_2$CO$_3$ (1164).
        By ammonolysis of 4-(MeO$_2$CCH$_2$NHCO)C$_6$H$_4$AsO (1164).
<u>Props.:</u> Chars >285° (1164).

4,6-DIAMINO-6-p-ARSENOSOANILINO-s-TRIAZINE $C_9H_9AsN_6O$

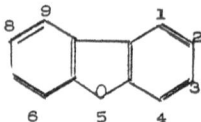

Prepn.: by reducing N-(4,6-diamino-s-triazin-2-yl)-arsanilic acid with SO_2 in HCl in the presence of HI; 78% yield (38); 48% yield (416, 438).
Props.: Crystallizes as dihydrate, which over P_2O_5 in vacuo loses one mol. of H_2O and at 135° becomes anhydrous (38). With alk. thioglycolates in alk. solns. forms stable solns. (448).
Rxns. with: HCl → the dichloroarsino deriv. (38).
HSR → the -As(SR)$_2$ derivs. (38, 434, 448).

6-ARSENOSO-4-HYDROXY-2-METHYLQUINOLINE $C_{10}H_8AsNO_2$

Prepn.: by reducing 4-hydroxy-methyl-6-methylquinolinearsonic acid with SO_2 in dild. HCl contg. KI followed by neutralization with NH_4OH; 75.1% yield (1319).
Props.: M 10° (1319).

Dibenzofuran

3-ARSENOSO-8-NITRODIBENZOFURAN $C_{12}H_2AsNO_4$
Prepn.: by boiling 3-dichloroarsino-8-nitrodibenzofuran with H_2O; theoret. yield (1145).
Props.: Light orange solid (1145).

2-ARSENOSODIBENZOFURAN $C_{12}H_7AsO_2$
Prepn.: by refluxing 2-dibenzofuranarsonic acid with PCl_3 in AcOH and boiling the resulting dichloroarsine with H_2O; 80% yield (1145).
Props.: M. 250° (1145).
Rxns. with: Hg(OAc.)$_2$, followed by $CaCl_2$ in AcOH → 2-dibenzyfurylmercury chloride (1145).

3-ARSENOSODIBENZOFURAN $C_{12}H_7AsO_2$
Prepn.: by refluxing 3-dichloroarsinodibenzofuran with H_2O; theoret. yield (1145).
Props.: M. 250° (1145).

7-ARSENOSO-2-DIBENZOFURANSULFONIC ACID $C_{12}H_7AsO_5S$
Prepn.: by reducing the arsonic acid with SO_2 in dild. HCl contg. KI;
Props.: M. 275° (1145).

4-ARSENOSO-4'-SULFANILANILIDE $C_{12}H_{13}AsN_2O_5$ 4-(4-$H_2NC_6H_4SO_2NH$)C_6H_4AsO
Prepn.: By hydrolysis of 4-(4-$AcNHC_6H_4SO_2NH$)C_6H_4AsO with 25% MeOH-HCl (388).
By reducing 4-(4-$H_2NC_6H_4SO_2NH$)$C_6H_4AsO_3H_2$ with $NaHSO_3$ in aq. HCl contg. KI (750) or with SO_2 in aq. NaOH (1257).
Props.: Orange ppt. (750); m. 205° (1257).

Rxns. with: HSCH$_2$SO$_2$Na → NaO$_2$SCH$_2$NHC$_6$H$_4$SO$_2$NHC$_6$H$_4$AsO
(-CH$_2$CO)$_2$O → NaO$_2$C(CH$_2$)$_2$CONHC$_6$H$_4$SO$_2$NHC$_6$H$_4$AsO (750).

N-(4-ARSENOSOHIPPURYL)GLYCINE METHYL ESTER C$_{12}$H$_{13}$AsN$_2$O$_5$ 4-(MeO$_2$CCH$_2$NHCOCH$_2$NHCO)-
C$_6$H$_4$AsO
Prepn.: From 4-(AgO$_2$CCH$_2$NHCOCH$_2$NHCO)C$_6$H$_4$AsO; 30% yield (1164).
Props.: Monohydrate, amorphous, chars >240° (1164).
Rxns. with NH$_4$OH → the amide (1164).

4-ARSENOSO-9-FLUORENONE C$_{13}$H$_7$AsO$_2$

Prepn.: by reducing the corresponding arsonic acid with SO$_2$ in 6N HCl contg. KI;
theoret. yield (1049).
Props.: Fluffy ppt. (1049).

1-ARSENOSO-9-FLUORENONE C$_{13}$H$_7$AsO$_2$

Prepn.: by reducing 9-fluorenone-1-arsonic acid with SO$_2$ in 6N HCl contg. KI;
85% yield (1049).
Props.: Yellow crystals, m. 138 140° (1049).

4-GUANYL-4'-ARSENOSOBENZENESULFONANILIDE C$_{13}$H$_{12}$AsN$_3$O$_3$S p-H$_2$NC(:NH)C$_6$H$_4$SO$_2$-
NHC$_6$H$_4$AsO
Prepn.: by treating p-NCC$_6$H$_4$SO$_2$NHC$_6$H$_4$AsO with HCl followed by NH$_3$ in EtOH-Et$_2$O
(274).
Props.: Monohydrate, yellow ppt. (274).

4-ARSENOSO-4'-CARBAMYLBENZANILIDE C$_{14}$H$_{11}$AsN$_2$O$_3$ 4-(4-H$_2$NOCC$_6$H$_4$CONH)C$_6$H$_4$AsO
Prepn.: by oxidation of p-NCC$_6$H$_4$CONH-p-C$_6$H$_4$AsO(OH)$_2$ with H$_2$O$_2$ followed by reduc-
tion with SO$_2$; 100% yield (288).
Props.: Amorphous, m. 319° (288).

2-(5-ARSENOSO-2-HYDROXYPHENYLAZO)-1-NAPHTHOL-4,8-DISULFONIC ACID
C$_{16}$H$_{11}$AsN$_2$O$_9$S$_2$

Prepn.: by coupling 2,5-HO(OAs)C$_6$H$_3$N$_2$Cl with 1-naphthol-4,8-disulfonic acid.
(428).
Rxns. with: Zn in alk. soln. → the corresponding hydrazino deriv. (428).

8-AMINO-2-(p-ARSENOSOPHENYLAZO)-1-NAPHTHOL-3,6-DISULFONIC ACID C$_{16}$H$_{12}$AsN$_3$O$_8$S$_2$

Propn.: by reducing the corresponding arsonic acid with SO_2 in HCl; 35% yield
(239).
Props.: Red. powder, m. 360° (289).

2-[x-Arsenosophenyl)hydrazino]-1-naphthol-n-(sulfonic acid)
(n = mono-, di-, or tri-)

2-[2-(5-ARSENOSO-2-HYDROXYPHENYL)HYDRAZINO]-1-NAPHTHOL-4,8-DISULFONIC ACID
$C_{16}H_{13}AsN_2O_9S_2$

Prepn.: By reducing the corresponding azo deriv. with Zn dust in alk. soln. (428).
By coupling $4\text{-}HOC_6H_4AsO$ with diazotized 2-amino-1-naphthol-4,8-disulfonic
acid, reducing the diazo deriv. with Zn dust in N NaOH or with H_2S in aq. NH_4OH,
and treating the thioarsenoso deriv. with aq. NaOH in H (428, 432).
Rxn. with Ac_2O → the corresponding N-acetyl and N,N —diacetyl derivs. (428).

2-[2-(2-HYDROXY-5-THIOARSENOSOPHENYL)HYDRAZINO]-1 NAPHTHOL-4,6,8-TRISULFONIC
ACID $C_{16}H_{13}AsN_2O_{11}S_4$

Prepn.: by coupling $1,4,6,8\text{-}HO(HO_3S)_3C_{10}H_3\text{-}2\text{-}N_2Cl$ with $4\text{-}HOC_6H_4AsO$ or $2,4\text{-}HO\text{-}$
$(OAs)C_6H_3N_2Cl$ with $4,6,8\text{-}(HO_3S)C_{10}H_4\text{-}1\text{-}OH$ and treating the rxn. prod. with H_2S
in 20% NH_3 (428, 432).

2-[2-(5-ARSENOSO-2-HYDROXYPHENYL)HYDRAZINO]-1-NAPHTHOL-4,6,8-TRISULFONIC ACID
$C_{16}H_{13}AsN_2O_{12}S_3$

Prepn.: by coupling $1,4,6,8\text{-}HO(HO_3S)_3C_{10}H_3\text{-}2\text{-}N_2Cl$ with $4\text{-}HOC_6H_4AsO$ in a non-
oxidizing atm. (432).

1,4-BIS(4-ARSENOSOPHENYL)PIPERAZINE $C_{16}H_{16}As_2N_2O_2$ $OAsC_6H_4NC_4H_8NC_6H_4AsO$

Prepn.: by reducing 1,4-bis(4-arsonophenyl)piperazine with PCl_3 in EtOAc (982).
Props.: Brown solid insol. in dil. acids, m. 270° (928).

2-[1-ACETYL-2-(5-ARSENOSO-2-HYDROXYPHENYL)HYDRAZINO]-1-NAPHTHOL-4,8-DISULFONIC
ACID $C_{18}H_{15}AsN_2O_{10}S_2$

Prepn.: by treating 2-[2-(5-arsenoso-2-hydroxyphenyl)hydrazino]-1-naphthol-4,8-di-
sulfonic acid with Ac_2O at 15° (428, 432).
Rxns. with: $Na_2S_2O_4$ → the corresponding arsenophenol deriv. (428).

2-[1,2-DIACETYL-2-(5-ARSENOSO-2-HYDROXYPHENYL)HYDRAZINO]-1-NAPHTHOL-4,8-DISUL-
FONIC ACID $C_{20}H_{17}AsN_2O_{11}S_2$

Prepn.: by treating 2-[2-(5-arsenoso-2-hydroxyphenyl)hydrazino]-1-naphthol-4,8-

sulfonic acid with Ac$_2$O at 80-100° (428, 432).

1,4-BIS[(4-ARSENOSOPHENYLCARBAMOYL)METHYL]PIPERAZINE   C$_{20}$H$_{22}$As$_2$N$_4$O$_4$ OAsC$_6$H$_4$-
NHCOCH$_2$NC$_4$H$_8$NCH$_2$CONHC$_6$H$_4$AsO

Prepn.:   by reducing the corresponding arsonic acid with PCl$_3$ in EtOAc (982).
Props.:   m. 280° (982).

1,4-BIS[N-(4-ARSENOSOPHENYL)GLYCYL]PIPERAZINE   C$_{20}$H$_{22}$As$_2$N$_4$O$_4$ OAsC$_6$H$_4$NHCH$_2$CO-
NC$_4$H$_8$NCH$_2$COC$_6$H$_4$AsO

Prepn.:   by reducing the corresponding diarsonic acid with PCl$_3$ in EtOAc (982).

2-[2-(p-ARSENOSOPHENYL)-1,2-DIBENZOYLHYDRAZINO]-1-NAPHTHOL-4,8-DISULFONIC ACID
BENZOATE   C$_{37}$H$_{25}$AsN$_2$O$_{11}$S$_2$

Prepn.:   by reducing the rxn. prod. of α-naphthol-4,8-disulfonic acid and p-
OAsC$_6$H$_4$N$_2$Cl with Fe in alk. soln. and treating the resulting hydrazino deriv.
with PhCOCl (428).
Props.:   Yellowish ppt. (428).

TABLE X
ARSENOSO COMPOUNDS

RAsO	Prepd. from	Yield %	Props., Rxns., Remarks	Ref.
C₃				
ClCH:C(Me)AsO (?)	MeC:CCO₂H + AsCl₃ at 120-130°, followed by N KOH			857
EtCH(AsO)₂	As₂O₃ + PrCOCl + AlCl₃		m. 76°	1024
C₄				
OCH:CHCH:CAsO	OCH:CHCH:CAsCl₂ + alk.	78	Colorless crystals, m. 148-150°; rxns.[1-4]	346 347 857
MeC(:CCO₂H)AsO (?)	MeC:CCO₂H + AsCl₃ at 110-115°, followed by KOH		Light yellow oil	857
MeCH₂C(:CHCl)AsO (?)	MeCH₂C CCO₂H + AsCl₃ at 150-155°, followed by KOH		Brownish oil	857
BuAsO		91	Rxn. with ROH → BuAs(OR)₂	718
C₅				
CH:CHCH:NCH:CAsO	RAsCl₂ + NH₄OH		m. 118-119°, rxns.[2,5]	85, 589 87
			Forms complex with heavy metals	
CH:CHCONHCH:CAsO	RAsO₃H₂ + SO₂ in H₂SO₄		m. 256-260°	1048a
	RAsCl₂ + NH₄OH		Forms complexes with heavy metals	85
C₆				
2,4,6-Br₃C₆H₂AsO	RN₂CO₂K + AsCl₃ in Me₂CO	36	m. 219-220°, rxn.[6]	1062
4,3,5-HO(O₂N)₂C₆H₂AsO	RAsO₃H₂ + Na₂SO₃·NaHSO₃ in H₂O			652
2,1,4-O₂NC₆H₃(AsO)₂	R(AsO₃H₂)₂ + NaHSO₃		m. 340° (dec.), rxns.[7,8]	79
4,2-Br(ClHg)C₆H₃AsO	4,2-Br(AcOHg)C₆H₃AsO + NaCl		Softens 249-252°, decomp. 274-277°	550
			On heating decomp. → 4-Br-C₆H₄HgCl	552
4,2-Cl(ClHg)C₆H₃AsO	4,2-Cl(AcOHg)C₆H₃AsO + NaCl		m. 260-263°	550
			On heating decomp. → 4-Cl-C₆H₄HgCl + A₂O₃	552
2,4-Cl₂C₆H₃AsO	RAsO₃H₂ + SO₂		Cryst. powder, m. 190-191°, rxn.[9]	550

98

Compound	Preparation	Yield (%)	Properties	Ref.
4-BrC$_6$H$_4$AsO	RAsO$_3$H$_2$ + SO$_2$		m. 254°, rxn[9]	550
			m. 239-240°	1062
4-ClHgC$_6$H$_4$AsO	RN$_2$CO$_2$K + AsCl$_3$	47	Unstable at elevated temp.	549
2-ClC$_6$H$_4$AsO	4-AcOHgC$_6$H$_4$AsO + NaCl		White powder, softens 208°	283
3-ClC$_6$H$_4$AsO	RAsO$_3$H$_2$ + SO$_2$		m. 140-150°	550
	RAsCl$_2$ + NaHCO$_3$		White powder in the pres., sirup in the absence of H$_2$O	283
4-ClC$_6$H$_4$AsO	RAsO$_3$H$_2$ + SO$_2$		White powder, m. 180-182°	550
	RAsCl$_2$ + NH$_4$OH	30	m. 184°, rxn.[9]	1336
3,4-Cl(HO)C$_6$H$_3$AsO	RAsO$_3$H$_2$ + SO$_2$		White powder, softens 263°, m. 267°	286
2-IC$_6$H$_4$AsO	RAsO$_3$H$_2$ + SO$_2$			283
2-O$_2$NC$_6$H$_4$AsO	RAsO$_3$H$_2$ + SO$_2$		Light yellow powder	283
3-O$_2$NC$_6$H$_4$AsO	RAsO$_3$H$_2$ + SO$_2$	72	m. 184.5-187.5°	865
4-O$_2$NC$_6$H$_4$AsO	RAsO$_3$H$_2$ + SO$_2$		White solid	283
3,2-HO(O$_2$N)C$_6$H$_3$AsO	RAsO$_3$H$_2$ + SO$_2$	87	Yellow powder, m. 220-223° (dec.), rxn.[9]	82
4,3-HO(O$_2$N)C$_6$H$_3$AsO	RAsO$_3$H$_2$ + Na$_2$SO$_3$ + NaHSO$_3$		Yellow crystals, sol. in EtOH with discoloration, rxn.[10]	652
1,4-C$_6$H$_4$(AsO)$_2$			Rxn. with H$_2$S in Me$_2$CO → ppt.	989
3,4-H$_2$N(Cl)C$_6$H$_3$AsO	RAsO$_3$H$_2$ + SO$_2$	88	Crystals, m. 182-184° rxn.?	286
2-HOC$_6$H$_4$AsO	RAsCl$_2$ + NH$_4$OH		Sirupy dihydrate loses H$_2$O at 150° to form brittle mass	283
3-HOC$_6$H$_4$AsO	RAsCl$_2$ + Na$_2$CO$_3$			283
3,4-(HO)$_2$C$_6$H$_3$AsO	RAsO$_2$H$_2$ + SO$_2$	10		286
2-HO$_3$SC$_6$H$_4$AsO	RAsBr$_2$ + NaOH			283
4-HO$_3$SC$_6$H$_4$AsO	RAsBr$_2$ + NaOH		White amorph. solid, softens 62°	955
3-H$_2$NC$_6$H$_4$AsO	RAsCl$_2$ + NH$_4$OH			283
3,2-H$_2$N(HO)C$_6$H$_3$AsO	RAsO$_3$H$_2$ + SO$_2$	78		286

1. Rxn. with alkali → unstable (OCH:CHCH:C)HAsO$_2$H (346).
2. Rxn. with H$_2$O$_2$ → the corresponding arsonic acid (85, 346, 347, 956).
3. Rxn. with Ac$_2$O → the anhydride with acetic acid, RAs(OAc)$_2$ (346, 347).
4. Rxn. with H$_3$PO$_2$ → the corresponding arseno deriv. (62, 346, 347, 1117).
5. Rxn. with 3-dichloroarsinopyridine (3:1) → chlorodi-3-pyridylarsine (589).
6. Rxn. with HgCl$_2$ in EtOH, followed by NaOH → cleavage of the As-C bond (1062).
7. Rxn. with Cl$_2$ in H$_2$O → the diarsonic acid (79).
8. Rxn. with Cl$_2$ in CHCl$_3$ → 2,1,4-O$_2$NC$_6$H$_3$(AsCl$_2$)$_2$ (79).
9. Rxn. with Hg(OAc)$_2$ in AcOH → substn. of one H atom by the AcOHg group (82, 549, 550, 552).
10. Rxn. with HSR' → the corresponding RAs(SR')$_2$ (39, 355, 356, 429, 434, 506, 1078, 1243).

TABLE X (cont'd.)
ARSENOSO COMPOUNDS

RAsO	Prepd. from	Yield %	Props., Rxns., Remarks	Ref.
4,2-H2N(HO)C6H3AsO	RAsO3H2 + SO2	52	m. 300°, salts[11]	37
4,3-H2N(HO)C6H3AsO	RAsO3H2 + SO2		Rxn. 12, salts[13]	362
5,2-H2N(HO)C6H3AsO	RAsO3H2 + SO2	84		286
5,3-H2N(HO)C6H3AsO	RAsO3H2 + SO2	21		286
4-H2NO2SC6H4AsO	RAsCl2 or RAsBr2 + NH4OH		Rxns. 2,14	956
3-H2NO2SC6H4AsO	RAsO3H2 + SO2			285
C7				
4-NCC6H4AsO	RAsO3H2 + SO2	52	Amorph., m. 195.5-197.5°, rxn.15	288
3,4-O2N(HO2C)C6H3AsO	RAsO3H2 + SO2	31		286
2,4-O2N(H2NOC)C6H3AsO	2,4-O2N(KO2C)C6H3AsO3H2 + PCl5 in CHCl3, followed by NH4OH	60	Hexagonal prisms, m. 162-163° (dec.), rxn.2	288
3,4-HO(H2NOC)C6H3AsO	RAsO3H2 + SO2	72	Cubes, m. >360°	288
4,3-HO(H2NOC)C6H3AsO	RAsO3H2 + SO2	72	Needles, m. 222-223°	288
3,4-O2N(MeO)C6H3AsO	RAsO3H2 + SO2	94		286
2,3-H2N(HO2C)C6H3AsO	RAsCl2 + NaHCO3	70	Hydrate, amorph., m. 188-189°	290
2,4-H2N(HO2C)C6H3AsO	RAsCl2 + NaHCO3	98	Hydrate, amorph., m. 225-226°	290
2,5-H2N(HO2C)C6H3AsO	RAsCl2 + NaHCO3	77	Hydrate, amorph., m. 221-222°	290
2,6-H2N(HO2C)C6H3AsO	RAsCl2 + NaHCO3	78	Needles, m. 151-152°	290
3,2-H2N(HO2C)C6H3AsO	RAsCl2 + NaHCO3	81	Hydrate, amorph., m. 214°	286
3,4-H2N(HO2C)C6H3AsO	RAsCl2 + NaHCO3	85		290
4,2-H2N(HO2C)C6H3AsO	RAsCl2 + NaHCO3	95	Amorph. m. 260°	290
		42	Hydrate, rectangular plates, m. 239-240° (dec.)	290
5,2-H2N(HO2C)C6H3AsO	RAsCl2 + NaHCO3	85	Hydrate, rectangular plates, m. 167°	290
3,4-O2N(MeO)C6H3AsO	RAsO3H2 + SO2	61.3	m. 247-248°	93
	RAsO3H2 + Na2SO3 + NaHSO3		Rxn. 10	652
4,3-HO(NaO2SCH2NH)C6H3AsO	4,3-HO(H2N)C6H3AsO + HOCH2-SO2Na			684
	4,3-HO(H2N)C6H3AsO + HOCHO + NaHSO3	90		48
4-H2NCONHC6H4AsO	RAsO3H2 + SO2	97	Rxn. 10	651

Compound	Reaction	Yield (%)	Properties	Ref.
3,4-H₂N(H₂NOC)C₆H₃AsO	RAsO₃H₂ + SO₂	40	Hexagonal prisms, m. ⊥(7-178°, rxn.[9] rxn.[16]	288
2-MeC₆H₄AsO	RAsO₃H₂ + SO₂		Decomp. 264-265° Rxn.[17]	549
4-HOCH₂C₆H₄AsO	RAsO₃H₂ + SO₂			1128
4-H₂NCH₂C₆H₄AsO	RAsO₃H₂ + SO₂			284, 389
3,4-H₂N(MeO)C₆H₃AsO	RAsCl₂ + alkali	45		286
4-MeHNO₂SC₆H₄AsO	4-ClO₂SC₆H₄AsCl₂ + MeNH₂ in C₆H₆ + H₂O			285
3,4-H₂N(H₂NCOHN)C₆H₃AsO	RAsCl₂ + alkali	90		286
C₈				
2,4,6-Cl₂(AcOHg)C₆H₂AsO	2,4-Cl₂C₆H₃AsO + Hg(OAc)₂		Decompn.[18]	550
3,5-(HO₂C)₂C₆H₃AsO	RAsO₃H₂ + PCl₃ + PCl₅, followed by NH₄OH	39	Amorphous, m. 224-225°	288
4,2-Br(AcOHg)C₆H₃AsO	4-BrC₆H₄AsO + Hg(OAc)₂		Plates, m. 232° Stability, rxn.[19]	550 552
4,2-Cl(AcOHg)C₆H₃AsO	4-ClC₆H₄AsO + Hg(OAc)₂		White crystals, m. 204°, decompn.[18] rxn.[19] m. 208-220° (dec.), rxn.[19]	550 1128
2,4-Cl(AcOHg)C₆H₃AsO	2-ClC₆H₄AsO + Hg(OAc)₂		m. 218-220° (dec.)	288
4-NCCH₂C₆H₄AsO	RAsCl₂ + Na₂CO₃	43	Hexagonal needles, m. 236.5-237.5°	
2,4-HO₂C(H₂NOC)C₆H₃AsO	RAsO₃H₂ + SO₂		Rectangular prisms, m. 221.5-222.5°	288
2,5-HO₂C(H₂NOC)C₆H₃AsO	RAsO₃H₂ + SO₂	67	Decomp. 179-180°, soly. data	
4,3-HO(ClCH₂CONH)C₆H₃AsO	3,4-H₂N(HO)C₆H₃AsO + ClCH₂COCl	98	Decompn.[18] rxn.[19-21]	113
4-AcOHgC₆H₄AsO	PhAsO + Hg(OAc)₂			549,552

11. Sulfate, m. >225° (dec.); Na salt light gray, m. >250° (37).
12. Rxn. with RAsO₃HNa + SO₂ → 4,3-H₂N(HO)C₆H₃As:AsR (362).
13. Salts with: glycine, phenylalanine, glutamic, and anthranilic acids (1238).
14. Rxn. with 10% HCl → the corresponding dichloroarsine, RAsCl₂ (39, 956).
15. Rxn. with HCl gas and EtOH in Et₂O → 4-Cl₂AsC₆H₄C(:NH)OEt.HCl.H₂O (288).
16. Rxn. with AcCl → the corresponding AcNH- deriv. (288).
17. Rxn. with RCl → the corresponding RNH- deriv. (391).
18. Decomp. at high temps. → cleavage of the C-As bond (549, 552).
19. Rxn. with HCl or HCl + NaCl → displacement of the AcO group by Cl (550, 552).
20. Rxn. with H₂O → hydrolysis of the Hg group (549).
21. Rxn. with KCN → PhAsO (552).

TABLE X (cont'd.)
ARSENOSO COMPOUNDS

RAsO	Prepd. from	Yield %	Props., Rxns., Remarks	Ref.
$3,5\text{-}(H_2NOC)_2C_6H_3AsO$	$3,5\text{-}(MeO_2C)_2C_6H_3AsO + NH_4OH$	81	Monohydrate, glass, sinters 75°	288
$4\text{-}H_2NCONHOCC_6H_4AsO$	$RAsO_3H_2 + SO_2$	75	m. 270-271°	1164
$4\text{-}AcC_6H_4AsO$	$RAsO_3H_2 + SO_2$		m. 209°	1118
$4\text{-}AcOC_6H_4AsO$	$RAsO_3H_2 + SO_2$	66		284
$4\text{-}HO_2CCH_2C_6H_4AsO$	$RAsO_3H_2 + SO_2$			285
$4\text{-}HOCH_2COC_6H_4AsO$	$RAsO_3H_2 + SO_2$		Yellow	231
$2,5\text{-}HO(Ac)C_6H_3AsO$	$RAsO_3H_2 + SO_2$	50	Monohydrate, m. 104°	36
$4\text{-}HO_2CCH_2OC_6H_4AsO$	$RAsO_3H_2 + SO_2$			285
$3,4\text{-}Cl(HOCH_2CH_2O)C_6H_3AsO$	$RAsO_3H_2 + SO_2$	62	m. >250°	502
$4\text{-}H_2NOCCH_2C_6H_4AsO$	$4\text{-}ClOCCH_2C_6H_4AsCl_2 + NH_4OH$			285
$3\text{-}AcNHC_6H_4AsO$	$RAsO_3H_2 + SO_2$	70	Needles, m. 139-140° Rxns.10,22	284
$4\text{-}AcNHC_6H_4AsO$			No method given Rxn.10	64
$4\text{-}MeC(:NOH)C_6H_4AsO$		56		284
$2,5\text{-}HO(AcNH)C_6H_3AsO$	$3,4\text{-}H_2N(HO)C_6H_3AsO + Ac_2O + AcCl$ in Et_2O	34.5	m. 185-186°, sol. in warm Na-OAc soln., rxns.10,23	1134
$4,3\text{-}HO(AcNH)C_6H_3AsO$				48
$2,3\text{-}O_2N(HOCH_2CH_2O)C_6H_3AsO$	$RAsO_3H_2 + SO_2$	70	Yellow amorph. powder, m. >270°, rxn.9	82
$4,3\text{-}O_2N(HOCH_2CH_2O)C_6H_3AsO$	$RAsO_3H_2 + SO_2$	87	Yellow amorph. powder, m. >270°, rxn.9	82
$4\text{-}H_2NCONHCH_2C_6H_4AsO$	$4\text{-}HCl \cdot H_2NCH_2C_6H_4AsO + KOCN$ in acid soln.	85	m. >300°	392
$4\text{-}(H_2NCOCH_2NHO_2S)C_6H_4AsO$	$4\text{-}ClO_2SC_6H_4AsCl_2 + H_2NCH_2CO\text{-}NH_2 \cdot HCl$	55	Amorph., m. 193-195° (dec.)	1164
$2,4\text{-}Me_2C_6H_3AsO$	$RAsO_3H_2 + SO_2$	60	Rxn.9	550
$2,6\text{-}Me_2C_6H_3AsO$	$RAsO_3H_2 + SO_2$			286
$3\text{-}HOCH_2CH_2OC_6H_4AsO$	$RAsO_3H_2 + SO_2$	55	Amorph. powder	82
$4\text{-}HOCH_2CH_2OC_6H_4AsO$	$RAsO_3H_2 + SO_2$	90	White ppt.; Na salt white rods	502
$4\text{-}MeNC_6H_4AsO$	$RAsO_3H_2 + SO_2$	74		284
$3,4\text{-}H_2N(EtO)C_6H_3AsO$	$RAsO_3H_2 + SO_2$			369
$3,4\text{-}H_2N(HOCH_2CH_2)C_6H_3AsO$	$RAsO_3H_2 + SO_2$	31		286
$4,2\text{-}H_2N(HOCH_2CH_2O)C_6H_3AsO$	$RAsO_3H_2 + SO_2$	60		37
$4\text{-}Me_2NO_2SC_6H_4AsO$	$RAs(HO)_2$ at 135-140°/10 mm.		Monohydrate, m.~100°	958
$4\text{-}EtNHO_2SC_6H_4AsO$	$4\text{-}ClO_2SC_6H_4AsCl_2 + EtNH_2$			285

Compound	Preparation	Yield (%)	Properties	Ref.
4-(HOCH₂CH₂NHO₂S)C₆H₄AsO	4-ClO₂SC₆H₄AsCl₂ + HOCH₂CH₂-NH₂			285
C₉				
4-Hydroxyquinoline-6-AsO	RAsO₃H₂ + SO₂	53.2	m. 317°	1319
4-[N:C(NH₂)N:C(Cl)N:CNH]-C₆H₃AsO	RAsO₃H₂ + SO₂			417, 419-24
4-(NCCH₂NHOC)C₆H₄AsO	4-ClOCC₆H₄AsCl₂ + H₂NCH₂CN	87	Amorph., chars >265°, rxn.24	288
3-(HO₂CCH:CH)C₆H₄AsO	RAsO₃H₂ + SO₂	97	Amorph., m. >360°	393
4-MeO₂CCH₂OC₆H₄AsO	RAsO₃H₂ + SO₂			285
4-(H₂NOCCH:CH)C₆H₄AsO	4-(ClOCCH:CH)C₆H₄AsCl₂ + NH₃			285
3,4-HO₂C(AcNH)C₆H₃AsO	RAsO₃H₂ + SO₂	84	Rxn. 25	1164
4-(H₂NCONHCONHCO)C₆H₄AsO			m. >360°	423
4-[N:C(NH₂)N:C(NH₂)N:CS]-C₆H₄AsO			Rxn. with HgO → displacement of AsO	
3,4-AcNH(H₂NOC)C₆H₃AsO	3,4-H₂N(H₂NOC)C₆H₃AsO + AcCl	35	Amorph., m. 263-264°	288
4-(H₂NCONHCOCH₂)C₆H₄AsO	4-ClOCCH₂C₆H₄AsCl₂ + H₂NCONH₂	45	Chars >272°	1164
4-(SCH:CHN:CNHO₂S)C₆H₄AsO	RAsO₃H₂ + SO₂			285
4,3-HO[N:C(NH₂)N:C(NH₂)N:CHN]C₆H₃AsO	RAsCl₂ + alkali		Rxn.10	417, 419-21 / 429
4-(AcCH₂O)C₆H₄AsO	RAsO₃H₂ + SO₂	80	Amorph. white powder, m. 206-209°, rxn.10	501
4-(HO₂CCH₂CH₂)C₆H₄AsO	RAsO₃H₂ + SO₂			285
3-(EtO₂C)C₆H₄AsO	4-ClOCC₆H₄AsCl₂ + EtOH, followed by Na₂CO₃			285
5,2-Ac(MeO)C₆H₃AsO	RAsO₃H₂ + SO₂		Monohydrate, m. 204° (dec.) Rxn.10	36
4-(MeO₂CCH₂O)C₆H₄AsO	RAsO₃H₂ + SO₂			355
2,4,5-MeO(HO)(Ac)C₆H₂AsO	RAsO₃H₂ + SO₂		Monohydrate, m. 260° (dec.)	35
4-(AcNHCH₂)C₆H₄AsO	RAsO₃H₂ + SO₂	60	m. 224-226° / m. 223-226°	284 / 389
4-(Me₂NOC)C₆H₄AsO	4-ClOCC₆H₄AsCl₂ + Me₂NH	65	Amorph., m. 184.5-185°	285
4-[EtOC(:NH)]C₆H₄AsO	RAsCl₂ + NaHCO₃			288
4-[EtOC(:NH.HCl)]C₆H₄AsO	4-NCC₆H₄As(OH)₂ + dry HCl in Et₂O-EtOH	95	m. 152° (dec.), rxn.26	738
4-[HO₂CCH(NH₂)CH₂]C₆H₄AsO	4-[HO₂CCH(NHBz)CH₂]C₆H₄AsO + aq. HCl	89	HCl salt, m. 282-284° (dec.)	123

22. Rxn. with R′N₂Cl → the corresponding RR′AsO₂H (64).
23. Rxn. with HSCH₂CHSHCH₂OH → 4,3-HO(AcNH)C₆H₃AsSCH₂CH(CH₂OH)S (441).
24. Rxn. with I in NaHCO₃ → the corresponding arsonic acid (288).
25. Rxn. with RAs(SR′CO₂H) → transesterification (1004).
26. Rxn. with alc. NH₃ → H₂NC(:NH)C₆H₄As(OH)₂ (738).

TABLE X (cont'd.)
ARSENOSO COMPOUNDS

RAsO	Prepd. from	Yield %	Props., Rxns., Remarks	Ref.
2,4-HO(EtO₂CNH)C₆H₃AsO	RAsO₃H₂ + PCl₃	78	m. 241°	37
			m. 159°	71
5,2-O₂N(MeCHOHCH₂O)C₆H₃AsO	RAsO₃H₂ + SO₂	90	Amorph., m. 152-154°	83
2,4-O₂N(HOCH₂CHOHCH₂O)C₆H₃-AsO	RAsO₃H₂ + SO₂	71	Yellow powder, m. 167-168°, rxn.4	62
4-(H₂NCONHCOCH₂NH)C₆H₄AsO	RAsO₃H₂ + SO₂	75	m. 166-168° (dec.)	1164
4-(HOCH₂CH₂NHCONH)C₆H₄AsO	RAsO₃H₂ + SO₂	67	Amorph.	1164
4-(H₂NCOCH₂NHCH₂)C₆H₄AsO	RAsO₃H₂ + SO₂		HCl salt, white microcrystals, m. 300°	391
	4-H₂NCH₂C₆H₄AsO + ClCH₂CONH₂			391
2-(MeCHOHCH₂O)C₆H₄AsO	RAsO₃H₂ + SO₂	55	Amorph., m. 115-120°	83
4-(HOCH₂CHOHCH₂O)C₆H₄AsO	RAsO₃H₂ + SO₂	82	White granules, m. 122-123°	62
4-(MeEtN)C₆H₄AsO	MeEtPhN + AsCl₃		m. 74-75°, rxn.27	1256
	RAsCl₂ + Na₂CO₃			1256
2,4-Me(Me₂N)C₆H₃AsO	3-MeC₆H₄NMe₂ + AsCl₃		m. 108°	1256
5,2-Me(Me₂N)C₆H₃AsO	4-MeC₆H₄NMe₂ + AsCl₃		m. 63-65°	1256
4,2-H₂N(MeCHOHCH₂O)C₆H₃AsO	RAsO₃H₂ + SO₂	45	Hydrate, m.~90°	37
5,2-H₂N(MeCHOHCH₂O)C₆H₃AsO	RAsO₃H₂ + SO₂	50	Amorph., m. 125-128°	83
3,4-H₂N(HOCH₂CHOHCH₂O)C₆H₃AsO	RAsO₃H₂ + SO₂	20	White powder, m. > 250°	62
4-[N:C(NHNH₂)N:C(NHNH₂)N:CNH-NH]C₆H₄AsO	RAsCl₂ + NH₄OH		White powder	418
C₁₀				
4-(SCH:CHN:CNHCO)C₆H₄AsO	4-ClOCC₆H₄AsCl₂ + 2-amino-thiazole	90	Amorph. powder, decomp. > 220°	828
1-C₁₀H₇AsO	RAsO₃H₂ + SO₂		m. 245°	549
2-C₁₀H₇AsO	RAsO₃H₂ + SO₂		White powder	287
4-[N:C(NH₂)N:C(Cl)C:CNH]C₆-H₄AsO	RAsO₃H₂ + SO₂			8a
4-(SCH:CHNHCNHCO)C₆H₄AsO	4-ClOCC₆H₄AsCl₂ + 2-amino-thiazoline			828
4-[N:C(NH₂)N:CHCH:CNH]C₆H₄-AsO	RAsO₃H₂ + SO₂		2HCl.0.5H₂O, m. 185-186°	8a, 39
	RAsCl₃ + alkali		Monohydrate, white solid, rxns.10,14	506

Compound	Preparation	Yield (%)	Properties	Ref.
3,4-HO[N̄:C(NH2)N]N:CHCH:C̄NH]-C6H3AsO	RAsO3H2 + SO2	86	Rxn.[14]	39
2,4-HO[N̄:C(NH2)N]N:CHCH:C̄NH]-C6H3AsO	RAsO3H2 + SO2			39
3,5-(MeO2C)2C6H3AsO	3,5-(HO2C)2C6H3AsO + MeOH		Rectangular prisms, m. 255°, rxn.[28]	288
4-(CH2:CHCH2NHOC)C6H4AsO	4-ClOCC6H4AsCl2 + CH2:CHCH2NH2	85		766
4-(HO2CCH2CH2CONH)C6H4AsO	RAsO3H2 + SO2			285
4,3-HO(HO2CCH2CH2CONH)C6-H3AsO				48
4-[N̄:C(NH2)N:C(NH2)CH:C̄NH]-C6H3AsO	RAsO3H2 + SO2 ; 4-N2NC6H4AsO + N̄:C(NH2)N:C-(NH2)CH:CCl		H2CO3.2H2O, m. 273-275° Isethionate.2H2O, m. 108° (dec.)	8, 8a
2,4,6-Me2(AcOHg)C6H2AsO	2,4-Me2C6H3AsO + Hg(OAc)2	45	M. 250° (dec.).	550
3,4-(AcNH)2C6H3AsO	RAsO3H2 + SO2			286
4-(H2NOCCH2CH2CONH)C6H4AsO	RAsO3H2 + SO2			285
4-(H2NOCCH2NHOCCH2)C6H4AsO	4-ClOCC6H4AsCl2 + H2NCH2-CONH2.HCl	50	2H2O, needles, m. 133° (dec.)	1164
4-(H2NOCCH2CH2NHOC)C6H4AsO	4-ClOCC6H4AsCl2 + H2NCH2CH2CO-NH2	50	Needles, m. 283-285° (dec.)	1164
3,4-H2N[N̄:C(Me)C(Me):NN̄]C6-H3AsO	RAsO3H2 + SO2	90	m. >250°	239
4-[MeCH2CH(CO2H)]C6H4AsO	RAsO3H2 + SO2	13	Amorph., softens 258°	393
4-[HO2C(CH2)3]C6H4AsO	RAsO3H2 + SO2			285
4-[H2NOC(CH2)3]C6H4AsO	4-[ClOC(CH2)3]C6H4AsCl2 + NH3			285
4-(MeCHOHCH2NHOC)C6H4AsO	4-ClOCC6H4AsCl2 + MeCHOHCH2-NH2			766
2,4-HO(PrCONH)C6H3AsO	RAsO3H2 + PCl3 in Et2O followed by NH4OH	61	m. 209°	37
		79	m. 198°	71
4-(HOCH2CHOHCH2NHOC)C6H4AsO	4-ClOCC6H4AsCl2 + H2NCH2CHOH-CH2OH	50	Amorph., chars >250°	288
2,4-MeO(EtO2CNH)C6H3AsO	RAsO3H2 + SO2 in H2SO4	47	m. 147°	37
4-[O(CH2CH2)2NO2S]C6H4AsO	RAsI2 + NH4OH		m. 196-197°	1276
3,4-H2N(Me2COHCH2O)C6H3AsO	RAsO3H2 + SO2		m. 123-124°	556
4,5,2-HO(Me)(Me2CH)C6H3AsO	5-carvacrolarsonic acid + SO2		Rxn.[2]	381
C11				
SC(COPh):CHCH:CAsO	RAsCl2 + 2N NaOH	91	White powder, rxn.[2]	1280
6-H2NOC10H6AsO	6-HO2CC10H6AsO3H2 + PCl3 + PCl5 + NH4OH		White powder	287

27. Rxn. with H2S → the corresponding thioarsenoso deriv. (1256).

TABLE X (cont'd.)
ARSENOSO COMPOUNDS

RAsO	Prepd. from	Yield %	Props., Rxns., Remarks	Ref.
4-[N:CHCH:CHN:CNHOC]C$_6$H$_4$AsO	4-ClOCC$_6$H$_4$AsCl$_2$ + 2-amino-pyrimidine			828
4-[S̄CH:C(Me)N:C̄NHCO]C$_6$H$_4$AsO	4-ClOCC$_6$H$_4$AsCl$_2$ + S̄CH:C(Me)-N:C̄NH$_2$			828
4-(N̄:CHCH:CHCH:C̄NHO$_2$S)C$_6$H$_4$-AsO	RAsO$_3$H$_2$ + SO$_2$			285
4-[N̄:CHC(NH$_2$):CHCH:C̄NH]C$_6$-H$_4$AsO	RAsO$_3$H$_2$ + SO$_2$	75	White powder, sinters at 122°, decomp. 132°, rxn.10	505
4-(HO$_2$CCH$_2$NHOCCH$_2$NHOC)C$_6$H$_4$-AsO	4-ClOCC$_6$H$_4$AsCl$_2$ + H$_2$NCH$_2$CONH-CH$_2$CO$_2$H		Hydrate, amorph., chars > 220°	1164
4-[N̄:C(NH$_2$)N:C(Me)CH:C̄NH]-C$_6$H$_4$AsO	RAsO$_3$H$_2$ + SO$_2$		m. 200-201°	8a
4-[N̄:CHN:C(NHMe)CH:C̄NH]C$_6$H$_4$-AsO	RAsO$_3$H$_2$ + SO$_2$		2H$_2$O, m. 110°	8a
4-[HO$_2$CCH(NHAc)CH$_2$]C$_6$H$_4$AsO	RAsO$_3$H$_2$ + SO$_2$		m. 250-252°	123
4-[H$_2$N(COCH$_2$NH)$_2$CO]C$_6$H$_4$AsO	4-(MeO$_2$CCH$_2$NHCOCH$_2$NHCO)C$_6$H$_4$-AsO + NH$_4$CH	50	Hydrate, rectangular prisms, chars >240°	1164
4-[N̄:C(NH$_2$)N:C(NHMe)CH:C̄NH]-C$_6$H$_4$AsO	RAsO$_3$H$_2$ + SO$_2$		Hydrate, m. 100°	8a
4-[N̄:C(NHMe)N:C(NH$_2$)CH:C̄NH]-C$_6$H$_4$AsO	RAsO$_3$H$_2$ + SO$_2$		Hydrate, m. 195-200°	8a
4-(AcNHCH$_2$CH$_2$NHOC)C$_6$H$_4$AsO	4-ClOCC$_6$H$_4$AsCl$_2$ + AcNHCH$_2$CH$_2$-NH$_2$	85	Amorph., m. 270-272° (dec.)	288
4-[HO$_2$CCH$_2$CH$_2$CMe$_2$)C$_6$H$_4$AsO	RAsO$_3$H$_2$ + SO$_2$		Hydroscopic	248
4-[HO$_2$C(CH$_2$)$_4$]C$_6$H$_4$AsO	RAsO$_3$H$_2$ + SO$_2$		Amorph., m. > 360°	393
4-Et$_2$NOCC$_6$H$_4$AsO	4-ClOCC$_6$H$_4$AsCl$_2$ + Et$_2$NH	53		285
4-[(CH$_2$(CH$_2$)$_4$NO$_2$S]C$_6$H$_4$AsO	RAsI$_2$ + NH$_4$OH	76	m. 194-195.5°	1276
2,4-(HOCH$_2$CH$_2$O)(Et$_2$O$_2$NH)C$_6$-H$_3$AsO	RAsO$_3$H$_2$ + SO$_2$	79	m. 222°	37
4-[HO(CH$_2$CH$_2$NH)$_2$OC]C$_6$H$_4$AsO	4-ClOCC$_6$H$_4$AsCl$_2$ + HOCH$_2$CH$_2$-NHCH$_2$CH$_2$NH$_2$			766
4,3-HO[HOCH$_2$CH$_2$(HOCH$_2$CHOHC-H$_2$)N]C$_6$H$_3$AsO	RAsO$_3$H$_2$ + SO$_2$		Rxn.29	362
3,4-HO[HOCH$_2$CH$_2$(HOCH$_2$CHOHC-H$_2$)N]C$_6$H$_3$AsO	RAsO$_3$H$_2$ + SO$_2$			368

Compound	Reaction	Yield, %	Description	Ref.
4-[(H2NCH2CH2)2NOC]C6H4AsO	4-ClOCC6H4AsCl2 + HN(CH2CH2-NH2)2		m. 181-182°	766
C12				
O(C6H4AsO-o)2	2-HOC6H4AsO3H2 + SO2			93
4-[N:CHCH:CHCH:CNHOC)C6-H4AsO	4-ClOCC6H4AsCl2 + 2-pyridyl-amine			285, 828
4-PhN2C6H4AsO	RAsO3H2 + SO2	74	Orange powder, m. 214-214.5°	289
4-(4-HOC6H4N2)C6H4AsO	RAsO3H2 + SO2	100	Orange powder, m. >360°	289
4,3-HO(PhN2)C6H3AsO	RAsO3H2 + SO2	94	Orange powder, m. 215°	289
4-(4-HO3SC6H4)C6H4AsO	RAsI2 + NH4OH	100	White needles, m. 221-222°	959
4-(4-H2NC6H4)C6H4AsO	RAsO3H2 + SO2	71	White needles, m. 272°	287
4,2-AcNHC10H6AsO	RAsO3H2 + SO2	65	White needles, m. 256.5°	287
2,1-AcNHC10H6AsO	RAsO3H2 + SO2			287
4-(4-H2NO2SC6H4)C6H4AsO	RAsI2 + NH4OH			959
4-PhNHO2SC6H4AsO	RAsI2 + NH4OH			958
4-[N:C(Me)CH:CHN:CNHOC]C6-H4AsO	4-ClOCC6H4AsCl2 + 2-amino-4-methylpyrimidine	86		828
4,3-HO(4-H2NC6H4SO2NH)C6-H3AsO	RAsO3H2 + SO2		Rxn. 4	1117
3,4-H2N(4-H2NC6H4SO3)C6-H3AsO	RAsO3H2 + SO2		m. 135-136° (dec.)	1258
4-Me2NC10H6AsO	1-C10H7NMe2 + AsCl3 at 100°		2H2O, m. <150°	388
4-[N:C(NMe)N:C(NMe)CH:CNH]C6H4AsO	RAsO3H2 + SO2		m. 98-100°, 0.5H2O, m. 130°	1256, 8a
4-[HO2C(CH2)5]C6H4AsO	RAsO3H2 + SO2	66	Amorph., m. 360°	393
2,4-(MeCHOHCH2O)(EtO2CNH)-C6H3AsO	RAsO3H2 + SO2	64	m. 175°	37
4,3-HO[(HOCH2CHOHCH2)2N]-C6H3AsO	RAsO3H2 + SO2			363
3,4-HO[(HOCH2CHOHCH2)2N]-C6H3AsO	RAsO3H2 + SO2		Rxn. 29	362
C13				
4-(4-NCC6H4SO2NH)C6H4AsO	4-NCC6H4SO2Cl + 4-H2NC6H4AsO	84	Needles, m. 163-166°	274
4-BzNHC6H4AsO	RAsO3H2 + SO2	90		284
4-(4-H2NOCC6H4)C6H4AsO	4-(4-HO2CC6H4)C6H4AsO3H2 + PCl3-PCl5 + NH4OH	85	White powder, m. 271-273°	287

28. Rxn. with NH4OH at 100° → 5-arsenosoisophthalimide (288).

29. Rxns. with R'AsH2 → RAs:AsR' (362).

TABLE X (cont'd.)
ARSENOSO COMPOUNDS

RAsO	Prepd. from	Yield %	Props., Rxns., Remarks	Ref.
4,3-HO(BzNH)C6H3AsO	3,4-H2N(HO)C6H3AsO + BzCl in EtOH in AcOH + NaOAc	32	m. 239°	1134 1134
3,4-O2N(PhCH2O)C6H3AsO	RAsO3H2 + Na2SO3-NaHSO3	57	rxn.10	652
4-(4-H2NOCC6H4N2)C6H4AsO	4-(4-HO2CC6H4N2)C6H4AsO3H2 + PCl3-PCl5 + NH4OH	88	Orange powder, chars > 260°	289
4-(4-H2NC6H4CONH)C6H4AsO	RAsO3H2 + SO2		White powder	284
4-[4-H2NC(:NH)C6H4SO3]C6H4-AsO	RAsO3H2 + SO2			275
4-(MePhNO2S)C6H4AsO	RAsI2 + NH4OH	66		1276
4-[N:C(Me)CH:C(Me)N:CNHCO]-C6H4AsO	4-ClOCC6H4AsCl2 + N:C(Me)CH:C(Me)N:CNH2			828
4-(4-H2NC6H4SO2NHCH2)C6H4AsO	RAsO3H2 + PhNHNH2 in Me2CO		White powder, m. 122-124°	389
4-[CH2(CH2)3NCH2CH2CONH]C6-H4AsO	RAsO3H2 + SO2			376
C14				
2,4-HO(BzCONH)C6H4AsO (-NHCOC6H4AsO)2	RAsO3H2 + PCl3 in Et2O + NH4OH	54	m. 224°	37
	4-ClOCC6H4AsCl2 + N2H4.H2O in C5H5N-C6H6	60	Amorph., m. > 360°	288
4,3-HO[3,4-HO(MeO)C6H3CH:N]-C6H3AsO		85		48
4-(4-AcNHC6H4)C6H4AsO	RAsO3H2 + SO2	100	White powder, m. 297.5-298.5°	287
4-(PhCH2NHOC)C6H4AsO	4-ClOCC6H4AsCl2 + PhCH2NH2		Amorph., m. 217°	285
2,4-HO(PhCH2O2CNH)C6H3AsO	RAsO3H2 + PCl3 in Et2O, followed by NH4OH	92		71
4-(4-AcNHC6H4SO2NH)C6H4AsO	4-H2NC6H4AsO + 4-AcNHC6H4-SO2Cl in Me2CO + C5H5N		2H2O, m. 198°, rxn.30	388
4,3-HO(AcNHC6H4SO2NH)C6H3AsO	RAsO3H2 + SO2		Rxn.4	1117
3,4-H2N(4-AcNHC6H4SO3)C6H3-AsO	RAsO3H2 + PhNHNH2 in MeOH		m. 171-176°	388
C15				
4-(4-AcNHC6H4CONH)C6H4AsO	RAsO3H2 + SO2			284
4-(4-AcNHC6H4NHOC)C6H4AsO	4-ClOCC6H4AsCl2 + 4-AcNHC6-H4NH2			285

Compound	Reaction	Yield	Description	Ref
4-(4-HOCH₂CH₂NHOCC₆H₄N₂)C₆-H₄AsO	4-(4-HO₂CC₆H₄N₂)C₆H₄AsO₃H₂ + PCl₃-PCl₅ in CHCl₃ + aq. H₂NCH₂CH₂OH	74	Orange powder, chars 275°	289
4-[BzNHCH(NH₂)CH₂]C₆H₄AsO	RAsO₃H₂ + SO₂ in HCl		White powder, m. 270-272°, rxn.31	123
C₁₆				
(-CH₂NHOCC₆H₄AsO-p)₂	4-ClOCC₆H₄AsO + (CH₂NH₂)₂ in Na₂CO₃	80	Amorph., m. 320°	288
2,4-HOCH₂CH₂O(PhCH₂O₂CNH)-C₆H₃AsO	RAsO₃H₂ + in H₂SO₄	59	m. 250°	37
4-(4-AcNHC₆H₄SO₂NHCH₂)C₆H₄-AsO	RAsO₃H₂ + PhNHNH₂ in MeOH		White powder, m. 215 219°, rxn.30	389
4-[EtO₂CN(CH₂CH₂)₂NCH₂CONH]-C₆H₄AsO	RAsO₃H₂ + PCl₃ in EtOAC		m. 120° (dec.)	982
C₁₇				
EtO₂CN(COC₆H₄AsO-p)₂	4-ClOCC₆H₄AsCl₂ + NaHNCO₂Et in C₆H₆ or C₅H₅N			1164
C₁₈				
4-(PhC₆H₄NHO₂S)C₆H₄AsO	RAsI₂ + NH₄OH	80		1276

30. Rxn. with 25% MeOH-HCl → hydrolysis of the AcNH-group (388, 389).
31. Rxn. with aq. HCl → (4-arsenosophenyl)alanine.HCl (123).

THIOARSENOSO DERIVATIVES

Thioarsenoso compounds are prepared or formed by:

1. Treating organodihaloarsines, $RAsX_2$, or arsenoso compounds, RAsO, with hydrogen sulfide in chloroform or ethanol.
2. Treating primary arsines, $RAsH_2$, with thionyl chloride in benzene to form RAsS and $RAsCl_2$ compounds or with N-sulfinylaniline, PhNSO, to form RAsS, RAsO, and $PhNH_2$ compounds
3. Cleaving arseno compounds, RAs:AsR, with sulfurdioxide to form a mixture of arsenoso and thioarsenoso compounds, RAsO + RAsS.
4. Dehydration of thioarsonous acids in vacuo.

TABLE XI
THIOARSENOSO COMPOUNDS

RAsS	Prepd. from	Props.	Ref.
C_1			
$CH_2(AsS)_2$	$CH_2(AsCl_2)_2 + H_2S$	m. 122-124° (dec.)	1213
C_4			
OCH:CHCH:CAsS	$RAsCl_2 + H_2S$	m. 120-160°[1]	347
C_6			
$3-O_2NC_6H_4AsS$	$RAsCl_2 + H_2S$	m. 96-98°, rxn.[2]	552
$4-O_2NC_6H_4AsS$	$RAsCl_2 + H_2S$	m. 141°, rxn.[2]	552
PhAsS	$RAsH_2 + SOCl_2$	Tetramer, m. 175-176°	16
	$RAsCl_2 + H_2S$	m. 151°, rxn.[2]	552
$3,4-H_2N(HO)C_6H_3AsS$	$(RAs:)_2 + SO_2$	Rxns.[3,4]	694
	RAs(SH)OH at 37°/0.5 mm.	Oligomeric aggregate	48
C_7			
$4-MeC_6H_4AsS$	$RAsH_2 + PhNSO$ in MeOH	m. 159-160°	16
C_8			
$4,2-O_2N(AcOHg)C_6H_3AsS$	$4-O_2NC_6H_4AsS + Hg(OAc)_2$ in AcOH	Crystals, m. 253°	552
$3,4-O_2N(AcOHg)C_6H_3AsS$	$3-O_2NC_6H_4AsS + Hg(OAc)_2$ in AcOH		552
$4-AcOHgC_6H_4AsS$	PhAsS + $Hg(OAc)_2$ in AcOH	m. 239-242°	552
C_9			
$4-(MeEtN)C_6H_4AsS$	$RAsO + H_2S$	m. 157°	1256

1. Crystallizes in the form of yellowish needles from EtOH or of octahedra from C_6H_6; turns yellow in light, changing to an insol. and non-fusible solid (347).
2. Rxn. with $Hg(OAc)_2$ in AcOH → substitution of one H atom in the ring by the AcOHg group via intermediate complex formation of the composition AcOHg-RAsS.$(RAsS)_2$ and AcOHgRAsS.RAsS (552).
3. Rxn. with I in an acid medium → the corresponding arsonic acid (694).
4. Rxn. with primary arsines, $R'AsH_2$ → RAs:AsR' (694).

TABLE XI (cont'd.)
THIOARSENOSO COMPOUNDS

RAsS	Prepd. from	Props.	Ref.
2,4-Me(Me$_2$N)C$_6$H$_3$AsS	RAsO + H$_2$S	m. 137°	1256
5,2-Me(Me$_2$N)C$_6$H$_3$AsS	RAsO + H$_2$S	m. 68°	1256
C$_{12}$			
4-Me$_2$NC$_{10}$H$_6$AsS	RAsO + H$_2$S	m. 144°	1256

ARSONOUS ACIDS

With the exception of 2-chloro-5-pyridine-, α-camphor-, and 3-carbazole-arsonous acids, all other arsonous acids reported to date contain only aromatic groups as the organic moiety of the molecule, however, the existance of ClCH:CHAs(OH)$_2$ has been detected by potentiometric titration of ClCH:CHAsCl$_2$ (1275). Being closely related to the arsenoso compounds, they are prepared by similar methods as the latter compounds, but they exist as monomers.

Arsonous acids are prepared or formed by:

1. Reducing arsonic acids, RAsO$_3$H$_2$, with sulfur dioxide in a hydrohalic acid, in the presence of catalytic amounts of potassium or sodium iodide, and neutralizing the intermediate organodihaloarsine with aqueous alkali or ammonia, or in diluted sulfuric acid at temperatures not higher than 10°. When the reduction is carried out in sulfuric acid, arsonous acids are formed directly during reduction.
2. Hydrolysis of organodihaloarsines with aqueous alkali, ammonia or amines.
3. Hydration of arsenoso compounds by recrystallization from 60% aqueous ethanol.
4. Condensing arsanilic acid with acylanilides in the presence of phosphorus trichloride or phosphorus oxychloride to form 1-iminoalkyl(or aryl)-aminobenzenearsonous acids.
5. Reacting arsenic trihalides with arylazocarboxylic salts, RN$_2$CO$_2$M, in acetone, treating the reaction product with hydrochloric acid, followed by water and potassium hydroxide.
6. Reacting N-acylarsanilic acid with aniline in the presence of phosphorus oxychloride at 125-130°, decomposing the reaction product with ice water and alkali, and neutralizing the solution.

3-AMINO-4-HYDROXYBENZENEARSONOUS ACID C$_6$H$_8$AsNO$_3$ 3,4-H$_2$N(HO)C$_6$H$_3$As(OH)$_2$
Prepn.: by reducing 3,4-H$_2$N(HO)C$_6$H$_3$AsO$_3$H$_2$ with SO$_2$ in aq. HCl in the presence of KI at temp. below 5° and neutralizing the solution with (NH$_4$)$_2$CO$_3$ to pH 6.5; 90% yield (43).
Props.: Colorless cryst. solid; soly. data; dehydration by azeotropic distn. with EtOH and C$_6$H$_6$ gave 2-amino-4-arsenosophenol (43, 1233). The cryst. arsonous acid was converted into an amorph. colorless substance by dissolving it in

H_2O contg. HCl, neutralizing the soln. with NH_4OH, sepg. the prod. at a low temp.
and grinding under ether (43). Neutralization curve (1232).
Rxns. with: H_2S at pH 6 → $3,4-H_2N(HO)C_6H_3As(SH)OH$ (48).
$(NH_4)_2S$ in dil. HCl → $3,4-H_2N(HO)C_6H_3As(SH)_2$ (48).
$NaNO_2$, followed by HCl → 4-hydroxy-3-diazoniumbenzenetrichloroarsonite (48).
$HS-R-CO_2R'$ → $3,4-H_2N(HO)C_6H_3As(S-RCO_2R')_2$ (1183, 1206).
HCl (1:1) → the HCl salt (43, 1233).

3-AMINO-4-HYDROXYBENZENEARSONOUS ACID HYDROCHLORIDE $C_6H_8AsNO_3.HCl$ $4,3-HO(HCl.$
$H_2N)C_6H_3As(OH)_2$

Prepn.: By adding one equiv. of HCl to aq. soln. of the arsonous acid and freezing the salt in vacuo. a soluble salt was obtd., while with 10% excess of HCl and
freezing in vacuo at somewhat higher temp. the salt obtd. was insol. in water (43).
      By treating the arsonous acid with one equiv. of HCl in abs. alc. and
pptg. with ether (1233).
Props.: White solid, occurs in two forms, one is sol. and the other insol. in
H_2O. The following structure was assigned to the insol salt:

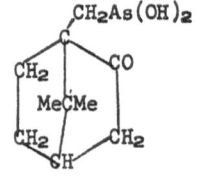

The sol. salt is dehydrated in vacuo over P_2O_5 to soluble 2-amino-4-arsenosophenol
hydrochloride, while the insol. salt is dehydrated to insol. $3,4-H_2N(HO)C_6H_3As-$
$(HO)Cl$. The dehydration of the sol. salt over P_2O_5 occurs step-wise and at r. t.
$[4,3-HO(HCl.H_2N)C_6H_3AsOH]_2O$ and at 70° $4,3-HO(HCl.H_2N)C_6H_3AsO$ are formed (43).
Rxn. with EtOH under reflux → hemialcoholate (43).

4-GUANYLBENZENEARSONOUS ACID $C_7H_9AsN_2O_2$ $4-H_2NC(:NH)C_6H_4As(OH)_2$

Prepn.: by reducing $4-NCC_6H_4AsO_3H_2$ with $NaHSO_3$ in 5% NaOH, then. treating the
arsenoso compd. with dry HCl in $Et_2O-EtOH$ and reacting the imido-chloroester
with alc. NH_3 at 65-70° (977).
Props.: m. 310° (977).

β-CAMPHORARSONOUS ACID $C_{10}H_{17}AsO_3$

$$CH_2As(OH)_2$$

Prepn.: by treating β-camphoryldichloroarsine with cond. NaOH (751).
Props.: Monohydrate, m. 100° (decomp.); anhyd., m. ~190° (751).
Rxn. with Cl_2 in H_2O or with H_2O_2 → the $RAsO_3H_2$ acid (751).

4-[1-(4-GUANYLPHENYLIMINO)ETHYLAMINO]BENZENEARSONOUS ACID $C_{15}H_{17}AsN_4O_2$
$H_2NC(:NH)C_6H_4N:C(Me)NHC_6H_4As(OH$

Prepn.: by treating $NCC_6H_4N:C(Me)NHC_6H_4As(OH)_2$ with dry HCl in $PhNO_2$ and hydrolyzing the imido ester with alc. NH_3 at 110-115° under pressure (980).
Props.: Brownish solid, m. 250° (980).

Rxns. with: H_3PO_2 in 5 HCl contg. KI → the arsenobenzene deriv. (980).

4-(4-GUANYL-α-PHENYLIMINOBENZYLAMINO)BENZENEARSONOUS ACID $C_{20}H_{19}AsN_4O_2$
$$H_2N(:NH)C_6H_4C(:NPh)NHC_6H_4As(OH)_2$$
Prepn.: by reacting $NCC_6H_4C:(NPh)NHC_6H_4As(OH)_2$ with dry HCl in $PhNO_2$ at 0° and treating the prod. with alc. NH_3 at 110-115° under pressure (980).
Props.: Yellow solid, m. 302° (980).

For other arsonous acids see Table XII on the following page.

TABLE XII
ARSONOUS ACIDS

RAs(OH)$_2$	Prepd. from	Yield %	Props.	Ref.
C$_5$				
CH:CHC(Cl):NCHCAs(OH)$_2$	RAsO$_3$H$_2$ + SO$_2$			590
C$_6$				
3,4-O$_2$N(Cl)C$_6$H$_3$As(OH)$_2$	RAsO$_3$H$_2$ + SO$_2$	80		286
3,4-O$_2$N(NaO$_3$S)C$_6$H$_3$As(OH)$_2$	RAsO$_3$H$_2$ + SO$_2$	44		286
2-O$_2$NC$_6$H$_4$As(OH)$_2$	RAsO$_3$H$_2$ + SO$_2$		m. 242°, rxn.[1]	551
			Unstable	552
3-O$_2$NC$_6$H$_4$As(OH)$_2$	RAsO$_3$H$_2$ + SO$_2$		m. 147-151°, rxn.[2]	551
			Stable compd.	283,
				552
4-O$_2$NC$_6$H$_4$As(OH)$_2$	RN$_2$CO$_2$K + AsCl$_3$ in Me$_2$O	64	Decomp. on heating	1062
	RAsO$_3$H$_2$ + SO$_2$		m. 234°, rxn.[2]	551
PhAs(OH)$_2$			Rxn. with RBr in alc. NaOH →	629
			RPhAsBr	
			Rxn.[3]	47
4-H$_2$NC$_6$H$_4$As(OH)$_2$				285
4-H$_2$NO$_2$SC$_6$H$_4$As(OH)$_2$	RAsO$_3$H$_2$ + SO$_2$			285
2-H$_2$NO$_2$SC$_6$H$_4$As(OH)$_2$	2-ClO$_2$SC$_6$H$_4$AsCl$_2$ + NH$_3$			285
4-H$_2$NNHC$_6$H$_4$As(OH)$_2$	RAsO$_3$H$_2$ + SO$_2$	32		47
C$_7$				
2,4-HO$_2$C(ClHg)C$_6$H$_3$As(OH)$_2$	2,4-HO$_2$C(AcOHg)C$_6$H$_3$As(OH)$_2$ + NaCl			551
4-NCC$_6$H$_4$As(OH)$_2$	RAsO$_3$H$_2$ + SO$_2$	85	Pale yellow-brown, m. 234°	738
	4-(H$_2$O.OAs)C$_6$H$_4$N$_2$Cl + CuCN		Rxn.[4]	738
4-NCNHC$_6$H$_4$As(OH)$_2$	RAsO$_3$H$_2$ + SO$_2$	46		47, 51
NHCONHC:CCH:CHCH:CAs(OH)$_2$	RAsO$_3$H$_2$ + SO$_2$	71		286
3-HO$_2$CC$_6$H$_4$As(OH)$_2$	RAsO$_3$H$_2$ + SO$_2$	83	Rxn.[2]	551
4-HO$_2$CC$_6$H$_4$As(OH)$_2$	2-ClOCC$_6$H$_4$AsCl$_2$ + NH$_3$		Rxn.[2]	1003
2-H$_2$NOCC$_6$H$_4$As(OH)$_2$	RAsO$_3$H$_2$ + SO$_2$			285
2,5-Me(O$_2$N)C$_6$H$_3$As(OH)$_2$	RAsO$_3$H$_2$ + SO$_2$	86	Light yellow solid, m. 240°, rxn.[2]	551
4-NaO$_3$SCH$_2$NHC$_6$H$_4$As(OH)$_2$	4-H$_2$NC$_6$H$_4$As(OH)$_2$ + HCHO~ SO$_3$HNa			47
4-[H$_2$NC(:NH)]C$_6$H$_4$As(OH)$_2$	4-[EtOC(:NH.HCl)]C$_6$H$_4$AsO + alc. NH$_3$ at 60°	80	HCl salt, hydroscopic, decomp. 210°	738

Compound	Preparation	Yield	Properties / Notes	Ref.
4-H$_2$NCONHC$_6$H$_4$As(OH)$_2$	RAsO$_3$H$_2$ + SO$_2$		rxn.5	285
	RAsO$_3$H$_2$ + NaHSO$_3$		m. 300°	108
4-MeSO$_2$C$_6$H$_4$As(OH)$_2$	RAsO$_3$H$_2$ + SO$_2$		m. 142.5°, rxn.6	285
4-H$_2$NCH$_2$C$_6$H$_4$As(OH)$_2$	RAsO$_3$H$_2$ + SO$_2$		HCl salt, microcryst., m.>300°	389
				389
C$_8$				
4,x-O$_2$N(AcOHg)C$_6$H$_3$As(OH)$_2$	4-O$_2$NC$_6$H$_4$As(OH)$_2$ + Hg(OAc)$_2$ in AcOH		m. 261° (dec.), rxn.7	551
3,x-O$_2$N(AcOHg)C$_6$H$_3$As(OH)$_2$	3-O$_2$NC$_6$H$_4$As(OH)$_2$ + Hg(OAc)$_2$ in AcOH		Stable to hydrolysis	552
			Explodes on direct heating, doesn't m. at 300°, rxns.7,8	551
4-H$_2$NOCCH$_2$OC$_6$H$_4$As(OH)$_2$	RAsO$_3$H$_2$ + SO$_2$			285
3,4-(H$_2$NCONH)$_2$C$_6$H$_3$As(OH)$_2$	RAsO$_3$H$_2$ + SO$_2$	73		286
4-Me$_2$NO$_2$SC$_6$H$_4$As(OH)$_2$	4-ClO$_2$SC$_6$H$_4$AsCl$_2$ + Me$_2$NH			285
	RAsI$_2$ + NH$_4$OH	88	At 135-140°/10 mm → the arsenoso deriv.	958
C$_9$				
4-NCCHClCH$_2$C$_6$H$_4$As(OH)$_2$	RAsO$_3$H$_2$ + SO$_2$	80	m. 120° (dec.)	657
3,4-HO$_2$C(AcOHg)C$_6$H$_3$As(OH)$_2$	3-HO$_2$CC$_6$H$_4$As(OH)$_2$ + Hg(OAc)$_2$ in AcOH		Shiny leaflets, m. 250° (dec.), rxns.7,8	551
4,2-HO$_2$C(AcOHg)C$_6$H$_3$As(OH)$_2$	4-HO$_2$CC$_6$H$_4$As(OH)$_2$ + Hg(OAc)$_2$ in AcOH		Doesn't m. 300°	551
			Stability, rxns.7,8	552
2,4-HO$_2$C(AcOHg)C$_6$H$_3$As(OH)$_2$	2-HO$_2$CC$_6$H$_4$As(OH)$_2$ + Hg(OAc)$_2$		Doesn't m. 300°	551
			Very stable, rxns.7,8	552
4-HO$_2$CCH:CHC$_6$H$_4$As(OH)$_2$	RAsO$_3$H$_2$ + SO$_2$			285
2,4,5-Me(AcOHg)(O$_2$N)C$_6$H$_2$-As(OH)$_2$	2,5-Me(O$_2$N)C$_6$H$_3$As(OH)$_2$ + Hg-(OAc)		m. 220-230°	551
4-(HO$_2$CCH$_2$NHOC)C$_6$H$_4$As(OH)$_2$	4-ClOC$_6$H$_4$AsCl$_2$ + H$_2$NCH$_2$CO$_2$H	85		285
3,4-(NaO$_3$SCH$_2$NH)(HOCH$_2$CH$_2$-O)C$_6$H$_3$As(OH)$_2$	RAsO$_3$H$_2$ + SO$_2$			48
C$_{10}$				
2-C$_{10}$H$_7$As(OH)$_2$	2-RN$_2$CO$_2$K + AsCl$_3$	30	m. 140°	1062

1. Rxn. with Hg(OAc)$_2$ in AcOH → 2-O$_2$NC$_6$H$_4$HgOAc (551).
2. Rxn. with Hg(OAc)$_2$ in AcOH → substitution of one H atom in the aromatic ring by the -HgOAc group (551).
3. Rxn. with HCHO:SO$_2$HNa → the NaO$_2$SCH$_2$NH- deriv. (47).
4. Rxn. with HCl in EtOH-Et$_2$O at 0° → 4-OAsC$_6$H$_4$C(:NH)Cl (738).
5. Rxn. with HCl, followed by diazotization and coupling with dimethyl-1-naphthylamine → the corresponding diazo deriv. (108).
6. Rxn. with KOCN → H$_2$NCONHCH$_2$C$_6$H$_4$AsO (392).

TABLE XII (cont'd.)
ARSONOUS ACIDS

$RAs(OH)_2$	Prepd. from	Yield %	Props.	Ref.
$3,4\text{-}(MeO_2C)_2C_6H_3As(OH)_2$	$RAsO_3H_2 + SO_2$	74		286
$4\text{-}(HOCH_2CH_2NHCOCH_2O)C_6H_4\text{-}As(OH)_2$	$RAsO_3H_2 + SO_2$	73		47
$4\text{-}(HOCH_2CH_2NHOCCH_2NH)C_6H_4\text{-}As(OH)_2$	$4\text{-}MeO_2CCH_2OC_6H_4As(OH)_2 + HOCH_2CH_2NH_2$; $RAsO_3H_2$	64		47, 47
$4\text{-}Et_2NO_2SC_6H_4As(OH)_2$	$4\text{-}ClO_2SC_6H_4AsCl_2 + Et_2NH$		m. 267-269°	285
C_{12} $HNC_6H_4C_6H_3As(OH)_2$	$RAsCl_2 + NH_4OH$	78		1127
$4\text{-}(4\text{-}H_2NO_2SC_6H_4NHO_2S)C_6H_4\text{-}As(OH)_2$	$RAsI_2 + NH_4OH$			958
$4\text{-}[(HOCH_2)_3CNHOCCH_2O]C_6H_4\text{-}As(OH)_2$	$RAsO_2H_2 + SO_2$	55		47
$4\text{-}[(HOCH_2CH_2)_2NOCCH_2O]C_6\text{-}H_4As(OH)_2$	$RAsO_3H_2 + SO_2$	62		47
$4,3\text{-}HO(C_6H_{11}O_5NH)C_6H_3As\text{-}(OH)_2$	$RAsO_3H_2 + SO_2$	54	$C_6H_{11}O_5$ = glucosyl	48
$4,3\text{-}HO(NaO_3SC_6H_{11}O_5)C_6H_3As\text{-}(OH)_2$	$RAsO_3H_2 + SO_2$	55	$NaO_3SC_6H_{11}O_5$ = Na glucose-bisulfite	48
$4\text{-}[(HOCH_2CH_2)_2NOCCH_2NH]\text{-}C_6H_4As(OH)_2$	$RAsO_3H_2 + SO_2$	72		47
C_{13} $4\text{-}PhNHOCC_6H_4As(OH)_2$	$4\text{-}ClOCC_6H_4AsCl_2 + PhNH_2$			285
$4\text{-}(4\text{-}HO_2CC_6H_4NHO_2S)C_6H_4\text{-}As(OH)_2$	$RAsI_2 + NH_4OH$	67		958
$CO[3,4\text{-}NH(Cl)C_6H_3As(OH)_2]_2$	$3,4\text{-}H_2N(Cl)C_6H_3As(OH)_2 + COCl_2$ in alk.			233, 712
$4\text{-}(MePhNO_2S)C_6H_4As(OH)_2$	$RAsI_2 + NH_4OH$		Rxn.[9]	1276
$CO[4\text{-}NHC_6H_4As(OH)_2]_2$	$4\text{-}H_2NC_6H_4As(OH)_2 + COCl_2$			712
$CO[3,4\text{-}NH(HO)C_6H_3As(OH)_2]_2$	$3,4\text{-}H_2N(HO)C_6H_3As(OH)_2 + COCl_2$			712
$CO[3,4\text{-}NH(HO)C_6H_3As(OH)_2]_2$	$3,4\text{-}H_2N(HO)C_6H_3AsCl_2 + COCl_2$ in aq. soln.		Rxn.[9]	233

C₁₄			
4-(4-H$_2$NOCC$_6$H$_4$NHOC)C$_6$H$_4$-As(OH)$_2$	4-ClOCC$_6$H$_4$AsCl$_2$ + 4-H$_2$NOCC$_6$-H$_4$NH$_2$	Colorless solid, m. 155°	285
4-(4-ClC$_6$H$_4$N:CMeNH)C$_6$H$_4$-As(OH)$_2$	4-H$_2$NC$_6$H$_4$AsO$_3$H$_2$ + 4-ClC$_6$H$_4$-NHAc + POCl$_3$		979
	4-AcNHC$_6$H$_4$AsO$_3$H$_2$ + 4-ClC$_6$-H$_4$NH$_2$ + POCl$_3$		979
4-(4-O$_2$NC$_6$H$_4$N:CMeNH)C$_6$H$_4$-As(OH)$_2$	4-H$_2$NC$_6$H$_4$AsO$_3$H$_2$ + 4-O$_2$NC$_6$-H$_4$NHAc + ·POCl$_3$	Yellow solid, m. 175°	979
	4-AcNHC$_6$H$_4$AsO$_3$H$_2$ + 4-O$_2$NC$_6$-H$_4$NH$_2$ + POCl$_3$		979
4-(PhN:CMeNH)C$_6$H$_4$As(OH)$_2$	4-H$_2$NC$_6$H$_4$AsO$_3$H$_2$ + PhNHAc + POCl$_3$	m. 155-156°, rxn.[10]	978
C₁₅			
4-(4-NCC$_6$H$_4$N:CMeNH)C$_6$H$_4$-As(OH)$_2$	4-H$_2$NC$_6$H$_4$AsO$_3$H$_2$ + 4-NCC$_6$H$_4$-NHAc + POCl$_3$	m. 210°, rxn.[11]	980
4-(4-MeOC$_6$H$_4$N:CMeNH)C$_6$H$_4$-As(OH)$_2$	4-H$_2$NC$_6$H$_4$AsO$_3$H$_2$ + 4-MeOC$_6$H$_4$-NHAc + POCl$_3$	Colorless solid, m. 156-157°	979
	4-MeOC$_6$H$_4$NH$_2$ + 4-AcNHC$_6$H$_4$As-O$_3$H$_2$ + POCl$_3$	Rxn.[10]	979
C₁₆			
4-[4-EtOC$_6$H$_4$N:C(Me)NH]C$_6$-H$_4$As(OH)$_2$	4-EtOC$_6$H$_4$NHAc + 4-H$_2$NC$_6$H$_4$As-O$_3$H$_2$ or 4-H$_2$NC$_6$H$_4$AsO + PCl$_3$	Colorless solid, m. 134-135°, rxn.[10]	978
C₁₉			
4-[Ph(PhCH$_2$)NO$_2$S]C$_6$H$_4$As-(OH)$_2$	RAsO recryst. from 60 EtOH[60]		1276
4-[NCC$_6$H$_4$C(:NPh)NH]C$_6$H$_4$-As(OH)$_2$	4-H$_2$NC$_6$H$_4$AsO$_3$H$_2$ + 4-NCC$_6$H$_4$-CONPh + POCl$_3$	m. 270°, rxn.[11]	980

7. Rxn. with HCl → hydrolysis of the AcOHg-group and formation of the ClHg-AcCl$_2$ deriv. (551, 552).
8. Rxn. with H$_2$O → partial hydrolysis of the Hg group (551, 552).
9. Rxn. with HSCH$_2$CO$_2$H → RAs(SCH$_2$CO$_2$H)$_2$ (233, 712).
10. Rxn. with H$_3$PO$_2$ → the corresponding arseno deriv. (978, 979).
11. Rxn. with dry HCl in PhNO$_2$ at 0°, followed by alc. NH$_3$ → conversion of the CN group into H$_2$NC-(:NH)-group (980).

117

BIS(ORGANOHYDROXYARSENIC) OXIDES

The oxides, [RAs(OH)]$_2$O, are hemihydrates of the corresponding arsenoso compounds or hemianhydrides of arsonous acids. They are formed by a controlled hydration of arsenoso compounds or by a controlled dehydration of arsonous acids. The compounds are fairly stable, and when the organo group contain functional groups,the latter enter ractions with suitable reagent, without changing the hydroxy oxide structure.

TABLE XIII

BIS(ORGANOHYDROXYARSENIC) OXIDES

[RAs(OH)]$_2$O	Prepd. from	Props.	Ref.
C_{12}			
[3,4-H$_2$N(HO)C$_6$H$_3$As(OH)]$_2$O	[3,4-H$_2$N(HO)C$_6$H$_3$AsO]n + 0.5 equiv. of H$_2$O	White powder, soly. data	43
[4,3-HO(HCl.H$_2$N)C$_6$H$_3$As-(OH)]$_2$O	[4,3-HO(HCl.H$_2$N)C$_6$H$_3$-AsO]n		43
	4,3-HO(HCl.H$_2$N)C$_6$H$_3$As-(OH)$_2$ over P$_2$O$_5$		43
C_{13}			
CS[NHC$_6$H$_4$As(OH)]$_2$O	CS(NHC$_6$H$_4$AsO$_3$H$_2$)$_2$ + PCl$_3$ + NH$_4$OH and drying at 60° in vacuo		712
CO[NHC$_6$H$_4$As(OH)]$_2$O	[4-H$_2$NC$_6$H$_4$As(OH)]$_2$O + COCl$_2$		712
C_{14}			
4,3-HO(HCl.H$_2$N)C$_6$H$_3$As-(OH)OAs(OEt)C$_6$H$_3$(NH$_2$.HCl)OH	[4,3-HO(HCl.H$_2$N)C$_6$H$_3$-AsO]n + EtOH	White solid	43
C_{15}			
CO[NHCH$_2$C$_6$H$_4$As(OH)]$_2$O	4-H$_2$NCH$_2$C$_6$H$_4$As(OH)$_2$ + COCl$_2$ + drying at 60° in vacuo		712

ARSONOUS ACID ESTERS

The esters, RAs(OR')$_2$ or RAsO-Y-O, are prepared from organodihaloarsines, RAsX$_2$, or from arsenoso compounds, RAsO, by:

1. Reacting organodichloroarsines with alkali metal alkoxides in the corresponding alcohol.
2. Reacting organodihaloarsines with alcohols or glycols in ether or in a hydrocarbon solvent in the presence of amines such as pyridine or triethylamine. When glycols are reacted, cyclic esters are formed, which are compiled on page 386.

3. Reacting organodihaloarsines with alcohols in the presence of potassium cyanide.
4. Reacting organodihaloarsines with ethylene oxide to form the corresponding arsonous acid di-(2-chloroethyl) esters.
5. Heating under reflux arsenoso compounds with alcohols in the presence of copper sulfate.
6. Heating arsenoso compounds or arsonous acids with glycols under atmospheric or reduced pressure or in ether in the presence of an amine, such as pyridine, to form cyclic esters which are compiled on page 386.
7. Transesterification of arsonous acid esters with higher-boiling alcohols.

TABLE XIV
ARSONOUS ACID ESTERS

$RAs(OR')_2$	Prepd. from	Yield	Props.	Ref.
C_3 $MeAs(OMe)_2$	$RAsCl_2$ + NaOMe		b. 108-110°, d_4^{20} 1.322, rxn.[1] b. 111°/760 mm.	677 1275
	$RAsI_2$ + MeOH + C_5H_5N	76	b. 109-110°/760 mm, d_{20} 1.3412, n_D^{20} 1.4608	646, 646a
C_4 ClCH:CHAs$(OMe)_2$	$RAsCl_2$ + MeOH + KCN		Pale yellow oil, b. 73°/13 mm.	125
	$RAsCl_2$ + NaOMe		b. 59°/6 mm.	1275
$EtAs(OMe)_2$	$RAsCl_2$ + NaOMe		b. 132-133°, d_{20}^{20} 1.286, rxn.[1] b. 129°/760 mm.	677 1275
C_5 $PrAs(OMe)_2$	$RAsI_2$ + MeOH + C_5H_5N	31	b. 62-63.5°/35-38 mm., d_{20} 1.2267, n_D^{20} 1.4583	646
$MeAs(OEt)_2$	$RAsCl_2$ + NaOEt		b. 136-137°, d_4^{20} 1.204, rxn.[1] b. 137°/760 mm.	677, 646a 1275
C_6 ClCH:CHAs$(OEt)_2$	$RAsCl_2$ + NaOEt		b. 71°/3 mm.	1275
$EtAs(OEt)_2$	$RAsCl_2$ + NaOEt		b. 159-160°, d_4^{20} 1.198, rxn.[1] b. 90-91°/89 mm., d_{20} 1.1596, n_D^{20} 1.4454, rxn.[2]	677 650
$BuAs(OMe)_2$	$RAsO$ + MeOH + $CuSO_4$	40.3	b. 65-67°/15 mm., d^{20} 1.1869, n_D^{20} 1.4555	718

1. Rxn. with SeO_2 under anhydrous conditions → $RAsO(OR')_2$ (677).

TABLE XIV (cont'd.)
ARSONOUS ACID ESTERS

RAs(OR)$_2$	Prepd. from	Yield %	Props.	Ref.
C$_7$				
PrAs(OEt)$_2$	RAsI$_2$ + NaOEt + C$_5$H$_5$N	66	b. 80.5-82.5°/30 mm. d$_{20}$ 1.1400, n$_D^{20}$ 1.4569	646, 646a
MeAs(OCHMe$_2$)$_2$	RAsCl$_2$ + Me$_2$CHOH + Et$_3$N	62	b. 160°/760 mm., d$_{25}$ 1.008, n$_D^{25}$ 1.4374	1283
C$_8$				
PhAs(OMe)$_2$	RAsCl$_2$ + NaOMe	50.4	b. 119-120°/13 mm, d$_0^{17}$ 1.5417, n$_D^{18}$ 1.6085	625
ClCH:CHAs(OCH$_2$CH:CH$_2$)$_2$	RAsCl$_2$ + NaOCH$_2$CH:CH$_2$		b. 99°/4 mm.	1275
EtAs(OCH$_2$CH:CH$_2$)$_2$	RAs(OEt)$_2$ + CH$_2$:CHCHOH	83.7	b. 77-79°/11 mm., d$_{20}$ 1.1741, n$_D^{20}$ 1.4762, rxn.[3]	650
ClCH:CHAs(OCHMe$_2$)$_2$	RAsCl$_2$ + NaOCHMe$_2$		b. 80°/3 mm., b. 88°/6 mm.	1275
EtAs(OPr)$_2$	RAsCl$_2$ + NaOPr	71	b. 86-90°/18 mm. b. 93°/25 mm., d$_4^{20}$ 1.106, rxn.[1]	44 677
BuAs(OEt)$_2$	RAsO + EtOH + CuSO$_4$	37.1	b. 81-83°/15 mm., d^{20} 1.1072, n$_D^{20}$ 1.4495	718
C$_9$				
MeAs(OBu)$_2$	RAsI$_2$ + BuOH + C$_5$H$_5$N	63	b. 108-109°/25 mm., d$_{20}$ 1.0938, n$_D^{20}$ 1.4522	646
PrAs(OPr)$_2$	RAsI + PrOH+C$_5$H$_5$N	67	b. 106-107°/35 mm., d$_{20}$ 1.0829, n$_D^{20}$ 1.4509	646
C$_{10}$				
PhAs(OCH$_2$CH$_2$Cl)$_2$	RAsCl$_2$ + $\overline{CH_2CH_2O}$	85.6	b. 190-193°/15 mm., d^{18} 1.42, rxn.[4,5]	787
ClCH:CHAs(OCH$_2$CH$_2$OEt)$_2$	RAsCl$_2$ + NaOCH$_2$CH$_2$OEt		b. 150°/3 mm.	1275
PhAs(OEt)$_2$	RAsCl$_2$ + NaOEt	61.3	b. 117-118°/10 mm., d$_0^{23}$ 1.3621, n$_D^{22}$ 1.5537	625

2. Rxn. with CH$_2$:CHCHOH at 110-120° → EtAs(OCH$_2$CH:CH$_2$)$_2$ (650).
3. Copolymerization with methyl acrylate (Bz$_2$O$_2$) → a solid copolymer (650).
4. Rxn. with H$_2$O → PhAsO + ClCH$_2$CH$_2$OH (787).
5. Rxn. with PCl$_5$ → PhAsCl$_2$ + (CH$_2$Cl)$_2$ (787).

TABLE XIV (cont'd.)
ARSONOUS ACID ESTERS

$RAs(OR)_2$	Prepd. from	Yield %	Props.	Ref.
$PhAs(OEt)_2$	$RAsCl_2$ + $(EtO)_3Sb$ at 140°		d_{20} 1.3959, n_D^{28} 1.5485	625
$BuAs(OPr)_2$	$RAsO$ + $PrOH$ + $CuSO_4$	64.4	b. 104-106°/12.5 mm., d^{20} 1.0617, n^{20} 1.4515	718
$EtAs(OBu)_2$	$RAsI_2$ + $BuOH$ + C_5H_5N	70	b. 104-106°/11 mm., d_{20} 1.1016, n_D^{20} 1.4580	646, 646a
C_{11} $PrAs(OBu)_2$	$RAsI_2$ + $BuOH$ + C_5H_5N	84	b. 113.5-115.5°/10-12 mm., d_{20} 1.0667, n_D^{20} 1.4559	646
C_{12} $PhAs(OCH_2CH:CH_2)_2$	$RAsCl_2$ + $NaOCH_2CH:CH_2$	74.8	b. 136-137°/11 mm., d_{20}, 1.2386, n_D^{20} 1.5450	650
	$RAsO$ + CH_2CHCH_2OH			650
$2\text{-}O_2NC_6H_4As(OCHMe_2)_2$	$RAsCl_2$ + Me_2CHOH + Et_3N	14.6	b. 94°/0.04 mm.	1283
$PhAs(OPr)_2$	$RAsCl_2$ + $NaOPr$	57.2	b. 139-140°/12 mm., d_0^0 1.3524, n_D^{16} 1.5190	625
$PhAs(OCHMe_2)_2$	$RAsCl_2$ + $NaOCHMe_2$		b. 118-119°/11 mm., d_0^{17} 1.3229, n_D^{17} 1.5109	625
$ClCH:CHAs(OC_5H_{11})_2$	$RAsCl_2$ + $Me_2CHCH_2CH_2ONA$		b. 130-138°/3 mm.	1275
$BuAs(OBu)_2$	$RAsO$ + $BuOH$ + $CuSO_4$	76.6	b. 132-134°/14 mm. d_0^{20} 1.0389, n_D^{20} 1.4540	718
C_{13} $MeAs(OC_6H_{11})_2$	$RAsCl_2$ + cyclo-$C_6H_{11}OH$	78	b. 120°/0.01 mm., d_{25} 1.182, n_D^{25} 1.5005	1283
C_{14} $PhAs(OBu)_2$	$RAsCl_2$ + $NaOBu$	50.6	b. 147-148°/10 mm. d_0^{25} 1.3268, n_D^{25} 1.5511, d^{20} 1.1424, n^{20} 1.5102	625 636
$PhAs(OCH_2CHMe_2)_2$	$RAsCl_2$ + Me_2CHCH_2ONa		b. 144-144.5°/12 mm. d_0^0 1.3287, n_D^{15} 1.5105	625

TABLE XVI (cont'd.)
ARSONOUS ACID ESTERS

RAs(OR)$_2$	Prepd. from	Yield %	Props.	Ref.
ClCH:CHAs(OC$_6$H$_{11}$)$_2$	RAsCl$_2$ + cyclo-C$_6$H$_{11}$OH in Et$_3$N	35.9	b. 102°/0.005 mm. d$_{25}$ 1.250, n$_D^{25}$ 1.5221	1283
C$_{16}$ PhAs(OC$_5$H$_{11}$)$_2$	RAsCl$_2$ + Me$_2$CH(CH$_2$)$_2$ONa	53.4	b. 153-154.5°/11 mm. d$_8$ 1.2969, n$_D^{25}$ 1.5492	625
C$_{20}$ BuAsPOC$_8$H$_{17}$)$_2$	RAsO + 'n-C$_8$H$_{17}$OH + CuSO$_4$	86.2	b. 193-195°/6 mm. d$_0^{20}$ 0.9777, n$_D^{20}$ 1.4603	718

THIO- AND DITHIOARSONOUS ACIDS

3,4-AMINOHYDROXYTHIOBENZENEARSONOUS ACID C$_6$H$_8$AsNO$_2$S 3,4-H$_2$N(HO)C$_6$H$_3$As(SH)OH
Prepn.: by neutralizing H$_2$N(HO)C$_6$H$_3$As(OH)$_2$ with HCl to pH 6 and satg. with
H$_2$S; 78% yield. In the presence of 2 equivs. of HCl, the HCl salt seps. (48).
Props.: at 37°/0.5 mm. loses H$_2$O and forms H$_2$N(HO)C$_6$H$_3$AsS (48).

3,4-AMINOHYDROXYDITHIOBENZENEARSONOUS ACID C$_6$H$_8$AsNOS$_2$ 3,4-H$_2$N(HO)C$_6$H$_3$As(SH)$_2$
Prepn.: by adding (NH$_4$)$_2$S to a soln. of H$_2$N(HO)C$_6$H$_3$As(OH)$_2$ in dil. HCl; 67%
yield (48).

ESTERS OF DITHIOARSONOUS ACIDS WITH MERCAPTANS

The esters are prepared or formed by:
1. Heating organodihaloarsines with mercaptans, mercaptocarboxylic acids, their esters or salts, dimercaptans, or arenedithiols in aqueous solutions or in alcohol (when the mercaptan is insoluble in water) in an inert atmosphere. Dimercaptans and arenedithiols yield cyclic 1,3,2-dithioarsenolane derivatives, which are compiled on page 389.
2. Reacting organodihaloarsines with mercaptans in ether in the presence of pyridine.
3. Reacting organodihaloarsines with sodium mercaptides in ether.
4. Heating arsenoso compounds or arsonous acids with mercaptans in aqueous solutions or in alcohols (when the mercaptan is insoluble in water) in an inert atmosphere.
5. Reacting arsonic acids with excess of ammonium or alkali metal mercaptides or dithioglycolates in aqueous or alkaline solutions at elevated temperatures. The reduction of the arsonic group to the trivalent state and esterification are usually carried out in one reaction, but the two steps can also be carried out separately, using a mercaptan such as ammonium mercaptoacetate or sodium mercaptomalate for the reduction step and then esterifying the reduced product with another mercaptan to form the desired ester.

TABLE XV

ESTERS OF DITHIOARSONOUS ACIDS WITH MERCAPTANS

RAs(SR)₂	Prepd. from	Yield %	Props.	Ref.
C₃				
MeAs(SMe)₂	RAsPOH)₂ + NaSMe		b.75-77°/2 mm.	1275
C₆				
EtAs(SEt)₂	RAsCl₂ + EtSH in CO₂	30.4		645, 647
	RAsCl₂ + EtSH + C₅H₅N	35.5	b. 80-82°/2 mm.	647
	RAsCl₂ + EtSNa in Et₂O	41.9	b. 75-77°/1.5 mm.	645, 647
EtAs(SCH₂CH:CH₂)₂	RAsCl₂ + CH₂:CHCH₂SNa in Et₂O	38.6	b. 101-102°/2 mm.	647
EtAs(SPr)₂	RAsCl₂ + PrSH in CO₂	40.6	b. 101-102°/2 mm.	645, 647
C₁₀				
3,4-O₂N(HO)C₆H₃As(SCH₂CO₂H)₂	RAsO + HSCH₂CO₂H		m. 255°	652, 1243
4-NCNHC₆H₄As(SCH₂CO₂H)₂	RAs(OH)₂ + HSCH₂CO₂H			51
3,4-H₂N(HO)C₆H₃As(SCH₂CO₂H)₂	RAs(OH)₂ + HSCH₂CO₂H	92		48
3,4-H₂N(HO)C₆H₃As(SCH₂CONH₂)₂	RAsO₃HNa + HSCH₂CONH₂	92		48
3,4-H₂N(HO)C₆H₃As(SCH₂CO₂NH₄)₂	RAsO₃HNa + HSCH₂CO₂NH₄			442
EtAs(SBu)₂	RAsCl₂ + BuSH in CO₂	54.7	b. 122-123°/2 mm.	645, 647
	RAsCl₂ + BuSH + C₅H₅N	43.5	b. 120-123°/2 mm., rxns.[1-4]	645, 647
C₁₁				
4-H₂NOCC₆H₄As(SCH₂CO₂H)₂	RAsO + HSCH₂CO₂H in H₂O at 100°			819
4,3-HO(HO₂SCH₂NH)C₆H₃As(SCH₂-CO₂H)₂	RAs(OH)₂ + HSCH₂CO₂H	70		48
4,3-HO(HO₃SCH₂NH)C₆H₃As(SCH₂-CO₂H)₂	RAs(OH)₂ + HSCH₂CO₂H	85		48
C₁₂				
3,4-O₂N(HO)C₆H₃As[SCH₂CH-(NH₂)CO₂H]₂	RAsO + HSCH₂CH(NH₂)CO₂H in aq. NaOH		m. 182°	652
3,4-H₂N(HO)C₆H₃As(SCH₂Ac)₂	RAs(OH)₂ + HSCH₂Ac	65		48
3,4-H₂N(HOCH₂CH₂O)C₆H₃As-(SCH₂CO₂H)₂	RAs(OH)₂ + HSCH₂CO₂H	82		48, 1183
3,4-H₂N(HOCH₂CH₂O)C₆H₃As-(SCH₂CONH₂)₂	RAs(OH)₂ + HSCH₂CONH₂	82		48, 1183

1. Rxn. with MeI → Me₂BuSI (645, 647).

TABLE XV (cont'd.)

ESTERS OF DITHIOARSONOUS ACIDS WITH MERCAPTANS

$RAs(SR)_2$	Prepd. from	Yield %	Props.	Ref.
EtAs(SCH₂CHSHCH₂OH)₂	RAsCl₂ + HSCH₂CHSHCH₂OH RAsCl₂ + Me₂CHCH₂CH₂SH in CO₂	43.2	b. 135-136°/2 mm.	443, 645, 647
C₁₃				
4-[N:C(NH₂)N:C(NH₂)N:CNH]-C₆H₄As(SCH₂CO₂K)₂ MeAs(SPh)₂	RAsCl₂ + HSCH₂CO₂K			429, 434
4-[N:C(NH₂)N:C(NH₂)N:CO]C₆-H₄As(SCH₂CO₂H)₂	RAs(OH)₂ + NaSPh RAsO₃H₂ + HSCH₂CO₂Na		b. 195-198°/1 mm. m. 226-228°	1275 1306
4-[N:C(NH₂)N:C(Cl)N:CNH]C₆-H₄As(SCH₂CONH₂)₂	RAsO₃H₂ + HSCH₂CONH₂ RAsO + HSCH₂CO₂H in 28% NH₄OH			429, 434, 38
4-(MeCOCH₂O)C₆H₄As(SCH₂CO₂-H)₂	RAsCl₂ + HSCH₂CO₂H			501
4-NCNHC₆H₄As[SCH₂CH(NH₂)CO₂-H]₂	RAs(OH)₂ + HSCH₂CH(NH₂)-CO₂Na			51
3,4-O₂N(MeO)C₆H₃As[SCH₂CH-(NH₂)CO₂H]₂	RAsO + HSCH₂CH(NH₂)CO₂H		m. 191-193°	652
3,4-(NaO₃SCH₂NH)(HOCH₂CH₂-O)C₆H₃As(SCH₂CONH₂)₂	RAs(OH)₂ + HSCH₂CONH₂	67		48
3,4-(NaO₃SCH₂NH)(HOCH₂CH₂O)-C₆H₃As(SCH₂CO₂Na)₂	RAs(OH)₂ + HSCH₂CO₂Na in alk. soln.	75		48
C₁₄				
3,4-O₂N(HO)C₆H₃As[SCH(CO₂-Na)CH₂CO₂Na]₂ ClCH:CHAs(SPh)₂	RAsO + HO₂CCH₂CHSHCO₂H in alk. soln. RAsCl₂ + PhSH		b. 180-190°/2 mm. (dec).	652 1275
3,4-MeO[N:CClN:CClN:CNH]C₆-H₃As(SCH₂CO₂K)₂	RAsCl₂ + HSCH₂CO₂K			429, 434
3-[SCH:C(Me)N:CNH]C₆H₄As-(SCH₂CO₂H)₂	RAsO₃H₂ + HSCH₂CO₂Na	60	m. 205-206°	250
4-[N:C(NH₂)N:CHCH₂NH]C₆H₄-As(SCH₂CO₂H)₂	RAsO + HSCH₂CO₂Na			39, 506
3,4-H₂N(HO)C₆H₃As[SCH(CO₂H)-CH₂CO₂H]₂	RAs(OH)₂ + HSCH(CO₂H)CH₂-CO₂H	68		48

Compound	Method	%	Properties	Ref.
$3,4\text{-}H_2N(HOCH_2CH_2O)C_6H_3As(SCH_2CO_2Me)_2$	$RAs(OH)_2 + HSCH_2CO_2Me$ in EtOH			1183
$EtAs(SC_6H_{13})_2$	$RAsCl_2 + n\text{-}C_8H_{13}SH + C_5H_5N$ in Et_2O	31.3	b. 163-164°/1.5 mm.	645, 647
C_{15}				
$4\text{-}[CH:CHC(NH_2):CHN:CNH]C_6H_4As(SCH_2CO_2H)_2$	$RAsO + HSCH_2CO_2Na$			505
$MeAs(SC_6H_4Me\text{-}p)_2$	$RAsCl_2 + 4\text{-}MeC_6H_4SH$		b. 180-190°/2 mm.	1275, 8a
$4\text{-}[N:C(NH_2)N:C(Me)CH:CNH]C_6H_4As(SCH_2CO_2H)_2$	$RAsO + HSCH_2CO_2H$		1.H_2O, m. 248°	
$4\text{-}[N:C(NH_2)N:C(NH_2)N:CNH]C_6H_4As[SCH_2CH(NH_2)CO_2H]_2$	$RAsO + HSCH_2CH(NH_2)CO_2H$			38
$4,3\text{-}HO[NH:C(NH_2)N:C(NH_2)N:CNH]C_6H_3As[SCH_2CH(NH_2.HCl)CO_2H]_2$	$RAsO +$ cystine.HCl			429, 434
C_{16}				
$4\text{-}[N:C(NH_2)N:CHCH:CNH]C_6H_4As[SCH_2CH(NH_2)CO_2H]_2$	$RAsO +$ cystine			506
$3,4\text{-}H_2N(HOCH_2CH_2O)C_6H_3As(SCH_2CO_2Et)_2$	$RAs(OH)_2 + HSCH_2CO_2Et$ in EtOH	68		1183
$4,3\text{-}HO(NaO_3SC_6H_{12}O_5\text{-}NH)C_6H_3As(SCH_2CO_2Na)_2$	$RAs(OH)_2 + HSCH_2CO_2Na$ in alk. soln.			48
C_{17}				
$4\text{-}(2\text{-}5C_6H_4N:CNH)C_6H_4As(SC_6H_4CO_2H)_2$	$RAsO_3H_2 + NaSCH_2CO_2Na$	56	m. 196.5-197.5°	250
$CH:NC(NHC_6H_4OEt):CHCHCAs(SCH_2CO_2H)_2$	$RAsO_3H_2 + NaSCH_2CO_2Na$	85	m. 200-201°	250
$4\text{-}[N:C(NHEt)N:C(NHEt)N:CO]C_6H_4As(SCH_2CO_2H)_2$	$RAsO_3H_2 + HSCH_2CO_2Na$ in H_2O		m. 170-173°	1306
$4\text{-}[N:C(NH_2)N:C(NMe_2)N:CNH]C_6H_4As(SCH_2CO_2H)_2$	$RAsO_3H_2 + HSCH_2CO_2H$			429, 434
$3,4\text{-}O_2N(MeO)C_6H_3As[SCH_2CH(NH_2)CO_2Et]_2$	$RAsO + HSCH_2CH(NH_2)CO_2Et$		m. 191-194° (dec.)	1243

2. Rxn. with AcCl → $EtAsCl_2 + AcSBu$ (645, 647).
3. Rxn. with Ac2O → $EtAs(OAc)_2 + AcSBu$ (645, 647).
4. Rxn. with H_2O on steam bath → $EtAsO_3H_2$ (647).

TABLE XV (cont'd.)

ESTERS OF DITHIOARSONOUS ACIDS WITH MERCAPTANS

RAs(SR')2	Prepd. from	Yield %	Props.	Ref.
C18				
3,4-H2N(HOCH2CH2O)C6H3As(SCH2-CO2CHMe2)2	RAs(OH)2 + HSCH2CO2CHMe2			1183
C19				
4-[···,N:C(NH2)N:CNH]-C6H4As(SC:CHCH:CHCH:N)2	RAsO + 2-mercaptopyridine			429, 434
3,4-O2N(PhCH2O)C6H3As[SCH2-CH(NH2)CO2H]2	RAsO + HSCH2CH(NH2)CO2H			652
C20				
4,3,5-HO(O2N)2C6H2As(SC6H4CO2-H-o)2	RAsO + 2-HSC6H4CO2H			652
3,4-O2N(HO)C6H3As(SC6H4CO2H-o)2	RAsO + 2-HSC6H4CO2H		m. 160-162°	652, 1243
3,4-Cl(H2N)C6H3As(SC6H4CO2-H-o)2	RAsO + 2-HSC6H4CO2H			429
3,4-H2N(HO)C6H3As(SC6H4CO2H-o)2	RAsO + 2-HSC6H4CO2H		m. 152-155°	832
C21				
3,4-O2N(MeO)C6H3As(SC6H4CO2H)2	RAsO + 2-HSC6H4CO2H			652
4-H2NCONHC6H4As(SC6H4CO2H-o)2	RAsO + 2-HSC6H4CO2H			1078
4-[N:C(NH2)N:C(NH2)N:CNH]C6-H4As(SC6H4SO3H-p)2	RAsO + 2-HSC6H4SO3H			429, 434
4,3-HO[N:C(NH2)N:C(NH2)N:CNH]C6H4As(SPh)2	RAsO3H2 + PhSH			429, 434
CO[3,4-NH(HO)C6H3As(SCH2CO2-H]2]2	CO[RAs(OH)2]2 + HSCH2CO2H			233, 712
CO[3,4-NH(Cl)C6H3As(SCH2CO2-H]2]2	CO[RAs(OH)2]2 + HSCH2CO2H			712
4-[N:C(NH2)N:C(NH2)N:CNH]C6-H4As(SC6H11O5)2	RAsO + mercaptoglucose			38
CO[4-NHC6H4As(SCH2CO2H)2]2	CO[RAs(OH)2]2 + HSCH2CO2H			712

Compound	Preparation	Yield (%)	M.p.	Ref.
C_{22}				
$2,4\text{-}HO(AcNH)C_6H_3As(SC_6H_4CO_2\text{-}H\text{-}o)_2$			m. 186°	1004
$4\text{-}AcNHC_6H_4As(SCH_2CO_2C_5H_{11})_2$	$RAsO + HSCH_2CO_2C_5H_{11}$			355-57
$2,5\text{-}HO(AcNH)C_6H_3As(SCHMeCO_2\text{-}Bu)_2$	$RAsO + MeCHSHCO_2Bu$			355-57
$3,4\text{-}H_2N(HO)C_6H_3As(SCH_2CO_2C_6\text{-}H_{13})_2$	$RAs(OH)_2 + HSCH_2CO_2C_6H_{13}$			1206
C_{23}				
$3,4\text{-}Cl[N{:}C(Cl)N{:}C(Cl)N{:}CNH]C_6\text{-}H_3As(SC_6H_4CO_2H\text{-}o)_2$	$3,4\text{-}Cl(H_2N)C_6H_3AsO + 2\text{-}HSC_6\text{-}H_4CO_2H + (ClCN)_3$	85		429
$2,4\text{-}AcNH[N{:}C(Cl)N{:}C(Cl)N{:}CNH]C_6H_3As(SPh)_2$	$RAsO_3H_3 + PhSH$			429
$4\text{-}[N{:}C(NH_2)N{:}C(NH_2)N{:}CNH]C_6\text{-}H_4As(SC_6H_4CO_2Na)_2$	$RAsCl_2 + 2\text{-}HSC_6H_4CO_2Na$	70		429, 434
$3\text{-}(Benzo[f]quinolyl\text{-}4\text{-}NH)C_6\text{-}H_4As(SCH_2CO_2Na)_2$	$RAsO_3H_2 + NaSCH_2CO_2Na$		m. 250°	250
$4\text{-}(Benzo[f]quinolyl\text{-}4\text{-}NH)C_6\text{-}H_4As(SCH_2CO_2H)_2$	$RAsO_3H_2 + NaSCH_2CO_2Na$		m. 237-238°	250
$4\text{-}[N{:}C(NH_2)N{:}C(NH_2)N{:}CNH]C_6\text{-}H_4As(SCH_2CO_2H)_2$	$RAsO + 4\text{-}mercaptopyridine$			429, 434
$4,2\text{-}Me_2N[N{:}C(NH_2)N{:}C(NH_2)N{:}CNH]C_6H_3As(SPh)_2$	$RAsO_3H_2 + PhSH$			429, 434
$3,4\text{-}O_2N(PhCH_2O)C_6H_3As[SCH_2\text{-}CH(NH_2)CO_2Et]_2$	$RAsO + HSCH_2CH(NH_2)CO_2Et$			1243
C_{24}				
$3,4\text{-}Me[N{:}C(Cl)N{:}C(Cl)N{:}CNH]\text{-}C_6H_3As(SC_6H_4CO_2H\text{-}o)_2$	$3,4\text{-}Me(H_2N)C_6H_3AsO + 2\text{-}HSC_6\text{-}H_4CO_2H + (ClCN)_2$		m. 208°	429
$2,4\text{-}HO(AcNH)C_6H_3As[SC_6H_3\text{-}4,3\text{-}(OMe)CO_2H]_2$				1004
$3,4\text{-}H_2N(HO)C_6H_3As(SCH_2CO_2\text{-}CHEtBu)_2$	$RAs(OH)_2 + HSCH_2CO_2CHEtBu$			1206
C_{25}				
$4\text{-}[N{:}C(NH_2)N{:}C(NH_2)N{:}CNH]\text{-}C_6H_4As(SC_6H_4CO_2H\text{-}p)_2$	$RAsO + 4\text{-}HSC_6H_4CO_2H$			429, 434
C_{26}				
$3,4\text{-}H_2N(HO)C_6H_3As(SCH_2CO_2\text{-}C_8H_{17})_2$	$RAs(OH)_2 + HSCH_2CO_2C_8H_{17}$	72		48, 1206

TABLE XV (cont'd.)

ESTERS OF DITHIOARSONOUS ACIDS WITH MERCAPTANS

RAs(SR)₂	Prepd. from	Yield %	Props.	Ref.
3,4-H₂N(HO)C₆H₃As(SCH₂CO₂C₈-H₁₇)₂	RAs(OH)₂ + HSCH₂CO₂CHMeC₆H₁₃			1206
C₂₇				
4-[N:C(NH₂)N:C(NH₂)N:CNH]C₆-H₄As(S-C₉H₆BrN)₂	RAsO + 2-mercapto-6-bromoquin-oline			434
3,4-O₂N(PhCH₂O)C₆H₃As(SC₆-H₄CO₂H-o)₂	RAsO + 2-HSC₆H₄CO₂H		m. 198°	652
4-[N:C(NH₂)N:C(NH₂)N:CNH]-As(S-C₉H₇N)₂	RAsO + 2-mercaptoquinoline			429, 434
4-[N:C(NH₂)N:C(NH₂)N:CNH]-C₆H₄As(S-C₉H₇N)₂	RAsO + 8-mercaptoquinoline			429, 434
C₂₈				
3,4-H₂N(HOCH₂CH₂O)C₆H₃As(SC-H₂CO₂C₈H₁₇)₂	RAs(OH)₂ + HSCH₂CO₂C₈H₁₇			48, 1183
3,4-H₂N(HOCH₂CH₂O)C₆H₃As(SC-H₂CO₂CHEtBu)₂	RAs(OH)₂ + HSCH₂CO₂CHEtBu			1183
3,4-H₂N(HOCH₂CH₂O)C₆H₃As(SC-H₂CO₂CHMeC₆H₁₃)₂	RAs(OH)₂ + HSCH₂CO₂CHMeC₆H₁₃			1183
C₂₉				
4-[N:C(NH₂)N:C(NH₂)N:CNH]-C₆H₄As(SC₁₀H₈N)₂	RAsO + 2-mercapto-4-methyl-quinoline			434
4-[N:C(NH₂)N:C(NH₂)N:CNH]-C₆H₄As(SCH₂CH₂NCOC₆H₄CO)₂	RAsO + COC₆H₄CONCH₂CH₂SH			429, 434
4,3-HO[N:C(NH₂)N:C(NH₂)N:CN-H] C₆H₃As(S-C₁₀H₁₆N₃)₂	RAsO + glutathione			429, 434
C₃₀				
4-[N:C(NH₂)N:C(NH₂)N:CNH]-C₆H₄As(S-C₁₀H₁₆N₃)₂	RAsO + glutathione			506
5,2-H₂N(HO)C₆H₃As[SCH(CO₂-Bu)CH₂CO₂Bu]₂	RAsCl₂ + HSCH(CO₂Bu)CH₂-CO₂Bu			355-6
	RAsO + HSCH(CO₂Bu)CH₂CO₂Bu			355, 357

2,5-HO(AcNH)C$_6$H$_3$As[SCH(Me)CO$_2$C$_8$H$_{17}$]$_2$	RAsO + MeCHSHCO$_2$C$_8$H$_{17}$	357
3,4-H$_2$N(HO)C$_6$H$_3$As(SCH$_2$CC$_2$C$_{10}$H$_{21}$)$_2$	RAs(OH)$_2$ + HSCH$_2$CO$_2$C$_{10}$H$_{21}$	1206
C$_{32}$		
4,3-HO(AcNH)C$_6$H$_3$As[SCH(CO$_2$-Bu)CH$_2$CO$_2$Bu]$_2$	RAsO + HSCH(CO$_2$Bu)CH$_2$CO$_2$Bu	355-7
2,4-HO(AcNH)C$_6$H$_3$As[SCH(CO$_2$-Bu)CH$_2$CO$_2$Bu]$_2$	RAsO + HSCH(CO$_2$Bu)CH$_2$CO$_2$Bu	355-7
2,5-HO(AcNH)C$_6$H$_3$As[SCH(CO$_2$-Bu)CH$_2$CO$_2$Bu]$_2$	RAsO + HSCH(CO$_2$Bu)CH$_2$CO$_2$Bu	355-7
3,4-H$_2$N(HOCH$_2$CH$_2$O)C$_6$H$_3$As-(SCH$_2$CO$_2$C$_{10}$H$_{21}$)$_2$	RAs(OH)$_2$ + HSCH$_2$CO$_2$C$_{10}$H$_{21}$	1183
C$_{33}$		
4-MeO$_2$CCH$_2$OC$_6$H$_4$As[SCH(CO$_2$-Bu)CH$_2$CO$_2$Bu]$_2$	RAsO + HSCH(CO$_2$Bu)CH$_2$CO$_2$Bu	355
C$_{34}$		
3,4-H$_2$N(HO)C$_6$H$_3$As(SCH$_2$CO$_2$-C$_{12}$H$_{25}$)$_2$	RAsO + HSCH$_2$CO$_2$C$_{12}$H$_{25}$	1206

ARSONOUS ANHYDRIDES AND ANHYDROSULFIDES WITH CARBOXYLIC AND CARBOTHIOLIC ACIDS

Arsonous anhydrides with carboxylic acids are prepared by:

1. Reacting arsenoso compounds, RAsO, with carboxylic anhydrides at elevated temperatures.
2. Reacting organodihaloarsines with silver carboxylates in benzene under reflux.
3. Treating arsonous acid esters or dithioesters with carboxylic anhydrides at elevated temperatures.

The anhydrides are listed in Table XVI.

Arsonous anhydrosulfides with carbothiolic acids are prepared by reacting organo-dihaloarsines with potassium carbothiolates or arsenoso compounds with carbothio-lic acids in aqueous solutions. The anhydrosulfides are compiled in Table XVII.

TABLE XVI
ARSONOUS ANHYDRIDES WITH CARBOXYLIC ACIDS

Anhydride	Prepd. from	Props.	Ref.
$EtAs(OAc)_2$	$EtAsOCH_2CH_2O + Ac_2O$ above $180°$	b. $126-127°/10$ mm., d_{20} 1.3300, n_D^{20} 1.4690	641
	$EtAs(SBu)_2 + Ac_2O$		645, 647
$PrAs(OAc)_2$	$PrAsCl_2 + AgOAc$	b. $120-123°/13$ mm.	44
$OCH:CHCH:CAs(OAc)_2$	$RAsO + Ac_2O$ at $80-90°$	b. $129-130°/2$ mm., d_4^{20} 1.526, n_D^{20} 1.5245; hydrolytically unstable	346-7
$PhAs(OAc)_2$	$PhAsOCH_2CH_2O$ or $PhAs-OC_6H_4O + Ac_2O$ (reflux)	b. $128-129°/2$ mm., d_{20} 1.4268, n_D^{20} 1.5484, hydrolytically unstable	639

TABLE XVII
ARSONOUS ANHYDROSULFIDES WITH CARBOTHIOLIC ACIDS

Anhydrosulfide	Prepd. from	Props.	Ref.
$MeAs(SCOMe)_2$	$MeAsO + MeCOSH$	b. $148-150°/14$ mm.	1249
	$MeAsCl_2 + MeCOSK$		1249
$MeAs(SCOC_6Cl_5)_2$	$MeAsO + C_6Cl_5COSH$	m. $161-162°$	1249
	$MeAsCl_2 + C_6Cl_5COSK$		1249
$MeAs(SCOC_6H_4Cl-p)_2$	$MeAsO + 4-ClC_6H_4COSH$	m. $122°$	1249
	$MeAsCl_2 + 4-ClC_6H_4COSK$		1249
$MeAs(SCOPh)_2$	$MeAsCl_2 + PhCOSK$	m. $62°$	1249
$PhAs(SCOPh)_2$	$PhAsO + PhCOSH$	m. $131-132°$	1249
	$PhAsCl_2 + PhCOSK$		

ARSONOUS ANHYDROSULFIDES WITH DITHIOCARBAMIC ACID DERIVATIVES

Arsonous anhydrosulfides with dithiocarbamic acids are prepared by reacting arsenoso compounds or organodihaloarsines with sodium or piperidinium dithiocarbamates in diluted hydrochloric acid. When sodium ethylenebis(dithiocarbamate) is used, 1,-3,4,8,2-dithiadiazarsenanonane derivatives are formed. Moreover, instead of dithiocarbamates, mixtures of secondary amines with carbondisulfide can be employed.

TABLE XVIII

ARSONOUS ANHYDROSULFIDES WITH DITHIOCARBAMIC ACIDS

Anhydrosulfide	Prepd. from	Props.	Ref.
MeAs[SC(S)NH$_2$]$_2$	RAsO + NaSC(S)NH$_2$	m. 138° (dec.)	1248
MeAs[SC(S)NHCH$_2$-]$_2$	RAsCl$_2$ + [CH$_2$NHC(S)SNa]$_2$	m. 96-100°	548
ClCH:CHAs[SC(S)NHCH$_2$-]$_2$	RAsCl$_2$ + [CH$_2$NHC(S)SNa]$_2$	m. 102-115°	548
EtAs[SC(S)NHCH$_2$-]$_2$	RAsCl$_2$ + [CH$_2$NHC(S)SNa]$_2$	m. 85-90°	548
MeAs[SC(S)NMe$_2$]$_2$	RAsO + Me$_2$NC(S)SNa	m. 144°	1248
BuAs[SC(S)NHCH$_2$-]$_2$	RAsCl$_2$ + [-CH$_2$NHC(S)SNa]$_2$	m. 102-104°	548
2,4-Cl(O$_2$N)C$_6$H$_3$As[SC(S)NH-CH$_2$-]$_2$	RAsCl$_2$ + [-CH$_2$NHC(S)SNa]$_2$	m. 122-124°	548
3-ClC$_6$H$_4$As[SC(S)NHCH$_2$-]$_2$	RAsCl$_2$ + [-CH$_2$NHC(S)SNa]$_2$	m. 95-98°	548
4-ClC$_6$H$_4$As[SC(S)NHCH$_2$-]$_2$	RAsCl$_2$ + [-CH$_2$NHC(S)SNa]$_2$	m. 101-103°	548
4-O$_2$NC$_6$H$_4$As[SC(S)NHCH$_2$-]$_2$	RAsCl$_2$ + [-CH$_2$NHC(S)SNa]$_2$	m. 172-182°	548
PhAs[SC(S)NHCH$_2$-]$_2$	RAsCl$_2$ + [-CH$_2$NHC(S)SNa]$_2$	m. 114-115°	548
2-MeC$_6$H$_4$As[SC(SNHCH$_2$-]$_2$	RAsCl$_2$ + [-CH$_2$NHC(S)SNa]$_2$	m. 107-108°	548
4-MeC$_6$H$_4$As[SC(S)NHCH$_2$-]$_2$	RAsCl$_2$ + [-CH$_2$NHC(S)SNa]$_2$	m. 101-103°	548
MeAs[SC(S)NEt$_2$]$_2$	RAsCl$_2$ + Et$_2$NH + CS$_2$	m. 102°	110
2-HO$_2$CC$_6$H$_4$As[SC(S)NHCH$_2$-]$_2$	RAsCl$_2$ + [-CH$_2$NHC(S)SNa]$_2$	m. 80-86°	548
ClCH:CHAs[SC(S)NEt$_2$]$_2$	RAsCl$_2$ + Et$_2$NH + CS$_2$	m. 85°	110
	RAsCl$_2$ + Et$_2$NC(S)SNa	m. 87°	1275
MeAs[SC(S)N(CH$_2$CH$_2$)$_2$O]$_2$	RAsCl$_2$ + $\overline{CH_2CH_2OCH_2CH_2N}$C-(S)SNa	m. 178°	1275
	RAsCl$_2$ + $\overline{CH_2CH_2OCH_2CH_2N}$C-(S)SNa	m. 174°	1248
MeAs[SC(S)N̅(CH$_2$)$_3$CH$_2$]$_2$	RAsO + $\overline{CH_2(CH_2)_3N}$C(S)SNa	m. 178°	1248
PhAs[SC(SNMe$_2$]$_2$	RAsO + Me$_2$NC(S)SNa	m. 221°	1248
ClCH:CHAs[SC(S)N(CH$_2$CH$_2$)$_2$-O]$_2$	RAsCl$_2$ + $\overline{CH_2CH_2OCH_2CH_2N}$C-(S)SNa	m. 161°	1275
4-H$_2$NC$_6$H$_4$As[SC(S)NMe$_2$]$_2$	RAsO + Me$_2$NC(S)SNa		1248
EtAs[SC(S)N(CH$_2$CH$_2$)$_2$O]$_2$	RAsCl$_2$ + $\overline{CH_2CH_2OCH_2CH_2N}$C-(S)SNa	m. 154°	1275
MeAs[SC(S)N̅(CH$_2$)$_4$CH$_2$]$_2$	RAsO + $\overline{CH_2(CH_2)_4N}$C(S)SNa	m. 157 °	1248
1-C$_{10}$H$_7$As[SC(S)NHCH$_2$-]$_2$	RAsCl$_2$ + [-CH$_2$NHC(S)SNa]$_2$	m. 108-110	548
PhAs[SC(S)N(CH$_2$CH$_2$)$_2$O]$_2$	RAsCl$_2$ + $\overline{CH_2CH_2OCH_2CH_2N}$C-(S)SNa	m. 218-219°	1275
2-MeOC$_6$H$_4$As[SC(S)NEt$_2$]$_2$	RAsCl$_2$ + Et$_2$NH + CS$_2$	m. 144°	110

TABLE XVIII (cont'd.)
ARSONOUS ANHYDROSULFIDES WITH DITHIOCARBAMIC ACIDS

Anhydrosulfide	Prepd. from	Props.	Ref
$4\text{-MeOC}_6\text{H}_4\text{As}[\text{SC}(\text{S})\text{NEt}_2]_2$	$\text{RAsCl}_2 + \text{Et}_2\text{NH} + \text{CS}_2$	m. 168°	110
$\text{MeAs}[\text{SC}(\text{S})\text{NMePh}]_2$	$\text{RAsO} + \text{MePhNC}(\text{S})\text{SNa}$	m. 191°	1248
$\overline{\text{CH}_2(\text{CH}_2)_4\text{NH}}.\text{HO}_3\text{SC}_6\text{H}_4\text{As-}$ $[\text{SC}(\text{S})\underline{\text{N}(\text{CH}_2)_4}\overline{\text{CH}_2}]_2$	$4\text{-NaO}_3\text{SC}_6\text{H}_4\text{AsI}_2 + \overline{\text{CH}_2\text{-}}$ $\underline{(\text{CH}_2)_4\text{NH}}.\text{HSC}(\text{S})\underline{\text{N}(\text{CH}_2)_4\text{-}}$ $\overline{\text{CH}_2}$ in not alc.	Colorless needles, de-comp. 230-232°	955
$\{\overline{\text{-CH}_2\text{As}[\text{SC}(\text{S})\underline{\text{N}(\text{CH}_2)_4\text{-}}}$ $\overline{\text{CH}_2}]_2\}_2$	$(\overline{\text{CH}_2\text{AsI}_2})_3 + \overline{\text{CH}_2(\text{CH}_2)_4\text{NH}}.\text{-}$ $\text{HSC}(\text{S})\underline{\text{N}(\text{CH}_2)_4\overline{\text{CH}_2}}$	m. 198-199° (effervesc.)	609

ARSONOUS ANHYDROSULFIDES WITH XANTHOGENIC AND PHOSPHORODITHIOIC ACID DERIVATIVES

The anhydrosulfides are prepared by reacting organodihaloarsines with potassium xantogenates and ethyl sodium phosphorodithioate, respectively.

TABLE XIX
ARSONOUS ANHYDROSULFIDES WITH XANTHOGENATES AND PHOSPHORODITHIOATES

Anhydrosulfide	Prepd. from	Props.	Ref.
$4\text{-ClC}_6\text{H}_4\text{As}[\text{SC}(\text{S})\text{OEt}]_2$	$\text{RAsCl}_2 + \text{KSC}(\text{S})\text{OEt}$	m. 59-60°	988
$4\text{-O}_2\text{NC}_6\text{H}_4\text{As}[\text{SC}(\text{S})\text{OEt}]_2$	$\text{RAsCl}_2 + \text{KSC}(\text{S})\text{OEt}$	m. 78-79°	988
$\text{PhAs}[\text{SC}(\text{S})\text{OEt}]_2$	$\text{RAsCl}_2 + \text{KSC}(\text{S})\text{OEt}$	m. 47-48°	988
$\text{PhAs}[\text{SC}(\text{S})\text{OCHMe}_2]_2$	$\text{RAsCl}_2 + \text{KSC}(\text{S})\text{OCHMe}_2$	m. 106-107°	988
$\text{PhAs}[\text{SC}(\text{S})\text{OC}_5\text{H}_{11}]_2$	$\text{RAsCl}_2 + \text{KSC}(\text{S})\text{OC}_5\text{H}_{11}\text{-n}$	Oil	988
$\text{PhAs}[\text{SP}(\text{S})(\text{OEt})_2]_2$	$\text{RAsCl}_2 + \text{NaSP}(\text{S})(\text{OEt})_2$	Oil	988

ARSINOUS ACIDS, ESTERS, ANHYDRIDES, AND THIO ANALOGS

ARSINOUS ACIDS

The acids having the general formula R_2AsOH or $\text{RR}'\text{AsOH}$ are prepared or formed by:
1. Hydrolyzing bis(diorganoarsenic) oxides, $(\text{R}_2\text{As})_2\text{O}$ or $(\text{RR}'\text{As})_2\text{O}$, with aqueous alkali (Method I).
2. Hydrolyzing organohaloarsines, R_2AsX or $\text{RR}'\text{AsX}$, with aqueous ammonia or alkali (Method II).
3. Reducing arsinic acids, $\text{R}_2\text{AsO}_2\text{H}$ or $\text{RR}'\text{AsO}_2\text{H}$, with sulfur dioxide in 10% sulfuric acid or in hydrochloric acid in the presence of potassium iodide. When hydrochloric acid is used as the solvent for the reduction, the reaction mixture is neutralized with aqueous alkali, and the product is precipitated with hydrochloric acid (Method III).

TABLE XX
ARSINOUS ACIDS

R₂AsOH or RR'AsOH	Prepd. from	Method	Props. and Remarks	Ref.
$(CF_3)_2AsOH$			$[(CF_3)_2AsO]_2Hg$, m.>310°	333
$(4-ClC_6H_4)[4,3-HO(O_2N)-C_6H_3]AsOH$	$RR'AsO_2H + SO_2$ in H_2SO_4	III	Powder, m. 198-202°	1189, 1191
$(4-O_2NC_6H_4)_2AsOH$	$R_2AsO_2H + SO_2$ in H_2SO_4	III	Granules, decomp. 120-130°	1189
			Potentiometric neutra-lization	1186
$Ph(2-O_2NC_6H_4)AsOH$	$RR'AsO_2H + SO_2$ in H_2SO_4	III	Columns, decomp. 135°	1189
$Ph(4-O_2NC_6H_4)AsOH$	$RR'AsO_2H + SO_2$ in H_2SO_4	III	Needles, m. 154°	1189
$Ph[3,4-O_2N(HO)C_6H_3]AsOH$	$RR'AsO_2H + SO_2$ in H_2SO_4	III	m. 140-142°	1189
			m. 160-176°	1191
			m. 116-117°, rxn.[1]	1242
$(4-HOC_6H_4)_2AsOH$	$R_2AsO_2H + SO_2$ in H_2SO_4	III	Decomp. 85°	1191
$Ph(2-H_2NC_6H_4)AsOH$	$RR'AsO_2H + SO_2$ in H_2SO_4	III	Powder, decomp. 140°	1191
$Ph(4-H_2NC_6H_4)AsOH$	$RR'AsO_2H + SO_2$ in H_2SO_4	III	Powder, m.113°, rxn.[1]	1189, 1191
$Ph[3,4-H_2N(HO)C_6H_3]AsOH$	$RR'AsO_2H + SO_2$ in H_2SO_4	III	HCl salt, powder de-comp. 160°, rxn.[1]	1189-1 1242
$(4-H_2NC_6H_4)_2AsOH$	$(4-HCl.H_2NC_6H_4)_2AsCl$	II	Amorphous solid, softens at 70-72°	64
	$R_2AsO_2H + SO_2$ in H_2SO_4	III	Powder, m. 97°	1189
			Potentiometric titra-tion	1186
$(4-O_2NC_6H_4)(4-HO_2CC_6H_4)-AsOH$	$RR'AsO_2H + SO_2$ in H_2SO_4	III	Powder, decomp. 217-220°	1189 1191
$Ph(2-HO_2CC_6H_4)AsOH$	$RR'AsO_2H + SO_2$ in H_2SO_4	III	Needles, m. 133°	1189
$Ph(4-HO_2CC_6H_4)AsOH$	$RR'AsO_2H + SO_2$ in H_2SO_4	III	Powder, decomp. 120°	1189, 1191
			Potentiometric titra-tion	1186
$(4-HOC_6H_4)(2-HO_2CC_6H_4)-AsOH$	$RR'AsO_2H + SO_2$ in H_2SO_4	III	Needles, m. 180-182°	1189
$(2-HO_2CC_6H_4)_2AsOH$	$R_2AsO_2H + SO_2$ in H_2SO_4	III	Prisms, decomp. 85°	1189
$(4-HO_2CC_6H_4)_2AsOH$	$R_2AsO_2H + SO_2$ in H_2SO_4	III	Powder, m.>300°, rxn.[1]	1189
			Potentiometric titra-tion	1186
$(4-AcNHC_6H_4)(4-O_2NC_6H_4)-AsOH$	$(RR'As)_2O + NaOH$	I	Colorless crystals, soft.>70°	64
$Ph[4-(\overline{N:CClN:CClN:CNH})-C_6H_4]AsOH$	$RR'AsO_2H + SO_2$ in HCl	III		658
$Ph[4-(\overline{N:CClN:COHN:CNH})-C_6H_4]AsOH$	$RR'AsO_2H + SO_2$ in HCl	III		658

1. Rxn. with mercaptocarboxylic acids $(HS-X-CO_2H) \rightarrow$ the corresponding $RR'AsS-X-CO_2H$ (1190, 1191).

TABLE XX (cont'd.)
ARSINOUS ACIDS

R$_2$AsOH or RR AsOH	Prepd. from	Method	Props., Rxns., Remarks	Ref.
[4-(N:C(OH)N:C(OH)N:C-NH)C$_6$H$_4$]$_2$AsOH	R$_2$AsO$_2$H + SO$_2$ in H$_2$-SO$_4$	III		1245
[4-(N:C(NH$_2$)N:C(OH)N:C-NH)C$_6$H$_4$]$_2$AsOH	R$_2$AsO$_2$H + SO$_2$ in H$_2$S-O$_4$	III		1245
[4-(N:C(NH$_2$)N:C(NH$_2$)N:CNH)C$_6$H$_4$]$_2$AsOH	R$_2$AsO$_2$H + SO$_2$ in H$_2$S-O$_4$	III		1245

ESTERS AND THIOESTERS OF ARSINOUS ACIDS

Esters of arsinous acids, R$_2$AsOR and RR'AsOR, are prepared by:
1. Reacting diorganohaloarsines, R$_2$AsX or RR'AsX, with alkali metal alkox-ides, MOR, in ether or in the corresponding alcohol, or with alcohols in ether in the presence of amines (pyridine) in an inert atmosphere.
2. Heating bis(diorganoarsenic) oxides, (R$_2$As)$_2$O, with alcohols under reflux, in the presence or absence of calcium carbide or copper sulfate, in an inert atmosphere.

Thioesters of arsinous acids, R$_2$AsSR and RR'AsSR, are prepared by:
1. Treating arsinic acids, R$_2$AsO$_2$H or RR'AsO$_2$H, with mercaptans.
2. Reacting arsinous acids, R$_2$AsOH or RR'AsOH, with mercaptans in hot water or in ethanol.
3. Reacting bis(diorganoarsenic) oxides, (R$_2$As)$_2$O or (RR'As)$_2$O, with mer-captans in hot water.

The esters of arsinous acids are compiled in Table XXI and the thioesters in Table XXII

TABLE XXI
ARSINOUS ACID ESTERS

Ester	Prepd. from	Yield	B. °/mm.; d_0^{20}; n_D^{20}, Rxns.	Ref.
Et₂AsOEt	Et₂AsI + NaOEt in Et₂O	76.2	141-142°/10; 1.1114, 1.4621; rxn.[2]	648
Et₂AsOCH₂CH:CH₂	Et₂AsCl + CH₂:CHCH₂OH + pyridine	44.5	80-82°/77; 1.1141; 1.4731; rxn.	650
MeBuAsOEt	MeBuAsI + NaOEt	55.9	51.2-52°/10; 1.0826; 1.4607; rxn.[2]	648
Et₂AsOPr	(Et₂As)₂O + PrOH	71	50-52°/10; 1.0859; 1.4613; rxn.[1,2]	251
MePhAsOMe	MePhAsCl + NaOMe (MePhAs)₂O + MeOH	24	101-102°/17 95-96°/17; 1.2975; 1.5628; rxn.[3]	636, 1160
Et₂AsOBu	(Et₂As)₂O + BuOH + CaC₂	77.7	72-72.5°/13; 1.0688; 1.4608; rxn.[1,2]	648
Pr₂AsOEt	(Pr₂As)₂O + EtOH + CaC₂	61.3	66-68°/12; 1.0636; 1.4620, rxn.[2]	648
MeBuAsOPr	(MeBuAs)₂O + PrOH + CaC₂	33	83°/10; 1.0470; 1.4603; rxn.[2]	648
MePhAsOEt	(MePhAs)₂O + EtOH		102-103°/20; 1.2473; 1.5559	636, 1160
Bu₂AsOMe	(Bu₂As)₂O + MeOH + CuSO₄	35.1	85-87°/12; 1.0593; 1.4653	719
EtPhAsOMe	(EtPhAs)₂O + MeOH	20	92-93°/10 Rxn.[1]	1160
MeBuAsOBu				649
(ClCH:CH)PhAsOEt	(ClCH:CH)PhAsCl + NaOEt		Yellow oil, 165-170°/3	263
MePhAsOPr	(MePhAs)₂O + PrOH	63	114-114.5°/18; 1.1987; 1.5403; rxn.[4]	636, 1160
EtPhAsOEt	(EtPhAs)₂O + EtOH	35	105-106°/12; 1.1980; 1.5430	636, 1160
PrPhAsOMe	(PrPhAs)₂O + MeOH	26	102-103°/10	1160
Bu₂AsOEt	(Bu₂As)₂O + EtOH + CuSO₄	64.8	89-91°/10; 1.0318; 1.4613	719
MePhAsOBu	(MePhAs)₂O + BuOH	85.6	122-123°/18; 1.1675; 1.5338; 1.1645; 1.5338	636, 1160
EtPhAsOPr	(EtPhAs)₂O + PrOH	61.5	110-111°/10; 1.1692; 1.5346	636, 1160
PrPhAsOEt	(PrPhAs)₂O + EtOH	30	101-103°/7; 1.1700; 1.5360	636, 1160
BuPhAsOMe	(BuPhAs)₂O + MeOH	31	117-118°/12; 1.1747; 1.5380	636, 1160
EtPhAsOBu	(EtPhAs)₂O + BuOH	79	122-123°/10; 1.1448; 1.5295	636, 1160

1. Rxn. with SeO₂ → mixture of the corresponding arsinic acid and arsinic acid ester (649).
2. Rxn. with H₂O in air → the corresponding arsinic acid (648, 650, 719).
3. Rxn. with Cu₂Cl₂ → MePhAsOMe.CuCl, m. 70-72° (1160).
4. Rxn. with MeI at 20-40° → Me₂PhAs(OMe)I, m. 109-110° (1160).

TABLE XXI (cont'd.)
ARSINOUS ACID ESTERS

Ester	Prepd. from	Yield %	B. °/mm.; d^{20}_0; n^{20}_D, Rxns.	Ref.
PrPhAsOPr	(PrPhAs)₂O + PrOH	81.5	126.5-127°/12.5; 1.1397; 1.5280 / 1.1397; 1.5260	1160 / 636
BuPhAsOEt	(BuPhAs)₂O + EtOH	38	127-128°/12; 1.1454; 15318	636, 1160
Bu₂AsOBu	(Bu₂As)₂O + BuOH + CuSO₄	81.4	117-118°/10; 1.0083; 1.4627	719
Ph₂AsOMe	Ph₂AsCl + NaOMe		138°/2	1275
PrPhAsOBu	(PrPhAs)₂O + BuOH	85	120-121°/4; 1.1253; 1.5260 / 1.1253; 1.5270	1160
BuPhAsOPr	(BuPhAs)₂O + PrOH	68	139-140°/10; 1.1158; 1.5220	636, 1160
Ph₂AsOEt	Ph₂AsCl + NaOEt	54	166.5-167°; d^{15}_0 1.5220; n^{16}_D 1.6025; complex⁵	625
BuPhAsOBu	(BuPhAs)₂O + BuOH	80	154-155°/16.5; 1.1038; 1.5198	636, 1160
Ph₂AsOPr	Ph₂AsCl + NaOPr		174-175°/10; d^{16}_0 1.2248; n^{17}_D 1.5925 complex⁶	625
Ph(4-MeC₆H₄)AsOEt	Ph(4-MeC₆H₄)AsCl + NaOEt	65.3	178-180°/9; d^{15}_0 1.2690; n^{15}_D 1.6120	625
Ph(4-MeC₆H₄)AsOPr	Ph(4-MeC₆H₄)AsCl + NaOPr		188-189°/11; d^{12}_0 1.2008; n^{12}_D 1.5865	625
Bu₂AsOC₈H₁₇	(Bu₂As)₂O + n-C₈H₇OH + CuSO₄	72.3	167-169°/11; 0.9780; 1.4650; rxn.²	719
Ph₂AsOPh			Diamagnetic anomaly	975

5. Complex: Ph₂AsOEt.CuI, m. 160-162° (dec.) (625).
6. Complex: Ph₂AsOPr.CuI, m. 140-142° (625).

TABLE XXII
ARSINOUS ACID THIOESTERS

Thioester	Prepd. from	Props.	Ref.
Me₂AsSCH₂CH(NH₂)CO₂H	R₂AsO₂H + cysteine.HCl	m. 219-220°	667
Ph₂AsSCH₂CO₂H	(R₂As)₂O + HSCH₂CO₂H	Needles, m. 94°	1190-1
Ph(4-HOC₆H₄)AsSCH₂CO₂H	RR'AsOH + HSCH₂CO₂H	Powder, m. 83-85°	1190-1
Ph[3,4-H₂N(HO)C₆H₃]AsSCH₂CO₂H	RR'AsOH + HSCH₂CO₂H	Needles, m. 53-55°	1190
Ph₂AsSCH₂CH(NH₂)CO₂H	(R₂As)₂O + cystine.HCl + AcONa in H₂O	Powder, m. 143-144°	1190 / 1191
	R₂AsO₂H + cystine.HCl + AcONa	m. 143-144° (dec.)	1244

136

TABLE XXII (cont'd.)
ARSINOUS ACID THIOESTERS

Thioester	Prepd. from	Props.	Ref.
Ph(4-HOC$_6$H$_4$)AsSCH$_2$CH(NH$_2$)CO$_2$H	RR'AsOH + cystine.HCl + AcONa in H$_2$O	Needles, decomp. 210°	1190
Ph(3,4 H$_2$N(HO)C$_6$H$_3$]AsSCH$_2$CH-(NH$_2$)CO$_2$H	RR'AsOH + cystine HCl + AcONa in H$_2$O	Powder, decomp. 158-162	1190-1
(4-HO$_2$CC$_6$H$_4$)$_2$AsSCH$_2$CO$_2$H	R$_2$AsOH + HSCH$_2$CO$_2$H	Mono-Na salt, needles	1191, 1241
Ph$_2$AsSCH(CO$_2$H)CH$_2$CO$_2$H	(R$_2$As)$_2$O + mercaptosuccinic acid	Powder, m. 135-136°	1190-1
Ph(4-HOC$_6$H$_4$)AsSCH(CO$_2$H)CH$_2$CO$_2$H	RR'AsOH + mercaptosuccinic acid	Plates, m. 93-94°	1190
Ph[3,4-H$_2$N(HO)C$_6$H$_3$]AsSCH(CO$_2$H)-CH$_2$CO$_2$H	RR'AsOH + mercaptosuccinic acid	Powder, m. 89-91°	1190-1
	RR'AsO$_2$H + mercaptosuccinic acid	m. 135-136°	1244
Ph[3,4-O$_2$N(HO)C$_6$H$_3$]AsSC$_6$H$_4$-CO$_2$H-o	RR'AsOH + 2-HSC$_6$H$_4$CO$_2$H	Powder, m. 88-90°	1190
2-(Ph$_2$AsS)C$_6$H$_4$CO$_2$H	(Ph$_2$As)$_2$O + 2-HSC$_6$H$_4$CO$_2$H	Plates or needles, m. 121-122°	1190-1
2-[(4-HOC$_6$H$_4$)$_2$AsS]C$_6$H$_4$CO$_2$H	R$_2$AsOH + 2-HSC$_6$H$_4$CO$_2$H	Powder	1241
2-[Ph(4-H$_2$NC$_6$H$_4$)AsS]C$_6$H$_4$CO$_2$H	RR'AsOH + 2-HSC$_6$H$_4$CO$_2$H	Powder, m. 172-173°	1190
2-[Ph(4-HOC$_6$H$_4$AsS]C$_6$H$_4$CO$_2$H	RR'AsOH + 2-HSC$_6$H$_4$CO$_2$H	Powder, decomp. 120°	1190
	RR'AsO$_2$H + 2-HSC$_6$H$_4$CO$_2$H	m. 120°	1244
Ph[3,4-H$_2$N(HO)C$_6$H$_3$]AsSC$_6$H$_4$CO$_2$H	RR'AsOH + 2-HSC$_6$H$_4$CO$_2$H	Powder, decomp. 96-98°	1190
	RR'AsO$_2$H + 2-HSC$_6$H$_4$CO$_2$H	m. 94°	1244
[3,4-H$_2$N(HO)C$_6$H$_3$]$_2$AsSC$_6$H$_4$CO$_2$H	R$_2$AsOH + 2-HSC$_6$H$_4$CO$_2$H	Brown needles	1241
2-[(2-HO$_2$CC$_6$H$_4$)$_2$AsS]C$_6$H$_4$CO$_2$H	R$_2$AsOH + 2-HSC$_6$H$_4$CO$_2$H	Powder	1241
2-[(4-HO$_2$CC$_6$H$_4$)$_2$AsS]C$_6$H$_4$CO$_2$H	R$_2$AsOH + 2-HSC$_6$H$_4$CO$_2$H	Powder, decomp. 242-244°	1190

137

ANHYDRIDES AND ANHYDROSULFIDES OF ARSINOUS ACIDS WITH CARBOXYLIC, CARBOTHIOLIC, XANTOGENIC, PHOSPHORODITHIOIC, AND DITHIOCARBAMIC ACID DERIVATIVES

The anhydrides and anhydrosulfides are prepared or formed by:

1. Reacting organomonhaloarsines, R_2AsX, with silver salts of carboxylic acids at 60° or with alkali metal salts of carboxylic acids in methanol.
2. Reacting bis(diorganoarsinic) oxides, $(R_2As)_2O$, with carboxylic anhydrides at 80-90°
3. Reacting organomonohaloarsines, R_2AsX, with alkali metal salts of thiocarboxylic acids.
4. Reacting organohaloarsines, R_2AsX, with alkali metal xanthates or their homologs or with dialkyldithiophosphates in acetone under reflux.
5. Reacting mixtures of carbon disulfide and secondary amines, R_2NH, with diorganohaloarsines, R_2AsX.

TABLE XXIII

ANHYDRIDES AND ANHYDROSULFIDES OF ARSINOUS ACIDS

Anhydr. or Anhydrosulf.	Prepd. from	Yield	Props.	Ref.
$Me_2AsOOCCF_3$	$Me_2AsCl + AgO_2CCF_3$	32	b. 136-137°, decomp. at 205° → Me_2AsCF_3, CHF_3, CO_2, and SiF_4	256
$MePhAsOAc$	$MePhAsCl + NaOAc$	36	Pale yellow liq., b. 140-142°/14 mm.	251
$(\overline{OCH:CHCH:C})_2AsOAc$	$(R_2As)_2O + Ac_2O$	85	Colorless, hydroscopic crystals, m. 48-50°, b. 161-162°/10 mm.	346, 347
$Ph_2AsSCOC_6H_4Cl-p$	$Ph_2AsCl + 4-ClC_6H_4COSK$		m. 87-88°	1249
$Me_2AsSC(S)OEt$	$Me_2AsCl + KSC(S)OEt$		b. 137-138°/18 mm.	977, 988
$Me_2AsSC(S)OCH_2CH:CH_2$	$Me_2AsCl + KSC(S)OC_3H_5$		b. 90-95°/1 mm.	988
$Me_2AsSC(S)OCHMe_2$	$Me_2AsCl + KSC(S)OCHMe_2$		b. 106-108°/3 mm.	988
$Me_2AsSC(S)OC_5H_{11}-n$	$Me_2AsCl + KSC(S)OC_5H_{11}$		b. 104°/0.2 mm.	988
$Me_2AsSC(S)OC_6H_{11}$				
$Me_2AsSC(S)OC_6H_{11}$	$Me_2AsCl + KSC(S)OC_6H_{11}$		b. 153-154°/2.5 mm. (dec.)	988
$Me_2AsSC(S)OC_{12}H_{25}$	$Me_2AsCl + KSC(S)OC_{12}H_{25}$		b. 198-201°/2 mm.	988
$Ph_2AsSC(S)OEt$	$Ph_2AsCl + KSC(S)OEt$		oil	988
$Ph_2AsSC(S)OCHMe_2$	$Ph_2AsCl + KSC(S)OCHMe_2$		oil	988
$Me_2AsSP(S)(OMe)_2$	$Me_2AsCl + HSP(S)(OMe)_2 + K_2CO_3$		b. 125-130°/2 mm.	988
$Me_2AsSP(S)(OEt)_2$	$Me_2AsCl + HSP(S)(OEt)_2 + K_2CO_3$		b. 116°/1.5 mm.	988
$Ph_2AsSP(S)(OEt)_2$	$Ph_2AsCl + NaSP(S)(OEt)_2$		Oil	988

TABLE XXIII (cont'd.)
ANHYDRIDES AND ANHYDROSULFIDES OF ARSINOUS ACIDS

Anhydr. or Anhydrosul.	Prepd. from	Yield	Props.	Ref.
$Me_2AsSC(S)NMe_2$	$Me_2AsCl + Me_2NH + CS_2$		m. 43°	110
$Me_2AsSC(S)NCH_2CH_2O\text{-}$ CH_2CH_2	$Me_2AsCl + O(CH_2CH_2)_2\text{-}$ $NH + CS_2$		m. 80°	110
$Me_2AsSC(S)NEt_2$	$Me_2AsCl + Et_2NH + CS_2$		m. 45°	110
$Me_2AsSC(S)N(CH_2)_4CH_2$	$Me_2AsCl + CH_2(CH_2)_4NH$ $+ CS_2$		m. 64°	110
$Ph_2AsSC(S)NEt_2$	$Ph_2AsCl + Et_2NH + CS_2$		m. 95°	110
$HN(C_6H_4)_2AsSC(S)NEt_2$				110

BIS(DIORGANOARSENIC) OXIDES

Bis(diorganoarsenic) oxides and sulfides, $(R_2As)_2O$ and $(R_2As)_2$, are anhydrides and anhydrosulfides of the corresponding arsinous and thioarsinous acids, R_2AsOH and R_2AsSH, respectively.

The oxides are prepared by:

1. Heating arsenic trioxide with acetic or propionic acid and potassium or cerium carbonate on pumice at 350° to form $(Me_2As)_2O$ and $(Et_2As)_2O$, respectively, in mixtures with the corresponding tetraorganodiarsines.
2. Fusion of arsenic trioxide with potassium or sodium acetate (Cadet method) to form a mixture containing cacodyl and cacodyl oxide.
3. Heating at 300° diarsenosomethane, $CH_2(AsO)_2$, with acetic anhydride and potassium acetate to form cacodyl oxide.
4. Distillation of diorganohaloarsines, R_2AsX or $RR'AsX$, with sodium hydroxide in a stream of carbon dioxide or by hydrolysis of diorganohaloarsines with aqueous or alcoholic alkali, aqueous ammonia, or alcoholic sodium alkoxides. The diorganohaloarsines can be formed in situ from arsinic acids by reduction with sulfur dioxide.
5. Metathesis of diorganohaloarsines, R_2AsX or $RR'AsX$, with mercury oxide.
6. Reacting arsenic trichloride or organodihaloarsines with aryldiazonium double salts with zinc or iron chloride, $RN_2Cl.MCl_x$, in acetone in the presence of iron or zinc and neutralizing the reaction product with aqueous alkali.
7. Reacting potassium arylazocarbonates, RN_2CO_2K, with arsenic trichloride in acetone or ethyl acetate, then threating the reaction product with aqueous hydrochloric acid, followed by 40% potassium hydroxide solution at an elevated temperature to form mixtures of bis(diarylarsenic) oxides with arsenosoarenes or arenearsonous acids.
8. Reducing arsonic acids, $RAsO_3H_2$, or arsenic acid, H_3AsO_4, with phenylhydrazine in the presence or absence of copper(I) oxide.
9. Reacting aryldiazonium chloride-double salt with arsenic trichloride, $RN_2ClAsCl_3$, with phenylhydrazine in aqueous-ethanolic hydrochloric acid in the presence of copper(I) chloride and iron(III) chloride and neutralizing the product with alkali.

10. Treating tertiary arsines, R_3As, with lithium in tetrahydrofuran, and hydrolyzing the lithium derivatives with water.
11. Oxidation of secondary arsines, R_2AsH, with atmospheric oxygen.

BIS(DIMETHYLARSENIC) OXIDE, CACODYL OXIDE $C_4H_{12}As_2O$ $(Me_2As)_2O$

Prep.: By passing As_2O_3 and 75% AcOH over K_2CO_3 on pumice at 350°, 66-77% yield of a mixture of $(Me_2As)_2O$ with $(Me_2As)_2$ (446, 447, 1308).
By heating As_2O_3 with KOAc or NaOAc at 300-400° (1224, 1307).
By distn. of Me_2AsCl with NaOH in CO_2 (602).
By heating at 300° $CH_2(AsO)_2$ or intermediate polymers formed from As-$(OAc)_3$, Ac_2O, and KOAc; isolated as $(Me_2As)_2O.2HgCl_2$ (1214).
By treating $Me_2AsCH \cdot CHCl$ with alc. KOH, 59.2% yield (674).
Props.: B. 118-120°, d_{15} 1.4603 (674).
Rxns. with: $HCl-FeCl_3$ or $HgCl_2 \rightarrow Me_2AsCl$ (447).
Atm. O or HgO $\rightarrow Me_2AsO_2H$ (1224).
Electrolytic redn. $\rightarrow (Me_2As)_2$ (1308).
Use: Stabilizer for vinyl ester monomers (305).
Complex compds.: $(Me_2As)_2O.2HgCl_2$, prisms, m. 210-212° (1214).
$[Pt[(Me_2As)_2O.H_2O]Cl_2]$, colorless needles, loses H_2O at 140-160° to form lemon-yellow crystals; dipole moment and mol. structure (602).
$[Pt[(Me_2As)_2O]Br_2]$, dipole moment (602).
$[Pt[Me_2As)_2O]I_2]$, dipole moment (602).

BIS(DIETHYLARSENIC) OXIDE $C_8H_{20}As)_2O$ $(Et_2As)_2O$

Prepn.: by passing As_2O_3 and $EtCO_2H$ over K_2CO_3 on pumice at 350° a mixt. of $(Et_2As)_2$ with $(Et_2As)_2O$ is formed. (446).
Props.: B. 107.5-110°/15 mm., d_{20} 1.3036, n_D^{20} 1.5123 (648).
Rxns. with: $HCl-FeCl_3 \rightarrow Et_2AsCl$ (447).
RONa $\rightarrow Et_2AsOR$ (648).

BIS(METHYLPHENYLARSENIC) OXIDE $C_{14}H_{16}As_2O$ $(MePhAs)_2O$

Prepn.: by treating $MePhAsI$ with cold 10 N NaOH; 80-90% yield (626).
Props.: B. 154-155°/3mm., d_0^{20} 1.4381, n_D^{20} 1.6252 (1160); d_0^{18} 1.4409; n_D^{20} 1.6250 (626); d_0^{20} 1.4381; n_D^{20} 1.6252; R at. 11.30 (636).
Rxns. with: $HNO_3 \rightarrow MePhAsOOH$ (640).
ROH $\rightarrow MePhAsOR$ (1160).

BIS[BIS(2,4,6-TRIBROMOPHENYL)ARSENIC] OXIDE $C_{24}H_8As_2Br_{12}O$
$[(2,4,6-Br_3C_6H_2As]_2O$

Prepn.: by reacting $2,4,6-Br_3C_6H_4N_2Cl.ZnCl_2$ with $AsCl_3$ in Me_2CO in the presence of Fe and treating the prod. with HCl followed by aq. KOH; 66% yield (1066).
Props.: M. 171-173° (1066).

BIS(4-BROMOPHENYLPHENYLARSENIC) OXIDE $C_{24}HL_6As_2Br_2O$ $[Ph(4-BrC_6H_4)As]_2O$

Prepn.: by adding $PhAsCl_2$ to $4-BrC_6H_4N_2Cl.FeCl_3$ in Me_2CO at 0°, followed by NaI, and hydrolyzing the prod. with 40% KOH (908).
Rxn. with HCl $\rightarrow Ph(4-BrC_6H_4)AsCl$ (908).

BIS(4-NITRODIPHENYLARSENIC) OXIDE $C_{24}H_{18}As_2N_2O_5$ $[Ph(4-O_2NC_6H_4)As]_2O$

Prepn.: by adding $PhAsCl_2$ to $4-O_2NC_6H_4N_2Cl.FeCl_3$ in Me_2CO at 0°, followed by NaI, and treating the prod. with 40% KOH; 51.6% yield along with $4-O_2NC_6H_4AsO_3H_2$(908).
Rxn. with: HCl $\rightarrow Ph(4-O_2NC_6H_4)AsCl$ (908).

BIS(DIPHENYLARSENIC) OXIDE $C_{24}H_{20}As_2O$ $(Ph_2As)_2O$

<u>Prepn.:</u> By reacting $PhNHNH_2$ with H_3AsO_4 in the presence of Cu_2O (56).

By reacting $PhAsO_3H_2$ with $PhNHNH_2$ at 70-75° in the absence of Cu; 40% yield (129).

By treating $PhN_2Cl.AsCl_3$ with $PhNHNH_2.HCl$ in aq. HCl-EtOH in the presence of $CuCl_2$ and $FeCl_3$ and neutralizing the rxn. mixt.; 42.5% yield along with 15.5% PhAsO (136).

By thermal decompn. of $Ph_2(HO_2CCH_2CH_2)AsO$; 98% yield (243).

By reacting $PhN_2Cl.ZnCl_2$ with $PhAsI_2$ in Me_2CO at 0° and treating the prod. with 40% KOH; 71% yield. With $PhAsCl_2$ instead of $PhAsI_2$ 42% yield was obtd. (908).

By treating Ph_2AsF with KOH (1100).

By reacting $PhN_2Cl.FeCl_3$ with $AsCl_3$ (1:1) in Me_2CO in the presence of Fe and treating the prod. with HCl followed by KOH; 38% yield (1066).

By-prod. in the rxn. of PhN_2CO_2K with $AsCl_3$ in Me_2CO, followed by a treatment with HCl, H_2O, and KOH; 21% yield (1062).

By reducing Ph_2AsO_2H with SO_2 in concd. HCl contg. KI, neutralizing, and dehydrating the prod. (1191).

By treating Ph_5As with Li in C_4H_8O and hydrolyzing the lithium derivative (1309).

<u>Props.:</u> M. 91° (908); m. 88-90° (1062); m. 92-93° (1066, 1100); m. 91-92° (1191); m. 92-93.5° (1309).

<u>Rxns. with:</u> HCN → Ph_2AsCN (56).

PhBr + Na-Hg → Ph_3As (96).

Na-Hg → Ph_2AsNa + Ph_2AsONa (?); Ph_2AsNa + O_2 + H_2O → Ph_2AsOOH (96).

SO_2 in HCl → Ph_2AsCl (136).

Cl_2 in $CHCl_3$ → $(Ph_2AsCl_2)_2O$ (145).

Oxidizing agents → Ph_2AsOOH (243).

$Hg(OAc)_2$ in AcOH → $[(4-AcOHgC_6H_4)_2As]_2O$ (552).

$HS-R-CO_2H$ → $Ph_2AsS-R-CO_2H$ (1190, 1191).

AcOK + Ac_2O → $CH_2(AsPh_2)_2$ (1215).

K_2CO_3 + Ac_2O → Ph_2AsMe (1215).

RCO_2K + $(RCO)_2O$ → Ph_2RAs (1215).

TABLE XXIV

BIS(DIORGANOARSENIC) OXIDES

(R$_2$As)$_2$O or (RR'As)$_2$O	Prepd. from	Yield %	Props.	Ref.
C$_4$				
[(CF$_3$)$_2$As]$_2$O	R$_2$AsI + HgO		b. 100° (calcd.) n_D^{20} 1.354, rxn.[1]	333-4, 1266
C$_6$				
[Me(ClCH:CH)As]$_2$O	RR'AsCl + NaOEt		Silky needles	261
(MeEtAs)$_2$O	RR'AsI + 10 N NaOH		b. 73-74°/11 mm., rxn[2]	626
[Me(HOCH$_2$CH$_2$)As]$_2$O	RR'AsO$_2$H + SO$_2$ in H$_2$O		Oil, rxn.[3]	715
C$_{10}$				
[Et(Me$_2$CH)As]$_2$O	RR'AsI + 10 N NaOH	76	Colorless liq. of unpleasant odor, b. 130-132°/17 mm.	626
(MeBuAs)$_2$O	RR'AsI + 10 N NaOH		b. 126.5-127.5°/11 mm., d_{20} 1.2290, n_D^{20} 1.5017	648
C$_{12}$				
(Pr$_2$As)$_2$O	R$_2$AsI + 10 N NaOH		b. 139-142°/12 mm., d_{20} 1.1855, n_D^{20} 1.500	648
C$_{16}$				
[(OCH:CHCH:C)$_2$As]$_2$O	R$_2$AsCl + NaOH		Oil, rxns.[4-7]	346-7, 626
(EtPhAs)$_2$O	RR'AsI + 10 N NaOH	66	d^{18} 1.3656, n_D^{18} 1.6102, d_0^{20} 1.3564, n_D^{20} 1.6062, b. 191-192°/9 mm., rxns.[8,9]	636, 1160
[Me(4-MeC$_6$H$_4$)As]$_2$O			Rxn.[10]	144
(Bu$_2$As)$_2$O			Rxn.	719
C$_{18}$				
[Et(PhCH$_2$)As]$_2$O	RR'AsI + 10 N NaOH	78	b. 174-175°/16 mm., d_0^{17} 1.2988, n_D^{15} 1.5960	626
(PrPhAs)$_2$O			b. 214-215°/10 mm., rxn.[9], d^{20} 1.2925, n_D^{20} 1.5900	1160, 636, 1160
[Et(4-MeC$_6$H$_4$)As]$_2$O	RR'AsI + 10 N NaOH		b. 231-232°/11 mm., n_D^{12} 1.5991	627

Compound	Preparation	Yield (%)	Properties	References
C20 (BuPhAs)$_2$O			b. 231-232°/18 mm., rxns. 8,9 d^{20} 1.2460, n$^{20}_D$ 1.5746	1160 636 1160
C22 [Me(1-C$_{10}$H$_7$)As]$_2$O	RR'AsCl + KOH in aq. MeOH		m. 58.5-59°, rxn.[11]	1155-6
C24 [(4-BrC$_6$H$_4$)$_2$As]$_2$O	RN$_2$CO$_2$K + AsCl$_3$ in Me$_2$CO	34	m. 157-159°	1062
[Ph(4-ClC$_6$H$_4$)As]$_2$O	4-ClC$_6$H$_4$N$_2$Cl.ZnCl$_2$ + Ph-AsI$_2$ in Me$_2$CO, followed by KOH	39	Rxn.[11]	908
[Ph(2-O$_2$NC$_6$H$_4$)As]$_2$O	2-O$_2$NC$_6$H$_4$N$_2$Cl.FeCl$_3$ + Ph-AsCl$_2$ in Me$_2$O, followed by NaI and KOH		Rxn.[12]	908
[Ph(2-H$_2$NC$_6$H$_4$)As]$_2$O	RR'AsCl + NH$_4$OH RR'AsH + atm. O		Semisolid mass, rxn.[13]	93 94
[Ph(4-H$_2$NO$_2$SC$_6$H$_4$)As]$_2$O [(3,4-H$_2$N(HO)C$_6$H$_3$]$_2$As]$_2$O	RR'AsBr + NH$_4$OH R$_2$AsCl + NaOH		m. 152-155° (dec.), rxn.[5]	957 95
C26 [Ph(4-MeC$_6$H$_4$)As]$_2$O	RR'AsI + NaOH 4-MeC$_6$H$_4$N$_2$Cl.FeCl$_3$ + PhAs-Cl$_2$ in Me$_2$CO + NaI + KOH soln.	71	m. 78-79° rxn.[11]	626 908
C28 [(2-HO$_2$CC$_6$H$_4$)$_2$As]$_2$O	R$_2$AsO$_2$H + SO$_2$, followed by neutralization and dehydration		m. 180-182°	1191

1. Rxn. with aq. NaOH → CHF$_3$ (353).
2. Rxn. with ClCH$_2$CO$_2$H in NaOH, followed by acidification and redn. with SO$_2$ → non-crystallizable oil (626).
3. Rxn. with SOCl$_2$ → Me(ClCH:CH)AsCl (715).
4. Rxn. with H$_2$O$_2$ → the corresponding R$_2$AsO$_2$H or RR'AsO$_2$H (346, 347, 908).
5. Rxn. with H$_3$PO$_2$ → the corresponding (R$_2$As)$_2$ (346).
6. Rxn. with Ac$_2$O → R$_2$AsOAc (347).
7. Rxn. with SO$_2$ → R$_2$AsOH (?) (347).
8. Rxn. with HNO$_3$ → the corresponding RR'AsO$_2$H (640).
9. Rxn. with R"OH → the corresponding R$_2$AsOR" or RR'AsOR" (719, 1160).

TABLE XXIV (cont'd.)

BIS(DIORGANOARSENIC) OXIDES

$(R_2As)_2O$ or $(RR'As)_2O$	Prepd. from	Yield %	Props.	Ref.
$[(4\text{-}AcNHC_6H_4)(4\text{-}O_2NC_6H_4)As]_2O$	$RR'AsO_2H + SO_2$, followed by neutralization		Crystallizes with 2 C_6H_6 as colorless needles, m. 133°[14] rxn.[15]	64
$[(4\text{-}MeC_6H_4)_2As]_2O$	$RN_2CO_2K + AsCl_3$ in Me_2CO	20	m. 103-105°	1062
$[Ph(2\text{-}EtOC_6H_4)As]_2O$	$2\text{-}EtOC_6H_4N_2Cl.FeCl_3 + Ph\text{-}AsCl_2$ in Me_2CO + NaI + KOH	64	Rxn.[4]	908
$[(4\text{-}MeOC_6H_4)_2As]_2O$	$R_2AsBr + NH_4OH$		m. 132-134°	95
C_{32} $[Ph(2\text{-}C_{10}H_7)As]_2O$	$2\text{-}C_{10}H_7N_2Cl.FeCl_3 + PhAs\text{-}Cl_2$ in Me_2CO + NaI + KOH	63	Rxn.[4]	908
$[(4\text{-}AcOHgC_6H_4)_2As]_2O$	$(Ph_2As)_2O + Hg(OAc)_2$		Shiny leaflets, decomp. 275-300°, rxns.[11,16]	551-2

10. Rxn. with $PbCl_4$ at -5° → $Me(4\text{-}MeC_6H_4)As(OH)_2Cl$; at -5 to 10° → $4\text{-}MeC_6H_4AsO_3H_2$ (144).
11. Rxn. with HX (X = Cl, Br, or F) → the corresponding $RR'AsX$ or R_2AsX (908, 1156).
12. Rxn. with HCl, followed by H_2O_2 → the corresponding $RR'AsCl$ → $RR'AsO_2H$ (908).
13. Acetylation → $[Ph(AcNHC_6H_4)As]_2O\cdot 1.5AcOH$, m. 180-181° (93).
14. Loses one mol. of C_6H_6 at 98° and the second at 180° (64).
15. Rxn. with NaOH soln. → $RR'AsOH$ (64).
16. Rxn. with KCN soln. → Ph_2AsO_2H (552).

BIS(DIORGANOARSENIC) SULFIDES

Bis(diorganoarsenic) sulfides are prepared by treating diorganohaloarsines with hydrogen sulfide in absolute ethanol.

BIS[(2-CHLOROVINYL)METHYLARSENIC] SULFIDE $C_6H_4As_2Cl_2S$ [Me(ClCH:CH)As]$_2$S
Prepn.: by passing H_2S through alc. soln. of Me(ClCH:CH)AsCl (261).
Props.: Light yellow oil (261).

BIS[(2-CHLOROVINYL)PHENYLARSENIC] SULFIDE $C_{16}HL_4As_2Cl_2S$ [Ph(ClCH:CH)As]$_2$S
Prepn.: by treating alc. Ph(ClCH:CH)AsCl with dry H_2S (263).
Props.: m. 141° (263).

BIS(DIPHENYLARSENIC) SULFIDE $C_{24}H_{21}As_2S$ (Ph$_2$As)$_2$S
Use: Stabilizer for synthetic rubber (10).

DIARSINES

Tetramethyldiarsine, which was named Cacodyl by its discoverer because of a bad odor, was the first synthetic organometallic compound.

The first two homologs of the series, tetramethyl- and tetraethyldiarsine, are formed by heating at 350° mixtures of As_2O_3 with AcOH and $EtCO_2H$, respectively, in the presence of K_2CO_3 or Ce_2CO_3 on pumice. The reaction products are mixtures, of the corresponding tetraalkyldiarsines, (R$_2$As)$_2$, with bis(dialkylarsenic) oxides, (R$_2$As)$_2$O, (Method I).

Other tetraorganodiarsines, (R$_2$As)$_2$ and (RR'As)$_2$, were prepared as follows:
By reducing arsinic acids, R_2AsO_2H or $RR'AsO_2H$, with 50% hypophosphorous acid, H_3PO_2, optionally in a solvent such as ethanol, in the presence or absence of KI, in an inert atmosphere (Method II).
By reducing bis(diorganoarsenic) oxides, (R$_2$As)$_2$O or (RR'As)$_2$O, with 50% hypophosphorous acid in the presence of KI (Method III).
By reacting diorganoarsines, R$_2$AsH or RR'AsH, with the corresponding bis-(diorganoarsenic) oxides, (R$_2$As)$_2$O or (RR'As)$_2$O, in ethanol (Method IV).
By reducing diorganohaloarsines, R$_2$AsX, with mercury, or zinc, or zinc-amalgam with or without hydrochloric acid (Method V).
By reacting arsenohydrocarbons, RAs:AsR, with metallic sodium or lithium and treating the di-sodium derivatives, (RAsNa)$_2$, with methyl iodide or dimethyl sulfate (Method VI).
By pyrolysis of arsenohydrocarbon polymers, (RAs)x, (Method VII).
By electrolytic reduction of bis(diorganoarsenic) oxide (Method VIII).

The compounds prepared during the period covered by the compilation are listed in Table XXV.

TABLE XXV
DIARSINES

Diarsine	Prepd. from	Method	Yield %	Props., Remarks	Ref.
[(CF$_3$)$_2$As]$_2$	R$_2$AsI + Hg	V		b. 106-107°, n$_D^{20}$ 1.372, rxns.1-4	117, 254, 334
(Me$_2$As)$_2$	R$_2$AsI + Zn + HCl	V		Rxn. 5	333
	R$_2$AsCl + Zn at 100°	V			254
	As$_2$O$_3$ + AcOH	I	66-77	Obtd. in mixt. with (Me$_2$-As)$_2$O	446, 447
	R$_2$AsO$_2$H + H$_3$PO$_2$	II		Rxns.6,7	1307, 993
	(RAs:)$_n$	VII			993
	(R$_2$As)$_2$O	VIII			1308
(Et$_2$As)$_2$	As$_2$O$_3$ + EtCO$_2$H	I		Thermodynamic data Obtd. in mixt. with (Et$_2$-As)$_2$O	868, 446, 447
[(CH$_2$:CHCH$_2$)As]$_2$ (?)	RMgBr+ AsCl$_3$, by-prod.	I		Explodes on contact with 0 Rxn.8	1262
(MePhAs)$_2$	(PhAs:)$_n$ + Na + MeI	VI	87	m. 81.5-82°	1056
	RR'AsO$_2$H + H$_3$PO$_2$	II	63		1056
Me[3,4-H$_2$N(HO)C$_6$H$_3$]As]$_2$	RR'AsO$_2$H + Zn-Hg + HCl (PhAs:)$_n$ + Li + MeI	VI		m. 70°, fumes in air 71-74°, rxn.9 m. 184-185°, 2 HCl, m. 168-170°	811
	RR'AsO$_2$H + H$_3$PO$_2$				1310, 94
[(OCH:CHCH:C)$_2$As]$_2$	(R$_2$As)$_2$O + H$_3$PO$_2$	III	90	colorless crystals, m. 107-108°, rxn.9	346-7
[(2,4,6-Br$_3$C$_6$H$_2$)$_2$As]$_2$	R$_2$AsO$_2$H + H$_3$PO$_2$ RN$_2$Cl. FeCl$_3$ + AsCl$_3$ in Me$_2$CO + Fe	II	11	m. 169-171°, by-prod.	346-7, 1066
(Ph$_2$As)$_2$	Ph$_2$AsH + CH$_2$:CHCN	II		by-prod.; rxn.10	242
[(4-H$_2$NC$_6$H$_4$)(4-O$_2$NC$_6$H$_4$)-As]$_2$	RR'AsO$_2$H + H$_3$PO$_2$	II		Orange powder	64
[Ph(3-HOC$_6$H$_4$)As]$_2$	RR'AsO$_2$H + H$_3$PO$_2$	II		m. 134-136°, rxn.11	94
[Ph'(2-H$_2$NC$_6$H$_4$)As]$_2$	Ph$_2$AsH + [Ph(2-H$_2$NC$_6$H$_4$)-As]$_2$O in EtOH	II		m. 133-134°, rxn.12	94
[Ph(3-H$_2$NC$_6$H$_4$)As]$_2$	RR'AsH + (RR'As)$_2$O (RR'As)$_2$O + H$_3$PO$_2$	III		m. 146-148°, rxn.12	94, 94

$[(4\text{-}H_2NC_6H_4)_2As]_2$	$R_2AsO_2H + H_3PO_2$	II	Colorless powder, m. 155-160°, rxns.13-15	64
$[4\text{-}H_2N(HO)C_6H_3)_2As]_2$	$R_2AsO_2H + H_3PO_2$ $(R_2As)_2O + H_3PO_2$	II III	m. 193-194°, 4.HCl salt, m. 170-172°; 4.H₃PO₂ salt, m. 202° (dec.), rxn.16	95 95
$[Ph(3\text{-}MeOC_6H_4)As]_2$	$[Ph(3\text{-}NaOC_6H_4)As]_2 + Me_2SO_4$		m. 98-99°	94
$[(4\text{-}AcNHC_6H_4)(4\text{-}O_2NC_6H_4)\text{-}As]_2$	$RR'AsO_2H + H_3PO_2$	II	Yellow crystals, soly. data	64
$[(4\text{-}AcNHC_6H_4)(4\text{-}H_2NC_6H_4)\text{-}As]_2$	$RR'AsO_2H + H_3PO_2$	II	Amorph. solid, m. 147-152°	64

1. Rxn. with aq. NaOH → HCF₃ and F ion, probably via (CF₃)₂AsH and (CF₃)₂AsOH (117, 333, 334).
2. Rxn. with NH₃ at 20° → HCF₃ (253).
3. Rxn. with CF₃I → (CF₃)₃As and (CF₃)₂AsI (254).
4. Rxn. with MeI → (CF₃)₂AsMe and (CF₃)₂AsI (254).
5. Rxn. with CF₃I → Me₂AsCF₃ (254).
6. Rxn. with I in MeOH → Me₂AsI (868).
7. Rxn. with H₂SO₄ → an Asⱽ compd. (993).
8. Rxn. with CH₂:CHCH₂Br, followed by Na picrate → tetrallylarsonium picrate (1262).
9. Rxn. with atm. O → the corresponding bis(diorganoarsenic) oxide and arsinic acid (346, 347, 1056).
10. Rxn. with PhBr + Na-Hg → Ph₃As (96).
11. Rxn. with Me₂SO₄ → [Ph(3-MeOC₆H₄)As]₂ (94).
12. Rxn. with Ac₂O, followed by I → Ph(H₂NC₆H₄)AsI (94).
13. Rxn. with acids → hydrolytic cleavage of the C-As bonds (964).
14. Rxn. with H₂O → hydrolytic cleavage of the C-As bonds (964).
15. Rxn. with alkali → hydrolytic cleavage of the C-As bonds (964).
16. Rxn. with 4-MeOC₆H₄C:CHC(:S)SS → compd. m. 200° (dec.) (109).

ARSENO COMPOUNDS

The elemental composition of organoarsenic derivatives formed from arsonic acids or from arsenoso compounds by reduction with hypophosphorous acid or other reducing agents corresponds to the empirical formula $(RAs{<})n$. Because many of the compounds have yellow color, similar to that of azo compounds, the presence of two arsenic atoms in the molecule with an As:As double bond was accepted.

The arseno compounds occur in two modifications; one is colorless or almost colorless and the other intensively colored.

Molecular weight determination of arsenomethane by Steinkopf et al. (Ber. 59, 1468 (1926)) led to a conclusion that the compound is a five-membered ring of trivalent arsenic atoms with five methyl groups:

$$
\begin{array}{c}
\text{Me} \\
\text{As} \\
\text{Me-As} \qquad \text{As-Me} \\
\text{As} \text{------} \text{As} \\
\text{Me} \qquad \text{Me}
\end{array}
$$

(I)

Recent investigations of Kraft et al. (Doklady Akad. Nauk S. S. S. R. 65, 509-12 (1949); ibid. 131, 1074-6 and 1342-4(1960) indicate that the colorless arseno compounds, particularly crystalline ones, exist as five- or six- membered arsenic rings, while the colored arseno derivatives, which occur as amorphous substances, are polymeric chains containing from about five to about fifty $RAs{<}$ units.

$$
X\text{-As-}\left[\text{-As-}\right]_x\text{-AsX}, \quad \text{where } X = OH, I, \text{ or } Cl.
$$
$$
\begin{array}{ccc} R & R & R \end{array}
$$

(II)

Electronic density measurements of colorless crystalline arsenobenzene by Kraft et al. showed that its molecule is a six-membered, chair-shaped ring of six arsenic atoms with six phenyl groups.

$$
\begin{array}{c}
\text{Ph} \\
\text{As} \\
\text{Ph-As} \qquad \text{As-Ph} \\
\text{Ph-As} \qquad \text{As-Ph} \\
\text{As} \\
\text{Ph}
\end{array}
$$

(III)

According to Kraft, the yellow color of arseno compounds should be attributed to a chain of unsaturated trivalent arsenic atoms (II). Colorless arsenobenzene, on melting, turns yellow, thus it is converted into a chain polymer.

Discovery of the chemotherapeutic action of 3,3'-diamino-4,4'dihydroxyarsenobenzene (Salvarsan) against Spirochaeta pallida led to intensive research on organoarsenicals, particularly on the arsenobenzene derivatives. Since many arseno compounds are insoluble in water and therefore unsuitable for medicinal use, various solubilizing groups were introduced into the molecules. Further, physiological action of certain arseno compounds was modified by introduction of sulfur, as sulfonyl or sulfino group, into the arsenoarene molecules.

The class of arseno compounds can be subdivided into symmetrical derivatives containing identical RAs< units and asymmetrical ones containing two different RAs< units.

SYMMETRICAL ARSENO COMPOUNDS

The symmetrical arseno compounds containing identical RAs< units are prepared in a non-oxidizing atmosphere as follows:
1. From arsonic acids by reduction with:
   (a) Hypophosphorous acid in the presence of hydrogen iodide in water at an elevated temperature, or in an acid medium, such as ascetic acid containing sodium acetate, or aqueous hydrochloric or sulfuric acid, or in ethanol containing small amounts of hydroquinone (Method Ia).
   (b) Sodium hydrosulfite in alkaline solution in the presence of magnesium chloride or in methanolic hydrogen chloride solution containing hydrogen iodide (Method Ib).
   (c) Sodium bisulfite in alkaline solution in the presence of magnesium chloride (Method Ic).
   (d) Tin(II) chloride in concentrated hydrochloric acid containing catalytic amounts of hydrogen iodide (Method Id).
   (e) Zinc powder or zinc alloys with lead or tin in hydrochloric or sulfuric acid in the presence of hydrogen iodide (Method Ie).
   (f) Formaldehyde-sodium sulfoxylate (Method If).
   (g) Electrolylically in hydrochloric or sulfuric acid at 80° (Method Ig).
2. From arsenoso compounds by reduction with:
   (a) Hypophosphorous acid in ethanol in the presence of hydroquinone (Method IIa).
   (b) Formaldehyde-sodium sulfoxylate in aqueous solution (Method IIb).
   (c) Electrolytically in hydrochloric or sulfuric acid at 80° (Method IIc).
3. From arsonous acids by reduction with hypophosphorous acid in an acid solution in the presence of hydrogen iodide (Method III).
4. From organodihaloarsines by reduction with:
   (a) Hypophosphorous acid in an acid medium in the presence of hydrogen iodide (Method IVa).
   (b) Metallic sodium or zinc in an anhydrous solvent (Method IVb).
5. By condensation of primary arsines, $RAsH_2$, with arsenoso compounds, RAsO, or with arsonic acids, $RAsO_3H_2$, in aqueous or alcoholic solution or in alcoholic hydrochloric acid (Method V).
6. By oxidation of primary arsines, $RAsH_2$, with atmospheric oxygen or with sulfuryl chloride (Method VI).
7. By metathesis of arseno compounds, (RAs:)n, with arsenoso compounds, RAsO, in hydrochloric acid (Method VII).

8. Arseno compounds containing functional groups, particularly amino groups, can be converted into N-mono- and N,N-disubstituted arseno derivatives by reacting: (a) with formaldehyde-sodium sulfoxylate or with formaldehyde-sodium hydrosulfite to yield N-sulfinomethyl derivatives; (b) with formaldehyde-sodium bisulfite to yield N-sulfomethyl derivatives; (c) with alkylene oxides, such as ethylene oxide, glycide, and epichlorohydrin, to yield N-2-hydroxyalkyl and N,N-di(2-hydroxyalkyl) derivatives; (d) with cyanuric chloride or chloro-s-triazine derivatives to yield N-triazinyl derivatives; (e) with sulfonyl chlorides to yield N-sulfonyl derivatives; (f) with phosgene to yield urea derivatives. The reactions lead to symmetrical and asymmetrical arseno derivatives; depending on the reaction conditions in certain reactions, one or both hydrogen atoms of every or every other amino group can be replaced by a substituent or substituents. Moreover, arseno compounds containing acylamido groups can be converted into amino derivatives by hydrolysis.

ARSENOMETHANE (MeAs)$_2$ and (MeAs)$_n$

Prepn.: By reducing $MeAsO_3H_2$ with H_3PO_2 in HCl (993).
By reacting [3,4-H_2N (HO)C_6H_3As:]n with MeAsO (1:2) in aqueous soln. and distg. the prod. with steam (693).
By treating MeAsO with [3,4-H_2N(HO)C_6H_3As:]n in 3% HCl, a polymer is formed (693).
Props.: The compound occurs in two forms, an almost colorless liquid and a dark polymer. The liquid form, b. 190°/15 mm., has a structure of a five-membered ring MeAsAs(Me)As(Me)As(Me)AsMe, which on heating changes into an intensively colored polymer with a linear structure (705, 706). If the polymerization is carried out with HCl, the polymer contains two terminal Cl atoms (693). The colorless arsenomethane on treating with I in C_6H_6 gave a polymer with two terminal I atoms and mol. wts. ranging from 2693 to 4349 (696).
The cyclic pentamer decomp. at an elevated temp. into (Me$_2$As)$_2$ and Me$_3$As (993).
Crystal structure of the cyclic pentamer (141). Vapor density (1271).
Rxn. with H_2SO_4 at r. t. → (MeAsO)n (993).

TETRAARSENODIACETIC ACID [(-As-AsCH$_2$CO$_2$H)$_2$]n

Prepn.: by adding As$_2$O$_3$ to concd. aq. NaOH, followed by ClCH$_2$CO$_2$H, then acidifying the rxn. mixt. with concd. H$_2$SO$_4$ and reducing with solid NaH$_2$PO$_2$ (823).
Props.: Decomp. 180° (823).
Rxn. with Br in CCl$_4$ → cleavage of the As-As bonds (825).

ARSENOBENZENE (PhAs:)n

Prepn.: By reacting PhAsH$_2$ with SO$_2$Cl$_2$ (17).
By reducing PhAsO with H$_3$PO$_2$ (706).
By reducing PhAsCl$_2$ with Na in Et$_2$O, 49% yield (756).
By reducing PhAsO$_3$H$_2$ with H$_3$PO$_2$ at 50-60°, 80% yield (1056).
Props.: The compound was isolated in various forms. Colorless crystals, m. 210-211°; crystallographic and electron density data indicate a six-membered, chair-shaped As ring (706). At the m. temp. turns yellow and polymerizes to a chain polymer (705). Pale yellow, monoclinic prismatic crystals, m. 213-214° (756). Yellow solid, m. 204-208 (1056). Mol. wts. of 917 and 881 indicate various sizes

of the molecules (756). Crystallographic data (756). Dipole moment (729). Magnetic props. (1029).

Rxns. with: SO_2Cl_2 → $PhAsCl_2$ (17).
Na → arsenobenzene-disodium and -tetrasodium derivs., which were concerted into (MePhAs)$_2$ and Me$_2$PhAs, resp., by reaction with MeI (1056).
Li in C_4H_8O, followed by MeI → (MePhAs)$_2$ (1310).
Li in C_4H_8O, followed by BzOH → PhAs(Li)As(H)Ph (1310).
Li in C_4H_8O, followed by HCl → (PhAs:)n + PhAsH$_2$ (1310).

4,4'-ARSENODIANILINE (4-$H_2NC_6H_4As$:)n
Prepn.: By oxidation of 4-$H_2NC_6H_4AsH_2$ by atm. O (242). Purification (664).
  By electrolytic redn. of 4-$H_2NC_6H_4AsO_3H_2$ in dild. H_2SO_4 at 80° (888).
  By reducing 4-$H_2NC_6H_4AsO_3H_2$ with Zn in HCl (1246).
Props.: Yellow crystals, m. 224-226° (242). Soly. data (76). Magnetic props. (1029).
Salts.: Sulfate, m. 168-170° (888). Dihydrochloride, magnetic props. (1029).
Rxns. with: H_2O, acids, or alkali at 180° → hydrolytic cleavage of the C-As bond (964).
Urea → (4-$H_2NCONHC_6H_4As$:)n (1246).

3,3'-DIAMINO-4,4'-DIHYDROXYARSENOBENZENE, SALVARSAN, ARSPHENAMINE [3,4-$H_2N(HO)$- C_6H_3As:]n
Prepn.: By reducing 3,4-$H_2N(HO)C_6H_3AsO_3H_2$ with $Na_2S_2O_4$ in aq. alkali at 60° (342, 343, 702). The degree of polymerization increases up to 25 RAs: units with increasing excess of the reducing agent, but when a certain excess ratio is attained, the polymerization degree decreases (702).
  By reducing 3,4-$H_2N(HO)C_6H_3AsO_3H_2$ with Zn (1:2) in H_2SO_4 at 58-60°, 93.4% yield of prod. with the same mol. wt. as that obtd. in the redn. with H_3PO_2. A lower degree of polymerization is obtd. when Zn-Pb or Zn-Sn is used instead of Zn (700).
  By reducing 3,4-$H_2N(HO)C_6H_3AsCl_2$ or its HCl salt with Zn in the presence of an anhydrous solvent (752).
  By reacting 3,4-$H_2N(HO)C_6H_3AsH_2$ with 3,4-$H_2N(HO)C_6H_3AsO$ or 3,4-$H_2N(HO)C_6$- $H_3AsO_3H_2$ in HCl (1259).
  By electrolytic redn. of 3,4-$H_2N(HO)C_6H_3AsO_3H_2$ on a Pb cathode in H_2SO_4 contg. KI at 50-55°; the mol. wt. of the prod. increases with increasing current density (701).
  By electrolytic redn. of 3,4-$H_2N(HO)C_6H_3AsO$ in 3 N HCl on a Hg cathode (1259).
Props.: Polymeric structure for the compound has been postulated because it is colloidal, and in reactions with oxidizing agents it requires less than a theoret. amt. of the reagents. The size of the mol. depends on the method of prepn. Thus, prod. prepd. with hypophosphoric acid is highly polymeric and insol. in H_2O. Redn. on a Pb cathode gives a H_2O-sol. prod., while redn. on Hg cathode and rxn. of 3,4-$H_2N(HO)C_6H_3AsH_2$ with the corresponding arsenoso derivative give a slowly dissolving prods. Iodometric titration indicated a degree of polymerization ranging from 7.3 to 20 RAs: units (698). The viscosity of the prod. increases during a treatment with NaOH solns., followed by pptn. with HCl, without any change in the chain length; the viscosity increase might be caused by the formation of O-bridges (703, 704). An increase in viscosity and mol. wt. was also observed when (3,4-$H_2N(HO)C_6$-

H₃As)n.sulfate was kept under N or its hydrochloric acid soln. under CO_2; the change is attributed to the formation of As-O-As bridges (704). The yellow color is attributed to the following As-chain structure: $HOAs(R)-[As(R)]_n-As(R)OH$ (705). Two interconvertible forms differing by their soly. in alc. and in Me_2CO were isolated. Form A, sol. in alc. and Me_2CO, was obtd. by the pptn. of form B from a NaOH soln. by AcOH, and form B can be obtd. from form A dissolved in an acid by adding excess NaOAc (1251). Magnetic props. (1029). Effect of red light upon solns. (1161).

Rxns. with: $CH_2OH.SO_3Na$ → the N,N'-NaO₃SCH₂- deriv.(322, 1251).
$CH_2OH.SO_2Na$ → the N-NaO₂SCH₂- deriv. (343, 671, 752).
$CH_2O + Na_2S_2O_5$ → $[4,3-HO(NaO_3SCH_2NH)C_6H_3As:]n$ (457).
Dextrose + $NaHSO_3$ → the N,N'-di(dextrose-Na sulfonate) deriv. (1205).
HNO_2 → $3,4-H_2N(HO)C_6H_3AsO + 3,4-H_2N(HO)C_6H_3AsO_3H_2$ (379).
$3,4-H_2N(HO)C_6H_3AsO_3H_2$ (1:2) → $3,4-H_2N(HO)C_6H_3AsO$ (693).
$4-HOC_6H_4AsO$ (1:2) → $(4-HOC_6H_4As:)_n + 3,4-H_2N(HO)C_6H_3AsO$ (693).
MeAsO (1:2) → $(MeAs:)n + 3,4-H_2N(HO)C_6H_3AsO$ (693).
SO_2 → $3,4-H_2N(HO)C_6H_3AsS + 3,4-H_2N(HO)C_6H_3AsO$ (694).
Atm. O → the arsenoso deriv. (698, 699, 1216).
RSO_2Cl → the N,N'-di(RSO₂)- deriv. (750, 1258).
$Me_2CO + H_3PO_2$ → $[4,3-HO[Me_2C(PH(O)OH)NH]C_6H_3As:]n$ (1107).

5,5´-ARSENO-2,2´-DIHYDROXYDIPHENYLUREA $C_{13}H_{10}As_2N_2O_3$ $[CO[3,4-NH(HO)C_6H_3As]_2?]$
Prepn.: by treating $3,4-H_2N(HO)C_6H_3AsO_3H_2$ dissolved in aq. alk. with theoret. amt. of $COCl_2$ in $CHCl_3$ and reducing the prod. with Na_2SO_4 in alk. soln. at 55° (970).
Props.: Green powder (970).

3-AMINO-4,4´-DIHYDROXY-3´-(SULFINOMETHYLAMINO)ARSENOBENZENE SODIUM SALT, NEO-SALVARSAN, NEOARSPHENAMINE $C_{13}H_{13}As_2N_2NaO_4S$ $[4,3-HO(NaO_2SCH_2NH)C_6H_3As-AsC_6-H_3-3,4-(NH_2)OH]n$
Prepn.: By adding slowly $CH_2OH.SO_2Na$ to a soln. of $(3,4-H_2N(HO)C_6H_3As:)n$ in an org. solvent (338, 343, 671, 752, 1176, 1251).
By reducing $4,3-HO(O_2N)C_6H_3AsO_3H_2$ with $Na_2S_2O_4.2H_2O$ in aq. NaOH at 60°, septg. the prod. under N, dissolving it in alc., and adding portion-wise slightly more than one equiv. of $CH_2OH.SO_2Na$ (242).
By simultaneous redn. and condensation of $3,4-H_2N(HO)C_6H_3AsO_3H_2$ with $CH_2OH.SO_2Na$ in H_2O-EtOH (976).
Props.: The compd. occurs in two forms, one is toxic and the other non-toxic. The toxic form is obtd. from the sol. and the other from the insol. $[3,4-H_2N(HO)C_6H_3As$ (1251). Sensitivity to atm. O (342, 343). Effect of aging and moisture (1038). Pptn. from aq. soln. by Me_2CHOH (227). UV spectrum (116). Analytical method (785). Compn. of com. prods. (786).
Rxns. with: OCH_2CHCH_2Cl → $[4,3-HO[(CH_2OHCHOHCH_2)H]C_6H_3As-AsC_6H_3-3,4-(NHCH_2SO_2Na OH]n$ (360).
Atm. O → the corresponding arsenoso derivs. (342, 343, 699).

5,5´-ARSENOBIS(2-BENZOXAZOLYLMERCAPTOACETIC ACID) $C_{18}H_{12}As_2N_2O_6S_2$
$[OC(SCH_2CO_2H):NC_6H_3As]n$
Prepn.: by reducing the corresponding arsono derivative in aq. soln. at 60-75° with H_3PO_2 in the presence of KI (372).

4,4'-BIS(2-THIAZINYLSULFAMYL)ARSENOBENZENE $C_{18}H_{14}As_2N_4O_4S_4$ [4-($\overline{SCH:CNH:C}$-
NHSO$_2$)C$_6$H$_4$As:]n

Prepn.: by reducing 4-(2-thiazolylsulfamyl)benzenearsonic acid with H_3PO_2 in aq.
soln. at 50° (680).

2,2'-ARSENOBIS(4-VINYLANISOLE) POLYMER $C_{18}H_{18}As_2O_2$ {[2,5-MeO(CH$_2$:CH)C$_6$H$_3$As:]n}n
Prepn.: by-prod. from catalytic reduction of 5-acetyl-2-methoxybenzenearsonic acid
oxime over Raney Ni, followed acidification; 60% yield (36).
Props.: Brown powder, m. 300° (36).

7,7'-ARSENOBIS[3-(4-CHLORO-2-BENZOXAZOLYLMERCAPTO)PROPIONIC ACID] $C_{20}H_{14}As_2$-
Cl$_2$N$_2$O$_6$S$_2$

Prepn.: by reducing 3(7-arsono-4-chlorobenzoxazolylmercapto)propionic acid with
H_3PO_2 in H_2O at 65° (372).
Props.: Yellow powder (372).

3,3'-ARSENOBIS(6-HYDROXYSUCCINANILIC ACID) DILACTONE $C_{20}H_{16}As_2N_2O_6$
[$\overline{OC(:O)CH_2CH_2C(:O)NHC}$$_6H_3$As:]n
Prepn.: by condensing 3,4-H$_2$N(HO)C$_6$H$_3$AsO$_3$H$_2$ with equal wt. of succinic anhyd.
and reducing the prod. with Na$_2$S$_2$O$_4$ in alk. soln. (970).
Props.: Greenish yellow, insol. powder (970).

6,6'-ARSENO[3-(2-BENZOXAZOLYLMERCAPTO)PROPIONIC ACID] $C_{20}H_{16}As_2N_2O_6S_2$
[$\overline{OC(SCH_2CH_2CO_2H):NC}$$_6H_3$As:]n
Prepn.: by reducing 3-(6-arsono-2-benzoxazolylmercapto)propionic acid with H_3PO_2
in H_2O at 65° (372).
Props.: Yellow powder (372).

4,4'-DIHYDROXY-3,3'-DISULFANILAMIDOARSENOBENZENE $C_{24}H_{32}As_2N_4O_6S_2$ [4,3-HO(H$_2$-
NC$_6$H$_4$SO$_2$NH)C$_6$H$_3$As:]n
Prepn.: By reacting 4-AcNHC$_6$H$_4$SO$_2$Cl with [3,4-H$_2$N(HO)C$_6$H$_3$As:]$_2$ in pyridine and
hydrolyzing the prod. with alc. HCl (750).
By reacting 4-O$_2$NC$_6$H$_4$SO$_2$Cl with [3,4-H$_2$N(HO)C$_6$H$_3$As:]$_2$ and reducing the
nitro group with Na$_2$S$_2$O$_4$ (750).
By reducing 4,3-HO(H$_2$NC$_6$H$_4$SO$_2$NH)C$_6$H$_3$AsO.   with 30% H$_3$PO$_2$ + KI in H$_2$O
at 60° (1117).
By reducing 4,3-HO(H$_2$NC$_6$H$_4$SO$_2$NH)C$_6$H$_3$AsO$_3$Na$_2$ with Na$_2$S$_2$O$_4$ in aq. soln. at
55° (1258).
Props.: M. 180-181° (decompn.) (1258).
Rxn. with CH$_2$OH.SO$_2$Na (1:1) → mono-N-sulfinomethyl deriv.(1158).

4,4'-[ARSENOBIS(PHENYLENEIMINO)]BIS(2-AMINO-1,6-DIMETHYLPYRIMIDINIUM PHOSPHATE)
$C_{24}H_{32}As_2N_8O_4P_2$

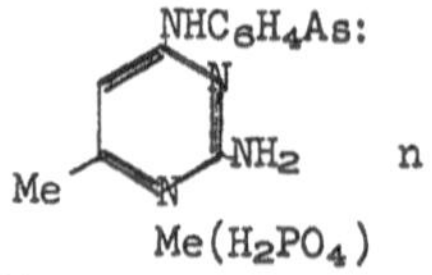

$$NHC_6H_4As:$$

Me — (ring) — NH_2 n

$Me(H_2PO_4)$

Prepn.: by reducing the corresponding arsonic acid methochloride with H_3PO_2 in MeOH contg. HI at 70-75° (8a).
Props.: $1.5H_2O$, m. 249-250° (decompn.) (8a).

7,7'-DIAMINO-2,2'-ARSENO-9-FLUORENONE $C_{26}H_{16}As_2N_2O_2$ [H_2N (structure) -As:]$_2$

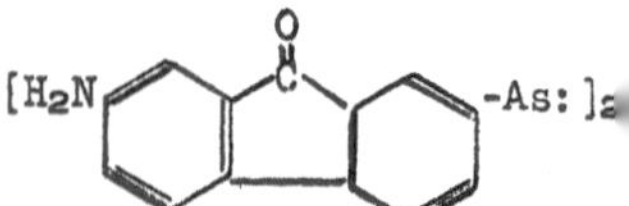

Prepn.: by reducing 7-amino-9-oxo-2-fluorenearsonic acid with H_3PO_2 in AcOH at 85-90°; 68% yield (1049).
Props.: Buff-orange ppt. (1049).

TETRANITRODIARSENOBISPHENYLENEMETHANE $C_{26}H_{16}As_4N_4O_8$

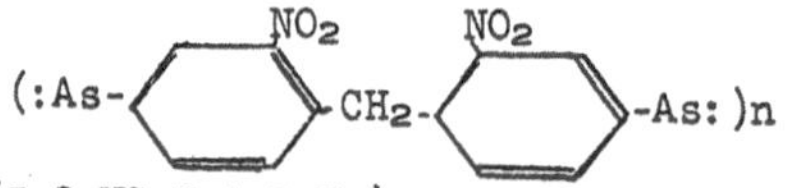

$(:As-$ (ring with NO_2) $-CH_2-$ (ring with NO_2) $-As:)n$

Prepn.: by heating $4,4'-CH_2(3-O_2NC_6H_3AsO_3H_2)_2$ with excess of H_3PO_2 in 30% H_2SO_4 (1095, 1332).
Props.: Amorph., orange yellow powder; soly. data (1332).

α,α-(2,2'-DINITRO-4,4'-ARSENODI-p-PHENYLENE)BIS[3-NITRO-p-TOLUENEARSONIC ACID]
$C_{28}H_{20}As_4N_4O_{14}$ {$3,4-O_2N[_2,4-O_2N(H_2O_3As)C_6H_3CH_2]C_6H_3As:)n$
Prepn.: by reducing $4,4'-CH_2(3-O_2NC_6H_3AsO_3H_2)_2$ with calcd. amt. of H_3PO_2 in 30% H_2SO_4 at boiling temp.; quantitat. yield (1095, 1332).
Props.: Amorph., greenish yellow, infusible powder; soly. data (1332).
Rxn. with concd. $HNO_3 \rightarrow 4,4'-CH_2(3-O_2NC_6H_3AsO_3H_2)_2$ (1332).

4,4'-BIS(5-HYDROXY-2-METHYL-3-CARBETHOXY-4-PYRRYLAZO)ARSENOBENZENE $C_{28}H_{28}As_2N_6O_6$
{$4-[EtO_2CC:C(Me)NHC(OH):CN:N]C_6H_4As:)n$
Prepn.: by reducing the corresponding arsonic acid with H_3PO_2 in the presence of KI under N (385).
Props.: Red, microcryst. powder, darkens at 220° (385).

3,3-BIS(5-HYDROXY-2-METHYL-3-CARBETHOXY-4-PYRRYLAZO)-p-ARSENOPHENOL
$C_{28}H_{28}As_2N_6O_8$ {$4,3-HO[EtO_2CC:C(Me)NHC(OH):CN:N]C_6H_3As:)n$
Prepn.: by reducing the corresponding arsonic acid with H_3PO_2 in the presence of KI (385).
Props.: Dark red powder, darkens at 210° (385).

3,3'-DIHYDROXY-4,4'-BIS(5-HYDROXY-2-METHYL-3-CARBETHOXY-4-PYRRYLAZO)ARSENOBENZENE
$C_{28}H_{28}As_2N_6O_8$ $\{3,4\text{-HO}[EtO_2C\overset{\shortmid}{C}:C(Me)NHC(OH):\overset{\shortmid}{C}N:N]C_6H_3As:\}n$

Prepn.: by reducing the corresponding arsonic acid with H_3PO_2 in the presence of KI under N (385).
Props.: Dark-red, microcryst. powder, darkens at 208°, m. 300° (385).

6,6'-ARSENOBIS(2-PHENYLQUINOLINE) $C_{30}H_{20}As_2N_2$

Prepn.: by reducing 2-phenylquinolenearsonic acid with H_3PO_2 in HCl in the presence of KI (33).
Props.: Hydrochloride, yellow powder, turns brown at 150°, becomes viscous at 220-222°, decomp. 250°, soly. data (33).

2,2'-BIS[2-(4-AMINO-6-p-SULFOANILINO-s-TRIAZIN-2-YL)HYDRAZONO]-p-ARSENOPHENOL
$C_{30}H_{28}As_2N_{14}O_8S_2$ $\{4,3\text{-HO}[N:C(NHC_6H_4SO_3H\text{-}p)N:C(NH_2)N:\overset{\shortmid}{C}NHNH]C_6H_3As:\}n$

Prepn.: by reducing the corresponding arsonic acid with $SnCl_2$ in HCl contg. HI (418).
Props.: Yellow ppt. (418).

3,3'-BIS(1-HYDROXY-4,8-DISULFO-2-NAPHTHYLAZO)ARSENOPHENOL $C_{32}H_{22}As_2N_4O_{16}S_4$

Prepn.: by reducing the corresponding arsonic acid with H_3PO_2 in the presence of HI (409, 412, 1040).
Props.: Stable to air dyestuff (409).

2,2'-BIS(4,8-DISULFO-1-HYDROXY-2-NAPHTHYLHYDRAZINO)-4-ARSENOPHENOL
$C_{23}H_{26}As_2N_4O_{16}S_4$

Prepn.: by reducing 2,2'-bis(4,8-disulfo-1-hydroxy-2-naphthylazo)-p-arseno-phenol with Zn dust in dild. $NaHCO_3$ (428, 432).

2,2'-{1,1'-DIACETYL-2,2'-[5,5'-ARSENOBIS(2-HYDROXYPHENYL)]HYDRAZO}DI-1-NAPHTHOL-
4,8-DISULFONIC ACID $C_{36}H_{30}As_2N_4O_{18}S_4$

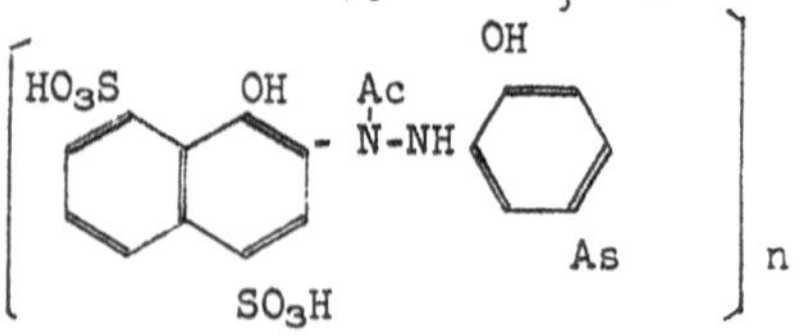

<u>Prepn.</u>: by reducing 2-[1-acetyl-2-(5-arsenoso-2-hydroxyphenyl)hydrazino]-1-
naphthol-4,8-disulfonic acid with $Na_2S_2O_4$ (428).

2,2'-[ARSENOBIS(p-PHENYLENEDIPROPIONYLHYDRAZO)]DI-1-NAPHTHOL-4,8-DISULFONIC
ACID DIPROPIONATE $C_{50}H_{50}As_2N_4O_{26}S_4$

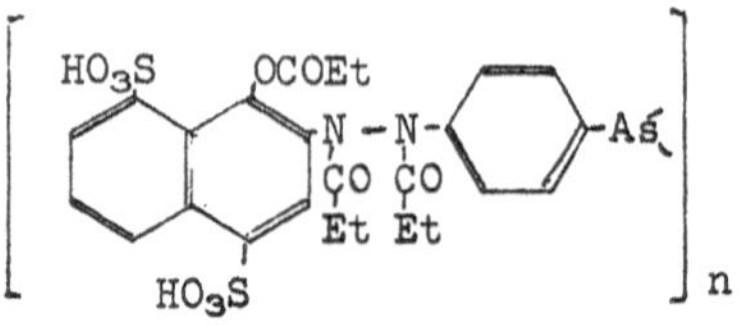

<u>Prepn.</u>: by reducing the corresponding arsonic acid with $Na_2S_2O_4$ (428, 432).

TABLE XXV

SYMMETRICAL ARSENO COMPOUNDS

(RAs:)n	Prepd. from	Method	Props. and Remarks	Ref.
C_2				
$(HO_2CCH_2As:)n$	$RAsO_3H_2$	Ia	Yellow needles, decomp. $\sim205°$ rxn.[1]	823, 825
C_3				
$[(HO_2C)_2(H_2O_3As)CAs:]n$	$(HO_2C)_2C(AsO_3H_2)_2$	Ia	Rxn.[2]	827
C_4				
$(\overline{OCH:CHCH:CAs:})n$	$RAsO_3H_2$ / $RAsO$	Ia / IIa	Yellow crystals, m. 87° Rxn.[3]	346-7 / 346-7
C_5				
$(\overline{CH:NCH:CHCHCAs:})n$	$RAsCl_2$	IVa	80% yield, rxn.[4]	85
C_6				
$(3,4\text{-}H_2N(AuS)C_6H_3As:)n$	$\{3,4\text{-}H_2N(HS)C_6H_3As:\}n + KAu\text{-}Br_4 + Na_2SO_3$			148
$[3,4\text{-}H_2N(F)C_6H_3As:]n$	$RAsO_3H_2$	Ic	2HCl salt, m. 195.2-196.4°	114
$[4\text{-}HOC_6H_4As:]n$	$\{3,4\text{-}H_2N(HO)C_6H_3As:\}n + 4\text{-}HOC_6H_4AsO$	VII		693
$(3\text{-}H_2NC_6H_4As:)n$	$RAsO_3H_2$	Ia	Rxn.[5]	964
$[3,4\text{-}H_2N(HS)C_6H_3As:]n$	$RAsO_3H_2$	Ib	Colorless, amorph. powder Rxn.[6]	147
C_7				
$(4\text{-}NCC_6H_4As:)n$	$RAsO_3H_2$	Ic	43% yield, rxn.[7] m.185.5-186.5°	737
$(4\text{-}MeC_6H_4As:)n$	$RAsCl_2 + Na$	IVb	Magnetic props.	756 / 1029

1. Rxn. with chlorine in Et_2O → cleavage of the As-As bonds and formation of the corresponding organodichloroarsine (825).
2. Rxn. with the H_2O_2 → $(HO_2C)_2C(AsO_3H_2)_2$ (827).
3. Rxn. with atm. O → $\overline{OCH:CHCH:CAsO}$, probably via $(\overline{OCH:CHCH:CAs})_2O$, m. 182-185° (?) (346, 347).
4. Rxn. with 3% H_2O_2 in cond. HCl → dichloroarsenopyridine (85).
5. Rxn. with H_2O, acids, or alkali → hydrolytic cleavage of the C-As bond (964).
6. Rxn. with $KAuBr_4 + Na_2SO_3$ → the S,S'-digold deriv. (148).
7. Rxn. with HCl in Et_2O-EtOH, followed by alc. NH_3 → 4-guanyl-4'-cyanoarsenobenzene (737).

TABLE XXV (cont'd.)
SYMMETRICAL ARSENO COMPOUNDS

$(RAs:)_n$	Prepd. from	Method	Props. and Remarks	Ref.
$[4,3\text{-HO}(NaO_3SCH_2NH)C_6H_3As:]_n$	$[4,3\text{-HO}(H_2N)C_6H_3As:]_n + CH_2\text{-OH.SO}_3Na$		Rxn.8	322
$[3,4\text{-HO}(NaO_3SCH_2NH)C_6H_3As:]_n$	$[4,3\text{-HO}(H_2N)C_6H_3As:]_n + CH_2\text{-O} + Na_2S_2O_5$; $3,4\text{-HO}(H_2N)C_6H_3AsO_3H_2$ followed by $CH_2O + NaHSO_3$	Ib	Pptd. from H_2O with Me_2CHOH	457, 227, 368
$[4\text{-H}_2NC(:NH)C_6H_3As:]_n$	$RAsO_3H_2$	Ic	$2HCl.4H_2O$, yellow plates, m. 240° (dec)	737
$(4\text{-H}_2NCONHC_6H_4As:)_n$	$(4\text{-HCl.H}_2NC_6H_4As:)_2 + CO(NH_2)_2$		m. 250° (dec.), rxn.9	1246
C_8				
$[2,5\text{-HO}(Ac)C_6H_3As:]_n$	$RAsO_3H_2$	Ia	Dark yellow, m. 193-198° (dec.)	36
$[3,4\text{-H}_2N(HOCH_2CH_2O)C_6H_3As:]_2$	$RAsO_3H_2$	Ia	Yellow ppt., m. 159-160°, decomp. ~225°, rxn.10	502
$[4\text{-H}_2NC(:NH)NHC(:NH)NHC_6H_4As:]_n$	$RAsO_3H_2$	Ia	Dark yellow solid	767, 1083-4
C_9				
$\{3,4\text{-H}_2N[N:C(Cl)N:C(Cl)N:C\text{-NH}]C_6H_3As:\}_n$	$4,3\text{-H}_2N(O_2N)C_6H_3AsO_3H_2 + (ClCN)_3$, followed by $SnCl_2$ in HCl	Id	Yellow ppt.	424
$\{4\text{-}[N:C(NH_2)N:C(Cl)N:CNH]\text{-}C_6H_3As:\}_n$	$RAsO_3H_2$	Id		424
$\{3,4\text{-O}_2N[O_2NOCH_2CH(ONO_2)CH_2\text{-O}]C_6H_3As:\}_n$	$RAsO_3H_2$	Ia	Light tan powder, m. 98-99°	62
$\{4\text{-}[N:C(NH_2)N:C(NH_2)N:CS]C_6H_4As:\}_n$	$RAsO_3H_2$	Id	Yellow ppt.; HCl salt, white crystals	426
$\{4\text{-}[N:C(NH_2)N:C(NH_2)N:CNH]\text{-}C_6H_4As:\}_n$	$RAsO_3H_2$	Ia; Id	Yellow ppt., chars at 250°	416, 424, 439
$\{4,2\text{-HO}[N:C(NH_2)N:C(NH_2)N:CNH]C_6H_3As:\}_n$	$RAsO_3H_2$	Id		417, 419, 421-2, 424, 429

[2,5-MeO(Ac)C6H3As:]n	RAsO3H2	Ia	Light yellow, m. 168° (dec.)	36
[4-AcCH2OC6H4As:]n	RAsO3H2	Ia	Yellow ppt.	501
[4,2,5-HO(MeO)(Ac)C6H2As:]n	RAsO3H2	Ia	Yellow, m. 228°	35
[4-[HO2CCH(NH2)CH2]C6H4As:]n	[4-[HO2CCH(NHBz)CH2]C6H4As]n acid hydrolysis		2HCl salt, m. 282-285° (dec.)	123
[3,4-O2N(HOCH2CHOHCH2O)C6H3-As:]n	RAsO3H2	Ia	Yellow powder, m. 197-198° (dec.)	62
[4-[N:C(NH2)N:C(NH2)N:CNH-NH]C6H4As:]n	RAsO3H2	Ib	Yellow powder	418
[4,3-HO[N:C(NHNH2)N:C(NHNH2)N:CNHNH]C6H3As:]n	RAsO3H2	Ib	Yellow ppt.	418
	RAsO3H2	Id		420
[2-(MeCHOHCH2O)C6H4As:]n	RAsO3H2	Ia	Powder, m. 121-124°	83
[2,5-MeO(MeCHOH)C6H3As:]n	RAsO3H2	Ia	Yellow powder, m. 245-250°, diacetate, m. 268° (dec.)	36
[4-(HOCH2CHOHCH2O)C6H4As:]n	RAsO3H2	Ia	Orange powder, m. 164-165°	62
[3,4-H2N(MeCHOHCH2O)C6H3-As:]n	RAsO3H2	Ia	Yellow solid, rxn.11	504
[3,4-H2N(HOCH2CH2CH2O)C6H3-As:]n	RAsO3H2	Ia	Rxn.11,12	504
[3,4-H2N(HOCH2CHOHCH2O)C6-H3As:]n	RAsO3H2	Ia	Yellow powder, m. 170-173°	62
	RAsO3H2	Ia	Rxn.11	504
[3,4-NaO3SCH2NH(HOCH2CH2O)-C6H3As:]n	[3,4-H2N(HOCH2CH2O)C6H3As:]n + CH2OH.SO3Na	Ia	Isolated as Na salt	1180
[HO[Me2C[PH(O)OH]NH]C6H3-As:]n	[HO(H2N)C6H3As:]n + H3PO2 + Me2CO		Isolated as Na salt	1107
C10 [3,4-Me[N:C(Cl)N:C(Cl)N:C-NH]C6H3As:]n	RAsO3H2	Id		424
[3,4-MeO[N:C(Cl)N:C(Cl)N:CNH]C6H3As:]n	3,4-MeO(H2N)C6H3AsO3H2 + (ClCN)3, Na2S2O4			424
[4-[N:C(NH2)N:C(Cl)CH:CNH]-C6H4As:]n	RAsO3H2	Ia	1.5H2O, m. 224-227° (dec.)	8a
[4-[N:C(NH2)N:CHCH:CNH]C6H4-As:]n	RAsO3H2		2HCl·3.5H2O, m. 260-262°	8a

8. Rxn. with CH2O in H2O, followed by NaHSO3 → 4,3-HO(NaO3SCH2NH)C6H3As-AsC6H3-3,4-[N(CH2SO3Na)2]OH (322).

9. Rxn. with H2O2 → the corresponding arsonic acid (1246).

10. Rxn. with dextrose + NaHSO3 in aq. soln. under CO2 → N,N'-di(dextrose-Na sulfonate) (1205).

11. Rxn. with CH2OH.SO2Na → the -NHCH2SO2Na and -NHCH2SO2Na-N'HCH2SO2Na derivs. (540).

12. Rxn. with CH2OH.SO3Na or CH2O + NaHSO3 → the N,N'-di(sulfcmethyleneamino) deriv. (504).

TABLE XXV (cont'd.)

SYMMETRICAL ARSENO COMPOUNDS

(RAs:)n	Prepd. from	Method	Props. and Remarks	Ref.
[3,4-AcNH(HO₂CCH₂O)C₆H₃As:]n	RAsO₃H₂	Ib		1047-8
{4-[N:C(NH₂)N:C(NH₂)CH:CNH]-C₆H₄As:]n			1.5H₂O, m. 286-287° (dec.)	8a
[3,4-(H₂NOCCH₂NH)(HO₂CCH₂-O)C₆H₃As:]n	RAsO₃H₂	Ib		1047-8
[3,4-HO₂CCH₂NH(HOCH₂CH₂O)-C₆H₃As:]n	RAsO₃H₂	Ia		1181
[3,4-NaO₂SCH₂NH(HOCH₂CH₂O)C₆H₃As:]n	[3,4-H₂N(HOCH₂CH₂CH₂O)C₆H₃-As:]n + CH₂OH.SO₃Na			504
	3,4-H₂N(HOCH₂CH₂CH₂O)C₆H₃-AsO₃H₂	If		504
[3,4-NaO₂SCH₂NH(MeCHOHCH₂-O)C₆H₃As:]n	3,4-H₂N(MeCHOHCH₂O)C₆H₃-AsO₃H₂	If		504
	3,4-H₂N(MeCHOHCH₂O)C₆H₃-AsO	If		504
[3,4-NaO₃SCH₂NH(HOCH₂CH₂-O)C₆H₃As:]n	[3,4-H₂N(HOCH₂CH₂CH₂O)-C₆H₃As:]n + CH₂OH:SO₃Na		Rxn.¹³	504
[3,4-NaO₃SCH₂NH(MeCHOHCH₂-O)C₆H₃As:]n	[3,4-H₂N(MeCHOHCH₂O)C₆H₃-As:]n+ CH₂OHSO₃Na			504
[4-MeCOHCH₂OC₆H₄As:]n	RAsO₃H₂	Ia	Yellow, m. 135-140°	556
[2,5-BuO(H₂N)C₆H₃As:]n	2,5-BuO(O₂N)C₆H₃AsO₃H₂ + H₂/Ni + H₃PO₂		m. 175°	818
3,4-H₂N(BuO)C₆H₃As:]n	3,4-O₂N(BuO)C₆H₃AsO₃H₂ + H₂/Ni + H₃PO₂		m. 121°	818
[3,4-H₂N(Me₂COHCH₂O)C₆H₃As:]n	RAsO₃H₂	Ia	Yellow, m. 125-130°	556
C₁₁				
{4-[N:C(NH₂)CH:C(Me)N:CNH]-C₆H₄As:]n	RAsO₃H₂	Ia	2HCl.3.5H₂O, m. 270-275° (dec.)	8a
{3-[HO₂CCH(NHAc)CH₂]C₆H₄-As:]n	RAsO₃H₂	Ib	m. 272-277° (dec.)	123
{4-[N:C(NHMe)N:C(NHMe)N:C-HN]C₆H₄As:]n	RAsO₃H₂	Ib		424
[2,4-HOCH₂CH₂O(EtO₂CNH)C₆-H₃As:]n	RAsO₃H₂	Ia	Needles, m. 222°; 90% yield	71

Compound	R / Synthesis		Properties	Ref.
$[5,2\text{-}H_2N(Me_2CHCH_2CH_2O)C_6H_3\text{-}As:]_n$	$5,2\text{-}O_2N(Me_2CHCH_2CH_2O)C_6H_3\text{-}AsO_3H_2 + H_2/Ni + H_3PO_2$	Ia	M. 153°	818
$[3,4\text{-}H_2N(Me_2CHCH_2CH_2O)C_6H_3\text{-}As:]_n$	$3,4\text{-}O_2N(Me_2CHCH_2CH_2O)C_6H_3\text{-}AsO_3H_2 + H_2/Ni + H_2PO_3$	Ib	m. 171°	818
$[4,3\text{-}HO[HOCH_2CH_2(HOCH_2CHOHCH_2)N]C_6H_3As:]_n$	$RAsO_3H_2$	Ia	Yellow ppt., sol. in H_2O	367
	$RAsO_3H_2$	Ib		367
C_{12}				
$[4\text{-}(PhNHO_2S)C_6H_4As:]_n$	$RAsO_3H_2$	Ib	Yellow powder	680
$[4\text{-}(4\text{-}H_2NC_6H_4SO_2NH)C_6H_4\text{-}As:]_n$	$RAsO_3H_2$	Ib	m. 158-160°	680, 750, 1257
$\{4\text{-}[3,4\text{-}H_2N(HO)C_6H_3SO_2NH]\text{-}C_6H_4As:\}_n$	$3,4\text{-}O_2N(HO)C_6H_3SO_2NHC_6H_4AsO_3H_2$	Ib		750
$[3,4\text{-}H_2N(4\text{-}H_2NC_6H_4SO_3)C_6H_3\text{-}As:]_n$	$RAsO_3H_2$	Ib	$\cdot H_2O$, m. 139.5°	388
$\{4,3\text{-}HO[(HOCH_2CHOHCH_2)_2N]\text{-}C_6H_3As:\}_n$	$RAsO_3H_2$	Ib	Yellow powder, sol. in H_2O	367, 579
C_{13}				
$[3,4\text{-}H_2N(PhCH_2O)C_6H_3As:]_n$	$RAsO_3H_2$	Ib	$2HCl\cdot H_2O$, yellow powder, m. 155-159°	388
$\{4\text{-}[4\text{-}H_2NC(:NH)C_6H_4SO_2NH]\text{-}C_6H_4As:\}_n$	$RAsO_3H_2$	Ib	Yellow powder	274
$[4\text{-}(4\text{-}H_2O_3AsC_6H_4CH_2)C_6H_4\text{-}As:]_n$	$CH_2(C_6H_4AsO_3H_2)_2$	Ib	Amorph., orange-yellow infusible powder	1095, 1332, 1333
$[4\text{-}(4\text{-}H_2O_3AsC_6H_4CO)C_6H_4\text{-}As:]_n$	$4,4'\text{-}CO(C_6H_4AsO_3H_2)_2$	Ia	Lemon-yellow powder, sol. in H_2SO_4 (red color), rxn.14	389
$[4\text{-}(4\text{-}H_2NC_6H_4SO_2NHCH_2)C_6H_3\text{-}As:]_n$	$RAsO_3H_2$	Ib	Yellow, amorph. ppt.	389
$\{4\text{-}[3,4\text{-}H_2N(MeO)C_6H_3SO_2NH]\text{-}C_6H_4As:\}_n$	$4\text{-}[3,4\text{-}O_2N(MeO)C_6H_3SO_2NH]C_6H_4AsO_3H_2$	Ib		750
$\{4\text{-}[\overline{CH_2(CH_2)_3NCH_2CH_2CONH}]\text{-}C_6H_4As:\}_n$	$RAsO_3H_2$	Ia		376
C_{14}				
$\{4\text{-}[4\text{-}ClC_6H_4NHC(:NH)NHC(:NH)NH]C_6H_4As:\}_n$	$RAsO_3H_2$	Ia	HCl salt, m. 228° (dec.)	1083-4

13. Rxn. with aq. HCl → hydrolysis of <u>one</u> NaO_3SCH_2NH-group to H_2N- (504).

14. With HNO_3 → violent rxn. (1333).

TABLE XXV (cont'd.)
SYMMETRICAL ARSENO COMPOUNDS

(RAs:)n	Prepd. from	Method	Props. and Remarks	Ref.
$[4-[PhN:C(Me)NH]C_6H_4As:]n$	$RAs(OH)_2$	III	m. 168-170°	978
	$RAsO$	IIa		1117
$[4,3-HO(4-AcNHC_6H_4SO_2NH)C_6H_3As:]n$	$[4,3-HO\{HCl.H_2N\}C_6H_3As:]n +$ $4-AcNHC_6H_4SO_2Cl$	Ib	Decomp. 230-231°, rxn.15	1258
$[3,4-H_2N(4-AcNHC_6H_4SO_3)C_6H_3As:]n$	$RAsO_3H_2$		Yellow powder, m. 177-183°	388
$\{3,4,5-AcNH(HO)(HOCH_2CHOH-CH_2)_2N\}C_6H_2As:]n$	$RAsO_3H_2 + SO_2$, followed by Na in AcOH	V	Yellow ppt., sol. in H_2O	367, 579
	$RAsO + RAsH_2$			367, 579
C₁₅				
$\{4-[4-MeOC_6H_4N:C(Me)NH]C_6H_4-As:\}n$	$RAs(OH)_2$	III	m. 300°	979
$[4-(4-AcNHC_6H_4SO_2NHCH_2)C_6H_4As:]n$	$RAsO_3H_2$	Ib	Yellow powder	389
$\{4-[H_2NC(:NH)C_6H_4N:C(Me)-NH]C_6H_4As:\}n$	$RAs(OH)_2$	III	Brown solid, m. 256-258°	980
$[4-[4-MeC_6H_4NHC(:NH)NHC(:N-H)NH]C_6H_4As:]n$	$RAsO_3H_2$	Ia		1083-4
$[4-[4-MeOC_6H_4NHC(:NH)NHC(:NH)NH]C_6H_4As:]n$	$RAsO_3H_2$	Ia	HCl salt, m. 225° (dec.)	1083-4
$\{4-[HOCH_2(CHOH)_4CH_2NMeCOCH_2-NH]C_6H_4As:\}n$	$RAsO_3H_2$	Ib	Light yellow powder	102
$\{4-[4-HO_2CCH(NHBz)CH_2]C_6H_4-As:\}n$	$RAsO_3H_2$	Ib	m. 294-295°, rxn.16	123
C₁₆				
$[4,3-EtO(PhNHCOCH_2NH)C_6H_3-As:]n$	$RAsO_3H_2$	Ib		1047-8
$[4,3-HO_2CCH_2O(PhNHCOCH_2NH)-C_6H_3As:]n$	$RAsO_3H_2$	Ib		1047-8
$[4,3-HO_2CCH_2O(4-HOC_6H_4NHCO-CH_2NH)C_6H_3As:]n$	$RAsO_3H_2$	Ib		1047-8
$\{4-[4-EtOC_6H_4N:C(Me)NH]C_6H_4-As:\}n$	$RAs(OH)_2$	III	m. 178-180°	978

			1083-4
[4-[4-AcNHC$_6$H$_4$NHC(:NH)NH-C(:NH)NH]C$_6$H$_4$As:]n	RAsO$_3$H$_2$	Ia	
C$_{17}$ [4,3-HOCCH$_2$O(4-MeC$_6$H$_4$NHCO-CH$_2$NH)C$_6$H$_3$As:]n	RAsO$_3$H$_2$	Ib	1047-8

15. Rxn. with acids or alkali → hydrolysis of the AcNH group (1258).
16. Rxn. with aq. HCl → hydrolysis of the BzNH-group (123).

ASYMMETRICAL ARSENO COMPOUNDS

The asymmetrical arseno compounds contain two different arseno units RAs< and R'As< in the molecule. They are formed from equivalent amounts of two different organoarsenic compounds (e.g., $RAsO_3H_2 + R'AsO_3H_2$, $RAsO + R'AsO$, $RAsH_2 + R'AsO$) or by asymmetrization of symmetrical arseno compounds containing functional groups in the aromatic rings by reacting some but not all of the functional groups with suitable reagents (introduction or cleavage of N-substituents).

The following methods have been employed for the preparation of these compounds:
1. Reduction of equimolar mixtures of two different arsonic acids with:
   (a) Hypophosphorous acid in water or in aqueous or alcoholic acids in the presence of hydrogen iodide (Method Ia).
   (b) Sodium hydrosulfite or pyrosulfite in alkaline medium in the presence or absence of magnesium chloride (Method Ib).
2. Reduction of equimolar mixtures of two different arsenoso compounds with hypophosphorous acid or with sodium hydrosulfite or pyrosulfite (Method II).
3. Condensation of a primary arsine, $RAsH_2$, with an arsenoso compound, $R'AsO$, or an arsonic acid, $R'AsO_3H_2$, which has an organic group different from that of the primary arsine. The reaction is carried out in alcoholic hydrochloric acid (Method IIIa).
4. Metathesis of two different arseno compounds or of an arseno compound with an arsenoso compound containing a different organic group. The reactions are carried out in aqueous solutions at an elevated temperature or in concentrated hydrochloric acid (Method IV).
5. Amino-substituted arsenoarenes are converted into N-sulfinomethyl or N-sulfomethyl derivatives by reacting the aminoarsenoarenes with sodium formaldehyde-sulfoxylate and sodium formaldehydebisulfite, respectively. Instead of the formaldehyde complexes, aqueous formaldehyde and sodium hydrosulfite, pyrosulfite, and bisulfite, respectively, can be used. Under properly controlled conditions, with a proper ratio of the reactants, one or both hydrogen atoms of the amino groups or only of every other amino group can be replaced by sulfoxymethyl and sulfomethyl groups, respectively. Moreover, since formaldehyde-sodium hydrosulfite is a reducing agent, it is possible to prepare N-substituted aminoarsenoarenes directly from arsonic acids or arsenoso compounds in one step, using proper amount of the reagent for the reduction and hydrogen substitution.

Reactions of amino-substituted arsenoarenes with cyanuric chloride or a chloro-s-triazine derivative lead to N-triazinyl arsenoarene compounds. Similarly arylsulfonyl chlorides yield the corresponding arylsulfonamid-arsenoarenes.

Alkylene oxides, such as ethylene oxide, epichlorohydrin , and glycide react with the amino groups of arsenoarenes to form the corresponding 2-hydroxyalkylamino derivatives. With proper ratios of the reagents it is possible to replace one or both hydrogen atoms of the amino groups or only of every other amino group. Moreover, introduction of two different hydroxy-alkyl groups is also possible. Phosgene, when used in a ratio of 1:2 with respect to the amino groups, leads to the urea type arseno compounds (<u>Method V</u>).

6. Cyano-substituted arseno compounds are converted partly or completely into the corresponding guanylarseno derivatives by treating the cyano derivatives with ethanolic hydrogen chloride and reacting the resulting imido ester with alcoholic ammonia with gentle warming (<u>Method VI</u>).

4-CYANO-4'-GUANYLARSENOBENZENE $C_{14}H_{11}As_2N_3$ [$H_2NC(:NH)C_6H_4As-AsC_6H_4CN$]n
<u>Prepn.</u>: by treating ($NCC_6H_4As::$)n with H^Cl in Et_2O-EtOH, followed by alc. NH_3 at 40°; 40% yield (737).
<u>Props.</u>: HCl salt.EtOH, pale yellow, m. 234° (737).

ASYMMETRICAL ARSENOBENZENE POLYMERS
Reduction of mixtures of arsonous acids, treated with magnesium chloride in aqueous alkali, by sodium hydrosulfite gave asymmetrical arsenobenzene polymers. Thus a mixture of 4-aminobenzenearsonous acid with 4-hydroxyarsinous acids in a molar ratio of 2:1 gave a product of the composition $-[As(C_6H_4NH_2)]n-[As(C_6H_4OH)]_{\overline{m}}$ with an As:N ratio of 3:2 and 1:1. Similarly a product obtained from $4-HOC_6H_4As(OH)_2$ and $3,4-H_2N(HO)C_6H_3As(OH)_2$ in ratio of 3:2 had an As:N ratio of 3:1 (697).

3-[8-ACETAMIDO-3-HYDROXY-1,4-BENZISOXAZINE-6-ARSENO)-4-HYDROXYANILINOMETHYLENE-
  SULFONIC ACID SODIUM SALT $C_{17}H_{16}As_2N_3NaO_7S$

<u>Prepn.</u>: by reducing a soln. of equiv. amts. of 8-acetamido-3-hydroxy-1,4-benzis-oxazine-6-arsonic acid and $2,5-HO(H_2N)C_6H_3AsO_3H_2$ with $Na_2S_2O_4$ soln. contg. $MgCl_2$ at 55°, and reacting the arseno deriv. with HCHO and $NaHSO_3$ (354).

N-[4-(8-ACETAMIDO-3-HYDROXY-1,4-BENZISOXAZIN-6-YLARSENO)PHENYL]GLYCINE
  $C_{17}H_{18}As_2N_3O_5$

<u>Prepn.</u>: by reducing aqueous soln. of equimol. amts. of 8-acetamido-3-hydroxy-1,4-benzisoxazine-6-arsonic acid and 4-arsonoanilinoacetic acid with sodium hydrosulfite soln. contg. $MgCl_2$ at 55-60° (354).

[5-(3-ACETAMIDO-4-HYDROXYPHENYLARSENO)-1-METHYLBENZIMIDAZOL-2-YLOXY]ACETIC
ACID $C_{18}H_{17}As_2N_3O_5$ [4,3-HO(AcNH)C$_6$H$_3$As-As-

$$-\text{OCH}_2\text{CO}_2\text{H}$$

Me

Prepn.: by reducing equimol. mixt. of the corresponding arsonic acid with H_3PO_2
in AcOH in the presence of KI and NaOAc (1175).

3,4'-DIHYDROXY-4-BIS(2-HYDROXYETHYL)AMINO-3'-(N-METHYL-N-SULFINOMETHYL)AMINO-
ARSENOBENZENE SODIUM SALT $C_{18}H_{24}As_2NO_6S$ {3,4-HO[(HOCH$_2$CH$_2$)$_2$N]C$_6$H$_3$As-
AsC$_6$H$_3$-3,4-[NMe(CH$_2$SO$_2$N)]OH}n
Prepn.: By reducing at 45° an equimolar mixt. of 4,3-HO(MeNH)C$_6$H$_3$AsO$_3$H$_2$ and
3,4-HO[(HOCH$_2$CH$_2$)$_2$N]C$_6$H$_3$AsO$_3$H$_2$ with H$_3$PO$_2$ in 10% HCl contg. KI and treating the
arseno deriv. with HOCH$_2$SO$_2$Na (361).
          By reacting[4,3-HO(NaO$_2$SCH$_2$NMe)C$_6$H$_3$As:AsC$_6$H$_3$-3,4-(HO)NH$_2$] with $\overline{CH_2CH_2O}$
in aq. soln. (361).

4,4'-DIHYDROXY-3-[(HYDROXYETHYL)(DIHYDROXYPROPYL)AMINO]-3'-(SULFINOMETHYLAMINO)-
ARSENOBENZENE SODIUM SALT $C_{18}H_{23}As_2N_2NaO_7S$ {4,3-HO[(HOCH$_2$CH$_2$)(HOCH$_2$CHOHCH$_2$)-
N]C$_6$H$_3$AsAsC$_6$H$_3$(NHCH$_2$SO$_2$Na)OH-3,4}n
Prepn.: By reacting [4,3-HO(H$_2$N)C$_6$H$_3$As-AsC$_6$H$_3$-3,4-(NHCH$_2$SO$_2$Na)OH]n with epichloro-
hydrin and ethylene oxide in H$_2$O (360).
          By reacting {4,3-HO[(HOCH$_2$CH$_2$)(HOCH$_2$CHOHCH$_2$N]C$_6$H$_3$As:)n hydrochloride with
[4,3-HO(HCl.H$_2$N)C$_6$H$_3$As:]n in H$_2$O at 80° and treating the asymmetrical arseno
compd. with CH$_2$OH.SO$_2$Na (364).
          By condensing 4,3-HO[(HOCH$_2$CH$_2$)(HOCH$_2$CHOHCH$_2$)N]C$_6$H$_3$AsO with 4,3-HO(Na-
O$_2$SCH$_2$NH)C$_6$H$_3$AsH$_2$ in water (363).
          By condensing 3,4-H$_2$N(HO)C$_6$H$_3$AsO with 4,3-HO(NaO$_2$SCH$_2$NH)C$_6$H$_3$AsH$_2$ and
treating the prod. with CH$_2$CH$_2$O and glycide (1:1:1) (576).

3,4'-DIHYDROXY-3'-(2-HYDROXYETHYL-2,3-DIHYDROXYPROPYLAMINO)-4-(SULFOMETHYL-
AMINO)-ARSENOBENZENE SODIUM SALT $C_{18}H_{23}As_2N_2NaO_8S$ {4,3-HO[(HOCH$_2$CH$_2$)-
(HOCH$_2$CH$_2$OHCH$_2$)N]C$_6$H$_3$As-AsC$_6$H$_3$-4,3-(NHCH$_2$SO$_3$Na)OH}n
Prepn.: By reducing equimol. mixt. of 4,3-HO[(HOCH$_2$CH$_2$)(HOCH$_2$CHOHCH$_2$)N]C$_6$H$_3$As-
O$_3$H$_2$ + 3,4-HO(H$_2$N)C$_6$H$_3$AsO$_3$H$_2$ with H$_3$PO$_3$ in 19% HCl contg. KI and treating the
arsenobenzene HCl dissolved in MeOH with NaHSO$_3$ followed by HCHO (365; 368, 578).
          By metathesis of [4,3-HO[HOCH$_2$CHOHCH$_2$)N]C$_6$H$_3$As:)n with [3,4-HO(H$_2$N)-
C$_6$H$_3$As:]n and treating the asymmetrical arseno compd. with CH$_2$OH.SO$_3$Na (578).

8-ACETAMIDO-8'-SULFINOMETHYLAMINO-6,6'-ARSENOBIS(3-HYDROXYBENZISOXAZINE) SODIUM
SALT $C_{18}H_{17}As_2N_4NaO_7S$

Prepn.: by dissolving equimol. amts. of 8-acetamido-3-hydroxy-1,4-benzisoxazine-

6-arsonic acid and 8-amino-8-hydroxy-1,4-benzisoxazine-6-arsonic acid in N
Na_2CO_3 soln., adding the soln. to an aq. soln. of $Na_2S_2O_4$ contg. Na_2CO_3 and
$MgCl_2$, heating at 55-60°C, and treating the prod. with $HOCH_2OSONa$ at 45-50°(354).

5-[4-(HYDROXYACETAMIDO)-2-METHYLPHENYLARSENO]-1-METHYLBENZIMIDAZOL-2-YLOXYACETIC
ACID $C_{19}H_{19}As_2N_3O_5$

Prepn.: by reducing a mixt. of the corresponding arsonic acids with H_3PO_2 in
AcOH + KI + NaOAc (1175).

5-(8-ACETAMIDO-3-HYDROXYBENZISOXAZIN-6-YLARSENO)-3-ACETAMIDO-2-HYDROXYANILINO-
METHYLENESULFONIC ACID SODIUM SALT $C_{19}H_{20}As_2N_4O_8S$

Prepn.: by reducing a neutral aq. soln. of equimolar quantities of 8-acetamido-
3-hydroxy-1,4-benzisoxazine-6-arsonic acid and $3,4,5-H_2N(HO)(AcNH)C_6H_2AsO_3H_2$ with
aq. $Na_2S_2O_4$ at 55-60° and treating the arseno deriv. HCl salt with HCHO and Na-
HSO_3 (354).

4-(4-AMINO-2,3-DIMETHYL-5-OXO-1-PYRAZOLINYL)-4'-(2-HYDROXYACETAMIDO)ARSENO-
BENZENE $C_{19}H_{20}As_2N_4O_3$ {4-[$\overline{N}$(Me)C(Me):C(NH_2)C(:O)$\overline{N}$]$C_6H_4As.AsC_6H_4$-4-
NHCOCH_2OH}n
Prepn.: by reacting 4-amino-1-(4-arsenosophenyl)-2,3-dimethylpyrazolinone with
4-(hydroxyacetamido)phenylarsine in alc. HCl (1173).

4,4'-DI(2-HYDROXYPROPOXY)-3-AMINO-3'-(SULFINOMETHYLAMINO)ARSENOBENZENE SODIUM
SALT $C_{19}H_{25}As_2N_2NaO_6S$ [$3,4-H_2N(MeCHOHCH_2O)C_6H_3As.AsC_6H_3-3,4-(NHCH_2SO_2Na)$-
OCH_2CHOHMe]n
Prepn.: By treating [$3,4-H_2N(MeCHOHCH_2O)C_6H_3As$:]$_2$ with $HOCH_2SO_2Na$ in aq. soln.
at 65-70° (504, 1180).
        By reducing $3,4-H_2N(MeCHOHCH_2O)C_6H_3AsO_3H_2$ or $3,4-H_2N(MeCHOHCH_2O)C_6H_3$-
AsO with $CH_2OH.SO_2Na$ in aq. soln. at 60°, with exclusion of oxygen (504).

3-AMINO-4,4'-DI(3-HYDROXYPROPOXY)-3'-SULFINOMETHYLENEAMINOARSENOBENZENE
SODIUM SALT $C_{19}H_{25}As_2N_2NaO_6S$ [$3,4-H_2N(HOCH_2CH_2CH_2O)C_6H_3As.AsC_6H_3-3,4$-
(NHCH_2SO_2Na)OCH_2CH_2CH_2OH]n
Prepn.: By treating [$3,4-H_2N(HOCH_2CH_2CH_2O)C_6H_3As$:]$_2$ with $HOCH_2SO_2Na$ (1:1) (504).
        By treating $3,4-H_2N(HOCH_2CH_2CH_2O)C_6H_3AsO_3H_2$ or $3,4-H_2N(HOCH_2CH_2CH_2O)$-
C_6H_3AsO with a calcd. amt. of $HOCH_2SO_2Na$ (504).

3-AMINO-4,4'-BIS(3-HYDROXYPROPOXY)-3'-SULFOMETHYLENEAMINOARSENOBENZENE
SODIUM SALT $C_{19}H_{25}As_2N_2NaO_7S$ [3,4-H_2N(HOCH$_2$CH$_2$O)C$_6$H$_3$As.AsC$_6$H$_3$-3,4-
(NHCH$_2$SO$_3$Na)OCH$_2$CH$_2$CH$_2$OH]n

Prepn.: by acidifying [3,4-NaO$_3$SCH$_2$NH(HOCH$_2$CH$_2$CH$_2$O)C$_6$H$_3$As:]n with dild. HCl
(504).

3-AMINO-4,4'-BIS(2-HYDROXYPROPOXY)-3'-SULFOMETHYLENEAMINOARSENOBENZENE SODIUM
SALT $C_{19}H_{25}As_2N_2NaO_7S$ [3,4-H_2N(MeCHOHCH$_2$O)C$_6$H$_3$As.AsC$_6$H$_3$-3,4-(NHCH$_2$SO$_3$Na)-
OCH$_2$CHOHMe]n

Prepn.: by acidifying aq. soln. of [3,4-(NaO$_3$SCH$_2$NH)(MeCHOHCH$_2$O)C$_6$H$_3$As:]$_2$ (504).
Props.: Orange yellow solid (504).

3,4'-DIHYDROXY-4-(2-HYDROXYETHYL-2,3-DIHYDROXYPROPYLAMINO)-3'-(METHYLSULFOXY-
METHYLAMINO)ARSENOBENZENE SODIUM SALT $C_{10}H_{25}As_2N_2NaO_7S$

Prepn.: by reducing equimol. mixt. of 3,4-HO(H_2N)C$_6$H$_3$AsO and 4,3-HO(MeNH)C$_6$-
H$_3$AsO$_3$HNa with Na$_2$S$_2$O$_5$, and treating the resulting arseno deriv. with HOCH$_2$-
SO$_2$Na, followed by one equiv. ethylene oxide in the cold and one equiv. of
glycide in the warm (362).
Props.: Yellow ppt. (362).

4,4'-DIHYDROXY-3-(2-HYDROXYETHYL-2,3-DIHYDROXYPROPYLAMINO)-3'-(N-METHYL-N-
SULFINOMETHYLAMINO)ARSENOBENZENE $C_{19}H_{26}As_2N_2O_7S$

Prepn.: By reducing equimol. mixt. of 4,3-HO[HOCH$_2$CH$_2$(HOCH$_2$CHOHCH$_2$)N]C$_6$H$_3$As-
O$_3$H$_2$ + 4,3-HO(MeNH)C$_6$H$_3$AsO$_3$H$_2$ with H$_3$PO$_2$ in 10% HCl contg. KI and treating
the arseno deriv. with CH$_2$OH.SO$_2$Na (361).
        By reducing equimol. mixt. of 4,3-HO(H_2N)C$_6$H$_3$AsO$_3$H$_2$ + 4,3-HO(MeNH)-
C$_6$H$_3$AsO$_3$H$_2$ with H$_3$PO$_2$ in 10% HCl contg. KI and treating the arseno deriv. with
1 equiv. of CH$_2$OH.SO$_2$Na, followed by 1 equiv. of CH$_2$CHO in the cold and then by
1 equiv. of glycide with warming (361).
        By reacting 4,3-HO[HOCH$_2$CH$_2$(HOCH$_2$CHOHCH$_2$)N]C$_6$H$_3$AsO with 4,3-HO(NaO$_2$S-
CH$_2$NMe)C$_6$H$_3$AsH$_2$ in aq. soln. (362).
Props.: Dark-yellow ppt. (361).

4,4'-DIHYDROXY-3-BIS(2,3-DIHYDROXYPROPYL)AMINO-3'-(SULFINOMETHYLAMINO)ARSENO-
BENZENE SODIUM SALT $C_{19}H_{25}As_2N_2NaO_8S$

$(HOCH_2CHOHCH_2)_2N$ ... HO- ... -As-As- ... -OH ... $NHCH_2SO_2Na$]n

Prepn.: By reacting equimol. mixt. of 4,4'-dihydroxy-3,3'-bis[bis(2,3-dihydroxy-
propyl)amino]arsenobenzene dihydrochloride and 3,3'-diamino-4,4-dihydroxyarsenoben-
zene dihydrochloride in H_2O at 80° and treating the prod. with $HOCH_2SO_2Na$ at 27°
(364).
                By reacting 4,3-$HO(H_2N)C_6H_3As:AsC_6H_3$-3,4-$(NHCH_2SO_2Na)OH$ with $ClCH_2$-
$CHCH_2O$ in H_2O (360).
                By condensing 4,3-$HO[(HOCH_2CHOHCH_2)_2N]C_6H_3AsO$ with 4,3-$HO[(HOCH_2$-
$CHOHCH_2)_2N]C_6H_3AsO$ with 4,3-$HO(NaOSOCH_2NH)C_6H_3AsH_2$ in aq. soln. (363, 576).
Props.: Yellow ppt. (364).

4,4'-DIHYDROXY-3-(2,3-DIHYDROXYPROPYL-2-HYDROXYETHYLAMINO)-5-METHOXY-3'-(SUL-
FOMETHYLAMINO)ARSENOBENZENE SODIUM SALT $C_{19}H_{25}As_2N_2NaO_9S$

$HOCH_2CHOHCH_2NCH_2CH_2OH$ ... HO- ... -As-As- ... OMe ... -OH ... $NHCH_2SO_3Na$]n

Prepn.: by reacting {4,3-$HO[HOCH_2CHOHCH_2N(CH_2CH_2OH)]C_6H_3As:$}$_2$ with [3,4,5-MeO(HO)-
$(H_2N)C_6H_2As:$]$_2$ in aq. soln. at elevated temp. and treating the asymmetric arseno
compound with $CH_2OH.SO_3Na$ (366, 368).
Props.: Yellow powder (366).

3,4'-DIHYDROXY-4-[BIS(2,3-DIHYDROXYPROPYL)AMINO]-3'-[N-METHYL-N-(SULFINOMETHYL)-
AMINO]ARSENOBENZENE SODIUM SALT $C_{20}H_{28}As_2N_2O_8S$ β,4-$HO[(HOCH_2CHOHCH_2)_2N]$-
$C_6H_3As-AsC_6H_3$-3,4-$[N(Me)CH_2SO_2Na]OH$n
Prepn.: By reducing at 30° equi mol. mixt. of 3,4-$HO[(HOCH_2CHOHCH_2)_2N]C_6H_3AsO_3H_2$
+ 4,3-$HO(MeNH)C_6H_3AsO_3H_2$ with H_3PO_2 in 10% HCl contg. KI and treating the rxn.
prod. with $CH_2OH.SO_2Na$ in aq. MeOH (361, 575).
                By reducing equimol. mixt. of 3,4-$HO(H_2N)C_6H_3AsO_3H_2$ and 4,3-$HO(MeNH)C_6H_3$-
AsO_3H_2 with H_3PO_2 in 10% HCl contg. KI, and treating the rxn. prod. with $CH_2OH.SO_2$-
Na (1:1), followed by $HOCH_2CHCH_2O$ (1:2) (361).
                By reacting 3,4-$HO[(HOCH_2CHOHCH_2)_2N]C_6H_4AsO$ with 4,3-$HO(MeNH)C_6H_3AsH_2$ in
aq. HCl and treating the rxn. prod. with $CH_2OH.SO_2Na$ (362).

2-ACETAMIDO-4-[2-(2-CARBOXYETHYLTHIO)-5-BENZOXAZOLYLARSENO]PHENOXYACETIC ACID
$C_{21}H_{20}As_2N_2O_7S$ [3,4-$AcNH(HO_2CCH_2CH_2O)C_6H_3As-As$

$-S(CH_2)_2CO_2H$] n

Prepn.: by reducing equimol. mixt. of 3,4-$AcNH(HO_2CCH_2CH_2O)C_6H_3AsO_3H_2$ and 3-(5-
arsono-2-benzoxazolylmercapto)propionic acid, dissolved in MeOH acidified with
HCl, with H_3PO_2 (372).
Props.: Na salt, yellow powder (372).

5-[4-(4-AMINO-2,3-DIMETHYL-5-OXO-1-PYRAZOLINYL)PHENYLARSENO]-1-METHYL-2-BENZI-
   MIDAZOLYLOXYACETIC ACID $C_{21}H_{21}As_2N_5O_4$

Prepn.: by reducing mixt. of the corresponding arsonic acids with H_3PO_2 in AcOH
+ NaOAC (1145).

   4-CARBOXYMETHOXY-3-METHYL-4'-(2,3-DIMETHYL-5-OXO-4-SULFOMETHYLENEAMINO-1-PYRA-
      ZOLINYL)ARSENOBENZENE $C_{21}H_{23}As_2N_3O_7S$ [3,4-Me(HO_2CCH_2O)$C_6H_3As.AsC_6H_4$-4-NN(Me).
   C(Me):C(NHCH_2SO_3H)C:O] n
Prepn.: by reducing equimol. mixt. of 3,4-Me(HO_2CCH_2O)$C_6H_3AsO_3H_2$ and 4-(4-amino-
2,3-dimethylpyrazolinyl)benzenearsonic acid with $Na_2S_2O_4$ in alk. soln. at 65° and
treating the prod. with $CH_2OH.SO_3Na$ in aq. soln.; isolated as di-Na salt (1171).

   4-CARBOXYMETHOXY-2-METHYL-4'-(2,3-DIMETHYL-5-OXO-4-SULFOMETHYLENEAMINO-1-PYRA-
      ZOLINYL)ARSENOBENZENE SODIUM SALT $C_{21}H_{23}As_2N_3O_7S$ [2,4-Me(HO_2CCH_2O)$C_6H_3As.As$-
   C_6H_4-4-NN(Me)C(Me):C(NHCH_2SO_3Na)C:O] n
Prepn.: by reducing equimol. mixt. of 4-$HO_2CCH_2OC_6H_4AsO_3H_2$
   and 4-(4-amino-2,3-dimethyl-5-pyrazolone-1-)benzenearsonic acid with $Na_2S_2O_4$ in
alk. soln. at 65° and treating the prod. with $CH_2OH.SO_3Na$ in aq. soln. (1171).

   3-ACETAMIDO-4,4'-DIHYDROXY-5-[BIS(2,3-DIHYDROXYPROPYL)AMINO]-3'-(SULFINOMETHYL-
      AMINO)ARSENOBENZENE SODIUM SALT $C_{21}H_{29}As_2N_3O_9S$ {3,4,5-AcNH(HO)[(HOCH_2CHOH-
   CH_2)$_2$N]C_6H_2As-AsC_6H_4-3,4-(NHCH_2SO_2Na)OH}n
Prepn.: By reducing a mixt. of 3,4,5-AcNH(HO)[(HOCH_2CHOHCH_2)$_2$N]$C_6H_2AsO_3H_2$ and 3,-
4-H_2N(HO)$C_6H_3AsO_3H_2$ with H_3PO_2 at 80° in aq. soln. contg. KI and treating the prod
with $CH_2OH.SO_2Na$ in aq. MeOH at 27° (370).
     By reacting [3,4,5-AcNH(HO)(H_2N)C_6H_2As:]n with [4,3-HO(NaO_2SCH_2NH)C_6H_3-
As:]n (1:1) in aq. soln. at an elevated temp. and treating the prod.. with glycide
at 65° (370).
     By reacting {3,4,5-AcNH(HO)[(HOCH_2CHOHCH_2)$_2$N]C_6H_2As:}n with [4,3-HO(NaO_2-
SCH_2NH)C_6H_3As:]n in aq. soln. at an elevated temp. (370).
     By reducing a mixt. of 3,4,5-AcNH(HO)(H_2N)$C_6H_2AsO_3H_2$ and 4,3-HO(NaO_2SCH_2-
NH)$C_6H_3AsO_3H_2$ with H_3PO_2 in aq. soln. contg. KI and treating the prod. with two
equivs. of glycide (370).
     By reacting {3,4,5-AcNH(HO)[(HOCH_2CHOHCH_2)$_2$N]C_6H_2As:}n with [3,4-H_2N-
(HO)C_6H_3As:]n in H_2O at 80° and treating the prod. with $CH_2OH.SO_2Na$ (370).
     By condensing 3,4,5-AcNH(HO)[(HOCH_2CHOHCH_2)$_2$N]C_6H_2AsO or 3,4,5-AcNH(HO)-
(H_2N)C_6H_2AsO (I) with 3,4-HO(NaO_2SCH_2NH)$C_6H_3AsH_2$ in H_2O and, if I is used, treat-
ing the prod. with glycide (370).

3-ACETAMIDO-4-CARBOXYMETHOXY-4'-[-2,3-DIMETHYL-5-OXO-4-(SULFOMETHYLENEAMINO)-1-PYRAZOLINYL]ARSENOBENZENE $C_{22}H_{24}As_2N_4O_8S$ [4-[$\overline{NN}$(Me)C(Me):C(NHCH$_2$SO$_3$H)$\bar{C}$-(:O)]C$_6$H$_4$As.AsC$_6$H$_3$-3,4-(NHAc)OCH$_2$CO$_2$H]n

Prepn.: by reducing equimol. mixt. of 3,4-AcNH(HO$_2$CCH$_2$O)C$_6$H$_3$AsO$_3$H$_2$ and 4-(4-amino-2,3-dimethyl-5-pyrazolone-1-)benzenearsonic acid with Na$_2$S$_2$O$_4$ in alk. soln. at 65° isolating the arseno deriv., and treating with HOCH$_2$OSO$_2$Na in aq. soln.; isolated as di-Na salt (1171).
Salt: Bi salt, prepd. by treating the corresponding Na salt with Bi(NO$_3$)$_3$ in H$_2$O-glycerol and neutralizing the soln. (1172).

5-{4-{3-[BIS(2,3-DIHYDROXYPROPYL)AMINO]-4-HYDROXYPHENYLARSENO}-2-BENZOXAZOLYL-THIO }PROPIONIC ACID $C_{22}H_{26}As_2N_2O_8S$

Prepn.: By reducing equimol. mixt. of 4,3-HO[(HOCH$_2$CHOHCH$_2$)N]C$_6$H$_3$AsO$_3$H$_2$ and 5-arsono-2-benzoxazolylmercaptopropionic acid with NaHSO$_3$, H$_3$PO$_3$, or H$_3$PO$_2$, suitably in the presence of HI or SnCl$_2$ (546).
By heating in aq. NaOH at 60° a mixt. of 3,3'-bis[di(2,3-dihydroxypropyl)-amino]-4,4'-dihydroxyarsenobenzene with 5,5'-arsenobis(2-benzoxazolylmercaptopropionic acid) (546).

4,4'-DIHYDROXY-N,N,N'-TRIS(2,3-DIHYDROXYPROPYL)-N-SULFINOMETHYL-3,3'-DIAMINO-ARSENOBENZENE SODIUM SALT $C_{22}H_{31}As_2N_2NaO_{10}S$

Prepn.: By reacting {4,3-HO[(CH$_2$OHCHOHCH$_2$)$_2$N]C$_6$H$_3$As-AsC$_6$H$_3$-3,4-(NHCH$_2$SO$_2$Na)OH}n with CH$_2$OHCHCH$_2$O in aq. soln. (361, 575).
By reacting [4,3-HO(NH$_2$.HCl)C$_6$H$_3$As:]n with one equiv. of CH$_2$OH.SO$_2$Na in aq. MeOH at 30° and treating the prod. with glycide in EtOH-Et$_2$O mixt. (361, 1176).
By condensing 4,3-HO(HOCH$_2$CHOHCH$_2$NH)C$_6$H$_3$AsH$_2$ with 4,3-HO(H$_2$N)C$_6$H$_3$AsO in dild. HCl, treating the arseno deriv. with HOCH$_2$SO$_2$Na, followed by 2 equivs. of glycide (362).
Props.: Yellow ppt. (361).

4,4'-DIHYDROXY-3-BIS(2,3-DIHYDROXYPROPYL)AMINO-3'-(2,3-DIHYDROXYPROPYLSULFO-METHYLAMINO)ARSENOBENZENE SODIUM SALT $C_{22}H_{31}As_2N_2NaO_{11}S$

Prepn.: by reducing a mixt. (1:1) of 4,3-HO[(HOCH$_2$CHOHCH$_2$)$_2$N]C$_6$H$_3$AsO$_3$H$_2$ and

4,3-HO(H_2N)$C_6H_3AsO_3H_2$ with H_3PO_2 in 10% HCl contg. KI and treating the prod. with glycide, followed by CH_2O and $NaHSO_3$ (365, 368).
Props.: Dark yellow powder sol. in H_2O (368).

5-[4-(4-ACETAMIDO-2,3-DIMETHYL-5-OXO-1-PYRAZOLINYL)PHENYLARSENO]-1-METHYLBENZ-
   IMIDAZOLYLOXYACETIC ACID $C_{23}H_{23}As_2N_5O_5$

Prepn.: by reducing equimol. mixt. of 4-(4-acetamido-2,3-dimethyl-5-oxo-1-pyra-
zolinyl)benzenearsonic acid and (5-arsono-1-methyl-2-benzimidazolyloxy)acetic
acid with H_3PO_2 in AcOH in the presence of KI and NaOAC (1175).

3-[5-(3-ACETAMIDO-5-[BIS(DIHYDROXYPROPYL)AMINO]-4-HYDROXYPHENYLARSENO)-2-BENZOX-
   AZOLYLMERCAPTO]PROPIONIC ACID $C_{24}H_{29}As_2N_3O_9S$

Prepn.: by reducing a mixt. of 3,4,5-AcNH(HO)[(HOCH$_2$CHOHCH$_2$)$_2$N]$C_6H_2AsO_3H_2$ and
3-(5-arsono-2-benzoxazolylmercapto)propionic acid with H_3PO_2 in dild. MeOH contg.
HCl (372).
Props.: Na salt, yellow powder (372).

N4-(HYDROXYMETHYL)-3',3'''-ARSENOBIS[6'-HYDROXYSULFANILANILANILIDE].SULFOXYLATE
   SODIUM SALT $C_{25}H_{24}As_2N_4O_8S_3$

Prepn.: by treating [4,3-HO($H_2NC_6H_4SO_2NH$)C_6H_3As:]$_2$ with calcd. amt. of CH_2OH.-
SO_2Na in dild. HCl at 50-60° (1258).
Props.: Decomp. 246-248° (1258).

TABLE XXVI

ASYMMETRICAL ARSENO COMPOUNDS

(RAs.AsR')n	Prepd. from	Method	Props.	Ref.
C$_{14}$				
[4-H$_2$NC$_6$H$_4$As-AsC$_6$H$_4$-4-NHOCCH$_2$OH]n	RAsO + R'AsH$_2$	IIIa	Light brown powder	1173
[3,4-H$_2$N(HO)C$_6$H$_3$As-AsC$_6$H$_4$-4-NHOCCH$_2$OH]n	RAsO + R'AsH$_2$	IIIa		1173
	RAsH$_2$ + R'AsO	IIIa		1173
C$_{15}$				
[4-(SCH:CHN:CNHO$_2$S)C$_6$H$_4$As-AsC$_6$H$_3$-3,4-(NH$_2$)-OH]n	RAsO$_3$H$_2$ + R'AsO$_3$H$_2$	Ib		680
[3,4-H$_2$N(HO)C$_6$H$_3$As-AsC$_6$H$_3$-3,4-(Me)OCH$_2$CO$_2$-H]n	RAsO$_3$H$_2$ + R'AsO$_3$H$_2$	Ia		1175
[4,3-HO(NaO$_3$SCH$_2$NH)C$_6$H$_3$As-AsC$_6$H$_3$-3,4-[N-(CH$_2$SO$_3$Na)$_2$]OH]n	[4,3-HO(NaO$_3$SCH$_2$NH)-C$_6$H$_3$As:]n + CH$_2$O + NaHSO$_3$			322
	[4,3-HO(H$_2$N)C$_6$H$_3$As:]n + CH$_2$O + NaHSO$_3$			322
[2,4-Me(HOCH$_2$CONH)C$_6$H$_3$As-AsC$_6$H$_3$-3,4-(NH$_2$)-OH]n	RAsH$_2$ + R'AsO.HCl	IIIa	Light yellow powder	1173
C$_{16}$				
[4-AcNHC$_6$H$_4$As-AsC$_6$H$_3$-3,4-(NHOCCH$_2$OH)OH]n	RAsH$_2$ + R'AsO	IIIa	Light yellow powder	1173
[4-H$_2$NCONHC$_6$H$_4$As-AsC$_6$H$_3$-2,4-(Me)OCH$_2$CO$_2$H]n	RAsO$_3$H$_2$ + R'AsO$_3$H$_2$	Ia		951, 366, 368
[4,3-HO(HOCH$_2$CHOHCH$_2$NH)C$_6$H$_3$As-AsC$_6$H$_3$-3,4-(NHCH$_2$SO$_3$Na)OH]n	[3,4-H$_2$N(HO)C$_6$H$_3$As:]n + [4,3-HO(HOCH$_2$CHOH-CH$_2$NH)C$_6$H$_3$As:]n + CH$_2$OH.SO$_3$Na.			
	RAsO$_3$H$_2$ + 4,3-HO(H$_2$N)C$_6$H$_3$AsO$_3$H$_2$ + CH$_2$OH-SO$_3$Na	Ia or Ib		356, 368
C$_{17}$				
[3,4-HO(H$_2$NCONH)C$_6$H$_3$As-AsC$_6$H$_3$-2,4-(NHAc)O-CH$_2$CO$_2$H]n	RAsO$_3$H$_2$ + R'AsO$_3$H$_2$	Ia		373
[2,4-Me(HO$_2$CCH$_2$O)C$_6$H$_3$As-AsC$_6$H$_3$-3,4-(NHAc)O-H]n	RAsO$_3$H$_2$ + R'AsO$_3$H$_2$	Ia		1175

TABLE XXVI (cont'd.)

ASYMMETRICAL ARSENO COMPOUNDS

(RAs.AsR')n	Prepd. from	Method	Props.	Ref.
[3,4-Me(HO₂CCH₂O)C₆H₃As-AsC₆H₃-4,2-(CH:NN-HCONH₂)OH]n	RAsO₂H₂ + R'AsO₃H₂	Ia		1175
[3,4-HO(H₂NCONH)C₆H₃As-AsC₆H₃-3,4-(NHAc)O-CH₂CO₂H]n	RAsO₃H₂ + R'AsO₃H₂	Ia		373, 951
[4,3-HO(H₂NCONH)C₆H₃As-AsC₆H₃-4,2-(NHAc)OCH₂CO₂H]n	RAsO₃H₂ + R'AsO₃H₂	Ib		373
[3,4-HO(H₂NCONH)C₆H₃As-AsC₆H₃-4,2-(NHAc)OC-H₂CO₂H]n	RAsO₃H₂ + R'AsO₃H₂	Ia		951
[2,5-HO(H₂NCONH)C₆H₃As-AsC₆H₃-4,2-(NHAc)OC-H₂CO₂H]n	RAsO₃H₂ + R'AsO₃H₂	Ia		373, 951
[2,5-HO(H₂NCONH)C₆H₃As-AsC₆H₃-3,4-(NHAc)OC-H₂CO₂H]n	RAsO₃H₂ + R'AsO₃H₂	Ia		373, 951
[4,3-HO(H₂NCONH)C₆H₃As-AsC₆H₃-3,4-(NHAc)OC-H₂CO₂H]n	RAsO₃H₂ + R'AsO₃H₂	Ia		373, 951
[3,4-HO(H₂NCONH)C₆H₃As-AsC₆H₃-5,2-(NHAc)O-CH₂CO₂H]n	4,3-HO(H₂N)C₆H₃AsO₃H₂ + R'AsO₃H₂, followed by KCNO	Ia		373, 951
[3,4-HO(H₂NCONH)C₆H₃As-AsC₆H₃-4,3-(NHAc)O-CH₂CO₂H]n	RAsH₂ + R'AsO₃H₂ (RAs:)n + [3,4-HO(Ac-NH)C₆H₃As:AsC₆H₃-4,3-(NHAc)OCH₂CO₂H]n	IIIb		373, 951 373
[4,3-Me(H₂NCONH)C₆H₃As-AsC₆H₃-2,4-(Me)OCH₂-CO₂H]n	RAsO₃H₂ + R'AsO₃H₂	Ia		951
[2,4-Me(HOCH₂CONH)C₆H₃As-AsC₆H₃-3,4-(NHAc)-OH]n	RAsH₂ + R'AsO	IIIa	Yellow ppt.	1173
[3,4-HO[(HOCH₂CH₂)N]C₆H₃As-AsC₆H₃-4,3-(NH-CH₂SO₂Na)OH]n	(RAs:AsR)n + [3,4-HO-(H₂N)C₆H₃As:]n + CH₂OH.SO₂Na	IV		364
C₁₈				
[3,4-H₂N(HO)C₆H₃As-AsC₆H₄-4-SO₂NHPh]n	RAsO₃H₂ + R'AsO₃H₂	Ib		680
[4-H₂NC₆H₄As-AsC₆H₄-4-NHO₂SC₆H₄NH₂]n	RAsO₃H₂ + R'AsO₃H₂	Ib		680
[3,4-H₂N(HO)C₆H₃As-AsC₆H₄-4-NHO₂SC₆H₄NH₂]n	RAsO₃H₂ + R'AsO₃H₂	Ib		680

Compound	Reaction	Method	Properties	References
$[4,2\text{-}AcNH(HO_2CCH_2O)C_6H_3As\text{-}AsC_6H_3\text{-}3,4\text{-}(NHAc)\text{-}OH]n$:	$RAsO_3H_2 + R'AsO_3H_2$	Ia	Bi salt, yellow	1175
$[3,4\text{-}AcNH(HO_2CCH_2O)C_6H_3As\text{-}AsC_6H_3\text{-}3,4\text{-}(NHAc)\text{-}OH]n$	$RAsO_3H_2 + R'AsO_3H_2$	Ia		1172
$[4\text{-}[HOCH_2CH_2(HOCH_2CHOHCH_2)N]C_6H_4As\text{-}AsC_6H_3\text{-}3,4\text{-}(NHCH_2SO_2Na)OH]n$	$RAsO + R'AsH_2$			1175, 363.
$[3,4\text{-}HO[HOCH_2CH_2(HOCH_2COHCH_2)N]C_6H_3As\text{-}AsC_6H_3\text{-}3,4\text{-}(NHCH_2SO_3Na)OH]n$	$RAsO + 3,4\text{-}H_2N(HO)\text{-}C_6H_3AsH_2$, $+ CH_2O + NaHSO_3$	IIIa	Dark yellow powder, sol. in H_2O	365, 368
C_{19}				
$[3,4\text{-}Me(HO_2CCH_2O)C_6H_3As\text{-}AsC_6H_3\text{-}3,4\text{-}(NHAc)\text{-}OCH_2CO_2H]n$	$RAsO_3H_2 + R'AsO_3H_2$	Ia		1175
$[3,4\text{-}HO[(HOCH_2CHOHCH_2)_2N]C_6H_3As\text{-}AsC_6H_3\text{-}4,3\text{-}(NHCH_2SO_2Na)OH]n$	$[3,4\text{-}HO(H_2N)C_6H_3As\text{-}As\text{-}C_6H_3\text{-}4,3\text{-}(NHCH_2\text{-}SO_2Na)OH]n + ClCH_2\text{-}\overline{CHCH_2O}$	IIIc		360
C_{20}				
$[4\text{-}[MeNC(Me):C(NH_2)C(:O)\overline{N}]C_6H_4As\text{-}AsC_6H_3\text{-}4,3\text{-}(OCH_2CO_2H)OH]n$	$(RAs:)n + (R'As:)n$	Ia		1175
$[4,3\text{-}HO[(HOCH_2CHOHCH_2)_2N]C_6H_3As\text{-}AsC_6H_2\text{-}3,4,5\text{-}(NHCH_2SO_2Na)(OH)OMe]n$	$RAsO + R'AsH_2$	IIIc		363
C_{21}				
$[3,4,5\text{-}AcNH(HO)[(HOCH_2CHOHCH_2)_2N]C_6H_2As\text{-}AsC_6H_3\text{-}4,3\text{-}(NHCH_2SO_3Na)OH]n$	$RAsO_3H_2 + 3,4\text{-}HO(H_2N)C_6H_3AsO_3H_2$, followed by $CH_2O + NaHSO_3$ $(RAs:)n + (R'As:)n$	Ia	Dark-yellow ppt.	366, 368
$[3,4,5\text{-}AcNH(HO)[(HOCH_2CHOHCH_2)_2N]C_6H_2As\text{-}AsC_6H_3\text{-}3,4\text{-}(NHCH_2SO_3Na)OH]n$	$RAsO_3H_2 + 4,3\text{-}HO(H_2N)C_6H_3AsO_3H_2$, followed by $CH_2OH.SO_3Na$		Dark-yellow ppt.	368, 368
C_{22}				
$[2,4\text{-}Me(HO_2CCH_2O)C_6H_3As\text{-}AsC_6H_4\text{-}4\text{-}\overline{NC(:O)C\text{-}(NHAc)}]:C(Me)\overline{N}Me]n$	$RAsO_3H_2 + R'AsO_3H_2$	Ia		1175
C_{23}				
$[3,4\text{-}AcNH(HO_2CCH_2O)C_6H_3As\text{-}AsC_6H_4\text{-}4\text{-}\overline{NC(:O)\text{-}C(NHAc):C(}\overline{NHAc)}:C(Me)\overline{N}Me]n$	$RAsO_3H_2 + R'AsO_3H_2$	Ia		1175

COMPOUNDS WITH THE AS-N BOND

Aminoorganoarsenic derivatives of trivalent arsenic are prepared by reacting organo-dihaloarsines, diorganohaloarsines, arsenic trichloride, or tertiary arsines with primary or secondary amines or with ammonia. The reactions proceed according to the following scheme:

$R_2AsX + HNR'R'' \rightarrow R_2AsNR'R''$ (R' and R'' may be H).
$RAsX_2 + H_2NR' \rightarrow RAs:NR'$
$RAsX_2 + HNR'R'' \rightarrow RAs(NR'R'')_2$
$AsX_3 + HNR'R'' \rightarrow R'R''NAsX_2$, $(R'R''N)_2AsX$, and $(R'R''N)_3As$.
$R_3As + NH_3 \rightarrow R_2AsNH_2$.

The aminoarsines are hydrolytically unstable.

AMINOBIS(TRIFLUOROMETHYL)ARSINE $C_2H_2AsF_6N$ $(CF_3)_2AsNH_2$
Prepn.: By-prod. in the rxn. of $(CF_3)_2AsCl$ with NH_3 in the gas phase (253).
    • By reacting $(CF_3)_3As$ with NH_3 at -64° (253).
    By-prod. in the rxn. of $(CF_3)_3As$ with NH_3 at 20° (253).
Props.: B. 89±1°, thermodynamic data; IR spectrum (253).

DICHLORODIMETHYLAMINOARSINE $C_2H_3AsCl_2N$ Me_2NAsCl_2
Prepn.: by reacting $AsCl_3$ with Me_2NH in Et_2O; 67.1% yield (637).
Props.: B. 73-75°/25 mm., d_{20} 1.6560, n_D^{20} 1.5532 (637).
Rxns. with: $Me_2NH \rightarrow (Me_2N)_2AsCl$ (637).

METHYLAMINOBIS(TRIFLUOROMETHYL)ARSINE $C_3H_4AsF_6H$ $(CF_3)_2AsNHMe$
Prepn.: by reacting $(CF_3)_2AsCl$ with $MeNH_2$ in the gase phase at 20° (253).
Props.: B. 84°; on heating at 120° for 15 hrs. decomp. $\rightarrow$ $(CF_3)_3As$, CHF_3, and As; thermodynamic data; IR spectrum (253).

IMINOBIS[BIS(TRIFLUOROMETHYL)ARSINE] $C_4HAs_2F_{12}N$ $[(CF_3)_2As]_2NH$
Prepn.: By reacting $(CF_3)_2AsCl$ with dry NH_3 at -64°; when the reaction is carried out in the gas phase $(CF_3)_2AsNH_2$ is also formed (253).
    By reacting $(CF_3)_3As$ with NH_3 at 20°; $(CF_3)_2AsNH_2$ is also formed (253).
Props.: B. 126.5°; thermodynamic data; IR spectrum (253).

ETHYLAMINOBIS(TRIFLUOROMETHYL)ARSINE $C_4H_6AsF_6N$ $(CF_3)_2AsNHEt$
Prepn.: by reacting $(CF_3)_2AsCl$ with $EtNH_2$ in the gase phase (253).
Props.: B. 98.5°; thermodynamic data; IR spectrum (253).

DIMETHYLAMINOBIS(TRIFLUOROMETHYL)ARSINE $C_4H_6AsF_6N$ $(CF_3)_2AsNMe_2$
Prepn.: by reacting $(CF_3)_2AsCl$ with Me_2NH in the gas phase (253).
Props.: B. 89°; on heating at 250-280° decomp. $\rightarrow$ CHF_3 and an As-contg. solid; thermodynamic data; IR spectrum (253).
Rxn. with $H_2O \rightarrow H_3AsO_3$, Me_2NH, and CHF_3 (253).

DICHLORO(DIETHYLAMINO)ARSINE $C_4H_{10}AsCl_2N$ Et_2NAsCl_2
Prepn.: by reacting $AsCl_3$ with Et_2NH in Et_2O; 70.9% yield (637); 91% yield (638).
Props.: B. 83-84°/13 mm., d_{20} 1.4727; n_D^{20} 1.5335; hydrolytically unstable (637, 638).

(ETHYLIMINO)ETHYLARSINE $C_4H_{10}AsN$ EtAs:NEt
Prepn.: by reacting $EtAsCl_2$ with $EtNH_2$ in $CHCl_3$ at -5°; 37% yield (44).
Props.: Light yellow liquid, b. 165-175°/3.5 mm. (44).

(DIMETHYLAMINO)DIMETHYLARSINE $C_4H_{12}AsN$ Me_2AsNMe_2
Prepn.: by reacting Me_2NH with Me_2AsCl at 46° (253) or with Me_2AsBr below -50° (851).
Props.: B. 112° (256), m. -110°, b. 110°; thermodynamic data; IR spectrum (253, 256, 851).
Rxn. with HCl → Me_2AsCl (851).

CHLOROBIS(DIMETHYLAMINO)ARSINE $C_4H_{12}AsClN_2$ $(Me_2N)_2AsCl$
Prepn.: by reacting Me_2NAsCl_2 with Me_2NH in Et_2O; 61.1% yield (637).
Props.: B. 42-43°/13 mm.; d_{20} 1.3460; n^{20}_{D} 1.5255 (637).
Rxn. with Me_2NH → $(Me_2N)_3As$ (637).

METHYLBIS(DIMETHYLAMINO)ARSINE $C_5H_{15}AsN_2$ $MeAs(NMe_2)_2$
Prepn.: by condensing $MeAsI_2$ with Me_2NH at a temp. < -40° (851).
Props.: M. -62°; b. 148°; thermodynamic data; IR spectrum (851).
Rxn. with HCl → $MeAsCl_2$ (851).

TRIS(DIMETHYLAMINO)ARSINE $C_6H_{18}AsN$ $(Me_2N)_3As$
Prepn.: By reacting $(Me_2N)_2AsCl$ with Me_2NH 52.3% yield (637).
    By condensing $AsCl_3$ with excess of Me_2NH below -78° (851).
Props.: B. 42-43°/3.5 mm; d_{20} 1.1248; n^{20}_{D} 1.4848 (637); m. -53°; b. 55-57°/10 mm.; thermodynamic data; IR spectrum (851).
Rxns. with: H_2O → Me_2NH + As_2O_3 (637).
HCl → $AsCl_3$ + Me_2NH (851).

CHLOROBIS(DIETHYLAMINO)ARSINE $C_8H_{20}AsClN$ $(Et_2N)_2AsCl$
Prepn.: by reacting Et_2NAsCl_2 with Et_2NH in Et_2O; 55.4% yield (637).
Props.: B. 118-120°/17 mm; d_{20} 1.2225; n^{20}_{D} 1.5098; hydrolytically unstable (637, 638).

TRIS(DIETHYLAMINO)ARSINE $C_{12}H_{30}AsN_3$ $(Et_2N)_3As$
Prepn.: by reacting $(Et_2N)_2AsCl$ with Et_2NH in Et_2O; 40.3% yield (637).
Props.: B. 91-93°/2 mm.; d_{20} 1.4825; fumes in air and is readily oxidized and hydrolyzed (637).

COMPOUNDS WITH THE As-P BOND

This class of compounds comprizes diorganoarsinophosphonic acid derivatives, R_2-$AsP(O)(OR')_2$, diorganoarsinophonium halides, $R_2AsP(OR')_3X$, and diorganoarsinohalo-triorganophosphoranes, $R_2AsPR'_3X$.

Diorganoarsinophosphonic esters are prepared by heating under reflux diorganohalo-arsines, R_2AsX or $RR'AsX$, with trialkyl phosphites or with alkali metal dialkyl phosphites. The reaction proceeds via Arbuzov rearrangement (transition of trivalent phosphorus to pentavalent state).

Diorganoarsinophosphonium halides are formed by addition of alkyl halides to diorganoarsinophosphonic esters:

$R_2AsP(O)(OR')_2 + RX \quad R_2AsP(OR')_3X$.

Diorganoarsinohalotriorganophosphoranes, R_2AsPR_3X, are prepared by reacting diorganohaloarsines with tertiary phosphines in cyclohexane. They are regarded as
arsino-phosphonium salts.

The arsenic-phosphorus bond in the arsinophosphonic acid derivatives is readily
cleaved by alkali, hydrohalic acids, and atmospheric oxygen.

TABLE XXVII

DIORGANOARSINOPHOSPHONIC ESTERS

Arsinophosphonic Esters	Prepd. from	Yield %	Props.	Ref.
Me$_2$AsP(O)(OMe)$_2$	Me$_2$AsBr + (MeO)$_3$P	60.5	b. 76.5°/1 mm., d^0 1.4011, rxn.[1]	634
Me$_2$AsP(O)(OEt)$_2$	Me$_2$AsBr + (EtO)$_3$P	74.3	b. 83°/1 mm., d^0 1.3036, n_D^0 1.2932, rxn.[1]	634
Et$_2$AsP(O)(OMe)$_2$	Et$_2$AsCl+ (MeO)$_3$P	29.8	b. 98.5-99.5°/1 mm., d^0 1.3205, rxn.[1]	634
Me$_2$AsP(O)(OPr)$_2$	Me$_2$AsBr + (PrO)$_3$P	64.9	b. 101.5°/1 mm.; d^0 1.2343, n_D^0 1.2242, rxns.[1,2]	634
Me$_2$AsP(O)(OCHMe$_2$)$_2$	Me$_2$AsBr + (Me$_2$CHO)$_3$P	64.4	b. 82-83°/1 mm., d^0 1.2112, n_D^0 1.2015, rxn.[1,2]	634
Et$_2$AsP(O)(OEt)$_2$	Et$_2$AsCl + (EtO)$_3$P	32.2	b. 105.5-106.5°/1 mm., d^{15} 1.2129, n_D^{15} 1.4878, rxns.[1,2]	634
MeEtAsP(O)(OCHMe$_2$)$_2$	MeEtAsI + NaOP(OCHMe$_2$)$_2$		b. 99-99.5°/2 mm., d_O^{18} 1.1733, n_D^{18} 1.4761, rxn.[1]	631
Me$_2$AsP(O)(OBu)$_2$	Me$_2$AsBr + (BuO)$_3$P	54.1	b. 122-123°/1 mm., d^0 1.933, rxns.[1,2]	634
Et$_2$AsP(O)(OPr)$_2$	Et$_2$AsBr + (PrO)$_3$P	36.2	b. 124-125°/1 mm., d^{15} 1.1675, n_D^{15} 1.4901, rxn.[1,2]	634
Et$_2$AsP(O)(OCHMe$_2$)$_2$	Et$_2$AsCl + (Me$_2$CHO)$_3$P	30.2	b. 106-107°/1 mm., d^{15} 1.1529, n_D^{15} 1.4782, rxns.[1,2]	634
EtBuAsP(O)(OEt)$_2$	EtBuAsI + NaOP(OEt)$_2$ in Et$_2$O		b. 112-113°/1 mm., d_O^{14} 1.1865, rxn.	629
MeEtAsP(O)(OBu)$_2$	MeEtAsI + NaOP(OBu)$_2$	56	b. 127-128°/3 mm., d_O^{20} 1.0710, rxn.[1,2]	631
Et(Me$_2$CHCH$_2$CH$_2$)AsP(O)-(OEt)$_2$	Et(Me$_2$CHCH$_2$CH$_2$)AsI + NaOP-(OEt)$_2$ in Et$_2$O	52	b. 118-120°/1 mm., d_O^{14} 1.2718, rxn.[1]	621
EtPhAsP(O)(OEt)$_2$	EtPhAsI + NaOP(OEt)$_2$ in Et$_2$O		b. 144-145°/1 mm., d_O^{14} 1.2734, rxn.[1]	629
Et$_2$AsP(O)(OBu)$_2$ (CH$_2$:CHCH$_2$)PhAsP(O)(OEt)$_2$	Et$_2$AsCl + (BuO)$_3$P (CH$_2$:CHCH$_2$)PhAsI + (EtO)$_3$P or NaOP(OEt)$_2$	16.9	b. 144-145°/1 mm., rxns.[1,2] b. 142-143°/1 mm., d^0 1.2568, rxn.[1]	634 629
EtPhAsP(O)(OPr)$_2$	EtPhAsI + NaOP(OPr)$_2$	50	b. 165-166°/2 mm., d_O^{24} 1.2427, n_D^{24} 1.5375	631
BuPhAsP(O)(OEt)$_2$	BuPhAsBr + NaOP(OEt)$_2$		b. 162-163°/1 mm., d_O^{17} 1.2345, rxn.[1]	629

1. Rxn. with dild. hydrohalic acids or alkali → cleavage of the As-P bond with formation of bis-(diorganoarsenic) oxide.

2. Rxn. with atm. O → cleavage of the As-P bond with formation of bis(diorganoarsenic) oxide.

TABLE XXVII (cont'd.)

DIORGANOARSINOPHOSPHONIC ESTERS

Arsinophosphonic Esters	Prepd. from	Yield %	Props.	Ref.
EtBuAsP(O)(OBu)$_2$	EtBuAsI + NaOP(OBu)$_2$	52	b. 146-147°/2 mm., d_0^{20} 1.1048, n_D^{20} 1.4775	631
Et(Me$_2$CHCH$_2$)AsP(O)(OBu)$_2$	Et(Me$_2$CHCH$_2$)AsI + NaOP(OBu)$_2$	54	b. 138.5-140°/2 mm., d_0^{20} 1.1087, n_D^{20} 1.4738, rxn.[1]	631
Ph$_2$AsP(O)(OEt)$_2$	Ph$_2$AsCl + NaOP(OEt)$_2$		b. 176-177°/1 mm., d_0^{18} 1.2845, rxn.[1]	629
EtPhAsP(O)(OBu)$_2$	EtPhAsI + NaOP(OBu)$_2$	60.7	b. 176-176.5°/2 mm., d_0^{13} 1.1875, n_D^{18} 1.5304, rxns.[1,2]	631

BUTYLETHYLARSINOTRIETHOXYPHOSPHONIUM IODIDE $C_{12}H_{29}AsIO_3P$ [EtBuAsP(OEt)$_3$]I (?)
<u>Prepn.:</u> by reacting EtBuAsI with (EtO)$_3$P at r.t.; 100% yield (629).
<u>Props.:</u> m. 182-183° (629).

TABLE XXVIII
DIORGANOARSINOHALOTRIORGANOPHOSPHORANES

[R$_2$AsPR'$_3$]X	Prepd. from	Props.	Ref.
(Me$_2$AsPEt$_3$)Cl	Me$_2$AsCl + Et$_3$P	Hydroscopic solid	237
Me$_2$AsPEt$_3$)I	Me$_2$AsI + Et$_3$P	m. 66°	237
	(Me$_2$AsPEt$_3$)Cl + NaI in		
	Me$_2$CO		237
(Me$_2$AsPMe$_2$Ph)Cl	Me$_2$AsCl + Me$_2$PhAs	m. 116°	237
(Me$_2$As(Me$_2$Ph)I	Me$_2$AsI + Me$_2$PhAs	m. 147-147.5°	237
	(Me$_2$AsPMe$_2$Ph)Cl + NaI		
	in H$_2$O		237

COMPOUNDS WITH THE As-Mg BOND

DIMETHYLARSINOMAGNESIUM BROMIDE Me$_2$AsMgBr
<u>Prepn.:</u> by reacting Me$_2$AsH with EtMgBr in Et$_2$O (9)
<u>Props.:</u> Insol. in Et$_2$O (9).
<u>Rxns. with:</u> H$_2$O → Me$_2$AsH (9).
EtBr → Me$_2$AsEt (9).
AcCl → Me$_2$AsAc (9).

PHENYLARSINOBIS(MAGNESIUM BROMIDE) PhAs(MgBr)$_2$
<u>Prepn.:</u> by reacting PhAsH$_2$ with PhMgBr or EtMgBr in C$_6$H$_6$ (68, 70, 244).
<u>Props.:</u> Sol. in benzene (68).
<u>Rxns. with:</u> SOCl$_2$ → PhAsSO (17).
o-C$_6$H$_4$(CH$_2$Br)$_2$ → 2-phenylisoarsindoline (68).
o-BrCH$_2$C$_6$H$_4$CH$_2$CH$_2$Br → 1,2,3,4-tetrahydro-2-phenylisoarsinoline (68).
1,8-BrCH$_2$C$_{10}$H$_6$CH$_2$Br → 2-phenylarsaperinaphthane (68).
BrCH$_2$CN → (PhAs:)n and PhAsBr$_2$ (68).
BrCH$_2$CO$_2$Et → (PhAs:)n, PhAsBr$_2$, and PhAs(CH$_2$CO$_2$Et)$_2$ (68).
BrCH$_2$C(OEt)$_2$Me → (PhAs:)n and PhAsBr$_2$ (68).
(o-BrCH$_2$C$_6$H$_4$-)$_2$ → 6H-5,7-dihydro-6-phenyldibenz[c,e]arsepin (68).
(BrCH$_2$CH$_2$)$_2$O → tetrahydro-4-phenyl-1,4-oxarsine (70).
o-PhC$_6$H$_4$CH$_2$Br→ PhAs(CH$_2$C$_6$H$_4$Ph-o) (244).

COMPOUNDS WITH THE As-B BOND

DIMETHYLARSINOBORINE TRIMER AND TETRAMER (Me$_2$AsBH$_2$)x (x = 3 and 4)
<u>Prepn.:</u> The compds. were obtd. by heating Me$_2$AsH.BH$_3$ at 100° (1169).
<u>Props.:</u> Trimer, m. 49.7-50.6°, logp_{mm} = 8.752-3074/T, b. (calcd.) 250°, stable on
heating at 150°, hydrolysis by 12 N HCl yields CH$_4$ (1169).

Tetramer, less volatile than the trimer, m. 149.5-150.5°, is formed from the trimer at 180°, log $p_{mm} = 18.635-7692/T$ (1169).

MISCELLANEOUS DERIVATIVES OF TRIVALENT ARSENIC

METHYLARSINOCARBINE HYDROBROMIDE   2(MeAs:C).3HBr (?)
Prepn.: by reacting MeAsH$_2$ with CHBr$_3$ in C$_5$H$_5$N (9).
Props.: Hydroscopic, crystalline solid, for which the $[MeAs:C{<}_{Br}^{H}]_2$.HBr formula was proposed (9).
Rxn. with dild. HCl → HCO$_2$H (9).

BIS [CHLORO(2-CHLOROVINYL)ARSENIC] OXIDE   C$_4$H$_4$As$_2$Cl$_4$O   [(ClCH:CH)AsCl]$_2$O
Rxn. with iodo-Cu(I) reagent (from NaI + CuSO$_4$) → ppt. (300).

PHENYLSULFINYLARSINE C$_6$H$_5$AsOS   PhAsSO
Prepn.: by reacting PhAs(MgBr)$_2$ with SO$_2$Cl$_2$ (17).

2-AMINO-4-(CHLOROHYDROXYARSINO)PHENOL C$_6$H$_7$AsClNO$_2$ 4,3-HO(H$_2$N)C$_6$H$_3$As(OH)Cl
Prepn.: by adding one equiv. of HCl to 4,3-HO(H$_2$N)C$_6$H$_3$As(OH)$_2$ in H$_2$O, colling the soln. to 0°, adding 10% excess HCl, freezing out the prod. and drying in vacuo over P$_2$O$_5$; 82% yield (43).
Props.: White, crystalline powder; slightly sol. in H$_2$O and 95% EtOH, insol. in abs. EtOH (43).

CHLORO(2,2'-DICHLOROISOPROPOXY)PHENYLARSINE C$_9$H$_{10}$AsCl$_3$O (ClCH$_2$)$_2$CHOAsPhCl
Prepn.: by satg. PhAsCl$_2$ with OCH$_2$CHCH$_2$Cl at r. t. (787).
Props.: Viscous liquid of pungent odor, b. 218-220°/5 mm., decomp. on heating under atm. pressure (787).
Rxn. with H$_2$O → hydrolysis (787).

CHLORO(2-CHLOROPROPOXY)PHENYLARSINE C$_9$H$_{11}$AsCl$_2$O MeCHClCH$_2$OAsPhCl
Prepn.: by reacting PhAsCl$_2$ with MeCHCH$_2$O at r. t. (787).
Props.: Viscous, almost odorless liquid, b. 190-195°/10 mm., d$_4^{17}$ 1.89 (787).
Rxn. with PCl$_5$ → MeCHClCH$_2$Cl + PhAsCl$_2$ (787).

2. COMPOUNDS OF PENTAVALENT ARSENIC

ALKANE-, ALKYLENE-, CYCLOALKANE-, AND ARYLALKANEARSONIC ACIDS

The aliphatic, cycloaliphatic, and arylaliphatic arsonic acids are prepared by:
1. Reacting alkyl or aralkyl halides, dissolved in alcohol, or alkyl sulfates with arsenic trioxide in alkaline solution or with alkalimetal arsenites in aqueous solution at elevated temperatures (Rosenmund Reaction).
2. Oxidative hydrolysis of alkyl- or aralkyldihaloarsines, using hydrogen peroxide in aqueous or acetic acid solution, chlorine in water, or nitric acid.
3. Reacting alkylene oxides, e.g., ethylene oxide, with alkali metal arsenites in aqueous solution.
4. Oxidation of arsonous acids with hydrogen peroxide in acetic acid or with chlorine in water.
5. Condensation of haloalkanearsonic acids with alkyl- or aralkylamines to form amino-substituted alkanearsonic acids.

METHANEARSONIC ACID $C_1H_5AsO_3$ $MeAsO_3H_2$
Prepn.: By reacting Na_3AsO_3 or K_3AsO_3 with Me_2SO_4 in aq. soln. under reflux (44, 744), or in alk. soln. at 72-78° (111a).
By treating aq. Na_3AsO_3 soln. with MeCl at 60°/60 psi. (845, 846).
By heating As_2O_3 with $MeNaSO_4$ in the presence of NaOH at 130-135° (1051).
Props.: M. 159.8° (44). Raman spectrum (994). Magnetic study (1029).
Rxns. with: $SO_2.HX$ (X = Cl or Br) contg. KI → $MeAsX_2$ (660, 774).
H_2SO_4 at 315° → $As(OH)_3$ (decompn. mechanism) (992, 993).
Na_3AsO_4 + H_3PO_2 → yellow to brown ppt. contg. organically bound As(995).
AcCl → $MeAsCl_2$ + $CH_2(AsMeCl)_2$ (1024).
Na_2SO_3 + H_3PO_2 in an acid medium → [MeAs(S)=]n (655).
Glycols in C_6H_6 (azeotrop. distn.) → arsenaspiro deriv. (1097).
Metal ions → pptn. (in quant. analysis) (1017).
$ClP(S)(OEt)_2$ in alk. soln. → $MeAs(O)[P(S)(OEt)_2]_2$ (1102).
Me_2SiCl_2 in C_6H_6 → $MeAs(O)(OSiMe_2Cl)_2$ (655a, 656).
$MeSiCl_3$ in C_6H_6 → $MeAs(O)(OSiMeCl_2)_2$ (655a, 656).
Salts: Ag salt occurs in two crystalline   forms, fine rhombohedral needles and hexagonal crystals (1028).
Di-sodium salt, Arrhenal, hexahydrate, prepd. by reacting As_2O_3, MeI, and NaOH in aq. alc. or Na_3AsO_3 with Me_2SO_4 in aq. soln. (1224).
UV spectrum (116). Raman spectrum (498). Cryoscopy of mixts. with Na molybdate (1154). Pptn. by $AgNO_3$ (1028).
Use: as herbicide (111a, 1158a), particularly as crabgrass killer (1201).

TABLE XXIX
ALIPHATIC ARSONIC ACIDS

$RAsO_3H_2$	Prepd. from	Props.	Ref.
C_1			
$CF_3AsO_3H_2$	CF_3AsI_2	Ionization, dehydration, salts[1]	334
$ClCH_2AsO_3H_2$	$ClCH_2AsCl_2 + H_2O_2$	Crystals, shrink 133-135°, partly decomp. 140° and remain unchanged up to 250°	121
C_2			
$CF_3CF_2AsO_3H_2$	$C_2F_5AsCl_2 + H_2O_2$	Microcryst. powder; di-Ag salt, scales, rxn.[2]	135
$HO_2CCH_2AsO_3H_2$	$ClCH_2CO_2H + As_2O_3$	White, hexagon. crystals, m. 148°(coarse) and 152°(fine crystals)	823
$ClCH:CHAsO_3H_2$	$BrCH_2CO_2H + K_3AsO_3$ cis- or trans-$ClCH:CHAsCl_2$ + Cl_2 in H_2O or H_2O_2	m. 152°, d_4^{20} 2.425, rxns.[3,4] cis-isomer, m. 90-91°, sol. in H_2O; trans-isomer, m. 130-131°, rxns.[2,5]	24 538
$EtAsO_3H_2$	$EtI + As_2O_3$ $EtBr + As_2O_3$ $EtNaSO_4 + As_2O_3$ + excess Na-OH at 130-5°	[87% yield] m. 99.6° Rxns.[2,6]	655, 744 44 1051
$HOCH_2CH_2AsO_3H_2$	$\overline{CH_2CH_2O} + K_3AsO_3$ or Na_2HAsO_3	Rxn.[7]	218
C_3			
$CH_2:C(CO_2H)AsO_3H_2$	$CH_2:CBrCO_2H + K_3AsO_3$	m. 159°, d^{20} 2.167	24
$HO_2CCH_2CH:As\ O_3H_2$	$Br(CH_2)_2CO_2H + K_3AsO_3$	m. 159°, d_4^{20} 2.167	24
$CH_2:CHCH_2AsO_3H_2$	$CH_2:CHCH_2Br + As_2O_3$	m. 128-129°; use[9]	4
$MeCH(CO_2H)AsO_3H_2$	$MeCHBrCO_2H + K_3AsO_3$	m. 134°, d 2.072, optical resolution[10]	24
$Cl(CH_2)_3AsO_3H_2$		Rxn.[8]	667
$PrAsO_3H_2$ (64% yield)	$PrBr + As_2O_3$	m. 134.6-135.2°	44
$HO(CH_2)_3AsO_3H_2$	$HO(CH_2)_3Cl + Na_3AsO_3$	Magnetic data	1029
$MeOCH_2CH_2AsO_3H_2$	$MeOCH_2CH_2Br + As_2O_3$	Rxn.[12] S...m	55 44

		Crystallographic data	
C₄			
$MeCH{:}C(CO_2H)AsO_3H_2$	$MeCH{:}CBrCO_2H + K_3AsO_3$	Isomers[11] Rxn.[2]	26
$CH_2{:}CMeCH_2AsO_3H_2$	$MeCHBrCH_2Br + K_3AsO_3$	Sublimes at 81°	44
	$CH_2{:}CMeCH_2Cl + As_2O_3$		24
$MeCH_2CH(CO_2H)AsO_3H_2$	$MeCH_2CHBrCO_2H + K_3AsO_3$	m. 127°, d^{24} 1.891, resolution[10]	44
$BuAsO_3H_2$ (96% yield)	$BuBr + As_2O_3$	m. 159.5-160°	44
	$BuAsCl_2 + H_2O_2$	m. 158-159°, rxns.[6,13]	674
$EtOCH_2CH_2AsO_3H_2$	$EtOCH_2CH_2Br + As_2O_3$	Sirup	44
$EtSCH_2CH_2AsO_3H_2$	$EtSCH_2CH_2Br + As_2O_3$	m. 109-110°	44
$O(CH_2CH_2AsO_3H_2)_2$	$O(CH_2CH_2Br)_2 + As_2O_3$	m. 173-174°	44
$SO_2(CH_2CH_2AsO_3H_2)_2$	$SO_2(CH_2CH_2Br)_2 + As_2O_3$	m. 215°	44
C₅			
$CH_2{:}CH(CH_2)_3AsO_3H_2$	$CH_2{:}CH(CH_2)_3Br + As_2O_3$	m. 144-6°	44
$MeCH_2CH(CO_2H)AsO_3H_2$	$MeCH_2CHBrCO_2H + K_3AsO_3$	m. 116-117° Optical resolution	24 25
$Me(CH_2)_4AsO_3H_2$	$n\text{-}C_5H_{11}Br + As_2O_3$	m. 162-163°, rxn.[14]	44
	$iso\text{-}C_5H_{11}AsCl_2 + HNO_3$	m. 194°	674
$Me_2CH(CH_2)_2AsO_3H_2$	$iso\text{-}C_5H_{11}Br + As_2O_3$	m. 192-194°	44
$MeEtCHCH_2AsO_3H_2$	$MeEtCHCH_2Br + As_2O_3$	m. 171-172°	44
$EtNH(CH_2)_3AsO_3H_2$	$Cl(CH_3)_2AsO_3H_2 + EtNH_2$	$HCl \cdot 0.5H_2O$ salt, m. 100-120°	667
C₆			
$Me(CH_2)_3CH(CO_2H)AsO_3H_2$	$Me(CH_2)_4CHBrCO_2H + K_3AsO_3$	m. 96° Optical resolution	24 25
$n\text{-}C_6H_{13}AsO_3H_2$	$n\text{-}C_6H_{13}Br + As_2O_3$	m. 165-166°	44

1. Ionization const. (334, 335, 337). Dehydration at 33°/0.01 mm. over $P_2O_5 \rightarrow [CF_3As(O)OH]_2O$, which at 73°/0.01 mm. $\rightarrow (CF_3AsO_2)n$; both are converted to $MeAsO_3H_2$ (334, 335). Forms mono- and di-Ag salt (335).
2. Redn. with SO_2 in HCl contg. HI $\rightarrow RAsCl_2$ (26, 135, 538, 744).
3. Redn. with $H_3PO_2 \rightarrow$ the arseno deriv. $(RAs{=})n$ (823, 825).
4. Rxn. with PCl_3 in $CCl_4 < 45° \rightarrow HO_2CCH_2AsCl_2$ (825).
5. Redn. of the trans-isomer with $POCl_3 \rightarrow$ mixt. of cis- and trans-$ClCH{:}CHAsCl_2$ (538).
6. Rxn. with H_2SO_4 at an elevated temp. $\rightarrow As(OH)_3$ (decompn. mechan.ism) (993).
7. Rxn. with SO_2 in HCl contg. KI $\rightarrow ClCH_2CH_2AsCl_2$ (218).
8. Rxn. with $RNH_2 \rightarrow$ the corresponding RNH-alkanearsonic acid (667).
9. Use: stabilizer for motor fuels and lubricating oils (1217).
10. Optical activity data and resolution (25).
11. The compd. occurs in cis and trans form (26).
12. Rxn. with SO_2 in HCl and $SOCl_2 \rightarrow Cl(CH_2)_3AsCl_2$ (55).

TABLE XXIX (cont'd.)
ALIPHATIC ARSONIC ACIDS

$RAsO_3H_2$	Prepd. from	Props.	Ref.
$PrNH(CH_2)_3AsO_3H_2$	$Cl(CH_2)_3AsO_3H_2 + PrNH_2$	HCl salt, m. 210°	667
C_7			
$PhCH_2AsO_3H_2$	$PhCH_2Br + As_2O_3$	m. 196-197°, rxn.[13]	667
$Me(CH_2)_4\text{-}CH(CO_2H)AsO_3H_2$	$Me(CH_2)_4CHBrCO_2H + K_3AsO_3$	m. 82-83°	24
$n\text{-}C_7H_{15}AsO_3H_2$	$n\text{-}C_7H_{15}Br + As_2O_3$	Optical resolution	25
		m. 156-157°	44
$BuNH(CH_2)_3AsO_3H_2$	$Cl(CH_2)_3AsO_3H_2 + BuNH_2$	HCl salt, m. 232°	667
C_8			
$PhCH:CHAsO_3H_2$	$PhCH:CHBr + As_2O_3$	m. 153°	265
$PhCH_2CH_2AsO_3H_2$	$PhCH_2CH_2Br + As_2O_3$	m. 167.5-168°	667
$PhOCH_2CH_2AsO_3H_2$	$PhOCH_2CH_2Br + As_2O_3$	m. 100-101°, decomp. 175° →	44
$Me(CH_2)_5CCl:CHAsO_3H_2$	$Me(CH_2)_5CCl:CHAsCl_2 + H_2O_2$	$n\text{-}C_6H_{13}CCl:CH_2 + As_2O_3$, forms mono- and dibasic salts	445
$Me(CH_2)_5CH(CO_2H)AsO_3H_2$	$Me(CH_2)_5CHBrCO_2H + K_3AsO_3$	m. 69°	24
		Optical resolution	25
$PhCH(CO_2H)AsO_3H_2$	$PhCHBrCO_2H + K_3AsO_3$	m. 110°	24
		Optical resolution	25
C_9			
$Me(CH_2)_6CH(CO_2H)AsO_3H_2$	$Me(CH_2)_6CHBrCO_2H + K_3AsO_3$	m. 115°	24
		Optical resolution	25
$C_6H_{11}NH(CH_2)_3AsO_3H_2$	$Cl(CH_2)_3AsO_3H_2 + cyclo\text{-}C_6\text{-}H_{11}NH_2$	HCl salt, m. 172-173°	667
C_{10}			
$PhCH_2NH(CH_2)_3AsO_3H_2$	$Cl(CH_2)_3AsO_3H_2 + PhCH_2NH_2$	HCl salt, m. 240° (dec.)	667
β-Camphorarsonic acid	β-camphorarsonous acid + Cl_2 in H_2O or H_2O_2	m. 210°	751
C_{11}			
$1\text{-}C_{10}H_7CH_2AsO_3H_2$	$1\text{-}C_{10}H_7CH_2Cl + As_2O_3$	Colorless needles, m. 142-144°	726
$2\text{-}C_{10}H_7CH_2AsO_3H_2$	$1\text{-}C_{10}H_7CH_2Br + As_2O_3$	Colorless needles, m. 159-161°	726
$Ph(CH_2)_2NH(CH_2)_3AsO_3H_2$	$Cl(CH_2)_3AsO_3H_2 + Ph_2CH_2CH_2NH_2$	HCl salt, m. 247-248°	667

13. Rxn. with glycols or α-hydroxycarboxylic acids in C_6H_6 with azeotropic distn. → spiroarsena derivs. (1097).

14. Redn. with Zn-Hg in HCl → $n\text{-}C_5H_{11}AsH_2$ (44).

AR ENEARSONIC ACIDS

The preparation of numerous arenearsonic acids is based primarily on the classical Bart method of coupling aryldiazonium salts with compounds of trivalent arsenic. Moreover, arenearsonic acids can also be prepared by oxidation of organic derivatives of trivalent arsenic, such as arsenoso compounds, primary arsines, arsenic-hetero-cycles with trivalent arsenic, and bis(diorganoarsenic) oxides. Aniline and phenol undergo direct arsonation with arsenic acid at elevated temperature.

In the Bart method aryldiazonium chlorides are reacted with alkali metal arsenites or metaarsenites in water or with arsenic trioxide in alkaline solution in the presence of copper salts or of copper or silver powder (Method I). According to the Schmidt modification of the Bart method, the Bart-Schmidt method, aryldiazonium salts are coupled with alkali metal arsenites in a neutral solution in the absence of catalysts (Method II).

Further modification of the method by Scheller, the Bart-Scheller method, brought a considerable improvement of yields. The diazotization of aromatic amines and coupling are carried out in one step; a calculated amount of aqueous sodium nitrite solution is added at a low temperature to a solution of an aromatic amine in alcohol containing sulfuric or hydrochloric acid or in acetic acid in the presence of arsenic trichloride and catalytic amounts of copper(I) chloride, the solvent is removed, after the nitrogen evolution is over, and the residue is treated with water and sodium hydrosulfite. (Method III). The method has been successfully applied for the preparation of arsonic acids from anilines substituted in the ring by nitro, carboxy, and sulfamido groups, particularly in the meta position.

According to the Doak modification of the Bart-Scheller method, the catalyst is added to the reaction mixture after completed diazotization of aromatic amines in alcoholic solution in the presence of arsenic trichloride (Method IV).

Since aryldiazonium fluoroborates and aryldiazonium double salts with metal halides, such as iron(III) and zinc chlorides, are more stable than aryldiazonium chlorides, they have been used for coupling with arsenic compounds. Thus arenearsonic acids are prepared by reacting aryldiazonium fluoroborates with sodium metaarsenite in aqueous alkali at room temperature in the presence of copper(I) chloride and completing the reaction at 65° (Method V). Coupling of aryldiazonium fluoroborates or of the metal halide double salts with arsenic trichloride in absolute alcohol in the presence of copper(I) chloride or bromide, followed by hydrolysis, yields arenearsonic acids along with different amounts of the corresponding diarylarsinic acids (Method VIa). When 80% alcohol is used instead of absolute alcohol, the ratio of the arsinic acids increases at the expense of the arsonic acids. Higher yields are obtained with copper(I) chloride than with the bromide (Method VIb).

Several arenearsonic acids were prepared by reacting arsenic acid, H_3AsO_4, with arylhydrazines. The reaction proceeds through intermediate reduction of arsenic acid to arsenious acid with simultaneous oxidation of the arylhydrazine to the aryldiazonium base which then interacts with the arsenious acid (Method VII).

Arenearsonic acids are also obtained by oxidation of aryldihaloarsines with hydrogen peroxide, or with chlorine in water, or in carbon tetrachloride. In the latter case the arylarsenic tetrachloride is then hydrolyzed by water. Similarly arsenoso-arenes, $(RAsO)n$, are oxidized by iodine in sodium bicarbonate solution to the corresponding arenearsonic acids (Method VIII).

Primary arylarsines, $RAsH_2$, are oxidized by atmospheric oxygen to arenearsonic acids.

Fused-ring, aromatic arsenic heterocycles are oxidized by nitric acid to arenearsonic acids, with cleavage of the As-ring.

Reaction of bis(diorganoarsenic) oxides with lead tetrachloride leads to organoarsenic tetrachlorides that can be converted into the corresponding arsonic acids by hydrolysis.

Direct arsonation of aniline and of phenol by heating with arsenic acid, H_3AsO_4, yields p-arsanilic acid and p-hydroxybenzenearsonic acid, respectively.

Arsanilic acid was also obtained in a yield of 42% by heating aniline arsenate, $PhNH_2 \cdot H_3AsO_4$, with aniline and diphenylurea for four hours at 155-160°, isolating the excess of aniline at 100° by alkalizing the reaction mixture, and precipitating the product by acidification.

Ring-substituted arenearsonic acids are prepared from ring-substituted aromatic amines by the methods described above or by treating arenearsonic acids with reagents such as halogen, nitric acid, and sulfonating agents. Arenearsonic acids substituted in the ring by functional groups, such as amino, nitro, hydroxy, sulfo, halo, cyano, and their homologs are readily converted into the corresponding arenearsonic acid derivatives by the known reactions of the particular functionally substituted aromatic compounds.

Reduction of arsonic acids to organodihaloarsines or arsenoso compounds and to arseno derivatives is the reaction common to all arsonic acids.

BENZENEARSONIC ACID $C_6H_7AsO_3$ $PhAsO_3H_2$

Prepn.: From aniline by the Bart method at 8° in the presence of $CuSO_4$, 80% yield along with a small amt. of $4-PhC_6H_4AsO_3H_2$ (58). A higher yield has been claimed using proper dilution of Na_3AsO_3 and maintaining the pH of the rxn. mixture constant (89).

By reacting arsenic acid with phenylhydrazine; rxn. mechansim elucidated (131).

By coupling PhN_2BF_4 with $NaAsO_2$ in aq. alkali in the presence of Cu_2Cl_2 at r. t., followed by heating to 65°, 58% yield (1091, 1092).

Props.: M. 154-156° (58); m. 158.5° (830); m. 158° (1092); $d_{calcd.}$ 1.830, $d_{exptl.}$ 1.760 (138). Monomeric structure by mol. wt. detn. (397). Dissocn. const. (1025; 1034). IR spectrum and H bonding (119a). UV spectrum (598). Crystal structure (138, 1129, 1130). Magnetic props. (1029). Polarography (830). Potentiometric titration (1236).

Rxns. with: SO_2 in HCl contg. KI → $PhAsCl_2$ (57) which after treatment with NaOH soln. and acidification to Congo red → $PhAs(OH)_2$ (1103).

SO_2 in H_2O at 85° → PhAsO (651).
$POCl_3$, followed by NaOH and HCl → PhAsO (981).
$NaHSO_3$ in the presence of Cl^-, I^-, or SO_4^{--} → PhAsO (58, 549).
H_2 in the presence of $CuSO_4$ → PhAsO (90).
AcCl + $AlCl_3$ → cleavage of the C-As bond (788).
$PhAsO(ONa)_2$ + $NaNH_2$ in piperidine → N-phenylpiperidine (126).
Metal ions → pptn. (quantitative analyses) (1013, 1025).
$PhNHNH_2$ → $(Ph_2As)_2O$ and Ph_3As (129).
HNO_3 → the m-nitro along with some o-nitro deriv. (398).
Ph_2SiCl_2 → $PhAs(O)(OSiPh_2Cl)_2$ (655a, 656).
$PhSiCl_3$ → $PhAs(O)(OSiPhCl_2)_2$ (655a, 656).
H_2SO_4 → $As(OH)_3$ (993).
AcCl → $CH_2(AsPhCl)_2$ + $PhAsCl_2$ (1024).
Glycols or α-hydroxycarboxylic acids in Ac_2O → spiroarsena derivs. (1097).
SF_4 → $PhAsF_4$ (1147a).
Uses: Reagent for pptn. of: Zr^{4+} (198); Bi (966); Pb (775); Nb and Ta (776); Zr^{4+} and Np^{4+} (1159); Pb (1284).
Stabilizer for motor fuels and lubricating oils contaminated with Cu (1217).
Turbidometric titration of Sn in Cu (196a).

TABLE XXX
HALOBENZENEARSONIC ACIDS

$RAsO_3H_2$	Prepd. from	Method	Yield %	Props.	Ref.
2-$BrC_6H_4AsO_3H_2$	RN_2BF_4	VIa	46	m. 173-175°	396
		VIb	29	Mg salt insol.	396
4-$BrC_6H_4AsO_3H_2$	RNH_2	I	80	m. > 300°	93, 480
		I	71	Rxns.[1,2]; salts[3]	850
	$RAsCl_2$ + H_2O_2			Rxns.[4,5]	1336
				Ionization const.	1034
	$(RN_2)_2ZnCl_4$	VIa	35		394
	RN_2BF_4	VIa	35	Hemihydrate, m. > 300°	396
2-$ClC_6H_4AsO_3H_2$	RN_2BF_4	VIb	19	Rxns.[2,6]; salts[7]	295
	RN_2BF_4	VIa	45	Use[9]	295
	RN_2BF_4	V	52	m. 182°	1091-2
				m. 180°	550
3-$ClC_6H_4AsO_3H_2$	RN_2BF_4	VIa	62	Rxn.[6], salts[8]	295
	RN_2BF_4	VIb	39		295

1. Rxn. with FCl_3 in AcOH → $RAsCl_2$ (480).
2. Rxn. with SO_2 in HCl contg. KI, followed by H_2O → RAsO (283, 550).
3. Salts with methylglucamine, ethylenediamine (diarsonate), and ethanolamine (831).
4. Rxn. with SO_2 in HX (X = halogen) contg. KI → $RAsX_2$ (99).
5. Rxn. with trihydrazino-s-triazine → 4-(4,6-dihydrazino-s-triazin-2-ylhydrazino)-benzenearsonic acid (418).
6. Rxns. with metal ions → pptn. (quant. analysis) (1020).
7. Salts: with methanolamine, m. 154-155° (831), with ethylenediamine, m. 217-218° (831).
8. Salts: with ethanolamine, m. 132-133° (831), with methylglucamine, m. 182-183° (831); with dodecylamine (1).

TABLE XXX (cont'd.)
HALOBENZENEARSONIC ACIDS

$RAsO_3H_2$	Prepd. from	Method	Yield %	Props.	Ref.
$4\text{-}ClC_6H_4AsO_3H_2$	RNH_2	I	79	White needles, m. > 300°	550, 850
	$RAsCl_2 + Cl_2$ in H_2O	VIII	99		896
	RN_2BF_4	V	63		1091, 1092
	$RAsH_2 + atm.\ O$			m. 383-385° (dec.) with	242
	RN_2BF_4	VIa	62	loss of H_2O	295
		Vb	23	Rxns.[2,6,10,11],salts[12]	295
				Ionization const.	1034
$2\text{-}IC_6H_4AsO_3H_2$				Rxn.[2]	
$4\text{-}IC_6H_4AsO_3H_2$	$4\text{-}IC_6H_4NH_2$	I	13		850

9. Wood preservative (241).
10. Rxn. with $HNO_3\text{-}H_2SO_4$ or alkali nitrate in $H_2SO_4 \rightarrow 4,3\text{-}Cl(O_2N)C_6H_3AsO_3H_2$ (594, 896).
11. Rxn. with Zn + HCl in MeOH in the presence of $HgCl_2 \rightarrow 4\text{-}ClC_6H_4AsH_2$ (242).
12. Salts with: dodecyl-, tetradecyl-, hexadecyl-, octadecyl-, and undecenyl-amine (1); with methylglucamine, m. 131-132° (831); with methanolamine, m. 135-136° (831); and with ethylenediamine, m. 218-129° (831).

TABLE XXXI
NITROSO- AND NITROBENZENEARSONIC ACIDS

$RAsO_3H_2$	Prepd. from	Method	Yield %	Props. and Remarks	Ref.
$4\text{-}ONC_6H_4AsO_3H_2$	$4\text{-}H_2NC_6H_4AsO_3HNa + H_2SO_5$			Rxn. with $RNH_2 \rightarrow RN:NC_6\text{-}H_4AsO_3H_2$	984
$2\text{-}O_2NC_6H_4AsO_3\text{-}H_2$	RNH_2	I		m. 194-196°	820
	$PhAsO_3H_2 + 100\%$ HNO_3			By-prod.; m. 225-229°, UV spectrum	398
	RNH_2	I	85	Prepd. at pH 7-9	551, 1088
	RN_2BF_4	V	67	m. 232-235° (dec.)	1091, 1092
				Light yellow needles, m. 235°, rxns.[1,2]; salts[4] other data[5]	551
$3\text{-}O_2NC_6H_4AsO_3\text{-}H_2$	$PhAsO_3H_2 + HNO_3$			m. 177-180°, m. 196-200°, resp.	398, 551

1. Rxn. with SO_2 in dild. HCl contg. KI, followed by hydrolysis $\rightarrow RAs(OH)_2$ (283, 551) or $(RAsO)n$ (283, 865).
2. Redn. with $FeCl_2$ in alk. soln. $\rightarrow$ the amino deriv. (820).
3. Redn. with Fe in aq. NaCl $\rightarrow$ the amino-arsonic acid and some diazobenzene-arsonic acid (1089).

TABLE XXIX (cont'd.)
NITROSO- AND NITROBENZENEARSONIC ACIDS

$RAsO_3H_2$	Prepd. from	Method	Yield	Props. and Remarks	Ref.
$3-O_2NC_6H_4AsO_3-H_2$	RNH_2	I	28		584, 883
	RNH_2	IV	54	Rxns. [6,7,8,9,10]	282
	RN_2BF_4	VIa	42	Other props.[11]	295
	RN_2BF_4	VIb	21		295
	RN_2BF_4	V	47	m. 182°	1091, 1092
$4-O_2NC_6H_4AsO_3-H_2$	RN_2BF_4	VIa	54		295
	RN_2BF_4	VIb	19	Rxns.[1,9,13]	295
	$(RN_2)_2ZnCl_4$	VIa	21	Other props.[12,13]	394
	RNH_2	III	45		394
	RNH_2	I	62		883, 986
	RN_2BF_4	V	79	Yellow crystals, decomp. 298-300°	1092
	$RN_2Cl.FeCl_3$ + PhAs-Cl_2 + NaI			By-prod.; decomp. 300°	908

4. Dodecylamine salt (1).
5. Polarography (1267a). Dissocn. const. (1034). Use as wood preservative (241).
6. Rxn. with SO_2 in dild. HCl contg. KI → $RAsCl_2$ (93, 283, 584).
7. Catalytic redn. over Raney Ni → the amino-arsonic acid (250).
8. Rxn. with Fe in aq. NaCl under reflux → the amino-arsonic acid (811).
9. Rxn. with heavy metals → ppt. (quant. analysis) (986, 1010, 1025).
10. Rxn. with As_2O_3 in aq. NaOH → $O(NC_6H_4AsO_3H_2)_2$ (865).
11. Dissocn. const. (1034). Use as wood preservative (241).
12. Dissocn.. const. (124, 1025, 1034).
13. Rxn. with $ClP(S)(OEt_2)$ → $RAs(O)[OP(S)(OEt)_2]_2$ (1102).

TABLE XXXI

BENZENEARSONIC ACIDS CONTAINING SUBSTITUENTS LINKED BY ALIPHATIC CARBON ATOMS

$RAsO_3H_2$	Prepd. from	Method*	Yield %	Props. and Remarks	Ref.
C₇					
4-$ClCOC_6H_4AsO_3H_2$	$RAsCl_2$ +·Cl_2 in CCl_4	VIII		m. 137-138°	1164
3-$F_3CC_6H_4AsO_3H_2$	RNH_2	I	51	Di-Na.H_2O salt, rxns. 1-4	475
4-$NCC_6H_4AsO_3H_2$	4-$H_2NC_6H_4AsO_3H_2$ + $NaNO_2$, followed by KCN + $NiCl_2$ in H_2O	I	12	Other props.5, using $CuSO_4$ instead of $NiCl_2$	284
			82		737
2-$HO_2CC_6H_4AsO_3H_2$	RNH_2	I		Shiny creamy plates	54
		I		White crystals	551
	1-chloroarsindole + HNO_3				262
	RN_2BF_4	V	65	Rxns. 6,7	1091, 1092
3-$HO_2CC_6H_4AsO_3H_2$	RNH_2	IV	76	Rxn.7	282
4-$HO_2CC_6H_4AsO_3H_2$	RNH_2	I		Rxns. 6,7,8, other props.9	551
	RNH_2	I			179
	RNH_2BF_4	V	67	Decomp. 232°	1091, 1092
				Shiny leaflets, m. 283° (dec.)	551
4-$ClCH_2C_6H_4AsO_3H_2$	$RAsCl_2$ + H_2O_2				1128
4-$N_2NC(:NH)C_6H_4AsO_3H_2$	4-$NCC_6H_4AsO_3H_2$ + alc. HCl at 0° + alc. NH_3 at 60°		62	Small scales, HCl salt, m. 280°, rxn.2	737
4-$H_2NNHCOC_6H_4AsO_3H_2$	4-$EtO_2CC_6H_4AsO_3H_2$ + H_2NNH_2		41	m. 159-160°	999, 1091, 1092
2-$MeC_6H_4AsO_3H_2$	RN_2BF_4	V	63		
3-$MeC_6H_4AsO_3H_2$	RNH_2	I	50	Rxns. 10,11, other props.9,12	549, 1091, 1092
	RN_2BF_4	V	54	m. 150°; other props.9	
4-$MeC_6H_4AsO_3H_2$	RN_2BF_4	V	73	Decomp. 300°	1091, 1092
4-$HOCH_2C_6H_4AsO_3H_2$	RNH_2		82	Rxns. 6,10,11; other props.13-5	549
	$(MeRAs)_2O$ + $PbCl_4$			m. 355°	144
4-$H_2NCH_2C_6H_4AsO_3H_2$	RNH_2	III		m. 165-171° (dec.), rxn.6,11	1128
	4-$AcNHCH_2C_6H_4AsO_3H_2$ by hydrolysis		47	Rxns. 11, 16-18	284, 389

* See pages 187-8.

Compound	Preparation			Properties	Refs.
C_8					
$3\text{-}ClCH_2COC_6H_4AsO_3H_2$	$3\text{-}AcC_6H_4AsO_3H_2 + Cl_2$ in AcOH in UV light		38	m. 202-203° (dec.), rxn.[6]	189
$4\text{-}ClCH_2COC_6H_4AsO_3H_2$	$4\text{-}AcC_6H_4AsO_3H_2 + Cl_2$ in EtOH + H_2O_2		76	m. 204-205° and 189-190°, resp., rxns. & derivs[6]. [19-22]	231, 1118
$4\text{-}NCCH_2C_6H_4AsO_3H_2$	RNH_2	III	7	m. > 280°, rxn.[23]	1128
$4\text{-}H_2NCONHCOC_6H_4AsO_3H_2$	RNH_2	I	17	Amorph., m. 326.5°, rxn.[11]	1164
$2\text{-}CH_2\!:\!CHC_6H_4AsO_3H_2$	$2\text{-}(HO_2CCHBrCH_2)C_6H_4AsO_3H_2 + 5\%\ Na_2CO_3$			Rxn.[8]	264
$(4\text{-}CH_2\text{-}CHC_6H_4AsO_3H_2)n$	Poly-4-aminostyrene	I		Resin contg. 14-16% As, absorbs cations at pH 3 Rxn.[24]	974
$3\text{-}AcC_6H_4AsO_3H_2$	RN_2BF_4	V	70	m. 175°, rxns.[6,25]	189, 1091, 1092
$4\text{-}AcC_6H_4AsO_3H_2$					

1. Rxn. with alc. HCl, followed by 10% alc. $NH_3 \rightarrow H_2NC(:NH)C_6H_4AsO_3H_2$ (737).
2. Rxn. with $NaHSO_3$ in alk. soln. contg. $MgCl_2 \rightarrow (RAs:)n$ (737).
3. Rxn. with SO_2 in dild. H_2SO_4 contg. $KI \rightarrow RAs(OH)_2$(737).
4. Rxn. with $NaHSO_3$ in 5% NaOH, followed by HCl gas in alc. and by alc. $NH_3 \rightarrow H_2NC(:NH)C_6H_4As(OH)_2$ (977).
5. Magnetic props. (1029).
6. Rxn. with SO_2 in $HCl \rightarrow RAsCl_2$ (54, 179, 189, 264, 393, 1118, 1128, 1188).
7. Rxn. with heavy metals → pptn. (quant. analysis) (1021).
8. Rxn. with SO_2 in HCl, followed by neutralization to Congo $\rightarrow RAs(OH)_2$ (551, 1003).
9. Ionization consts. (1034).
10. Rxn. with $POCl_3$, followed by alk. and neutralization $\rightarrow (RAsO)n$ (657, 981).
11. Rxn. with SO_2 in HCl contg. KI, followed by neutralization $\rightarrow (RAsO)n$ (123, 231, 248, 284, 389, 391, 393, 545, 1129, 1164).
12. Forms insol. Mg salt (398).
13. Dissocn. consts. (424, 1025, 1034).
14. Redn. potential (124).
15. Magnetic props. (873, 1029).
16. Rxns. with $RSO_2Cl \rightarrow$ the RSO_2NH- derivs. (389).
17. Rxn. with RCl → the RNH- derivs. (390).
18. Rxn. with $CH_2\!:\!CHCN \rightarrow$ the $NCCH_2CH_2NH$- deriv. (390).
19. Rxns. with RNH_2 and $RRNH \rightarrow$ the $RNHCH_2CO$- and $RRNHCH_2CO$- derivs., resp. (231).
20. Rxn. with $PhNHNH_2 \rightarrow CH_2N(Ph)N\!:\!CC_6H_4AsO_3H_2$ (231).
21. Rxn. with $HCO_2K + H_2O$ in MeOH → hydrolysis of the Cl atom (231).
22. Semicarbazone darkens at 210°; oxime, m. 173-173.5° (231).
23. Rxn. with SO_2 in cold HCl contg. I $\rightarrow 4\text{-}NCCH_2C_6H_4AsCl_2$; if the redn. is carried out at r. t.. 4-$H_2NCOCH_2C_6H_4AsCl_2$ is formed. (189).
24. Rxn. with Cl_2 in AcOH in UV light $\rightarrow 3\text{-}ClCH_2COC_6H_4AsO_3H_2$ (189).

193

TABLE XXXI (cont'd.)

BENZENEARSONIC ACIDS CONTAINING SUBSTITUENTS LINKED BY ALIPHATIC CARBON ATOMS

$RAsO_3H_2$	Prepd. from	Method	Yield %	Props. and Remarks	Ref.
4-$HO_2CCH_2C_6H_4AsO_3H_2$	RNH_2	III	35	Hexagonal plates, rxn.[26]	285
4-$HOCH_2COC_6H_4AsO_3H_2$	4-$BrCH_2COC_6H_4AsO_3H_2$ + H_2O under reflux		81	Colorless prisms contg. 1 mol. H_2O, rxns.[27,28]	1278
	4-$ClCH_2COC_6H_4AsO_3H_2$ + HCO_2-K in MeOH + H_2O			Yellow substance, rxn.[11]	231
4-$HO_2CCH(HO)C_6H_4AsO_3H_2$	RNH_2	I	43	Forms di-Na salt, rxn.[6]	393
4-$H_2NOCH_2C_6H_4AsO_3H_2$	4-$MeO_2CCH_2C_6H_4AsO_3H_2$ + NH_4-OH			Prisms	1268
4-$MeO_2CCH_2C_6H_4AsO_3H_2$	$RAsCl_2$ + H_2O_2			m. >280°	1128
C₉					
4-$NCCHClCH_2C_6H_4AsO_3H_2$	4-(H_2O_3As)$C_6H_4N_2Cl$ + CH_2:CHCN in Me_2CO		50	m. >300°, rxn.[11]	657
4-$NCCH_2NHOCC_6H_4AsO_3H_2$	RAsO + I in $NaHCO_3$		100	Rectangular prisms, m. 251-252° (dec.)	288
2-HO_2CCH:$CHC_6H_4AsO_3H_2$	2-EtO_2CCH:$CHC_6H_4N_4NH_2$	I		m. 205-206°, rxn.[29]	264
3-HO_2CCH:$CHC_6H_4AsO_3H_2$	RNH_2	III	44	m. >360°, rxn.[11]	393
4-HO_2CCH:$CHC_6H_4AsO_3H_2$	RNH_2	III	30	Needles, rxn.[30]	285
2-$HO_2CCHBrCH_2C_6H_4AsO_3H_2$	2-HO_2CCH:$CHC_6H_4AsO_3H_2$ + HBr	I	17	Yellow, m. 185°, rxn.[31]	264
4-$H_2NCO(NHCO)_2C_6H_4AsO_3H_2$	RNH_2	I	5	Rxn.[11]	1164
4-$H_2NOCH_2NHCOC_6H_4AsO_3H_2$	4-$HO_2CCH_2C_6H_4AsO_3H_2$ + HCl			m. 211-213° (dec.) rxn.[11]	1264
4-$MeO_2CCH_2C_6H_4AsO_3H_2$	4-$MeO_2CCH_2C_6H_4AsO_3H_2$ + HCl in MeOH			Na salt, needles; mono-hydrate, leaflets	1268
4-$EtO_2CC_6H_4AsO_3H_2$	RN_2BF_4	V	60	m. 260°	1091, 1092
4-$HO_2CCH_2CH_2C_6H_4AsO_3H_2$	RNH_2				1269
	4-HO_2CCH:$CHC_6H_4AsO_3H_2$ + H_2/Ni	I	69	Prisms, rxn.[26]	285
4-$MeNHCOCH_2C_6H_4AsO_3H_2$	4-$HO_2CCH_2C_6H_4AsO_3H_2$ + $MeNH_2$			Pale yellow, hexagonal prisms	1268
4-$H_2NCO(CH_2)_2C_6H_4AsO_3H_2$	4-$MeO_2C(CH_2)_2C_6H_4AsO_3H_2$ + NH_4OH			Hexagonal leaflets	1269
4-$AcNHCH_2C_6H_4AsO_3H_2$	RNH_2	III	62	m. >300°, rxns.[11,32]	284
	RNH_2	II			389

Product	Preparation (reactants)		Yield %	Properties	References
$4\text{-}HO_2CCH_2NHCH_2C_6H_4AsO_3H_2$	$4\text{-}H_2NCH_2C_6H_4AsO_3H_2 + Cl\text{-}CH_2CO_2H$			White powder, m. 142-143°	390
$4\text{-}HO_2CCH(NH_2{\cdot}HCl)CH_2C_6H_4AsO_3H_2 + HCl$	$4\text{-}HO_2CCH(NHBz)CH_2C_6H_4AsO_3H_2 + HCl$		87	m. 265° (dec.)	123
$4\text{-}H_2NCOCH_2NHCH_2C_6H_4AsO_3H_2$	$4\text{-}H_2NCH_2C_6H_4AsO_3H_2 + ClCH_2CONH_2$; RNH_2	I		Lustrous colorless plates, m. 236°, rxn.11, salts[33]; m. 238-238.5° (dec.), rxn. 11	390, 391, 1164
$4\text{-}HOCH_2CH_2NHCONHC_6H_4AsO_3H_2$	RNH_2				390
$4\text{-}HOCH_2CH_2NHCH_2C_6H_4AsO_3H_2$	$4\text{-}H_2NCH_2C_6H_4AsO_3H_2 + CH_2CH_2O$	I	50	White powder, m. 251-2°	390
C₁₀					
$4\text{-}HO_2CCH_2CH_2COC_6H_4AsO_3H_2$				Rxn.[34]	1270
$4\text{-}NCCH_2CH_2NHCH_2C_6H_4AsO_3H_2$	$4\text{-}H_2NCH_2C_6H_4AsO_3H_2 + CH_2{:}CHCN$	I		White needles, m. >300°, rxn.[35]	390
$4\text{-}MeO_2C(CH_2)_2C_6H_4AsO_3H_2$	$4\text{-}HO_2C(CH_2)_2C_6H_4AsO_3H_2 + HCl$ in MeOH ; RNH_2			Prisms ; Rxn.[36]	1269
$4\text{-}HO_2C(CH_2)_3C_6H_4AsO_3H_2$	$4\text{-}MeO_2C(CH_2)_2C_6H_4AsO_3H_2 + MeNH_2$; RNH_2	I	75	m. 125.5-126.5° ; Octagonal leaflets	285, 1269
$4\text{-}MeNHCO(CH_2)_2C_6H_4AsO_3H_2$	RNH_2				
$4\text{-}EtCH(CO_2H)C_6H_4AsO_3H_2$	$4\text{-}MeO_2CCH_2C_6H_4AsO_3H_2 + Me_2NH$	I	25	m. 148-149°, rxn. 11 ; Leaflets	393, 1268
$4\text{-}Me_2NOCCH_2C_6H_4AsO_3H_2$	$4\text{-}MeO_2CCH_2C_6H_4AsO_3H_2 + EtNH_2$			Prisms	1268
$4\text{-}EtNHOCCH_2C_6H_4AsO_3H_2$	$4\text{-}NCCH_2CH_2NHCH_2C_6H_4AsO_3H_2 + HCl$; $4\text{-}H_2NOCCH_2CH_2NHCH_2C_6H_4AsO_3H_2 + HCl$			White flakes, m. 235-236°	390
$4\text{-}HO_2CCH_2CH_2NHCH_2C_6H_4AsO_3H_2$	$4\text{-}NCCH_2CH_2NHCH_2C_6H_4AsO_3H_2$			White needles, m. >300°, rxn.[37]	390

25. Rxn. with SO_2 in $H_2O \rightarrow (RAsO)n$ (118).
26. Esterification with MeOH in the presence of HCl → the $MeO_2C\text{-}$ deriv (1268, 1269).
27. Rxn. with $RCHO + NH_4OH$ in the presence of CuOAc → $HNCR{:}NCH{:}CC_6H_4AsO_3H_2$ (1278).
28. Phenylhydrazone, yellow needles, m. >400° (1278).
29. Rxn. with HBr → $2\text{-}HO_2CCHBrCH_2C_6H_4AsO_3H_2$ (264).
30. Catalytic redn. over Raney Ni → reduction of the double bond (285).
31. Rxn. with ½ $Na_2CO_3 \rightarrow 2\text{-}CH_2{:}CHC_6H_4AsO_3H_2$ (264).
32. Rxn. with ½ HCl or with 6N NaOH → hydrolysis of the AcNH group (285, 299, 389).
33. Mono-Na salt, white crystals, m. >300°; di-Na salt (390). HCl salt, colorless needles, m. 266° (390).

TABLE XXXI (cont'd.)

BENZENEARSONIC ACIDS CONTAINING SUBSTITUENTS LINKED BY ALIPHATIC CARBON ATOMS

RAsO$_3$H$_2$	Prepd. from	Method	Yield %	Props. and Remarks	Ref.
4-Cl$_2$CHCONHCH(CH$_2$OH)CH(OH)-C$_6$H$_4$AsO$_3$H$_2$	RNH$_2$	I	16	Occurs in the l- and d-threo form, [α]$^{22}_{5?}$-7.8°; [α]$^{22}_D$ 8.6°	1126
4-HO$_2$CCH(NHAc)CH$_2$C$_6$H$_4$AsO$_3$H$_2$	RNH$_2$	I	52	Grayish-white solid, m. 242-245°, rxn.11,38	123
4-HO$_2$C(CH$_2$)$_4$C$_6$H$_4$AsO$_3$H$_2$	RNH$_2$	I	62	m. 204°, rxn.11	393
4-HO$_2$CCH$_2$CMe$_2$C$_6$H$_4$AsO$_3$H$_2$	RNH$_2$	I	75	Rxn.11	248
4-PrNHCOCH$_2$C$_6$H$_4$AsO$_3$H$_2$	4-MeO$_2$CCH$_2$C$_6$H$_4$AsO$_3$H$_2$ + PrNH$_2$			Needles; Na salt, rhombic plates	1268
4-Me$_2$NOC(CH$_2$)$_2$C$_6$H$_4$AsO$_3$H$_2$	4-MeO$_2$C(CH$_2$)$_2$C$_6$H$_4$AsO$_3$H$_2$ + Me$_2$NH			Hexagonal tablets	1269
4-EtNHOC(CH$_2$)$_2$C$_6$H$_4$AsO$_3$H$_2$	4-MeO$_2$C(CH$_2$)$_2$C$_6$H$_4$AsO$_3$H$_2$ + EtNH$_2$			Octagonal leaflets	1269
C$_{12}$					
$\overline{SCH:CHCH:CCH:CH}C_6H_4AsO_3H_2$	4-H$_2$O$_3$AsC$_6$H$_4$N$_2$Cl + $\overline{SCH:CH-CH:CCH:CH}CO_2$H		30	m. > 300°	402
$\overline{OCH:CHCH:CCH:CH}C_6H_4AsO_3H_2$	4-H$_2$O$_3$AsC$_6$H$_4$N$_2$Cl + $\overline{OCH:CH-CH:CCH:CH}CO_2$H				400, 404
4-[O(CH$_2$CH$_2$)$_2$NCH$_2$CO]C$_6$H$_4$AsO$_3$H$_2$	4-ClCH$_2$COC$_6$H$_4$AsO$_3$H$_2$ + morpholine	I	63	HCl.salt, m. 172-173°	231
4-HO$_2$C(CH$_2$)$_5$C$_6$H$_4$AsO$_3$H$_2$	RNH$_2$		38	m. 146-150°, rxn.11	393
4-Et$_2$NCH$_2$COC$_6$H$_4$AsO$_3$H$_2$	4-ClCH$_2$COC$_6$H$_4$AsO$_3$H$_2$ + Et$_2$-NH		33	HCl salt, m. 186-187	231
4-PrNHCO(CH$_2$)$_2$C$_6$H$_4$AsO$_3$H$_2$	4-MeO$_2$C(CH$_2$)$_2$C$_6$H$_4$AsO$_3$H$_2$ + PrNH$_2$			Needles	1269
C$_{13}$ 4,4'-CO(C$_6$H$_4$AsO$_3$H$_2$)$_2$	Cl$_2$C:C(C$_6$H$_4$AsO$_3$H$_2$)$_2$ + CrO$_3$ in acid CO(C$_6$H$_4$NH$_2$)$_2$ CH$_2$(C$_6$H$_4$AsO$_3$H$_2$)$_2$+ CrO$_3$ or KMnO$_4$	I		Small white needles, soly. data, rxn.39	963, 1069, 1333, 1333
2-PhCH$_2$C$_6$H$_4$AsO$_3$H$_2$	RNH$_2$	I	48	Colorless crystals, m. 155-156°, rxn.40	537
	RNH$_2$	IV	74		537

Compound	Preparation		Yield (%)	Properties	Ref.
4,4'-CH₂(C₆H₄AsO₃H₂)₂	CH₂(C₆H₄NH₂)₂	I		Square-based needles, de-comp. 250°, soly. data, basicity; rxn.39,41, 42	1095, 1332
4-(4-H₂NC₆H₄SO₂NHCH₂)C₆H₄-AsO₃H₂	4-AcNHC₆H₄SO₂NHCH₂C₆H₄AsO₃-H₂ + alc. HCl			Decomp. 307-308°, rxn.43	389
4-[CH₂(CH₂)₄NCOCH₂]C₆H₄AsO₃-H₂	4-MeO₂CCH₂C₆H₄AsO₃H₂ + piperidine			Cream, hexagonal prisms	1268
4-[CH₂(CH₂)₄NCH₂CO]C₆H₄As-O₃H₂	4-ClCH₂COC₆H₄AsO₃H₂ + piperidine		61	HCl salt, m. 186-187°	231
C₁₄ 4,4'-(Cl₂C:C)C₆H₄AsO₃H₂)₂	Cl₃CCH(C₆H₄AsO₃H₂)₂ + alc. KOH	I		Nacreous blades, rxn.42	963
4-[4,3-HO(O₂N)C₆H₃CH:CH]-C₆H₄AsO₃H₂	Cl₂C:C(C₆H₄NH₂)₂ 4-H₂O₃AsC₆H₄N₂Cl + 3,4-HO-(O₂N)C₆H₃CH:CHCO₂H		2	Yellow ppt.	1096, 401
3-(PhCH:CH)C₆H₄AsO₃H₂	3-H₂O₃AsC₆H₄N₂Cl + PhCH:CH-CO₂H + CuCl₂			Pale yellow plates, m. 300°	403, 404
4-(PhCH:CH)C₆H₄AsO₃H₂	4-H₂O₃AsC₆H₄N₂Cl + PhCH:CH-CO₂H + CuCl₂		35	Buff, m.>300°	399, 400, 404
4-(4-HOC₆H₄CH:CH)C₆H₄AsO₃H₂	4-H₂O₃AsC₆H₄N₂Cl + 4-HOC₆-H₄CH:CHCO₂H + CuCl₂	I		Pale yellow, m. > 300°	399, 400
3,3'-Cl₃CCH:(C₆H₄AsO₃H₂)₂	Cl₃CCH:(C₆H₄NH₂)₂			Cream-colored prisms, rxn.44	963
4-(4-H₂NC₆H₄CH:CH)C₆H₄AsO₃H₂	AcNH-deriv. + acid		80	Yellowish crystals, salts45	1096, 299
4-(PhNHCOCH₂)C₆H₄AsO₃H₂	4-MeO₂CCH₂C₆H₄AsO₃H₂ + PhNH₂			Needles	1268

34. Rxn. with N₂H₄·H₂O → the pyridazone deriv. (1270).
35. Rxn. with concd. HCl at r. t., followed by NaOAc → the H₂NOC-deriv. (390).
36. Rxn. with NH₄OH or amines → the corresp. carbamido deriv. (1269).
37. Rxn. with concd. HCl, followed by H₂O + NaOAc → the carboxy deriv. (390).
38. Rxn. with Na₂S₂O₄ + MgCl₂ in aq. alkali → the arseno deriv. (123, 389).
39. Redn. with H₃PO₂ → the arseno deriv. (1095, 1332, 1333).
40. Rxn. with H₂SO₄ → acridarsinic acid (537).
41. Rxn. with HNO₃-H₂SO₄ → the 3,3'-dinitro deriv. (1095, 1332).
42. Rxn. with CrO₃ or KMnO₄ → CO(C₆H₄AsO₃H₂)₂ (963, 1333).
43. Redn. with PhNHNH₂ in MeOH → the arsenoso deriv. (389).
44. Rx. with alc. KOH → Cl₂C:C((C₆H₄AsO₃H₂)₂ (963).
45. Hydrochloride, colorless needles; monoethanolamine salt (299).

TABLE XXXI (cont'd.)

BENZENEARSONIC ACIDS CONTAINING SUBSTITUENTS LINKED BY ALIPHATIC CARBON ATOMS

$RAsO_3H_2$	Prepd. from	Method	Yield	Props. and Remarks	Ref.
$4-[CH_2(CH_2)_4NCOCH_2CH_2]C_6H_4$ AsO_3H_2	$4-MeO_2C(CH_2)_2C_6H_4AsO_3H_2$ + piperidine				1269
C_{15}					
$4-[PhCH:C(CN)]C_6H_4AsO_3H_2$	$4-H_2O_3AsC_6H_4N_2Cl$ + PhCH:-CHCN + $CuCl_2$		20	Pale yellow, m. $> 300°$	399 400 404
$4-(4-MeC_6H_4CH:CH)C_6H_4AsO_3-$ H_2	$4-H_2O_3AsC_6H_4N_2Cl$ + 4-Me-$C_6H_4CH:CHCO_2H$ + $CuCl_2$		30	Pale yellow, m. $> 300°$	399
$4-(PhMeCHN:CH)C_6H_4AsO_3H_2$	$4-OCHC_6H_4AsO_3H_2$ + PhMeCH-NH_2			m. 79°, optical rotation data	1202
$4-[PhNHOC(CH_2)_2]C_6H_4AsO_3H_2$	$4-MeO_2C(CH_2)_2C_6H_4AsO_3H_2$ + $PhNH_2$			Hexagonal needles	1269
$4-(4-AcNHC_6H_4SO_2NHCH_2)C_6-$ $H_4AsO_3H_2$	$4-H_2NCH_2C_6H_4AsO_3H_2$ + $AcNHC_6H_4SO_2Cl$			m. $> 300°$ (dec.), rxn.38,43	389
C_{16}					
$4-[4-MeOC_6H_4CH:C(CO_2H)]-$ $C_6H_4AsO_3H_2$	$4-H_2O_3AsC_6H_4N_2Cl$ + 4 MeO-$C_6H_4CH:CHCO_2H$ + $CuCl_2$		20	m. $> 300°$	399
$4-(4-AcNHC_6H_4CH:CH)C_6H_4As-$ O_3H_2	$4-H_2O_3AsC_6H_4N_2Cl$ + 4-AcN-$HC_6H_4CH:CHCO_2H$ + $CuCl_2$		25	Rxn.32	299
$4-[HO_2CCH(NHBz)CH_2]C_6H_4As-$ O_3H_2	RNH_2	I		Shiny needles, m. 258 260° rxn. 11,38,46	123
$4-[PhCONHCH(CH_2OH)CH(OH)-]$ $C_6H_4AsO_3H_2$	RNH_2	I	22	l - threo-, $[\alpha]_D^{22}$ 89°, decomp. 142-143°	1125

46. Acid hydrolysis → 4-arsonophenylalanine hydrochloride (123).

2-AMINOBENZENEARSONIC ACID, o-ARSANILIC ACID $C_6H_8AsNO_3$ $2-H_2NC_6H_4AsO_3H_2$

repn.: By reducing $2-O_2NC_6H_4AsO_3H_2$ with Fe contg. $FeCl_2$ in H_2O, 70% yield (250, 20), or with Fe filings in aq. NaCl under reflux, 83% yield, along with $(2-H_2O_3-sC_6H_4N:)_2$ (1089).

rops.: M. 144° (830); m. 152° (1089). Stable in alkali at 160-180° (964). Polargraphy (830). Ionization const. (1034).

xns. with: $NaNO_2$, followed by $AcCH_2CO_2Et$ → $2-[EtO_2CC(Ac):NNH]C_6H_4AsO_3H_2$ (240).
aNO_2, followed by Na_3AsO_3 → $o-C_6H_4(AsO_3H_2)_2$ (327).
aNO_2, followed by 2-naphthol-3,6-disulfonic acid → hydroxydisulfonaphthyldiazoenzenearsonic acid (820).
N_2Cl in alk. soln. → $2,?-H_2N(RN_2)C_6H_3AsO_3H_2$ (222)
n + HCl in the presence of $HgCl_2$ → $2-H_2NC_6H_4AsH_2$ (242).
$_2NC(:NH)NHCN$ → $2-[H_2NC(:NH)NHC(:NH)NH]C_6H_4AsO_3H_2$ (1101).
aloheterocyclic compds. → N-heterocyclylarsanilic acid (250, 332).
Furaldehyde → $OCH:CHCH:CCH:NC_6H_4AsO_3H_2$ and $H_2O_3AsC_6H_4NHCH:CHCH:C(OH)CH:NC_6H_4As-O_3H_2$ (720).
,4,5-Triamino-6-hydroxypyrimidine + 1,1,3-tribromopropane → N-[(2-amino-4-hydroxy-6-pteridyl)methyl]arsanilic acid (452).
,2-Naphthoquinone-4,8-disulfonic acid → the 4-(2 arsonoanilino) deriv. (1247).
$_2NCN + HCl$ → $H_2NC(:NH)NHC_6H_4AsO_3H_2$ (1101).
se: Quant. detn. of heavy metals (219, 1009).

TABLE XXXII
N-SUBSTITUTED o-ARSANILIC ACIDS

2-(RNH)C₆H₄AsO₃H₂ in which R is:	Prepd. by reacting 2-H₂NC₆H₄AsO₃H₂ with	Yield %	Props. and Remarks	Ref.
7 $_2NC(:NH)-$	H_2NCN		Darkens at 280°	1101
8 $_2NC(:NH)NHCONH-$	Hydrolysis of H₂NC-(:NH)NHC(:NH)NH-		m. 210°	1101
$_2NC(:NH)NHC(:NH)NH-$	$H_2NC(:NH)NHCN$		colorless needles, m. 260°	1101
11 CH:CHCH:CCH:	Furfuraldehyde in aq. soln.[1]		m. 166-167°	720
H:CHC(NO₂):CHN:C-	2 chloro 5 nitropyridine	52	m. 236-237°, rxn.[2]	250
H:CHC(NH₂):CHN:C-	Rxn.[2]	29	m. 230-231° (dec.)	250

. In concd. solns. $H_2O_3AsC_6H_4NHCH:CHCH:C(OH)CH:NC_3H_4AsO_3H_2$ is formed as by prod. (720).
. Catalytic redn. over Raney Ni → the 5-amino-2-pyridyl deriv. (250).

TABLE XXXII (cont'd.)
N-SUBSTITUTED o-ARSANILIC ACDS

2-(RNH·)$C_6H_4AsO_3H_2$ in which R is:	Prepd. by reacting 2-$H_2NC_6H_4AsO_3H_2$ with	Yield %	Props. and Remarks	Ref.
C_{13} 2-HOC_6H_4CH=	Salicyl aldehyde		Colorimetric reagent for Se, Ti(III), and Fe(III)	754b
2-Amino-4-hydroxy-6-pteridylmethyl	2,4,5-Triamino-6-hydroxypyrimidine. H_2SO_4 + Br_2CHCOCH$_2$-Br at pH 2		UV spectrum	15, 452
C_{15} 5-Nitro-1-isoquinolinyl	5-Nitro-1-chloroisoquinoline in H_2O + NaOH	13	Decomp. 242-245°	332
1-Isoquinolyl	1-Chloroisoquinoline at 175-185°	13	Decomp. 195-200°	332
C_{16} 3,4-Dioxo-5-sulfo-1-naphthyl	1,2-Naphthoquinone-4,8-disulfonic acid			1151, 1247
C_{17} -CH:CHCH:C(OH)CH:[3]	2-Furfuraldehyde		Colored salts with Sn^{++}, Ti, Zr, Hf, Cb, Ta, Th, Sb, Bi, Cr^3, Mo, W and U	720

3. $H_2O_3AsC_6H_4NHCH:CHCH:C(OH)CH:NHC_6H_4AsO_3H_2$.

m-ARSANILIC ACID AND N-SUBSTITUTED DERIVATIVES

m-ARSANILIC ACID $C_6H_8AsNO_3$ m-$H_2NC_6H_4AsO_3H_2$

Prepn.: By catalytic redn. of m-$O_2NC_6H_4As(O)(ONa)_2$ in the presence of Raney Ni; 96% yield (250).

By reducing m-$O_2NC_6H_4AsO_3H_2$ with Fe filings in aq. NaCl under reflux; 85% yield (811).

Props.: M. 210-215° (811). Stable in alkali at 160-180° (964). Ionization const (1034).

Rxns. with: $NaNO_2$, followed by AcCH$_2$CO$_2$Et → EtO$_2$CC(Ac):NNHC$_6$H$_4$AsO(OH)$_2$ (240).
Haloheterocycles → N-heterocyclyl-m-arsanilic acid (250, 333).
SO_2 in concd. HCl cont. KI → m-$H_2NC_6H_4AsCl_2$ (283).
AcCl in the presence of NaOAc → m-AcNHC$_6$H$_4$AsO(OH)$_2$ (284).
$NaNO_2$, followed by amino-isoquinolines → amino-isoquinolinazo-m-$C_6H_4AsO(OH)_2$ (332).

NaNO$_2$, followed by aromatic amines → the corresponding aminodiazobenzenearsonic acids (380).
NaNO$_2$, followed by As$_2$O$_3$, (Bart method) → m-C$_6$H$_4$(AsO$_3$H$_2$)$_2$ (882).
NaNO$_2$, followed by RCH:CHCOR + CuCl$_2$ → RCH:CHC$_6$H$_4$-3-AsO(OH)$_2$ (403).
NaNO$_2$, followed by 4-amino-6-chloro-s-triazin-2-ylthiol → 3-(4-amino-6-chloro-s-triazin-2-ylthio)benzenearsonic acid (426).
NaNO$_2$, followed by 1-naphthylamine-4,8-disulfonic acid, Fe redn. in N NaOH, and methylation with Me$_2$SO$_4$ → 2-[2-(m-arsonophenyl)-1,2-dimethylhydrazino]-1-methyl-aminonaphthalene-4,8-disulfonic acid (428).
RN$_2$Cl in alk. soln. → 3,?-H$_2$N(RN$_2$)C$_6$H$_3$AsO$_3$H$_2$ (222).
NaNO$_2$, followed by hydrolysis → 3-HOC$_6$H$_4$AsO$_3$H$_2$ (811).
Heavy metals → ppts. (quant. analysis) (1009).

TABLE XXXIII
N-SUBSTITUTED m-ARSANILIC ACIDS

3-(RNH)C$_6$H$_4$AsO$_3$H$_2$ where R is:	Prepd. by reacting 3-H$_2$NC$_6$H$_4$AsO$_3$H$_2$ with (or)	Yield %	Props. and Remark	Ref.
C$_8$ Ac–	AcCl	50	m. 208-209°, rxns[1,2]	284
C$_9$ SCH:CHN:C –	SCH:CHN:CCl	13	m. 204-206°	250
C$_{10}$ N:C(NH$_2$)N:C(Cl)CH:C –	N:C(NH$_2$)N:C(Cl)CH:CCl	54	m. 183-184°	14
SCH:CMeN:C –	SCH:CMeN:CCl	65	m. > 250°, rxn.[3]	250
N:C(NH$_2$)N:CHCH:C –	N:C(NH$_2$)N:CHCH:CCl	49	m. 222-223°	14
C$_{11}$ N:CHC(NO$_2$):CHCH:C –	N:CHC(NO$_2$):CHCH:CCl	73	m. > 250°	250
	RNHC$_6$H$_4$NH$_2$ by the Bart method			505
N:CHCH:CHCH:C –	N:CHCH:CHCH:CCl	21	m. 124.5-125.5°	250
N:C(NH$_2$)N:CMeCH:C –	N:C(NH$_2$)N:CMeCH:CCl	55		14
C$_{12}$ MeCH:CHCOCH$_2$CO –	3-(ClCH$_2$CONH)C$_6$H$_4$AsO$_3$-H$_2$ + Na crotonate		m. 230-232°	1174
C$_{13}$ N:C(SEt)N:CMeCH:C –	N:C(SEt)N:CMeCH:CCl	61	Rxn.[4]	14
C$_{15}$ 4-Br-1-isoquinolyl-	4-Br-1-Cl-isoquioline	57	Decomp. 215-218°	332

1. Rxn. with SO$_2$ in dild. HCl contg. KI → the arsenoso deriv. (284).
2. Polagraphic redn. (1077).
3. Rxn. with NaSCH$_2$CO$_2$Na → the RNHC$_6$H$_4$(SCH$_2$CO$_2$Na)$_2$ (250).
4. Rxn. with H$_2$O$_2$ in alk. soln. → cleavage of the EtS group and fromation of the 2-hydroxy-6-methylpyrimidyl deriv. (14).

TABLE XXXIII (cont'd.)
N- SUBSTITUTED m-ARSANILIC ACIDS

3-(RNH)$C_6H_4AsO_3$ where R is:	Prepd. by reacting 3 $H_2NC_6H_4AsO_3H_2$ with (or)	Yield %	Props. and Remark	Ref.
5-O_2N-1-isoquinolyl-	1-Cl-5 O_2N-isoquino-line	11	Decomp. 220-221°	332
1-Isoquinolyl-	1-Cl-isoquinoline	52	Decomp. 195 196°	332
C_{19} 1-Benzo[f]quinolyl-	1-Cl-benzo[f]quino-line	86	m. 250°, rxn3	250
4-Benzo[h]quinolyl-	4-Cl-benzo[h]quino-line	37	m. 350°	250

p-ARSANILIC ACID AND N-SUBSTITUTED DERIVATIVES

ARSANILIC ACID, 4-AMINOBENZENEARSONIC ACID $C_6H_8AsNO_3$ 4 $H_2NC_6H_4AsO_3H_2$

Prepn.: By heating $PhNH_2$ with H_3AsO_4 (d$_4^{15}$ 1.781) (1:1:1) at 155-160° for four hrs., cooling the rxn. mixt. and reheating until all amorphous matl. dissolves, then filtering, and washing with H_2O at 0°; 82.4-90.5% yield (66, 288). In another process the prod. was isolated by sepn. from unchanged aniline with alkali and pptn. by acidification to pH 3 (986).

By heating $PhNH_2 \cdot H_3AsO_4$ with $PhNH_2$ and $(PhNH)_2CO$ (1:1:1) at 155-160° for 4 hrs.; 42.4% yield (588). The prod. was isolated as monosodium salt (343).

By adding dropwise 80% H_3AsO_4 to $PhNH_2$ at 130-140° and heating the mixt. at 160-170° for 5 hrs. (1234).

Props.: M. 231 232° (274, 830). Stable in 2N alk. at 160 180° under pressure; max decompn. with 0.4 0.6 N NaOH (964). Dissocn. const. (124, 1034). Redn. potential (124). Polarograph (830). Magentic susceptibility (873, 1029). UV spectrum (116). Effect on barley germination (351). Potentiometric titration (1236).

Rxns. with: $NaNO_2$ in acid medium and coupling with proteins, e.g. bovine serum, thyroxine, and proteins → the corresponding 4-(proteinoazo)benzenearsonic acids (28, 52, 378, 453, 564, 653, 835).

$NaNO_2$ in acid medium and coupling with aromatic amines → the corresponding amino arylazobenzenearsonic acids (4, 78, 332, 380, 617, 1135, 1305).

$NaNO_2$ in acid medium and coupling with 2,5 $H_2N(HO_3S)C_6H_3CO_2H$ → 2,4 $(CO_2H)(SO_3H)$ $C_6H_3HN:NC_6H_4AsO_3H_2$ (378).

$NaNO_2$ in acid meidum and coupling with $AcCH_2CO_2Et$ → $AcC(CO_2Et):NNHC_6H_4AsO_3H_2$ (240, 525, 1123).

$NaNO_2$ in acid medium and coupling with phenols and naphthols → the corresponding hydroxyphenyl- and hydroxynaphthylazobenzenearsonic acids (289, 378, 411, 413, 768, 984, 376a).

$NaNO_2$ in acid medium and coupling with $RCH:CHCOR$ → $RCH:CHC_6H_4AsO_3H_2$ or $RCH:C-(COR)C_6H_4AsO_3H_2$ (399, 400, 401, 402).

$NaNO_2$ in acid medium and coupling with pyrrole derivs. → the corresponding pyrrylazobenzenearsonic acids (870, 871).

aNO$_2$ in acid medium and coupling with furan and thiophene → 2-furyl- and 2-thienyl-
benzenearsonic acid (400, 402, 404).

aNO$_2$ in acid medium and coupling with CH$_2$:CHCN → NCCHClCH$_2$C$_6$H$_4$AsO$_3$H$_2$ (657).

aNO$_2$ in acid medium and coupling with thioammeline → 4-(4,6-diamino-s-triazin-2-
ylthio)benzenearsonic acid (426).

aNO$_2$, followed by redn. with Na$_2$S$_2$O$_4$ → 4-H$_2$NNHC$_6$H$_4$AsO$_3$H$_2$ (418).

aNO$_2$ in acid medium, followed by KCN → 4-NCC$_6$H$_4$AsO$_3$H$_2$ (284, 737).

aNO$_2$ in acid medium and coupling with As$_2$O$_3$ by the Bart method → p-C$_6$H$_4$(AsO$_3$H$_2$)$_2$
(883).

aNO$_2$ in acid medium and coupling with 2-aminobenzothiazole → benzothiazolylamino-
azobenzene- or dihydro-2-imino-3-benzothiazolylazobenzenearsonic acid (1264).

liphatic or arom. RCl → RNHC$_6$H$_4$C$_6$H$_4$AsO$_3$H$_2$ (102, 107a, 302, 577, 761, 764, 769,
879, 982, 1075, 1319).

aloheterocyclic compds. → N-heterocyclylarsanilic acids (8a, 250, 332, 418, 505,
506, 1043, 1319).

AcCH$_2$-)$_2$ → MeC:CHCH:C(Me)NC$_6$H$_4$AsO$_3$H$_2$ (474).

RCHO → RCH:NC$_6$H$_4$AsO$_3$H$_2$;HCHO gave (HOCH$_2$)$_2$NC$_6$H$_4$AsO$_3$H$_2$ (?) (217).

RCHO + H$_2$/Raney Ni → RCH$_2$NHC$_6$H$_4$AsO$_3$H$_2$ (284).

PhCH:CHCHO in HCl → 2-phenyl-6-quinolinearsonic acid (33).

(AcH)$_3$ in the presence of HCl → 2-methyl-6-quinolinearsonic acid (459).

AcCH$_2$CO$_2$Et → MeC(:CHCO$_2$Et)NHC$_6$H$_4$AsO$_3$H$_2$ (34).

CH$_2$(CN)CO$_2$Et → NCCH$_2$CONHC$_6$H$_4$AsO$_3$H$_2$ (216).

RCOCl → RCONHC$_6$H$_4$AsO$_3$H$_2$ (78, 288, 725, 833, 862, 863, 872, 887, 894, 895).

RSO$_2$Cl → RSO$_2$NHC$_6$H$_4$AsO$_3$H$_2$ (27, 274, 458, 707, 794, 1257).

RN$_2$Cl in alk. soln. → 4,?-H$_2$N(RN$_2$)C$_6$H$_3$AsO$_3$H$_2$ (222, 1041).

Br$_2$ in AcOH → 3,4-Br(H$_2$N)C$_6$H$_3$AsO$_3$H$_2$ (452).

PhNH$_2$ → (4-H$_2$NC$_6$H$_4$)AsO$_2$H (484).

RHNCN → RNHC(:NH)NHC$_6$H$_4$AsO$_3$H$_2$ (497, 1101).

(ClCN)$_3$ → N-(4,6-dichloro-s-triazin-2-yl)arsanilic acid (406, 407, 410, 414, 415,
416, 437, 1039).

N:C(NH$_2$)N:C(NH$_2$)N:CCl → N-(4,6-dimino-s-triazin-2-yl)arsanilic acid (38, 408, 410,
1039).

2,4,5-Triamino-6-hydroxypyrimidine + Br$_2$CHCOCH$_2$B4 → N-[(2-amino-4-hydroxy-6-pteridyl)-
methyl]arsanilic acid (452, 457, 457a).

BrCN → NCNHC$_6$H$_4$AsO$_3$H$_2$ (41, 51).

RHNC(:NH)NHCN → RHNC(:NH)NHC(:NH)NHC$_6$H$_4$AsO$_3$H$_2$ (41, 1083, 1084, 1101).

RNCS → RNHCSNHC$_6$H$_4$AsO$_3$H$_2$ (324, 326).

2,5-HO(O$_2$N)C$_6$H$_3$CH$_2$OH → 4-[2,4-HO(O$_2$N)C$_6$H$_3$CH$_2$NH]C$_6$H$_4$AsO$_3$H$_2$ (1277).

9-Phenoxyacridines + PhOH → the corresponding 9-acridylarsanilic acids (302).

1,2-Naphthoquinone-4,x,x-polysulfonic acids → the 4-(4-arsonophenylamino) deriv.
(1149, 1150, 1151, 1247).

Dihydroresorcinols → the corresponding N-cyclohexenonylarsanilic acid (554).

NaOCOCH$_2$SCN.H$_2$O → H$_2$NOCSCH$_2$CONHC$_6$H$_4$AsO$_3$H$_2$ (1279).

COCl$_2$ + NH$_4$OH → H$_2$NCONHC$_6$H$_4$AsO$_3$H$_2$ (887).

H(electrolytic redn. in H$_2$SO$_4$) → [:AsC$_6$H$_4$NH$_2$]$_2$.H$_2$SO$_4$ (888).

Rxn. prod. from urea + KOH at 160° treated with H$_2$NC$_6$H$_4$AsO$_3$HNa in AcOH → H$_2$NCONH-
C$_6$H$_4$AsO$_3$H$_2$ (259).

CO(NH$_2$)$_2$ at 110-120° in MePh → H$_2$NCONHC$_6$H$_4$AsO$_3$H$_2$ (889).

RNHAc + PCl$_3$ or POCl$_3$ → RN:C(Me)NHC$_6$H$_4$AsO$_3$H$_2$ (978, 979, 980).

RNO → RN:NC$_6$H$_4$AsO$_3$H$_2$ (984).
o-C$_6$H$_4$(CO)$_2$O → o-C$_6$H$_4$(CO)$_2$NC$_6$H$_4$AsO$_3$H$_2$ (1255).
(CH$_2$CO$_2$H)$_2$ or (CH$_2$CO)$_2$O → HO$_2$C(CH$_2$)$_2$CONHC$_6$H$_4$AsO$_3$H$_2$ (1005).
(CO$_2$H)$_2$ at 150° → HO$_2$CCONHC$_6$H$_4$AsO$_3$H$_2$ (343).
Ni-Al alloy in aq. alk. → PhNH$_2$ (1110).
H$_2$SO$_5$ → 4-ONC$_6$H$_4$AsO$_3$H$_2$ (984).
Zn + HCl in the presence of HgCl$_2$ → 4-H$_2$NC$_6$H$_4$AsH$_2$ (242).
Zn + concd. H^{C}l → (4-HCl.H$_2$NC$_6$H$_4$As:)$_2$ (1246).
Heavy metal ions → ppts. (quant. analyses) (876, 877, 986, 1008, 1009, 1011, 1012
Salts: Monosodium salt, Atoxyl, 4 H$_2$NC$_6$H$_4$AsO$_3$HNa, forms mono , tri , penta , and
hexahydrates (485).
Crystal structure (485). With Br it gives tribromoarsanilic acid (11).
With NaOBr, red color is formed (11). On heating with concd. HCl gives purple col
(186).
4-H$_2$NC$_6$H$_4$As(O)O$_2$SnBu$_2$, decomp. 175-185° (1267).
Decylamine salt, m. 140-141° (1).
2,2'-Methylenebis(1-naphthalenesulfonate) (170).
Ru and Sc salts form mono- and trihydrates; crystal structure (485).
Ba, Sr, and Mg, salts form pentahydrates; crystal structure (486).
Methanolamine salt (831).
HCl salt, magnetic props. (1029).
1,1'-Ethylidenebis(2-naphthalenesulfonate) (1036, 1037).

DERIVATIVES CONTAINING SUBSTITUENTS LINKED BY ALIPHATIC CARBON ATOMS

p-CARBAMIDOBENZENEARSONIC ACID, CARBARSONE C$_7$H$_9$AsN$_2$O$_4$ H$_2$NCONHC$_6$H$_4$AsO$_3$H$_2$
Prepn.: By heating urea with KOH at 160°, dissolving the solid mass in H$_2$O and
adding p-H$_2$NC$_6$H$_4$AsO(OH)(ONa) and glacial AcOH (259).
By dissolving arsanilic acid in excess NH$_4$OH and treating the mixt. with COCl$_2$
80-85% yield (887, 887a).
By heating arsanilic acid with urea in MePh at 110-120°; 33% yield (889).
By treating alk. soln. of (4-H$_2$NCONHC$_6$H$_4$As:)$_2$ with H$_2$O$_2$ in H$_2$O-C$_6$H$_6$ (1246).
Props.: M. 172° (887); m. 169-170° (889); m. 168-170° (decompn.) (1246).
Potentiometric titration (1236). Crystal structure (483).
Rxns. with: NaHSO$_3$ → H$_2$NCONHC$_6$H$_4$AsO(OH)$_2$ (108).
SO$_2$ + HI → (p H$_2$NCONHC$_6$H$_4$)$_3$As (456).
SO$_2$ in H$_2$O at 55° → H$_2$NCONHC$_6$H$_4$AsO (651).
Salts with: 7-(3-Octylaminopropylamino)benz[c]acridine, mono- and di-arsonate (861

4-ARSONOÖXANILIC ACID C$_8$H$_8$AsNO$_6$ 4-HO$_2$CCONHC$_6$H$_4$AsO$_3$H$_2$
Prepn.: by heating Na arsanilate with (CO$_2$H)$_2$ at 150° (343).
Props.: Crystallographic data (482).
Rxns. with: HNO$_3$-H$_2$SO$_4$, followed by hydrolysis → 4,3-HO(O$_2$N)C$_6$H$_3$AsO$_3$H$_2$ (343).
ROH + H$_2$SO$_4$ → RO$_2$CCONHC$_6$H$_4$AsO$_3$H$_2$ (863).
Salts: Hemi , mono , and dianiline salt (863).
Amides: The following amides were prepd. respectively by ammonolysis and aminolys
of   MeO$_2$CCONHC$_6$H$_4$AsO$_3$H$_2$: amide, prisms, forms Na salt, prisms; methylamide,
leaflets, forms Na salt dihydrate, needles; dimethylamide, rhombic leaflets, forms I

salt monohydrate; ethylamide, microneedles, forms Na salt monohydrate, hexagonal
leaflets; propylamide, rectangular plates, forms Na salt, needles; piperidide, needles
forms Na salt monohydrate; anilide, microneedles, forms di-Na salt dihydrate, needles
(363).

N-ACETYLARSANILIC ACID $C_8H_{10}AsNO_4$ 4-AcNHC$_6$H$_4$AsO$_3$H$_2$

Prepn.: Radioactive (As^{76}) compd. obtd. by reacting a mixt. of p-AcNHC$_6$H$_4$NH$_2$.HCl,
As^{76}Cl$_3$, and Cu$_2$Cl$_2$ with NaNO$_2$ (157).
  By acetylation of arsanilic acid (872).
Props.: Dissocn. const. (124, 1025). Redn. potential (124).
Rxns. with: Aq. NaOH → 4-H$_2$NC$_6$H$_4$AsO$_3$H$_2$ (157).
HNO$_3$ → 3,4-O$_2$N(AcNH)C$_6$H$_3$AsO$_3$H$_2$ (872).
Cl$_2$^{36} in AcOH under reflux, followed by HI → 2- and 3- Cl^{36}C$_6$H$_4$NH$_2$ (918).
RNH$_2$ + POCl$_3$ → RN:C(Me)NHC$_6$H$_4$As(OH)$_2$ (979).
Metal ions → ppts. (quant. analyses) (1025, 1026).
Salts: Na acetylarsanilate, Arsacetin, 4-AcNHC$_6$H$_4$AsO(OH)ONa, UV spectrum (116).

N-HYDROXYACETYL-p-ARSANILIC ACID $C_8H_{10}AsNO_5$ 4-HOCH$_2$CONHC$_6$H$_4$AsO$_3$H$_2$

Rxn. with Zn + HCl in MeOH → 4-HOCH$_2$CONHC$_6$H$_4$AsH$_2$ (1173).
Salts: 7-(3-Octylaminopropylamino)benz[c]acridine salt (861).
With 8-hydroxyquinoline, m. 152-162°; with·5-chloro-8-hydroxy-7-iodoquinoline, m.
  177-178°; and with 8-hydroxy-5,7-diioquinoline, m. 191-192° (277).
Piperazine salt with 2 mols. of the acid has no definite m. temp.; piperazine mono-
arsonate hydrochloride decomp. 200-300° (278).
Chloroquine tri(N-hydroxyacetyl-p-arsanilate), m. 158-160°, and di(n-hydroxyacetyl-
  p-arsanilate), m. 158-161° (279).

N-(4-ARSONOPHENYL)GLYCINE $C_8H_{10}AsNO_5$ 4-HO$_2$CCH$_2$NHC$_6$H$_4$AsO$_3$H$_2$

Rxns. with: Glycols → the corresponding monoesters (762, 763).
2,4,5-Triamino-6-hydroxypyrimidine + Br$_2$CHCOCHBr → N[(2-amino-4-hydroxy-6-pteridyl)-
  methyl]-N-(carboxymethyl)arsanilic acid (452).

4-GUANYLCARBAMIDO- or 4-CARBAMYLGUANIDOBENZENEARSONIC ACID (?) $C_8H_{11}AsN_4O_4$
H_2NCONHC(:NH)NH-C$_6$H$_4$AsO$_3$H$_2$.HCl or HN:C(NH$_2$)NHCONH-C$_6$H$_4$AsO$_3$H$_2$.HCl

Prepn.: by refluxing 4-biguanidobenzenearsonic acid with 4N HCl (41, 1101).
Props.: Free acid, m. 210° (1101).

N-(AMIDINOAMIDINO)ARSANILIC ACID $C_8H_{12}AsN_5O_3$ H_2NC(:NH)NHC(:NH)NHC$_6$H$_4$AsO$_3$H$_2$

Prepn.: by reacting arsanilic acid dissolved in H$_2$O or dild. EtOH at pH 4.8 with
dicyandiamide under reflux (41, 765, 1083, 1084, 1101).
Props.: Granular crystals, sol. in acids and alkali (41). Colorless needles, m.
260° (1101).
Rxns. with: 4 N HCl → 4-carbamylguanido- or 4-guanylcarbamidobenzenearsonic acid
  hydrochloride (41, 1101)
HCHO → a compd. C$_9$H$_{12}$AsN$_5$O$_3$, m. > 300° (767).
H$_3$PO$_2$ → the arseno deriv. (767, 1083, 1084).
Ac$_2$CH$_2$ → N-[(4,6-dimethyl-2-pyrimidinyl)amidino]arsanilic acid (767).

N-(4-ARSONOPHENYL)GLYCINE 3-AMINO-2-HYDROXYPROPYL ESTER $C_{11}H_{17}AsN_2O_6$
    $H_2NCH_2CHOHCH_2O_2CCH_2NHC_6H_4AsO_3H_2$
Prepn.: by heating N-(4-arsonophenyl)glycine with $H_2NCH_2CHOHCH_2OH$ and PhOH at
160-170° under reduced pressure (762).

N-[3-(5-NITROFURFURYLIDENE)CARBAZOLYL]ARSANILIC ACID $C_{12}H_{11}AsN_4O_7$

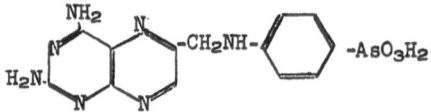

O_2N⟩-CH:NNHCONH-⟨⟩-AsO_3H_2

Prepn.: by reacting 5-nitrofuraldehyde dissolved in EtOH with aq. soln. of H_2N-
$NHCONHC_6H_4AsO_3H_2$ (1265).
Props.: M. 215° (decompn.) (1265).

N-[(2-AMINO-4-HYDROXY-6-PTERIDYL)METHYL]ARSANILIC ACID $C_{13}H_{13}AsN_6O_4$

Prepn.: By reacting arsanilic acid in H_2O with 2,4,5-triamino-6-hydroxypyrimidine
sulfate and $Br_2CHCOCH_2Br$ at 80° at pH 2.0-3.5 and treating the rxn. prod. with Ca-
Cl_2 and $ZnCl_2$ in aq. NaOH (15, 452).
    By reacting arsanilic acid with 2,3-dibromopropion aldehyde and condensing the
product with 2,4,5-triamino-6-hydroxypyrimidine and dehydrogenating the pteridyl
deriv. (457, 457a).
Props.: UV spectrum (452, 457, 457a).

N-[(2,4-DIAMINO-6-PTERIDYL)METHYL]ARSANILIC ACID $C_{13}H_{14}AsN_7O_3$

Prepn.: by reacting arsanilic acid with 2,3-dibromopropionaldehyde followed by
2,4,5,6-tetraaminopyrimidine (15, 456, 457a).
Props.: UV spectrum (456, 457a).

N,N'-(THIOCARBONYL)DIARSANILIC ACID $C_{13}H_{14}As_2N_2O_6S$ $SC(NHC_6H_4AsO_3H_2)_2$
Prepn.: by treating arsanilic acid (neutralized with NaOH in satd. NaOAc soln.)
with $CSCl_2$ dissolved in MePh (712).
Rxns. with: PCl_3, followed by NH_4OH → $CS(NHC_6H_4AsOH)_2O$ (712).

N,N'-CARBONYLDIARSANILIC ACID $C_{13}H_{14}As_2N_2O_7$ $CO(NHC_6H_4AsO_3H_2)_2$
Prepn.: by treating arsanilic acid (neutralized with NaOH in satd. NaOAC soln.)
with $COCl_2$ in PhMe; 93% yield (712, 984).

N-(4,6-DIMETHYL-2-PYRIMIDYLCARBAMIDO)ARSANILIC ACID $C_{13}H_{16}AsN_5O_3$

Prepn.: by reacting $H_2NC(:NH)NHC(:NH)NHC_6H_4AsO_3H_2$ with Ac_2CH_2 in aq. soln. at r.
t.; 78% yield (767).
Props.: Hydroscopic solid, m. 300° (767).

4-PHTHALIMIDOBENZENEARSONIC ACID $C_{14}H_{10}AsNO_5$ $4-[o-C_6H_4(CO)_2N]C_6H_4AsO_3H_2$
Prepn.: by reacting arsanilic acid with phthalic anhydride (1:1.5-2) in AcOH
(1255).
Props.: M. 383° (decompn.) (1255).

N,N'-OXALYLDIARSANILIC ACID $C_{14}H_{14}As_2N_2O_8$ $(-CONHC_6H_4AsO_3H_2)_2$
Prepn.: by reacting arsanilic acid with oxalyl chloride (984).

N,N'-ETHYLENEBIS(p-ARSANILIC ACID) $C_{14}H_{18}As_2N_2O_6$ $(-CH_2NHC_6H_4AsO_3H_2)_2$
Props.: Analytical method (858a, 1261a).
Salts with: 7-(3-Octylaminopropylamino)benz[c]acridine (861).
Tetracycline (448a).
Spiramycin, $[\alpha]^{29}_D$ -56° (1151a).

N-(4-ARSONOPHENYL)GLYCINE 6-ESTER WITH 1,5-MANNITAN $C_{14}H_{20}AsNO_9$
$OCH_2CH(OH)CH(OH)CHCH(OH)CH_2O_2CCH_2NHC_6H_4AsO_3H_2$
Prepn.: by heating N-(4-arsonophenyl)glycine with mannitan in the presence of
PhOH under reduced pressure; isolated as Ca salt (762).

N-[(2-AMINO-4-HYDROXY-6-PTERIDYL)METHYL]-N-CARBOXYMETHYLARSANILIC ACID
$C_{15}H_{15}AsN_6O_6$

Prepn.: by refluxing arsanilic acid with $ClCH_2CO_2H$ (1:1) in N NaOH and reacting
the resulting p-arsonophenylglycine with 2,4,5-triamino-6-hydroxypyrimidine sulfate and 1,1,3-tribromopropanone in H_2O at 80° (15, 452).
Props.: UV spectrum (452).

N-[(2-AMINO-4-HYDROXY-6-PTERIDYL)METHYL]-N-(CARBAMYLMETHYL)ARSANILIC ACID
$C_{15}H_{16}AsN_4O_5$

Prepn.: by reacting N-(carbamylmethyl)arsanilic acid with 2,4,5-triamino-6-hydroxy-

pyrimidine sulfate and 1,1,3-tribromopropanone in H_2O at 80° (15, 42).
<u>Props.:</u> UV spectrum (452).

N-[(1,2,3,6-TETRAHYDRO-1,3,7-TRIMETHYL-2,6-DIOXYPURIN-8-YL)METHYL]ARSANILIC ACID
$C_{15}H_{18}AsN_5O_5$

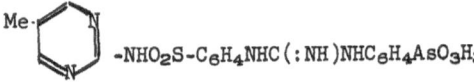

<u>Prepn.:</u> by warming 8-chloromethylcaffeine with arsanilic acid in EtOH in the presence of NaOH; 25-30% yield (107a).

1,3-BIS(4-ARSONOPHENYLAMINO)PROPANE $C_{15}H_{20}As_2N_2O_6$ $CH_2(CH_2NHC_6H_4AsO_3H_2)_2$
<u>Prepn.:</u> by refluxing arsanilic acid with $Br(CH_2)_3Br$ in aq. alk. (764).

N-(N'-GLUCONYL-N'-METHYLCARBAMYLMETHYL)-p-ARSANILIC ACID $C_{15}H_{25}AsN_2O_9$
$HOCH_2(CHOH)_4CH_2NMeCOCH_2NHC_6H_4AsO(OH)_2$
<u>Prepn.:</u> by heating a mixt. of p-arsanilic acid and aq. NaOH with $ClCH_2CON(Me)$-$CH_2(CHOH)_4CH_2OH$ (102, 577) or treating $4-H_2O_3AsC_6H_4NHCH_2CO_2Me$ with methylglucosamine (577).
<u>Props.:</u> White powder, sol. in H_2O (102).
<u>Rxns. with:</u> $Na_2S_2O_4.2H_2O \rightarrow$ the corresponding arsenobenzene (102).

N,N'-SULCINYLDIARSANILIC ACID $C_{16}H_{18}As_2N_2O_8$ $(-CH_2CONHC_6H_4AsO_3H_2)_2$
<u>Prepn.:</u> by reacting arsanilic acid with succinic anhydride (984).

N-[p-(6-METHYL-2-PYRIMIDYLSULFAMYL)PHENYLGUANYL]ARSANILIC ACID $C_{18}H_{19}AsN_6O_4S$

Me· [pyrimidine ring] $-NHO_2S-C_6H_4NHC(:NH)NHC_6H_4AsO_3H_2$

<u>Prepn.:</u> by reacting arsanilic acid with p-(6-methyl-2-pyrimidylsulfamyl)phenyl cyanimide in pyridine (497).
<u>Props.:</u> Turns black > 300° (497).

N,N'-ADIPYLDIARSANILIC ACID $C_{18}H_{22}As_2N_2O_8$ $(-CH_2CH_2CONHC_6H_4AsO_3H_2)_2$
<u>Prepn.:</u> by heating $H_2NC_6H_4AsO_3HNa$ with adipyl chloride in C_6H_6 (519, 984).

N,N'-PHTHALOYLDIARSANILIC ACID $C_{20}H_{18}As_2N_2O_8$ [benzene ring] $CONHC_6H_4AsO_3H_2$ / $CONHC_6H_4AsO_3H_2$
<u>Prepn.:</u> by reacting arsanilic acid with phthalyl chloride (984).

N,N'-ISOPHTHALOYLDIARSANILIC ACID $C_{20}H_{18}As_2N_2O_8$ $CONHC_6H_4AsO_3H_2$ [benzene ring] $CONHC_6H_4AsO_3H_2$

<u>Prepn.:</u> by reacting arsanilic acid with isophthalyl chloride (984).

N,N'-TEREPHTHALOYLDIARSANILIC ACID $C_{20}H_{18}As_2N_2O_8$ $p\text{-}C_6H_4[CONH\text{-}p\text{-}C_6H_4AsO(OH)_2]_2$
Prepn.: by condensing terephthaloyl chloride with arsanilic acid; 25% yield (288, 984), or with atoxyl in benzene (519).
Props.: Amorphous, m. >360° (288, 519).

1,4-BIS[N-(4-ARSONOPHENYL)GLYCYL]PIPERAZINE $C_{20}H_{26}As_2N_4O_8$ $H_2O_3AsC_6H_4NHCH_2CON\text{-}C_4H_8NCOCH_2NHC_6H_4AsO_3H_2$
Prepn.: by refluxing 1,4-bis(chloroacetyl)piperazine with $H_2NC_6H_4AsO_3Na_2$ (982).
Rxns. with: $PCl_3 \rightarrow$ the diarsenoso deriv. (982).

1,4-BIS[(4-ARSONOPHENYLCARBAMOYL)METHYL]PIPERAZINE $C_{20}H_{26}As_2N_4O_8$
$H_2O_3AsC_6H_4NHCOCH_2NC_4H_8NCH_2CONHC_6H_4AsO_3H_2$
Prepn.: by refluxing piperazine with $ClCH_2CONHC_6H_4AsO_3H_2$ in alk. soln. (982).
Rxns. with: PCl_3 in EtOAc $\rightarrow$ the diarsenoso deriv. (982).

N-(N-7-SULFO-3-ACRIDYLGLYCYL)ARSANILIC ACID $C_{21}H_{18}AsN_3O_7S$

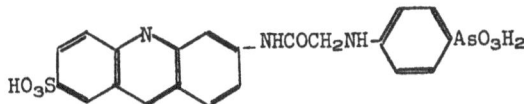

Prepn.: by reacting N-(chloroacetyl)arsanilic acid with 7-aminoacridyl-2-sulfonic acid (833).
Props.: Dark red prisms, turning to tar at 340-360° (833).

N,N'-AZELAYLDIARSANILIC ACID $C_{21}H_{26}As_2N_2O_8$ $CH_2[(CH_2)_3CONHC_6H_4\text{-}4\text{-}AsO_3H_2]_2$
Prepn.: by reacting Na arsanilate with azelayl dichloride in aq. alk. (862).
Props.: Amorphous solid (862).

N,N'-SEBACYLDIARSANILIC ACID $C_{22}H_{30}As_2N_2O_8$ $[-(CH_2)_4CONHC_6H_4\text{-}4\text{-}AsO_3H_2]_2$
Prepn.: reacting Na arsanilate with $ClOC(CH_2)_8COCl$ in aq. Na_2CO_3 (862, 984).
Props.: Amorphous solid, forms tri-Na salt (862).

N,N'-(HEXADECAMETHYLENEDICARBONYL)DIARSANILIC ACID (?) $C_{30}H_{46}As_2N_2O_8$
$[-(CH_2)_8CONHC_6H_4AsO_3H_2]_2$
Prepn.: by-prod. in the rxn. of Na arsanilate with $ClOC(CH_2)_{16}COCl$ (863).
Props.: Brown, gelatinous solid (863).

TABLE XXXIV

N-SUBSTITUTED p-ARSANILIC ACIDS WITH SUBSTITUENTS LINKED BY ALIPHATIC CARBON ATOMS

$RHNC_6H_4AsO_3H_2$ or $RRNC_6H_4AsO_3H_2$, where R is (are)	Prepd. by reacting $4\text{-}H_2NC_6H_4AsO_3H_2$ (A) with	Yield %	Props. and Remarks	Ref.
C7				
NC-	$BrCN$ + A + NaOH	70	Colorless crystals, soly. data, rxns.[1,2]	41, 51
$H_2NC(:NH)$-	H_2NCN + A + NH_4OH		Monohydrate[3]	105
	H_2NCN + A in acid		Colorless prisms	1101
C8				
Cl_3CCO-	Cl_3CCOCl + A in MePh		Fine needles	519
Cl_2CHCO-	$Cl_2CHCOCl$ + A in MePh		Shiny leaflets	519
$ClCH_2CO$-	$ClCH_2COCl$ + A		m. 295-296°, rxns.[4,5]	833
ICH_2CO-	ICH_2COCl + A		m. 195-196°, rxn.[6]	985
H_2NCOCH_2-			Na salt, crystal structure, rxns.[7,8]	482
Me_2-	$Me_2NC_6H_4NH_2$ by the Bart-Scheller method	24	Rxns.[2], Zr complex, analytical reagent for F^-	284, 622a
Et-	$MeCHO$ + A + H_2/Ni	5		284
$HOCH_2NHCO$-	$MeO_2CNHC_6H_4AsO_3H_2$ + $HOCH_2$-NH_2	52	Rxn.[23]	47
$HOCH_2CH_2$-	$\overline{CH_2CH_2O}$ + A		Soly.	675
C9				
$Cl_2C:CClCO$-	$Cl_2C:CClCOCl$ + A	63	Clusters of needles Rxns.[9,10]	519
$NCCH_2CO$-	$NCCH_2CO_2Et$ + A			216
MeO_2CCO-	$HO_2CCONHC_6H_4AsO_3H_2$ + MeOH		Rhombic plates, unstable in neutral or alk. soln., rxn[11]	863
H_2NCSCH_2CO-	$NCCH_2CONHC_6H_4AsO_3H_2$ + H_2S in $NaHCO_3$	43		216
$H_2NOC(S)CH_2CO$-	$NaOCOCH_2SCN\cdot H_2O$ + A		Rxn.[11]	1279
MeO_2CCH_2-				47, 577
$HOCH_2NHCOCH_2$-	$MeO_2CCH_2NHC_6H_4AsO_3H_2$ + $HO\text{-}CH_2NH_2$		Rxn.[22]	47
$HO_2CCH_2CH_2$-	$BrCH_2CH_2CO_2H$ + A	41		769
$MeCH(CO_2H)$-	$MeCHBrCO_2H$ + A	65		769
EtO_2C-			Rxn.[11]	534

Group	Reaction / Method	Description	Ref.
...2NH(:NH)CH₂CO-	HCl + NH₃ + etc.		210
ClCH₂CHOHCH₂-	(ClCH₂)₂CHOH or ŌCH₂CHCH₂Cl + A		764
HOCH₂CH₂NHCO-	Bart method		536, 580
	EtOCONHC₆H₄AsO₃H₂ + HOCH₂CH₂-NH₂ + A	White cryst. powder, decomp. 200°	534, 536, 580
EtMe- MeCHOHCH₂-	EtMeNC₆H₄AsO by oxidation MeCHCH₂O + A	Sol. in H₂O	1256 675
C₁₀			
HO₂C(CH₂)₂CO-	(CH₂CO)₂O or (CH₂CO₂H)₂ + A		1005
AcOCH₂CO-	ClCH₂CONHC₆H₄AsO₃H₂ + AcONa		1174
EtO₂CCO-	EtO₂CCOCl + A		863
Me(HOCH₂CH₂NHCO)-	Me(EtOCO)NC₆H₄AsO₃H₂ + HOCH₂-CH₂NH₂	White powder	534
HOCH₂CHOHCH₂NHCO-	EtOCONHC₆H₄AsO₃H₂ + H₂NCH₂C-HOHCH₂OH	White powder	534
HOCH₂CH₂OCH₂CH₂- Me₂CHNHC(:NH)- HOCH₂CH₂NHCOCH₂-	ClCH₂CH₂OCH₂CH₂OH Me₂CHN:C:NH + A MeO₂CCH₂NHC₆H₄AsO₃H₂ + HOCH₂- 61 CH₂NH₂	Rxn.21	764 497 47
C₁₁			
ClCH₂CHOHCH₂O₂CCH₂- HOCH₂CHOHCH₂NHCOCH₂-	MeO₂CCHNHC₆H₄AsO₃H₂ + H₂NCH₂-CHOHCH₂OH	Hydroscopic, m. 220°	762 102

1. Rxn. with aq. alkali → H₂NCONHC₆H₄AsO₃H₂ (41, 51).
2. Rxn. with SO₂ in HCl contg. KI → the arsenoso deriv. (47, 51, 284, 288, 376, 683, 862).
3. Monohydrate, lustrous, transparent, colorless crystals, loses H₂O in vacuo and becomes white and opaque; darkens <350° without melting (105).
4. Rxns. with primary or secondary amines → the corresponding glycylarsanilic acids (84, 376, 833, 982).
5. Rxns. with RCO₂Na → RCO₂CH₂CONHC₆H₄AsO₃H₂ (1174).
6. Rxn. with bovine serum (985).
7. Rxn. with Br → white ppt. tribromoarsanilic acid (11).
8. Rxn. with NaOBr → red color. (11).
9. Rxn. with alc. HCl in Et₂O, followed by NH₃ → H₂NC(:NH)CH₂CONH- deriv. (216).
10. Rxn. with H₂S in 1 equiv. of aq. NaHCO₃ → H₂NCSCH₂CONH- deriv. (216).

TABLE XXXIV (cont'd.)

N-SUBSTITUTED p-ARSANILIC ACIDS WITH SUBSTITUENTS LINKED BY ALIPHATIC CARBON ATOMS

$RHNC_6H_4AsO_3H_2$ or $RRNC_6H_4AsO_3H_2$, where R is (are)	Prepd. by reacting $4\text{-}H_2NC_6H_4AsO_3H_2$ (A) with	Yield %	Props. and Remarks	Ref.
$HOCH_2CHOHCH_2O_2CCH_2-$			Hydroscopic, m. 185°	762
C₁₂				
$MeCH{:}CBrCO_2CH_2CO-$	$ClCH_2CONHC_6H_4AsO_3H_2 + MeCH{:}CBrCO_2K$		m. 268-270°	1174
$MeCH{:}CHCO_2CHCO-$	$4\text{-}ClCH_2CONHC_6H_4AsO_3H_2 + MeCH{:}CHCO_2Na$		m. >270°	1174
$EtO_2CC{:}CMe-$	$AcCH_2CO_2Et + A$		m. >300°	34
$CH_2(CH_2)_3NCH_2CO-$	$4\text{-}ClCH_2CONHC_6H_4AsO_3H_2 + $ pyrrolidine			376
$(HOCH_2CH_2)_2NCOCH_2-$	$MeO_2CCH_2NHC_6H_4AsO_3H_2 + (HOCH_2CH_2)_2NH$	76	Rxn.[26]	47
C₁₃				
$3,5\text{-}(O_2N)_2C_6H_3CO-$	$3,4\text{-}(O_2N)_2C_6H_3COCl + A$		Rxn.[12]	725
$3,5\text{-}H_2N(O_2N)C_6H_3CO-$	$4\text{-}[3,4\text{-}(O_2N)_2C_6H_3CONH]\text{-}C_6H_4\text{-}AsO_3H_2 + (NH_4)_2S$		Needles, rxn.[13]	725
$4\text{-}BrC_6H_4NHC(:NH)-$	$4\text{-}BrC_6H_4N{:}C{:}NH + A$		m. 125° (dec.)	497
$4\text{-}ClC_6H_4NHC(:NH)-$	$4\text{-}ClC_6H_4N{:}C{:}NH + A$		m. 180° (dec.)	497
$4\text{-}H_2NC_6H_4CO-$	not reported		Rxn.[2]	284
	$4\text{-}EtO_2CNHC_6H_4COCl + A$	100	Condensation and hydrolysis	984
$2,5\text{-}HO(O_2N)C_6H_3CH_2-$	$2,4\text{-}HO(O_2N)C_6H_3CH_2OH + A$		m. 245°	1277
$PhNHC(:NH)-$	$PhN{:}C{:}NH + A$		m. 135° (dec.)	497
$4\text{-}H_2NO_2SC_6H_4NHCS$	$4\text{-}H_2NO_2SC_6H_4NCS + A$			326
$Me_2CHCH_2CO_2CH_2CO-$	$4\text{-}ClCH_2CONHC_6H_4AsO_3H_2 + Me_2\-CHCH_2CO_2CH_2CO-$		m. >270°	1174
$CH_2(CH_2)_3NCH_2CH_2CO-$	$4\text{-}Cl(CH_2)_2CONHC_6H_4AsO_3H_2 + $ pyrrolidine		Rxns.[2,14,15]	376
C₁₄				
$2,3,4,5,6\text{-}Cl_4(HO_2C)C_6CO-$	Tetrachlorophthalic anhyd. + A	35	Amorph., rxn.[16]	519
$4\text{-}NCC_6H_4CO-$	$4\text{-}NCC_6H_4COCl + A$	5	Rosets	288
$4\text{-}HO_2CC_6H_4CO-$	$p\text{-}C_6H_4(COCl)_2 + A$	95	Needles, rxn.[2]	288
$4\text{-}H_2NOCC_6H_4CO-$	$4\text{-}NCC_6H_4CONHC_6H_4AsO_3H_2 + H_2O_2$			288

4-MeC₆H₄CO-	4-MeC₆H₄COCl + A	Needles, m. > 360°; Insol. Mg salt	100	288 398
4-O₂NC₆H₄SO₂NHCOCH₂-	4-O₂NC₆H₄SO₂NHCOCH₂Cl + A	Rxn.17		1075
4-ClC₆H₄NHC(:NH)NHC(:NH)-	4-ClC₆H₄NHC(:NH)NHCN + A	Rxn.15		1083, 1084
4-MeC₆H₄NHC(S)-	4-MeC₆H₄NCS + A	Decomp.~ 262°		164
4-MeC₆H₄NHC(:NH)-	4-MeC₆H₄N:C:NH + A	m. 160° (dec.)		497
4-H₂NC₅H₄SO₂NHCOCH₂-	Redn. of the corresp. nitro deriv. with Na₂S₂O₄	White solid		1075
4-H₂NC(:NH)NHO₂SC₆H₄NHC(:NH)-	H₂NC(:NH)NHO₂SC₆H₄N:C:NH + A	m. 150° (dec.)		497
4-Et₂NCH₂CHOHCH₂NHCO-	Et₂NCH₂CHOHCH₂NH₂ + EtO₂CNH-C₆H₄AsO₃H₂	White cryst. powder		534
C₁₅ PhCO₂CH₂CO-	ClCH₂CONHC₆H₄AsO₃H₂ + PhCO₂-Na	m. > 270°		1174
4-AcNHC₆H₄CO-	Method not given	Rxn.2	100	284
4-MeC₆H₄NHC(:NH)NHC(:NH)-	4-MeC₆H₄NHC(:NH)NHCN + A	Rxn.15		1083, 1084
4-MeOC₆H₄NHC(:NH)NHC(:NH)-	4-MeOC₆H₄NHC(:NH)NHCN + A	Rxn.15		1083, 1084
CH₂(CH₂-)₂	Br(CH₂)₃Br + A	m. >250°		764
HOCH(CH₂-)₂	HOCH(CH₂Cl)₂ + A	Glistening leaflets, rxn.²		764
HO₂C(CH₂)₇CO-	Hydrolysis of p-arsenoazelanic acid esters	m. 210° (dec.), rxn.¹⁸		862
EtO₂CN(CH₂CH₂)₂NCH₂CO-	4-ClCH₂CONHC₆H₄AsO₃H₂ + EtO₂-CN(CH₂CH₂)₂NH			982
H₂NOC(CH₂)₇CO-	Ammonolysis of the corresp. Me ester in a sealed tube	Branching needles		862

11. Rxn. with aq. NH₃ or amines → the corresp. carbamyl deriv. (47, 534, 577, 863).
12. Redn. with (NH₄)₂S → the 3-amino deriv. (725).
13. Rxn. with (CH₂CO)₂O → the 5-nitro-3-succinoylamino deriv. (725).
14. Rxn. with Na₂S₂O₄ in aq. NaOH + MgCl₂ → the arseno deriv. (274, 376).
15. Rxn. with H₃PO₂ in the presence of KI → the arseno deriv. (376, 1083, 1084).
16. Rxn. with H₂O₂ → H₂NOCC₆H₄CONHC₆H₄AsO₃H₂ (288).
17. Redn. with Na₂S₂O₄ in Na₂CO₃ soln. + MgCl₂ → the amino deriv. (1075).
18. Rxn. with PCl₃ → the arseno deriv. (982).
19. Rxn. with aq. alkali → hydrolysis of the ester group (862).
20. Rxns. with NH₃OH or RR'NH → ammonolysis and aminolysis, resp., of the ester group (862).
21. Redn. with SO₂ → the corresponding arsonous acid (47).
22. Rxn. with HCHO → HOCH₂NHCOCH₂N(CH₂OH)C₆H₄AsO₃H₂ (47).

TABLE XXXIV (cont'd.)

N-SUBSTITUTED p-ARSANILIC ACIDS WITH SUBSTITUENTS LINKED BY ALIPHATIC CARBON ATOMS

$RHNC_6H_4AsO_3H_2$ or $RRNC_6H_4-AsO_3H_2$, where R is (are)	Prepd. by reacting $4-H_2NC_6H_4AsO_3H_2$ (A) with	Yield %	Props. and Remarks	Ref.
C$_{16}$				
$SCH:CHN:CHNHO_2SC_6H_4NHC(:NH)-$ $AcNHC_6H_4SO_2NHCOCH_2-$ $4-AcNHC_6H_4NHC(:NH)NHC(:NH)-$	$SCH:CHN:CHNHO_2SC_6H_4NHCN + A$ $AcNHC_6H_4SO_2NHCOCH_2Cl + A$ $4-AcNHC_6H_4NHC(:NH)NHCN + A$		Darkens at 223° White powder Rxn.[15]	497 1075 1083, 1084
$MeO_2C(CH_2)_7CO-$	$MeO_2C(CH_2)_7COCl + A$		Hair-like needles, forms mono-Na salt, rxns.[19,20]	862
$HO_2C(CH_2)_8CO-$	Hydrolysis of the corresp. carboxylic esters		Needles, forms Na salts, rxn.[2]	862
$MeNHOC(CH_2)_7CO-$	Aminolysis of $MeO_2C(CH_2)_7CO-$ $NHC_6H_4AsO_3H_2$		Amorph. solid; Na salts, leaflets	862
$H_2NOC(CH_2)_8CO-$	Ammonolysis of $MeO_2C(CH_2)_8-$ $CONHC_6H_4AsO_3H_2$		Jagged needles; Na salt, leaflets	862
$HOCH_2(CHOH)_4CH_2N(CH_2CH_2OH)-$ $COCH_2-$	$ClCH_2CON(CH_2CH_2OH)CH_2(CHOH)_4-$ $CH_2OH + A$			102
C$_{17}$				
$3,5-O_2N(H_2N)C_6H_3(HO_2CCH_2CH_2CONH)C_6-$ H_3CO-	$3,5-O_2N(H_2N)C_6H_3CONHC_6H_4AsO_3-$ $H_2 + (CH_2CO)_2O$		Rxn. with Zn + HCl in EtOH → redn. of the NO_2 group	725
$3,5-H_2N(HO_2CCH_2CH_2CONH)-$ C_6H_3CO-	Redn. of the corresp. 3-nitro deriv.			725
$BrCH_2CHBr(CH_2)_8CO-$	$BrCH_2CHBr(CH_2)_8COCl + A$			519
$Me[Me(CH_2)_7]CHCO-$	$Me[Me(CH_2)_7]CHCOCl + A$			519
$EtO_2C(CH_2)_7CO-$	$EtO_2C(CH_2)_7COCl + A$		Leaflets; Na-salt, plates	862
$MeO_2C(CH_2)_8CO-$	$MeO_2C(CH_2)_8COCl + A$		Plates; Na salt, glistening leaflets, rxns.[19,20]	862
$EtNHOC(CH_2)_7CO-$	$MeO_2C(CH_2)_7CONHC_6H_4AsO_3H_2 +$ $EtNH_2$		Plates; Na salt, leaflets	862
$MeNHOC(CH_2)_8CO-$	$MeO_2C(CH_2)_8CONHC_6H_4AsO_3H_2 +$ $MeNH_2$		Minute needles; Na salt, leaflets	862
$Me(CH_2)_9CO-$ $HOCH_2(CHOH)_4CH_2N(Me)CO-$ $CH_2-(HOCH_2CH_2)$	$Me(CH_2)_9COCl + A$ $HOCH_2CH_2NHC_6H_4AsO_3H_2 + ClCH_2C-$ $ON(Me)CH_2(CHOH)_4CH_2OH$			519 102

C_{18} $EtO_2C(CH_2)_8CO-$ $Me_2NOC(CH_2)_8CO-$	$EtO_2C(CH_2)_8COCl + A$ $MeO_2C(CH_2)_8CONHC_6H_4AsO_3-$ $H_2 + Me_2NH$	Prisms Feathery needles; Na salt, rectangular prisms	862 862
C_{20} $HOCH_2(CHOH)_4CH_2N(Me)COCH(Me)-$	$MeCHBrCON(Me)CH_2(CHOH)_4CH_2-$ $OH + A$		102
C_{21} $(4-BrC_6H_4NHCO)_2CH-$	$(4-BrC_6H_4NHCO)_2CHBr + A$	m. 251-253°	879
C_{22} $PhNHOC(CH_2)_8CO-$	$PhNHOC(CH_2)_8COCl + A$	White gelatinous solid; Na salt, white needles	862
$n-C_{15}H_{31}CO-$	$n-C_{15}H_{31}COCl + A$		78
C_{23} $(PhCH_2NHCO)_2CH-$ $(4-MeC_6H_4NHCO)_2CH-$ $Me(CH_2)_{10}CO_2CH_2CH(OH)CH_2O_2-$ CCH_2-	$(PhCH_2NHCO)_2CHBr + A$ $(4-MeC_6H_4NHCO)_2CHBr + A$ 4-Arsonophenylglycine + lauryl monoglyceride	m. 266° m. 233° (dec.)	879 879 763
C_{24} $HO_2C(CH_2)_{16}CO-$ $Me(CH_2)_{16}CO-$	$ClOC(CH_2)_{16}COCl + A$ in aq. alkali $Me(CH_2)_{16}COCl + A$	Rxn. with MeOH + $H_2SO_4 \rightarrow$ the carbomethoxy deriv.	863 78
C_{25} $MeO_2C(CH_2)_{16}CO-$	$MeO_2C(CH_2)_6COCl + A$ $HO_2C(CH_2)_{16}CONHC_6H_4AsO_3H_2 +$ $MeOH + H_2SO_4$		863 863
$Me(CH_2)_{12}CO_2CH_2CH(OH)CH_2-$ O_2CCH_2-	4-Arsonophenylglycine + myristoyl monoglyceride		763
C_{29} $Me(CH_2)_{16}CO_2CH_2CH(OH)CH_2O_2-$ CH_2-	4-Arsonophenylglycine + stearyl monoglyceride		763

23. Rxn. with HCHO $\rightarrow$ HOCH$_2$NHCON(CH$_2$OH)C$_6$H$_4$AsO$_3$H$_2$ (47).

TABLE XXXV

N-ARALKYLIDENE-p-ARSANILIC ACIDS *

(Prepared by reacting p-arsanilic acid with aldehydes in diluted alcohol**)

RCH:NC$_6$H$_4$AsO$_3$H$_2$	Starting Aldehyde	Props. and Remarks	Ref
PhCH:NC$_6$H$_4$AsO$_3$H$_2$	PhCHO	Mono-Na salt, colorless crystals	217
4-MeOC$_6$H$_4$CH:NC$_6$H$_4$AsO$_3$H$_2$	4-MeOC$_6$H$_4$CHO	Mono-Na salt, colorless needles	217
4-Me$_2$NC$_6$H$_4$CH:NC$_6$H$_4$AsO$_3$-H$_2$	4-Me$_2$NC$_6$H$_4$CHO	Mono-Na salt, colorless needles	217
2-HOC$_6$H$_4$CH:NC$_6$H$_4$AsO$_3$H$_2$	2-HOC$_6$H$_4$CHO	Mono-Na salt, yellow shiny needles	217
PhCH:CHCH:NC$_6$H$_4$AsO$_3$H$_2$	PhCH:CHCHO	Mono-Na salt, yellow flakes	217

*The Schiff bases form bisulfite derivatives, RCH(SO$_3$Na)NHC$_6$H$_4$AsO(ON)ONa. (217).
**Formaldehyde yielded (HOCH$_2$)$_2$NC$_6$H$_4$AsO$_3$H$_2$, colorless hydroscopic solid (217).

TABLE XXXVI

N-ARYL-p-ARSANILIC ACIDS

4-(RNH)C$_6$H$_4$AsO$_3$H$_2$ where R is	Prepd. from	Yield %	Props. and Remarks	Ref.
C$_{12}$ 2,4-(HO$_2$N)$_2$C$_6$H$_3$-	RCl + 4-H$_2$NC$_6$H$_4$AsO$_3$H$_2$	41	Rxn. with FeSO$_4$ in alk. soln. → redn. of the NO$_2$ group	761
4-H$_2$NC$_6$H$_4$-	RNH$_2$ + 4-ClC$_6$H$_4$AsO$_3$H$_2$		White crystals, rxn. with SO$_2$ in HCl → the arsenoso deriv.	683
2,4-(H$_2$N)$_2$C$_6$H$_3$-	4-[2,4-(O$_2$N)$_2$C$_6$H$_3$NH]-C$_6$H$_4$AsO$_3$H$_2$ + FeSO$_4$ in aq. NaOH	36	Dark red	761
C$_{13}$ 2-HO$_2$CC$_6$H$_4$-	RCl + 4-H$_2$NC$_6$H$_4$AsO$_3$H$_2$	73	m. 278°, cyclization in H$_2$SO$_4$ → 9 oxo 2 acridinearsonic acid	302
C$_{14}$ 4-Me$_2$NC$_6$H$_4$-	RNH$_2$ + 4-ClC$_6$H$_4$AsO$_3$-H$_2$	41		683
C$_{16}$ 4-Et$_2$NC$_6$H$_4$	RNH$_2$ + 4-ClC$_6$H$_4$AsO$_3$-H$_2$	35		683

N-(5,5-DIMETHYL - 3-OXO-1-CYCLOHEXEN-1-YL)ARSANILIC ACID $C_{14}H_{18}AsNO_4$
   $CH_2CMe_2CH_2C(:O)CH:CNHC_6H_4AsO_3H_2$
Prepn.: by reacting 5,5-dimethyldihydroresorcinol with $4-H_2NC_6H_4AsO_3H_2$ (554).
Props.: m. 209° (554).

N-(3-CHLORO-1,4-DIHYDRO-1,4-DIOXO-2-NAPHTHYL)ARSANILIC ACID $C_{16}H_{11}AsClNO_5$

Prepn.: by condensing 2,3-dichloro-1,4-naphthoquinone with excess arsanilic acid
in alc. in the presence of NaOAc (162).
Props.: Orange-brown powder, m. > 360° (162).

4-(2-HYDROXY-1,4-NAPHTHOQUINOIMINO)BENZENEARSONIC ACID $C_{16}H_{12}AsNO_5$

Prepn.: by reacting Na 1,2-naphthoquinone-4-sulfonate with arsanilic acid in aq.
alk. at 75° (1090).
Props.: Tetrahydrate, m. > 365° (1090).

N-(3,4-Dihydro-3,4-dioxo-x-sulfonaphthyl)arsanilic Acid

N-(3,4-DIHYDRO-3,4-DIOXO-5-SULFONAPHTHYL)ARSANILIC ACID $C_{16}H_{12}AsNO_8S$
Prepn.: by reacting potassium 1,2-naphthoquinone-4,8-disulfonate with potassium
arsanilate in the presence of an oxidizing agent (1149, 1150, 1247).

N-(3,4-DIHYDRO-3,4-DIOXO-6-SULFONAPHTHYL)ARSANILIC ACID $C_{16}H_{12}AsNO_8S$
Prepn.: by reacting 1,2-naphthoquinone-4,7-disulfonic acid with arsanilic acid
in the presence of alkali and H_2O_2 (1151, 1247).

N-(3,4-DIHYDRO-3,4-DIOXO-7-SULFONAPHTHYL)ARSANILIC ACID $C_{16}H_{12}AsNO_8S$
Prepn.: by reacting 1,2-naphthoquinone-4,6-disulfonic acid with arsanilic acid
in the presence of alkali (1150, 1151).

N-(3,4-DIHYDRO-3,4-DIOXO-5,7-DISULFONAPHTHYL)ARSANILIC ACID
Prepn.: by reacting 1,2-naphthoquinone-4,6,8-trisulfonic acid with arsanilic acid
in the presence of alkali and H_2O_2 (1151, 1247).

217

N,N'-(3,6-DICHLORO-2,5-p-BENZOQUINONYLENE)DIARSANILIC ACID $C_{18}H_{14}As_2Cl_2N_2O_8$

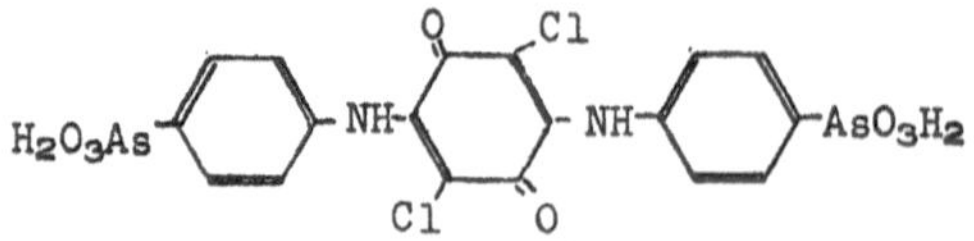

Prepn.: by condensing 2,3,5,6-tetrachloro-p-quinone with arsanilic acid in alc. in the presence of NaOAc (162).
Props.: Chars at 335° (162).

N-(4-PHENYL-3-OXO-1-CYCLOHEXEN-1-YL)ARSANILIC ACID $C_{18}H_{18}AsNO_4$

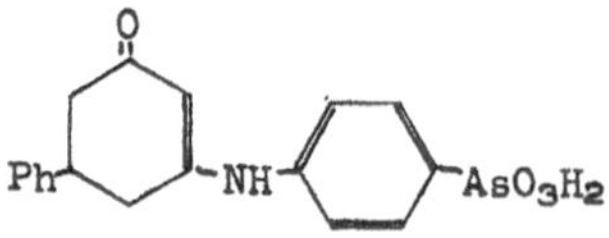

Prepn.: by treating Na arsanilate with 5-phenyldihydroresorcinol (554).
Props.: m. 264° (554).

N-(4-BENZYL-3-OXO-1-CYCLOHEXEN-1-YL)ARSANILIC ACID $C_{19}H_{20}AsNO_4$

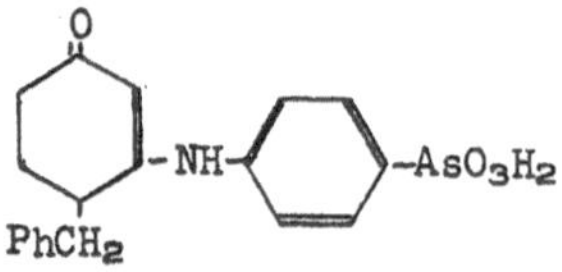

Prepn.: by treating Na arsanilate with 4-benzyldihydroresorcinol (554).
Props.: m. 205° (554).

N-SULFONYL DERIVATIVES

N-SULFANILYARSANILIC ACID $C_{12}H_{13}AsN_2O_5S$ $4-(H_2NC_6H_4SO_2NH)C_6H_4AsO_3H_2$
Prepn.: by reacting $AcNHC_6H_4SO_2Cl$ with p-arsanilic acid in a neutral medium and hydrolyzing the prod. (27, 707, 794, 1257).
Props.: White crystalline powder, discolors at 220° (27); m. 244-245° (794).
Rxns. with: BzH → the Schiff base (707, 795).
$Na_2S_2O_4$ + $MgCl_2$ in alk. soln. → the arsenobenzene deriv. (750, 1257).
$NaHSO_3$ in dild. HCl contg. KI → the arsenoso deriv. (750).
$NaNO_2$, followed by 2,6-diaminopyridine → 2,6-diamino-3-pyridyl-azobenzenesulfonyl-arsonic acid (794).
SO_2 in NaOH soln. → the arsenoso deriv. (1257).

N-(4-CYANOPHENYLSULFONYL)ARSANILIC ACID $C_{13}H_{11}AsN_2O_5S$ $4-(4-NCC_6H_4SO_2NH)C_6H_4-$
AsO_3H_2
Prepn.: by reacting $4-NCC_6H_4SO_2Cl$ with $4-H_2NC_6H_4AsO_3H_2$ (274).
Props.: Brownish crystals (274).
Rxn. with alc. HCl followed by alc. NH_3 → conversion of the CN group into H_2NC-
(:NH)- group (274).

218

N-(4-METHOXY-3-NITROPHENYLSULFONYL)ARSANILIC ACID $C_{13}H_{13}AsN_2O_8S$
    4-[3,4-O_2N(MeO)$C_6H_3SO_2NH$]$C_6H_4AsO_3H_2$
<u>Prepn.</u>: by reacting 3,4-O_2N(MeO)$C_6H_4SO_2Cl$ with 4-$H_2NC_6H_4AsO_3H_2$ (750).
<u>Rxn.</u> with $Na_2S_2O_4$ → redn. of the O_2N and AsO_3H_2 groups to form the diamino-arseno-
benzene deriv. (750).

N-(4-AMIDINOPHENYLSULFONYL)ARSANILIC ACID $C_{13}H_{14}AsN_3O_5S$
    4-[4-H_2NC(:NH)$C_6H_4SO_2NH$]$C_6H_4AsO_3H_2$
<u>Prepn.</u>: by treating 4-(4-$NCC_6H_4SO_2NH$)$C_6H_4AsO_3H_2$ with alc. HCl, followed by alc.
NH_3 (274).
<u>Props.</u>: Crystals; the compd. probably forms internal salt (274).
<u>Rxn.</u> with $Na_2S_2O_4$ in aq. NaOH + $MgCl_2$ → the arseno deriv. (274).

N-(N-ACETYLSUFANILYL)ARSANILIC ACID $C_{14}H_{15}AsN_2O_6S$ 4-(4-$AcNHC_6H_4SO_2NH$)$C_6H_4AsO_3H_2$
<u>Prepn.</u>: by reacting 4-$AcNHC_6H_4SO_2Cl$ with 4-$H_2NC_6H_4AsO_3H_2$; 35% yield (27).
<u>Props.</u>: Shiny leaflets, discolor at 230° (27). Fine colorless needles, m. > 300°
(274). Decomp. 290° (1257).
<u>Rxn.</u> with aq. NaOH or aq. HCl → hydrolysis of the AcNH group (27, 1257).

p-(8-ETHOXY-5-QUINOLYLSULFONAMIDO)BENZENEARSONIC ACID $C_{17}H_{17}AsN_2O_6S$

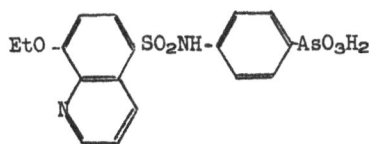

<u>Prepn.</u>: by reacting 8-ethoxy-5-quinolylsulfonyl chloride with arsanilic acid in
aq. Na_2CO_3 soln. (458).
<u>Props.</u>: Colorless cryst. powder, m. 242-244° (dec.) (458).

N-[p-[2,6-DIAMINO-3-PYRIDYLAZO)PHENYLSULFONYL]ARSANILIC ACID $C_{17}H_{17}AsN_6O_5S$
    4-{4-[H_2NC:NH(NH_2):CHCH:CN_2]$C_6H_4SO_2NH$}$C_6H_4AsO_3H_2$
<u>Prepn.</u>: by coupling ClN:$NC_6H_4SO_2NHC_6H_4AsO_3H_2$ with diaminopyridine (794).
<u>Props.</u>: Dull red, microcryst. powder, m. 275-277° (794).

N-(N-BENZOYLSULFANILYL)ARSANILIC ACID $C_{18}H_{17}AsN_2O_6S$
    4-(4-$BzNHC_6H_4SO_2NH$)$C_6H_4AsO_3H_2$
<u>Prepn.</u>: by reacting 4-(4-$H_2NC_6H_4SO_2NH$)$C_6H_4AsO_3H_2$ with BzCl (795).
<u>Props.</u>: Plates, m. 205.5-206.5° (795).

N-HETEROCYCLYL DERIVATIVES

N-(4,6-DICHLORO-s-TRIAZIN-2-YL)ARSANILIC ACID $C_9H_7AsCl_2N_4O_3$
    4-(N:$CClN$:$CClN$:CNH)$C_6H_4AsO_3H_2$
<u>Prepn.</u>: by reacting arsanilic acid or its salts with $(ClCN)_3$; up to 85% yield
(406, 407, 410, 414, 415, 416, 437, 1039).
<u>Props.</u>: White powder; soly. data (1039).

Rxns. with: 10% aq. NH_3 at 45° → N-(4-amino-6-chloro-s-triazin-2-yl)arsanilic acid (410, 414, 415, 416, 419, 420, 421, 424, 1039).
25% aq. NH_3 at 100-130° → the 4,6-diamino-s-triazinyl deriv. (406, 410, 414, 1039).
Aq. primary or secondary amines → the corresponding 4,6-diamino-s-triazinyl deriv. (410, 414, 416, 1039).
10% NaOH at 40° → the 4-chloro-6-hydroxy-s-triazinyl deriv., while at 120° under pressure the 4,6-dihydroxy-s-triazinyl deriv. was obtd. (416).
20% H_2NNH_2 → the 4-chloro-6-hydrazino-s-triazinyl deriv. (418).
75% H_2NNH_2 → the 4,6-dihydrazino-s-triazinyl deriv. (418).
p-$H_2NC_6H_4AsO_3H_2$ in 2-NaOH → 2,4-bis(p-arsonoanilino)-6-chloro-s-triazine (416).
4-$H_2NC_6H_4SO_3H_2$ → the 2-chloro-4-sulfanilyl-s-triazinyl deriv. (416).
1-Amino-8-hydroxy-3,6-naphthalenedisulfonic acid → the 4-chloro-6-(3,6-disulfo-8-1-naphthylamino)-s-triazinyl deriv. (416).

N-(4,6-DIAMINO-s-TRIAZIN-2-YL)ARSANILIC ACID $C_9H_{11}AsN_6O_3$
4-[N:C(NH_2)N:C(NH_2)N:CNH]C_6H_4AsO_3H_2

Prepn.: By reacting 2,4-diamino-6-chloro-s-triazine with arsanilic acid in aq. HCl under reflux; up to 96% yield (4, 14, 38, 40, 408, 1039).
By reacting N-(4,6-dichloro-s-triazin-2-yl)arsanilic acid with 25-28% NH_4OH at 110-130°; up to 57% yield (406, 410, 414, 416, 1039).
By heating a mixt. of arsanilic acid with NHNHC(:NH)NKCN in H_2O at 97° and acidbying the rxn. mixt. (621).
By heating 4-arsonophenylisomelamine in aliphatic acls. or ethers at 100° in the presence of a basic catalyst (1204).
By reacting 4-(4,6-diamino-s-triazin-2-ylamino)phenyldiazonium chloride with Na_3AsO_3 in the presence of $CuSO_4$; 55% yield (1306).
Props.: M. > 320°, sol. in hot H_2O and cold alkali (621).
Rxns. with: SO_2 in aq. HCl cong. HI → the arsenoso deriv. (38, 416, 438).
H_3PO_2 + HI in MeOH → the arsenobenzene deriv. (416, 439).
$SnCl_2$ in HCl contg. HI → the arsenobenzene deriv. (416).
RSH (1:4) → the corresponding -As(SR)_2 deriv. (429, 1036).
HNO_3-H_2SO_4 → the 3-nitro deriv. (1306).
Salts: HCl salt, white crystals; picrate, fine needles; chloroplatinate, yellow crystals (406, 1039). Di-Na salt crystallizes with 3 mol. H_2O and loses H_2O at 135° in vacuo; the anhyd. salt forms tetrahydrate on exposure to air (38). Mono-Na salt crystallizes with 4 mols. H_2O and is stepwise dehydrated at 315° in vacuo to trihydrate and anhydrate (38).

N-(4-AMINO-6-HYDRAZINO-s-TRIAZIN-2-YL)ARSANILIC ACID $C_9H_{12}AsN_7O_3$
4-[N:C(NHNH_2)N:C(NH_2)N:CNH]C_6H_4AsO_3H_2

Prepn.: by treating the 4-chloro-6-hydrazino deriv. with liq. NH_3 at 50° under pressure (418).
Props.: White ppt. (418).
Rxns. with: $AgNO_3$ → Ag. (418)
H_3PO_2 in HCl-HI → yellow ppt. (418).
BzH → white ppt. (418).
PhCH:CHCHO → the 6-cinnamylidenehydrazino deriv. (418).
$AcCH_2CO_2Et$ → the 6-(3-methyl-5-oxo-1-pyrazolyl) deriv. (418).

N-(2-AMINO-4-PYRIMIDYL)ARSANILIC ACID $C_{10}H_{11}AsN_4O_3$ 4-[$\overline{N:C(NH_2)N:CHCH:CNH}$]$C_6H_4AsO_3$
Prepn.:   By refluxing arsanilic acid with 2-amino 4 chloropyrimidine in aq. sus-
pension or in Me_2CO-H_2O mixt. (39, 40) or in H_2O contg. HCl (506).
   By coupling $\overline{N:C(NH_2)N:CHCH:CNH}C_6H_4N_2Cl$ with Na_3AsO_3 in the presence of a Cu
salt (506).
Props.:   Dihydrate, white ppt., m. 250° (506).
Rxn. with SO_2 in HCl contg. HI, followed by neutralization → the corresp. arsenoso
compd. (39).
Salts: HCl salt, white crystals, m. 250°; di Na salt, white crystals (506).

   N (2 BENZOTHIAZOLYL)-p-ARSANILIC ACID $C_{13}H_{11}AsN_2O_3S$

Prepn.:   by refluxing p-arsanilic acid with 2-chlorobenzothiazole in dild. HCl;
87% yield (250).
Props.:   m. > 250° (250).
Rxns. with: $NaSCH_2CO_2Na$ → the corresponding bis(carboxymethylmercapto)arsino
deriv. (250).

   N-[4 AMINO-6-(3-METHYL-5-PYRAZOLON-1-YL)-s-TRIAZIN-2-YL)ARSANILIC ACID
      $C_{13}H_{13}AsN_7O_4$

Prepn.:   by reacting the 4-amino-6-hydrazino deriv. with $AcCH_2CO_2Et$ in 50% AcOH
(418).
Props.:   White microcryst. ppt. (418).

   N-(4-QUINAZOLYL)ARSANILIC ACID $C_{14}H_{12}AsN_3O_3$

Prepn.:   by reacting 4-chloroquinazoline with arsanilic acid in $HCONMe_2$ at 50°;
58% yield (1319).
Props.:   M. > 320° (1319).

N-(4 BROMO-1-ISOQUINOLYL)ARSANILIC ACID $C_{15}H_{12}AsBrN_2O_3$

Prepn.:   by refluxing 4-bromo-1-chloroisoquinoline with arsanilic acid (1:1) in H_2O contg. 1 equiv. HCl; 46% yield (332).
Props.:   Decomp. 203-205° (332).

N-(5-NITRO-1-ISOQUINOLYL)ARSANILIC ACID $C_{15}H_{12}AsN_3O_5$

Prepn.:   by refluxing 5-nitro-1-chloroisoquinoline with arsanilic acid (1:1) in H_2O contg. 5 equiv NaOH; 11% yield (332).
Props.:   Decomp. 225-226° (332).

N-(8-NITRO-4-QUINOLYL)ARSANILIC ACID $C_{15}H_{12}AsN_3O_5$

Prepn.:   by reacting 4-chloro-8-nitroquinoline with arsanilic acid at 50° in $HCONMe_2$; 30.2% yield (1319).
Props.:   M. 288° (1319).

N-[2 CHLORO-4-(p-SULFANILYL)-s-TRIAZIN-6-YL]ARSANILIC ACID $C_{15}H_{13}AsClN_5O_6S$

Prepn.:   by reacting N-(4,6-dichloro-s-triazi-2-yl)arsanilic acid with p $H_2NC_6H_4$-SO_3H in 2% NaOH; 34% yield (416).

N-1-ISOQUINOLYLARSANILIC ACID $C_{15}H_{13}AsN_2O_3$

Prepn.:   by refluxing 1 chloroisoquinoline with arsanilic acid (1:1) in dlld. HCl; 58% yield (332).
Props.:   Decomp. 300° (332).

222

2-CHLORO-4,6-BIS(p-ARSONOANILINO)-s-TRIAZINE

$$C1$$

Prepn.: by reacting 4,6-dichloro-1,3,5-triazin-2-ylarsanilic acid with arsanilic acid in 2% NaOH at 28°; 61% yield (416).
Rxns. with: 20% NH$_4$OH at 100° → the 6-amino deriv. (416).

2-AMINO-4,6-BIS(p-ARSONOANILINO)-s-TRIAZINE C$_{15}$H$_{16}$As$_2$N$_6$O$_6$

Prepn.: by treating the 2-chloro deriv. with 20% NH$_4$OH at 110°; 46% yield (416).

N-[4,6-BIS(2-HYDROXYPROPYLAMINO)-s-TRRIAZI-2-YL]ARSANILIC ACID C$_{15}$H$_{23}$AsN$_6$O$_5$

NHCH$_2$CHOHMe

Prepn.: by refluxing 2-chloro-4,6-bis(2-hydroxypropylamine)-s-triazine with p-arsanilic acid in H$_2$O (46).

N-(2-METHYL-6-NITRO-4-QUINOLYL)ARSANILIC ACID C$_{16}$H$_{14}$AsN$_3$O$_5$

Prepn.: by reacting 4-chloro-2-methyl-6-nitroquinoline with arsanilic acid in HCONMe$_2$ at 50°; 50.1% yield (1319).
Pr_ps.: M. 288° (1319).

N-(2-METHYL-4-QUINOLYL)ARSANILIC ACID $C_{16}H_{15}AsN_2O_3$

Prepn.: by reacting 4-chloro-2-methylquinoline with arsanilic acid in HCONMe$_2$
at 50°; 33.2% yield (1319).
Props.: m. 285° (1319).

N-(4,6-BIS(DIETHYLAMINO)-1,3,5-TRIAZIN-2-YL)ARSANILIC ACID $C_{17}H_{27}AsN_6O_3$

Prepn.: by treating the 4,6-dichloro deriv. with aq. Et$_2$NH at 110°, 58% yield
(416).

N-[4-AMINO-6-(CINNAMYLIDENEHYDRAZINO)-s-TRIAZIN-2-YL)-ARSANILIC ACID
$C_{18}H_{18}AsN_7O_3$

Prepn.: by reacting the 4-amino-6-hydrazino deriv. with cinnamaldehyde in dild.
HCl (418).
Props.: Yellow crystals (418).

N-(7-CHLORO-2-NITRO-9-ACRIDINYL)ARSANILIC ACID $C_{19}H_{13}AsClN_3O_5$

Prepn.: by condensing 9-amino-7-chloro-2-nitroacridine with 4-ClC$_6$H$_4$AsO$_3$H$_2$(6).
Props.: M. 330° (dec.) (6).

N-[4-CHLORO-6-[3,6-DISULFO-8-HYDROXY-1-NAPHTHYLAMINO)-s-TRIAZIN-2-YL]ARSANILIC
ACID $C_{19}H_{15}AsClN_5O_{10}S_2$

Prepn.: by reacting the 4,6-dichloro-s-tr izinyl acid with 8,3,6-HO(HO$_3$S)$_2$—

$C_{10}H_4NH_2$ in 2% NaOH at 35°; 29% yield (416).

N-9-ACRIDYLARSANILIC ACID $C_{19}H_{15}AsN_2O_3$

Prepn.: by digesting arsanilic acid with 9-phenoxyacridine and PhOH; 92% yield
(302).
Props.: m. 264-265° (302).

N-(4-BENZO[h]QUINOLYL)-p-ARSANILIC ACID $C_{19}H_{15}AsN_2O_3$

Prepn.: by refluxing p-arsanilic acid with 4-chlorobenzo[h]quinoline in dild.
HCl; 41% yield (250).
Props.: m. >250° (250).

N-(1-BENZO[f]QUINOLYL)-p-ARSANILIC ACID $C_{19}H_{15}AsN_2O_3$

Prepn.: by refluxing p-arsanilic acid with 1-chlorobenzo[f]quinoline in dild.
HCl; 60% yield (250).
Props.: m. >250° (250).
Rxns. with: $NaSCH_2CO_2H$ → the corresponding bis(carboxymethylmercapto)arsino
deriv. (250).

N,N-BIS(6-NITRO-2-BENZOTHIAZOLYL)ARSANILIC ACID $C_{20}H_{12}AsN_5O_7S_2$

Prepn.: by reacting arsanilic acid with 2-chloro-6-nitrobenzothiazole (303).
Props.: m. 63° (303).

N-(7-CHLORO-2-METHOXY-9-ACRIDYL)ARSANILIC ACID $C_{20}H_{16}AsClN_2O_4$

NH—⟨ ⟩—AsO$_3$H$_2$

Cl

OMe

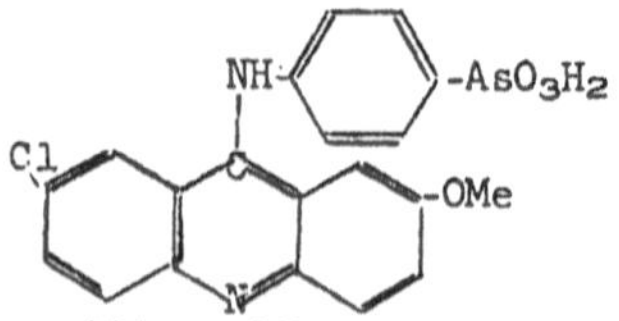

Prepn.: by reacting 7,8-dichloro-2-methoxyacridine with arsanilic acid in PhOH at 100-110° (1144).
Props.: M. 218-220° (1144).

N-(7-CHLORO-4-METHOXY-9-ACRIDYL)ARSANILIC ACID $C_{20}H_{16}AsClN_2O_4$

NH—⟨ ⟩—AsO$_3$H$_2$

Cl

OMe

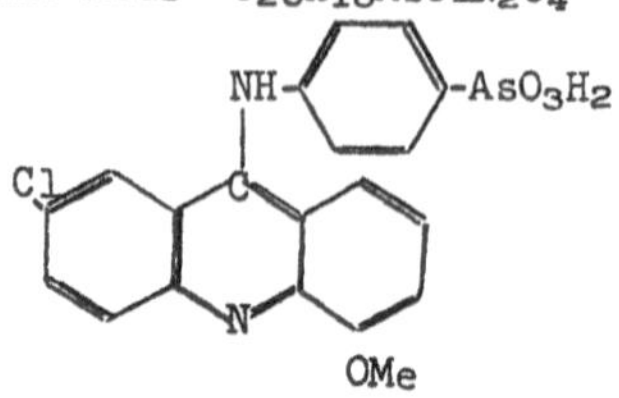

Prepn.: by reacting 4-methoxy-7,9-dichloroacridine with arsanilic acid in PhOH at 100-110° (1143).
Props.: M. 220-222° (decompn.) (1143).

N-(5-CHLORO-2-METHOXY-9-ACRIDYL)ARSANILIC ACID $C_{20}H_{16}AsClN_2O_4$

NH—⟨ ⟩—AsO$_3$H$_2$

OMe

Cl

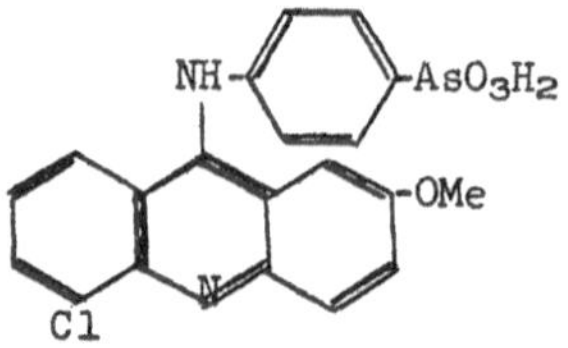

Prepn.: by reacting 5,9-dichloro-2-methoxyacridine with arsanilic acid in PhOH at 100-110° (1142).
Props.: M. 257-259° (1142).

N-(2-METHYL-9-ACRIDYL)ARSANILIC ACID $C_{20}H_{17}AsN_2O_3$

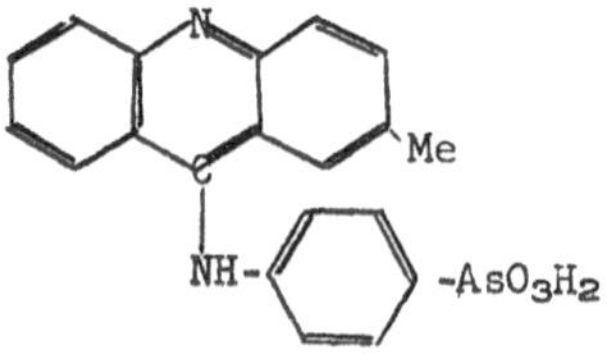

Me

NH—⟨ ⟩—AsO$_3$H$_2$

Prepn.: by digesting arsanilic acid with 2-methyl-9-phenoxyacridine and PhOH on a steam bath; 95% yield (302).
Props.: M. 268-269° (302).

226

N-(2-METHOXY-9-ACRIDYL)ARSANILIC ACID $C_{20}H_{17}AsN_2O_4$

Prepn.: by digesting arsanilic acid with 2-methoxy-9-phenoxyacridine and PhOH; 70% yield (302).
Props.: M. 245-248° (302).

TABLE XXXVII

N-HETEROCYCLYL-p-ARSANILIC ACIDS

$RNHC_6H_4AsO_3H_3$ where R is:	Prepd. from: (Ars stands for $-HNC_6H_4AsO_3H_2$)	Yield %	Props. and Remarks	Ref.
C_9				
$N{:}C(OH)N{:}C(Cl)N{:}C-$	$N{:}C(Cl)N{:}C(Cl)N{:}C-Ars + 10\%$ NaOH at 40°	76	Rxn.[1]	416
$N{:}C(NH_2)N{:}C(Cl)N{:}C-$	$N{:}C(Cl)N{:}C(Cl)N{:}C-Ars + 10\%$ NH_4OH	69	Rxns.[2,3]	410, 414-6 420-1 437, 1039
$SCH{:}CNH{:}C-$	2-Chlorothiazole + HArs	30	Small white crystals	250
$\cdot N{:}C(OH)N{:}C(OH)N{:}C-$	$N{:}C(Cl)N{:}C(Cl)N{:}C-Ars + 10\%$ NH_4OH at 120°/pressure	60	m. 238-240°	416
$N{:}C(NHNH_2)N{:}C(Cl)N{:}C-$	$N{:}C(Cl)N{:}C(Cl)N{:}C-Ars + 20\%$ H_2NNH_2		Rxns.[4,5,6]	418
$N{:}C(NH_2)N{:}C(OH)N{:}C-$	$N{:}C(NHNH_2)N{:}C(Cl)N{:}CCl + HArs$ $N{:}C(NH_2)N{:}C(OH) N{:}CCl + HArs$	51		418 46
$N{:}C(NHNH_2)N{:}C(NH_2)N{:}C-$	$N{:}C(Cl)N{:}C(OH)N{:}CArs + 28\%$ + 2 equivs. NaOH $N{:}C(Cl)N{:}C(OH)N{:}CArs + 10\%$ NH_4OH at 120° $N{:}C(NHNH_2)N{:}C(Cl)N{:}CArs +$ liq. NH_3 at 50°/pressure	62	White ppt.	416 418
$N{:}C(NHNH_2)N{:}C(NHNH_2)N{:}C-$	$N{:}C(NHNH_2)N{:}C(Cl)N{:}CArs$ or $N{:}C(Cl)N{:}C(Cl)N{:}CArs + 75\%$ H_2NNH_2		White cryst. ppt., rxns.[7,8]	418
C_{10}				
$N{:}CHC(NO_2)CHN{:}C-$	$N{:}CHC(NO_2)CHN{:}CCl + HArs$ in dild. HCl	22		14
$CH{:}C(Cl)N{:}C(NH_2)N{:}C-$	$CH{:}C(Cl)N{:}C(NH_2)N{:}CCl + HArs$ in dild. HCl	49	m. 293°, rxns.[3,9]	8a, 14
$SCH{:}C(Me)N{:}C-$	$SCH{:}C(Me)N{:}CCl + HArs$	21	m. > 250°, rxn.[10]	250
$CH{:}CHN{:}C(NH_2)N{:}C-$	$CH{:}CHN{:}C(NH_2)N{:}CCl + HArs$	61	Rxns.[3,9]	8a, 14
$N{:}C(NH_2)CH{:}CHN{:}C-$	$N{:}C(NH_2)CH{:}CHN{:}CCl + HArs$		HCl·H_2O salt, m. 331° Rxn.[3a]	14 39
$N{:}C(NH_2)N{:}C(NH_2)CH{:}C-$	$N{:}C(NH_2)N{:}C(NH_2)CH{:}CCl + HArs$		HCl·H_2O salt, m. 278° (dec.) rxns.[3,9]	8a

MeNH₂ at 110°

Compound	Reaction	Yield %	Properties	Ref.
C₁₁				
ĊH:CHC(NO₂):CHN:Ċ-	CH:CHC(NO₂):CHN:ĊCl + HArs in dild. HCl	55	m. >250°, rxn.11	250, 505
ĊH:CHCH:CHN:Ċ-	CH:CHCH:CHN:ĊCl + HArs	71	m. 220-221°	250
ĊH:CHC(NH₂):CHN:Ċ-	CH:CHC(NO₂):CHN:ĊArs + H₂/Raney Ni or Fe(OH) + alkali		White crystals, m. 240° (dec.), turns lavender in air, rxn.3	505
ĊH:C(Me)N:C(NH₂)N:Ċ-	CH:C(Me)N:C(NH₂)N:ĊCl + HArs	48	HCl.5H₂O salt, m. 301°, rxn.9	8a
ĊH:C(NHMe)N:CHN:Ċ-	CH:C(NHMe)N:CHN:ĊCl + HArs		HCl.H₂O, salt, m. 268°, rxn.3	14
ĊH:C(NHMe)N:C(NH₂)N:Ċ-	CH:C(NHMe)N:C(NH₂)N:ĊCl + HArs		HCl salt, m. 280°, rxn.3	8a
Cl.MeN:C(NH₂)N:C(NH₂)CH:Ċ-	ClMeN:C(NH₂)N:C(NH₂)CH:ĊCl + HArs		Hydrate, m. 289° (dec.).	8a
ĊH:C(NH₂)N(Me.Cl):C(NH₂)N:Ċ-	CH:C(NH₂)N(Me.Cl):C(NH₂)ĊCl + HArs	52	m. 265° (dec.)	8a
N:C(NHMe)N:C(NHMe)N:Ċ-	N:C(Cl)N:C(Cl)N:ĊArs + 17% MeNH₂ at 110°		Rxn.12	410, 414, 416, 1039
C₁₂				
N:C(Me)CH:C(Me)N:Ċ-	N:C(Me)CH:C(Me)N:ĊCl + HArs		HCl salt, m.246° (dec.)	8a
ĊH:C(Me)N(Me.Cl):C(NH₂)N:Ċ-	CH:C(Me)N(Me.Cl):C(NH₂)N:ĊCl + HArs		Hydrate, m. 236-237° (dec.). rxn.9	8a
ĊH:C(NHMe)N:C(NHMe)N:Ċ-	CH:C(NHMe)N:C(NHMe)N:ĊCl + HArs		HCl salt, m. 283-284°, rxn.3	8a

1. Rxn. with 28% NH₄OH → the 4-amino-6-hydroxy-s-triazinyl deriv. (416).
2. Rxn. with MeNH₂ at 110° → the 4-amino-6-methylamino-s-triazinyl deriv. (416).
3. Rxn. with SO₂ in HCl followed by neutralization → the arsenoso deriv. (8a, 417, 419-422, 424).
3a.Rxn. with SO₂ in concd. HCl → the dichloroarsino deriv. (506).
4. Rxn.with H₂NNH₂ → the 4,6-dihydrazino-s-triazinyl deriv. (418).
5. Rxn. with liq. NH₃ at 50°/pressure → the 4-amino-6-hydrazino-s-triazinyl deriv. (418).
6. Rxn. with SnCl₂ in HCl contg. HI → the corresp. arsenobenzene deriv. (424).
7. Rxn. with PhCH:CHCHO in HCl → yellow ppt. (418).
8. Rxn. with SnCl₂ in HCl → yellow ppt. (418).
9. Rxn. with H₃PO₂ + HI in MeOH → the arsenobenzene deriv. (8a).
10.Rxn. with NaSCH₂CO₂Na → the corresp. bis(carboxymethylmercapto)arsino deriv. (250).

TABLE XXXVII (cont'd.)

N-HETEROCYCLYL-p-ARSANILIC ACIDS

RNHC$_6$H$_4$AsO$_3$H$_3$ where R is:	Prepd. from: (Ars stands for-HNC$_6$H$_4$AsO$_3$H$_2$)	Yield %	Props. and Remarks	Ref.
C$_{13}$ CH:C(Me)N:C(SEt)N:C- N:C(NMe$_2$)N:C(NMe$_2$)N:C- N:C(NHCH$_2$CH$_2$OH)N:C- (NHCH$_2$CH$_2$OH)N:C-	CH:C(Me)N:C(SEt)N:CCl + HArs N:C(Cl)N:C(Cl)N:CArs + Me$_2$NH N:C(NHCH$_2$CH$_2$OH)N:C- (NHCH$_2$CH$_2$OH)N:CCl + HArs	76	Rxn.^{13}	14 407 46

11. Catalytic redn. over Raney Ni → the 5-amino deriv. (505).

12. Redn. with Na$_2$S$_2$O$_4$ in aq. alkali → the arsenobenzene deriv. (424).

13. Rxn. with H$_2$O$_2$ in alk. soln. → cleavage of the EtS group to from the 2-hydroxy-6-methyl-4-pyrimidyl deriv. (14).

4-HYDRAZINOBENZENEARSONIC ACID $C_6H_9AsN_2O_3$ $4-H_2NNHC_6H_4AsO_3H_2$

__Prepn.:__ by diazotiazation of p-arsanilic acid and reducing the diazo compd.
with $Na_2S_2O_4$, 50% yield (47, 418) or with $SnCl_2$ + HCl, 43% yield (565).
__Props.:__ M. 198° (565).
__Rxns. with:__ $(ClCN)_3 \rightarrow$ 4-(4,6-dichloro-s-triazin-2-ylhydrazino)benzenearsonic
acid (418).
Reducing agents $\rightarrow$ the corresponding arsonous acid (47).
1,2-Naphthoquinone-4,8-disulfonic acid $\rightarrow$ 4-(1-hydroxy-4,8-disulfo-2-naphthylazo)ben-
zenearsonic acid (411).

4-[2-(4,6-DICHLORO-s-TRIAZIN-2-YL)HYDRAZINO]BENZENEARSONIC ACID $C_9H_8AsCl_2N_5O_3$
4-[$\overline{N:C(Cl)N:C(Cl)N:\dot{C}NHNH}$]$C_6H_4AsO_3H_2$
__Prepn.:__ by reacting $4-H_2NNHC_6H_4AsO_3H_2$ with $(ClCN)_3$ in aq. $NaHCO_3$ at 2° (418).
__Rxns. with:__ $SnCl_2$ in HCl-HI $\rightarrow$ yellow ppt. (418).
10% H_2NNH_2 at 30° $\rightarrow$ the 4-chloro-6-hydrazino-s-triazin-2-ylhydrazino deriv. (418).
75% H_2NNH_2 $\rightarrow$ the 4,6-dihydrazino-s-triazin-2-ylhydrazino deriv. (418).
15% NH_4OH at 30° $\rightarrow$ the 4-amino-6-chloro-s-triazin-2-ylhydrazino deriv. (418).
Liq. NH_3 at -40 to 25° $\rightarrow$ the 4,6-diamino deriv. (418).
Hot concd. NH_4OH $\rightarrow$ the 4,6-diamino deriv. (418).

4-[2-(4-AMINO-6-CHLORO-s-TRIAZIN-2-YL)HYDRAZINO]BENZENEARSONIC ACID $C_9H_{10}AsClN_6O_3$
4-[$\overline{N:C(NH_2)N:C(Cl)N:\dot{C}NHNH}$]$C_6H_4AsO_3H_2$
__Prepn.:__ by reacting 4,6-dichloro-s-triazin-2-ylhydrazinobenzenearsonic acid with
15% aq. NH_3 at 30° (418).
__Props.:__ White ppt. (418).
__Rxn. with 75% aq. H_2NNH_2__ $\rightarrow$ the 4-amino-6-hydrazino-s-triazin-2-ylhydrazino deriv.
(418).

4-(4-CHLORO-6-HYDRAZINO-s-TRIAZIN-2-YLHYDRAZINO)BENZENEARSONIC ACID $C_9H_{11}AsClN_7O_3$
$\overline{N:C(NHNH_2)N:C(Cl)N:\dot{C}NHNH}C_6H_4AsO_3H_2$
__Prepn.:__ by reacting the 4,6-dichloro-s-triazin-2-ylhydrazino deriv. with 10% H_2N-
NH_2 at 30° (418).
__Props.:__ White substance (418).
__Rxn. with PhCH:CHCHO in HCl__ $\rightarrow$ yellow ppt. (418).

4-[2-(4,6-DIAMINO-s-TRIAZIN-2-YL)HYDRAZINO]BENZENEARSONIC ACID $C_9H_{12}AsN_7O_3$
$\overline{N:C(NH_2)N:C(NH_2)N:\dot{C}NHNH}C_6H_4AsO_3H_2$
__Prepn.:__ by treating 4,6-dichloro-s-triazin-2-ylhydrazinobenzenearsonic acid with
liq. NH_3 at -40 to -25° or with hot concd. NH_4OH (418).
__Rxn. with $Na_2S_2O_4$ in MeOH contg. HCl + HI__ $\rightarrow$ the corresp. arsenobenzene deriv.
(418).

p-(ISOPROPYLIDENEHYDRAZINO)BENZENEARSONIC ACID $C_9H_{13}AsN_2O_3$ $Me_2C:NNHC_6H_4AsO_3H_2$
__Prepn.:__ by warming 4-hydrazinobenzenearsonic acid with Me_2CO; 42% yield (47).
__Props.:__ Almost white crystalls. (47).

4-[2-(4-AMINO-6-HYDRAZINO-s-TRIAZIN-2-YL)HYDRAZINO]BENZENEARSONIC ACID
$C_9H_{13}AsN_8O_3$ $\bar{N}:C(NHNH_2)N:C(NH_2)N:CNHNHC_6H_4AsO_3H_2$
Prepn.: by treating 4-amino-6-chloro-s-triazin-2-ylhydrazinobenzenearsonic acid
with 75% aq. H_2NNH_2 (418).
Props.: White powder (418).

4-(4,6-DIHYDRAZINO-s-TRIAZIN-2-YLHYDRAZINO)BENZENEARSONIC ACID $C_9H_{14}AsN_9O_3$
$\bar{N}:C(NHNH_2)N:C(NHNH_2)N:CNHNHC_6H_4AsO_3H_2$
Prepn.: By reacting the 4,6-dichloro deriv. with 75% aq. H_2NNH_2 (418).
By reacting trihydrazino-s-triazine with $4-BrC_6H_4AsO(ONa)_2$ in PhCl in the
presence of Cu-bronze at 100° (418).
Props.: White microcryst. ppt. (418).
Rxns. with: $SnCl_2$ in HCl → yellow ppt. (418).
PhCH:CHCHO → yellow ppt. (418).
1,2-Naphthoquinone → yellow ppt. (418).
$AcCH_2CO_2Et$ → the pyrazolone deriv. (418).
SO_2 in concd. HCl contg. HI → the dichloroarsino deriv. (418).

α,β-DIOXOBUTYTIC ACID 4-ARSONOPHENYLHYDRAZONE $C_{10}H_{11}AsN_2O_6$ 4-[$HO_2CC(Ac):NNH$]-
$C_6H_4AsO_3H_2$
Prepn.: by diazoti zation of p-arsanilic acid and condensation with $AcCHCO_2Et$,
followed by hydrolysis with $Ba(OH)_2$; 82% yield (240).
Props.: M. 240-241° (dec.) (240).

CHLOROGLYOXYLIC ACID ETHYL ESTER 4-ARSONOPHENYLHYDRAZONE $C_{10}H_{12}AsClN_2O_5$
$EtO_2CC(Cl):NNHC_6H_4AsO_3H_2$
Prepn.: by coupling $4-H_2O_3AsC_6H_4N_2Cl$ with $ClCH_2CO_2Et$ in EtOH in the presence of
AcONa; 34% yield (1123).

γ-BROMO-α,β-DIOXOBUTYRIC ACID ETHYL ESTER α-(4-ARSONOPHENYLHYDRAZONE)
$C_{12}H_{14}AsBrN_2O_6$ $BrCH_2COC(CO_2Et):NNHC_6H_4AsO_3H_2$
Prepn.: By bromination of $MeCOC(CO_2Et):NNHC_6H_4AsO_3H_2$ in AcOH; 71% yield (1123).
By coupling $4-H_2O_3AsC_6H_4N_2Cl$ with $BrCH_2COCH_2CO_2Et$ in aq. EtOH in the presence
of AcONa; 9% yield (1123).
Rxn. with AcOK in EtOH → pyrazole deriv. formation by a ring closure (1123).

α,β-DIOXOBUTYRIC ACID ETHYL ESTER 2-ARSONOPHENYLHYDRAZONE $C_{12}H_{15}AsN_2O_6$
$2-[EtO_2CC(Ac):NNH]C_6H_4AsO_3H_2$
Prepn.: by diazotization of o-arsanilic acid and treating the diazonium deriv.
with ethyl acetoacetate in EtOH; 90% yield (240).
Props.: M. 182.5-183.5° (240).

α,β-DIOXOBUTYRIC ACID ETHYL ESTER 3-ARSONOPHENYLHYDRAZONE $C_{12}H_{15}AsN_2O_6$
$3-(EtO_2CC(Ac):NNH]C_6H_4AsO_3H_2$
Prepn.: by diazotization of m-arsanilic acid and treating the diazonium deriv.
with $AcCH_2CO_2Et$; 92% yield (240).
Props.: M. 197-198° (240).

α,β-DIOXYBUTYRIC ACID ETHYL ESTER α-(4-ARSONOPHENYLHYDRAZONE) $C_{12}H_{15}AsN_2O_6$

$MeCOC(CO_2Et):NNHC_6H_4AsO_3H_2$

Prepn.: by diazotization of arsanilic acid in alk. soln., treating the rxn. prod.
with HCl, and coupling with $AcCH_2CO_2Et$ in aq. EtOH contg. AcONa; 92% yield (1123).

Rxns. with: Br (1:1) in AcOH-AcONa at 20° → the γ-bromo deriv. (1123).

Br (1:2) in AcOH-AcONa at 40° → $BrC:C(OH)C(CO_2Et):NN-C_6H_4AsO_3H_2$ (1123).

2-[2-(p-ARSONOPHENYL)HYDRAZINO]-1-NAPHTHOL-4,8-DISULFONIC ACID $C_{16}H_{15}AsN_2O_{10}S_2$

Prepn.: by reducing the corresponding azo compd. with Zn dust in an alk. soln.
(428, 432).

Props.: Yellowish cryst. powder (428).

Rxns. with: $EtOCOCl$ → 2-[2-(p-arsonophenyl)-1,2-diethoxycarbonylhydrazino]-1-
naphthol-4,8-disulfonic acid ethylcarbonate (428).

BzCl → the 1-benzoylhydrazino and the 1,2-dibenzoylhydrazino benzoate deriv. (428,
432).

$CH_2(COCl)_2$ → the N,N'-malonyl deriv. (428).

MeCHO → the -N-N-CHMe deriv. (428).

HCHO → coupling of two molecules by means of two methylene bridges between the
N atoms and formation of a tetrazine ring (428).

PhNCO → the N,N'-dicarbanilyl deriv. (428).

EtCOCl → the 1,2-dipropionylhydrazino propionate deriv. (432).

3-[2-(1-METHYLAMINO-4,8-DISULFO-2-NAPHTHYL)-1,2-DIMETHYLHYDRAZINO]BENZENEARSONIC
ACID $C_{19}H_{21}AsN_3O_{10}S_2$

Prepn.: by coupling diazotized m-arsanilic acid with α-naphthylamine, reducing
the azo group with Fe powder in N NaOH and methylating with Me_2SO_4 (428, 432).

4-[2-(1-HYDROXY-4,8-DISULFO-2-NAPHTHYL)-2-BENZOYLHYDRAZINO]BENZENEARSONIC
ACID $C_{23}H_{19}AsN_2O_{11}S_2$

Prepn.: by treating the di-Na salt of 2-[2-(p-arsonophenyl)-hydrazino]-1-naphthol-
4,8-disulfonic acid with BzCl in air-free water under H in the presence of $CaCO_3$
(428).

4-[2-(1-HYDROXY-4,8-DISULFO-2-NAPHTHYL)-1,2-DIPROPIONYLHYDRAZINO]BENZENEARSONIC
ACID PROPIONATE $C_{25}H_{27}AsN_2O_{13}S_2$

Prepn.: by treating 2-[2-(-p-arsonophenyl)hydrazino)-1-naphthol-4,8-disulfonic
acid with Et COCl in dild. NaOH under H (428, 432).
Props.: Na salt, white crystals. (428).
Rxns. with $Na_2S_2O_4$ → the corresponding arseno deriv. (428).

4-[2-(1-HYDROXY-4,8-DISULFO-2-NAPHTHYL)-1,2-BIS(PHENYLCARBAMYL)HYDRAZINO]BEN-
ZENEARSONIC ACID $C_{30}H_{25}AsN_4O_{12}S_2$

Prepn.: by treating 2-[2-(p-arsonophenyl)hydrazino]-1-naphthol-4,8-disulfonic acid
with PhNCO in dild. $NaHCO_3$ soln. (428, 432).

4-[2-(1-HYDROXY-4,8-DISULFO-2-NAPHTHYL)-1,2-DIBENZOYLHYDRAZINO]BENZENEARSONIC
ACID BENZOATE $C_{37}H_{27}AsN_2O_{13}S_2$

Prepn.: by treating the corresponding 1-naphthol hydrazino deriv. with BzCl (432).

AZOBENZENEARSONIC ACIDS

3-(5-AMINO-4-BROMO-8-ISOQUINOLYLAZO)BENZENEARSONIC ACID $C_{15}H_{12}AsBrN_4O_3$

Prepn.: by reacting $3-H_2O_3AsC_6H_4N_2Cl$ with 5-amino-4-bromoisoquinoline in dil.
AcOH; 73% yield (332).
Props.: Decomp. 180-182° (332).

4-(5-AMINO-4-BROMO-8-ISOQUINOLYLAZO)BENZENEARSONIC ACID $C_{15}H_{12}AsBrN_4O_3$

Prepn.: by coupling $4-H_2O_3AsC_6H_4N_2Cl$ with 5-amino-4-bromoisoquinoline in dild. AcOH; 47% yield (332).
Props.: Decomp. 219-221° (332).

4-(8-AMINO-4-BROMO-5-ISOQUINOLYLAZO)BENZENEARSONIC ACID $C_{15}H_{12}AsBrN_4O_3$

Prepn.: by $4-H_2O_3AsC_6H_4N_2Cl$ with 8-amino-4-bromoisoquinoline in dil. AcOH; 66% yield (332).
Props.: Decomp. 206-207° (332).

4-(5-AMINO-1-CHLORO-8-ISOQUINOLYLAZO)BENZENEARSONIC ACID $C_{15}H_{12}AsClN_4O_3$

Prepn.: by reacting $4-H_2O_3AsC_6H_4N_2Cl$ with 5-amino-1-chloroisoquinoline in dil. AcOH; 82% yield (332).
Props.: Decomp. 260° (332).

4-(8-HYDROXY-5-QUINOLYLAZO)BENZENEARSONIC ACID $C_{13}H_{12}AsN_3O_4$

Prepn.: by coupling $4-H_2O_3AsC_6H_4N_2Cl$ with 8-quinolinol (1135).
Props.: M. 235.4° (1135).
Salts: HCl, m. 219.8° (1135).

3-(5-AMINO-8-ISOQUINOLYLAZO)BENZENEARSONIC ACID $C_{15}H_{13}AsN_4O_3$

Prepn.: by reacting 3-$H_2O_3AsC_6H_4N_2Cl$ with 5-aminoisoquinoline in dild. AcOH; 46% yield (332).
Props.: Decomp. 180-182° (332).

4-(5-AMINO-8-ISOQUINOLYLAZO)BENZENEARSONIC ACID $C_{15}H_{13}AsN_4O_3$

Prepn.: by coupling 4-$H_2O_3AsC_6H_4N_2Cl$ with 5-aminoisoquinoline in dild. AcOH; 95% yield (332).
Props.: Decomp. 300° (332).

2-(2-HYDROXY-3,6-DISULFO-1-NAPHTHYLAZO)BENZENEARSONIC ACID $C_{16}H_{13}AsN_2O_{10}S_2$

Prepn.: by coupling 2-$H_2O_3AsC_6H_4N_2Cl$ with 2,3,6-$HOC_{10}H_5(SO_3Na)_2$ (820).
Use: Reagent for thorium (820).

4-(1-HYDROXY-4,8-DISULFO-2-NAPHTHYLAZO)BENZENEARSONIC ACID $C_{16}H_{13}AsN_2O_{10}S_2$

Prepn.: by coupling p-$H_2O_3AsC_6H_4N_2Cl$ with 1-naphthol-4,8-disulfonic acid or p-hydrazinobenzenearsonic acid with 1,2-naphthoquinone-4,8-disulfonic acid (411, 41?

Rxns. with: Zn in aq. NaOH → the hydrazino deriv. (428).

2-(2-HYDROXY-3,6-DISULFO-1-NAPHTHYLAZO)BENZENEARSONIC ACID $C_{16}H_{13}AsN_2O_{10}S_2$

Prepn.: by reacting 2-hydroxy-3,6-naphthalenedisulfonic acid with $2\text{-}H_2O_3AsC_6H_4\text{-}N_2Cl$ in 10% NaOH (709).
Props.: Polarography (1267a).
Rxns. with: Sn(IV), U(VI), Ta, Nb, rare earths, and Zr → orange-yellow ppts. (1045).
Th and Bi → red, sol. complexes (colorimetric detn.) (1045).
Use: Photometric detn. of Th (110a, 234, 834, 965a, 1042a).

4-(2-HYDROXY-6,8-DISULFO-7-NAPHTHYLAZO)BENZENEARSONIC ACID $C_{16}H_{13}AsN_2O_{10}S_2$

Prepn.: by treating 6,8-disulfo-2-naphthol with $4\text{-}H_2O_3AsC_6H_4N_2Cl$ in 10% NaOH (709).
Use: Reagent for titration of rare earths and F^- (725a).

2-(1,8-DIHYDROXY-3,6-DISULFO-2-NAPHTHYLAZO)BENZENEARSONIC ACID $C_{16}H_{13}AsN_2O_{11}S_2$

Prepn.: by coupling $2\text{-}H_2O_3AsC_6H_4N_2Cl$ with chromotropic acid (586).
Props.: Absorption spectrum (608). Dissocn. const. (673a). Polarography (1267a).
Rxns. with: U, Al, and Ca ions → Complex compds. (608).
Pu(IV) → Pu complex (953).
Use: Reagent for detn. of: Th (586); Rare earth metals (721); Al (722); Zr (723); Be (1128a); Pu(IV) (953); Rare earths in U (51a); and U-ethylenedinitrilotetraacetic acid (444c).
Spectrophotometric detn. of γ-amts. of La (1128b).

Detection of small amts. of In (723a).
Photometric titration of rare earths in ethylenedinitrilotetraacetic acid (725a).

4-(1-AMINO-3,6,8-TRISULFO-2-NAPHTHYLAZO)BENZENEARSONIC ACID $C_{16}H_{13}AsN_2O_{13}S_3$

Prepn.: by reacting 8-aminonaphthalene-1,3,6-trisulfonic acid with 4-$H_2O_3AsC_6$-H_4N_2Cl in 10% NaOH (709).

4-(2-AMINO-8-HYDROXY-3,6-DISULFO-1-NAPHTHYLAZO)BENZENEARSONIC ACID $C_{16}H_{14}AsN_3O_1$

Prepn.: by reacting 2-amino-8-naphthol-3,6-disulfonic acid with diazotized arsanilic acid in alk. soln. (493).
Props.: K salt, red-black leaves (493).
Rxns. with tetraazotized [3,4-I(H_2N)C_6H_3]$_2CH_2$ →

4-(1-AMINO-8-HYDROXY-3,6-DISULFO-7-NAPHTHYLAZO)BENZENEARSONIC ACID
   $C_{16}H_{14}AsN_3O_{10}S_2$

Prepn.: by treating 1-amino-8-hydroxynaphthalene-3,6-disulfonic acid with 4-$H_2O_3AsC_6H_4N_2Cl$ in 10% NaOH (709).

p-(4-HYDROXY-2-METHYLPYRIMIDO[1,2-a]BENZIMIDAZOL-3-YLAZO)BENZENEARSONIC ACID
$C_{17}H_{14}AsN_5O_4$

Prepn.: by coupling $4-H_2O_3AsC_6H_4N_2Cl$ with $AcCH_2CO_2Et$ in the presence of two
equivs. of NaOH and reacting the diazo compd. with 2-aminobenzimidazole in AcOH;
70% yield (525).

  4-(1-AMINO-2-METHYL-3-NAPHTHYLAZO)BENZENEARSONIC ACID $C_{17}H_{16}AsN_3O_3$

Prepn.: by coupling $4-H_2O_3AsC_6H_4N_2Cl$ with 2-methyl-1-naphthylamine (1305).

  p-PHENYLENEDI(4-AZOBENZENEARSONIC ACID) $C_{18}H_{16}As_2N_4O_6$ $1,4-C_6H_4(N:NC_6H_4AsO_3H_2)_2$
Prepn.: by condensing $4-ONC_6H_4AsO_3H_2$ with $1,4-C_6H_4(NH_2)_2$ (984).

  4,4'-(4-HYDROXY-m-PHENYLENEBISAZO)DIBENZENEARSONIC ACID $C_{18}H_{16}As_2N_4O_7$

Prepn.: by coupling $4-H_2O_3AsC_6H_4N_2Cl$ with $4-(4-HOC_6H_4N_2)C_6H_4AsO_3H_2$ (378).
Props.: Brown amorph. powder (378).

  4,4'-(2,6-DIHYDROXY-p-PHENYLENEBISAZO)DIBENZENEARSONIC ACID $C_{18}H_{16}As_2N_4O_8$

Prepn.: by coupling $4-H_2O_3AsC_6H_4N_2Cl$ with resorcinol (2:1) (376a).

  4,4'-(4-HYDROXY-5-METHYL-m-PHENYLENEBISAZO)DIBENZENEARSONIC ACID $C_{19}H_{18}As_2N_4O_7$

Prepn.: by coupling $H_2O_3AsC_6H_4N_2Cl$ with o-cresol (984).

3-(4-[3-(p-ARSONOPHENYLAZO)-4-HYDROXY-5-IODOPHENOXY]-3,5-DIIODOPHENYL) ALANINE
$C_{21}H_{17}AsI_3N_3O_7$

$$HO_2CCH(NH_2)CH_2- \text{(ring with I, I)} -O- \text{(ring with I, OH)} -N:N- \text{(ring)} -AsO_3H_2$$

<u>Prepn.</u>: by coupling $4-H_2O_3AsC_6H_4N_2Cl$ with thyroxine (52).

3-(4-ARSONOBENZENEAZO)-N-(4-AMINOBENZENESULFONYL)TYROSINE $C_{21}H_{21}AsN_4O_8S$

$$H_2N- \text{(ring)} -SO_2NHCH(CO_2H)CH_2- \text{(ring, OH)} -N:N \text{(ring)} AsO_3H_2$$

<u>Prepn.</u>: by coupling N-(p-aminobenzenesulfonyl)tyrosine with $4-H_2O_3AsC_6H_4N_2Cl$
(835).
<u>Props.</u>:   m. >300° (835).

BIS(4-ARSONOPHENYLAZO)1-TYROSINE $C_{21}H_{21}As_2N_5O_9$

$$\text{CH}_2\text{CH(NH}_2)\text{CO}_2\text{H} \quad \text{(ring, OH)} \quad (N:NC_6H_4AsO_3H_2)_2$$

<u>Prepn.</u>: by reacting $4-H_2O_3AsC_6H_4N_2Cl$ with 1-tyrosine (681).

8-AMINO-2,7-BIS(4-ARSONOPHENYLAZO)-1-NAPHTHOL-5-SULFONIC ACID $C_{22}H_{19}As_2N_2O_{10}S$

$$H_2O_3As- \text{(ring)} -N:N- \text{(naphthol, SO_3H, OH, NH_2)} -N:N- \text{(ring)} -AsO_3H_2$$

<u>Prepn.</u>: by coupling 1-amino-8-naphthol-4-sulfonic acid with $4-H_2O_3AsC_6H_4N_2Cl$
(984).

4,4'-(4,4'-DIHYDROXY-3,3'-BIPHENYLENEDIAZO)DIBENZENEARSONIC ACID
$C_{24}H_{20}As_2N_4O_8$

$$\left[-\text{(ring, -OH)} N:N- \text{(ring)} - AsO_3H_2 \right]_2$$

<u>Prepn.</u>: by coupling p-biphenol with $4-H_2O_3AsC_6H_4N_2Cl$ (984).

240

2,4,6-TRIS(p-ARSONOPHENYLAZO)RESORCINOL $C_{24}H_{21}As_3N_6O_{11}$

Prepn.: by coupling resorcinol with $4\text{-}H_2O_3AsC_6H_4N_2Cl$ (984).

2,4,6-TRIS(p-ARSONOPHENYLAZO)PHLOROGLUCINOL $C_{24}H_{21}As_3N_6O_{12}$

Prepn.: by coupling phloroglucinol with $4\text{-}H_2O_3AsC_6H_4N_2Cl$ (984).

4-(2-HYDROXY-1-NAPHTHYLAZO)BENZENEARSONIC ACID DECANOATE $C_{26}H_{31}AsN_2O_5$

Prepn.: by heating 4-(2-hydroxy-1-naphthylazo)benzenearsonic acid with $n\text{-}C_9\text{-}C_{19}COCl$ in C_6H_6 at 100° (1218).
Props.: Red-brown, m. 235-236° (1218).

4-(2-DECANOYLAMIDO-1-NAPHTHYLAZO)BENZENEARSONIC ACID $C_{26}H_{32}AsN_3O_4$

Prepn.: by heating 2-aminonaphthylazcbenzenearsonic acid with $C_9H_{09}COCl$ in C_6H_6 at 80-100° (1218).
Props.: Red-brown, m. 127° (1218).

4,6-BIS[4-(4-ARSONOPHENYLAZO)PHENYLAZO]o-CRESOL $C_{31}H_{26}AsN_8O_7$

Prepn.: by coupling o-cresol with $H_2O_3AsC_6H_4N_2C_6H_4N_2Cl$ (984).

4-(2-PALMITOYLAMIDO-1-NAPHTHYLAZO)BENZENEARSONIC ACID $C_{32}H_{44}AsN_3O_4$

Props.:   Decomp. 294° (78).

3,3;5,5'-TETRAKIS(4-ARSONOPHENYLAZO)-2,4,4'-TRIHYDROXYAZOBENZENE $C_{36}H_{30}As_4N_{10}O$

Prepn.:   by coupling 2,4,4'-trihydroxyazobenzene with $4\text{-}H_2O_3AsC_6H_4N_2Cl$ (984).

2,4,6-TRIS[4-(4-ARSONOPHENYLAZO)PHENYLAZO]PHLOROGLUCINOL $C_{42}H_{33}As_3N_{12}O_{12}$

Prepn.:   by coupling phloroglucinol with $H_2O_3AsC_6H_4N_2C_6H_4N_2Cl$ (984).

BIS{4-[7-AMINO-8-(p-ARSONOPHENYLAZO)-1-HYDROXY-3,6-DISULFO-2-NAPHTHYLAZO]-3-
    IODOPHENYL}METHANE $C_{45}H_{34}As_2I_2N_{10}O_{20}S_4$

Prepn.:   by reacting 4-(2-amino-8-hydroxy-3,6-disulfo-1-naphthylazo)benzenearsonic
acid with tetraazotized di(4-amino-3-iodophenyl)methane (493).
Props.:   K salt, dark powder (493).

4,4'-(p-BIPHENYLENEDISAZO)BIS{2,6-BIS[p-(p-ARSONOPHENYLAZO)PHENYLAZO]RESORCINOL}
$C_{72}H_{54}As_4N_{30}O_{16}$

$H_2O_3AsC_6H_4N:NC_6H_4N:N$ $\quad\quad\quad\quad\quad\quad\quad\quad\quad\quad$ $N:NC_6H_4N:NC_6H_4AsO_3H_2$

HO— $\quad$ —NN— $\quad$ —NN— $\quad$ —OH

$H_2O_3AsC_6H_4N:NC_6H_4N:N$ $\quad\quad\quad\quad\quad\quad\quad\quad\quad\quad$ $N:NC_6H_4N:NC_6H_4AsO_3H_2$

Prepn.: by coupling 4,4'-bis(2,4-dihydroxyphenylazo)biphenyl with $H_2O_3AsC_6H_4NN$-$C_6H_4N_2Cl$ (984).

TABLE XXXVIII
AZOBENZENEARSONIC ACID

X-(RN:N)C$_6$H$_4$AsO$_3$H$_2$ where R is:	Prepd. from (Ars = -C$_6$H$_4$AsO$_3$H$_2$)	Yield %	Props. and Remarks	Ref.
C$_{10}$				
2-[O:CON:C(Me)CH-	2-[AcC(CO$_2$Et):NNH]Ars + HON-H$_2$.HCl + NaHCO$_3$	28	m. 232-233°	240
3-[O:CON:C(Me)CH-	3-[AcC(CO$_2$Et):NNH]Ars + HON-H$_2$.HCl + NaHCO$_3$	55	m. >250°	240
4-[AcCH(CO$_2$H)-	4-ClN:NArs + AcCH$_2$CO$_2$Et		Rxn.1	525
2-[O:CNHN:C(Me)CH-	2-AcC(CO$_2$H):NNHArs + H$_2$NNH$_2$.HCl in alk. soln.	22	m. 268-269° (dec.)	240
3-[O:CNHN :C(Me)CH-	3-AcC(CO$_2$H):NNHArs + H$_2$NNH$_2$-HCl in alk. soln.	65	m. >250°	240
4-[O:CNHN:C(Me)CH-	4-AcC(CO$_2$H)C:NNHArs + H$_2$NN-H$_2$.HCl in alk. soln.	44	m. >250°	240
C$_{12}$				
4-PhN:NC$_6$H$_4$AsO$_3$H$_2$	PhN:NC$_6$H$_4$NH$_2$ by the Bart or Bart-Scheller method	24	Orange plates, m. 332.5-333.-5°, rxns.[2,3]	289
4-(4-HOC$_6$H$_4$-	PhOH + ArsN$_2$Cl in alk. soln.		Golden yellow plates, rxn.[2,4]	289, 378, 822, 984
4-[2,4-(HO)$_2$C$_6$H$_3$-	Resorcinol + ArsN$_2$Cl			984, 709, 984
4-(4-H$_2$NC$_6$H$_4$-	1,4-C$_6$H$_4$(NH$_2$)$_2$ + Ars-N$_2$Cl			617
	4-AcNHC$_6$H$_4$NH$_2$ + 4-ONArs in AcOH, followed by hydrolysis			984
	PhNH$_2$.MeSO$_3$H + ArsN$_2$Cl, followed by hydrolysis		Red powder	984
4-[3,4-H$_2$N(HO)C$_6$H$_3$-	4-[3,4-AcNH(HO)C$_6$H$_3$N$_2$]Ars + HCl in MeOH	70	HCl salt, red powder, rxn.[5]	289
4-[4,2-H$_2$N(HO)C$_6$H$_3$-	3-H$_2$NC$_6$H$_4$OH + ArsN$_2$Cl			617
4-(4-HO$_3$SC$_6$H$_4$NH-	4-H$_2$NC$_6$H$_4$SO$_3$H + ArsN$_2$Cl			380
4-[2,5-(HO$_3$S)$_2$C$_6$H$_3$NH-	2,5-(HO$_3$S)$_2$C$_6$H$_3$NH$_2$ + ArsN$_2$-Cl			380
(:NC$_6$H$_4$AsO$_3$H$_2$-o)$_2$	Redn. of 2-O$_2$NC$_6$H$_4$AsO$_3$H$_2$ with Fe in aq. NaCl		m. 269°	1089

...AsO₃H₂-m/2	...(NC₆H₄AsO₃H₂-m/2)... aq. NaCl		Orange crystals	013
(:NC₆H₄AsO₃H₂-p)	4-H₂NArs + 4-ONArs		White, microcryst. crystals, rxn.2	984
2-H₂O₃AsC₆H₄N:NC₆H₄-2-AsO₃H₂	2-ArsN₂Cl + 4-O₂NC₆H₄AsCl₂ in alk. soln.			611
O(:NC₆H₄AsO₃H₂-m)₂	3-O₂NC₆H₄AsO₃H₂ + As₂O₃ in aq. NaOH	90	Yellow, explodes on heating rxn.6,7	865
	Addn. of 3-O₂NC₆H₄AsCl₂, Na₂CO₃, and CuSO₄ to 2-MeC₆H₄-NH₂, HCl, and NaNO₂ in H₂O	48	Colorless powder	613
C₁₃				
4-HO₂CC₆H₄-	4-MeC₆H₄N₂Ars + KMnO₄	68	Orange needles, rxn.8	289
4-[4,3-HO(HO₂C)C₆H₃-	4-ArsN₂Cl + 2-HOC₆H₄CO₂H			709
2-SC₆H₄N:CNH- or HN:CSC₆H₄N-	4-ArsN₂Cl + 2-aminobenzothiazole		M. 176-178°	1264
4-[4,3-H₂N(HO₂C)C₆H₃-	2-H₂NC₆H₄CO₂H + 4-ArsN₂Cl			709
4-[2,4-HO₂C(HO₃S)C₆H₃NH-	2,5-H₂N(HO₃S)C₆H₃CO₂H + 4-ArsN₂Cl	84	Yellow needles, decomp. 210°	378
3-[2,5-HO₂C(HO₃S)C₆H₃NH-	2,4-H₂N(HO₃S)C₆H₃CO₂H + 3-ArsN₂Cl			380
4-(4-MeC₆H₄-	4-(4-MeC₆H₄N:N)C₆H₄NH₂. by the Bart-Scheller method	41	Orange needles, rxn.9	289
4-[2,5-HO(Me)C₆H₃-	4-ArsN₂Cl + p-cresol			709
4-[4,2-HO(Me)C₆H₃-	4-ArsN₂Cl + m-cresol			709
4-[4,3-HO(Me)C₆H₃-	4-ArsN₂Cl + o-cresol			709
C₁₄				
4-[4,3-HO(AcNH)C₆H₃-	4-ArsN₂Cl + 2-AcNHC₆H₄OH	75	Red powder, m. 224.8-225.2° rxn.10	289

1. Rxn. with 2-aminobenzimidazole → 4-(4-hydroxy-2-methylpyrimido[1,2-a]benzimidazol-3-ylazo)benzenearsonic acid (525).
2. Rxn. with SO₂ → the arsenoso deriv. (239, 611).
3. Catalytic redn. over Raney Ni → p-arsanilic acid (289).
4. Rxn. with 4-H₂O₃AsC₆H₄N₂Cl → H₂O₃AsC₆H₄N₂C₆H₃(OH)N₂C₆H₄AsO₃H₂ (378).
5. Redn. with SO₂ → red insol. prod. (239).
6. Redn. with SO₂ in HCl → [3-Cl₂AsC₆H₄N]₂O (865).
7. Redn. with Fe in 5% aq. NaCl → {NC₆H₄AsO₃H₂-m}₂ (613).
8. Rxn. with PCl₃ and FCl₅ in CHCl₃, followed by NH₄OH → OAsC₆H₄N:NC₆H₄CONH₂ (289).
9. Oxidation with KMnO₄ → HO₂CC₆H₄N:NC₆H₄AsO₃H₂ (289).
10. Hydrolysis with methanolic HCl → the 3-amino deriv. (289).

TABLE XXVIII (cont'd.)
AZOBENZENEARSONIC ACID

x-(RN:N)C6H4AsO3H2 where R is:	Prepd. from (Ars = -C6H4AsO3H2):	Yield %	Props. and Remarks	Ref.
4-(4-Me2NC6H4-	4-ArsN2Cl + PhNMe2	79	m. 310° (dec.)	4, 682
4-(4-EtNHC6H4-	4-ArsN2Cl + PhNHEt		m. 276° (dec.)	4
4-[EtO2CC:C(Me)NHC(OH):C-	4-ArsN2Cl + 3-carbethoxy-5-hydroxy-2-methylpyrrole		Dark yellow powder, m. 180° (dec.), stable in air	870
2-[CH2CH2C(Me)2CH2C(OH):C-	2-ArsN2Cl + methone			709
3-[CH2CH2C(Me)2CH2C(OH):C-	3-ArsN2Cl + methone			709
4-[CH2CH2C(Me)2CH2C(OH):C-	4-ArsN2Cl + methone			709
4-[HNC(Me):C(CO2Et)C(Me):C-	4-ArsN2Cl + 3-carbethoxy-2,4-dimethylpyrrole		Orange-yellow crystals, decomp. 210°, stable in air	871
4-[MeC:C(CO2Et)NHC(Me):C-	4-ArsN2Cl + 2-carbethoxy-3,5-dimethylpyrrole		Yellow-orange powder, m. 185° (dec.)	871
C15				
4-(4-PrNHC6H4-	4-ArsN2Cl + PhNHPr		m. 286° (dec.)	4
C16				
4-(2,3-HOC10H6-	4-ArsN2Cl + 2-naphthol		Brick red, cryst. leaflets, tautomerims 11, rxns.12	709
4-(4,1-HOC10H6-	4-ArsN2Cl + 1-naphthol	91		378
4-(4,1-H2NC10H6-	4-ArsN2Cl + 1-C10H7NH2			709
4-[2,1-H2NC10H6-	4-ArsN2Cl + 2-C10H7NH2		Rxns. 13,14	709
4-(4-Et2NC6H4-	4-ArsN2Cl + PhNEt2	37.6		682
4-(4-BuNHC6H4-	4-ArsN2Cl + PhNHBu			4
4-[4-(Me2CHCH2NH)C6H4-	4-ArsN2Cl + PhNHCH2CHMe2		Brick red, m. 303° (dec.) Crystallizes with 1 EtOH	4
4-[4-(MeEtCHNH)C6H4-	4-ArsN2Cl + PhNHCHMeEt			4
C17				
4-(4,1-MeOC10H6-	4-(4-HOC10H6N2)C6H4AsO3H2 + Me2SO4	72	Rxn. 15	378
4-[4-(n-C5H11NH)C6H4-	4-ArsN2Cl + PhNHC5H11			4
4-[4-(Me2CHCH2CH2NH)C6H4-	4-ArsN2Cl + PhNHCH2CH2CHMe2			4
4-[4-(sec-C5H11NH)C6H4-	4-ArsN2Cl + PhNH-sec-C5H11			4
4-[4-(MeBuN)C6H4-	4-ArsN2Cl + PhNMeBu		m. 245° (dec.)	4

C_{18}			
4-[4-(HOC$_6$H$_4$N:N)C$_6$H$_4$-	4-ArsN:NC$_6$H$_4$N$_2$Cl + PhOH	Brown red, m. 292° (dec.)	984
4-[4-(cyclo-C$_6$H$_{11}$NH)C$_6$H$_4$-	4-ArsN$_2$Cl + cyclo-C$_6$H$_{11}$NHPh	m. 270° (dec.)	4
4-[4-(n-C$_6$H$_{13}$NH)C$_6$H$_4$-	4-ArsN$_2$Cl + n-C$_6$H$_{13}$NHPh	m. 265° (dec.)	4
4-{4-[Me$_2$CH(CH$_2$)$_3$NH]C$_6$H$_4$-	4-ArsN$_2$Cl + Me$_2$CH(CH$_2$)$_3$NHPh		4
C_{19}			
4-[4-(PhCH$_2$NH)C$_6$H$_4$-	4-ArsN$_2$Cl + PhCH$_2$NHPh	m. 340° (dec.)	4
4-[4-(n-C$_7$H$_{15}$NH)C$_6$H$_4$-	4-ArsN$_2$Cl + n-C$_7$H$_{15}$NHPh		4
C_{20}			
4-[2,5-HO(Me$_3$CCH$_2$CMe$_2$)C$_6$H$_3$-	4-ArsN$_2$Cl + 4-(Me$_3$CCH$_2$CMe$_2$)-C$_6$H$_4$OH		768
4-[4-(n-C$_8$H$_{17}$NH)C$_6$H$_4$-	4-ArsN$_2$Cl + n-C$_8$H$_{17}$NHPh	Red substance, m. 155° (dec.)	78
C_{21}			
4-{4-[Et$_2$N(CH$_2$)$_3$CHMe]C$_6$H$_4$-	4-ArsN$_2$Cl + Et$_2$N(CH$_2$)$_3$CHMe-NHPh	Red powder, decomp. 125°	1196
C_{24}			
4-[2,1-(n-C$_8$H$_{17}$NH)C$_{10}$H$_6$-	4-ArsN$_2$Cl + 2-n-C$_8$H$_{17}$NHC$_{10}$H$_7$	m. 206° (dec.)	78
4-[4-(n-C$_{12}$H$_{25}$NH)C$_6$H$_4$-	4-ArsN$_2$Cl + n-C$_{12}$H$_{25}$NHPh		4
C_{26}			
4-[4-(n-C$_{14}$H$_{29}$NH)C$_6$H$_4$-	4-ArsN$_2$Cl + n-C$_{14}$H$_{29}$NHPh	Brown-red plates	4
C_{28}			
4-[4-(n-C$_{16}$H$_{33}$NH)C$_6$H$_4$-	4-ArsN$_2$Cl + n-C$_{16}$H$_{33}$NHPh	Red, m. 283° (dec.)	78
C_{30}			
4-[4-(n-C$_{18}$H$_{37}$NH)C$_6$H$_4$-	4-ArsN$_2$Cl + n-C$_{18}$H$_{37}$NHPh	Orange leaflets	4
C_{39}			
4-[4-(C$_{27}$H$_{45}$NH)C$_6$H$_4$-	4-ArsN$_2$Cl + cholesteryl-NHPh		4

11. In an acid soln. exists in the hydrazonaphthoquinone and in alkali in the azo-hydroxy form (378).
12. Rxn. with Me$_2$SO$_4$ → the 4-methoxynaphthylazo deriv. (378).
13. RCOCl → the RCONH- deriv. (78, 1218).
14. Cholesteryl chloroformate → compd. C$_{44}$H$_{58}$AsN$_3$O$_5$, m. 290° (78).
15. Rxn. with Na$_2$S$_2$O$_2$ → redn. of the N:N group (378).

HYDROXYBENZENEARSONIC ACIDS AND O-SUBSTITUTED DERIVATIVES

2-HYDROXYBENZENEARSONIC ACID $C_6H_7AsO_4$ $2-HOC_6H_4AsO_3H_2$
Prepn.: from $2-HOC_6H_4NH_2$ by the Bart method; 65% yield (883).
Rxns. with: PhN_2Cl at pH 8.9-9.1 → $2,5-HO(PhN_2)C_6H_3AsO(OH)_2$ (289).
$AcNHC_6H_4SO_2C_6H_4N_2Cl$ → $2,4-HO(AcNHC_6H_4SO_2C_6H_4N_2)C_6H_4AsO_3H_2$ (222).
Diazotized poly(aminostyrene), followed by condensation with HCHO → metal complex-
ing resin (273a).

3-HYDROXYBENZENEARSONIC ACID $C_6H_7AsO_4$ $3-HOC_6H_4AsO_3H_2$
Prepn.: By diazotization of m-arsanilic acid, followed by decompn. of the diazon-
ium compd.; 85% yield (811); can not be prepd. from $3-HOC_6H_4NH_2$ by the Bart
method (883).
Props.: M. 160-170° (dec.) (811).
Rxn. with Me_2SO_4 → $3-MeOC_6H_4AsO_3H_2$ (811).

4-HYDROXYBENZENEARSONIC ACID $C_6H_7AsO_4$ $4-HOC_6H_4AsO_3H_2$
Prepn.: From $4-HOC_6H_4NH_2$ by the Bart method; 61-70% yield (91, 883); isolated
as $ZnCl_2$ salt (1109).
    By heating PhOH with H_3AsO_4 at 146°; 32.5% yield (1234).
    By heating $4-HOC_6H_4SO_3H$ with 95% H_3AsO_4; 65% yield (rxn. mechanism suggested
(1235)); 90% yield (1240).
Props.: Dissocn. consts. (124, 1025, 1034). Redn. potential (124). Potentiometric titration (1236). Magnetic study (1029).
Rxns. with: ROH in NaOH soln. → $4-ROC_6H_4AsO_3H_2$ (62).
Ac_2O → $4-AcOC_6H_4AsO_3H_2$ (284).
PhN_2Cl → $4-PhN_2C_6H_4OH$, $4,3-HO(PhN_2)C_6H_3AsO_3H_2$, and $2,4-(PhN_2)_2C_6H_3OH$ (289).
$AcNHC_6H_4SO_2C_6H_4N_2Cl$ → $4,?-HO(AcNHC_6H_4SO_2C_6H_4N_2)C_6H_3AsO_3H_2$ (222).
RCl → $ROC_6H_4AsO_3H_2$ (501, 502, 765, 769, 971, 972, 1047).
$Br(CH_2)_nBr$ → $(CH_2)n-2(OC_6H_4AsO_3H_2)_2$ (999).
$RR'CCH_2O$ → $RR'C(OH)CH_2OC_6H_4AsO_3H_2$ (556).
RSO_2Cl in alk. soln. → $RSO_3C_6H_4AsO_3H_2$ (1082).
HCHO in H_2SO_4 → $CH_2[2,5-HOC_6H_3AsO_3H_2]_2$ (358).
HCHO in 20% NaOH with/without 1 equiv. PhOH → ion exchange resin (555).
Metal ions → ppts. (quant. analysis) (1025, 1027).
Use: Reagent for Zr^{++} (198). Stabilizer for motor fuels and lubricating oils
contaminated with Cu (1217).

4,4'-(ETHYLENEDIOXY)DIBENZENEARSONIC ACID $C_{14}H_{16}As_2O_8$ $(-CH_2OC_6H_4AsO_3H_2)_2$
Prepn.: by refluxing $4-HOC_6H_4AsO_3H_2$ with $(-CH_2Br)_2$ in 10% NaOH (999).

4,4'-(TRIMETHYLENEDIOXY)DIBENZENEARSONIC ACID $C_{15}H_{18}As_2O_8$ $CH_2(CH_2OC_6H_4AsO_3H_2)_2$
Prepn.: by refluxing $4-HOC_6H_4AsO_3H_2$ with $CH_2(CH_2Br)_2$ in 10% NaOH (999).

4,4'-(2-HYDROXYTRIMETHYLENEDIOXY)DIBENZENEARSONIC ACID $C_{15}H_{18}As_2O_9$ $HOCH(CH_2OC_6-H_4AsO_3H_2)_2$
Prepn.: by refluxing $4-HOC_6H_4AsO_3H_2$ with $HOCH(CH_2Cl)_2$ in 6 N NaOH and acidifying
the rxn. prod., 30% yield (62).
Props.: M. >270° (62).
Rxn. with $HNO_3-H_2SO_4$ at 0° → the 3,3'-dinitro derив. (62).

HYDROXYBENZENEARSONIC ACIDS AND O-SUBSTITUTED DERIVATIVES

$ROC_6H_4AsO_3H_2$ where R is:	Prepd. from (Ars stands for $-OC_6H_4AsO_3H_2$)	Yield %	Props. and Remarks	Ref.
C_7				
3-Me-	3-HArs + Me_2SO_4	75	Rxns.[1,2]	811
			m. 138-141°	613
4-Me-			Dissocn. const. and redn. pot-ential	123
C_8				
4-Ac-	4-HArs + Ac_2O	100	Rxns.[3]	284
4-Et-	4-HArs + EtBr		Rxn. with HNO_3-H_2SO_4 → the 3-NO_2 deriv.	1047
	4-$EtOC_6H_4N_2BF_4$ + $NaAsO_2$ + $CuCl_2$	73		1091
3-$HOCH_2CH_2$-	3-$HOCH_2CH_2OC_6H_4N_2Cl$ + Na_3AsO_3	53	m. 110°, rxns.[4,5]	82
4-$HOCH_2CH_2$-	4-$HOCH_2CH_2OC_6H_4N_2Cl$ + Na_3AsO_3		m. 146-147°, rxns. [1,6,7]	532
	4-HArs + $HOCH_2CH_2Cl$		White hexagonal plates, m. 128°	502, 971
C_9				
4-[N:C(NH_2)N:C(NH_2)N:C-	4-N:C(NH_2)N:C(NH_2)N:COC_6H_4-NH_2 by the Bart method	55	m. > 250°	1306
	N:C(NH_2)N:C(NH_2)N:CCl + 4-HArs			1306
4-$AcCH_2$-	$AcCH_2Cl$ + 4-HArs	9	White crystals, m. 172-173°	501, 972
4-MeO_2CCH_2-	$ClCH_2CO_2H$ + 4-HArs, followed by MeOH + H_2SO_4		Na salt, rxns.[3, 8, 9] m. 191°, rxn.[10]	501, 1153

1. Rxn. with SO_2 in HCl → the corresponding dichloroarsine (271, 284, 502, 613).
2. Redn. with Zn-Hg in HCl → 3-$MeOC_6H_4AsH_2$ (811).
3. Redn. with SO_2 in dild. HCl followed by neutralization → the arsenoso deriv. (62, 83, 275, 284, 501).
4. Rxn. with HNO_3 → 2,3-$O_2N(O_2NOCH_2CH_2O)C_6H_3AsO_3H_2$ + 4,3-$O_2N(O_2NOCH_2CH_2O)C_6H_3AsO_3H_2$ (82).
5. Rxn. with H_2SO_3 + HI → the arsenoso deriv. (82).
6. Rxn. with HNO_3 → the 3-NO_2 deriv. (502, 532).
7. Rxn. with HNO_3-H_2SO_4 on a steam bath → the 2,6-dinitro deriv. (502).
8. Redn. with H_3PO_2 → the arseno deriv. (62, 83, 501).
9. Rxn. with HNO_3-H_2SO_4 → the 3-nitro deriv. (62, 501, 556, 1047).
10. Rxns. with amines → the corresp. acetamide derivs. (47, 1153).

249

TABLE XXXIX (cont'd.)

HYDROXYBENZENEARSONIC ACIDS AND O-SUBSTITUTED DERIVATIVES

$ROC_6H_4AsO_3H_2$ where R is:	Prepd. from (Ars stands for $-OC_6H_4AsO_3H_2$)	Yield %	Props. and Remarks[k]	Ref.
$4-HO_2CCH_2CH_2-$	$ClCH_2CH_2COH + 4-HArs$	5		769
$4-HOCH_2NHCOCH_2-$	$4-MeO_2CCH_2Ars + HOCH_2NH_2$	32		47
$4-MeCH(CO_2H)-$	$MeCHClCO_2H + 4-HArs$	37		769, 1047
$4-Pr-$	$PrBr + 4-HArs$			1048
$4-HOCH_2CH_2CH_2-$	$ClCH_2CH_2CH_2OH + 4-HArs$	51	Bobbin-shaped crystals, m. 146°	502, 1048
$2-MeCHOHCH_2-$	$2-ROC_6H_4NH_2$ by the Bart method	15	Colorless plates, m. 167°; Na salt, white granules, rxns.3,8, 11	971, 83
$4-HOCH_2CHOHCH_2-$	$ClCH_2CHOHCH_2OH + 4-HArs$	60	White powder	62, 504
			Na salt, ‡ tiny white granules rxns.	62
C_{10}				
$4-\overset{\displaystyle .}{N}:C(NH_2)N:CHCH:C-$	$\overset{\displaystyle .}{N}:C(NH_2)N:CHCH:CCl + 4-HArs$	50	White needles, m. 227-228° Rxn.9	39
$4-HOCH_2CH_2NHCOCH_2-$	$4-MeO_2CCH_2Ars + HOCH_2CH_2NH_2$ at 100°		m. 161-164° Rxn. 9	47
$4-Bu-$	$BuBr + 4-HArs$			1153
$4-Me_2C(OH)CH_2-$	$Me_2\overset{.}{C}H_2O + 4-HArs$			1047
$4-HOCH_2CH_2OCH_2CH_2-$	$HOCH_2CH_2OCH_2CH_2Cl + 4-HArs$	65	m. 189-192°; Na salt, rxn. 9	556
			m. 127°, rxn. 12	765
C_{11}				
$4-MeCHOHCH_2NHCOCH_2-$	$4-MeO_2CCH_2Ars + MeCHOHCH_2NH_2$	97		47
$4-[\overset{.}{N}:C(NH_2)N:C(NHEt)N:C-$	$ROC_6H_4NH_2$ by the Bart method	25		1306
C_{12}				
$2-(4-BrC_6H_4-$	$ROC_6H_4NH_2$ by the Bart method	20	Needles, m. 183-184°	858
$4-(4-BrC_6H_4-$	$4-PhArs + Br_2$ in CCl_4		m. 354-355°	271
$4-(4-ClC_6H_4-$	$4-ROC_6H_4NH_2$ by the Bart method		m. 375°	271
$4-(4-ClC_6H_4SO_2-$	$4-ClC_6H_4SO_2Cl + 4-HArs$		m. >250°	1082
$4-Ph-$	$4-ROC_6H_4NH_2$ by the Bart method	26	Colorless plates, m. 365°, rxns.1, 13	271
$4-PhSO_2-$	$PhSO_2Cl + 4-HArs$		m. 170-172°	1082
$4-(4-H_2NC_6H_4SO_2-$	$4-H_2NC_6H_4SO_2Cl + 4-HArs$		m. 179-181°	1082

Compound	Preparation	%	Properties	Ref.
4-AcOCH₂CH₂OCH₂CH₂CH₂-	4-AcOCH₂CH₂OCH₂CH₂Ars + Ac₂O	45	m. 125°	765
4-(CH₂CH₂OCH₂CH₂NCOCH₂-	4-MeO₂CCH₂Ars + morpholine	43	Redn. with SO₂ → the arsonous acid	47
4-[(HOCH₂)₃CNHCOCH₂-	4-MeO₂CCH₂Ars + (HOCH₂)₃CNH₂	82		47
C₁₃				
4-(4-NCC₆H₄SO₂-	4-NCC₆H₄SO₂Cl + 4-HArs	30	Colorless needles, rxn.[14]	275
4-[4-H₂NC(:NH)C₆H₄SO₂-	4-NCC₆H₄SO₂Ars + alc. HCl + NH₃		Small colorless needles, rxn.[3]	275
2-(3-MeC₆H₄-	2-(3-MeC₆H₄O)C₆H₄NH₂ by the Bart method	40	Prisms, m. 193-194°	858
2-(2-MeC₆H₄-	2-(2-MeC₆H₄O)C₆H₄NH₂ by the Bart method	25	m. 184-185°	858
2-(4-MeOC₆H₄-	2-(4-MeOC₆H₄O)C₆H₄NH₂ Bart method	11	Octahedral crystals, m. 188-189°	858
4-(4-MeC₆H₄SO₂-	4-MeC₆H₄SO₂Cl + 4-HArs		m. 205-207° (dec.)	1082
4-[N:C(NHEt)N:C(NHEt)N:C-	4-ROC₆H₄NH₂ by the Bart method	9	Rxn.[15]	1306
C₁₄				
4-(4-NCC₆H₄CH₂-	4-BrCH₂C₆H₄CN + 4-HArs	66	Colorless needles, rxn.[14]	276
4-(PhCOCH₂-	PhCOCH₂Cl + 4-HArs		Transparent needles, m. 250° rxn.[8]	501, 972, 1047, 1048
4-(PhNHCOCH₂-			Rxn.[16]	1082
4-(4-AcNHC₆H₄SO₂-	4-AcNHC₆H₄SO₂Cl + 4-HArs	16	Na salt, colorless plates; Ag. salt, white ppt.	276
4-(4-H₂NC(:NH)C₆H₄CH₂-	4-NCC₆H₄CH₂Ars + alc. HCl, followed by NH₃	13	Long prisms, m. 184-185°	858
2-(2,4-Me₂C₆H₃-	2-ROC₆H₄NH₂ by the Bart method	12	Needles, m. 177.5-178°	858
2-(2,5-Me₂C₆H₃-	2-ROC₆H₄NH₂ by the Bart method		Prisms, 178-179°	858
2-(3,5-Me₂C₆H₃-	2-ROC₆H₄NH₂ by the Bart method	20	m. 69-70°	765
4-(PrCO₂CH₂CH₂CH₂CH₂-	4-HOCH₂CH₂CH₂CH₂Ars + (PrCO)₂O			
C₁₅				
4-HOCH₂(CHOH)₄CH₂NMeCOCH₂-	MeO₂CCH₂OArs + glucosylmethyl-amine			102
C₁₆				
4-(2-C₁₀H₇SO₂-	2-C₁₀H₇SO₂Cl + 4-HArs		m. 278-280° (dec.)	1082

11. Rxn. with HNO₃ at 0° → nitrate of the 5-nitro deriv. (83).
12. Rxn. with (RCO)₂O → RCOOCH₂CH₂OCH₂CH₂CH₂OC₆H₄AsO₃H₂ (765).
13. Rxn. with Br in CCl₄ → 4-(4-BrC₆H₄O)C₆H₄AsO₃H₂ (271).
14. Rxn. with alc. HCl, followed by NH₃ → conversion of the CN to H₂NC(:NH)- (275, 276).
15. Rxn. with HSCH₂CO₂Na → the -As(SCH₂CO₂H)₂ (1306).
16. Rxn. with HNO₃-H₂SO₄, followed by redn. → the 3-amino deriv. (1047, 1048).

TABLE XL

MERCAPTO-, SULFO-, AND SULFONYLBENZENEARSONIC ACIDS

$R\text{-}C_6H_4AsO_3H_2$ where R is:	Prepd. from	Yield %	Props. and Remarks	Ref.
C_6				
4-ClO_2S-	4-$ClO_2SC_6H_4AsCl_2$ + Cl_2, followed by hydrolysis		Rxns.[1]	956
4-HS-	4-$NCSC_6H_4AsO_3H_2$ + 10% NaOH	80	Orange, amorph. solid; salts[2]; rxns.[3,4]	864, 960
2-HO_3S-	2-$IC_6H_4AsO_3H_2$ + Na_2SO_3		Rxn.[5]	283
4-HO_3S-	4-$HO_3SC_6H_4NH_2$ by the Bart method	28	Rxns.[5,6]	955
3-H_2NO_2S-	3-$H_2NO_2SC_6H_4NH_2$ by the Bart-Scheller method	58	m. 218-129°	282
4-H_2NO_2S-	4-$H_2NO_2SC_6H_4NH_2$ by the Bart-Scheller method	52	Rxns.[6,8,16,17]	282
	by the Bart method	25	At 185-190° in vac. forms anhydride	956
	4-$H_2NO_2SC_6H_4AsO$ + H_2O_2			956
C_7				
4-NCS-	4-$ClN_2C_6H_4AsO_3H_2$ + CuSCN + KSCN	45	Rxn.[7]	864
2-MeS-	4-$ClN_2C_6H_4AsO_3H_2$ + CuCN	50	m. 154-154.5°, rxn.[8]	960
	2-$MeSC_6H_4NH_2$ by the Bart method			743
4-MeS-	4-$MeSC_6H_4NH_2$ by the Bart-Scheller method	50	Needles, rxn.[8]	285
4-$MeSO_2$-	$MeSC_6H_4AsO_3H_2$ + H_2O_2	100	Plates	285
C_8				
4-HO_2CCH_2S-	4-$HSC_6H_4AsO_3H_2$ + $ClCH_2CO_2H$	55	m. 192°, rxn.[9]	864
4-$HO_2CCH_2SO_2$-	4-$HO_2CCH_2SC_6H_4AsO_3H_2$ + H_2O_2	91	m. 188-189°	864
4-$HOCH_2CH_2S$-	4-$ClCH_2CH_2SC_6H_4AsO_3H_2$ by hydrolysis		m. 118°(from H_2O), m. 132-133° (from 5% HCl)	960
	4-$HSC_6H_4AsO_3H_2$ + $ClCH_2CH_2OH$	91	Rxn.[9]	960
	$RC_6H_4NH_2$ by the Bart method	48	Leaflets, m. 120.5-121° or needles, m. 132-133°	864
4-$HOCH_2CH_2SO_2$-	4-$HOCH_2CH_2SC_6H_4AsO_3H_2$ + H_2O_2	59	m. 177°	864
4-Me_2NO_2S-	4-$ClO_2SC_6H_4AsO_3H_2$ + Me_2NH	57	m. 166-168°, rxn.[6]	958

		Yield %		Ref.
C9				
4-N:CC1N:CC1N:CS-	4-HS C6H4AsO3H2 + (C1CN)3		White ppt. Rxn.10	426
3-N:C(NH2)N:CC1N:CS-	N:C(NH2)N:CC1N:CSH + 3-C1N2-C6H4AsO3H2			426
4-SCH:CHN:CNHO2S-	4-RC6H4NH2 Bart method	55	Cubic crystals	285
4-N:C(NH2)N:C(NH2)N:CS-	N:C(NH2)N:C(NH2)N:CSH + 4-C1-N2C6H4AsO3H2		white cryst. ppt., rxn.8, 11	426
4-AcCH2S-	4-HSC6H4AsO3H2 + AcCH2C1	18	m. 172.5°, rxn.9	864
4-AcCH2O2S-	4-AcCH2SC6H4AsO3H2 + H2O2	30	m. 202.5-203.5°	864
4-HO(CH2)3S-	4-HSC6H4AsO3H2 + HO(CH2)3C1	31	m. 116.3-117.5°, rxn.9	864
4-HO(CH2)3O2S-	4-HO(CH2)3SC6H4AsO3H2 + H2O2	78	m. 160.5°	864
C10				
4-N:C(NH2)N:CHCH:CS-	N:C(NH2)N:CHCH:CC1 + 4-HSC6H4AsO3H2	39	m. 131.5-132°	864
4-EtO2CCH2S-	C1CH2CO2Et + 4-HSC6H4AsO3H2	19	m. 123°	864
4-EtO2CCH2O2S-	4-EtO2CCH2SC6H4AsO3H2 + H2O2	59	m. 165-166°	864
4-HOCH2CH2OCH2CH2O2S-	HOCH2CH2OCH2CH2C1 + 4-HSC6H4-AsO3H2	83	m. 180°	864
4-EtOCH2CH2S-	EtOCH2CH2C1 + 4-HSC6H4AsO3H2	52	m. 121-122°	864
4-HOCH2CH2OCH2CH2S-	HOCH2CH2OCH2CH2C1 + 4-HSC6H4-AsO3H2	54	m. > 250°	864
4-EtOCH2CH2O2S-	EtOCH2CH2C1 + 4-HSC6H4AsO3H2	89	m. 182.5-184.5°	864

1. Rxns. with RNH2 or R2NH → RNHO2SC6H4AsO3H2 and R2NO2SC6H4AsO3H2 (956, 958, 1276).
2. Au salt, prepd. by reacting 4-HSC6H4AsO3H2 with KAuBr4 + Na2SO3 in NaOH (148).
3. Rxns. with RC1 in alk. soln. → RSC6H4AsO3H2 (960).
4. Rxn. with I in aq. NaHCO3 → (-SC6H4AsO3H2)2 (960).
5. Redn. with SO2 in HBr → HO3SC6H4AsBr2 (283, 955).
6. Rxn. with HI in H2O → the corresponding diiodoarsine (RAsI2) (955, 956, 958).
7. Rxn. with 10% NaOH → 4-HSC6H4AsO3H2 (864, 960).
8. Redn. with SO2 in HC1 → the corresponding dichloroarsine (RAsC12) (426, 743).
9. Rxn. with H2O2 → the corresponding sulfonyl arsonic acid (285, 288, 864).
10. Rxn. with Et2NH → the 4-amino-6-diethylamino-triazinylthio deriv. (426).
11. Rxn. with SnCl2 in HC1 contg. KI → the arsenobenzene deriv. (426).
12. Catalytic redn. → the 4-aminophenylmercapto deriv. (288).
13. Redn. with FeSO4 in 10% NaOH → the amino deriv. (708).
14. Rxn. with CuCN by the Sandmeyer rxn. → 4-(4-NCC6H4S)C6H4AsO3H2 (288).
15. Redn. by SO2 in aq. HI → the diiodoarsino deriv. (RAsI2) (1276).
16. Redn. by SO2 in 30% HBr → 4-H2NO2SC6H4AsBr2 (956).
17. Rxn. with PC13 +PC15 → 4-C1O2SC6H4AsC12 (956).

TABLE XL (cont'd.)

MERCAPTO-, SULFO-, AND SULFONYLBENZENEARSONIC ACIDS

$R-C_6H_4AsO_3H_2$ where R is:	Prepd. from	Yield %	Props. and Remarks	Ref.
$4-[S(CH_2CH_2)_2NO_2S]-$	$4-[S-(CH_2CH_2)_2NO_2S]C_6H_4NH_2$ by the Bart-Scheller method	59	m. 250°	250
C11				
$4-N{:}CHCH{:}CHNO_2S-$	$4-[N{:}CHCH{:}CHCH{:}CHNO_2S)C_6H_4NH_2$ by the Bart-Scheller method	25	Prisms (285)	285
$4-CH_2(CH_2)_4NO_2S-$	$CH_2(CH_2)_4NH + 4-ClO_2SC_6H_4AsO_3H_2$	48		1276
C12				
$4-(4-O_2NC_6H_4S)-$	$4-(4-O_2NC_6H_4S)C_6H_4NH_2$ by the Bart-Scheller method	39	Yellow needles, m. 291-292° Rxn.[12]	288
$4-(4-O_2NC_6H_4O_2S)-$	$4-O_2NC_6H_4Cl + 4-HSC_6H_4AsO_3H_2$	52	m. 183°	864
	$4-(4-O_2NC_6H_4S)C_6H_4AsO_3H_2 + H_2O_2$	40		864
$4-(2-O_2NC_6H_4NHO_2S)-$	$4-(2-O_2NC_6H_4NHO_2S)C_6H_4NH_2$ by the Bart method		Rxn.[13]	708
$4-(3-O_2NC_6H_4NHO_2S)-$	$4-(3-O_2NC_6H_4NHO_2S)C_6H_4NH_2$ by the Bart method		Rxn.[13]	708
$4-(4-O_2NC_6H_4NHO_2S)-$	$4-(4-O_2NC_6H_4NHO_2S)C_6H_4NH_2$ by the Bart method		Rxn.[13]	708
$4-(4-H_2NC_6H_4S)-$	$4-(4-O_2NC_6H_4S)C_6H_4AsO_3H_2$ H_2/catalyst	67	Needles, decomp. 190°, rxn.[14]	288
$4-(4-H_2NC_6H_4O_2S)-$	$4-(4-H_2NC_6H_4S)C_6H_4AsO_3H_2 + H_2O_2$	31	m. 211.5° (dec.)	864
	$PhNH_2 + 4-ClO_2SC_6H_4AsO_3H_2$	52	m. 229-230° (dec.)	864
$4-PhNHO_2S-$		33	Rxn.[6]	956, 958
$4-(2-HOC_6H_4NHO_2S)-$	$4-(2-HOC_6H_4NHO_2S)C_6H_4NH_2$ by the Bart method			708
$4-(3-HOC_6H_4NHO_2S)-$	$4-(3-HOC_6H_4NHO_2S)C_6H_4NH_2$ by the Bart method			708
$4-(3-HOC_6H_4NHO_2S)-$	$4-(4-HOC_6H_4NHO_2S)C_6H_4NH_2$ by the Bart method			708
$4-(4-H_2O_3AsC_6H_4SS)-$	$4-HSC_6H_4AsO_3H_2 + I$ in $NaHCO_3$	92	m. 220° (dec.), rxn.[18]	960

Compound	Preparation	Yield %	Notes	Ref.
4-(3-H$_2$NC$_6$H$_4$NHO$_2$S)-	+ FeSO$_4$ in 10% NaOH			708
4-(4-H$_2$NC$_6$H$_4$NHO$_2$S)-	4-(3-O$_2$NC$_6$H$_4$NHO$_2$S)C$_6$H$_4$AsO$_3$H$_2$ + FeSO$_4$ in 10% NaOH			708
4-(4-H$_2$NO$_2$SC$_6$H$_4$NHO$_2$S)-	4-(4-O$_2$NC$_6$H$_4$NHO$_2$S)C$_6$H$_4$AsO$_3$H$_3$ + FeSO$_4$ in 10% NaOH 4-H$_2$NO$_2$SC$_6$H$_4$NH$_2$ + 4-ClO$_2$SC$_6$-H$_4$AsO$_3$H$_2$	47	Rxn.[6]	958
C$_{13}$				
4-(4-NCC$_6$H$_4$S)-	4-(4-H$_2$NC$_6$H$_4$S)C$_6$H$_4$AsO$_3$H$_2$ + CuCN by Sandmeyer rxn.	32	Yellow needles, decomp. > 200° rxn.[9]	288
4-(4-H$_2$NOCC$_6$H$_4$O$_2$S)-	4-(4-NCC$_6$H$_4$S)C$_6$H$_4$AsO$_3$H$_2$ + H$_2$O$_2$	28	Needles, m. 310.5°	288
4-(4-HO$_2$CC$_6$H$_4$NHO$_2$S)-	4-H$_2$NC$_6$H$_4$CO$_2$H + ClO$_2$SC$_6$H$_4$-AsO$_3$H$_2$	54	Rxn.[6]	958
4-(HO$_2$CC$_6$H$_4$NHO$_2$S)-	HO$_2$CC$_6$H$_4$NHO$_2$SC$_6$H$_4$NH$_2$ by the Bart method		o-, m-, and p- carboxy deriv. were prepd.	708
4-PhCH$_2$S-	4-PhCH$_2$C$_6$H$_4$NH$_2$ by the Bart method		Light yellow needles, decomp. ~250°	1184
4-(2-MeC$_6$H$_4$NHO$_2$S)-	4-(2-MeC$_6$H$_4$NHO$_2$S)C$_6$H$_4$NH$_2$ by the Bart method			708
4-(3-MeC$_6$H$_4$NHO$_2$S)-	4-(3-MeC$_6$H$_4$NHO$_2$S)C$_6$H$_4$NH$_2$ by the Bart method			708
4-(4-MeC$_6$H$_4$NHO$_2$S)-	4-(4-MeC$_6$H$_4$NHO$_2$S)C$_6$H$_4$NH$_2$ by the Bart method			708
4-(MePhNO$_2$S-	PhNHMe + 4-ClO$_2$SC$_6$H$_4$AsO$_3$H$_2$	82	Rxn.[15]	1276
4-(2-MeOC$_6$H$_4$NHO$_2$S)-	4-(2-MeOC$_6$H$_4$NHO$_2$S)C$_6$H$_4$NH$_2$ by the Bart method			708
4-(3-MeOC$_6$H$_4$NHO$_2$S)-	4-(3-MeOC$_6$H$_4$NHO$_2$S)C$_6$H$_4$NH$_2$ by the Bart method			708
4-(4-MeOC$_6$H$_4$NHO$_2$S)-	4-(4-MeOC$_6$H$_4$NHO$_2$S)C$_6$H$_4$NH$_2$ by the Bart method			708
4-[N:C(NH$_2$)N:C(NEt$_2$)N:CS]-	4-[N:C(NH$_2$)N:C(NEt$_2$)N:CS]C$_6$H$_4$AsO$_3$-H$_2$ + Et$_2$NH		White powder	426
C$_{14}$				
4-(4-AcNHC$_6$H$_4$O$_2$S)-	4-(4-AcNHC$_6$H$_4$O$_2$S)C$_6$H$_4$NH$_2$ by the Bart method		White powder, decomp. 230-238°	222

18. Rxn. with conc. HCl → (Cl$_2$OAsC$_6$H$_4$S-)$_2$ (?), which after treatment with SO$_2$ in conc. HCl-CHCl$_3$ contg. NaI → (Cl$_2$AsC$_6$H$_4$S-)$_2$ (960).

TABLE XL (cont'd.)

MERCAPTO-, SULFO-, AND SULFONYLBENZENEARSONIC ACIDS

R-$C_6H_4AsO_3H_2$ where R is:	Prepd. from	Yield %	Props. and Remarks	Ref.
C_{18} 4-(4-PhC$_6$H$_4$NHO$_2$S)-	4-PhC$_6$H$_4$NH$_2$ + 4-ClO$_2$SC$_6$H$_4$-AsO$_3$H$_2$	90	Rxn.[15]	1276
C_{19} 4-[Ph(PhCH$_2$)NO$_2$S]-	PhCH$_2$NHPh + 4-ClO$_2$SC$_6$H$_4$AsO$_3$H$_2$	80	Rxn.[15]	1276

DISUBSTITUTED BENZENEARSONIC ACIDS

4-CHLORO-3-NITROBENZENEARSONIC ACID $C_6H_5AsClNO_5$ $4,3-Cl(O_2N)C_6H_3AsO_3H_2$
Prepn.: by reacting $4,3-Cl(O_2N)C_6H_3N_2Cl$ with Na_3AsO_3 in the presence of $CuSO_4$ (Bart method) (92).
By treating $4-ClC_6H_4AsO_3H_2$ with excess alkali nitrate in H_2SO_4 at 70° (594, 896).
Rxns. with: H_2S → $4,3-Cl(O_2N)C_6H_3AsS_2$ (147).
RNH_2 → $3,4-O_2N(RNH)C_6H_3AsO_3H_2$ (595, 760, 761).
40% KOH → $3,4-O_2N(HO)C_6H_3AsO_3H_2$ (896).

4-HYDROXY-3-NITROBENZENEARSONIC ACID $C_6H_6AsNO_6$ $4,3-HO(O_2N)C_6H_3AsO_3H_2$
Prepn.: By hydrolyzing $4,3-Cl(O_2N)C_6H_3AsO_3H_2$ with 40% KOH on a water bath; 65% yield (896).
By treating $4-HO_2CCONHC_6H_4AsO_3H_2$ with HNO_3 in H_2SO_4 and hydrolyzing the prod. with strong alkali (343).
By treating $3,4-O_2N(H_2N)C_6H_3AsO_3H_2$ with aq. KOH at 80° (344).
Rxns. with: $Fe(OH)_2$ in alk. soln. → the 3-amino deriv. (1237).
Fe in HCl at 85-90° → the 3-amino deriv. (379).
$FeCl_2$ in NaOH soln. → the 3-amino deriv. (669).
Glucose in alk. soln. → the 3-amino deriv. (340, 669).
$PhCH_2Cl$ in NaOH soln. → $3,4-O_2N(PhCH_2O)C_6H_3AsO_3H_2$ (388).
$Na_2S_2O_4$ in alk. soln. → the diamino-arsenobenzene deriv. (342, 343).
RSO_2Cl → $3,4-O_2N(RSO_3)C_6H_3AsO_3H_2$ (1082).
AcCl → $3,4-O_2N(AcO)C_6H_4AsO_3H_2$ (1117).
Polarographic redn. (1077).
Salts: with dodecylamine (1); with methylglucamine (831).
Tin derivs.: $[3,4-O_2N(HO)C_6H_3As(O)(OH)O]_2SnBu_2$, decomp. 164-170° (1267).
$3,4-O_2N(HO)C_6H_3As(O)O_2SnBu_2$, decomp. 191° (1267).
Use: Analytic reagent for turbidimetric detn. of Sn(IV) (654) and Bi (724).

6-NITRO-m-ARSANILIC ACID $C_6H_7AsN_2O_5$ $3,6-H_2N(O_2N)C_6H_3AsO_3H_2$
Rxn. with 2,4-Diamino-6-chloro-s-triazine → N-(4,6-diamino-s-triazin-2-yl)-6-nitro-m-arsanilic acid (419).

3-NITRO-p-ARSANILIC ACID $C_6H_7AsN_2O_5$ $3,4-O_2N(H_2N)C_6H_3AsO_3H_2$
Prepn.: by nitration of $4-AcNHC_6H_4AsO_3H_2$, followed by hydrolysis (872).
Rxns. with: $NaNO_2$, followed by Na_3AsO_3 or As_2O_3 in alk. soln. (Bart method) → $3,1,4,-O_2NC_6H_3(AsO_3H_2)_2$ (79, 883).
$NaNO_2$,followed by $AcCHCO_2Et$→ $3,4-O_2N[EtO_2CC(Ac):NNH]C_6H_3AsO_3H_2$ (240).
$NaNO_2$ in aq. Na_2CO_3, followed by pyrrole deriv. → 3-hydroxy-4-pyrrylazobenzene-arsonic acid (870).
Aq.,alkali → $3,4-O_2N(HO)C_6H_3AsO_3H_2$ (344, 872).
2,4-Diamino-6-chloro-s-triazine → 4,6-diamino-s-triazinyl deriv. (420).
Sn derivs.: $[3,4-O_2N(H_2N)C_6H_3As(O)(OH)O]_2SnBu_2$, decomp. 189-193° (1266).
$3,4-O_2N(H_2N)C_6H_3As(O)O_2SnBu_2$, decomp. 262° (1266).

3-AMINO-4-MERCAPTOBENZENEARSONIC ACID $C_6H_8AsNO_3S$ $3,4-H_2N(HS)C_6H_3AsO_3H_2$
Prepn.: by treating $3,4-O_2N(Cl)C_6H_3AsS_2$ with Na_2S in aq alc. under reflux (147).

Props.: Amorphous colorless powder insol. in acids, sol. in alkali (147).
Rxns. with: O in alk. soln. → [2,4-H₂N(H₂O₃As)C₆H₃S-]₂ (147).

H_3PO_2 → the arsenobenzene deriv. (147).
$Na_2S_2O_4$ + $MgCl_2$ → the arsenobenzene deriv. (147).
$KAuBr_4$ + Na_2SO_3 → 3,4-H₂N(AuS)C₆H₃AsO₃H₂ (148).
Au deriv.: 3,4-H₂N(AuS)C₆H₃AsO₃H₂, m. 207°, polarography (830).

3-AMINO-4-HYDROXYBENZENEARSONIC ACID $C_6H_8AsNO_4$ 3,4-H₂N(HO)C₆H₃AsO₃H₂

Prepn.: by reducing 3,4-O₂N(HO)C₆H₃AsO₃H₂ with glucose in alk. soln. at 100-105°, 86% yield (340), or with Fe in HCl at 85-90°, 90% yield (379), or with $FeCl_2$ in NaOH soln. (669), or with $FeSO_4$ + Ca(OH)₂ in aq. NaOH (1237).
Props.: Potentiometric titration (1236).
Rxns. with: RCOCl or (RCO)₂O → the 3-RCONH- derivs. (149, 150, 970, 1007).
ROCOCl → the 3-ROCONH- derivs. (149).
SO_2 in aq. HCl contg. KI, followed by neutralization → the arsenoso deriv. (43, 37 1112).
RSO₂Cl + 2 N NaOH in Me₂CO → 3,4-H₂N(RSO₃)C₆H₃AsO₃H₂ (275, 388, 1082, 1117).
RSO₂Cl in aḷ. alkali at 30-40° → 3,4-RSO₂NH(HO)C₆H₃AsO₃H₂ (1258).
Haloheterocycles → the N-heterocyclyl deriv. (250, 332).
(ClCN)₃ in 2% NaOH, followed by NH₄OH → N-(4,6-diamino-s-triazin-2-ylamino)4-hydroxybenzenearsonic acid (416).
$NaNO_2$, followed by pyrrole derivs. → the corresp. 3-pyrrylazo deriv. (385, 871).
$NaNO_2$, followed by 1-naphthol-4,8-disulfonic acid → the 3-naphthylazo deriv. (409, 412).
$NaNO_2$, followed by $NaAsO_2$ → 4,1,3-HOC₆H₃(AsO₃H₂)₂ (880).
2,4-Dichloro-6-hydrazine → 3-(4-chloro-6-hydrazino-s-triazin-2-ylamino) deriv. (418).
HOCH₂CHSHCH₂SH → 3,4-H₂N(HO)C₆H₃AsSCH₂CH(CH₂OH)S (441).
HOCH₂SO₂Na → 4,3-HO(NaO₂SCH₂NH)C₆H₃AsO₃H₂ (1046).
HCl, followed by (MeCHO)₃ → 8-hydroxy-2-methyl-5-quinolinearsonic acid (459).
Zn + HCl (1:3:6)→ 3,4-H₂N(HO)C₆H₃AsH₂ (700).
Zn + HCl (1:2:4) → [3,4-H₂N(HO)C₆H₃As:]n (700).
$Na_2S_2O_4$ → [3,4-H₂N(HO)C₆H₃As:]n (702).
Electrolytic redn. → 3,4-H₂N(HO)C₆H₃AsO → [3,4-H₂N(HO)C₆H₃As:]n (701) → 3,4-H₂N-(HO)C₆H₃AsH₂ (1259).
[3,4-H₂N(HO)C₆H₃As:]n 2:1 → 3,4-H₂N(HO)C₆H₃AsO (693).
Polarographic redn. (1077).
1,2-Naphthoquinone-4,x-disulfonic acid → substitution of the 4-SO₃H group by 3,4-NH(HO)C₆H₃AsO₃H₂ (1243).

4-AMINO-2-HYDROXYBENZENEARSONIC ACID $C_6H_8AsNO_4$ 4,2-H₂N(HO)C₆H₃AsO₃H₂

Prepn.: by heating 3-(EtOCONH)C₆H₄OH with H_3AsO_4 and hydrolyzing the rxn. prod. (37).
By reducing 2,4-HO(O₂N)C₆H₃AsO₃H₂ with Fe in HCl in the presence of H_3AsO_4 or H_3AsO_3 (152).
Props.: M. 184° (37).
Rxns. with: SO_2 in 2 N H_2SO_4 in the presence of KI → the arsenoso deriv. (37).
(RCO)₂O → the 4-RCONH- deriv. (71, 149).
ROCOCl → the 4-ROCONH- deriv. (71).

258

RCHO → the 4-RCH:N- deriv. (149).
Haloheterocycles → the N-heterocyclyl derivs. (250).
(ClCN)$_3$, followed by NH$_4$OH → the N-(4,6-diamino-s-triazin-2-yl) deriv. (416).
Hydrohalic acids, H$_3$PO$_2$, H$_2$/Ni, PCl$_3$, or NaHSO$_3$ → dearsonation (37).
1,2-Naphthoquinone-4,x,x-polysulfonic acid → substitution of the 4-SO$_3$H by 4,3-NH(HO)C$_6$H$_3$AsO$_3$H$_2$ (1243).

4-ARSONO-2-NITROPHENOXYACETIC ACID C$_8$H$_8$AsNO$_8$ 3,4-O$_2$N(HO$_2$CCH$_2$O)C$_6$H$_3$AsO$_3$H$_2$
Rxns. with: Fe(OH)$_2$ in alk. soln., followed by ClCH$_2$CONH$_2$ → 3,4-H$_2$NϙCCH$_2$NH(HO$_2$C-CH$_2$O)C$_6$H$_3$AsO$_3$H$_2$ (1047).
Na$_2$S$_2$O$_4$ → the arsonobenzoxazine deriv. (1047).
Na$_2$S$_2$O$_4$, followed by ClCH$_2$CONHR → 3,4-RNHCOCH$_2$NH(HO$_2$CCH$_2$O)C$_6$H$_3$AsO$_3$H$_2$ (1047).

3-CARBOMETHOXY-4-HYDROXYBENZENEARSONIC ACID C$_8$H$_9$AsO$_6$ 4,3-HO(MeO$_2$C)C$_6$H$_3$AsO$_3$H$_2$
Prepn.: by esterifying 3,4-HO$_2$C(HC)C$_6$H$_3$AsO$_3$H$_2$ with MeOH, 59% yield (288).
Props.: Needles, soften at 193° (288).
Rxn. with NH$_3$ at 0° → 4,3-HO(H$_2$NOC)C$_6$H$_3$AsO$_3$H$_2$, while at 100° NH$_3$ cleaves As (288).

3-CHLORO-4-(2-HYDROXYETHOXY)BENZENEARSONIC ACID C$_8$H$_{10}$AsClO$_5$ 3,4-Cl(HOCH$_2$-CH$_2$O)C$_6$H$_3$AsO$_3$H$_2$
Prepn.: by adding neutral soln. of 3,4-O$_2$N(HOCH$_2$CH$_2$O)C$_6$H$_3$AsO$_3$H$_2$ and urea to Fe(OH)$_2$, shaking the mixture, and neutralizing the sol. fraction (502).
Props.: White rectangular plates, m. 141° (502).
Rxn. with SO$_2$, followed by neutralization → the arsenoso deriv. (502).

3-ACETAMIDO-4-HYDROXYBENZENEARSONIC ACID, ACETARSONE, STOVARSOL, SPIROCID, C$_8$H$_{10}$AsNO$_5$ 3,4-AcNH(HO)C$_6$H$_3$AsO$_3$H$_2$
Prepn.: by acetylation of crude 3,4-H$_2$N(HO)C$_6$H$_3$AsO$_3$H$_2$ and purification at pH 4.5-5.0 in the presence of Pb(OAc)$_2$ and C at 55° (341). Acetylation with Ac$_2$O at 5° gave 90% yield (1007).
Props.: Dissocn. const. (454). UV spectrum (116).
Rxns. with: HNO$_3$-H$_2$SO$_4$ → the 5-nitro deriv. (149).
Conc. HCl at 120° → red colored prod. (186).
6-Methoxy-8-quinolinediazonium salt → dark brick-red powder, m. 220-222° (decomp.) (488).
RSO$_2$Cl → 3,4-AcNH(RSO$_3$)C$_6$H$_3$AsO$_3$H$_2$ (1082).
Salts: H$_2$O-sol. di-alkali salts with Na and K, Na, and Li, and K and Li (73).
Na salt on heating with conc. HCl at 120° → red color (186). Vuzine salt (1141).
7-(3-Octylaminopropylamino)benz[c]acridine mono- and disalts (861). Diethylammonium salt on heating with conc. HCl gives red color (186). Diethyl-(2-hydroxyethyl)ammonium salt (Arsaminol) on heating with conc. HCl at 120° gives red color (186). 8-(3-Diethylaminopropyl)amino-6-methoxyquinoline salt, m. 100-120° (dec.) (487).

4-ACETAMIDO-2-HYDROXYBENZENEARSONIC ACID, ORSANINE, C$_8$H$_{10}$AsNO$_5$ 4,2-AcNH(HO)-C$_6$H$_3$AsO$_3$H$_2$
Prepn.: by heating 4,2-H$_2$N(HO)C$_6$H$_3$AsO$_3$H$_2$ with AcOH-Ac$_2$O, 88% yield (71)
Props.: Monoclinic prisms, decomp. 266° (71). Dissocn. const. (454).

Rxns. with: SO_2, followed by neutralization → the arsenoso deriv. (1003).
$\overline{HSCH_2CO_2Na}$ → $2,4\text{-}HO(AcNH)C_6H_3As(SCH_2CO_2H)_2$ (1003).

3-(2-HYDROXYETHOXY)-2-NITROBENZENEARSONIC ACID $C_8H_{10}AsNO_7$ $2,3\text{-}O_2N(HOCH_2CH_2\text{-}O)C_6H_3AsO_3H_2$

Prepn.: by nitrating $3\text{-}HOCH_2CH_2OC_6H_4AsO_3H_2$ and hydrolyzing the nitrate with HCl;
50% yield, along with the $4\text{-}NO_2$ deriv. (82).
Props.: Yellow plates, m. >270° (82).
Rxns. with: 6 N NaOH → $2,3\text{-}O_2N(OH)C_6H_3AsO_3H_2$ (82).
$\overline{H_2SO_3 + HI}$ → the arsenoso deriv. (82).

4-(2-HYDROXYETHOXY)-3-NITROBENZENEARSONIC ACID $3,4\text{-}O_2N(HOCH_2CH_2O)C_6H_3AsO_3H_2$
$C_8H_{10}AsNO_7$

Prepn.: by treating $4\text{-}HOCH_2CH_2OC_6H_3AsO_3H_2$ with HNO_3 at 20° and hydrolyzing the
prod. with 3 N HCl under reflux (502).
Rxns. with: $FeCl_2$ in aq. alkali → redn. of the $-NO_2$ group (502).
$\overline{Fe(OH)_2}$ in the presence of urea, followed by HCl → the 3-Cl deriv. (502).

3-(2-HYDROXYETHOXY)-4-NITROBENZENEARSONIC ACID $C_8H_{10}AsNO_7$ $4,3\text{-}O_2N(HOCH_2CH_2O)\text{-}$
$C_6H_3AsO_3H_2$

Prepn.: by-prod. from nitration of $3\text{-}HOCH_2CH_2OC_6H_4AsO_3H_2$, 30% yield (82).
Props.: Irregular prisms, m. 164° (82).
Rxns. with: 6 N NaOH → the 3-hydroxy deriv. (82).
$\overline{H_2SO_3 + HI}$ → the arsenoso deriv. (82).

4-HYDROXY-3-METHYLUREIDOBENZENEARSONIC ACID $C_8H_{11}AsNO_5$ $4,3\text{-}HO(MeNHCONH)C_6H_3As$
Prepn.: by heating at 100° a mixt. of 2-oxo-5-benzoxazolinearsonic acid with
$\overline{MeNH_2}$ (375).

4-(2-HYDROXYETHYLAMINO)-3-NITROBENZENEARSONIC ACID $C_8H_{11}AsN_2O_6$ $3,4\text{-}O_2N\text{-}$
$(HOCH_2CH_2NH)C_6H_3AsO_3H_2$

Prepn.: by heating $3,4\text{-}O_2N(MeO_2CCH_2O)C_6H_3AsO_3H_2$ or $3,4\text{-}O_2N(HO_2CCH_2O)C_6H_3AsO_3H_2$
or $3,4\text{-}O_2N(HOCH_2CH_2O)C_6H_3AsO_3H_2$, or $3,4\text{-}O_2N(MeO)C_6H_3AsO_3H_2$, or $3,4\text{-}O_2N(HO)C_6H_3\text{-}$
AsO_3H_2 with $H_2NCH_2CH_2OH$ at 80°; 76, 56, 42, 67, and 24% yield, resp. (1182).

4-(2-HYDROXYETHOXY)-m-ARSANILIC ACID $C_8H_{12}AsNO_5$ $3,4\text{-}H_2N(HOCH_2CH_2O)C_6H_3AsO_3H_2$
Prepn.: by reducing $3,4\text{-}O_2N(HOCH_2CH_2O)C_6H_3AsO_3H_2$ with $FeCl_2$ in aq. alkali (502).
Props.: White rods, m. 156-157°; HCl salt, granular solid, m. 169°; mono-Na
salt, white granular crystals (502).
Rxns. with: SO_2 in aq. HCl + KI → $3,4\text{-}(HCl.H_2N)(HOCH_2CH_2O)C_6H_3AsCl_2$ (502).
$\overline{H_3PO_2}$ → the arsenobenzene deriv. (502).
$H_2NC(:NH)NHCN$ → $4,3\text{-}HOCH_2CH_2O[H_2NC(:NH)NHC(:NH)NH]C_6H_3AsO_3H_2$ (1181)
RCl → the 3-RNH- derivs. (1181).
RCOCl → the 3-RCONH- derivs. (1181).
KCNO → the 3-H_2NCONH- deriv. (1181).

3-CHLORO-4-(4,6-DICHLORO-s-TRIAZIN-2-YLAMINO)BENZENEARSONIC ACID $C_9H_7AsCl_3N_4O_3$
$3,4\text{-}Cl(\overline{N:CClN:CClN:CNH})C_6H_3AsO_3H_2$

Prepn.: by reacting $3,4\text{-}Cl(H_2N)C_6H_3AsO_3H_2$ with $(ClCN)_3$ in alk. soln. (424).
Rxn. with NH_3, followed by $Na_2S_2O_4$ → $3,3''$-dichloro-4,4'-bis(4,6-diamino-s-triazin-
2-ylamino)arsenobenzene (424).

3-[2-(4,6-DICHLORO-s-TRIAZIN-2-YL)HYDRAZINO]-4-HYDROXYBENZENEARSONIC ACID

$C_9H_8AsCl_2N_5O_5$ 4,3-HO(N:CClN:CClN:CNHNH)$C_6H_3AsO_3H_2$

Prepn.: by reacting 4,3-HO(H_2NNH)$C_6H_3AsO_3H_2$ with (ClCN)$_3$ in aq. NaHCO$_3$ at 2° (418).
Rxn. with 4-H_2NC_6H_4SO$_3$Na, followed by NH$_4$OH → 3-[2-(4-amino-6-p-sulfoanilino-s-triazin-2-yl)hydrazino]-4-hydroxybenzenearsonic acid (418).

3-(4-CHLORO-6-HYDRAZINO-s-TRIAZIN-2-YLAMINO)-4-HYDROXYBENZENEARSONIC ACID

$C_9H_{10}AsClN_6O_4$ 4,3-HO[N:CClN:C(NHNH$_2$)N:CNH]$C_6H_3AsO_3H_2$

Prepn.: By reacting 4,3-HO(N:CClN:CClN:CNH)$C_6H_3AsO_3H_2$ + 20% H_2NNH$_2$ at 25° (418).
By reacting 4,3-HO(H_2N)$C_6H_3AsO_3H_2$ N:CClN:C(NHNH$_2$)N:CCl in aq. Na$_2$CO$_3$ at 25° (418).
Props.: White ppt. (418).
Rxns. with: Liq. NH$_3$ at -40 to +35° → the 4-amino-6-hydrazino-s-triazinyl deriv. (418).
Conc. H_2NNH$_2$.H_2O → the 4,6-dihydrazino-s-triazinyl deriv. (418).

4-(2,4-DIHYDROXYPROPOXY)-3-NITROBENZENEARSONIC ACID $C_9H_{10}AsN_3O_{12}$

3,4-O$_2$N[O$_2$NOCH$_2$CH(ONO$_2$)CH$_2$O]$C_6H_3AsO_3H_2$

Prepn.: by treating 4-(HOCH$_2$CHOHCH$_2$O)$C_6H_4AsO_3H_2$ with HNO$_3$-H_2SO$_4$; 75-80% yield (62, 504).
Props.: Yellow plates, m. 132-133° (62, 504).
Rxns. with: 2 N HCl or H_2SO$_4$ → hydrolysis of the nitrate groups (62, 504).
H_3PO$_2$ → the arseno deriv. (62).

3-(4,6-DIAMINO-s-TRIAZIN-2-YLAMINO)-4-HYDROXYBENZENEARSONIC ACID $C_9H_{11}AsN_6O_4$

4,3-HO[N:C(NH$_2$)N:C(NH$_2$)N:CNH]$C_6H_3AsO_3H_2$

Prepn.: by reacting 3,4-H_2N(HO)$C_6H_3AsO_3H_2$ with (ClCN)$_3$ in aq. NaOH and satg. the rxn. mixt. with NH$_3$; 53% yield (416, 149, 420).
Rxns. with: SO$_2$ in HCl in the presence of NaI → the dichloroarsino deriv. (417, 419, 420, 421, 424).
SnCl$_2$ in concd. HCl contg. HI → the arsenophenol deriv. (417, 419-422, 424, 429).

4-(4,6-DIAMINO-s-TRIAZIN-2-YLAMINO)-2-HYDROXYBENZENEARSONIC ACID $C_9H_{11}AsN_6O_4$

Prepn.: By refluxing 2,4-diamino-6-chloro-s-triazine with 2,4-HO(H_2N)$C_6H_3AsO_3H_2$ in H_2O (46).
By reacting 2,4-HO(H_2N)$C_6H_3AsO_3H_2$ with (ClCN)$_3$ in aq. NaOH and satg. the rxn. mixt. with NH$_3$; 49% yield (416).
Rxn. with SO$_2$ in conc. HCl cont. HI → the dichloroarsino deriv. (420, 429).

2-METHOXY-5-VINYLBENZENEARSONIC ACID POLYMER (C$_9H_{11}AsO_3$)n [2,5-MeO(CH$_2$:CH)-$C_6H_3AsO_3H_2$]n

Prepn.: by acidifying a soln. of 2,5-MeO(MeCHOH)$C_6H_3AsO_3H_2$ and evapg. the solution to dryness; 65% yield (36).
Props.: Light-brown solid, m. 295-320° (36).

3-CARBETHOXYAMINO-4-HYDROXYBENZENEARSONIC ACID $C_9H_{12}AsNO_6$ 4,3-HO(EtOCONH)-$C_6H_3AsO_3H_2$

Prepn.: by treating 3,4-H_2N(HO)$C_6H_3AsO_3H_2$ dissolved in aq. NaOH with EtOCOCl at 0° (149).

Props.: M. 210-211° (149).
Rxn. with $HOCH_2CH_2NH_2$ → 4,3-$HO(HOCH_2CH_2NHCONH)C_6H_3AsO_3H_2$ (534).

4-CARBETHOXYAMINO-2-HYDROXYBENZENEARSONIC ACID $C_9H_{12}AsNO_6$ 2,4-$HO(EtOCONH)$-$C_6H_3AsO_3H_2$

Prepn.: By treating a soln. of 2,4-$HO(H_2N)C_6H_3AsO_3H_2$ in aq. NaOH with $EtOCOCl$ at 0° (149).
   By heating m-(carbethoxyamino)phenol with 87% H_3AsO_4; 62% yield (37).
Props.: M. 220-221° (149); m. 231° (37).
Rxns. with: H_2SO_4-HNO_3 → 2,3,4-$HO(O_2N)(EtOCONH)C_6H_2AsO_3H_2$ (149).
3 N NaOH → 4,2-$H_2N(HO)C_6H_3AsO_3H_2$ (37).
Me_2SO_4 → 2-methoxy.deriv. (37).
PCl_3 in Et_2O, followed by hydrolysis → the arsenoso deriv. (37, 71).
CH_2CH_2O → 4,2-$EtOCONH(HOCH_2CH_2O)C_6H_3AsO_3H_2$ (71).

4-CARBETHOXYAMINO-3-HYDROXYBENZENEARSONIC ACID $C_9H_{12}AsNO_6$ 3,4-$HO(EtOCONH)$-$C_6H_3AsO_3H_2$

Rxn. with Ethanolamine → 3,4-$HO(HOCH_2CH_2NHCONH)C_6H_3AsO_3H_2$ (534).

3-(HYDROXYETHYLCARBAMINO)-4-HYDROXYBENZENEARSONIC ACID $C_9H_{13}AsN_2O_6$ 4,3-HO-$(HOCH_2CH_2NHCONH)C_6H_3AsO_3H_2$

Prepn.: By heating 4,3-$HO(EtOCONH)C_6H_3AsO_3H_2$ with ethanolamine at 110° (534).
   By heating 2-oxo-5-benzoxazolinearsonic acid with $HOCH_2CH_2NH_2$ (375).
Props.: Slightly colored cryst. powder (534).

2-(4-ARSONO-2-NITROPHENYL)-5-METHYL-2,1,3-TRIAZOLE-4-CARBOXYLIC ACID $C_{10}H_9AsN_4$ 3,4-$O_2N[N:C(CO_2H)C(Me):NN]C_6H_3AsO_3H_2$

Prepn.: by oxidation of 3,4-$O_2N(N:CMeCMe:NN)C_6H_3AsO_3H_2$ with alk. $KMnO_4$, 49% yield (239).
Props: M. > 250° (239).
Rxns. with: H_2/Raney Ni → redn. of the NO_2 group (239).
EtOH in the presence of H_2SO_4 → esterification of the CO_2H group (239).

4-HYDROXY-3-(4-SULFANILYLAMINO)BENZENEARSONIC ACID $C_{12}H_{13}AsN_2O_6S$ 4,3-$HO(4-H_2NC_6H_4SO_2NH)C_6H_3AsO_3H_2$

Prepn.: by hydrolyzing 4,3-$HO(4-AcNHC_6H_4SO_2NH)C_6H_3AsO_3H_2$ with 5 N HCl at 100° (11 1258) or with 18% HCl-MeOH at r. t. (1117).
Props.: Darkens at 207° (1117), m. 190-191° (dec.) (1258).
Rxns. with: SO_2 in HCl contg. KI, followed by neutralization → the arsenoso deriv. (1117, 1258).
$Na_2S_2O_4$ → the arseno deriv. (1258).

4,4'-DITHIOBIS(m-ARSANILIC ACID) $C_{12}H_{14}As_2N_2O_6S_2$ 4,3-$[-S(H_2N)C_6H_3AsO_3H_2]_2$

Prepn.: by oxidation of 3,4-$H_2N(HS)C_6H_3AsO_3H_2$ with air in alk. soln. (147).
Props.: Amorphous colorless powder, insol. in acids, sol. in alkali (147).
Rxn. with Ac_2O → 4,3-$[-S(AcNH)C_6H_3AsO_3H_2]_2$ (147).

4, -CARBONYLDI(3-NITROBENZENEARSONIC ACID) $C_{13}H_{10}As_2N_2O_{11}$ 4,4'-CO(3-O$_2$NC$_6$H$_3$-SO$_3$H$_2$)$_2$

Prepr   by oxidation of 4,4'-CH$_2$(3-O$_2$NC$_6$H$_3$AsO$_3$H$_2$)$_2$ with KMnO$_4$ in alk. soln.; 70-80 yield (1333).

Props   Yellow, amorph. powder; soly. data (1333).

N- (2-AMINO-4-HYDROXY-6-PTERIDYL)METHYL]-3-BROMOARSANILIC ACID $C_{13}H_{12}AsBrN_6O_4$

Prepn.: by treating arsanilic acid with Br$_2$ in AcOH and reacting the 3-bromo deriv. with 2,4,5-triamino-6-bromopyrimidine and 1,1,3-tribromopropanone in H$_2$O at pH 2-3 at 80° (452).

By reacting 3,4-Br(H$_2$N)C$_6$H$_3$AsO$_3$H$_2$ with 2,4,5-triamino-6-hydroxypyrimidine and 1,-1,3-bribromopropanone (15).

Props.: Microscopic, yellow scales, greasy to the touch; soly. data (1332).

Rxns. with: H$_3$PO$_2$ → the arseno deriv. (1332).

KMnO$_4$ in alk. soln. → 4,4'-CO(3-O$_2$NC$_6$H$_3$AsO$_3$H$_2$)$_2$ (1333).

N,N'-CARBONYLDI(4-CHLORO-m-ARSANILIC ACID) $C_{13}H_{12}As_2Cl_2N_2O_7$ CO[NHC$_6$H$_3$-2,5-(Cl)AsO$_3$H$_2$]$_2$

Prepn.: by treating 3,4-H$_2$N(Cl)C$_6$H$_3$AsO$_3$H$_2$ with COCl$_2$ in alk. soln. (712).

3,3'-METHYLENEBIS(4-HYDROXYBENZENEARSONIC ACID) $C_{13}H_{14}As_2O_8$ 3,3'-CH$_2$(4-HOC$_6$H$_3$AsO$_3$H$_2$)

Prepn.: by adding HCHO soln. to p-HOC$_6$H$_4$AsO(OH)$_2$ (1:2) in 92% H$_2$SO$_4$ at 0-5°; 22% yield (358).

Props.: m. > 300° (358).

3-(4-CARBETHOXY-2-HYDROXY-5-METHYL-3-PYRRYLAZO)-4-HYDROXYBENZENEARSONIC ACID
$C_{14}H_{16}AsN_3O_7$

EtO$_2$C-C———C-N:N-〔HO〕-AsO$_3$H$_2$
Me-C———C -OH
N
H

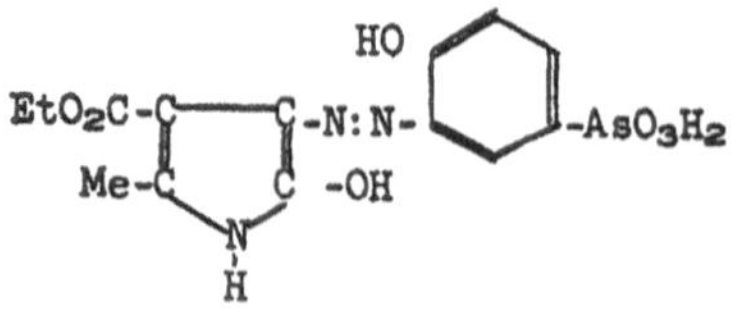

<u>Prepn.:</u> by reacting 2,5-HO(H$_2$O$_3$As)C$_6$H$_3$N$_2$OH with 3-carbethoxy-5-hydroxy-2-methyl-pyrrole in 96% EtOH (385).
<u>Props.:</u> Dark yellow crystals, darkens at 190° (385).
<u>Rxns. with:</u> H$_3$PO$_2$ + KI → the arsenophenol deriv. (385).

4-(3-CARBETHOXY-5-HYDROXY-2-METHYL-4-PYRRYLAZO)-3-HYDROXYBENZENEARSONIC ACID
$C_{14}H_{16}AsN_3O_7$

EtO$_2$C ——— -N:N- 〔 〕-AsO$_3$H$_2$
Me- N -OH OH
H

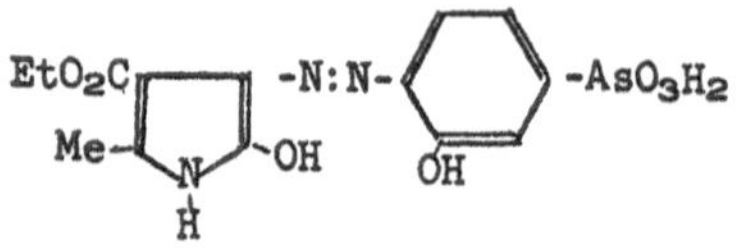

<u>Prepn.:</u> by diazotization of 3-nitroarsanilic acid in 2 N aq. Na$_2$CO$_3$, followed by adding the soln. to 2 N H$_2$SO$_4$, treating the diazo salt with NaOAc and adding the azo compd. to 3-carbethoxy-5-hydroxy-2-methylpyrrole in 96% EtOH (870).
<u>Props.:</u> Red powder, decomp. at 208-210° (870).

3-(3-CARBETHOXY-2,4-DIMETHYL-5-PYRRYLAZO)-4-HYDROXYBENZENEARSONIC ACID
$C_{14}H_{18}AsN_3O_6$

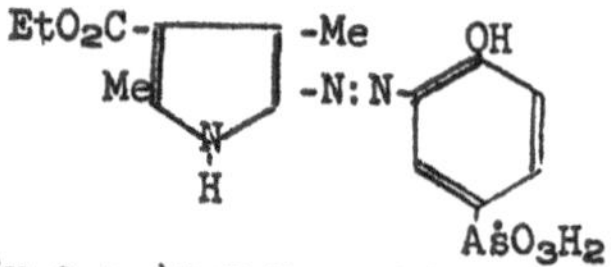

EtO$_2$C- ——— -Me OH
Me- -N:N- 〔 〕
N
H
AsO$_3$H$_2$

<u>Prepn.:</u> by coupling 2,5-HO(H$_2$O$_3$As)C$_6$H$_3$N$_2$-sulfate with 3-carbethoxy-2,4-dimethylp: role in abs. EtOH (871).
<u>Props.:</u> Yellow needles, decomp. 160°, stable in air under light (871).

4-(5-AMINO-8-ISOQUINOLYLAZO)-2-NITROBENZENEARSONIC ACID $C_{15}H_{12}AsN_5O_5$

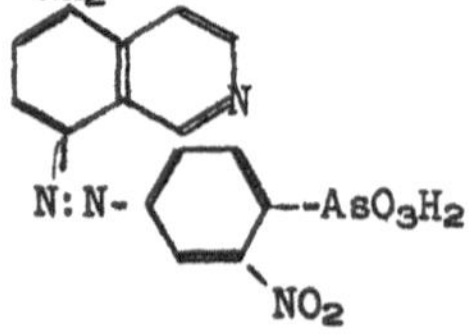

NH$_2$

N:N- 〔 〕--AsO$_3$H$_2$
NO$_2$

<u>Prepn.:</u> by reacting 3,4-O$_2$N(H$_2$O$_3$As)C$_6$H$_3$N$_2$Cl with 5-aminoisoquinoline in dild. Ac

264

79% yield (332).
Props.: Decomp. 300° (332).

4-HYDROXY -N-1-ISOQUINOLYL-m-ARSANILIC ACID $C_{15}H_{13}AsN_2O_4$

Prepn.: by refluxing 1-chlorosioquinoline with $4,3-HO(H_2N)C_6H_3AsO_3H_2$ (1:1) in H_2O contg. 1 equiv. HCl; 65% yield (332).
Props.: Decomp. 207-209° (332).

BIS(4-ARSONO-2-NITROPHENOXY)ISOPROPYL NITRATE $C_{15}H_{15}As_2N_3O_{15}$ $[2,4-O_2N(H_2O_3As)-C_6H_3OCH_2]_2CHONO_2$

Prepn.: by treating $(4-H_2O_3AsC_6H_4OCH_2)_2CHOH$ with $HNO_3-H_2SO_4$ at 0°, 92% yield (62).
Props.: Yellow compd. decomp. at 218° (62).
Rxn. with 2 N HCl under reflux → hydrolysis of the nitrate group (62).

N-[2-(4-AMINO-6-p-SULFOANILIND-s-TRIAZIN-2-YL)HYDRAZINO]-4-HYDROXY-m-ARSANILIC ACID $C_{15}H_{16}AsN_7O_7S$

Prepn.: by reacting the 3-[2-(4,6-dichloro-s-triazin-2-yl)hydrazino]-4-hydroxy-benzenearsonic acid with $4-H_2NC_6H_4SO_3Na$ in aq. alkali and treating the rxn. prod. with 30% NH_4OH at 100° under pressure (418).
Props.: White ppt. (418).
Rxn. with $SnCl_2$ in HCl contg. HI → the arsenophenol deriv. (418).

BIS(2-NITRO-4-ARSONOPHENOXY)ISOPROPANOL $C_{15}H_{16}As_2N_2O_{13}$ $[2,4-O_2N(H_2O_3As)C_6H_3-OCH_2]_2CHOH$

Prepn.: by hydrolysis of the corresponding isopropyl nitrate (62).
Props.: Yellow solid decompg. at 260° (62).
Rxn. with H_2 over Raney Ni → the corresponding 2-H_2N deriv. (62).

BIS(2-AMINO-4-ARSONOPHENOXY)ISOPROPANOL $C_{15}H_{18}As_2N_2O_9$ $[2,4-H_2N(H_2O_3As)C_6H_3OCH_2]_2CHOH$

Prepn.: by catalytic redn. of $[2,4-O_2N(H_2O_3As)C_6H_3As)C_6H_3OCH_2]_2CHOH$ over Raney Ni. 74% yield (62).
Props.: Decomp. 186° (62).

4-(1-AMINO-8-HYDROXY-3,6-DISULFO-2-NAPHTHYLAZO)-2,6-DIIODOBENZENEARSONIC ACID
$C_{16}H_{12}AsI_2N_3O_{10}S_2$

Prepn.: by reacting sodium 1-amino-8-naphthol-3,6-disulfonate with 3,5,4-I_2(H_2-O_3As)$C_6H_2N_2Cl$ in aq. NaOH (493).
Props.: Dark maroon powder (493).
Rxn. with [3,4-Me(ClN_2)C_6H_3-]$_2$ → 3,3'-dimethylbiphenyl-4,4'-bis[2-azo-8-amino-1-hydroxy-3,6-disulfo-7-naphthaleneazo-2',6'-diiodobenzenearsonic acid (493).

3-(3,4-DIHYDRO-3,4-DIOXO-5-SULFONAPHTHYLAMINO)-4-HYDROXYBENZENEARSONIC ACID
$C_{16}H_{12}AsNO_9S$

Prepn.: by reacting 1,2-naphthoquinone-4,8-disulfonic acid with 3,4-H_2N(HO)C_6H_3-AsO_3H_2 in the presence of alkali (1150, 1151).

4-(3,4-DIHYDRO-3,4-DIOXO-5-SULFO-1-NAPHTHYLAMINO)-3-HYDROXYBENZENEARSONIC ACID
$C_{16}H_{12}AsNO_9S$

Prepn.: by reacting 1,2-naphthoquinone-4,8-disulfonic acid with 3,4-HO(H_2N)C_6H_3-AsO_3H_2 in the presence of alkali and an oxidizing agent (1247).

3-(3,4-DIHYDRO-3,4-DIOXO-6-SULFO-1-NAPHTHYLAMINO)-4-HYDROXYBENZENEARSONIC ACID
$C_{16}H_{12}AsNO_9S$

Prepn.: by reacting 1,2-naphthoquinone-4,7-disulfonic acid with 3,4-H_2N(HO)C_6H_3-

AsO$_3$H$_2$ in the presence of alkali and an oxidizing agent (1247).

3-(3,4-DIHYDRO-3,4-DIOXO-6-SULFO-1-NAPHTHYLAMINO)-4-HYDROXYBENZENEARSONIC ACID
C$_{16}$H$_{12}$AsNO$_9$S

Prepn.: by reacting 1,2-naphthoquinone-4,7-disulfonic acid with 3,4-H$_2$N(HO)C$_6$H$_3$-AsO$_3$H$_2$ in the presence of alkali and H$_2$O$_2$ (1151, 1247).

3-(3,4-DIHYDRO-3,4-DIOXO-7-SULFONAPHTHYLAMINO)-4-HYDROXYBENZENEARSONIC ACID
C$_{16}$H$_{12}$AsNO$_9$S

Prepn.: by reacting potassium 1,2-naphthqouinone-4,6-disulfonate with 4,3-HO(H$_2$N)-C$_6$H$_3$AsO$_3$HNa in the presence of an oxidiazing agent (1149, 1247).

4-(3,4-DIHYDRO-3,4-DIOXO-5-SULFONAPHTHYLAMINO)-2-HYDROXYBENZENEARSONIC ACID
C$_{16}$H$_{12}$AsNO$_9$S

Prepn.: by reacting 1,2-naphthoquinone-4,8-disulfonic acid with 2,4-HO(H$_2$N)C$_6$H$_3$-AsO$_3$H$_2$ in the presence of alkali and H$_2$O$_2$ (1151,1247).

267

4-(3,4-DIHYDRO-3,4-DIOXO-5,7-DISULFO-1-NAPHTHYLAMINO)-2-HYDROXYBENZENEARSONIC
$C_{16}H_{12}AsNO_{12}S_2$

Prepn.: by reacting 1,2-naphthoquinone-4,6,8-trisulfonic acid with 2,4-HO(H_2N)-
$C_6H_3AsO_3H_2$ in the presence of alkali and H_2O_2 (1151, 1247).

3-AMINO-4-(3,4-DIHYDRO-3,4-DIOXO-5-SULFO-2-NAPHTHYLAMINO)BENZENEARSONIC ACID
$C_{16}H_{13}AsN_2O_8S$

Prepn.: by reacting 1,2-naphthoquinone-4,8-disulfonic acid with 3,4-(H_2N)$_2C_6H_3$-
AsO_3H_2 in the presence of alkali and H_2O_2 (1151, 1247).

4-HYDROXY-3-(1-HYDROXY-4,8-DISULFO-2-NAPHTHYLAZO)BENZENEARSONIC ACID
$C_{16}H_{13}AsN_2O_{11}S_2$

Prepn.: by coupling 2,5-HO(H_2O_3As)$C_6H_3N_2Cl$ with 1-naphthol-4,8-disulfonic acid
(409, 412, 1040).
Rxns. with: H_3PO_2 in a mineral acid contg. KI → the arsenobenzene deriv.(409, 4
1040).
Zn in aq. NaOH → the hydrazino deriv. (428).

4-HYDROXY-3-(1-HYDROXY-4,8-DISULFO-2-NAPHTHYLHYDRAZINO)BENZENEARSONIC ACID
$C_{17}H_{15}AsN_2O_{11}S_2$

Prepn.: by reducing the corresponding naphthylazo deriv. with Zn dust in aq.
alkali (428, 432).

268

4,4'-DITHIOBIS(N-ACETYL-m-ARSANILIC ACID) $C_{16}H_{18}As_2N_2O_8S_2$ $[2,4-AcNH(H_2O_3As)C_6H_3S-]_2$
Prepn.: by treating $[2,4-H_2N(H_2O_3As)C_6H_3S-]_2$ with Ac_2O (147).
Props.: Amorph. colorless powder with no m.p., insol.in acids, sol. in alk. (147).
Rxn. with $FeSO_4$ in alk. soln. → $3,4-AcNH(HS)C_6H_3AsO(OH)_2$ (147).

3-(8-ETHOXY-5-QUINOLYLSULFONAMIDO)-4-HYDROXYBENZENEARSONIC ACID $C_{17}H_{17}AsN_2O_7S$

Prepn.: by reacting 8-ethoxy-5-quinolylsulfonyl chloride with $3,4-H_2N(HO)C_6H_3AsO_3-$
H_2 in aq. Na_2CO_3.soln. (458).
Props.: M. 215-217° (decompn.) (458).

3,3'-DIARSONO-4,4'-DIMETHOXY-α,α'-DIMETHYLDIBENZYLAMINE $C_{18}H_{25}As_2NO_8$

Prepn.: by-prod. in the catalytic redn. of 5-acetyl-2-methoxybenzenearsonic acid
oxime (36).
Derivs.: Acetamide, white crystals m. 205° (36).

N-(1-BENZO[f]QUINOLYL)-4-HYDROXY-m-ARSANILIC ACID $C_{19}H_{25}AsN_2O_4$

Prepn.: by refluxing 4-hydroxy-m-arsanilic acid with 1-chlorobenzo[f]quinoline in
dild. HCl; 41% yield (250).
Props.: M. > 250° (250).

3,3',3"-TRIARSONO-4,4',4"-TRIMETHOXY-$\alpha,\alpha',\alpha"$-TRIMETHYLTRIBENZYLAMINE
   $C_{27}H_{36}As_3NO_{12}$

Prepn.: by-prod. in the catalytic redn. of 5-acetyl-2-methoxybenzenearsonic acid
oxime (36).
Props.: m. 205° (36).

3,3'-DIMETHYLBIPHENYL-4,4'-BIS[4-(2-AZO-8-AMINO-1-HYDROXY-3,6-DISULFO-7-NAPHTHAI
AZO)-2,6-DIIODOBENZENEARSONIC ACID $C_{46}H_{34}As_2I_4N_{10}O_{20}S_4$

Prepn.: by condensing tetrazotized o-tolidine with 4-(1-amino-8-hydroxy-3,6-disul
2-naphthylazo)-2,6-diiodobenzenearsonic acid (493).
Props.: Na salt, dark maroon leaves (493).

4,4'-(p-BIPHENYLENEDIAZO)BIS[2,6-BIS(p-ARSONOPHENYLAZO)RESORCINOL] $C_{48}H_{38}As_4N_1$

Prepn.: by coupling 4,4'-bis(2,4-dihydroxyphenylazo)biphenyl with 4-H_2O_3AsC_6H_4N_2-
Cl (984).

DISUBSTITUTED BENZENEARSONIC ACIDS

RR₁C₆H₃AsO₃H₂ where R and R₁ are:	Prepd. from	Yield %	Props. and Remarks	Ref.
C₆				
4,3-Br(O₂N)-			Rxn.[1]	250
2,4-Cl₂-	2,4-Cl₂C₆H₃NH₂ by the Bart method		Shiny leaflets, m. 204°, rxns.[2, 3, 4]; use[5]	550
4,3-F(O₂N)-	by the Bart-Scheller method			120
	4,3-F(O₂N)C₆H₃NH₂ by the Bart-Scheller method		m. 279.2-281.3°, rxns.[6, 7]	114
3,4-F₂-	3,4-F₂C₆H₃NH₂ by the Bart-Scheller method		m. >300°	529
2,4-HO(O₂N)-	2,4-(O₂N)₂C₆H₃NH₂ by the Bart method, followed by rxn with NH₄OH + NH₄Cl + MgCl₂		Rxns.[8, 9]	670
3,2-HO(O₂N)-	3,2-O₂N(HOCH₂CH₂O)C₆H₃AsO₃H₂ by alk. hydrolysis	40	Yellow plates, m. 208° (dec.), rxn.[2]	82
3,4-HO(O₂N)-	4,3-O₂N(HOCH₂CH₂O)C₆H₃AsO₃H₂ by alk. hydrolysis	60	Yellow plates, m. >270°, use[10]	82
3,4-O₂N(HO₃S)-	3,4-O₂N(Cl)C₆H₃AsO₃H₂ + Na₂SO₃	54	Rxn.[11]	286
3,4-Br(H₂N)-	4-H₂NC₆H₄AsO₃H₂ + Br₂		m. ~255°, rxn.[12]	452
3,4-H₂N(Cl)-	Method not given			286
3,4-H₂N(F)-	3,4-O₂N(F)C₆H₃AsO(OH)(ONa) + H₂/Ni	52	Very sensitive to air	114
2,4-O₂N(H₂N)-	2,4-O₂N(AcNH)C₆H₃NH₂ by the Bart method, followed by hydrolysis of the AcNH group	46	Orange-red needles, m. 255-258° (dec.), rxn.[13]	1023

1. Rxn. with amines → the 4-amino deriv. (250).
2. Redn. by SO₂ in HCl contg. KI, followed by neutralization → the arsenoso deriv. (36, 39, 62, 82, 83, 102, 288, 289, 362, 368, 369, 388, 550, 556, 1117).
3. Redn. by SO₂ in HCl → the dichloroarsino deriv. (120, 289, 290, 393, 418).
4. Dissocn. const. and redn. potential (241).
5. Used as wood preservative.
6. Rxn. with Na₂S₂O₄ → redn. of the NO₂ and AsO₃H₂ groups to the diamino arsenobenzene derivs. (114, 1047).
7. Catalytic redn. over Raney Ni → redn. of the -NO₂ group to -NH₂ (62, 83, 114, 239, 288, 290, 388, 506).

TABLE XLI (cont'd.)

DISUBSTITUTED BENZENEARSONIC ACIDS

$RR_1C_6H_3AsO_3H_2$ where R and R_1 are:	Prepd. from	Yield %	Props. and Remarks	Ref.
$2,4\text{-}O_2N(H_2NO_2S)\text{-}$	$2,4\text{-}O_2N(H_2NO_2S)C_6H_3NH_2$ by the Bart method		m. 220°	1225
$2,4\text{-}(HO)_2\text{-}$	Rxn. with HCHO → ion-exchange resin			1203
$2,5\text{-}HO(H_2N)\text{-}$	$2,5\text{-}HO(PhN_2)C_6H_3AsO_3H_2 + H_2$ over Raney Ni		Rxn.[3]	289
$5,2\text{-}HO(H_2N)\text{-}$	$5,2\text{-}HO(PhN_2)C_6H_3AsO_3H_2 + H_2$ / Raney Ni		Rxn.[3]	289
$4,3\text{-}HO(H_2NNH)\text{-}$	$4,3\text{-}HO(H_2N)C_6H_3AsO_3H_2$ by diazotization followed by redn.		Rxn.	418
$2,4\text{-}H_2N(H_2NO_2S)\text{-}$	$2,4\text{-}O_2N(H_2NO_2S)C_6H_3AsO_3H_2 + Fe(OH)_2$ in alk. soln. or H_2/Ni			1225
C_7				
$2,3\text{-}HO_2C(O_2N)\text{-}$	$2,3\text{-}H_2N(O_2N)C_6H_3CO_2H$ by the Bart method	29	Needles, m. 305° (dec.) rxn.[7]	290, 288
$2,4\text{-}HO_2C(O_2N)\text{-}$	$2,5\text{-}H_2N(O_2N)C_6H_3CO_2H$ by the Bart method; $2,4\text{-}Me(O_2N)C_6H_3AsO_3H_2$ by oxidation	63	Needles, m. 331° (dec.) rxn.[7]	288, 290, 288
$2,5\text{-}HO_2C(O_2N)\text{-}$	$2,5\text{-}H_2N(O_2N)C_6H_3CO_2$ by the Bart method	62	Needles, m. 344.5°, rxn.[7] Na salt, needles, m. >360°	290
$2,6\text{-}HO_2C(O_2N)\text{-}$	$2,6\text{-}Me(O_2N)C_6H_3AsO_3H_2 + KMnO_4$	59	Needles, explode at 280-295°	290
$3,2\text{-}HO_2C(O_2N)\text{-}$	$3,2\text{-}H_2N(O_2N)C_6H_3CO_2H$ by the Bart method	49	Na salt, rectangular plates m. 228°	290
$3,5\text{-}HO_2C(O_2N)\text{-}$	$3,5\text{-}Me(O_2N)C_6H_3AsO_3H_2 + KMnO_4$	52	K salt, prism, m. >360°	290
$4,2\text{-}HO_2C(O_2N)\text{-}$	$4,3\text{-}H_2N(O_2N)C_6H_3CO_2H$ by the Bart method	49	Hexagonal plates, m. 244-245°, rxn.[15]	290
$4,3\text{-}HO_2C(O_2N)\text{-}$	$4,2\text{-}H_2N(O_2N)C_6H_3CO_2H$ by the Bart method	42	K salt, rectangular prisms, sinter 319°, rxn.[7]	290
$5,2\text{-}HO_2C(O_2N)\text{-}$	$5,2\text{-}H_2N(O_2N)C_6H_3CO_2H$ by the Bart method	53	1.H_2O, cubes, m. 240.5-241.5°, rxn.[15]	290
$3,4\text{-}O_2N(H_2NOC)\text{-}$	$3,4\text{-}O_2N(H_2NOC)C_6H_3AsO + H_2O_2$	60	Rectangular prisms, chars 270°, rxn.[7]	288

Compound	Method	Yield	Form, m.p.	Ref.
	Ni			288, 290
2,4-HO₂C(H₂N)-	2,4-HO$_2$C(O$_2$N)C$_6$H$_3$AsO$_3$H$_2$ + H$_2$/Ni	72	Plates, m. >360°, rxn.3, 16	288, 290
2,5-HO₂C(H₂N)-	2,5-HO$_2$C(O$_2$N)C$_6$H$_3$AsO$_3$H$_2$ + H$_2$/Ni	83, 94	Needles, m. >360°, rxns.3, 16	290, 288
3,2-HO₂C(H₂N)-	3,2-HO$_2$C(O$_2$N)C$_6$H$_3$AsO$_3$H$_2$ + Fe(OH)$_2$ in alk. soln.	45	Na salt, rectangular plates or needles, m. 256.5°, rxn.3	290
3,4-HO₂C(H₂N)-	3,4-HO$_2$C(O$_2$N)C$_6$H$_3$AsO$_3$H$_2$ + H$_2$/Ni	75	Needles, m. 246-248° (dec.) rxn.3	290
4,2-HO₂C(H₂N)-	4,2-HO$_2$C(O$_2$N)C$_6$H$_3$AsO$_3$H$_2$ + Fe(OH)$_2$ in alk. soln.	57	Needles, m. >360°, rxn.3	290
4,3-HO₂C(H₂N)-	4,3 HO$_2$C(O$_2$N)C$_6$H$_3$AsO$_3$H$_2$ + H$_2$/Ni	88	Rectangular prisms, rnx.3	290
5,2-HO₂C(H₂N)-	5,2-HO$_2$C(O$_2$N)C$_6$H$_3$AsO$_3$H$_2$ + Fe(OH)$_2$ in alk. soln.	67	Needles, rxn.3	290
5,3-HO₂C(H₂N)-	5,3-HO$_2$C(O$_2$N)C$_6$H$_3$AsO$_3$H$_2$ + H$_2$/Ni	59	Needles, m. 240°, rxn.3	290
6,2-HO₂C(H₂N)-	6,2-HO$_2$C(O$_2$N)C$_6$H$_3$AsO$_3$H$_2$ + Fe(OH)$_2$ in alk. soln.	45	Rectangular prisms	290
2,4-Me(O₂N)-	2,4-Me(O$_2$N)C$_6$H$_3$NH$_2$ by the Bart-Scheller method	67	Needles, m. 240°	288
2,5-Me(O₂N)-	2,5-Me(O$_2$N)C$_6$H$_3$NH$_2$ by the Bart method		White crystals, m. 261°, rxn.f1	551
2,6-Me(O₂N)-	2,6-Me(O$_2$N)C$_6$H$_3$NH$_2$ by the Bart method	52	Needles, m. 235-236° m. 228-230°, rxns.15, 17	288, 351

8. Rxn. with Fe in HCl in the presence of H$_3$AsO$_4$ or H$_3$AsO$_3$ → redn. of the NO$_2$ group (152).
9. Redn. followed by acetylation → the acetamido deriv. (965).
10. Reagent for turbidometric titration of Sn in Cu (196a).
11. Redn. by SO$_2$ in HCl contg. KI, followed by neutralization → the arsonous acid, RAs(OH)$_2$,(286, 551).
12. Rxn. with 2,4,5-triamino-6-hydroxypyrimidine + Br$_2$CHCOCH$_2$Br → N-[(2-amino-4-hydroxy-6-pteridyl)methyl]-3-bromoarsanilic acid (452).
13. Rxn. with NaNO$_2$, followed by RCH:CHCOR → 2,4-O$_2$N(RCH:CH)C$_6$H$_3$AsO$_3$H$_2$ (401, 404).
14. Rxn. with (ClCN)$_3$ → the 2-(4,6-dichloro-s-triazin-2-ylhydrazino) deriv. (418).
15. Redn. with Fe^{++} in alk. soln. → redn. of the -NO$_2$ to-NH$_2$ (250, 290, 351, 393, 760, 1047, 1117).
16. Diazotization followed by CuCN under Sandmeyer condns. → the NC- deriv. (288).
17. Rxn. with alk. KMnO$_4$ → oxidation of the Me groups to -CO$_2$H (239, 288, 290).

TABLE XLII (cont'd.)
DISUBSTITUTED BENZENEARSONIC ACIDS

$RR_1C_6H_3AsO_3H_2$ where R and R_1 are:	Prepd. from	Yield %	Props. and Remarks	Ref.
3,5-Me(O_2N)-	3,5-Me(O_2N)$C_6H_3NH_2$ by the Bart method	11	Pale yellow needles, m. 307°, rxn.[17]	290
4,3-Me(O_2N)-	4,3-Me(O_2N)$C_6H_3NH_2$ by the Bart-Scheller method	40		282
4,3-HO(H_2NOC)-	4,3-HO(MeO_2C)$C_6H_3AsO_3H_2$ + NH_4OH	90	Needles, char at 330°, rxn.[2]	288
3,4-H_2N(H_2NOC)-	3,4-O_2N(H_2NOC)$C_6H_3AsO_3H_2$ + H_2/Ni	60	Plates, decomp. >230°, rxn.[2]	288
4,3-HO(H_2NCONH)-	3,4-HO(H_2N)$C_6H_3AsO_3H_2$ + KOCN or $H_2NCONHNO_2$		Rxn.[19]	374
3,4-HO(H_2NCONH)-			Rxn.[20]	533
2,6-H_2N(Me)-	2,6-O_2N(Me)$C_6H_3AsO_3H_2$ + Fe(OH)₂ in alk. soln.			351
4,3-HO(MeNH)-				362
4,3-HO(HO_2SCH_2NH)-	4,3-HO(H_2N)$C_6H_3AsO_3H_2$ + $HOCH_2$-SO_2Na		Rxn.[21]	1046
4,3-HO($H_2NNHCONH$)-	2-oxo-6-benzoxazolinearsonic acid + hydrazine			375

C_8

$RR_1C_6H_3AsO_3H_2$ where R and R_1 are:	Prepd. from	Yield %	Props. and Remarks	Ref.
2,5-HO_2C(NC)-	2,5-HO_2C(H_2N)$C_6H_3AsO_3H_2$ by the Sandmeyer method	41	Needles, decomp. >351°, rxn.[22]	288
2,4-HO_2C(NC)-	2,4-HO_2C(NC)$C_6H_3AsO_3H_2$ by the Sandmeyer method	23	White plates, rxn.[22]	288
3,5-(HO_2C)₂-	3,5-$Me_2C_6H_3AsO_3H_2$ + $KMnO_4$	75	K salt, white needles, rxn.[23]	288
2,4-HO_2C(H_2NOC)-	2,4-HO_2C(NC)$C_6H_3AsO_3H_2$ + H_2O_2	57	Rectangualr prisms, m. 347.5°, rxn.[2]	288
2,5-HO_2C(H_2NOC)-	2,5-HO_2C(NC)$C_6H_3AsO_3H_2$ + H_2O_2	63	Plates, rxn.[2]	288
3,4-O_2N(AcO)-	3,4-O_2N(HO)$C_6H_3AsO_3NH_2$ + AcCl		Isolated as Na.$2H_2O$ salt, rxns.[15, 24]	1117
4,3-F(AcNH)-	4,3-F(H_2N)$C_6H_3AsO_3H_2$ + Ac_2O		Brown, m. 217.1-218.2°	114
3,4-O_2N(H_2NOCCH_2O)-	3,4-O_2N(MeO_2CCH_2O)$C_6H_3AsO_3H_2$ + EtOH-NH_3	72	Pale yellow	1182
	or NH_4OH	61	Rxn.[25]	1182

3,4-O2N(O2NOC6H4CH2O)-	4-HOC6H2OC6H4AsO3H2 + HNO3		Long white needles, m. 217?, 218°, rxn.26	302
2,5-HO(Ac)-	2,5-HO(Ac)C6H3NH2 by the Bart method at 20°	35	m. 225°, rxns.2, 20	36
4,3-HS(AcNH)-	[2,4-AcNH(H2O3As)C6H3S-]2 + FeSO4 in alk. soln.		Amorph. colorless powder, Au deriv.27	147
3,4-O2N(HOCH2CH2)-	4-HOCH2CH2C6H4AsO3H2 by acetylation and nitration	84	m. 119-120°	286
3,4-O2N(EtO)-	4-HOC6H4AsO3H2 + EtBr in alk. soln. + HNO3-H2SO4		Rxn.6, 15	1047
2,4-Me2-	2,4-Me2C6H3NH2 by the Bart method		White needles, m. 210°, rxn.2	550
3,5-Me2-	3,5-Me2C6H3NH2 by the Bart method	18	White needles, m. 222-223°, rxn.17	288
3,4-Me2-	3,4-Me2C6H3NH2 by the Bart method		White needles, change form at 110°, rxn.28	1016
2,3-Me2-	2,3-Me2C6H3NH2 by the Bart method		Fine needles, sublimes at 200°, rxn.28	1016
2,6-Me2-	2,6-Me2C6H3NH2 by the Bart method	30	m. 207-208°	286
3,4-H2N(EtO)-	3,4-O2N(EtO)C6H3AsO3H2 + Na2S2O4 or Fe(OH)2 in aq. alk.		Rxn.2, 29	1047, 1048
4,2-H2N(HOCH2CH2O)-	4,2-PhCH2OCONH(HOCH2CH2O)-C6H3AsO3H2 by hydrolysis	68	Tetragonal prisms, m. 164°	71
3,4-O2N(H2NCH2CH2NH)-	3,4-O2N(MeO2CCH2O)C6H3AsO3H2 + (-CH2NH2)2	37		1182
4,3-HO[H2NC(:NH)NHC(:NH)NH]-	3,4-H2N(HO)C6H3AsO3H2 + H2NC(:NH)NHCN	78		1181

18. Rxn. with HNO3-H2SO4 → 4,3,5-Me(O2N)2C6H2AsO3H2 (865).
19. Rxn. with Hg(OAc)2 in aq. NaOH, followed by I + KI → the 5-iodo deriv. (374).
20. Redn. by H3PO2 → the arsenobenzene deriv. (36, 71, 424, 504, 556, 951, 1181).
21. Redn. by Zn in HCl → the primary arsine, RAsH2, (362).
22. Rxn. with H2O2 → oxidation of the -CN group to -CONH2 (288).
23. Rxn. with PCl3 + PCl5 in CHCl3 followed by NH4OH → the arsenoso deriv. (288).
24. Rxn. with hot H2O → hydrolysis of the AcO group (1117).
25. Catalytic redn. over Ni → 6-arsono-3-hydroxy-1,4,2H-benzoxazine (1182).
26. Aq. HCl → hydrolysis of the -ONO2 group (83, 502).
27. 3,4-AcNH(AuS)C6H3AsO3H2, prepd. by treating the mercapto deriv. with KAuBr4 + Na2SO3 in alk. soln. (148).
28. Rxns. with metal ions → quantitat. pKn. (1016).
29. Rxn. with Ac2O → the 3-AcNH deriv. (1047).

TABLE XLI (cont'd.)

DISUBSTITUTED BENZENEARSONIC ACIDS

$RR_1C_6H_3AsO_3H_2$ where R and R₁ are:	Prepd. from	Yield %	Props. and Remarks	Ref.
C₉				
4,3-HO[N:CClN:CClN:CNH]-			Rxn.30	418
3,4-O₂N[N:C(NH₂)N:C(NH₂)N:Cq-	4-[N:C(NH₂)N:C(NH₂)N:C0]C_6H_4-AsO_3H_2 + HNO_3-H_2SO_4	13	Yellow solid	1306
3,4-O₂N(MeCOCH₂O)-	4-MeCOCH₂O$C_6H_4AsO_3H_2$ + HNO_3-H_2SO_4 <5°	80	Long yellow needles, m. 190° rxn.31	501
3,4-O₂N(CH₂:CHCH₂NH)-	3,4-O₂N(Cl)$C_6H_3AsO_3H_2$ + CH₂:CHCH₂NH₂		Rxn.15	760
5,2-O₂N[MeCH(ONO₂)CH₂O]-	2-MeCHOHCH₂O$C_6H_4AsO_3H_2$ + NHO₃	78	Colorless needles, m. 186°, rxn.26	83
5,2-Ac(MeO)-	5,2-Ac(MeO)$C_6H_3NH_2$ by the Bart method	60	m. 212°, rxns.2, 20, 32	36
2,4-HO(EtCONH)-	2,4-HO(H₂N)$C_6H_3AsO_3H_2$ + (EtCO)₂O		Colorless powder, m. 232-233° (dec.)	149
2,5-MeO(MeC:NOH)-	2,5-MeO(Ac)$C_6H_3AsO_3H_2$ + HONH₂·HCl	90	m. 200°, rxn.33	36
5,2-O₂N(MeCHOHCH₂O)-	2-(MeCHOHCH₂O)$C_6H_4AsO_3H_2$ + HNO₃, followed by hydrolysis	66	m. 165-167°, rxn.2,7	83
3,4-O₂N(HOCH₂CHOHCH₂O)-	3,4-O₂N[O₂NOCH₂CH(ONO₂)CH₂O]-$C_6H_3AsO_3H_2$ by hydrolysis	68	Yellow powder, rxns.2, 7	62, 504, 418
4,3-HO[N:C(NH₂)N:C(NHNH₂)N:CNH]-	4,3-HO[N:CClN:C(NHNH₂)N:CNH]-$C_6H_3AsO_2H_2$ + liq. NH₃		White ppt., rxn.3	760
3,4-H₂N(CH₂:CHCH₂NH)-	3,4-O₂N(CH₂:CHCH₂NH)$C_6H_3AsO_3$-H_2 + FeSO₄ in aq. NaOH			750
3,4-O₂N(MeCHOHCH₂NH)-	3,4-O₂N(Cl)$C_6H_3AsO_3H_2$ + MeCH-OHCH₂NH₂	68		1181
3,4-H₂NCONH(HOCH₂CH₂O)-	3,4-H₂N(HOCH₂CH₂O)$C_6H_3AsO_3H_2$ + KCNO		White cryst. powder, decomp. 237° Rxn.34	534
3,4-HO(HOCH₂CH₂NHCONH)-	3,4-HO(EtOCONH)$C_6H_3AsO_3H_2$ + HOCH₂CH₂NH₂ at 110°			418
4,3-HO[N:C(NHNH₂)N:C(NHNH₂)N:CNH]-	4,3-HO[N:CClN:C(NHNH₂)N:CNH]-$C_6H_3AsO_3H_2$ + H₂NNH₂·H₂O		Na salt, m. >300° (dec.), rxns.20, 35, 36	36
dl-2,5-MeO(MeCHOH)-	2,5-MeO(Ac)$C_6H_3AsO_3H_2$ + H₂/Ni	98		36
2,4-Me(Me₂N)-	2,4-Me(Me₂N)C_6H_3AsO + H_2O_2		m. >250°	1256

		%		
2,5-MeO(MeCHNH₂)-	2,5-MeO(MeC:NOH)C₆H₃AsO₃H₂ + H₂/Ni		White crystals, m. 248° (dec.)	36
4,2-H₂N(MeCHOHCH₂O)-	4,2-PhCH₂OCONH(MeCHOHCH₂O)C₆-H₃AsO₃H₂ by hydrolysis	75	Hydroscopic powder, softens at 159°, rxn.37	71
5,2-H₂N(MeCHOHCH₂O)-	5,2-O₂N(MeCHOHCH₂O)C₆H₃AsO₃-H₂ + H₂/Ni	80	.2H₂O, colorless needles loses 2H₂O at 110°, m. 184°, rxn.2 Rxns.20, 38	83
3,4-H₂N(MeCHOHCH₂O)-	4,3-HO(H₂N)C₆H₃AsO₃NaH + HO-CH₂CHCH₂O		Rxn.20	504 504
3,4-H₂N(HOCH₂CH₂CH₂O)-				362, 368
4,3-HO(HOCH₂CHOHCH₂NH)-			Colorless powder, rxn.21	62, 504
5,4-H₂N(HOCH₂CHOHCH₂O)-	3,4-O₂N(HOCH₂CHOHCH₂O)C₆H₃As-O₃H₂ + H₂ /Ni	60	Light gray powder, m. 194-196°; Na salt, granules; rxns.62, 20	
3,4-O₂N[MeCH(NH₂)CH₂NH]-	3,4-O₂N(Cl)C₆H₃AsO₃H₂ + H₂NCH-MeCH₂NH₂		Rxn.15	760
3,4-O₂N(H₂NCH₂CHOHCH₂NH)-	3,4-O₂N(Cl)C₆H₃AsO₃H₂ + (H₂N-CH₂)₂CHOH			760
3,4-H₂N[MeCH(NH₂)CH₂NH]-	3,4-O₂N[MeCH(NH₂)CH₂NH]C₆H₃-AsO₃H₂ + FeSO₄ in aq. NaOH			760
C₁₀ 3,4-Me(N:CC1N:CC1N:CNH)-	3,4-Me(H₂N)C₆H₃AsO₃H₂ + (Cl-CN)₃			424
3,4-O₂N(N:CHCH:CHN:C NH)-	3,4-O₂N(Br)C₆H₃AsO₃H₂ + 2-aminopyridine		Yellow crystals, at 148-156°, liq. crystals, m. 160-163;rxn7	506
2,5-HO[N:C(NH₂)N:CC1CH:CNH]-	2,5-HO(H₂N)C₆H₃AsO₃H₂ + H₂NC:NCC1:CHCC1:N	23		14
3,4-O₂N(CONH:CMeCHN:N)-	3,4-O₂N[AcC(CO₂H):NNH]C₆H₃As-O₃H₂ + H₂NNH₂.HCl	32	m. > 250°	240

30. Rxn. with 20% aq. H₂NNH₂ → the 4-chloro-6-hydrazino-s-triazinyl deriv. (418).
31. Rxn. with FeSO₄ or Na₂S₂O₄ in alk. soln. → 3-methylbenzoxazine-6-arsonic acid (503).
32. Catalytic redn. → dl-5,2-MeHOH(MeO)C₆H₃AsO₃H₂ (36).
33. Catalytic redn. → a mixt. of 5,2-Ac(MeO)C₆H₃AsO₃H₂, [3,4-H₂O₃As(MeO)C₆H₃CHMe]₂NH, 2,5-MeO-(MeCHNH₂)C₆H₃AsO₃H₂, and [3,4-H₂O₃As(MeO)C₆H₃CHMe]₃N (36).
34. Rxn. with Na₂S₂O₄ in alk. soln. → the arseno deriv. (102, 388, 418, 1047).
35. Rxn. with Ac₂O → the acetate (36).
36. Rxn. with strong acids → dehydration of the MeCHOH group and polymerization (36).
37. Rxn. with EtOCOCl → the 4-EtOCONH deriv (71).
38. Rxn. with HOCH₂SO₂Na → 3,4-NaO₂SCH₂NH(MeCHOHCH₂O)C₆H₃AsO₃H₂ (504).

TABLE XLI (cont'd.)
DISUBSTITUTED BENZENEARSONIC ACIDS

$RR_1C_6H_3AsO_3H_2$ where R and R_1 are:	Prepd. from	Yield %	Props. and Remarks	Ref.
$3,4$-$O_2N[N{:}C(NH_2)N{:}C(OMe)N{:}CS]$-	$3,4$-$O_2N(ClN{:}N)C_6H_3AsO_3H_2$ + $N{:}C(NH_2)N{:}C(OMe)N{:}CSH$		Yellow cryst. powder	426
$3,4$-$H_2N(N{:}CHCH{:}CHN{:}CNH)$-	$3,4$-$O_2N(N{:}CHCH{:}CHN{:}CNH)C_6H_3AsO_3H_2$ + H_2/Ni		Brownish solid	506
$2,4$-$HO[CH{:}CHN{:}C(NH_2)N{:}CNH]$-	$2,4$-$HO(H_2N)C_6H_3AsO_3H_2$ + $CH{:}CHN{:}C(NH_2)N{:}CCl$		Rxn.[2]	39
$3,4$-$HO[N{:}C(NH_2)N{:}CHCH{:}CNH]$-	$3,4$-$HO(H_2N)C_6H_3AsO_3H_2$ + $N{:}C(NH_2)N{:}CHCH{:}CCl$		m. 247-249°	14
$4,3$-$HO[N{:}C(NH_2)N{:}CHCH{:}CNH]$-	$4,3$-$HO(H_2N)C_6H_3AsO_3H_2$ + $N{:}C(NH_2)N{:}CHCH{:}CCl$		Rxn.[2]	39
$2,5$-$HO[N{:}C(NH_2)N{:}CHCH{:}CNH]$-	$2,5$-$HO(H_2N)C_6H_3AsO_3H_2$ + $N{:}C(NH_2)N{:}CHCH{:}CCl$	46		14
$3,4$-$H_2N[N{:}C(CO_2H)CMe{:}NN]$-	$3,4$-$O_2N[N{:}C(CO_2H)CMe{:}NN]C_6H_3AsO_3H_2$ + H_2/Ni	54	m. > 250°	239
$3,4$-$O_2N(N{:}CMeCMe{:}NN)$-	4-$N{:}CMeCMe{:}NNC_6H_4AsO_3H_2$ + HNO_3-H_2SO_4	70	m. 187.5-188.5°, rxns.[7,17]	239
$3,4$-$O_2N[HO_2C(CH_2)_3]$-	4-$[HO_2C(CH_2)_3]C_6H_4AsO_3H_2$ + HNO_3-H_2SO_4	68	m. 262°, rxn.[15]	393
$3,4$-$ClCH_2CONH(HOCH_2CH_2O)$-	$3,4$-$H_2N(HOCH_2CH_2O)C_6H_3AsO_3H_2$ + $ClCH_2COCl$ in AcOH-NaOAc	60	Rxn.[40]	1181
$4,3$-$HO(CH_2{:}CHCH_2NHCONH)$-	2-oxo-5-benzoxazolinearsonic acid + allylamine			375
$3,4$-$O_2N[S(CH_2CH_2)_2N]$-	$3,4$-$O_2N(Br)C_6H_3AsO_3H_2$ + $S(CH_2CH_2)_2NH$ + K_2CO_3	96	Bright yellow powder, m. 187-189°, rxn.[15]	250
$3,4$-$O_2N[O(CH_2CH_2)_2N]$-	$3,4$-$O_2N(Br)C_6H_3AsO_3H_2$ + morpholine + K_2CO_3	91	Bright yellow substance, m. 203-204°, rxn.[15]	250
$3,4$-$HO[O(CH_2CH_2)_2NCOCH_2O]$-	$3,4$-$HO(MeO_2CCH_2O)C_6H_3AsO_3H_2$ + morpholine	46	Rxn.[34]	1182
$3,4$-$(H_2NCOCH_2NH)(HO_2CCH_2O)$-	$3,4$-$O_2N(HO_2CCH_2O)C_6H_3AsO_3H_2$ + $Fe(OH)_2$ in alk. soln. + $ClCH_2CONH_2$			1047
$3,4$-$H_2N(N{:}CMeCMe{:}NN)$-	$3,4$-$O_2N(N{:}CMeCMe{:}NN)C_6H_3AsO_3H_2$ + H_2/Ni	77	m. > 250°, rxn.[41]	239
$3,4$-$H_2N[HO_2C(CH_2)_3]$-	$3,4$-$O_2N[HO_2C(CH_2)_3]C_6H_3AsO_3H_2$ + $Fe(OH)_2$ in alk. soln.	81	m. 309.5° (dec.), rxn.[3]	393

Substituent	Synthesis	Yield %	Properties	Ref.
	…HO(Me2CHCONH)C6H3NH2 + Ac2O COCl		195°	150
3,4-AcNH(EtO)-	4,3-HO(Me2CHCONH)C6H3NH2 by the Bart method			149
2,4-MeO(EtOCONH)	4,3-HO(H2N)C6H3AsO3H2 + (Me2-CHCO)2O			1047, 37
3,4-HO(PrOCONH)-	3,4-H2N(EtO)C6H3AsO3H2 + Ac2O 2,4-HO(EtOCONH)C6H3AsO3H2 + Me2SO4	79	Forms dihydrate, rxn.41	71
2,4-HO(BuO)-	2,4-HO(H2N)C6H3NH2 by the Bart-Scheller method		Orthorhombic leaflets, de-comp. 220°, rxn.42	818
3,4-O2N(BuO)-	3,4-HO(BuO)C6H3NH2 by the Bart-Scheller method		m. 119°	1047, 1048
	4-BuOC6H4AsO3H2 + HNO3-H2SO4		Rxns.15,43	818
5,2-O2N(BuO)-	5,2-O2N(BuO)C6H3NH2 by the Bart-Scheller method		Rxn.43	556
3,4-O2N(Me2COHCH2O)-	4-(Me2COHCH2O)C6H4AsO3H2 + HN-O3-H2SO4		m. 210-215°, rxn.20,44	1181
3,4-HO2CCH2NH(HOCH2CH2O)-	3,4-H2N(HOCH2CH2O)C6H3AsO3H2 + ClCH2CO2H in aq. NaOH	55	Rxn.20	1181
3,4-HO3SCH2CONH(HOCH2CH2O)-	3,4-ClCH2CONH(HOCH2CH2O)C6H3-AsO3H2 + Na2SO3	76		250
3,4-H2N[S(CH2CH2)2N]-	3,4-O2N[S(CH2CH2)2N]C6H3AsO3-H2 + Fe(OH)2 in alk. soln.	22	m. 172° (dec.)	250
3,4-H2N[O(CH2CH2)2N]-	3,4-O2N[O(CH2CH2)2N]C6H3AsO3-H2 + Fe(OH)2 in alk. soln.	33	m. 256-258° (dec.)	760
3,4-H2N(HOCH2CMe2NH)-	3,4-O2N(Cl)C6H3AsO3H2 + Me2C-(NH2)CH2OH in alk. soln. at 160-170°		Rxn.15	1181
3,4-H2NOCCH2NH(HOCH2CH2O)-	3,4-H2N(HOCH2CH2O)C6H3AsO3H2 + ClCH2CONH2	78		375
4,3-HO[HO(CH2)3NHCONH]-	2-oxo-5-benzoxazolinearsonic acid + HO(CH2)3NH2			

39. Redn. with SnCl2 in HCl contg. HI → the arseno deriv. (424).
40. Rxn. with Na2SO3 → 3,4-(HO3SCH2CONH)(HOCH2CH2O)C6H3AsO3H2 (1181).
41. Redn. with SO2 in 6 N H2SO4 contg. KI → the arsenoso deriv. (37, 239).
42. Redn. by PCl3 in Et2O → the arsenoso deriv. (37, 71).
43. Catalytic redn. over Raney Ni, followed by H3PO2 → the diamino-arseno deriv. (818).
44. Redn. by FeCl3 in N NaOH or by H2/Ni → redn. of the NO2 group to -NH2 (556).

TABLE XLI (cont'd.)

DISUBSTITUTED BENZENEARSONIC ACIDS

$RR_1C_6H_3AsO_3H_2$ where R and R_1 are:	Prepd. from	Yield %	Props. and Remarks	Ref.
$3,4\text{-}O_2N[(HOCH_2CH_2)_2N]\text{-}$	$3,4\text{-}O_2N(MeO_2CCH_2O)C_6H_3AsO_3\text{-}$ $H_3 + (HOCH_2CH_2)_2NH$	55		1182
$3,4\text{-}H_2N(Me_2COHCH_2O)\text{-}$	$3,4\text{-}O_2N(Me_2COHCH_2O)C_6H_3As\text{-}$ $O_3H_2 + FeCl_2$ in N NaOH or H_2/Ni		m. 150-155°; monohydrate, m. 65-70°, rxns.2, 2o	556
$3,4\text{-}HO[(HOCH_2CH_2)_2N]\text{-}$	$3,4\text{-}HO(H_2N)C_6H_3AsO_3HNa +$ CH_2CH_2O		Colorless powder	361
$3,4\text{-}O_2N(HOCH_2CH_2NHCH_2CH_2NH)\text{-}$	$3,4\text{-}O_2N(Cl)C_6H_3AsO_3H_2 + HOCH_2\text{-}$ $CH_2NHCH_2CH_2NH_2$			760
$4,3\text{-}HOCH_2CH_2O[H_2NC(:NH)NHC(:N\text{-}$ $H)NH]\text{-}$	$4,3\text{-}HOCH_2CH_2O(H_2N)C_6H_3AsO_3H_2$ $+ H_2NC(:NH)NHCN$	86	Dihydrate crystals	1181
$3,4\text{-}H_2N(HOCH_2CMe_2NH)\text{-}$	$3,4\text{-}O_2N(HOCH_2CMe_2NH)C_6H_3AsO_3\text{-}$ $H_2 + FeSO_4$ in alk. soln.			760
$3,4\text{-}O_2N(H_2NCH_2CH_2NHCH_2CH_2NH)\text{-}$	$3,4\text{-}O_2N(Cl)C_6H_3AsO_3H_2 + (H_2\text{-}$ $NCH_2CH_2)_2NH$		Rxn.15	760
$3,4\text{-}H_2N[H_2N(CH_2CH_2NH)_2]\text{-}$	$3,4\text{-}O_2N(H_2NCH_2CH_2NHCH_2CH_2NH)\text{-}$ $C_6H_3AsO_3H_2 + FeSO_4$ in aq. NaOH			760
C_{11}				
$4,3\text{-}HO[\overline{N}:CHC(NO_2):CHCH:\overline{C}NH]\text{-}$	$4,3\text{-}HO(H_2N)C_6H_3AsO_3H_2 + \overline{C}H:\text{-}$ $CHC(NO_2):CHN:\overline{C}Cl$	51	m. > 250°	250
$2,4\text{-}HO[\overline{N}:CHC(NO_2):CHCH:\overline{C}NH]\text{-}$	$2,4\text{-}HO(H_2N)C_6H_3AsO_3H_2 + \overline{C}H:\text{-}$ $CHC(NO_2):CHN:\overline{C}Cl$	70	m. 176-178°	250
$2,5\text{-}HO[\overline{N}:C(NH_2)N:CMeCH:\overline{C}NH]\text{-}$	$2,5\text{-}HO(H_2N)C_6H_3AsO_3H_2 + \overline{N}:C\text{-}$ $(NH_2)N:CMeCH:\overline{C}Cl$	52		14
$3,4\text{-}HO[\overline{N}:C(NH_2)N:CMeCH:\overline{C}NH]\text{-}$	$3,4\text{-}HO(H_2N)C_6H_3AsO_3H_2 + \overline{N}:C\text{-}$ $(NH_2)N:CMeCH:\overline{C}Cl$	32		14
$2,4\text{-}Me(AcOCH_2CONH)\text{-}$	$2,4\text{-}Me(ClCH_2CONH)C_6H_3AsO_3H_2$ $+ AcONa$		m. 155-157°	1174
$4,3\text{-}HO[\overline{C}H_2(CH_2)_3\overline{N}CONH]\text{-}$	2-Oxo-5-benzoxazolinearsonic acid + pyrrolidine		m.~320° (dec.)	375
$2,5\text{-}MeO[AcOCH(Me)]\text{-}$	$2,5\text{-}MeO(MeCHOH)C_6H_3AsO_3H_2 +$ Ac_2O			36
$2,5\text{-}MeO[AcNHCH(Me)]\text{-}$	$2,5\text{-}MeO[MeCH(NH_2)]C_6H_3AsO_3H_2$ $+ Ac_2O$		White crystals, m. 205°	36
			m. 227-228°	149

		Yield (%)		
2,4-HO(Me₂CHCH₂CONH)-	2,4-HO(H₂N)C₆H₃AsO₃H₂ + (Me₂-CHCH₂CO)₂O		m. 232°	149
4,3-HO(BuCONH)-	4,3-HO(H₂N)C₆H₃AsO₃H₂ + (Bu-CO)₂O		m. 208-209°	149
4,3-HO(Me₂CHCH₂CONH)-	4,3,-HO(H₂N)C₆H₃AsO₃H₂ + (Me₂-CHCH₂CO)₂O		m. 226-227°	149
5,2-O₂N(Me₂CHCH₂CH₂O)-	5,2-O₂N(Me₂CHCH₂CH₂O)C₆H₃NH₂ by the Bart-Scheller method		Rxn.[43]	818
2,4-HOCH₂CH₂O(EtOCONH)-	2,4-HO(EtOCONH)C₆H₃AsO₃H₂ + CH₂CH₂O + KOH in EtOH	65	Tetragonal needles, m. 233° Rxn.20	71
4,2-Me₂N[N:C(NH₂)N:C(NH₂)-N:CNH]-	4,2-Me₂N(Cl)C₆H₃AsO₃H₂ + melamine		Rxn.20	424
4,3-HO(BuNHCONH)-	2-Oxo-5-benzoxazolinearsonic acid + BuNH₂			375
3,4-HO[(HOCH₂CH₂)₂NCONH]-	2-Oxo-6-benzoxazolinearsonic acid + (HOCH₂CH₂)₂NH			375
3,4-H₂N(iso-C₅H₁₀O)-	3,4-HO(H₂N)C₆H₃AsO₃H₂ + CH₂CH₂O + HOCH₂CHCH₂O		Rxn.[45] Rxn.2	368
3,4-HO[HOCH₂CHOHCH₂(HOCH₂CH₂)N]-				368
4,3-HO[HOCH₂CHOHCH₂(HOCH₂CH₂)N]-	4,3-HO(H₂N)C₆H₃AsO₃H₂ + CH₂-CH₂O + HOCH₂CHCH₂O		Colorless powder, rxn.2	361
C₁₂				
3,4-O₂N(4-ClC₆H₄SO₃)-	3,4-O₂N(HO)C₆H₃AsO₃H₂ + 4-Cl-C₆H₄SO₂Cl		m. 255-257° (dec.)	1082
3,4-H₂N(4-ClC₆H₄SO₃)-	3,4-H₂N(HO)C₆H₃AsO₃H₂ + 4-Cl-C₆H₄SO₂Cl		HCl salt, m. 242-243° (dec.)	1082
2,5-HO(PhN₂)-	2-HOC₆H₄AsO₃H₂ + PhN₂Cl	45	Orange needles, m. 257.3°, rxn.[48]	289
4,3-HO(PhN₂)-	4,3-HO(PhN₂)C₆H₃NH₂ by the Bart-Scheller method	40	Orange needles, m. 290°, rxn.2	289
	4-HOC₆H₄AsO₃H₂ + PhN₂Cl	5.9		289
5,2-HO(PhN₂)-	3-HOC₆H₄AsO₃H₂ + PhN₂Cl	65	Golden plates, m. 237.5°, rxn.[48]	289
3,4-O₂N(PhNH)-	3,4-O₂N(Cl)C₆H₃AsO₃HNa + Ph-NH₂	89	Yellow orange crystals, rxn.[47]	595
3,4-O₂N(4-HOC₆H₄NH)-	3,4-O₂N(Cl)C₆H₃AsO₃HNa + 4-HOC₆H₄NH₂		Orange-red crystals, rxn.[47]	595

45. Rxn. with ClCH₂CONHPh → 3,4-(PhNHCOCH₂NH)(iso-C₅H₁₁)C₆H₃AsO₃H₂ (1047).
46. Catalytic redn. over Raney Ni → HO(H₂N)C₆H₃AsO₃H₂ (289).
47. Rxn. with Na₂S₂O₄ in alk. soln. → redn. of the -NO₂ to -NH₂ (595).

TABLE XLI (cont'd.)

DISUBSTITUTED BENZENEARSONIC ACIDS

$RR_1C_6H_3AsO_3H_2$ where R and R_1 are:	Prepd. from	Yield %	Props. and Remarks	Ref.
$3,4\text{-}O_2N(3\text{-}HOC_6H_4NH)\text{-}$	$3,4\text{-}O_2N(Cl)C_6H_3AsO_3HNa$ + $3\text{-}HOC_6H_4NH_2$		Orange crystals	595
$3,4\text{-}H_2N(PhSO_3)\text{-}$	$3,4\text{-}H_2N(HO)C_6H_3AsO_3H_2$ + $PhSO_2Cl$		m. 195-197°	1082
$3,4\text{-}O_2N(4\text{-}H_2NC_6H_4SO_3)\text{-}$	$3,4\text{-}O_2N(HO)C_6H_3AsO_3H_2$ + $4\text{-}H_2NC_6H_4SO_2Cl$		HCl salt, m. 267-269°	1082
$4,3\text{-}HO[2,4\text{-}HO(H_2O_3As)C_6H_3N_2]\text{-}$	$2,4\text{-}HO(H_2O_3As)C_6H_3N_2Cl$ + $3\text{-}H_2NC_6H_4OH$		m. 190-192°	617
$4,3\text{-}Cl(MeCH:CHCO_2CH_2CONH)\text{-}$	$4,3\text{-}Cl(ClCH_2CONH)C_6H_3AsO_3H_2$ + $MeCH:CHCO_2Na$			1174
$3,4\text{-}H_2N(PhNH)\text{-}$	$3,4\text{-}O_2N(PhNH)C_6H_3AsO_3H_2$ + $Na_2S_2O_4$		Pinkish ppt.	595
$3,4\text{-}H_2N(4\text{-}HOC_6H_4NH)\text{-}$	$3,4\text{-}O_2N(4\text{-}HOC_6H_4NH)C_6H_3AsO_3H_2$ + $Na_2S_2O_4$		White cryst. platelets	595
$3,4\text{-}H_2N(4\text{-}H_2NC_6H_4SO_3)\text{-}$	$3,4\text{-}H_2N(4\text{-}AcNHC_6H_4SO_3)C_6H_3AsO_3H_2$ + 20% alc. HCl		Long needles, rxns. [2, 34]	388
$4,3\text{-}H_2N(4\text{-}H_2NO_2CC_6H_4N_2)\text{-}$	$4\text{-}H_2NC_6H_4AsO_3H_2$ + $4\text{-}H_2NO_2SC_6H_4N_2Cl$	60	Di-Na salt, light yellow solid, m. 240° (dec.).	709, 1041
$3,4\text{-}O_2N[\overline{N:C(CO_2Et)CMe:NN}]\text{-}$	$3,4\text{-}O_2N[\overline{N:C(CO_2H)CMe:NN}]C_6H_3AsO_3H_2$ + $EtOH(H_2SO_4)$		Sinters at 135-136°, m. 189-209° (dec.).	239
$3,4\text{-}O_2N[AcC(CO_2Et):NNH]\text{-}$	$3,4\text{-}O_2N(ClN:N)C_6H_3AsO_3H_2$ + $Ac\text{-}CHCO_2Et$	84	m. 251-252° (dec.)	240
$3,4\text{-}O_2N[\overline{C(Me):NC(NH_2):NC(Me):NCNH}]\text{-}$	$3,4\text{-}O_2N(Cl)C_6H_3AsO_3H_2$ + $MeC:N\text{-}C(NH_2):NC(Me)CNH_2$		Dihydrate, m. 180° (dec.)	8a
$4,3\text{-}HOCH_2CH_2O[\overline{N:C(NH_2)N:CH\text{-}CH:CNH}]\text{-}$	$4,3\text{-}HOCH_2CH_2O(H_2N)C_6H_3AsO_3H_2$ + $\overline{N:C(NH_2)N:CHCH:CCl}$		Rxn. [2]	39
$4,3\text{-}HO[(EtCO)_2N]\text{-}$	$4,3\text{-}HO(H_2N)C_6H_3AsO_3H_2$ + $(EtCO)_2O$		m. 182°	149
$2,4\text{-}HO[\overline{CH_2(CH_2)_3NCH_2CONH}]\text{-}$	$2,4\text{-}HO(ClCH_2CONH)C_6H_3AsO_3H_2$ + $\overline{CH_2(CH_2)_3NH}$			376
$3,4\text{-}HO[\overline{CH_2(CH_2)_3NCH_2CONH}]\text{-}$	$3,4\text{-}HO(ClCH_2CONH)C_6H_3AsO_3H_2$ + $\overline{CH_2(CH_2)_3NH}$			376
$4,3\text{-}HO[\overline{CH_2(CH_2)_3NCH_2CONH}]\text{-}$	$4,3\text{-}HO(ClCH_2CONH)C_6H_3AsO_3H_2$ + $\overline{CH_2(CH_2)_3NH}$			376

Product	Reaction	Yield	Description	Ref.
	+ ɛᴛᴄᴄᴄɪ			
3,4-O₂N[O(CH₂CH₂)₂NCH₂CH₂NH]-	3,4-O₂N(Cl)C₆H₃AsO₃H₂ + O(CH₂-CH₂)₂NCH₂CH₂NH₂		Colorless powder, rxn.[2]	760
3,4-HO[(HOCH₂CHOHCH₂)₂N]-	3,4-HO(H₂N)C₆H₃AsO₃H₂ + HOCH₂CHCH₂O			361
3,4-O₂N[H₂N(CH₂)₂NH(CH₂)₂NH(CH₂)₂NH]-	3,4-O₂N(Cl)C₆H₃AsO₃H₂ + (H₂N-CH₂CH₂NHCH₂-)₂		Rxn.[15]	760
3,4-H₂N[H₂N(CH₂)₂NH(CH₂)₂NH(CH₂)₂NH]-	3,4-H₂N[H₂N(CH₂)₂NH(CH₂)₂NH(CH₂)₂NH]C₆H₃AsO₃H₂ + FeSO₄ in aq. NaOH			760
C₁₃				
3,4-O₂N(4-NCC₆H₄SO₃)-	3,4-O₂N(Cl)C₆H₃AsO₃H₂ + 4-NC-C₆H₄SO₂Cl	26	Colorless platelets	275
3,4-H₂N(4-NCC₆H₄SO₃)-	3,4-H₂N(HO)C₆H₃AsO₃H₂ + 4-NC-C₆H₄SO₂Cl	72	Colorless plates	275
4,2-Cl(PhCH₂)-	4,2-Cl(PhCH₂)C₆H₃NH₂ by the Bart method	21.8	m. 189°, rxn.[48]	270a
5,2-Cl(4-MeC₆H₄O)-	5,2-Cl(4-MeC₆H₄O)C₆H₃NH₂ by the Bart method		m. 199-200°, rxn.[49]	731
2,4-HO(2-HOC₆H₄CH:N)-	2,4-HO(H₂N)C₆H₃AsO₃H₂ 2-HOC₆H₄CHO		Crystals, decomp. 210°	149
3,4-O₂N(PhCH₂O)-	3,4-O₂N(HO)C₆H₃AsO₃H₂ + PhCH₂Cl	74.5	Fine needles, m. 250°, rxn.[7]	388
3,4-O₂N(2-MeOC₆H₄NH)-	3,4-O₂N(Cl)C₆H₃AsO₃HNa + 2-Me-OC₆H₄NH₂		Red-orange crystals	595
3,4-O₂N(4-MeOC₆H₄NH)-	3,4-O₂N(Cl)C₆H₃AsO₃HNa + 4-Me-OC₆H₄NH₂		Orange-yellow crystals	595
3,4-H₂N(PhCH₂O)-	3,4-O₂N(PhCH₂O)C₆H₃AsO₃H₂ + H₂/Ni	78	Slightly tan needles, m. 248-249° (dec.), rxns.[2, 34]	388
3,4-H₂N(4-MeC₆H₄SO₃)-	3,4-H₂N(HO)C₆H₃AsO₃H₂ + 4-Me-C₆H₄SO₂Cl		m. 180-190° (dec.)	1082
2,4-Me(MeCH:CHCO₂CH₂CONH)-	2,4-Me(ClCH₂CONH)C₆H₃AsO₃H₂ + MeCH:CHCO₂Na		m. 205°	1174
3,4-HO[N:C(SEt)N:C(Me)CH:CN-H]-	3,4-HO(H₂N)C₆H₃AsO₃H₂ + N:C(S-Et)N:C(Me)CH:CCl	44		14
2,5-HO[N:C(SEt)N:C(Me)CH:CN-H]-	2,5-HO(H₂N)C₆H₃AsO₃H₂ + N:C(S-Et)N:C(Me)CH:CCl	59		14

48. Rxn. with H₂SO₄ → 2-chloroacridarsonic acid (270a).
49. Rxn. with H₂SO₄ → 8-chloro-2-methylphenoxarsinic acid (731).

TABLE XLI (cont'd.)

DISUBSTITUTED BENZENEARSONIC ACIDS

$RR_1C_6H_3AsO_3H_2$ where R and R_1 are:	Prepd. from	Yield %	Props. and Remarks	Ref.
C_{14}				
$2,4-O_2N(4-HOC_6H_4CH:CH)-$	$3,4-O_2N(H_2O_3As)C_6H_3N_2Cl + 4-HOC_6H_4CH:CHCO_2H$		Reddish ppt.	400, 401, 404
$3,4-AcNH(4-ClC_6H_4SO_3)-$	$3,4-AcNH(HO)C_6H_3AsO_3H_2 + 4-Cl\ C_6H_4SO_2Cl$		m. 180-185°	1082
$3,4-O_2N(4-AcNHC_6H_4SO_3)-$	$3,4-O_2N(HO)C_6H_3AsO_3H_2 + 4-Ac\ NHC_6H_4SO_2Cl$		m. 188-189° (dec.)	1082
$3,4-O_2N(PhNHCOCH_2O)-$	$4-(PhNHCOCH_2O)C_6H_4AsO_3H_2 + HNO_3-H_2SO_4$		Rxn.15	1047
$2,4-HO(PhCH_2OCONH)-$	$2,4-HO(H_2N)C_6H_3AsO_3H_2 + PhCH_2OCOCl$	82	Hexagonal needles; decomp. 223°, rxns.42, 50	71
$3,4-AcNH(PhSO_3)-$	$3,4-AcNH(HO)C_6H_3AsO_3H_2 + Ph-SO_2Cl$		m. 190-193°	1082
$3,4-O_2N(3-AcNHC_6H_4NH)-$	$3,4-O_2N(Cl)C_6H_3AsO_3HNa + 3-Ac\ NHC_6H_4NH_2.HCl$		Yellow ppt., rxn.47	595
$3,4-O_2N(4-AcNHC_6H_4NH)-$	$3,4-O_2N(Cl)C_6H_3AsO_3HNa + 4-Ac\ NHC_6H_4NH_2.HCl$	85	Orange crystals, rxn.47	595
$3,4-H_2N(PhNHCOCH_2O)-$	$4-(PhNHCOCH_2O)C_6H_4AsO_3H_2 + HN\ O_3-H_2SO_4,$ followed by $Fe(OH)_2$ or $Na_2S_2O_4$ in alk. soln.			1047, 1048
$3,4-O_2N(4-EtOC_6H_4NH)-$	$3,4-O_2N(4-HOC_6H_4NH)C_6H_3AsO_3H_2$ + EtOH		Rxn.47	595
$4,3-HO(4-AcNHC_6H_4SO_2NH)-$	$4,3-HO(H_2N)C_6H_3AsO_3Na_2 + 4-Ac\ NHC_6H_4SO_2Cl$		m. 165-166°, forms dibasic salts	1117, 1258
$3,4-H_2N(4-AcNHC_6H_4SO_3)-$	$3,4-H_2N(HO)C_6H_3AsO_3H_2 + 4-Ac\ NHC_6H_4SO_2Cl$		Fine needles, m. 250° (dec.) rxns.34, 51, 52	388
$4,2-Me(PhCH_2)-$	$4,2-Me(PhCH_2)C_6H_3NH_2$ by the Bart method	51	m. 178-179°, rxn53	270a
$3,4-H_2N(4-AcNHC_6H_4NH)-$	$3,4-O_2N(4-AcNHC_6H_4NH)C_6H_3AsO_3\ H_2 + Na_2S_2O_4$			595
$3,4-H_2N(3-AcNHC_6H_4NH)-$	$4,3-O_2N(3-AcNHC_6H_4NH)C_6H_3AsO_3\ H_2 + Na_2S_2O_4$		White needles turning pink in air	595
$3,4-O_2N(4-Me_2NC_6H_4NH)-$	$3,4-Cl(O_2N)C_6H_3AsO_3H_2 + 4-H_2-\ NC_6H_4NMe_2$	56	Red brown	761

Compound	Preparation	Properties	Ref.
...$NH]\text{-}$	...$3,4\text{-}\ldots As O_3H_2\ +\ ClCH_2\text{-}CONHCH_2(CHOH)_4CH_2OH$	Almost white powder, chars without melting, rxns.[2, 34]	102
$3,4\text{-}O_2N(H_2NCH_2CH_2NHCH_2CH_2NH\text{-}CH_2CH_2NHCH_2CH_2NH)\text{-}$	$3,4\text{-}O_2N(ClC_6H_3AsO_3H_2\ +\ H_2NCH_2\text{-}CH_2(NHCH_2CH_2)_2NHCH_2CH_2NH_2$		760
C15			
$4,3\text{-}Cl(BzOCH_2CONH)\text{-}$	$4,3\text{-}Cl(ClCH_2CONH)C_6H_3AsO_3H_2\ +\ BzCNa$	m. 186-188°	1174
$3,4\text{-}AcNH(4\text{-}NCC_6H_4SO_3)\text{-}$	$3,4\text{-}AcNH(HO)C_6H_3AsO_3H_2\ +\ 4\text{-}NC\text{-}C_6H_4SO_2Cl$ 31	Colorless needles, rxn.[54]	275
$3,4\text{-}AcNH(4\text{-}MeC_6H_4SO_3)\text{-}$	$3,4\text{-}AcNH(HO)C_6H_3AsO_3H_2\ +\ Me\text{-}C_6H_4SO_2Cl$	m. 203-205° (dec.)	1082
$3,4\text{-}AcNH[4\text{-}H_2NC(:NH)C_6H_4SO_3]\text{-}$	$3,4\text{-}AcNH(4\text{-}NCC_6H_4SO_3)C_6H_3As\text{-}O_3H_2$, see footnote [54]	Rxn.[55]	275
$3,4\text{-}Me(AcNHC_6H_4SO_2NH)\text{-}$	$3,4\text{-}Me(H_2N)C_6H_3AsO_3H_2\ +\ 4\text{-}Ac\text{-}NHC_6H_4SO_2Cl$		707
$4,3\text{-}HO[HOCH_2(CHOH)_4CH_2NMeCO\text{-}CH_2NH]\text{-}$	$4,3\text{-}HO(H_2N)C_6H_3AsO_3H_2\ +\ ClCH_2\text{-}CONMeCH_2(CHOH)_4CH_2OH$	Almost white powder, rxns.[2, 34]	102
C16			
$3,4\text{-}O_2N(2\text{-}C_{10}H_7SO_3)\text{-}$	$3,4\text{-}O_2N(HO)C_6H_3AsO_3H_2\ +\ 2\text{-}C_{10}H_7SO_2Cl$	m. 258-259° (dec.)	1082
$3,4\text{-}H_2N(2\text{-}C_{10}H_7SO_3)\text{-}$	$3,4\text{-}H_2N(HO)C_6H_3AsO_3H_2\ +\ 2\text{-}C_{10}H_7SO_2Cl$	HCl salt, m. 139-140° (dec.)	1082
$3,4\text{-}AcNH(PhNHCOCH_2O)\text{-}$	$4\text{-}(PhNHCOCH_2O)C_6H_4AsO_3H_2\ +\ HN\text{-}O_3\text{-}H_2SO_4$, followed by $Na_2S_2O_4$ and Ac_2O	Rxns.[34]	1047
$4,3\text{-}HO_2CCH_2O(PhNHCOCH_2NH)\text{-}$	$3,4\text{-}O_2N(HO_2CCH_2O)C_6H_3AsO_3H_2\ +\ Na_2S_2O_4$, followed by $PhNHCO\text{-}CH_2Cl$		1047, 1048
$2,4\text{-}HOCH_2CH_2O(PhCH_2OCONH)\text{-}$	$2,4\text{-}HO(PhCH_2O_2OCONH)C_6H_3AsO_3H_2\ +\ \overline{CH_2CH_2O}$ 79	Tetragonal needles, m. 235°, rxns.[56, 57]	71

50. Rxn. with $\overline{CH_2CH_2O}$ → $2\text{-}HOCH_2CH_2O(PhCH_2OCONH)C_6H_3AsO_3H_2$ (71).
51. Rxn. with 5 N HCl at 100° or with alc. HCl at r. t. → deacetylation (388, 1117, 1258).
52. Redn. with $PhNHNH_2$ in MeOH → the arsenoso deriv. (388).
53. Cyclization in H_2SO_2 on steam bath → 3-methylacridarsinic acid (270a).
54. Rxn. with alc. HCl in Et_2O, followed by alc. NH_3 → $3,4\text{-}AcNH[4\text{-}H_2NC(:NH)C_6H_4SO_3]C_6H_3AsO_3H_2$ (275).
55. Rxn. with aq. HCl → $4\text{-}H_2NCOC_6H_4SO_3H\ +\ 4,3\text{-}HO(H_2N)C_6H_3AsO_3H_2$ (275).
56. Rxn. with 0.5 N NaOH → hydrolysis of the carbamate group (71).
57. Rxn. with $Me\overline{CHCH_2O}$ + KOH → displacement of the $HOCH_2CH_2O$-group by $MeCHOHCH_2O\text{-}$ (71).

TABLE XLI (cont'd.)

DISUBSTITUTED BENZENEARSONIC ACIDS

$RR_1C_6H_3AsO_3H_2$ where R and R_1 are:	Prepd. from	Yield %	Props. and Remarks	Ref.
3,4-O_2N(4-$Et_2NC_6H_4NH$)-	3,4-O_2N(Cl)$C_6H_3AsO_3H_2$ + 4-$H_2NC_6H_4NEt_2$	60	Red-brown substance	761
2,4-Me[$HOCH_2$(CHOH)$_4CH_2N$(Me)-$COCH_2NH$]-	2,4-Me(H_2N)$C_6H_4AsO_3H_2$ + $ClCH_2CONMeCH_2$(CHOH)$_4CH_2OH$			102
3,4-MeO[$HOCH_2$(CHOH)$_4CH_2N$(Me)-$COCH_2NH$]-	3,4-MeO(H_2N)$C_6H_3AsO_3H_2$ + $ClCH_2CONMeCH_2$(CHOH)$_4CH_2OH$			102
C_{17}				
4,x-H_2N(N:CHCH:CHCH:CNHO$_2$S-$C_6H_4N_2$)-	4-$H_2NC_6H_4AsO_3H_2$ + 4-(2-pyridylsulfamyl)phenyldiazonium chloride			1041
2,4-$MeCHOHCH_2O$(PhCH$_2$OCONH)-	2,4-HO(PhCH$_2$OCONH)$C_6H_3AsO_3H_2$ + MeCHCH$_2$O		Tetragonal needles, m. 176°, rxn. 56	71
C_{18}				
5,2-O_2N(4-$HO_3SC_6H_4N$:NC_6H_4-NHN:N)-	2,4-H_2O_3As(O_2N)$C_6H_3N_2Cl$ + 4-$H_2NC_6H_4N$:$NC_6H_4SO_3H$		Colorimetric reagent for Pb	754a
3,4-AcNH(2-$C_{10}H_7SO_3$)-	3,4-AcNH(HO)$C_6H_3AsO_3H_2$ + 2-$C_{10}H_7SO_2Cl$		m. 203-205° (dec.)	1082
3,4-O_2N(4-$H_2NC_6H_4C_6H_4NH$)-	3,4-O_2N(Cl)$C_6H_3AsO_3H_2$ + benzidine		Exists in two isomeric forms, orange and red	595
C_{19}				
3,4-$PhNHCOCH_2NH$(iso-$C_5H_{11}O$)-	3,4-H_2N(iso-$C_5H_{11}O$)$C_6H_3AsO_3H_2$ + ClCH$_2$CONPh			1047, 1048
C_{20}				
4,x-HO(4-$AcNHC_6H_4SO_2C_6H_4N_2$)-	4-$HOC_6H_4AsO_3H_2$ + $AcNHC_6H_4SO_2$-$C_6H_4N_2Cl$			222
2,4-HO(4-$AcNHC_6H_4SO_2C_6H_4N_2$)-	2-$HOC_6H_4AsO_3H_2$ + $AcNHC_6H_4SO_2$-$C_6H_4N_2Cl$			222
2,x-H_2N(4-$AcNHC_6H_4SO_2C_6H_4N_2$)-	2-$H_2NC_6H_4AsO_3H_2$ + $AcNHC_6H_4SO_2$-$C_6H_4N_2Cl$		Yellow-red powder	222
3,x-H_2N(4-$AcNHC_6H_4SO_2C_6H_4N_2$)-	3-$H_2NC_6H_4AsO_3H_2$ + $AcNHC_6H_4SO_2$-$C_6H_4N_2Cl$			222
4,x-H_2N(4-$AcNHC_6H_4SO_2C_6H_4N_2$)-	4-$H_2NC_6H_4AsO_3H_2$ + $AcNHC_6H_4SO_2$-$C_6H_4N_2Cl$			222

TABLE XLII

TRI- AND TETRA-SUBSTITUTED BENZENEARSONIC ACIDS

Arsonic acid	Prepd. from	Yield %	Props. and Remarks	Ref.
C_6				
$4,3,5\text{-}HO(O_2N)_2C_6H_2AsO_3H_2$			On heating explodes with liberation of As and AsH_3	1006
$3,4,5\text{-}H_2N(HO)(I)C_6H_2As\text{-}O_3H_2$			Rxn.[1]	374
C_7				
$4,3,5\text{-}Me(O_2N)_2C_6H_2AsO_3H_2$	$4,3\text{-}Me(O_2N)C_6H_3AsO_3H_2 +$ $HNO_3\text{-}H_2SO_4$	64	Yellow	865
$3,4,5\text{-}H_2NCONH(HO)(I)C_6\text{-}H_2AsO_3H_2$	$3,4,5\text{-}H_2N(HO)(I)C_6H_2As\text{-}O_3H_2 + KOCN$, followed by AcOH		White powder, chars on heating	374
	$3,4\text{-}H_2NCONH(HO)C_6H_3AsO_3H_2$ $+ Hg(OAc)_2$ in AcOH $+ I_2\text{-}$ KI			374
	$3,4,5\text{-}H_2NCONH(HO)(I)C_6H_2N\text{-}H_2$ by the Bart method		Rxn. with $H_2SO_4 \rightarrow$ elimination of I	374
C_8				
$3,4,5\text{-}AcNH(HO)(O_2N)C_6\text{-}H_2AsO_3H_2$	$3,4\text{-}AcNH(HO)C_6H_3AsO_3H_2 +$ $HNO_3\text{-}H_2SO_4$ at -10 to 0°		m. 218-220°; rxn.[2]	149
$4,3,5\text{-}EtO(O_2N)_2C_6H_2As\text{-}O_3H_2$	$4,3\text{-}EtO(O_2N)C_6H_3AsO_3H_2$ $+ HNO_3\text{-}H_2SO_4$	64	Yellow	865
$4,2,6\text{-}HOCH_2CH_2O(O_2N)_2\text{-}C_6H_2AsO_3H_2$	$4\text{-}HOCH_2CH_2OC_6H_4AsO_3H_2 +$ $HNO_3\text{-}H_2SO_4$ on steam bath		Yellow, diamond-shaped plates, m. 212-215°, decomp. 228°	502
$3,4,5\text{-}AcNH(HO)(H_2N)C_6\text{-}H_2AsO_3H_2$	$3,4,5\text{-}AcNH(HO)(O_2N)C_6H_2\text{-}AsO_3H_2 + Na_2S_2O_4$		Crystals	149
C_9				
$4,3,5\text{-}\overline{N\text{:}CClN\text{:}CClN\text{:}C}NH\text{-}(O_2N)_2C_6H_3AsO_3H_2$	$4,3,5\text{-}H_2N(O_2N)_2C_6H_2AsO_3\text{-}H_2 + (ClCN)_3$		Rxns.[3-5]	419, 422

1. Rxn. with KOCN $\rightarrow$ $3,4,5\text{-}H_2NCONH(HO)(I)C_6H_3AsO_3H_2$ (374).
2. Rxn. with $Na_2S_2O_4$ in aq. NaOH $\rightarrow$ redn. of the NO_2 group (149).
3. Rxn. with aq. NH_3, followed by $SnCl_2$ in HCl $\rightarrow$ $4,3,5\text{-}\overline{N\text{:}C(NH_2)N\text{:}C(NH_2)N\text{:}C}NH\text{-}$ $(H_2N)_2C_6H_2AsCl_2$ (419).
4. Rxn. with Me_2NH, followed by $SnCl_2$ in HCl $\rightarrow$ $4,3,5\text{-}\overline{N\text{:}C(NMe_2)N\text{:}C(NMe_2)N\text{:}C}NH(H_2N)_2\text{-}C_6H_2AsCl_2$ (419).
5. Rxn. with $SnCl_2$ in HCl $\rightarrow$ redn. of the $O_2N\text{-}$ and AsO_3H_2 groups to NH_2 and $AsCl_2$, resp. (420, 421, 422).

TABLE XLII (cont'd.)
TRI- AND TETRA-SUBSTITUTED BENZENEARSONIC ACIDS

Arsonic acid	Prepd. from	Yield %	Props. and Remarks
$2,3,4$-HO(O_2N)(EtOCONH)- C$_6$H$_2$AsO$_3$H$_2$	$2,4$-HO(EtOCONH)C$_6$H$_3$AsO$_3$- H$_2$ + HNO$_3$-H$_2$SO$_4$ at -5°		m. 240° (dec.), rxn.[2]
$2,4,5$-OHC(MeO)$_2$C$_6$H$_2$As- O$_3$H$_2$	$2,4,5$-OCH(MeO)$_2$C$_6$H$_2$NH$_2$ by the Bart method	46.2	m. ~300°, stable in 15% HCl or 30% NaOH, decomp. in 25% HCl or 37% NaOH rxn.[6]
$3,4,6$-Ac(HO)(MeO)C$_6$H$_2$- AsO$_3$H$_2$	$3,4,6$-Ac(HO)(MeO)C$_6$H$_2$NH$_2$ by the Bart method	58	m. 225° (dec.), rxns.[7,8]
$2,4,5$-HO$_2$C(MeO)$_2$C$_6$H$_2$As- O$_3$H$_2$	$2,4,5$-OHC(MeO)$_2$C$_6$H$_2$AsO$_3$- H$_2$ + KMnO$_4$		m. >300°; Ba salt, needles
$2,3,4$-HO(H$_2$N)(EtOCONH)- C$_6$H$_2$AsO$_3$H$_2$	$2,3,4$-HO(O_2N)(EtOCONH)C$_6$- H$_2$AsO$_3$H$_2$ + Na$_2$S$_2$O$_4$ in NaOH		Decomp. 170°
C$_{10}$			
$2,4,5$-O$_2$N[$\overline{\text{N:C(NH}_2\text{)N:C-}}$ (NH$_2$)N:CNH](MeO)C$_6$H$_2$- AsO$_3$H$_2$	$2,4,5$-O$_2$N(H$_2$N)(MeO)C$_6$H$_2$- + N:C(NH$_2$)N:C(NH$_2$)N:CCl		Rxn.[5]
$2,4,5$-$\overline{\text{N:C(NH}_2\text{)N:C(NH}_2\text{)-}}$ N:CNH(MeNH)(O_2N)C$_6$H$_2$As- O$_3$H$_2$	$2,4,5$-Cl(MeHN)(O_2N)C$_6$H$_2$- AsO$_3$H$_2$ + $\overline{\text{N:C(NH}_2\text{)N:C(NH}_2\text{)}}$ N:CNH		Rxn.[9]
$2,4,5$-(MeO)$_2$(Ac)C$_6$H$_2$AsO$_3$ H$_2$	$2,4,5$-(MeO)$_2$(Ac)C$_6$H$_2$NH$_2$ by the Bart method $2,4$-(HO)$_2$C$_6$H$_3$Ac + H$_3$AsO$_4$ at 150-155° for 5 hrs., followed by methylation	7	m. 250°
$2,3,4,5$-Me(Br)(HO)(Me$_2$- CH)C$_6$HAsO$_3$H$_2$	$2,3,4,5$-Me(Br)(HO).(Me$_2$- CH)C$_6$HNH$_2$ by the Bart method		Plates, m. 234-235°
$2,4,5$-H$_2$NCONH:CH(MeO)$_2$- C$_6$H$_2$AsO$_3$H$_2$	$2,4,5$-OHC(MeO)$_2$C$_6$H$_2$AsO$_3$H$_2$ + H$_2$NCONHNH$_2$		Dihydrate, m. 256°
$2,4,5$-Me$_2$CH(HO)(Me)C$_6$H$_2$- AsO$_3$H$_2$	$2,4,5$-Me$_2$CH(HO)(Me)C$_6$H$_2$- NH$_2$ by the Bart method		Colorless prisms, decomp. 185°, rxn.[7]
C$_{11}$			
$2,4,5$-(MeO)$_2$(EtCO)C$_6$H$_2$- AsO$_3$H$_2$	$2,4$-(HO)$_2$C$_6$H$_3$COEt + H$_3$As- O$_3$ at 100°, followed by methylation	11	m. 243°

6. Rxn. with KMnO$_4$ → oxidation of the Me group to CO$_2$H (1122).
7. Redn. by SO$_2$ in an acid contg. NaI, followed by neutralization → the arsenoso deriv. (35, 381).
8. Redn. by H$_3$PO$_2$ → the arseno deriv. (35).
9. Redn. by SO$_2$ in dild. HCl contg. KI → redn. of the NO$_2$ and AsO$_3$H$_2$ groups to -NH$_2$ and -AsCl$_2$, resp. (419).

TABLE XLII. (cont'd.)
TRI- AND TETRA-SUBSTITUTED BENZENEARSONIC ACIDS

Arsonic acid	Prepd. from	Yield %	Props. and Remarks	Ref.
C_{14} 3,4,5-AcNH(HO)[(HOCH$_2$-CHOHCH$_2$)$_2$N]C$_6$H$_2$AsO$_3$H$_2$	3,4-AcNH(HO)C$_6$H$_3$AsO$_3$H$_2$ + HNO$_3$, followed by redn. of the NO$_2$ group and treating with glycide			368
C_{15} 3,4,5-(CH$_2$(CH$_2$)$_3$NCH$_2$-CH$_2$CONH)(HO)(AcNH)C$_6$-H$_2$AsO$_3$H$_2$	3,4,5-ClCH$_2$CH$_2$CONH(HO)-(AcNH)C$_6$H$_2$AsO$_3$H$_2$ + pyrrolidine			376
C_{18} 2,4,3,5-(HO)$_2$(PhN$_2$)$_2$-C$_6$HAsO$_3$H$_2$	2,4-(HO)$_2$C$_6$H$_3$AsO$_3$H$_2$ + Ph-N$_2$Cl at pH 7.6-7.9	84	Orange powder, m. 268°	289

ARYLBENZENE- AND FUSED-RING AROMATIC ARSONIC ACIDS

7-NITRO-9-OXO-1-FLUORENEARSONIC ACID C$_{13}$H$_8$AsNO$_6$

Prepn.: By adding 7-nitro-9-oxo-1-fluorenyldiazonium chloride to an alk. soln. of As$_2$O$_3$ contg. CuSO$_4$; 14% yield (450).
By nitration of 9-oxo-1-fluorenearsonic acid with HNO$_3$-H$_2$SO$_4$; 90% yield (1049).
Rxns. with: PCl$_3$ in AcOH → the 1-dichloroarsino deriv. (450, 1049).
Fe(OH)$_2$ in alk. soln. → the 7-amino deriv. (1049).

7-NITRO-9-OXO-2-FLUORENEARSONIC ACID C$_{13}$H$_8$AsNO$_6$

Prepn.: by nitration of 9-oxo-2-fluorenearsonic acid with HNO$_3$-H$_2$SO$_4$ at 20-30° (1049).
Rxn. with Fe(OH)$_2$ in alk. soln. → the 7-amino deriv. (1049).

7-NITRO-9-OXO-4-FLUORENEARSONIC ACID C$_{13}$H$_8$AsNO$_6$

Prepn.: By nitrating 9-oxo-4-fluorenearsonic acid with HNO$_3$-H$_2$SO$_4$ in ice-salt bath; 90% yield (1049).

By coupling 7-nitro-9-oxo-4-fluorenyldiazonium salt with AsCl$_3$ in alk. medium and decompg. the double salt in the presence of CuSO$_4$; 43% yield (1049).

Rxn. with PCl$_3$ in AcOH → dichloro(7-nitro-9-oxo-4-fluorenyl)arsine (1049).

9-OXO-1-FLUORENEARSONIC ACID C$_{13}$H$_9$AsO$_4$

Prepn.: from 1-amino-9-fluorenone by the Bart method; 45% yield (1049).
Props.: Pale yellow solid (1049).
Rxns. with: SO$_2$ in 6 N HCl contg. KI → the 1-arsenoso deriv. (1049).
HNO$_3$-H$_2$SO$_4$ → the 7-nitro deriv. (1049).

9-OXO-2-FLUORENONEARSONIC ACID C$_{13}$H$_9$AsO$_4$

Prepn.: by coupling 2-fluorenonediazonium sulfate with AsCl$_3$ in water and decompg the double salt with Cu$_2$Br$_2$; 40% yield (1049).
Rxn. with HNO$_3$-H$_2$SO$_4$ → the 7-nitro deriv. (1049).

9-OXO-4-FLUORNEONEARSONIC ACID C$_{13}$H$_9$AsO$_4$

Prepn.: from 4-amino-9-fluorenone by the Bart-Scheller method; 45% yield (1049).
Props.: Light yellow solid, m. 138-139° (1049).
Rxns. with: SO$_2$ in 6 N HCl contg. KI → the arsenoso deriv. (1049).
PCl$_3$ in AcOH → dichloro(9-oxo-4-fluroenyl)arsine (1049).
HNO$_3$-H$_2$SO$_4$ → the 7-nitro deriv. (1049).
Salt: Na salt, yellow orange crystals (1049).

1-AMINO-9-OXO-4-FLUORENEARSONIC ACID C$_{13}$H$_{10}$AsNO$_4$

Prepn.: by adding H$_3$AsO$_4$.H$_2$O in small portions to 1-amino-9-fluorenone at 145° and completing the rxn. at 160-165°; 11.7% yield (1049).
Props.: Orange-yellow ppt. (1049).

7-AMINO-9-OXO-1-FLUORENEARSONIC ACID $C_{13}H_{10}AsNO_4$

Prepn.: by reducing the 7-nitro deriv. with $Fe(OH)_2$ in alk. medium at 90°; 42% yield (1049).
Props.: Yellow-orange crystals (1049).
Rxn. with $ClCH_2CONH_2$ in alk. soln. → the 7-(carbamylmethylamino) deriv. (1049).

7-AMINO-9-OXO-2-FLUORENEARSONIC ACID $C_{13}H_{10}AsNO_4$

Prepn.: by reducing the 7-nitro deriv. with $Fe(OH)_2$ in alk. soln. (1049).
Rxns. with: $NaNO_2$ followed by K_3AsO_3 and $CuSO_4$·→ 9-oxo-2,7-fluorenediarsonic acid (1049).
H_3PO_2 in AcOH at 85-90° → 7,7'-diamino-2,2'-arseno-9-fluorenone (1049).

7-AMINO-9-OXO-4-FLUORENEARSONIC ACID $C_{13}H_{10}AsNO_4$

Prepn.: by reducing the 7-nitro derivatives with $Fe(OH)$ in alk. medium at 90°; 56% yield (1049).
Props.: Brown powder (1049).
Rxns. with: Ac_2O → the N-acetyl deriv. (1049).
$ClCH_2CONH_2$ in 2 N KOH contg. KI → the 7-(carbamylmethylamino)- deriv. (1049).

1-ANTHRAQUINONEARSONIC ACID $C_{14}H_9AsO_5$

Prepn.: by adding gradually 1-anthraquinonediazonium sulfate and Cu_2Br_2 to a soln. of $AsCl_3$ in EtOH at r. t., heating the rxn. mixt. at 60-65°, and hydrolyz-ing with aq. NaOH; 65% yield. The Bart rxn. with the diazonium chloride or fluoro-borate gives only 17 and 23% yield, respectively (753).

7-ACETAMIDO-9-OXO-4-FLUORENEARSONIC ACID $C_{15}H_{12}AsNO_5$

Prepn.: by treating 7-amino-9-oxo-4-fluorenearsonic acid with Ac_2O at 75-80°; theoret. yield (1049).

Props.: Bright orange powder (1049).

7-(CARBAMYLMETHYLAMINO)-9-OXO-1-FLUORENEARSONIC ACID $C_{15}H_{13}AsN_2O_5$

H_2NCOCH_2NH- [structure] AsO_3H_2

Prepn.: by reacting the 7-amino deriv. with ClCH_2CONH_2 in alk. soln.; 72% yield
(1049).
Props.: Red-purple substance (1049).

7-(CARBAMYLMETHYLAMINO)-9-OXO-4-FLUORENEARSONIC ACID $C_{15}H_{13}AsN_2O_5$

H_2NCOCH_2NH- [structure] AsO_3H_2

Prepn.: by reacting the 7-amino deriv. with ClCH_2CONH_2 in alk. soln. contg. KI
at 100°; 85% yield (1049).
Props.: Purple crystals (1049).

N-1-ISOQUINOLYL-4-AMINONAPHTHALENEARSONIC ACID $C_{19}H_{15}AsN_2O_3$

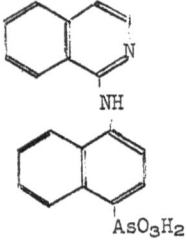

Prepn.: by heating 1-chloroisoquinoline with 4-aminonaphthalenearsonic acid (1:1)
at 175-185°; 20% yield (332).
Props.: Decomp. 202-205° (332).

3-(2-HYDROXY-1-NAPHTHYLAZO)-1-NAPHTHALENEARSONIC ACID $C_{20}H_{15}AsN_2O_4$

[structure] OH
N:N-
AsO_3H_2

Prepn.: by diazotization of 3,1-H_2NC_{10}H_6AsO_3H_2, followed by coupling with 2-naph

292

90% yield (553).
Props.: Crimson microneedles, decompg. on heating (553).

2-[2-HYDROXY-3,6-DISULFO-1-NAPHTYLAZO]NAPHTHALENEARSONIC ACID $C_{20}H_{15}AsN_2O_{10}S_2$

Use: Color reagent for Th, U, and La (327a).

p-9-ANTHRYLBENZENEARSONIC ACID $C_{20}H_{15}AsO_3$

Prepn.: by reacting anthracene with $p-H_2O_3AsC_6H_4N_2X$ in a mixt. of $AcOH-Ac_2O$-AcONa at 80° or in a mixt. of $Me_2CO-AcONa-CuCl_2$ at r. t. (350).
Props.: Needles (350).

TABLE XLIII

ARYLBENZENE- AND FUSED-RING AROMATIC ARSONIC ACIDS

Arenearsonic Acid	Prepd. from	Yield %	Props. and Remarks	Ref.
C10				
2-ClC$_{10}$H$_6$AsO$_3$H$_2$	2-Chloronaphthylamine Bart method	56	m. 296°, rxn.[1]	113
1-ClC$_{10}$H$_6$-2-AsO$_3$H$_2$	1-H$_2$NC$_{10}$H$_6$-2-AsO$_3$H$_2$ Sandmeyer rxn.	82	m. 317°, rxn.[1]	113
3-O$_2$NC$_{10}$H$_6$AsO$_3$H$_2$	3-O$_2$NC$_{10}$H$_6$-1-NH$_2$ Bart method	42	Pale-orange rods, decomp. on heating, rxn.[2] Dissocn. const.	553
4-O$_2$NC$_{10}$H$_6$AsO$_3$H$_2$ 1-C$_{10}$H$_7$AsO$_3$H$_2$	1-C$_{10}$H$_7$NH$_2$ Bart method	63	m. 197°, rxns.[3,4] Dissocn. const.	1034 549 1034
2-C$_{10}$H$_7$AsO$_3$H$_2$	2-C$_{10}$H$_7$N$_2$Cl + AsCl$_3$ in HCl, followed by NH$_4$OH	50	Rxns.[4,5]; use[6] Dissocn. const.	131, 136, 1034
4-HOC$_{10}$H$_6$AsO$_3$H$_2$	4-H$_2$NC$_{10}$H$_6$OBz Bart-Scheller method	2.5	White needles, m. >360°,	287
3,4-H$_2$N(HO)C$_{10}$H$_5$AsO$_3$H$_2$	4-BzOC$_{10}$H$_6$AsO$_3$H$_2$ + MeOH-HCl 3,4-PhN$_2$(HO)C$_6$H$_4$AsO$_3$H$_2$ catalytic redn.	57 52	Rxns.[7,8] White needles, decomp. in air, rxn.[27]	289 289
3-H$_2$NC$_{10}$H$_6$AsO$_3$H$_2$ 4-H$_2$NC$_{10}$H$_6$AsO$_3$H$_2$	3-O$_2$NC$_{10}$H$_6$AsO$_3$H$_2$ + Fe(OH)$_2$	53	m. 252° (dec.), rxn.[9] Dissocn. const.	553 1034
C11				
6-HO$_2$CC$_{10}$H$_6$-2-AsO$_3$H$_2$	6-H$_2$NC$_{10}$H$_6$AsO$_3$H$_2$ + Ni(CN)$_2$ Sandmeyer method, followed by hydrolysis	22	White powder, rxn.[10]	287
C12				
4,2-O$_2$N(4-O$_2$NC$_6$H$_4$)C$_6$H$_3$AsO$_3$H$_2$ 2-(4-O$_2$NC$_6$H$_4$)C$_6$H$_4$AsO$_3$H$_2$	4,2-O$_2$N(4-O$_2$NC$_6$H$_4$)C$_6$H$_3$NH$_2$	77	m. 220° Yellow crystals, m. 290° (dec.), rxns. 2, 11-13	377 377
4-(4-O$_2$NC$_6$H$_4$)C$_6$H$_4$AsO$_3$H$_2$	4-(4-O$_2$NC$_6$H$_4$)C$_6$H$_4$NH$_2$ by Bart-Scheller method	34	Pale yellow needles, rxn.[13]	287
4,2-O$_2$N(Ph)C$_6$H$_3$AsO$_3$H$_2$	4,2-O$_2$N(Ph)C$_6$H$_3$NH$_2$ Bart-Scheller method	50	Glistening plates, m. 228-229°, rxns. 2, 4, 14	377

Compound	Preparation	Yield (%)	Properties	Ref.
4-[3,4-O$_2$N(H$_2$N)C$_6$H$_3$]C$_6$H$_4$AsO$_3$H$_2$	4-[3,4-O$_2$N(H$_2$N)C$_6$H$_3$]C$_6$H$_4$NH$_2$ Bart method; O$_3$H$_2$ + 25% KOH under reflux	14.9	Brown powder, rxn.[15]	287
2-PhC$_6$H$_4$AsO$_3$H$_2$	2-PhC$_6$H$_4$NH$_2$ Bart method		m. 216-222°	179, 245
4-PhC$_6$H$_4$AsO$_3$H$_2$	Bart-Scheller method; 4-PhC$_6$H$_4$NH$_2$ by the Bart-Scheller method	40	Rxns.[5,14,16], salt[17] UV spectrum	377, 395, 959
4-(4-HO$_3$SC$_6$H$_4$)C$_6$H$_4$AsO$_3$H$_2$	4-PhC$_6$H$_4$AsO$_3$H$_2$ + 96% H$_2$SO$_4$ at 110-120° or ClSO$_3$H followed by hydrolysis		Forms Ba and Na salts, rxn.[18]	959, 959
4,2-O$_2$N(Ph)C$_6$H$_3$AsO$_3$H$_2$	4,2-O$_2$N(Ph)C$_6$H$_3$AsO$_3$H$_2$ by redn. with FeSO$_4$ or catalytically over Ni	66	Needles, m. 187°, rxn.[14]	377
2-(4-H$_2$NC$_6$H$_4$)C$_6$H$_4$AsO$_3$H$_2$	2-(4-O$_2$NC$_6$H$_4$)C$_6$H$_4$AsO$_3$H$_2$ by FeSO$_4$ or catalytically over Ni	92	Microcrystals, m. 212-213°, rxn.[14]	377

1. Rxn. with phenols in PhNO$_2$ in the presence of K$_2$CO$_3$ + Cu → the corresponding phenoxynaphthalenearsonic acid (113).
2. Rxn. with Fe(OH)$_2$ in alk. soln. → redn. of the -NO$_2$ group to -NH$_2$ (377, 553).
3. Redn. with SO$_2$ in HCl contg. KI, followed by hydrolysis → the arsenoso deriv. (549).
4. Rxn. with metal ions → ppt. (quant. detn.) (1015, 1016).
5. Redn. with SO$_2$ in HCl contg. KI → redn. of the -AsO$_3$H$_2$ to -AsCl$_2$ (136, 189, 245).
6. Wood preservative (241).
7. Rxn. with PhN$_2$Cl → bright red dye (287).
8. Rxn. with PhN$_2$Cl at pH 7.1-7.4 → 4,3-HO(PhN$_2$)C$_{10}$C$_5$AsO$_3$H$_2$ (289).
9. Rxn. with NaNO$_2$ followed by 2-C$_{10}$H$_7$OH → 3-(2-hydroxy-1-naphthylazo)-1-naphthalenearsonic acid (553).
10. Rxn. with PCl$_3$, followed by NH$_3$ → 6-H$_2$NOCC$_{10}$H$_6$AsO (287).
11. Rxn. with POCl$_3$ under reflux → [2-(4-O$_2$NC$_6$H$_4$)C$_6$H$_4$]$_2$AsO$_2$H (377).
12. Rxn. with PCl$_3$ in AcOH → redn. of the -AsO$_3$H$_2$ group to -AsCl$_2$ (449).
13. Hydrogenation over Raney Ni → redn. of the -NO$_2$ group to NH$_2$ (287, 377).
14. Heating with H$_2$SO$_4$ → the dibenzarsenolic acid deriv. (377).
15. Heating under reflux with 25% KOH → 4-[3,4-O$_2$N(HO)C$_6$H$_3$]C$_6$H$_4$AsO$_3$H$_2$ (287).
16. Rxn. with HNO$_3$ in H$_2$SO$_4$ → the dinitro deriv. (377).
17. Forms di-Na salt tetrahydrate, m. 202-206° (245).
18. Rxn. with HI → PhC$_6$H$_4$SO$_3$H (959).

TABLE XLIII(cont'd.)

ARYLBENZENE- AND FUSED-RING. AROMATIC ARSONIC ACIDS

Arenearsonic acid	Prepd. from	Yield %	Props. and Remarks	Ref
$4-(4-H_2NC_6H_4)C_6H_4AsO_3H_2$	$4-(4-O_2NC_6H_4)C_6H_4AsO_3H_2$ by redn. with $FeSO_4$ or catalytically over Ni	80	White needles, rxns.[19,20]	287
$3-AcNHC_{10}H_6AsO_3H_2$	$3-H_2NC_{10}H_6AsO_3H_2 + Ac_2O$		Needles, decomp. on heating Rxn.[21]	553 959
$4-(4-H_2NO_2SC_6H_4)C_6H_4AsO_3H_2$	$4-PhC_6H_4AsO_3H_2 + ClSO_3H$ followed by $10\% NH_4OH$			
$2-[PhAs(O_2H)]C_6H_4AsO_3H_2$	$4-(4-H_2NC_6H_4)C_6H_4AsO_3H_2$ Sandmeyer rxn. followed by hydrolysis		White needles, rxn.[5]	287
C_{13}				
$4-(4-HO_2CC_6H_4)C_6H_4AsO_3H_2$	$4-(4-H_2NC_6H_4)C_6H_4AsO_3H_2 + Ni(CN)_2$ Sandmeyer rxn., followed by hydrolysis		White needles, rxn.[22]	287
C_{14}				
$2-(4-ClCH_2COC_6H_4)C_6H_4AsO_3H_2$	$2-(4-AcC_6H_4)C_6H_4AsO_3H_2 + Cl_2$ in AcOH in UV under reduced pressure		Rxn.[15]	189
$2-(4-AcC_6H_4)C_6H_4AsO_3H_2$	$2-(4-AcC_6H_4)C_6H_4NH_2$ Bart-Scheller method		Yellow, m. 235-236°, rxn.[23]	189
$2-(2-AcNHC_6H_4)C_6H_4AsO_3H_2$	$2-(2-AcNHC_6H_4)C_6H_4NH_2$ Bart-Scheller method		Crystals, m. 122°, rxn.[14]	377
$4-(4-AcNHC_6H_4)C_6H_4AsO_3H_2$	$4-(4-H_2NC_6H_4)C_6H_4AsO_3H_2$ by acetylation		White needles	287
$4,2-Me(4-MeC_6H_4)C_6H_3AsO_3H_2$	$4,2-Me(4-MeC_6H_4)C_6H_3NH_2$ Bart-Scheller method		Needles, m. 237°	377
C_{15}				
$2-(4-EtCO_2NHC_6H_4)C_6H_4AsO_3H_2$		15	Crystals, m. 228°, rxn.[14]	377
C_{16}				
$2-(2-ClC_6H_4O)C_{10}H_6AsO_3H_2$	$2-ClC_{10}H_6AsO_3H_2 + 2-ClC_6H_4OH$	42	m. 230-231°, rxn.[24]	113
$2-(4-ClC_6H_4O)C_{10}H_6AsO_3H_2$	$2-ClC_{10}H_6AsO_3H_2 + 4-ClC_6H_4OH$		m. 217-219°, rxn.[24]	113

Compound	Method	Yield	Properties	Ref.
4,3-HO(PhN₂)C₁₀H₅AsO₃H₂	4-HOC₁₀H₆AsO₃H₂ + PhN₂Cl at pH 7.1-7.4	20	Red needles, m. 245°, rxn.[25]	289
1,2-PhOC₁₀H₆AsO₃H₂	1,2-ClC₁₀H₆AsO₃H₂ + PhOH	66.5	m. 322°, rxn.[24]	113
2-PhOC₁₀H₆AsO₃H₂	2-ClC₁₀H₆AsO₃H₂ + PhOH	51	m. 211-295°, rxn.[24]	113
C₁₇				
3-(4-O₂NC₆H₄CH:N)C₁₀H₆AsO₃H₂	3-H₂NC₁₀H₆AsO₃H₂ + 4-O₂NC₆H₄CHO		Yellow needles, decomp. on heating	553
4-PhCO₂C₁₀H₆AsO₃H₂	4-H₂NC₁₀H₆O₂CPh Bart-Scheller method	19	Plates, m. 199.8-200°, rxn.[24,26]	289
1,2-(2-MeC₆H₄O)C₁₀H₆AsO₃H₂	1,2-ClC₁₀H₆AsO₃H₂ + 2-MeC₆H₄OH	18	m. 181°, rxn.[24]	113
2-(2-MeC₆H₄O)C₁₀H₆AsO₃H₂	2-ClC₁₀H₆AsO₃H₂ + 2-MeC₆H₄OH	48	m. 295°, rxn.[24]	113
2-(4-MeC₆H₄O)C₁₀H₆AsO₃H₂	2-ClC₁₀H₆AsO₃H₂ + 4-MeC₆H₄OH	30	m. 225-273°	113

19. Rxn. with AcCl → the N-acetyl deriv. (287).
20. Rxn. with NaNO₂, followed by Ni(CN)₂ and hydrolysis → the 4'-carboxy deriv. (287).
21. Rxn. with 50% HI at 75-80° → redn. of the -AsO₃H₂ group to -AsI₂ (959).
22. Rxn. with PCl₃-PCl₅, followed by NH₄OH → 4-(4-H₂NOCC₆H₄)C₆H₄AsO (287).
23. Rxn. with Cl₂ in AcOH in UV light → 2-(4-ClCH₂COC₆H₄)C₆H₄AsO₃H₂ (189).
24. On boiling with aq. AcOH → the corresponding benzophenoxarsinic acid (113).
25. Catalytic redn. over Raney Ni → 3,4-H₂N(HO)C₁₀H₅AsO₃H₂ (289).
26. Methanolic HCl → 4-HOC₁₀H₆AsO₃H₂ (289).
27. Rxn. with ClCH₂COCl → 3-hydroxy-4-naphth[1,4]-p-oxazine-6-arsonic acid (289).

HETEROCYCLE-SUBSTITUTED BENZENEARSONIC ACIDS

4-[2-(2-FURYL)IMIDAZOLYL]BENZENEARSONIC ACID $C_{13}H_{11}AsN_2O_4$

Prepn.: by heating under reflux $4-H_2O_3AsC_6H_4COCH_2OH$ in 60% MeOH with Cu acetate, 25% NH_4OH, and furfurol, crystallizing the prod. from dil. HCl and liberating from HCl soln. by neutralization; 20% yield (1279).
Props.: Colorless conglomer. prisms, decomp. 297° (1278).
Hydrochloride crystallizes in colorless needles (1278).

4-(2-OXO-2H-1-BENZOPYRAN-3-YL)BENZENEARSONIC ACID $C_{15}H_{11}AsO_5$

Prepn.: by reacting $p-H_2O_3AsC_6H_4N_2Cl$ with coumarin in the presence of $CuCl_2$; 55% yield (399, 400, 404).
Props.: Pale yellow plates, m.>300° (399).

4-ARSONOCINCHOPHEN $C_{16}H_{12}AsNO_4$

Prepn.: by heating isatin and $4-AcC_6H_4AsO_3H_2$ with 33% KOH at 150-160° (1118).
Props.: Yellowish crystals (1118).

4-[(4-ARSONOPHENYL)2-IMIDAZOLYL]BENZOIC ACID $C_{16}H_{13}AsN_2O_6$

Prepn.: by he ting under reflux aq. soln. of $4-H_2O_3AsC_6H_4COCH_2OH$ with Cu acetate, NH_4OH, and 4,3 $HO(HO_2C)C_6H_3CHO$; 41% yield (1278).
Props.: Microcryst. prisms, decomp. >300°; HCl salt, needle clusters, m. 307° (dec.) (1278).

4-[2-(1-HYDROXY-4,8-DISULFO-2-NAPHTHYL)-3-METHYL-1-DIAZIRIDINYL]BENZENEARSONIC
ACID $C_{18}H_{17}AsN_2O_{10}S_2$

Prepn.: by reacting 2-[2-(p-arsonophenyl)hydrazino]-1-naphthol-4,8-disulfonic acid
with MeCHO in dild. NaHCO$_3$ soln. under H (428, 432).
Props.: White ppt. (428).

1-[2-(1-HYDROXY-4,8-DISULFO-2-NAPHTHYL)-3,5-DIOXO-1-PYRAZOLIDYL]BENZENEARSONIC
ACID $C_{19}H_{15}AsN_2O_{12}S_2$

Prepn.: by reacting di-Na salt of 2-[2-(p-arsonophenyl)-hydrazino]-1-naphthol-4,8-
disulfonic acid with malonyl dichloride (428, 432).

1,5-BIS(4-ARSONOPHENYL)HEXAHYDRO-2,4-BIS(1-HYDROXY-4,8-DISULFO-2-NAPHTHYL)-s-
TETRAZINE $C_{34}H_{30}As_2N_4O_{20}S_4$

Prepn.: by reacting the di-Na salt of 2-[2-(p-arsonophenyl)hydrazino]-1-naphthol-
4,8-disulfonic acid with formaldehyde (428, 432).

TABLE XLIV

HETEROCYCLE-SUBSTITUTED BENZENEARSONIC ACIDS

R-C6H4AsO3H2 where R is:	Prepd. from	Yield %	Props. and Remarks	Ref.
C8				
4-N:CHCH:NN-	4-NCHCH:NN-C6H4NH2 by the Bart method	27	m. >285°	1076
C9				
4-N:C(CO2H)CH:NN-	N:C(CO2H)C(CO2H):NNC6H4AsO3-H2 by decarboxylation	33	m. >250°	239
4-N:CHSCH:C-	4-N:CHSCH:CC6H4NH2 by the Bart method	60	Needles, decomp. 331°	948
4-HNCH:NCH:C-	4-HO2CCOC6H4AsO3H2 + HCHO + 25% NH4OH + CuOAc under reflux		Hydrate, m. 310°	1278
C10				
4-N:C(CO2H)C(CO2H):NN-	4-CMeCMe:NNC6H4AsO3H2 + KMnO4	62	m. >250°, rxn.[1]	239
4-N:C(CO2H)COH:CHN-	N:C(CO2Et)COH:CHNC6H4AsO3H2 by alk. hydrolysis	80		1123
4-SCH:CHCH:C-	4-ClN2C6H4AsO3H2 + thiophene + CuCl2	25	Yellow, m. >300°	400, 402, 404
4-OCH:CHCH:C-	4-ClN2C6H4AsO3H2 + furan + CuCl2		Yellow, m. >300°	400, 402, 404
4-N:C(CO2H)CMe:NN-	4-N:CMeCMe:NNC6H4AsO3H2 + KMnO4	86	m. >250°, rxn.[2]	239
4-HNCMe:NCH:C-	4-HOCH2COC6H4AsO3H2 + AcH + NH4OH + CuOAc	50	Chars >300°	1278
4-CH2CH2C(O)NHN:C-	4-HO2C(CH2)2COC6H4AsO3H2 + H2NNH2.H2O			1270
2-N:CMeCMe:NN-	2-N:CMeCMe:NNC6H4NH2 by the Bart method	24	m. 265-266°	239
4-N:CMeCMe:NN-	4-N:CMeCMe:NNC6H4NH2 by the Bart method	47	m. >250°, rxns.[3,4]	239
4-N:NC(CH2OH):C(CH2OH)N-	4-N3C6H4AsO3H2 + (:CCH2OH)2		Decomp. 180°	917

	Preparation	Yield (%)	Properties	Refs.
C₁₁				
4-HNCEt:NCH:C̄-	4-HOCH₂C₆H₄AsO₃H₂ + EtCHO + NH₄OH + CuOAc	54	Colorless prisms, m. 315° (dec.), HCl.H₂O salt, m. 275	1278, 1278
C₁₂				
4-N̄:C(CO₂H)C(O)CH:CHN-	4-[C̄(O)CH:CMeOC(O)CN:N]C₆H₄AsO₃H₂ + aq. NaOH under reflux		m. 252°	867
4-N̄:C(CO₂Et)COH:CBrN-	4-[AcC(CO₂Et):NNH]C₆H₄AsO₃H₂ + Br₂ in AcOH-AcONa		Rxn. 5	1123
4-N̄:C(CO₂Et)COH:CHN-	4-[BrCH₂COC(CO₂Et):NNH]C₆H₄AsO₃H₂ + AcOH in EtOH	57	Rxn. 5	1123
4-MeCH:CHCH:CMeN̄-	4-H₂NC₆H₄AsO₃H₂ + (AcCH₂-)₂	70	m. 230° (dec.).	474
4-N̄:C(CO₂Et)CMe:NN-	4-N̄:C(CO₂Et)CMe:NNC₆H₄AsO₃H₂ + EtOH	70	m. >250°	239
4-HN̄C(Pr):NCH:C̄-	4-HOCH₂C₆H₄AsO₃H₂ + PrCHO + NH₄OH + CuOAc in 50% EtOH	51	Colorless prisms, m. 250° (dec.), HBr salt, colorless flat prisms	1278
C₁₃				
4-[ŌC(CH:CHCO₂H):CHCH:C̄]-	4-ClN₂C₆H₄AsO₃H₂ + ŌCH:CHCH:- C̄CH:CHCO₂H + CuCl₂	38.8	m. >300°	400, 402
4-HN̄C(Bu):NCH:C̄-	4-HOCH₂COC₆H₄AsO₃H₂ + BuCHO + NH₄OH + CuOAc		Prisms, decomp. 270°	1278
C₁₄				
4-C̄H₂N(Ph)N:C-	4-ClCH₂COC₆H₄AsO₃H₂ + PhNHNH₂		Yellow solid, darkens at 210-215°	231
C₁₅				
4-[S̄C(:NC₆H₄Cl-o)NHC̄OCH]-	4-ClN₂C₆H₄AsO₃H₂ + 2-(S̄CH₂-CONHC̄:N)C₆H₄Cl	30		1043
4-[S̄C(:NC₆H₄Cl-p)NHC̄OCH]-	4-ClN₂C₆H₄AsO₃H₂+4-(S̄CH₂CONH-C̄:N)C₆H₄Cl	30		1043

1. Heating with Cu₂O in quinoline → elimination of one CO₂ group (239).
2. Esterification with EtOH → the carbethoxy deriv. (239).
3. Rxn. with HNO₃, H₂SO₄ → 4-triazolyl-3-nitrobenzenearsonic acid (239).
4. Rxn. with KMnO₄ → a mixt. of mono- and dicarboxy derivs. by oxidation of one and both Me groups at the tirazole ring (239).
5. Rxn. with a₄. NaOH → hydrolysis of the -CO₂Et group (1125).

TABLE XLIV (Cont'd.)

HETEROCYCLE-SUBSTITUTED BENZENEARSONIC ACIDS

$R\text{-}C_6H_4AsO_3H_2$ where R is:	Prepd. from	Yield %	Props. and Remarks	Ref.
$4\text{-}[H\overline{NC}(C_6H_4NO_2\text{-}p){:}NCH{:}\underline{C}]\text{-}$	$4\text{-}HOCH_2COC_6H_4AsO_3H_2 + 4\text{-}O_2N\text{-}C_6H_4CHO + NH_4OH + CuOAc$	31	Orange needles, decomp. 320-323°	1278
$4\text{-}[\overline{S}C(:NC_3H_4NO_2\text{-}p)NHCO\overline{C}H]\text{-}$	$4\text{-}ClN_2C_6H_4AsO_3H_2 + 4\text{-}(\overline{S}CH_2\text{-}CONHC{:}N)C_6H_4NO_2$	25		1043
$4\text{-}[\overline{S}C(:NC_6H_4NO_2\text{-}m)NHCO\overline{C}H]\text{-}$	$4\text{-}ClN_2C_6H_4AsO_3H_2 + 3\text{-}(\overline{S}CH_2CON\text{-}H\overline{C}{:}N)C_6H_4NO_2$	25		1043
$4\text{-}H\overline{NC}(Ph){:}NCH{:}\overline{C}\text{-}$	$4\text{-}HOCH_2COC_6H_4AsO_3H_2 + BzH + NH_4OH + CuOAc$	39	Prismatic stars, decomp. 330° HCl salt, pine needles, dec. 303°	1278 1278
$4\text{-}\overline{S}C(:NPh)NHCO\overline{C}H\text{-}$	$4\text{-}ClN_2C_6H_4AsO_3H_2 + PhN{:}\overline{C}NHCO\text{-}CH_2S + CuCl_2$	30	m. 300°	1043
$4\text{-}H\overline{NC}(n\text{-}C_6H_{13}){:}NCH{:}\overline{C}\text{-}$	$4\text{-}HOCH_2COC_6H_4AsO_3H_2 + n\text{-}C_6H_{13}\text{-}CHO + NH_4OH + CuOAc$	70	Voluminous needles changing to prisms above 27°6	1278
C_{16} $4\text{-}H\overline{NC}(C_6H_4CO_2H\text{-}p){:}NCH{:}\overline{C}\text{-}$	$4\text{-}HOCH_2C_6H_4AsO_3H_2 + 4\text{-}HO_2\text{-}CC_6H_4CHO + NH_4OH + CuOAc$	49	Microcryst. prisms, m. 320°, HCl salt, forms prisms and needles	1278
$4\text{-}\overline{S}C(:NC_6H_4CO_2H\text{-}o)NHC\overline{C}H\text{-}$	$4\text{-}ClN_2C_6H_4AsO_3H_2 + 2\text{-}(\overline{S}CH_2CON\text{-}H\overline{C}{:}N)C_6H_4CO_2H + CuCl_2$ in Me_2CO	25		1043
$4\text{-}\overline{S}C(:NC_6H_4CO_2H\text{-}m)NHCO\overline{C}H\text{-}$	$4\text{-}ClN_2C_6H_4AsO_3H_2 + 3\text{-}(\overline{S}CH_2CON\text{-}H\overline{C}{:}N)C_6H_4CO_2H + CuCl_2$ in Me_2CO	25		1043
$4\text{-}\overline{S}C(:NC_6H_4CO_2\text{-}p)NHCO\overline{C}H\text{-}$	$4\text{-}ClN_2C_6H_4AsO_3H_2 + 4\text{-}(\overline{S}CH_2CON\text{-}H\overline{C}{:}N)C_6H_4CO_2H + CuCl_2$ in Me_2CO	25		1043
$4\text{-}H\overline{NC}(C_6H_4OMe\text{-}p){:}NCH{:}\overline{C}\text{-}$	$4\text{-}HOCH_2COC_6H_4AsO_3H_2 + 4\text{-}MeO\text{-}C_6H_4CHO + NH_4OH + CuOAc$	35	Flat, colorless prisms, decomp. 310°, HCl salt, colorless platelets, decomp. 270°	1278
$4\text{-}\overline{S}C(:NC_6H_4Me\text{-}o)NHCO\overline{C}H\text{-}$	$4\text{-}ClN_2C_6H_4AsO_3H_2 + 2\text{-}(\overline{S}CH_2CO..NHC{:}N)C_6H_4Me + CuCl_2$ in $Me_2\text{-}CO$	28		1043

4-SC(:NC₆H₄Me-p)NHCOCH-	4-ClN₂C₆H₄AsO₃H₂ + 4-(SCN₂CO-NHC:N)C₆H₄Me-... CuCl₂ in Me₂-CO	29	1043
4-[4-H₂O₃AsC₆H₄N(CH₂CH₂)₂N]-	H₂NC₆H₄N(CH₂CH₂)₂NC₆H₄NF₂ by the Bart method	Light brown solid, m. > 280° rxn.⁷	982

6. The needles sinter at 190°, m. 195-197°; the prisms, m. 256-260° (1278).
7. Rxn. with PCl₃ in EtOAc → the arsenoso deriv. (982).

2-OXO-5-BENZOXAZOLINEARSONIC ACID $C_7H_6AsNO_5$

Rxn. with RR'NH → 4,3-HO(RR'NCONH)$C_6H_3AsO_3H_2$ (375).

2-OXO-6-BENZOXAZOLINEARSONIC ACID $C_7H_6AsNO_5$

Rxn. with RR'NH → 3,4-HO(RR'NCONH)$C_6H_3AsO_3H_2$ (375).

2-MERCAPTOBENZIMIDAZOLE-5-ARSONIC ACID $C_7H_7AsN_2O_3S$

Rxns. with: SO$_2$ in glycerol contg. NaI → the arsenoso deriv. (371, 535, 581).
PhNHNH$_2$ → the arsenoso deriv. (353, 581).

4-HYDROXY-5-QUINAZOLINEARSONIC ACID $C_8H_7AsN_2O_4$

Prepn.: from 5-amino-4-hydroxyquinazoline by the Bart method, 48.3% yield
(1319).
Props.: M. >320° (1319).

4-HYDROXY-6-QUINAZOLINEARSONIC ACID $C_8H_7AsN_2O_4$

Prepn.: from 6-amino-4-hydroxyquinazoline by the Bart method, 22.2% yield (1319)
Props.: M. > 320° (1319).

4-HYDROXY-7-QUINAZOLINEARSONIC ACID $C_8H_7AsN_2O_4$

Prepn.: from 7-amino-4-hydroxyquinazoline by the Bart method, 42.2% yield (1319)
Props.: M. 320° (1319).
Rxns. with: SO$_2$ in dild. HCl contg. KI, followed by neutralization → the arseno
deriv. (1319).

3,4-DIHYDRO-3-OXO-1,4,2H-BENZOTHIAZINE-6-ARSONIC ACID $C_8H_8AsNO_4S$

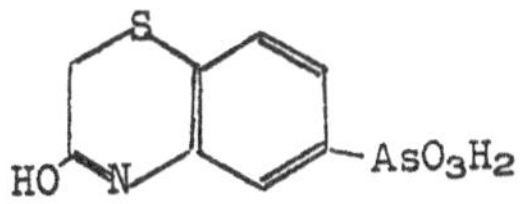

HO—N ... S ... AsO_3H_2

Prepn.: by reacting 3,4-dihydro-3-oxo-1,4,2H-benzothiazin-2-yldiazonium salt with NaAsO$_2$ in aq. alk. (770).
Props.: Pale yellow, decomp. >300° (770).

3-HYDROXY-1,4,2H-BENZOXAZINE-6 ARSONIC ACID $C_8H_8AsNO_5$

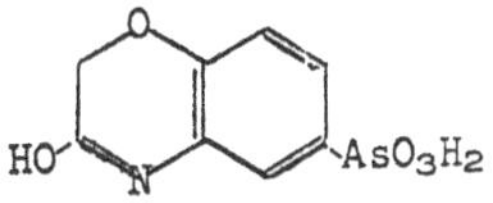

HO—N ... O ... AsO_3H_2

Prepn.: By catalytic hydrogenation of 3,4-$O_2N(H_2NCOCH_2O)C_6H_3AsO_3H_2$ in the presence of Ni (1182).
 By reducing 3,4-$O_2N(HO_2CCH_2O)C_6H_3AsO_3H_2$ with $Na_2S_2O_4$ and acidifying the rxn. mixt. (1047).

2,6,8-TRICHLORO-5-QUINOLINEARSONIC ACID $C_9H_5AsCl_3NO_2$

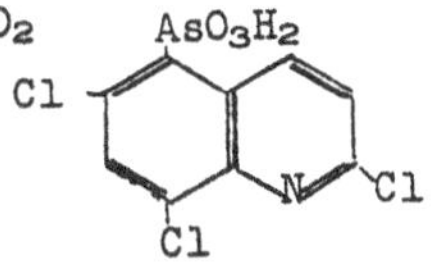

AsO_3H_2
Cl ... N—Cl
Cl

Prepn.: from 5-amino-2,6,8-trichloroquinoline by the Bart method (1103).
Props.: M. > 310° (1103).

6-BROMO-2-CHLORO-5-QUINOLINEARSONIC ACID $C_9H_6AsBrClNO_3$

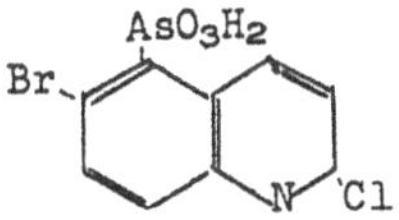

AsO_3H_2
Br ... N—Cl

Prepn.: from 5-amino-6-bromo-2-chloroquinoline by the Bart method; 7.7% yield (583).
Props.: m. >330° (583).

2,6-DICHLORO-5-QUINOLINEARSONIC ACID $C_9H_6AsCl_2NO_3$

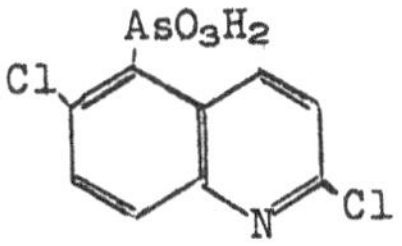

AsO_3H_2
Cl ... N—Cl

Prepn.: from 5-amino-2,6-dichloroquinoline by the Bart method; 6.7% yield (583).
Props.: m. 302-303° (583).

6,8-DICHLORO-5-QUINOLINEARSONIC ACID $C_9H_6AsCl_2O_3$

Prepn.: from 5-amino-6,8-dichloroquinoline by the Bart method (1103).
Props.: m. 292-293° (1103).

6-BROMO-2-HYDROXY-5-QUINOLINEARSONIC ACID $C_9H_7AsBrNO_4$

Prepn.: from 5-amino-6-bromo-2-hydroxyquinoline by the Bart method; 8.8% yield (583).
Props.: m. > 330° (583).

2-CHLORO-5-QUINOLINEARSONIC ACID $C_9H_7AsClNO_3$

Prepn.: from 5-amino-2-chloroquinoline by the Bart method, 24% yield (183).
Props.: Orthorhombic needles (183).
Rxn. with cellosolve in the presence of dissolved Na → 2-(2-ethoxyethyl)-5-quinolinearsonic acid (183).

2-CHLORO-6-QUINOLINEARSONIC ACID $C_0H_7AsClNO_3$

Prepn.: from 6-amino-2-chloroquinoline by the Bart method, 18% yield (183).
Props.: Orthorhombic needles (183).

2-CHLORO-8-QUINOLINEARSONIC ACID $C_9H_7AsClNO_3$

Prepn.: from 8-amino-2-chloroquinoline by the Bart method, 11% yield (183).
Props.: Orthorhombic needles, m. 273-276° (183).
Rxn. with aq. HCl → 2-hydroxy-8-quinolinearsonic acid (183).

6-CHLORO-2-HYDROXY-5-QUINOLINEARSONIC ACID $C_9H_7AsClNO_4$

Prepn.: from 5 amino 6 chloro 2 hydroxyquinoline by the Bart method; 6.5% yield (583).
Props.: m. 325-236° (583).

4-ARSONO-7-CHLORO-2-BENZOXAZOLYLMERCAPTOACETIC ACID $C_9H_7AsClNO_6S$

Prepn.: by refluxing 7-chloro-2 mercapto 5 benzoxazolearsonic acid with $ClCH_2$-CO_2H in aq. NaOH soln. (372).
Rxn. with H_3PO_2 → the arseno deriv. (372).

8-HYDROXY-7-IODO-5-QUINOLINEARSONIC ACID $C_9H_7AsINO_4$

Prepn.: by reacting 8-hydroxy-5-quinolinearsonic acid with calcd. amt. of KI and $KBrO_3$ in dil. H_2SO_4; isolated as the acetate (197).

8-HYDROXY-7-NITRO-5-QUINOLINEARSONIC ACID $C_9H_7AsN_2O_6$

Prepn.: by treating 8-hydroxy-5-quinolinearsonic acid with calcd. amt. of HNO_3 in concd. H_2SO_4; isolated as the sulfate (197).
Props.: Dark yellow powder (197).
Rxn. with $FeSO_4$ in dild H_2SO_4 → the 7-amino deriv. (197).

2-HYDROXY-5-QUINOLINEARSONIC ACID $C_9H_8AsNO_4$

Prepn.: from 5-amino-2 hydroxyquinoline by the Bart method, 23% yield (183).
Props.: Monohydrate crystallizes in tetragonal plates (183).

2-HYDROXY-7-QUINOLINEARSONIC ACID $C_9H_8AsNO_4$

H_2O_3As (structure)

Prepn.: from 7-amino-2-hydroxyquinoline by the Bart method, 13% yield (183).
Props.: Orthorhombic needles (183).

2-HYDROXY-8-QUINOLINEARSONIC ACID $C_9H_8AsNO_4$

AsO_3H_2 (structure)

Prepn.: from 8-amino-2-hydroxyquinoline by the Bart method; theoret. yield (183).
Props.: Orthorhombic needles (183).

4-HYDROXY-6-QUINOLINEARSONIC ACID $C_9H_8AsNO_4$

H_2O_3As (structure) OH

Prepn.: from 6-amino-4-hydroxyquinoline by the Bart method, 29.7% yield (1319).
Props.: m. 320° (1319).
Rxn. with SO_2 in HCl contg. KI, followed by neutralization → the arsenoso
deriv. (1319).

8-HYDROXY-5-QUINOLINEARSONIC ACID $C_9H_8AsNO_4$

AsO_3H_2 (structure) OH

Prepn.: from 5-amino-8-hydroxyquinoline by the Bart method (197).
Rxns. with: KI + $KBrO_3$ in H_2SO_4 → 8-hydroxy-7-iodo-5-quinolinearsonic acid (197)
HNO_3-H_2SO_4 → 8-hydroxy-7-nitro-5-quinolinearsonic acid (197).
Salt: Sulfate, red powder, slightly sol. in H_2O, EtOH, and Me_2CO (197).

5-ARSONO-2-BENZOXAZOLYLMERCAPTOACETIC ACID $C_9H_8AsNO_6S$

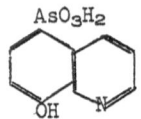
H_2O_3As—(structure)C–SCH_2CO_2H

Prepn.: by refluxing 2-mercapto-5-benzoxazolearsonic acid with $ClCH_2CO_2H$ in aq.
NaOH or methanolic KOH (372).
Rxn. with H_3PO_2 → the arseno deriv. (372).

7-AMINO-8-HYDROXY-5-QUINOLINEARSONIC ACID $C_9H_9AsN_2O_4$

Prepn.: by reducing 8-hydroxy-7-nitro-5-quinolinearsonic acid with a calcd.amt. of $FeSO_4$ in dild. H_2SO_4 (197).

3-METHYLBENZOXAZINE-6-ARSONIC ACID $C_9H_{10}AsNO_2$

Prepn.: by reducing $3,4-O_2N(MeCOCH_2O)C_6H_3AsO_3H_2$ with $FeSO_4$ or $Na_2S_2O_4$ in alk. soln. and acidifying the reduced prod. (503).
Props.: Light tan crystals (503).
Rxn. with SO_2 in HCl contg. KI, followed by neutralization → 6-arsenoso-3-methyl-benzoxazine (503).

8-BROMO-2-CHLORO-6-METHYL-5-QUINOLINEARSONIC ACID $C_{10}H_8AsBrClNO_3$

Prepn.: from 5-amino-8-bromo-2-chloro-6-methylquinoline by the Bart method, 2.3% yield (281).
Props.: m. 290-292° (281).

2,8-DICHLORO-6-METHYL-5-QUINOLINEARSONIC ACID $C_{10}H_8AsCl_2NO_3$

Prepn.: from 5-amino-2,8-dichloro-6-methylquinoline by the Bart method (892).
Props.: m. 307-308° (decompn.) (892).

6-BROMO-8-METHYL-5-QUINOLINEARSONIC ACID $C_{10}H_9AsBrNO_3$

Prepn.: from 5-amino-6-bromo-8-methylquinoline by the Bart method; 10.5% yield (583).
Props.: m. 244-245° (decompn.) (583).

8-BROMO-6-METHYL-5-QUINOLINEARSONIC ACID $C_{10}H_9AsBrNO_3$

AsO_3H_2

Me—

Br, N

Prepn.: from 5-amino-8-bromo-6-methylquinoline by the Bart method; 6.9% yield (28:
Props.: m. 276-277° (decomp.) (281).

8-BROMO-2-HYDROXY-6-METHYL-5-QUINOLINEARSONIC ACID $C_{10}H_9AsBrNO_4$

AsO_3H_2

Me—

Br, N, OH

Prepn.: from 5-amino-8-bromo-2-hydroxy-6-methylquinoline by the Bart method, 28%
yield (281),
Props.: m. >300° (281).

2-CHLORO-6-METHYL-5-QUINOLINEARSONIC ACID $C_{10}H_9AsClNO_3$

AsO_3H_2

Me—

N, Cl

Prepn.: from 5-amino-2-chloro-6-methylquinoline by the Bart method, 16% yield
(184).

2-CHLORO-8-METHYL-5-QUINOLINEARSONIC ACID $C_{10}H_9AsClNO_3$

AsO_3H_2

Me, N, Cl

Prepn.: from 5-amino-2-chloro-8-methylquinoline by the Bart method, 11.9% yield
(88).
Props.: m. >315° (88).

2-CHLORO-4-METHYL-7-QUINOLINEARSONIC ACID $C_{10}H_9AsClNO_3$

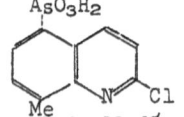

Me

H_2O_3As— N, Cl

Prepn.: from 7-amino-2-chloro-4-methylquinoline by the Bart method, 7% yield
(183).
Props.: Orthorhombic needles, m. 192° (183).
Rxn. with cellosolve in the presence of Na → 2-(2-ethoxyethyl)-4-methyl-7-quinolin
arsonic acid (183).

2-CHLORO-6-METHYL-8-QUINOLINEARSONIC ACID $C_{10}H_9AsClNO_3$

repn.: from 8-amino-2-chloro-6-methylquinoline by the Bart method, 13.1% yield (184).
rops.: Colorless, prismatic needles, m. 323-324° (184).
xn. with 25% H_2SO_4 at 173° → the 2-hydroxy deriv. (184).

2-CHLORO-7-METHYL-8-QUINOLINEARSONIC ACID $C_{10}H_9AsClNO_3$

repn.: from 8-amino-2-chloro-7-methylquinoline by the Bart method (185).
rops.: Decomp. 240° (185).

6-CHLORO-8-METHYL-5-QUINOLINEARSONIC ACID $C_{10}H_9AsClNO_3$

repn.: from 5-amino-6-chloro-8-methylquinoline by the Bart method; 17% yield (583).
rops.: m. 255-256° (583).

8-CHLORO-6-METHYL-5-QUINOLINEARSONIC ACID $C_{10}H_9AsClNO_3$

repn.: from 5-amino-8-chloro-6 methylquinoline by the Bart method; 8.9% yield (892).
rops.: m. >325° (892).

3-(7-ARSONO-4-CHLORO-2-BENZOXAZOLYLMERCAPTO)PROPIONIC ACID $C_{10}H_9AsClNO_6S$

repn.: by refluxing 4-chloro-2-mercapto-7-benzoxazolearsonic acid with 3-ClCH$_2$-H$_2$CO$_2$H in EtOH-KOH (372).
xn. with H_3PO_2 → the arseno deriv. (372).

3-(7-ARSONO-4-CHLORO-2-BENZOXAZOLYLMERCAPTO)PROPIONIC ACID $C_{10}H_9AsClNO_6S$

Prepn.: by refluxing 4-chloro-2-mercapto-7-benzoxazolearsonic acid with 3-ClCH$_2$-CH$_2$CO$_2$H in EtOH-KOH (372).
Rxn. with H$_3$PO$_2$ → the arseno deriv. (372).

8-HYDROXY-7-IODO-2-METHYL-5-QUINOLINEARSONIC ACID $C_{10}H_9AsINO_4$

Prepn.: by treating 8-hydroxy-2-methyl-5-quinolinearsonic acid with alk. ICl soln. (459).
Props.: Insol. in org. solvents, m. > 300° (459).
Rxn. with H$_2$SO$_4$ or HNO$_3$ → evolution of I (459).

2-METHYL-6-QUINOLINEARSONIC ACID $C_{10}H_{10}AsNO_3$

Prepn.: by treating arsanilic acid with HCl followed by paraldehyde (459).
Props.: Brown, m. > 300° (459).

6-METHYL-5-QUINOLINEARSONIC ACID $C_{10}H_{10}AsNO_3$

Prepn.: from 5-amino-6-methylquinoline by the Bart method, 9.85% yield (184).
Props.: m. 257° (dec.) (184).

6-METHYL-8-QUINOLINEARSONIC ACID $C_{10}H_{10}AsNO_3$

Prepn.: from 8-amino-6-methylquinoline by the Bart method, 11.8% yield (184).
Props.: Long, slender needles, darken at 270°, melt 280° (184).

6-METHYL-8-QUINOLINEARSONIC ACID $C_{10}H_{10}AsNO_3$

Prepn.: from 8-amino-6-methylquinoline by the Bart method, 11.8% yield (184).
Props.: Long, slender needles, darken at 270°, melt 280° (184).

7-METHYL-8-QUINOLINEARSONIC ACID $C_{10}H_{10}AsNO_3$

Prepn.: from 8-amino-7-methylquinoline by the Bart method (185).
Props.: Long, tan needles, (185).

8-METHYL-5-QUINOLINEARSONIC ACID $C_{10}H_{10}AsNO_3$

Prepn.: from 5-amino-8-methylquinoline by the Bart method, 12.8% yield (88).
Props.: m. 224-226° (88).

2-HYDROXY-4-METHYL-7-QUINOLINEARSONIC ACID $C_{10}H_{10}AsNO_4$

Prepn.: from 7-amino-2-hydroxy-4-methylquinoline by the Bart method, 31% yield (183).
Props.: Monohydrate crystallizes in orthorhombic plates (183).
Salt: Na salt dihydrate forms tetragonal granules (183).

2-HYDROXY-6-METHYL-5-QUINOLINEARSONIC ACID $C_{10}H_{10}AsNO_4$

Prepn.: from 5-amino-2-hydroxy-6-methylquinoline by the Bart method, 14.4% yield (184).
Props.: Pale tan needles, m. 250° (184)

2-HYDROXY-6-METHYL-8-QUINOLINEARSONIC ACID $C_{10}H_{10}AsNO_4$

Me —⬡⬡— OH
AsO₃H₂

$$Me—\text{[quinoline ring]}—OH,\ AsO_3H_2$$

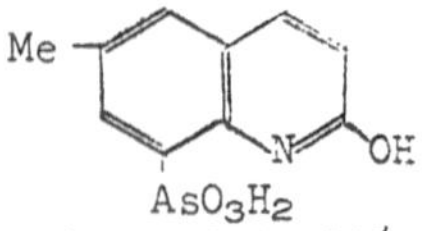

Prepn.: by hydrolyzing 2-chloro-6-methyl-8-quinolinearsonic acid in 25% H_2SO_4
at 173° in an autoclave (184).
Props.: m. 306° (184).

2-HYDROXY-7-METHYL-8-QUINOLINEARSONIC ACID $C_{10}H_{10}AsNO_4$

Me —[quinoline ring]— OH
AsO₃H₂

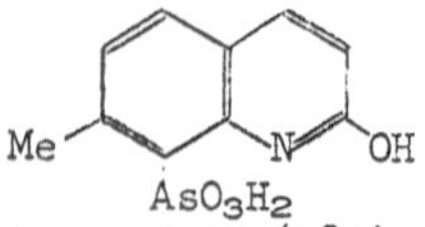

Prepn.: by refluxing the 2-chloro arsonic acid with NaOMe in MeOH (185).
Props.: Small tablet-like crystals, decomp. > 260° (185).

2-HYDROXY-8-METHYL-5-QUINOLINEARSONIC ACID $C_{10}H_{10}AsNO_4$ AsO₃H₂

[quinoline ring] OH
Me

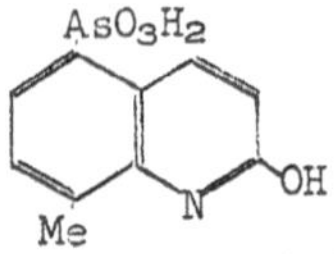

Prepn.: from 5-amino-2-hydroxy-8-methylquinoline by the Bart method, 14.8% yield
(88).
Props.: m. > 315° (88).

4-HYDROXY-2-METHYL-6-QUINOLINEARSONIC ACID $C_{10}H_{10}AsNO_4$

OH
H_2O_3As—[quinoline ring]— Me

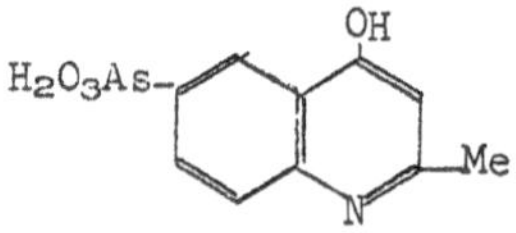

Prepn.: from 6-amino-4 hydroxy-2-methylquinoline by the Bart method, 31.6% yield
(1319).
Props.: m. >320° (1319).

8-HYDROXY-2-METHYL-5-QUINOLINEARSONIC ACID $C_{10}H_{10}AsNO_4$

AsO₃H₂
[quinoline ring] Me
OH

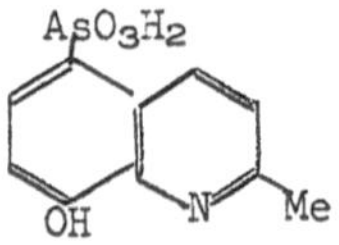

Prepn.: by treating 4,3-HO(H_2N)$C_6H_3AsO_3H_2$ with concd. HCl followed by paraldehyde
(459).
Rxn. with alk. ICl soln. → the 7-iodo deriv. (459).

3-(5-ARSONO-2-BENZOXAZOLYLMERCAPTO)PROPIONIC ACID $C_{10}H_{10}AsNO_6S$

Prepn.: by refluxing 2-mercapto-5-benzoxazolearsonic acid with $ClCH_2CH_2CO_2H$ in MeOH-KOH (372).
Rxn. with $H_3PO_2 \rightarrow$ the arseno deriv. (372).

3-(6-ARSONO-2-BENZOXAZOLYLMERCAPTO)PROPIONIC ACID $C_{10}H_{10}AsNO_6S$

Prepn.: by refluxing 2-mercapto-6-benzoxazolearsonic acid with $ClCH_2CH_2CO_2H$ in EtOH-KOH (372).
Rxn. with $H_3PO_2 \rightarrow$ the arseno deriv. (372).

(5-ARSONO-7-METHOXY-2-BENZOXAZOLYLMERCAPTO)ACETIC ACID $C_{10}H_{10}AsNO_7S$

Prepn.: by refluxing 2-mercapto-7-methoxy-5-benzoxazolearsonic acid with $ClCH_2$-CO_2H in aq. NaOH (372).
Rxn. with $H_3PO_2 \rightarrow$ the arseno deriv. (372).

5-BENZOYL-2-THIOPHENARSONIC ACID $C_{11}H_9AsO_4S$

Prepn.: by oxidation of the arsenoso. or dichloroarsino deriv. with 3% H_2O_2 (1280).
Props.: Pearly, lustrous flakes, m. 360° (decompn.), loses 1 mol. of H_2O below 140° (1280).

2-(5-ARSONO-2 BENZOXAZOLYLMERCAPTO)BUTYRIC ACID $C_{11}H_{12}AsNO_6S$

Prepn.: by refluxing 2-mercapto-5-benzoxazolearsonic acid with 2-bromo-butyric acid in the presence of solid KOH (372).
Rxn. with $H_3PO_2 \rightarrow$ the arseno deriv. (372).

2-(4-SULFAMYLANILINO)-5-PYRIDINEARSONIC ACID $C_{11}H_{12}AsN_3O_5S$

$$H_2NO_2S-\bigcirc-NH-\bigcirc_N-AsO_3H_2$$

Prepn.: by refluxing p-sulfamylaniline with 2-chloro-5-pyridinearsonic acid in di
HCl, 50% yield (250).
Props.: m. > 250° (250).

2-(4-ARSONOANILINE)-5-PYRIDINEARSONIC ACID $C_{11}H_{12}As_2N_2O_6$

$$H_2O_3As-\bigcirc-NH-\bigcirc_N-AsO_3H_2$$

Prepn.: by refluxing p-arsanilic acid with 2-chloro-5-pyridinearsonic acid in
dild. HCl, 90% yield (250).
Props.: m. > 250° (250).

8-NITRO-3-DIBENZOFURANARSONIC ACID $C_{12}H_8ANO_6$

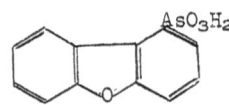

Prepn.: From 3-amino-8-nitrodibenzofuran by the Bart method, 35% yield (1145).
By treating 3-dibenzofuranarsonic acid with HNO_3, 70% yield (1145).
Props.: Yellow, m. > 280° (1145).
Rxns. with: PCl_3 in AcOH → the dichloroarsino deriv. (1145).
H_2/Ni → the 8-amino deriv. (1145).

1-DIBENZOFURANARSONIC ACID $C_{12}H_9AsO_4$

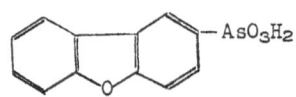

Prepn.: by heating 1-dibenzofuranmercuri chloride with $AsCl_3$ in C_6H_6 at 100°
(271).
Props.: Monohydrate, buff, m. 186-188° (decomp.) (271).

2-DIBENZOFURANARSONIC ACID $C_{12}H_9AsO$

$$\bigcirc\bigcirc_O\bigcirc-AsO_3H_2$$

Prepn.: From 2-aminodibenzofuran by the Bart method (271).
By heating dibenzofuran with $AsCl_3$ in $o\text{-}C_6H_4Cl_2$ in the presence of $AlCl_3$,
hydrolyzing the rxn. prod. with alkali, and acidifying with HCl (271).
By heating dibenzofuran with H_3AsO_4 at 175-220° (1145).
Props.: Monohydrate, buff, m. 213-214°; anhydride, m. ~360° (271).
Rxn. with PCl_3 in AcOH followed by H_2O → the arsenoso deriv. (1145).

3-DIBENZOFURANARSONIC ACID $C_{12}H_9AsO_4$

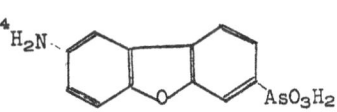

repn.: from 3-aminodibenzofuran by the Bart method (271, 1145).
rops.: Monohydrate, m. 385° (271).
alts: Ba salt, colorless crystals (271).
xns. with: PCl_3 in AcOH → the dichloroarsino deriv. (1145).
NO_3 → the 8-nitro deriv. (1145).
$_2SO_4$ → the 8-sulfo deriv. (1145).

8-SULFO-3-DIBENZOFURANARSONIC ACID (?) $C_{12}H_9AsO_7S$

repn.: by heating 3-dibenzofuranarsonic acid with H_2SO_4 on a steam bath (1145).
rops.: M. > 300° (1145).
xn. with SO_2 in dild. contg. KI → the arsenoso deriv. (1145).

3-CARBAZOLARSONIC ACID $C_{12}H_{10}AsNO_3$

repn.: from 3-aminocarbazole by the Bart method, 27% yield (1127).
rops.: m. 346-347° (1127).
xn. with SO_2 in concd. HCl contg. I → the dichloroarsino deriv. (1127).

8-AMINO-3 DIBENZOFURANARSONIC ACID $C_{12}H_{10}AsNO_4$

repn.: by catalytic redn. of the 8-nitro deriv. over Raney Ni; theoret. yield (1145).
rops.: White ppt., m. > 250° (1145).

3,4-DIHYDRO-3-OXO-2H-NAPHTHO[1,2-b]-1,4-THIAZINE-6-ARSONIC ACID $C_{12}H_{10}AsNO_4S$

repn.: from 6-amino-2,3-dihydro-3-oxo-2H-naphtho[1,2-b]-1,4-thiazine by the Bart
ethod, 30% yield (240a).
rops.: m. 265° (240a).

3-HYDROXY-4-NAPHTH[1,4]-p-OXAZINE-6-ARSONIC ACID $C_{12}H_{10}AsNO_5$

Prepn.: by reacting $3,4-H_2N(HO)C_{10}H_5-1-AsO_3H_2$ with $ClCH_2COCl$ (289).
Props.: Needles (289).

4-(5-ARSONO-2-PYRIDYLAMINO)BENZOIC ACID $C_{12}H_{11}AsN_2O_5$

Prepn.: by refluxing p-aminobenzoic acid with 2-chloro-5-pyridinearsonic acid in
dild. HCl, 80% yield (250).
Props.: m. 247-248° (250).

9-OXO-2-ACRIDANEARSONIC ACID $C_{13}H_{10}AsNO_4$

Prepn.: by cyclizing $4-(2 HO_2CC_6H_4NH)C_6H_4AsO_3H_2$ with concd. H_2SO_4 at 120-130°
(302)
Props.: m. > 300° (302).

1-(4-ARSONOPHENYLCARBAMYLMETHYL)-1,6-DIHYDRO-6-OXO-3-PYRIDINEARSONIC ACID
$C_{14}H_{14}As_2N_2O_4$

Prepn.: by refluxing 2-pyridono-5-arsonic acid with N-(chloroacetyl)arsanilic
acid in alc. alkali, 54.6% yield (84).
Props.: Fine needles, decomp. ~262° (84).

2-p-PHENETIDINO-5-PYRIDINEARSONIC ACID $C_{13}H_{15}AsN_2O_4$

Prepn.: by refluxing p-phenetidine with 2-chloro-5-pyridinearsonic acid in
dild. HCl; 77% yield (250).
Props.: m. 216-218° (250).

318

2-(2-ETHOXYETHOXY)-5-QUINOLINEARSONIC ACID $C_{13}H_{16}AsNO_5$

Prepn.:   by refluxing 2-chloro-5-quinolinearsonic acid with a soln. of metallic Na in ethylcellosolve and neutralizing the rxn. prod.; 90% yield (183).
Props.:   Flat prisms, m. 172° (183).

2-[2-(N-THIOMORPHOLINYL)-5-PYRIDYLAMINO]-5-PYRIDINEARSONIC ACID $C_{14}H_{17}AsN_4O_3S$

Prepn.:   by refluxing 2-chloro-5-pyridinearsonic acid with 2-(N-thiomorpholinyl)-5-aminopyridine in dild. HCl; 80% yield (250).
Props.:   m. 174-175° (250).

2-[2-(4-MORPHOLINYL)-5-(YRIDYLAMINO]-5-PYRIDINEARSONIC ACID $C_{14}H_{17}AsN_4O_4$

Prepn.:   by refluxing 2-chloro-5-pyridinearsonic acid with 5-amino-2-(4-morpholinyl)-pyridine in dild. HCl, 96% yield (250).
Props.:   m. 128-129° (250).

2-(2-ETHOXYETHOXY)-4-METHYL-7-QUINOLINEARSONIC ACID $C_{14}H_{18}AsNO_5$

Prepn.:   by refluxing 2-chloro-4-methyl-7-quinolinearsonic acid with a soln. of metallic Na in ethylcellosolve and neutralizing the rxn. prod., 59% yield (183).
Props.:   Monoclinic needles, m. 180° (183).

2-PHENYL-6-QUINOLINEARSONIC ACID $C_{15}H_{12}AsNO_3$

Prepn.: by condensing arsanilic acid with PhCH:CHCHO in the presence of HCl at 115° (33).
Props.: Colorless, cryst. powder, insol. in H_2O, m. >340° (33).
Rxns. with: H_3PO_2 in HCl in the presence of KI → 2,2'-diphenyl-6,6'-arsenoquino-line-2HCl (33).

8-HYDROXY-2-PHENYL-5-QUINOLINEARSONIC ACID $C_{15}H_{12}AsNO_4$

Prepn.: by condensing 3-amino-4-hydroxybenzenearsonic acid with PhCH:CHCHO in HCl at 115-120° and neutralizing the rxn. mixt. (33).
Props.: m. 330° (33).

8-(4-DIETHYLAMINO-1-METHYLBUTYLAMINO)-6-METHOXY-5-QUINOLINEARSONIC ACID (?)
   $C_{18}H_{30}AsN_2O_3$

Prepn.: by treating 8-amino-6-methoxyquinoline with $Et_2N(CH_2)_3CH(OH)Me$ and H_3AsO_4 (1327).

9-(DIMETHYLAMINOPHENYL)-2-ACRIDINEARSONIC ACID $C_{21}H_{19}AsN_2O_3$

Prepn.: by reacting 9-oxo-2-acridanearsonic acid with $POCl_3$ + $PhNMe_2$; 53% yield (302).
Props.: m. 230-232° (302).

320

HETEROCYCLE – ARSONIC ACIDS

$RAsO_3H_2$, where R is:	Prepd. from	Yield %	Props. and Remarks	Ref.
C_4 $OCH:CHCH:C-$	$RAsO + H_2O_2$	73	Colorless crystals, m. 133° decomp. > 50° and in aq. HCl	346, 347
C_5 $HC:NCCl:CHCH:C-$	$RNH_2 + AsCl_3$ in conc. HCl followed by $Cu_2Cl_2 + NaNO_2$	50	Rxns.[1,2]	85, 590
$HC:CHCH:NCH:C-$	$RAsO + H_2O_2$ $RAsCl_2 + H_2O_2$ $R_2AsCl.2HCl$ + alkali $R_2As + KCN$ or Aq CN	73–85 77	m. 156°, salt[3] by-prod. by-prod.	85 591 592
$CH:CHC(:O)NHCH:C-$	$RNH_2 + AsCl_3$ in conc. HCl followed by $Cu_2Cl_2 + NaNO_2$	40	Rxns.[4-6]	85, 86
C_7 $CH:NC(SEt):CHCH:C-$	RNH_2 Bart method		m. 145°	1185
$CH:NC(NHEt):CHCH:C-$	$CH:NC(Cl):CHCH:CAsO_3H_2 + Et-NH_2$ in MeOH		Purification[7]	84
C_8 $CH:CHC(NEt_2):NCH:C-$	$CH:CHC(Cl):NCH:CAsO_3H_2 + Et_2NH$ in EtOH		Purification[7]	84

1. Redn. by SO_2 in aq. HCl, followed by neutralization → the corresponding arsonous acid, $RAs(OH)_2$ (590).
2. Rxns. with primary or secondary amines → replacement of the Cl atom by an amino group (84, 250).
3. HCl salt, m. 195-197° (85).
4. Redn. by SO_2 in H_2SO_4 contg. KI → the arsenoso deriv., RAsO, (1048a).
5. Redn. by $H_3PO_2 + HI$, followed by oxidation with H_2O_2 in conc. HCl → $CH:CHC(:O)NHCH:CAsCl_2.HCl$ (85).
6. Rxn. with $ClCH_2CONHC_6H_4AsO_3H_2$ → $H_2O_3AsC_6H_4NHCOCH_2NCH:C(AsO_3H_2)CH:CHCO$ (85).
7. The compd. was purified by redn. with $H_3PO_2 + KI$ to the arseno deriv., followed by oxidation to the arsonic acid (84).

DIARSONIC ACIDS

METHANEDIARSONIC ACID $CH_6As_2O_6$ $CH_2(AsO_3H_2)_2$
Prepn.: by oxidation of $CH_2(AsO)_2$ with 30% H_2O_2 at an elevated temp. (1024, 1213).
Props.: Coarse crystals, m. 168° (dec.) (1024).
Rxn. with $Na_2S_2O_4 \rightarrow$ yellow ppt., $(:AsCH_2As:)x$ (?) (1024).

DIARSONOMALONIC ACID $C_3H_6As_2O_{10}$ $(H_2O_3As)_2C(CO_2H)_2$
Prepn.: by reacting $Br_2C(CO_2Et)_2$ with As_2O_3 in aq. KOH in exothermic rxn. at 70°
(824).
Props.: Colorless needles, m. 128° (824).
Salts: $(NaO_2C)_2C(AsO_3HNa)_2.14H_2O$, rhombic, colorless crystals, stable in air,
m. 61° (824).
Rxn. with $H_3PO_2 \rightarrow [H_2O_3AsC(CO_2H)_2As-AsC(CO_2H)_3AsO_3H_2]n$ and $\overline{(HO_2C)_2CAs:AsC(CO_2H)_2-}$
$\overline{As:As}$ (827).

DIARSONOSUCCINIC ACID $C_4H_8As_2O_{10}$ $[-CH(AsO_3H_2)CO_2H]_2$
Prepn.: by reacting $(-CHBrCO_2Et)_2$ with As_2O_3 in aq. NaOH, isolated as Ba, Ca,
and Na salts (826).

2-NITRO-p-BENZENEDIARSONIC ACID $C_6H_7As_2NO_8$ $2,1,4-O_2NC_6H_3(AsO_3H_2)_2$
Prepn.: by oxidation of $2,1,4-O_2NC_6H_3(AsO)_2$ with Cl_2 in H_2O (79).
Props.: Pale yellow powder, m. 238-240° (79).

3-NITRO-1,4-BENZENEDIARSONIC ACID $C_6H_7As_2NO_8$ $3,1,4-O_2NC_6H_3(AsO_3H_2)_2$
Prepn.: From $3,4-O_2N(H_2N)C_6H_3AsO_3H_2$ by the Bart method; 66% yield (883).

4-HYDROXY-5-NITRO-m-BENZENEARSONIC ACID $C_6H_7As_2NO_9$ $4,5,3,1-HO(O_2N)C_6H_2(AsO_3H_2$
Prepn.: from $3,4,5-H_2N(HO)(O_2N)C_6H_2AsO_3H_2$ by the Bart method; 40% yield (880).
Props.: Yellowish crystals (880).

o-BENZENEDIARSONIC ACID $C_6H_8As_2O_6$ $o-C_6H_4(AsO_3H_2)_2$
Prepn.: from $2-H_2NC_6H_4AsO_3H_2$ by the Bart method; 52.6% yield (327).
Props.: Monohydrate, m. > 360° (327).
Rxn. with SO_2 in concd. HCl contg. KI $\rightarrow o-C_6H_4(AsCl_2)_2$ (327).

m-BENZENEDIARSONIC ACID $C_6H_8As_2O_6$ $m-C_6H_4(AsO_3H_2)_2$
Prepn.: from $3-H_2NC_6H_4AsO_3H_2$ by the Bart method; 13% yield (383).

p-BENZENEDIARSONIC ACID $C_6H_8As_2O_6$ $p-C_6H_4(AsO_3H_2)_2$
Prepn.: from $4-H_2NC_6H_4AsO_3H_2$ by the Bart method; 24% yield (883).

4-HYDROXY-1,2-BENZENEDIARSONIC ACID $C_6H_8As_2O_7$ $4,1,2-HOC_6H_3(AsO_3H_2)_2$
Prepn.: From $2,4-H_2N(HO)C_6H_3AsO_3H_2$ by the Bart method; 24% yield (883).

4-HYDROXY-1,3-BENZENEDIARSONIC ACID $C_6H_8As_2O_7$ $4,1,3-HOC_6H_3(AsO_3H_2)_2$
Prepn.: from $3,4-H_2N(HO)C_6H_3AsO_3H_2$ by the Bart method; 25% yield (880).
Props.: Red-brown crystals (880).

3-NITRO-4,4'-BIPHENYLDIARSONIC ACID $C_{12}H_{11}As_2NO_8$

Prepn.: from 3-nitrobenzidine by the Bart method; 19.4% yield (237).
Props.: Yellow powder, m. 249.5-250.5° (287).

4,4'-BIPHENYLDIARSONIC ACID $C_{12}H_{12}As_2O_6$ $(4-C_6H_4AsO_3H_2)_2$
Prepn.: from benzidine by the Bart method; 3.4% yield (287).
Props.: White prisms (287).

9-OXO-2,7-FLUORENEDIARSONIC ACID $C_{13}H_{10}As_2O_7$

Prepn.: from 7-amino-9-oxofluorenearsonic acid by the Bart method; 67% yield (1049).

2,5-BIS(4-ARSONOPHENYL)QUINONE $C_{18}H_{14}As_2O_8$

Prepn.: by reacting $4-H_2O_3AsC_6H_4N_2Cl$ with quinone (400, 404).

4,4'-(9,10-ANTHRYLENE)DIBENZENEARSONIC ACID $C_{26}H_{20}As_2O_6$

Prepn.: by-prod. in the rxn. of anthracene with $4-H_2O_3AsC_6H_4N_2Cl$ in a mixt. of AcOH-Ac$_2$O-AcONa at 80° or in a mixt. of Me$_2$CO-AcONa-CuCl$_2$ at r. t. (350) or in AcOH in the presence of AcONa at 45° for 3 hrs. and at 80° for 6 hrs. (350a).
Props.: Yellow microcrystals (350).

3,3'-(4,4'-BIPHENYLENEBISAZO)BIS[4-AMINO-1-NAPHTHALENEARSONIC ACID] $C_{32}H_{25}As_2N$

Prepn.: By reacting $(-C_6H_4N_2Cl)_2$ with $4,1-H_2N-C_{10}H_6-AsO_3H_2$; 69% yield (682).

ARSONIC ACID ESTERS AND ANHYDRIDES

Methanearsonic Acid Esters and Anhydrides

$C_3H_9AsO_3$ MeAsO(OMe)$_2$, prepd. by reacting MeAsO(OAg)$_2$ with MeI, 55% yield; b. 97°/10 mm, d_4^{20} 1.562, n_D^{20} 1.469 (677).

$C_3H_9AsCl_4O_3Si_2$ MeAsO(OSiMeCl$_2$)$_2$, a solid, prepd. by treating MeAsO$_3$H$_2$ with MeSiCl in C_6H_6 at r. t.; hydrolysis by $H_2O \rightarrow$ poly(methylarsonomethylsiloxane) (655a, 656).

$C_5H_{13}AsO_3$ MeAsO(OEt)$_2$, prepd. by oxidation of MeAs(OEt)$_2$ with SeO$_2$ under anhydrou condns., 70-75% yield; b. 122-124°/19 mm., d_4^{20} 1.354, n_D^{20} 1.458 (677). On boiling under atm. pressure undergoes rearrangement to MeOAs(OEt)$_2$ (678).

$C_5H_{15}AsCl_2O_3Si_2$, MeAsO(OSiMe$_2$Cl)$_2$, pale yellow oil, prepd. by treating MeAsO$_3$H$_2$ with Me$_2$SiCl$_2$ in C_6H_6 at r. t.; hydrolysis by $H_2O \rightarrow$ poly(methylarsonodimethyl-siloxane) (655a, 656).

$C_7H_{17}AsO_3$ MeAsO(OPr)$_2$, prepd. by treating MeAsO(OAg)$_2$ with PrI in C_6H_6 (1097); b. 138-140°/18 mm., d_4^{20} 1.243, n_D^{20} 1.451 (677); b. 138-140°/13-15 mm., d_4^{20} 1.24175, n_D^{20} 1.45235 (1097).

$C_7H_{17}AsO_3$ MeAsO(OCHMe$_2$)$_2$, prepd. by oxidation of MeAs(OCHMe$_2$)$_2$ with SeO$_2$ in dry benzene under reflux or by refluxing MeAsO(OAg)$_2$ with Me$_2$CHCl; 83% yield; b. 82°/3 mm., d_{25} 1.2133, n_D^{25} 1.4420 (1283).

$C_9H_{21}AsO_3$ MeAsO(OBu)$_2$, prepd. by azeotrop. distn. of MeAsO$_3$H$_2$ with BuOH in dry C_6H_6 (1097); or by oxidation of MeAs(OBu)$_2$ with SeO$_2$ in dry C_6H_6, 49% yield (1283); or by refluxing MeAsO(OAg)$_2$ with BuCl; b. 105°/2 mm (1283); b. 160.3-161.3°/13 mm., d_4^{20} 1.1728, n_D^{20} 1.45240 (1097). Decomp. at 200-220° into BuOH and MeAs(OBu)$_2$ (649).

$C_9H_{23}AsO_7P_2S_2$ MeAsO[OP(S)(OEt)$_2$]$_2$, prepd. by refluxing MeAsO(ONa)$_2$ with ClP(S)-(OEt)$_2$ in alc.; b. 112-116°/17 mm (1102).

$C_{11}H_{17}AsCl_2O_3Si_2$ MeAsO[OSiClMe(C$_4$H$_4$N)]$_2$ (C$_4$H$_4$N = pyrryl (655a).

$C_{13}H_{25}AsCl_4O_3Si_2$ MeAsO(OSiCl$_2$C$_6$H$_{11}$)$_2$, white solid (655a).

$C_{17}H_{21}AsCl_4O_3Si_2$ MeAsO[OSiCl$_2$(CH$_2$CH$_2$Ph)]$_2$, yellow semisolid (655a).

Ethanearsonic Acid Esters

$C_4H_{11}AsO_3$ EtAsO(OMe)$_2$, prepd. by oxidation of EtAs(OMe)$_2$ with SeO$_2$ under anhydr. condns.; b. 107-109°/15 mm., d_4^{20} 1.432, n_D^{20} 1.457 (677). On heating under reflux/atm. pressure undergoes rearrangement to EtOAs(OMe)$_2$ (678).

$C_6H_{15}AsO_3$ EtAsO(OEt)$_2$, prepd. by oxidation of EtAs(OEt)$_2$ with SeO$_2$ under anhydr. conds., 70% yield; b. 123-125°/15 mm., d_4^{20} 1.297, n_D^{20} 1.451 (677).

$C_{10}H_{23}AsO_3$ EtAsO(OBu)$_2$, decomp. at 200-220° into BuOH, EtAs(OBu)$_2$, and an un-identified carbonyl compd. (649).

2-Propenearsonic Acid Ester

$C_9H_{15}AsCl_4O_3Si_2$ $CH_2:CHCH_2AsO[OSiCl_2(CH_2CH:CH_2)]_2$, rubbery solid, prepd. by adding $CH_2:CHCH_2SiCl_3$ to a suspension of $CH_2:CHCH_2AsO_3H_2$ in C_6H_6 and heating under reflux (655a).

Propanearsonic Acid Esters

$C_7H_{17}AsO_3$ $PrAsO(OEt)_2$, b. 121.5–124°/11 mm., d_{20} 1.2435, n^{20}_D 1.4520, at 200–220° undergoes rearrangement and $PrOAs(OEt)_2$ is formed (649).

$C_9H_{21}AsO_3$ $PrAsO(OPr)_2$, decomp. at 200–220° (649).

$C_{11}H_{25}AsO_3$ $PrAsO(OBu)_2$, decomp. at 200–220° into BuOH and $PrAs(OBu)_2$ (649).

Butanearsonic Acid Esters

$C_{10}H_{23}AsO_3$ $BuAsO(OPr)_2$, prepd. by azeotropic distn. of $BuAsO_3H_2$ with PrOH in C_6H_6; b. 151.8–152.2°/6 mm., d^{20}_4 1.1488, n^{20}_D 1.45401 (1097).

$C_{12}H_{27}AsO_3$ $BuAsO(OBu)_2$, prepd. by azeotropic distn. of $BuAsO_3H_2$ with BuOH in C_6H_6; b. 168–168.5°/7 mm., d^{20}_4 1.1040, n^{20}_D 1.45066 (1097).

Benzenearsonic Acid Esters and Anhydrides

$C_{18}H_{20}As_2Au_2O_6$ $PhAsO(OAuEt)O-]_2$, m. 128° (387).

$C_{18}H_{15}AsCl_4O_3Si_2$ $PhAsO(OSiCl_2Ph)_2$, amorphous solid, prepd. by reacting $PhAsO_3H_2$ with $PhSiCl_3$ in C_6H_6 at r. t. (655a, 656).

$C_{30}H_{25}AsCl_2O_3Si_2$ $PhAsO(OSiClPh_2)_2$, amorphous solid, prepd. by reacting $PhAsO_3H_2$ with Ph_2SiCl_2 in C_6H_6 at r. t. (655a, 656).

$C_{42}H_{35}AsO_3Si_2$ $PhAsO(OSiPh_3)_2$, waxy solid, prepd. from $PhAsO_3H_2$ and Ph_3SiCl in C_6H_6 under reflux (655a, 656).

4-Nitrobenzenearsonic Acid Anhydrides

$C_{12}H_{10}AsNO_5Sn$ $4-O_2NC_6H_4AsO(OH)OSnPh$, decomp. 297° (1267).

$C_{14}H_{22}AsNO_5Sn$ $4-O_2NC_6H_4As(O)O_2SnBu_2$, decomp. 296° (1267).

$C_{14}H_{24}AsO_9P_2S_2$ $4-O_2NC_6H_4As(O)[OP(S)POEt]_2]_2$, prepd. by refluxing $4-O_2NC_6H_4AsO-(ONa)_2$ with $ClP(S)(OEt)_2$ in alc., m. 63–64° (1102).

$C_{20}H_{28}As_2N_2O_{10}Sn$ $[4-O_2NC_6H_4As(O)(OH)O]_2SnBu_2$, decomp. 165–172° (1267).

ARSINIC ACIDS

Dialkyl- and diarylarsinic acids, R_2AsO_2H and $RR'AsO_2H$, are prepared or formed by the following methods.

1. From organic derivatives of trivalent arsenic by:
   (a) Oxidation of bis(diorganoarsenic) oxides, $(R_2As)_2O$ or $(RR'As)_2O$, by air, hydrogen peroxide, nitric acid, chlorine (followed by hydrolysis), or mercury oxide (Method Ia).
   (b) Oxidation of arsinous acid, R_2AsOH or $RR'AsOH$, by hydrogen peroxide. Arsinous acids can be prepared by reacting organodihaloarsines with aryldiazonium chloride-metal chloride double salts, $RN_2Cl.MCln$, in acetone at 0°, treating the reaction product with sodium iodide, and hydrolyzing by aqueous alkali (Method Ib).
   (c) Oxidation of arsinous acid esters, R_2AsOR or $RR'AsOR$, with selenium dioxide to form the corresponding arsinic acids (Method Ic).

(d) Oxidation of diorganohaloarsines, R_2AsX or $RR'AsX$, with nitric acid, aqueous hydrogen peroxide, or potassium permanganate in alkaline solution. Diorganohaloarsines can be obtained by reacting organodihaloarsines with aryldiazonium chloride-metal halide double salts, $RN_2Cl.MCln$, in acetone in the presence of metals such as iron or zinc. Diorganohalogenarsines are also prepared by treating bis(diorganoarsen oxides with hydrogen halides (Method Id).

(e) Oxidation of diorganohaloarsines, R_2AsX or $RR'AsX$, with halogens to diorganoarsenic trihalides, followed by hydrolysis of the latter, or treating diorganohaloarsines with chlorine in water (Method Ie).

(f) Treating diorganohaloarsines with 100% excess of alkali or with potassium or silver cyanide (Method If).

(g) Oxidation of secondary arsines, R_2AsH with oxygen in ether (Method Ig)

(h) Oxidation of tetraorganodiarsines, $(R_2As)_2$ or $(RR'As)_2$, with atmospheric oxygen (Method Ih).

(i) Treating organodihaloarsines, $RAsX_2$, or arsenoso compounds, $(RAsO)n$, with alkyl halides in aqueous or methanolic sodium hydroxide solutions or with dimethyl sulfate in aqueous sodium hydroxide (Method Ii).

(j) Reacting organodihaloarsines with aryldiazonium salts in alcohol, in the presence of copper or copper salts, and hydrolyzing the product The diazonium compounds can be prepared in the presence of organodihaloarsines as follows. Organodihaloarsines are added to a mixture of a primary aromatic amine salt and a copper salt in alcohol, the solution is treated at 3° with a nitrite, and the resulting double sal: $RN_2Cl.RAsCl_2$, or diorganoarsenic trichloride, R_2AsCl_3, is decomposed with aqueous alkali. (Method Ij).

(k) Reacting organodihaloarsines, $RAsX_2$, or arsenoso compounds, $(RAsO)n$, or arsonous acid salts, $RAs(ONa)_2$, with aryldiazonium chlorides in a neutral or alkaline medium in the presence or absence of copper or copper salts, and acidifying the reaction mixture after evolution of nitrogen is over (Method Ik).

(l) Reaction of arsenic trichloride with aryldiazonium fluoroborates or aryldiazonium double salts, $RN_2Cl.MCln$, in absolute or 80% ethanol, in presence of copper salts, followed by hydrolysis, to yield mixtures of arsonic and arsinic acids (Method Il).

(m) Reacting aryldiazonium chlorides with sodium arsenite in alkaline solution, acidifying the mixture to Congo, to precipitate arsonic acid, and then acidifying the fitrate with hydrochloric acid to precipitate arsinic acid (Method Im).

(n) Oxidation of tertiary arsines with hydrogen peroxide in alkaline medium or with nitric acid (Method In).

(o) Reaction of arsenoso compounds $(RAsO)n$, with alkylene oxides in aqueous solution (Method Io).

(p) Reacting trialkyl arsenites with Grignard compounds in ether at -70° to yield mixtures of arsonic and arsinic acids (Method Ip).

2. From organic derivatives of pentavalent arsenic by:

(a) Treating tertiary arsine oxides, R_3AsO, with 10 N sodium hydroxide at boiling temperature (Method IIa).

(b) Fusion of tertiary arsine oxides, R_3AsO, with sodium hydroxide, follow by neutralization (Method IIb).

(c) Decomposition of tertiary arsine dihalides, R_3AsX_2, with water (Method IIc).

(d) Hydrolysis of diorganohaloarsine dihydroxides, $RR'As(OH)_2Cl$ (Method IId).

(e) Disproportionation of arsonic acids, $RAsO_3H_2$, by boiling with phosphorus oxychloride (Method IIe).

(f) Heating arsonic acids $RAsO_3H_2$, with aromatic amines, to yield the corresponding amino-substituted arsinic acids, or with phenols to yield the corresponding hydroxy-substituted arsinic acids (Method IIf).

Diarylarsinic acids can be converted, by reacting with nitrating mixtures, into the corresponding nitro derivatives, which may be reduced catalytically or by ferrous hydroxide in alkaline medium to the corresponding aminodiaryl- or bis-(aminoaryl)-arsinic acids.

Amino-substituted diarylarsinic acids can be reacted with cyanuric chloride or chloro-s-triazines to yield s-triazinylamino derivatives; with acyl halides or carboxylic anhydrides to yield acylamido derivatives; with phosgene to yield urylene- or with phosgene and ammonia to yield ureido- derivatives; with sulfonyl chlorides to yield sulfonamido derivatives; with sodium or amyl nitrite to yield diazonium compounds which may be coupled with aromatic compounds or hydrolyzed to yield hydroxy derivatives.

Diarylarsinic acids containing 4,6-dichloro-s-triazin-2-ylamino group may be converted into the corresponding mono- and diamino-, mono- and dihydroxy-, and aminohydroxy-s-triazin-2-ylamino derivatives in step-wise reactions with ammonia, amines, and aqueous alkali, respectively.

Diarylarsinic acids containing phenolic groups are readily etherified by dimethyl sulfate.

POLYMETHYLENEARSINIC ACID $(CH_3AsO_2)x$ $(-CH_2AsO_2H)n$
Prepn.: By heating As_2O_3 with AcOK in Ac_2O, boiling the prod. with H_2O and treating the ppt. with H_2O_2; formation mechanism (1214).

BIS(TRIFLUOROMETHYL)ARSINIC ACID $C_2HAsF_6O_2$ $(CF_3)_2AsO_2H$
Prepn.: By reacting $(CF_3)_2AsI$ with H_2O in a sealed tube at r. t., 86% yield (333).
Props.: Hydrate crystallizes in the form of large, translucent, trigonal rhombs; loses H_2O in a desiccator and changes into a white powder; sublimes at 150°/ 0.001 mm. with decompn. to give needles (333). Ionization const. (335). Relative basicity (337).
Salt: Ag salt, glittering trigonal crystals (333).

DIMETHYLARSINIC ACID, CACODYLIC ACID $C_2H_7AsO_2$ Me_2AsO_2H
Prepn.: by oxidation of $(Me_2As)_2O$ with air or HgO (1224).
Props.: Basicity (665). Raman spectrum (994). Magnetic props. (1029).
Rxns. with: H_3PO_2 in HCl or HBr → Me_2AsCl and Me_2AsBr, resp. (660).
H_3PO_2 in the presence of EtI → Et_2Me_2AsI (993).

SO$_2$ in H$_2$SO$_4$ in the presence of KI → Me$_2$AsI (660).
H$_2$SO$_4$ at 320° → [Me$_2$As(OH)$_2$SO$_4$H] → [MeSO$_4$H+MeAs(OH)$_2$] → As(OH)$_3$ (991, 993).
Metal ions → pptn. (quant. analyses) (1017, 1018, 1019).
Neutron → As^{76} (1170).
Salts: Me$_2$AsO$_2$Na.3H$_2$O and (Me$_2$AsO$_2$)$_3$Fe (1224).
(Me$_2$AsO$_2$)$_2$Cd, forms decahydrate (silky crystals), heptahydrate, and dihydrate;
soly. and stability data (1209).
(Me$_2$ASO$_2$)$_2$Zn, prepd. from the acid and ZnO or from the Ba salt and ZnSO$_4$, forms
mono- and heptahydrate (1212). Crystal structure of the heptahydrate (1207).
Soly. and stability (1208).
(Me$_2$AsO$_2$)$_2$Ba, forms mono-, tri-, and nona-hydrate; soly. and stability (1210).
 Crystal structure (1211).
(Me$_2$AsO$_2$)$_2$Ca, forms mono-, nona-, and an-hydrate; soly. and stability (1210).
 Crystal structure (1211).
(Me$_2$AsO$_2$)$_2$Sr, forms mono-, tri-, and tridecahydrate; soly. and stability (1210).
 Crystal structure (1211).
Me$_2$AsO$_2$Ag, orthorhombic prisms and large needles (1028).
Esters: Me$_2$AsO[OP(S)(OEt)$_2$], b. 103-104°/17 mm. (1102)
Me$_2$AsO[OP(O)(OEt)$_2$], prepd. by refluxing Me$_2$AsO$_2$Na with ClP(O)(OEt)$_2$ in EtOH (1102)

DIETHYLARSINIC ACID C$_4$H$_{11}$AsO$_2$ Et$_2$AsO$_2$H
Prepn.: by-prod. in the oxidation of Et$_2$AsOEt with SeO$_2$; 1.5% yield (649).
Props.: M. 136-137° (649).
Rxn. with Penicillium brevicaule in bread cultures → Et$_2$AsMe (193).
Ester: Et$_2$AsO$_2$Bu, prepd. by oxidation of Et$_2$AsOBu with SeO$_2$; b. 130-130.5°/4 mm.,
d$^{20}_4$ 1.1922, n$^{20}_D$ 1.4721 (649).

BUTYLMETHYLARSINIC ACID C$_5$H$_{13}$AsO$_2$ MeBuAsO$_2$H
Esters: MeBuAsO$_2$Et, prepd. by oxidation of MeBuAsOEt with SeO$_2$, b. 110-111°/2 mm.
d$^{24}_4$ 1.2265; n$^{28}_D$ 1.4729 (649).
MeBuAsO$_2$Bu, prepd. by oxidation of MeBuAsOBu with SeO$_2$, b. 133-134°/3 mm., d$^{20}_4$
1.1507, n$^{28}_D$ 1.4676 (649).

METHYLPHENYLARSINIC ACID C$_7$H$_9$AsO$_2$ MePhAsOOH
Prepn.: By treating PhAsCl$_2$ dissolved in aq. NaOH with Me$_2$SO$_4$ (0.5:0.62); 97%
yield (251).
    By oxidation of (MePhAs)$_2$O with 5 N HNO$_3$ (640).
    By oxidation of (MePhAs-)$_2$ with atm. O (1056).
Props.: IR spectrum and H bonding (119a).
Salts: MePhAsOOAg, gave MePhAsOOMe when refluxed with MeI (640).
Rxns. with: SO$_2$ in HCl contg. KI → MePhAsCl (251).
HNO$_3$-H$_2$SO$_4$ → Me(3-O$_2$NC$_6$H$_4$)A.sOOH (251).
NH$_4$OH, followed by Ag NO$_3$ → MePhAsOOAg (640).
Zn-Hg + HCl → MePhAsH (811).
H$_3$PO$_2$ → (MePhAs-)$_2$ (1056).
Esters: MePhAsOOMe, b. 89-91°/14 mm; n$^{28}_D$ 1.5740; prepd. by refluxing MePhAsOOAg
with MeI (640).

BIS(4-NITROPHENYL)ARSINIC ACID $C_{12}H_9AsN_2O_6$ $(4-O_2NC_6H_4)_2AsO_2H$

Prepn.: By treating $4-O_2NC_6H_4As(ONa)_2$ with $4-O_2NC_6H_4N_2Cl$ in alk. soln. in the presence of copper or copper salts and neutralizing the rxn. mixture (1187).

By reacting $4-O_2NC_6H_4AsCl_2$ with $4-O_2NC_6H_4N_2Cl$ in an acid medium to form $(4-O_2NC_6H_4)_2AsCl_3$ and hydrolyzing the prod. (1188, 1228).

By-prod in the rxn. of $AsCl_3$ with $4-O_2NC_6H_4N_2BF_4$ in 80% EtOH in the presence of Cu_2Br_2, 21% yield (295).

By-prod. in the rxn. of $(4-O_2NC_6H_4N_2)_2ZnCl_4$ with $AsCl_3$ and Cu_2Br_2 in abs. EtOH, 3% yield; or when $4-O_2NC_6H_4N_2$-sulfate is used, 7% yield (394).

Props.: Yellow, m. 227° (1187); m. 278° (1188, 1228); m. 263°, polarography (829).

Rxns. with: SO_2 in 10% HCl + KI → the arsinous acid (1189).
H_2 over Pd/C → $(4-H_2NC_6H_4)AsO_2H$ (1228).

DIPHENYLARSINIC ACID $C_{12}H_{11}AsO$ Ph_2AsOOH

Prepn.: By reacting $PhAsCl_2$ with PhN_2Cl in the presence of $CuSO_4$, neutralizing the mixt. with NaOH, and acidifying after 12 hrs.; 61% yield (136, 1188).

By adding $PhNH_2.HCl$ in EtOH to $PhAsCl_2$, reacting the mixt. with amyl nitrite, treating the pink $PhN_2Cl.PhAsCl_2$ ppt. with $NaHCO_3$, and acidifying the soln.; 56.5% yield (136).

By oxidation of $(Ph_2As)_2O$ (243).

By reacting $PhAs(ONa)_2$ with PhN_2Cl in the presence of ammoniacal $CuSO_4$ (90, 1187).

By oxidation of Ph_2AsCl with H_2O_2 (674).

By oxidation of Ph_2AsF with HNO_3, isolated as $Ph_2AsO_2H.HNO_3$ (1100).

By fusion of Ph_3AsO with NaOH, followed by neutralization, 87% yield (562).

By treating $(Ph_2As)_2O$ with Na-Hg in C_6H_6 and hydrolyzing the Na deriv. (96).

By oxidation of $CH_2(AsPh_2)_2$ with concd. HNO_3 (1215).

By oxidation of Ph_2AsH with O in Et_2O (1293).

By-prod. in the rxn. of Ph_3As with Li in C_4H_8O, followed by Me_3SiCl and hydrolysis (1309).

Props.: M. 168-170° (96); m. 173° (136); m. 174° (562, 1215); m. 173-174° (674). m. 167-168° (829); m. 178° (1187, 1188); m. 172° (1293). Polarography (829). Potentiometric titration (1186).

Rxns.with: Zn-Hg + HCl → Ph_2AsH (242).

Metal ions → pptn. (quantitate. analysis) (1013, 1228).

$AcCl$ → Ph_2AsCl (1024).

SO_2 in HCl contg. KI, followed by hydrolysis → $(Ph_2As)_2O$ (1191).

$HS-R-CO_2H$ → $Ph_2AsS-R-CO_2H$ (1244).

Derivs.: S-benzylthiuronium salt, colorless needles, m. 147° (efferv.) (243). $Ph_2As(O)[OP(S)(OC_6Cl_5)_2]$, m. 217, prepd. by refluxing $Ph_2AsOONa$ with $ClP(S)(O-C_6Cl_5)_2$ in EtOH (1102).

2-(PHENYLARSINICO)BENZOIC ACID $C_{13}H_{11}AsO_4$ $Ph(2-HO_2CC_6H_4)AsO_2H$

Prepn.: by adding $PhAsCl_2$ to $2-MeO_2CC_6H_4N_2Cl.FeCl_3$ in Me_2CO at 0°, followed by NaI, hydrolyzing the prod. with 40% KOH and oxidizing with 30% H_2O_2; 50% yield along with 20% $Ph(2-MeO_2CC_6H_4)AsO_2H$ (908).

By treating 3-chloro-1-phenylarsindole with H_2O_2 (266).

By reacting $PhAsCl_2$ with $2-HO_2CC_6H_4N_2Cl$ to form $Ph(2-HO_2CC_6H_4)AsCl_3$ and hydrolyzing the prod. (1188).

Props.: m. > 300° (908); m. 166° (266); Decomp. 166° (1188).
Potentiometric neutralization (1186).
Rxn. with SO$_2$ in 10% H$_2$SO$_4$ + KI → Ph(2-HO$_2$CC$_6$H$_4$)AsOH (1189).

4-{N-[(2-AMINO-4-HYDROXY-6-PTERIDYL)METHYL]AMINO}PHENYLARSINICOACETIC ACID C$_{15}$H$_{15}$AsN$_6$O$_5$

Prepn.: by reacting p-aminophenylarsinicoacetic acid with 2,4,5-triamino-6-hydroxypyrimidine sulfate and 1,1,3-tribromopropanone in aq. soln. at pH 2 and 80° (15, 452).
Props.: UV spectrum (452).

(4-AMINOPHENYL)[4-(3-CARBETHOXY-5-HYDROXY-2-METHYL-4-PYRRYLAZO)PHENYL]ARSINIC ACID C$_{20}$H$_{21}$AsN$_4$O$_5$

Prepn.: by diazotizing (4-H$_2$NC$_6$H$_4$)$_2$AsO$_2$H in dild. HCl at 0-5°, satg. the diazonium salt with NaOAc, and adding the soln. to 3-carbethoxy-5-hydroxy-2-methylpyrrole in 96% EtOH (870).
Props.: Yellow powder (870).

BIS[4-(3-CARBETHOXY-2,4-DIMETHYL-5-PYRRYLAZO)PHENYL]ARSINIC ACID C$_{28}$H$_{33}$AsN$_6$O$_6$

Prepn.: by diazotization of (4-H$_2$NC$_6$H$_4$)$_2$AsO$_2$H in concd. HCl, followed by coupling with 3-carbethoxy-2,4-dimethylpyrrole in EtOH contg. NaOAc (871).
Props.: Dark orange microcyrstals, m.151° (decompn.). (871).

TABLE XLVI
ARSINIC ACIDS

R₂AsO₂H or RR'AsO₂H	Prepd. from	Method	Yield %	Props.	Ref.
C₂ (ClCH₂)₂AsO₂H	R₂AsCl + H₂O₂	Id		Sinters 102°, m. 117-126° (dec.)	121
C₃ Me(ClCH:CH)AsO₂H	RR'AsCl + H₂O₂	Id		m. 149-150°	261
Me(HOCH₂CH₂)AsO₂H	RAsO + $\overline{CH_2CH_2O}$	Io		Oil, decomp. at elevated temp., rxn.[1]	715
C₄ Et₂AsO₂H	R₂AsOEt + SeO₂	Ic	1.5	m. 136-137°, rxn.[3]	649
MePrAsO₂H				Rxn.[4]	193
C₅ EtPrAsO₂H				Rxn.[4]	193
C₆ (NCCH₂CH₂)₂AsO₂H	R₂AsCl + HNO₃	Id	80	m. 142° (dec.), R₂As(O)ONO₂, fine needles, m. 113-116°	243
C₇ Me(3-O₂NC₆H₄)AsO₂H	RAsCl₂ + 3-O₂NC₆H₄NH₂	Ij	50	Light yellow solid, m. 223-223.5°, rxn.[2]	251
	MePhO₂H + HNO₃-H₂SO₄		30		251
Me[3,4-H₂N(HO)C₆H₃]AsO₂H	Me[3,4-O₂N(HO)C₆H₃]AsO₂H + Fe(OH)₂		71	m. 233-234°, rxns.[2], 5	93
C₈ (SCE:CHCH:C)₂AsO₂H	R₃AsO + NaOH	IIa		m. 170° with decompn. to (R₂As)₂O, m. 315-316°	349

1. Rxn. with SO₂ in H₂O in the presence of KI → (RR'As)₂O (715).
2. Rxn. with SO₂ in concd. HX (X = Cl or Br) in the presence of KI → the corresponding RR'AsX (64, 93, 95, 96, 179, 199, 242, 245, 251, 455, 658, 805, 957, 1336).
3. Rxn. with ClP(S)(OEt)[OC₆H₃-3,4-(NO₂)Cl], using Et₂AsO₂Na → Et₂As(O){OP(S)(OEt)[OC₆H₃-4,3-(NO₂)Cl]} (1102).
4. Rxn. with Penicillium brevicaule in bread cultures → RR'AsMe (193).
5. Rxn. with H₃PO₂ in the presence of HI → the corresponding arseno deriv. (64, 94, 95).

TABLE XLVI (cont'd.)
ARSINIC ACIDS

R_2AsO_2H or $RR'AsO_2H$	Prepd. from	Method	Yield %	Props.	Ref.
$(QCH{:}CHCHC)_2AsO_2H$	$R_3AsO + NaOH$	IIa		Rxn.[5]	346
	$R_3As + H_2O_2$	In		Colorless needles, m. 141°	346-7
	$(R_2As)_2O + H_2O_2$	Ia			346-7
	$R_2AsCl + H_2O_2$	Id	53		347
$(ClCH{:}CH)PhAsO_2H$	$R_2AsCl + Cl_2$	Ie		m. 138°	753
$(HO_2CCH_2)(4\text{-}H_2NC_6H_4)AsO_2H$	$RR'AsCl + H_2O_2$	Id		m. 135°	263
				Rxn.[6]	452
$Me[3,4\text{-}O_2N(MeO)C_6H_3]AsO_2H$	$R'AsO + MeI$	Ii	97	m. 216-217°	93
$EtPhAsO_2H$	$(RR'As)_2O + HNO_3$	Ia		Rxns.[7,8]	640
$Me(4\text{-}MeC_6H_4)AsO_2H$	$RR'As(OH)_2Cl$	IId		m. 151°	144
C9					
$(CH_2{:}CHCH_2)PhAsO_2H$				Rxn.[9]	
C10					
$(CH{:}NCH{:}CHCHC)_2AsO_2H$	$R_2AsCl \cdot 2HCl$	If	83	Needles, m. 143-145.5°, rxn.[2]	951
$[HO_2C(CH_2)_3]PhAsO_2H$	$PhAsCl_2 + Br(CH_2)_3CN$	Ii	89	Colorless crystals, m. 201-202° (dec.); rxn.[2]	242
$[-CH_2(Bu)AsO_2H]_2$	$BuAsCl_2 + BrCH_2CH_2Br$	Ia			455
C11					
$Me(1\text{-}C_{10}H_7)AsO_2H$	$RR'AsCl + H_2O_2$	Id		m. 188.6°, rxn.[10]	1156
C12					
$(2\text{-}BrC_6H_4)_2AsO_2H$	$AsCl_3 + RN_2BF_4$ in 80% EtOH	Il	12	m. 275-278°	396
	in abs. EtOH	Il	5		396
$(4\text{-}BrC_6H_4)_2AsO_2H$	$R_2AsCl + H_2O_2$	Id		m. 186-187°, rxn.[2]	1336
	$AsCl_3 + RN_2BF_4$	Il	3	m. 189-190°	396
$(4\text{-}ClC_6H_4)(4\text{-}O_2NC_6H_4)AsO_2H$	$AsCl_3 + (RN_2)_2ZnCl_4$	Il	3		394
$(4\text{-}ClC_6H_4)(4\text{-}O_2NC_6H_4)AsO_2H$	$R'As(ONa)_2 + RN_2Cl$	Ik		Needles, decomp. 116-117°	1187
$(4\text{-}ClC_6H_4)[4,3\text{-}HO(O_2N)C_6H_3]AsO_2H$	$R'AsCl_2 + RN_2Cl$	Ij		Yellow needles, m. 221-222°, rxn.[11]	1188
$(2\text{-}ClC_6H_4)_2AsO_2H$	$AsCl_3 + RN_2BF_4$	Il	13	m. 245-248°	295
$(3\text{-}ClC_6H_4)_2AsO_2H$	$AsCl_3 + RN_2BF_4$	Il	18	m. 150-152°	295
$(4\text{-}ClC_6H_4)_2AsO_2H$	$AsCl_3 + RN_2BF_4$	Il	14	m. 177-179°	295
$(4\text{-}FC_6H_4)_2AsO_2H$	$R_2AsCl + H_2O_2$	Id		m. 136-138°	1336

Compound	Synthesis		Yield (%)	Properties	Ref.
$(2\text{-}O_2NC_6H_4)_2AsO_2H$	$AsCl_3 + RN_2BF_4$ $Na_3AsO_3 + RN_2X$	I1 Im	17	m. 218°; polarography, rxns. 12, 13	829
$(3\text{-}O_2NC_6H_4)_2AsO_2H$				m. 255° m. 244°; polarography	295 695 829
$(3\text{-}O_2NC_6H_4)(4\text{-}O_2NC_6H_4)AsO_2H$	$R'PhAsO_2H + HNO_3$		100	m. 230-232°, rxn. 12	93
$(2\text{-}O_2NC_6H_4)(4\text{-}O_2NC_6H_4)AsO_2H$				m. 230°, polarography	829
$(2\text{-}O_2NC_6H_4)(3\text{-}O_2NC_6H_4)AsO_2H$				m. 248-250°, polarography m. 257°, polarography	829 829
$(3\text{-}O_2NC_6H_4)[4,3\text{-}HO(O_2N)\text{-}C_6H_3]AsO_2H$	$Ph(4\text{-}HOC_6H_4)AsO_2H + HNO_3\text{-}H_2SO_4$		95	m. 195-196°	93
$(4\text{-}O_2NC_6H_4)[4,3\text{-}HO(O_2N)\text{-}C_6H_3]AsO_2H$	$R'AsCl_2 + RN_2Cl$	Ij		Yellow, m. 215°	1188
$[4,3\text{-}HO(O_2N)C_6H_3]_2AsO_2H$ $Ph(4\text{-}BrC_6H_4)AsO_2H$	$PhAsO + R'N_2Cl$ $PhAsCl_2 + R'N_2BF_4$ $RR'AsCl + H_2O_2$	Ik Ij Id	13	Rxns. 2, 12 m. 184-185° m. 191-198° Rxns. 2	93 396 908, 1066 908 1066
$Ph(2\text{-}ClC_6H_4)AsO_2H$	$RAs(ONa)_2 + R\,N_2Cl$	Ik		m. 176° m. 184-185° Columns, m. 188°	494, 1187
$Ph(4\text{-}ClC_6H_4)AsO_2H$	$RAs(ONa)_2 + R'N_2Cl$ $RR'AsCl + H_2O_2$	Ik Id	28	m. 161°, m. 162-163° m. 161-162° m. 189-190°	494, 805 908 1187

6. Rxn. with 2,4,5-triamino-6-hydroxypyrimidine + $Br_2CHCOCH_2Br \rightarrow 4\text{-}[N\text{-}[(2\text{-amino-4-hydroxy-6-}$ pteridyl)methyl]amino]phenylarsinicoacetic acid (452).

7. Rxn. with NH_4OH, followed by Ag $NO_3 \rightarrow EtPhAsO_2Ag$ (640).

8. Rxn. of $EtPhAsO_2Ag + EtI \rightarrow EtPhAsO_2Et$, b. 88-89°, d8 1.2091, n_D^{20} 1.5520 (640).

9. Rxn. with $AcCl \rightarrow PhAsCl_2 + CH_2{:}CHCH_2OAc$ (650).

10. Rxn. with concd. HCl or HBr $\rightarrow$ cleavage of $C_{10}H_8$ (1156).

11. Rxn. with SO_2 in 10% H_2SO_4 + KI $\rightarrow RR'AsOH$ (1189, 1191, 1242, 1245).

12. Redn. with $Fe(OH)_2$ in alk. soln. $\rightarrow$ redn. of the NO_2 group or groups (64, 95, 547, 1189).

13. Electrolytic redn. $\rightarrow$ the diamino arsinic acid (547).

TABLE XLVI (cont'd.)

ARSINIC ACIDS

R_2AsO_2H or $RR'AsO_2H$	Prepd. from	Method	Yield %	Props.	Ref.
$Ph(2\text{-}IC_6H_4)AsO_2H$	$RAs(ONa)_2 + R'N_2Cl$	Ik		Colorless crystals, m. 211-214°, salt[14]	243
$Ph(4\text{-}IC_6H_4)AsO_2H$	$RAs(ONa)_2 + R'N_2Cl$	Ik		Needles, decomp. 196°, Rxns.[2, 11, 12, 15]	1187
$Ph(2\text{-}O_2NC_6H_4)AsO_2H$	$(RR'As)_2O + HCl + H_2O$	Id		Needles, m. 199°	908
$Ph(3\text{-}O_2NC_6H_4)AsO_2H$	$RAsCl_2 + R'N_2Cl$	Ij		Needles, m. 154-155°, rxn.[12]	1188
$Ph(4\text{-}O_2NC_6H_4)AsO_2H$	$PhAsO + R'N_2Cl$	Ik	51	IR spectrum and H bonding	93
	$RR'AsCl + H_2O_2$	Id		m. 173°, rxn.[11, 12, 16]	119a
	$RAs(ONa)_2 + R'N_2Cl$	Ik		Yellow needles, m. 180°	908
	$RAsCl_2 + R'N_2Cl$	Ij		m. 181°	1187
$Ph[3,4\text{-}O_2N(HO)C_6H_3]AsO_2H$	$RAsCl_2 + R'N_2Cl$	Ij		Yellow, m. 167°, rxn.[11]	1188
$Ph(4\text{-}ClNHO_2SC_6H_4)AsO_2H$	$RAsCl_2 + R'N_2Cl$	Ij		m. 160-161° (dec.)	1188
	$Ph(4\text{-}H_2NO_2SC_6H_4)AsO_2H + NaOCl$				957
$(4\text{-}H_2NC_6H_4)(4\text{-}O_2NC_6H_4)AsO_2H$	$(4\text{-}AcNHC_6H_4)(4\text{-}O_2NC_6H_4)AsO_2H + 25\% HCl$		90	Yellow needles, m. 239°, rxn.[5]	64
$Ph(2\text{-}HOC_6H_4)AsO_2H$	$Ph(2\text{-}H_2NC_6H_4)AsO_2H$ by diazotization			m. 221-223°, rxn.[17]	93
$Ph(3\text{-}HOC_6H_4)AsO_2H$	$Ph(3\text{-}H_2NC_6H_4)AsO_2H$ by diazotization			m. 230-232°, rxns.[5, 17]	93
$Ph(4\text{-}HOC_6H_4)AsO_2H$	$RAs(ONa)_2 + R'N_2Cl$	Ik		Plates, decomp. 260-265°, rxns.[11, 18]	1187
$(2\text{-}HOC_6H_4)_2AsO_2H$	$PhOH + H_3AsO_4$	IIf		m. 209-210°	1325
$(4\text{-}HOC_6H_4)_2AsO_2H$	$Na_3AsO_3 + RN_2Cl$	Im		m. 238-239° (dec.), rxn.[11]	695
$(3\text{-}HOC_6H_4)(4\text{-}HOC_6H_4)AsO_2H$	$(3\text{-}H_2NC_6H_4)(4\text{-}H_2NC_6H_4)AsO_2H$ by diazotization			m. 210-211°	93
$Ph(4\text{-}HO_3SC_6H_4)AsO_2H$	$RAsCl_2 + R'N_2Cl$	Ij		m. 129-130°, rxns.[2, 11, 19, 20]	957
$Ph(2\text{-}H_2NC_6H_4)AsO_2H$	$Ph(2\text{-}O_2NC_6H_4)AsO_2H + Fe(OH)_2$			m. 132-133°	93
$Ph(3\text{-}H_2NC_6H_4)AsO_2H$	$Ph(3\text{-}O_2NC_6H_4)AsO_2H + Fe(OH)_2$			m. 210-212°, rxn.[2]	1189
$Ph(4\text{-}H_2NC_6H_4)AsO_2H$	$Ph(4\text{-}O_2NC_6H_4)AsO_2H + Fe(OH)_2$			Needles, m. 219-221°	93, 1189
	$Ph(4\text{-}O_2NC_6H_4)AsO_2H + H_2$ over Pd/C			m. 219-221°, rxns.[2, 21, 22]	1228
$Ph[2,4\text{-}H_2N(HO)C_6H_3]AsO_2H$	$Ph[2,4\text{-}O_2N(HO)C_6H_3]AsO_2H + Fe(OH)_2$			Powder, decomp. 203-206°	1189

Ph(4-H$_2$NO$_2$SC$_6$H$_4$)AsO$_2$H	Ik	RAsO + R'N$_2$Cl	rxns.11, 18 m. 229-231°, rxns.2,23-4	957
(3-H$_2$NC$_6$H$_4$)$_2$AsO$_2$H			m. 247°, polarography, rxns25	830
(3-H$_2$NC$_6$H$_4$)(4-H$_2$NC$_6$H$_4$)AsO$_2$-H		(3-O$_2$NC$_6$H$_4$)(4-O$_2$NC$_6$H$_4$)As-O$_2$H + Fe(OH)$_2$	m. 176-178°, rxn.19	93
		RAsO$_3$H$_2$ + RNH$_2$ at 220°	m. 182°; polarography	830
			Crystal structure	484
(4-H$_2$NC$_6$H$_4$)$_2$AsO$_2$H		(4-AcNHC$_6$H$_4$)(4-H$_2$NC$_6$H$_4$)AsO$_2$-H + 5N HCl	Colorless microprisms, m. 248°	64
	11		m. 223.5°; polarography Stability26;rxn2,5,11,22,27	830
			Needles, m. 248-249°,rxn.21	1228
(2-H$_2$NC$_6$H$_4$)$_2$AsO$_2$H		(4-O$_2$NC$_6$H$_4$)$_2$AsO$_2$H + H$_2$ over Pd/C (4-O$_2$NC$_6$H$_4$)$_2$AsO$_2$H + Fe(OH)$_2$ or electrolyt.		547
[3,4-H$_2$N(HO)C$_6$H$_3$]$_2$AsO$_2$H		[3,4-O$_2$N(HO)C$_6$H$_3$]$_2$AsO$_2$H + Fe(OH)$_2$	m. 154°, polarography	830
			Light brown crystals, rxns.2,5	95
Ph[HO$_2$C(CH$_2$)$_5$]AsO$_2$H	Ii	RAsCl$_2$ + Br(CH$_2$)$_5$CN	decomp. 218°, rxns.2,5	
			Colorless crystals, m. 101.5-103°	242

C$_{13}$

(4-O$_2$NC$_6$H$_4$)(4-HO$_2$CC$_6$H$_4$)AsOH	Ij	RAsCl$_2$ + R'N$_2$Cl	Decomp. 275°, rxn.11	1188
Ph(3-OHCC$_6$H$_4$)AsO$_2$H	Ik	RAsO + R'N$_2$Cl	Colorless prisms, m. 168°, deriv.28	494
Ph(3-HO$_2$CC$_6$H$_4$)AsO$_2$H	Ik	RAsO + R'N$_2$Cl	Yellow prisms, m. 215°	494

14. S-Benzylthi uronium salt, colorless crystals, m. 162°(243).
15. Rxn. with Zn + HgCl$_2$ in MeOH → Ph(2-H$_2$NC$_6$H$_4$)AsH (243).
16. Catalytic redn. over Pd/C in AcOH → the 4-amino deriv. (1228).
17. Rxn. with Me$_2$SO$_4$ → the methoxy deriv. (93).
18. Rxn. with HSR" → RR'AsSR" (1244).
19. Rxn. with NaNO$_2$, followed by hydrolysis → the corresponding hydroxy deriv. (93).
20. Rxn. with Zn-Hg in NaOH → the corresponding secondary arsine (94).
21. Rxn. with NaNO$_2$ in HCl, followed by coupling with PhNMe$_2$ → the corresponding 4-(4-dimethylamino-phenylazo)phenyl derivs. (1228).
22. Rxn. with (ClCN)$_3$ in aq. alk. → the N-dichloro-s-triazinyl deriv. (658, 1245).
23. Rxn. with HI in AcOH → RR'AsI (93, 957).
24. Rxn. with 1% NaOH + alk. NaOCl → Ph(ClNHO$_2$SC$_6$H$_4$)AsO$_2$H (957).
25. Rxn. with COCl$_2$ and NH$_4$OH → (3-H$_2$NCONHC$_6$H$_4$)$_2$AsO$_2$H (887a).
26. Stable at 160° in alk. soln., but is hydrolyzed by 0.4-0.8 N HCl (964).
27. Rxn. with NaNO$_2$, followed by pyrrole derivs. → bis(4-pyrrylazophenyl) derivs. (870, 871).

TABLE XLVI (cont'd.)
ARSINIC ACIDS

R₂AsO₂H or RR'AsO₂H	Prepd. from	Method	Yield %	Props.	Ref.
Ph(4-HO₂CC₆H₄)AsO₂H	RAsO + R'N₂Cl	Ik		Needles, m. 229-231°	494
	RAsCl₂ + R'N₂Cl	Ij		Columns, decomp. 237-239°	1188
	R(4-MeC₆H₄)AsO₂H + KMnO₄			White powder, rxn.11	1189
(2-ClC₆H₄)(2-MeC₆H₄)AsO₂H	R'AsO + RN₂Cl	Ik		Colorless prisms, m. 218-224°	494
(2-ClC₆H₄)(4-MeC₆H₄)AsO₂H	R'AsO + RN₂Cl	Ik		Colorless prisms, m. 170-172°	494
(3-ClC₆H₄)(4-MeC₆H₄)AsO₂H	R'AsO + RN₂Cl	Ik		Colorless prisms, m. 124-125°	494
(3-ClC₆H₄)(2-MeC₆H₄)AsO₂H	R'AsO + RN₂Cl	Ik		Colorless prisms, m. 138°	494
(4-ClC₆H₄)(2-MeC₆H₄)AsO₂H	R'AsO + RN₂Cl	Ik		Colorless prisms, m. 131-134°	494
(4-ClC₆H₄)(4-MeC₆H₄)AsO₂H	R'AsO + RN₂Cl	Ik		Colorless needles, m. 144°	494
(2-O₂NC₆H₄)(2-MeC₆H₄)AsO₂H	R'AsO + RN₂Cl	Ik		Yellow prisms, m. 220-221°	494
(3-O₂NC₆H₄)(2-MeC₆H₄)AsO₂H	R'AsO + RN₂Cl	Ik		Colorless prisms, m. 183-186°	494
(4-O₂NC₆H₄)(2-MeC₆H₄)AsO₂H	R'AsO + RN₂Cl	Ik		Colorless prisms, m. 169-171°	494
(4-O₂NC₆H₄)(4-MeC₆H₄)AsO₂H	R'AsO + RN₂Cl	Ik		Colorless needles, m. 142-144°	494
Ph[4,3-Me(O₂N)C₆H₃]AsO₂H	RAsCl₂ + R'N₂Cl	Ij		Yellow needles, m. 167°	1188
Ph(2-MeC₆H₄)AsO₂H	RAsCl₂ + R'N₂Cl	Ij		Granules, m. 168°	1188
	R'AsCl₂ + RN₂Cl	Ij			1188
Ph(3-MeC₆H₄)AsO₂H	RAsO + R'N₂Cl	Ik		Colorless prisms, m. 163°	494
Ph(4-MeC₆H₄)AsO₂H	RAsO + R'N₂Cl	Ik		Colorless prisms, m. 117-119°	494
Ph(4-MeC₆H₄)AsO₂H	RAsO + R'N₂Cl	Ik	40	m. 150-152°	805, 1187
	RR'AsCl + H₂O₂	Id		m. 162°, rxns.2, 29	908
	RAsCl₂ + R'N₂Cl	Ij		Needles, m. 159°	1188
Ph(PhCH₂)AsO₂H	Ph(2-HOC₆H₄)AsO₂H + MeSO₄			IR spectrum and H-bonding rxn.23	119a
Ph(2-MeOC₆H₄)AsO₂H	Ph(3-HOC₆H₄)AsO₂H + MeSO₄			m. 187-188°	93
Ph(3-MeOC₆H₄)AsO₂H				m. 120-121°	93
Ph[Me₂CHCH(CO₂H)NHCOCH₂]-AsO₂H	RAsCl₂ + R'Cl	Ii		m. 188° (dec.)	75

336

C14					
$(4\text{-}HO_2CC_6H_4)_2AsO_2H$	$(4\text{-}MeC_6H_4)_2AsO_2H + KMnO_4$			Anhydride, needles, de-comp. 240°, rxn.[11]	1189
$(4\text{-}AcNHC_6H_4)(4\text{-}O_2NC_6H_4)AsO_2H$	$RAsO + R'AsCl$	Ik	52	Potentiometric titration	1186
				Cream-colored needles, softens 245-259°, m. 262° (dec.), Na salt sparingly sol. in H_2O; rxns[5;12;30-1]	64
$[3,4\text{-}O_2N(MeO)C_6H_3]_2AsO_2H$	$(4\text{-}MeOC_6H_4)_2AsO_2H + HNO_3$			m. 321°, rxn.[12]	95
$Ph(2\text{-}MeO_2CC_6H_4)AsO_2H$	$RAsCl_2 + R'N_2Cl.FeCl_3$				908
$Ph(2\text{-}AcNHC_6H_4)AsO_2H$	$Ph(2\text{-}O_2NC_6H_4AsO_2H + Fe\text{-}(OH)_2 + Ac_2O\text{-}AcOH$	Id	20	m. 193-195°	494
$(4\text{-}H_2NC_6H_4)(4\text{-}AcNHC_6H_4)AsO_2H$	$(4\text{-}O_2NC_6H_4)(4\text{-}AcNHC_6H_4)AsO_2H + Fe(OH)_2$	Ik	83.6	Colorless crystals, m. 279° (dec.), rxn.[5,30]	64
$(3\text{-}H_2NCONHC_6H_4)_2AsO_2H$	$(3\text{-}H_2NC_6H_4)_2AsO_2H + COCl_2 + NH_4OH$			m. >250°	887a
$(2\text{-}MeC_6H_4)_2AsO_2H$	$RAsO + RN_2Cl$	Ik		Colorless prisms, m. 197°	494
	$RAsCl_2 + RN_2Cl$	Ij		Needles, m. 202°	1188
$(3\text{-}MeC_6H_4)_2AsO_2H$	$RAsO + RN_2Cl$	Ik		Colorless needles, m. 140-143°	494
$(2\text{-}MeC_6H_4)(3\text{-}MeC_6H_4)AsO_2H$	$RAsO + R'N_2Cl$	Ik		Colorless plates, m. 129-131°	494
$(4\text{-}MeC_6H_4)_2AsO_2H$	$RAsCl_2 + RN_2Cl$	Ij		Needles, m. 167°, rxn.[29]	1188
$(3\text{-}MeC_6H_4)(4\text{-}MeC_6H_4)AsO_2H$	$R'AsO + RN_2Cl$	Ik		Yellow prisms, m. 145-146°	494
$(PhCH_2)_2AsO_2H$	$RMgCl + (BuO)_3As$	Ip		m. 208-210°	727
$Ph(PhCH_2CH_2)AsO_2H$	$RAsCl_2 + R'Br$	Il		m. 142-143°, rxns.[32, 33]	614
$Et(2\text{-}PhC_6H_4)AsO_2H$	$R'AsCl_2 + RBr$	Il		Props.[34]; rxn.[2]; salt[35]	245
$Ph(2,4\text{-}Me_2C_6H_3)AsO_2H$	$RAsO + R'N_2Cl$	Ik		Colorless needles, m. 165-167°	494
$Ph(2\text{-}EtOC_6H_4)AsO_2H$	$(RR'As)_2O + H_2O_2$	Ia		m. 182°	908
$Ph(4\text{-}EtOC_6H_4)AsO_2H$	$RAsCl_2 + (R'N_2Cl)_2 ZnCl_2$	Id		m. 158°	1066

28. Dinitrophenylhydrazone, yellow, m. 270° (dec.) (494).
29. Rxn. with $KMnO_4$ → oxidation of the Me group to $-CO_2H$ (1189).
30. Rxn. with 25% HCl → hydrolysis of the AcNH group to NH_2 (64).
31. Rxn. with SO_2 in HCl-AcOH + HI → $(RR'As)_2O$ (64).
32. Rxn. with H_2SO_4 at 100°, followed by SO_2 in aq. HCl contg. KI → $\overline{C_6H_4CH_2CH_2AsPh}$ (614).
33. Rxn. with SO_2 in aq. HCl contg. KI → $PhAsCl_2 + Ph(PhCH_2CH_2)AsCl$ (614).
34. Hemihydrate crystallizes in two forms; α-modification, hexagonal plates, m. 160-161°, β-modification, long crystals, m. 151-152°; the two forms can be inversely converted into each other (245).
35. S-(p-Chlorobenzyl)thiuronium salt with 2.5 H_2O, m. 84-85° (245).

TABLE XLVI (cont'd.)
ARSINIC ACIDS

R_2AsO_2H or $RR'AsO_2H$	Prepd. from	Method	Yield %	Props.	Ref.
$(4-MeCC_6H_4)_2AsO_2H$	$R_2AsBr + H_2O_2$ or $KMnO_4$	Id		m. 190-191°, rxns.[2], 36	95
$[-CH_2(Ph)AsO_2H]_2$	$PhAsCl_2 + (-CH_2Br)_2$	Ii		m. 210-212°, rxn.[2]	199
	$[CH_2(Ph)AsO]_2O + acid$			m. 200° (effervesc.)	609
$[3,4-H_2N(MeO)C_6H_3]_2AsO_2H$	$[3,4-O_2N(MeO)C_6H_3]_2AsO_2H + Fe(OH)_2$			Crystals, m. 183-184°	95
C_{15}					
$Ph[4-N:C(Cl)N:C(Cl)N:CNHC_6H_4]AsO_2H$	$Ph(4-H_2NC_6H_4)AsO_2H + (Cl-CN)_3$			Rxns.[2], 37, 38	658, 1245
$Ph[4-N:C(Cl)N:C(OH)N:CNHC_6H_4]AsO_2H$	$Ph[N:C(Cl)N:C(Cl)N:CNHC_6H_4]-AsO_2H + 10\% NaOH$			Rxns.[2], 39	658, 1245
$Ph[4-N:C(NH_2)N:C(Cl)N:CNH-C_6H_4]AsO_2H$	$Ph[4-N:C(Cl)N:C(Cl)N:CNH-C_6H_4]AsO_2H + NH_4OH$				1245
$Ph[4-N:C(NH_2)N:C(OH)N:CNHC_6H_4]AsO_2H$	$Ph[4-N:C(Cl)N:C(OH)N:CNH-C_6H_4]AsO_2H + NH_4OH$ under pressure with heat			Rxn.[2]	1245
$(2-MeC_6H_4)(2,4-Me_2C_6H_3)As-O_2H$	$RAsO + R'N_2Cl$	Ik		Colorless needles, m. 184-185°	494
$(4-MeC_6H_4)(2,4-Me_2C_6H_3)As-O_2H$	$RAsO + R'N_2Cl$	Ik		Colorless needles, m. 138-140°	494
$Ph(PhCH_2OCH_2CH_2)AsO_2H$	$RAsCl_2 + R'Cl$	Ii	82	m. 109°; salts[40]	242
C_{16}					
$Ph(1-C_{10}H_7)AsO_2H$	$RAsO + R'N_2Cl$	Ik		m. 181-183°	494
$Ph(2-C_{10}H_7)AsO_2H$	$RAsO + R'N_2Cl$	Ik		m. 157-159°	494
	$(RR As)_2O + H_2O_2$	Ia		m. 182°	908
$(4-AcC_6H_4)_2AsO_2H$	$RAs(OH)_2 + RN_2Cl$	Ik		m. 185-186°, rxn.[2]	1118
$S[CH_2CH_2As(Ph)O_2H]_2$	$PhAsO + S(CH_2CH_2Br)_2$	Ii	64	m. 189°	44
$SO_2[CH_2CH_2As(Ph)O_2H]_2$	$PhAsO + SO_2(CH_2CH_2Br)_2$	Ii	87	m. 197°	44
C_{17}					
$(4-MeC_6H_4)(1-C_{10}H_7)AsO_2H$	$RAsO + R'N_2Cl$	Ik		Colorless plates, m. 198-200°	494
$(4-MeC_6H_4)(2-C_{10}H_7)AsO_2H$	$RAsO + R''N_2Cl$	Ik		Pink prisms, m. 160-162°	494
C_{18}					
$\{4-[N:C(Cl)N:C(Cl)N:CNH]-C_6H_4\}_2AsO_2H$	$(4-H_2NC_6H_4)_2AsO_2H + (Cl-CN)_3$			Rxns.[41, 42]	1245
$\{4-[N:C(Cl)N:C(OH)N:CNH]-C_6H_4\}_2AsO_2H$	$\{4-[N:C(Cl)N:C(Cl)N:CNH]-C_6H_4\}_2AsO_2H + NaOH$			Rxns.[42]	1245

338

Ph[2-(4-O2NC6H4)C6H4]AsO2H	RAsCl2 + R'N2Cl	Ik		m. 241-244[3]	179
	RAsCl2 + R'N2Cl	Ij	32	Rxn.[43]	179
{4-[N:C(NH2)N:C(Cl)N:CHN]-C6H4}2AsO2H	{4-[N:C(Cl)N:C(Cl)N:CNH]-C6H4}2AsO2H + NH4OH			See	1245
{4-[N:C(OH)N:C(Cl)N:CNH]-C6H4}2AsO2H	{4-[N:C(Cl)N:C(Cl)N:CNH]-C6H4}2AsO2H + NaOH			See[41], rxn. 11	1245
Ph(4-PhC6H4)AsO2H	RAsO. + R'N2Cl	Ik	12	Rxn.[2]	805
{4-[N:C(NH2)N:C(OH)N:CNH]-C6H4}2AsO2H	{4-[N:C(NH2)N:C(Cl)N:CNH]-C6H4}2AsO2H + NH4OH			See[42], rxn. 11	1245
{4-[N:C(NH2)N:C(Cl)N:CNH]-C6H4}2AsO2H	{4-[N:C(Cl)N:C(Cl)N:CNH]-C6H4}2AsO2H + NH4OH			See[42], rxn. 11	1245
C19 (4-ClC6H4)(2-PhC6H4CH2)AsO2H	RAsCl2 + R'Br	Ii	90	S-p-chlorobenzylthiuronium salt, m. 135°	244
Ph(2-PhC6H4CH2)AsO2H	RAsCl2 + R'Br	Ii	93	m. 138-140°; salt[44]	244
C20 (4-HO2CC6H4)[2-(4'-MeOC6H4)]C6H4]AsO2H	RAsCl2 + R'N2Cl	Ij		m. 280°; rxns.[2,45]	179
Ph[4-(4-Me2NC6H4N2)C6H4]AsO2H	Ph(4-H2NC6H4)AsO2H + NaNO2 + PhNMe2			Salt[46]	1228
C24 [2-(4-O2NC6H4)C6H4]2AsO2H	RAsO3H2 + POCl3	IIe		m. >260°	377
C28 [4-(4-Me2NC6H4N2)C6H4]2-AsO2H	(4-H2NC6H4)2AsO2H + NaNO2 + PhNMe2			m. >300°, salt[47]	1228

36. Rxn. with HNO3-H2SO4 → the 3,3'(O2N)2 deriv. (95).
37. Rxn. with 10% NaOH at 40° → hydrolysis of one Cl group (658, 1245).
38. Rxn. with NH4OH at 30° → the 4-chloro-6-hydroxy-s-triazinyl deriv. (1245).
39. Rxn. with NH4OH under pressure → the 4-amino-6-hydroxy-s-triazinyl deriv. (1245).
40. Na salt trihydrate, m. 77°, remelts 249° (dec.); hydrochloride, RR'As(OH)2Cl, plates, m. 94° (242).
41. Rxn. with aq. NaOH at 40° → the bis(4-chloro-6-hydroxy-s-triazinyl) deriv.; and with aq. NaOH under reflux → the bis(4,6-dihydroxy-s-triazinyl) deriv. (1245).
42. Rxn. with NH4OH at 30° → the bis(4-amino-6-chloro-s-triazinyl) deriv.; and with NH4OH under pressure with heating → the bis(4,6-diamino-s-triazinyl deriv. (1245).
43. Heating with H2SO4 at 140-150° → 3-nitro-5-phenyldibenzarsenole As-oxide (179).
44. S-p-Chlorobenzylthiuronium salt, m. 136-137° (244).
45. Rxn. with polyphosphoric acid, followed by SO2 → 4-(3-methoxy-5-dibenzarsenolyl)benzoic acid (179).
46. (RR'AsO2)4Zr.HCl (1228); useful in detection of fluorine (1229).
47. (R2AsO2)4Zr.2HCl (1228); useful in detection of fluorine (1229).

ORGANOARSENIC TETRAHALIDES

PHENYLARSENIC TETRACHLORIDE $C_6H_5AsCl_4$ $PhAsCl_4$
Rxn. with $LiAlH_4$ or $LiBH_4 \rightarrow PhAsH_2$ (1289).

PHENYLARSENIC TETRAFLUORIDE $C_6H_5AsF_4$ $PhAsF_4$
Prepn.: by reacting $PhAsO_3H_2$ with SF_4; high yield (1147a).
Props.: B. 52-53°/2 mm (1147a).

o-TOLYLARSENIC TETRACHLORIDE $C_7H_7AsCl_4$ $o\text{-}MeC_6H_4AsCl_4$
Prepn.: by reacting $o\text{-}MeC_6H_4AsCl_2$ with $PbCl_4$ (144).
Props.: viscous oil (144).

DIORGANOARSENIC TRIHALIDES

BIS(TRIFLUOROMETHYL)ARSENIC TRICHLORIDE $C_2AsCl_3F_6$ $(CF_3)_2AsCl_3$
Prepn.: by a month-long reaction of $(CF_3)_3As$ with Cl_2 (1:1) in a sealed tube at
r. t.; 30% yield (333).
Props.: b. 93-95°/722 mm; n_D^{20} 1.423 (333).
Rxn. with aq. NaOH $\rightarrow CHF_3$ (333).

DIPHENYLARSENIC TRIBROMIDE $C_{12}H_{10}AsBr_3$ Ph_2AsBr_3
Prepn.: by treating Ph_2AsH with Br_2 in Et_2O (1293).
Props.: Golden-yellow ppt., m. 128° (1293).

DIPHENYLARSENIC TRICHLORIDE $C_{12}H_{10}AsCl_3$ Ph_2AsCl_3
Props.: Inhibits corrosion of steel (30).
Rxn. with $LiAlH_4$ or $LiBH_4 \rightarrow Ph_2AsH$ (1289).

TRIORGANOARSINE DI- AND TETRAHALIDES, HYDROXIDE SALTS, DIHYDROXIDES, OXIDES, SULFIDES, AND SELENIDES

Triorganoarsine dihalides are prepared by:
1. Treating triorganoarsines with halogens in a sealed tube or in solvents, such as chloroform, carbon tetrachloride, diiodomethane, acetone, acetic acid or petroleum ether.
2. Treating triorganoarsines with lead tetrachloride.
3. Treating triorganoarsine oxides with concentrated hydrohalic acids.
4. Treating alkyltriarylarsonium iodides with chlorine to form triarylarsine chloride iodide.

The di- and tetrahalides are compiled in Table XLVII.

Triorganoarsine hydroxide salts, $R_3As(OH)X$, are prepared by:
1. Reacting tertiary arsines with nitric acid in ether to form triorganoarsine hydroxide nitrates.
2. Reacting triorganoarsine oxides with diluted nitric acid or with alcoholic or aqueous hydrochloric acid, or with sulfuric acid to form triorganoarsine hydroxide nitrates, chlorides, and sulfates, respectively.
3. Treating triorganoarsine dihydroxides with hydrochloric acid to form tri-

organoarsine hydroxide chlorides.

4. Treating triorganoarsine oxides with picric, styphnic, or picrolonic acid to form triorganoarsine hydroxide picrates, styphnates, and picrolonates, respectively.

The hydroxide salts are compiled in Table XLVIII.

Triorganoarsine dihydroxides, $R_3As(OH)_2$, are prepared by oxidation of tertiary arsines with halogens in aqueous alcohol or in anhydrous solvents, but in the latter case the triorganoarsine dihalides are hydrolyzed with concentrated alkali or with potassium permanganate in acetone. Some triorganoarsine oxides can be converted into the corresponding dihydroxides by hydration and, vice versa, the dihydroxides on drying in vacuo over a dehydration agent lose water and are converted into the oxides. Moreover, (4-ethylphenyl)phenyl-4-tolylarsine under those conditions yields the bis(triarylarsine hydroxide) oxide. The dihydroxides are compiled in Table XLIX.

Triorganoarsine oxides are prepared or formed by:

1. Direct and instantaneous oxidation of tertiary arsines with sulfuric acid at an elevated temperature.
2. Oxidation of tertiary arsines with halogens to tertiary arsine dihalides, followed by hydrolysis with alkali. Depending on the nature of substituents in the organic moieties, some triorganoarsine oxides crystallize from aqueous solutions, while others can be obtained by azeotropic removal of water or by dehydration of the corresponding triorganoarsine dihydroxides.
3. Oxidation of tertiary arsines by hydrogen peroxide at 100° in the absence of solvents or at a lower temperature in acetone, or by potassium permanganate or perbenzoic acid in acetone.
4. Reaction of diorganohaloarsines with alkyl halides in alkaline solutions.
5. Reaction of tertiary arsines with chloramine-T, followed by hydrolysis.
6. Oxidation of tertiary arsines with mercury oxide in ethanol or with selenium dioxide in benzene.
7. Metathesis of the so called "ylidene"-triorganoarsenic compounds, $R_3As:CR'R''$ with carbonyl compounds or by hydrolysis of the "ylidene" derivatives.
8. Oxidation of the $R_3AsAlCl_3$ complexes by air.

The oxides are compiled on page 347 and in Table L.

Triorganoarsenic sulfides are prepared by:

1. Treating triorganoarsine oxides with hydrogen sulfide.
2. Treating triorganoarsine hydroxide halides with hydrogen sulfide.
3. Treating triorganoarsine dihalides with sodium or hydrogen sulfide.

The sulfides are compiled in Table LI.

Triphenylarsine selenide, Ph_3AsSe, was prepared by treating triphenylarsine with selenium dioxide in benzene at room temperature to form a mixture of Ph_3AsO and Ph_3AsSe (see page 353).

TABLE XLVII

TRIORGANOARSINE DI- AND TETRAHALIDES

R_3AsX_2	Prepd. from	Props.	Ref.
C3			
$(CF_3)_3AsCl_2$	$(CF_3)_3As + Cl_2$ in sealed tube at r.t.	B. 98.5°, n_D^{20} 1.386, decompn. Rxns. 2-8	333
$(CF_3)_3AsF_2$	$(CF_3)_3AsCl_2 + AgF$	B. 57-58°, rxn.[4]	333
Me_3AsBr_2	$Me_3As + Br$ in Me_2CO	Hydroscop. needles, decomp. 190-192°, rxns.[9-14]	1314
C6			
$(CH_2:CH)_3AsI_2$	$R_3As + I$ in CCl_4	Decomp. on heating in vacuo → AsI_3, $(CH_2:CH)_2AsI$, $CH_2:CHI$	773
C7			
$Me(NCCH_2CH_2)_2AsBr_2$	$RR'_2As + Br$	Deliquescent, m. 76°	243
Et_3AsBr_2	$R_3As + Br$ in CCl_4	at 100° in vacuo → Et_2AsBu, rxn.[14]	516
C9			
$MeEtPhAsCl_2$	$RR'R''As + PbCl_4$	Hydroscop. needles, m. 83°	144
C10			
$Me_2(PHCH:CH)AsCl_2$	$R_2R'As + Cl$ in CCl_4	Decompn.[15]	265
C12			
$(OCH:CHCH:C)_3AsCl_2$	By-prod. from $R_2AsCl + Cl_2$; $R_3As + Cl_2$	m. 132°, rxn.[16]; Colorless crystals, m. 128-129°, fumes in air	753, 346
$Me_2(1-C_{10}H_7)AsBr_2$	$R_2R'As + Br$ in CCl_4	m. 105-106° (dec.); dry distn. → $Me(1-C_{10}H_7)AsBr$	1156
$Me_2(1-C_{10}H_7)AsCl_2$	$R_2R'As + Cl$ in CCl_4	Decomp. 128-130°; dry distn. → $Me(1-C_{10}H_7)AsCl$	1155
$(Me_3SiCH_2)_3AsBr_2$	$R_3As + Br$ in petr. ether	m. 118-120°	1121
$(Me_3SiCH_2)_3AsI_2$	$R_3As + I$ in CH_2I_2	m. 118-120° (dec.)	1121
C13			
$CF_3Ph_2AsBr_2$	$RR'_2As + Br$ in CCl_4	Colorless, hydroscop. rhombohedrons, m. 110°, decomp.	255

MePh₂AsBr₄	RR'₂As + Br in CHCl₃	m. 63-64°	146
MePh₂AsI₂	RR'₂As + I in CHCl₃	m. 104° (dec.)	146
C₁₄			
Me₂(4-PhOC₆H₄)AsI₂	R₂R'As + I in petr. ether	m. 134°	271
[HO₂C(CH₂)₃]₂PhAsCl₂	R₂R'As + concd. HCl	m. 152°	242
C₁₇			
Ph₂(N:CHCH:CHCH:C)AsBr₂	R₂R'As + Br in AcOH	Deliquescent crystals, rxn.[17]	806
C₁₈			
Ph₃AsCl₂[18]	R₃As + Cl in CCl₄	at 230° in vacuo → Ph₂AsCl; Dipole moment, rxn.[19]	516 603
(cyclo-C₆H₁₁)₃AsCl₂	R₃As + Cl in CCl₄	At 230° in vacuo → (C₆H₁₁)₂-AsCl	515
C₃₆			
(2-PhC₆H₄)₃AsClI	R₃MeAsI + Cl	Yellow, m. 172-174° (dec.)	1318

$Ph_2(N{:}CHCH{:}CHCH{:}C)AsBr_2$

Ph_3AsCl_2

$(cyclo\text{-}C_6H_{11})_3AsCl_2$

$(2\text{-}PhC_6H_4)_3AsClI$

1. Decomposes at 125° → (CF₃)₂AsCl + CF₃AsCl₂ (333).
2. Rxn. with Hg → (CF₃)₃As + HgCl₂ (333).
3. Rxn. with AgF → (CF₃)₃AsF₂ (333).
4. Rxn. with aq. NaOH → CHF₃ (333).
5. Rxn. with H₂O → (CF₃)₃As + (CF₃)₂AsO₂H (335).
6. Rxn. with Ag₂O → (CF₃)₂AsCl, CO₂, F, and (CF₃)₃As (335).
7. Rxn. with AcOH → (CF₃)₂AsCl + CF₃Cl (335).
8. Rxn. with EtOH → (CF₃)₃As (335).
9. Rxn. with MeLi in Et₂O, followed by Br → Me₃AsBr₂ (1314).
10. Rxn. with MeLi in Et₂O → evolution of CH₄ + C₂H₆ (1314).
11. Rxn. with MeLi in Et₂O at -50 to -60°, followed by Ph₃B → (Me₄As)(BMePh₃) (1314).
12. Rxn. with MeLi in Et₂O at -50 to -60°, followed by I in C₆H₆ → Me₃AsI₂ (1314).
13. Rxn. with MeLi in Et₂O at -50 to -60°, followed by HCl → Me₄AsCl (1314).
14. Rxn. with H₂O → R₃As(OH)Br (947).
15. Decomp. on heating → 1-methylarsindole [via Me(PhCH:CH)AsCl] (265).
16. Rxn. with H₂O → R₃AsO (346).
17. Rxn. with alkali in iild. AcOH → R₂RAsO (346).
18. The compd. is apparently formed in the rxn. of PhN₂Cl with As in Me₂CO in the presence of CaCO₃ (1274).
19. Rxn. with LiBH₄ or LiAlH₄ → Ph₃As (1289).

TABLE XLVIII

TRIORGANOARSINE HYDROXIDE SALTS, $R_3As(OH)X$

$RR'R''As(OH)X$	Prepd. from	Props. and Remarks	Ref.
C_3			
$Me_3As(OH)NO_3$	$Me_3As + HNO_3$ in Et_2O	m. 127° m. 128-129°; dissocn. const.	1053 947
C_8			
$Me_2PhAs(OH)Cl$	$Me_2PhAs + PbCl_4$ in $CHCl_3$	Formed by partial hydrolysis	144
C_{10}			
$Me(NCH_2CH_2)PhAs(OH)NO_3$		m. 118-120°	811
C_{12}			
$MeEtPrAs(OH) \cdot$ picrate	$MeEtPrAs + HNO_3$, followed by picric acid	m. 108-109°	193
C_{13}			
$MePh_2As(OH)Cl$	$MePh_2AsO + EtOH-HCl$	m. 147°	146
$MePh_2As(OH)NO_3$	$MePh_2AsO + HNO_3$	m. 106-107°	1053
C_{14}			
$MePh(2-HO_2CC_6H_4)As(OH)Cl$ ·		m. 226-227°	623
$MePh(3-HO_2CC_6H_4)As(OH)Cl$		m. 148-150°	623
$MePh(4-HO_2CC_6H_4)As(OH)Cl$		m. 150-152°	623
C_{18}			
$Ph_3As(OH)Cl$	$Ph_3As + HNO_3$	Dipole moment	603-4
$Ph_3As(OH)NO_3$		m. 124-126°	1053
C_{21}			
$(N:CHCH:CHCH:C)_3As(OH) \cdot$ picrate		m. 144-147°	806
$(2-MeC_6H_4)_3As(OH)Cl$	$(2-MeC_6H_4)_3As(OH)_2 + 5\%$ HCl	Colorless crystals	1198
	$(2-MeC_6H_4)_3As(OH)NHO_2SC_6H_4-$ NHAc $+ 5\%$ HCl	m. 224-226°	1198
C_{23}			
$(2-MeC_6H_4)As(OH)NHO_2SC_6H_4NHAc$	$R_3As + 4-AcNHC_6H_4SO_2NKCl$ in	m. 215-217° (dec.)	1198

C24 Ph[4-BrC6H4)(4-ClC6H4)As(OH)OC6H4-(NO2)3	RR'R''AsO + picric acid	Yellow, m. 184°	805
[3,4-O2N(EtO)C6H3]3As(OH)Cl	R3AsO + 20% HCl	Pale yellow, m. 170-170.5° (dec.).	1261
C25 Ph2As(O)CH2As(OH)(Ph)2ONO2	CH2(AsPh2)2 + HNO3 Ph2AsCH2As(O)Ph2 + concd. HNO3 at 20°	Colorless needles m. 176-178°	245 245
Ph2As(O)CH2As(OH)(Ph)2Br	CH2[As(O)Ph2]2 + aq. HBr	Colorless crystals, m. 149-151° Props.[1]	245
CH2[AsPh2(OH)Cl]2	CH2[As(O)Ph2]2 + aq. HCl		245
C27 Ph(4-MeC6H4)(4-EtC6H4)As(OH)-picrate	[RR'R''As(OH)]2O + picric acid	Yellow, m. 133°	803
C30 [MePh(HO)AsCH2-]2 dipicrate		m. 171-172° (dec.)	609
Ph[4-PhC6H4)(4-ClC6H4)As(OH) picrate	RR'R''AsO + picric acid	Greenish yellow, m. 146.5-148.5°	805
C31 Ph2AsCH2AsPh2(O)OC6H2(NO2)3	Ph2AsCH2AsPh2(O) + picric acid	Yellow needles, m. 147°	245
C35 Ph2AsCH2AsH2(OH)OC6H(OH)(NO2)3	Ph2AsCH2AsPh2(O) + styphnic acid	Pale yellow, needles, m. 142-143°	245
Ph2AsCH2AsPh2(OH)C10H7N4O5	Ph2AsCH2AsPh2(O) + picrolonic acid	Pale yellow, powder, m. 152°	245
C36 [Ph3As(OH)]2SO4		m. 195	1053

1. Forms sesquihydrate, colorless crystals, m. 127-130°, loses one mole of H2O and one HCl during recrystallization from a small volume of hot H2O to form Ph2As(O)CH2AsPh2(OH)Cl hemihydrate. The latter compd. can also be prepared by drying the bis(hydroxide chloride) over CaCl2 in vacuo or by adding Cl dissolved in CCl4 to Ph2AsCH2As(O)Ph2 in CHCl3; m. 151-152° (245).

TABLE XLIX
TRIORGANOARSINE DIHYDROXIDES

$R_3As(OH)_2$	Prepd. from	Props.	Ref.
$Me_2PhAs(OH)_2$	$R_2R'As + Br + NaOH$	m. 157-161° with dehydration	1053
$MePh_2As(OH)_2$	$RR'_2As + Br + NaOH$	Hydroscopic solid	1053
$(4\text{-}BrC_6H_4)Ph_2As(OH)_2$	$RR'_2As + KMnO_4$ in Me_2CO	m. 180-181°	96
$Ph_3As(OH)_2$	$R_3As + Br + NaOH$	m. 115-116°, rxn.[1]	1053, 1198
$Ph(4\text{-}BrC_6H_4)(2\text{-}MeO_2CC_6H_4)As(OH)_2$		m. 212-215°	908
$(2\text{-}MeC_6H_4)_3As(OH)_2$	$R_3As + Br + NaOH$	m. 125-126°, rxns.[1,2]	1198
$(4\text{-}MeC_6H_4)_3As(OH)_2$	$R_3As + Br + NaOH$	m. 93-95°, rxn.[1]	1198
$(4\text{-}MeOC_6H_4)_3As(OH)_2$	$R_3As + KMnO_4$ in Me_2CO $(4\text{-}HOC_6H_4)_3AsO + CH_2N_2$ in Et_2O	Colorless needles, m. 92-93°	96 96
$Ph_2(1\text{-}C_{10}H_7)As(OH)_2$	$R_2R'As + KMnO_4$ in Me_2CO	m. 190-191°	96
$[3,4\text{-}O_2N(EtO)C_6H_3]_3As(OH)_2$	$[3,4\text{-}O_2N(EtO)C_6H_3]_3AsO + Ac_2O$ contg. H_2SO_4 on steam bath followed by H_2O	Pale yellow, m. 119-121°	1261
$(3\text{-}AcNHC_6H_4)_3As(OH)_2$	$R_3As + KMnO_4$ in Me_2CO	White crystals, m. > 300°, rxn.[3]	96
$Ph(1\text{-}C_{10}H_7)_2As(OH)_2$	$RR'_2As + KMnO_4$ in Me_2CO	m. 242-243°	96
$(2\text{-}PhC_6H_4)_3As(OH)_2$	$R_3As + Cl$ or $Br + KOH$	m. 237-238°	1318

1. Rxn. with $RSO_2NH_2 \rightarrow R_3As(OH)NHO_2SR$ (1198).
2. Rxn. with 5% HCl $\rightarrow R_3As(OH)Cl$ (1189).
3. Rxn. with NaOH in aq. EtOH $\rightarrow R_3AsO$ (96).

TRIORGANOARSINE OXIDES

TRIPHENYLARSINE OXIDE $C_{18}H_{15}AsO$ Ph_3AsO
Prepn.: By oxidation of Ph_3As with $KMnO_4$ in Me_2CO-H_2O (606).
    By hydrolyzing fluorenylidenetriphenylarsenic with alc. NaOH under reflux (606).
    By reacting fluorenylidenetriphenylarsenic with aldehydes in $CHCl_3$ (606).
    By air oxidation of $Ph_3AsAlCl_3$ (755).
    By treating Ph_3As with SeO_2 in C_6H_6 (842)
    By-prod. from rxn. of $PhN_2Cl.ZnCl_2$ with $AsCl_3$ in Me_2CO in the presence of Fe;
11% yield along with Ph_3As (1066).
    By treating Ph_3As with 30% H_2O_2 in Me_2CO at 25-30° and removing H_2O by azeo-
trop. distn. with C_6H_6; 84-87% yield (1136).
Props.: Colorless needles, m. 193.5-195.5° (606); m. 170° (842); m. 189° (1136).
UV spectrum (598).
Dipole moment (640).
Rxns. with: $RMgX \rightarrow Ph_3RAsX$ (98).
K in $C_4H_8O \rightarrow$ intensively red-brown soln. and ppt. (527).

METHYLENEBIS(DIPHENYLARSINE) OXIDE $C_{25}H_{22}As_2O$ $Ph_2As(O)CH_2AsPh_2$
Prepn.: by reacting Ph_2AsCl with CH_2N_2 in Et_2O at 0°; 25% yield, along with
Ph_2AsOOH and $(Ph_2As)_2O$ (245).
Props.: m. 184-186° (245).
Rxns. with: Picric acid → the hydroxy picrate (245).
Picrolonic acid → the hydroxy picrolonate (245).
Styphnic acid → the hydroxy styphnate (245).
HCl in $CHCl_3$, followed by $H_2S \rightarrow Ph_2AsCH_2As(S)Ph_2$ (245).
SO_2 in $CHCl_3 \rightarrow Ph_2AsCH_2AsPh_2$ (245).
H_2O_2 in EtOH → $CH_2[As(O)Ph_2]_2$ (245).
Concd. $HNO_3 \rightarrow Ph_2As(O)CH_2As(OH)(NO_3)Ph_2$ (245).

TRIS[o-(2-CARBOXYVINYL)PHENYL]ARSINE OXIDE $C_{27}H_{21}AsO_7$ $(o-HO_2CCH:CHC_6H_4)_3AsO$
Prepn.: By adding $o-HO_2CCH:CHC_6H_4N_2$ salt soln. to a mixt. of Na_2CO_3, As_2O_3 and
$CuSO_4$ in H_2O (264).
Props.: m. > 300° (264).

3-(DIPHENYLARSINOMETHYL)CICHOPHEN OXIDE $C_{29}H_{22}AsNO_3$

Prepn.: by refluxing a mixt. of $Ph_2AsCH_2CH_2Bz$, isatin, and KOH in EtOH contg.
H_2O (243).
Props.: Colorless plates, m. 212-213° (dec.) (243).
Derivs.: Monopicrate, yellow crystals, m. 185-188° (dec.) (243).
Benzylammonium salt monohydrate, colorless crystals, m. 173-176° (243).

TABLE L

TRIORGANOARSINE OXIDES

R_3AsO	Prepd. from	Props. and Remarks	Ref.
C₃ Me_2AsO	$R_3As + H_2SO_4$ at 250° $R_3As + Br + alkali$	Rxn. with H_2SO_4 at 320° → $As(OH)_3$	990, 993, 1053
C₆ Et_3AsO		Decomp. on heating → Et_3As + EtOH	649
Me_2BuAsO		Decomp. on heating → Me_2BuAs + $MeBu_2As$	649
C₈ Me_2PhAsO	$R_2R'As(OH)_2$	$Cu(ClO_4)$ complex, pale cream colored crystals	1053, 494a
$(\dot{N}:CHCH:CHCH:\dot{C}CH_2)Me_2AsO$		$Cu(ClO_4)_2$, deep-blue, elongated crystals	494a
C₉ Pr_3AsO		Decomp. on heating → Pr_3As + PrOH	649
$MeBu_2AsO$		Decomp. on heating → $MeBu_2As$, Bu_3As, Bu_2AsOMe, BuOH, PrCHO	649
C₁₀ $(HO_2CCH_2)_2PhAsO$	$RR'AsCl + ClCH_2CO_2H$ $R_2R'As + H_2O_2$	m. 148°	119
$Me(HO_2CCH_2)PhAsO$		Rxn. with SO_2 in HCl → $R_2R'As$ m. 165°	119, 811
C₁₂ $(\dot{S}CH:CHCH:\dot{C})_2AsO$	$R_3As + H_2O_2$ at 100°	m. 160°, rxn. with aq. NaOH → R_2AsO_2H	349
$(\dot{O}CH:CHCH:\dot{C})_3AsO$	$R_3As + Cl$; followed by alkali	Colorless crystals, m. 137-138°, rxn. with NaOH → R_2AsO_2H	346-7, 346

C-13			
Me[4,3-Br(O$_2$N)C$_6$H$_3$]$_2$AsO	Me(4-BrC$_6$H$_4$)$_2$AsO + HNO$_3$-H$_2$SO$_4$	m. 213-215° (dec.); rxn.[1]	99
Me(4-BrC$_6$H$_4$)$_2$AsO	RR'$_2$As + KMnO$_4$	m. 221-223°, rxn.[2]	99
Me[4,3-HO(O$_2$N)C$_6$H$_3$]$_2$AsO	Me[4,3-Br(O$_2$N)C$_6$H$_3$]$_2$AsO + aq. KOH	Pale yellow crystals, m. 239-240°	99
MePh$_2$AsO	RR'$_2$As + KMnO$_4$	m. 154-155°; rxns.[3,4] complexes [5,7]	96
C$_{14}$			
MePh(2-HO$_2$CC$_6$H$_4$)AsO	MePh(2-MeC$_6$H$_4$)As + KMnO$_4$	Rxns. 8, 9	623
MePh(3-HO$_2$CC$_6$H$_4$)AsO	MePh(3-MeC$_6$H$_4$)As + KMnO$_4$	Rxns. 8, 9	623
MePh(4-HO$_2$CC$_6$H$_4$)AsO	MePh(4-MeC$_6$H$_4$)As + KMnO$_4$	Rxns. 8, 9, 10	623
Ph[HO$_2$C(CH$_2$)$_3$]$_2$AsO	RR'AsCl + Br(CH$_2$)$_3$CN in 40% NaOH	Crystals, m. 133-136°	242
	RR'$_2$As + H$_2$O$_2$	m. 131-132°, rxns. 10, 11	119
C$_{15}$			
(N:CHCH:CHCH:C)$_3$AsO	R$_3$As + chloroamine-T followed by hydrolysis	Picrate, m. 144-147°	806
Ph$_2$(HO$_2$CCH$_2$CH$_2$)AsO	R$_2$R'As + HgO	m. 152° (dec.), decomp. above m. → (Ph$_2$As)$_2$ + EtCO$_2$H	243
EtPh(4-HO$_2$CC$_6$H$_4$)AsO	RR'(4-MeC$_6$H$_4$)As + KMnO$_4$	m. 154-155° (dec.), rxns.9,10	624
Ph$_2$(HO$_2$CCH$_2$CH$_2$)AsO or Ph$_2$As$^+$–(CH$_2$CH$_2$CO$_2^-$)OH (?)	R$_2$R'As + H$_2$O$_2$	m. 158°	110
C$_{16}$			
Ph$_2$[MeCH(CO$_2$H)CH$_2$]AsO	R$_2$R'As + HgO	Colorless crystals, m. 157-160° (effersesc.), decomp. above m. → (Ph$_2$As)$_2$O + Me$_2$CHCO$_2$H	243
C$_{17}$			
Ph$_2$(N:CHCH:CHCH:C)AsO	R$_2$R'AsBr$_2$ + alkali	N-picrate, m. 144-145°	806

1. Rxn. with KOH in H$_2$O under reflux → Me[4,3-HO(O$_2$N)C$_6$H$_3$]$_2$AsO (99).
2. Rxn. with HNO$_3$-H$_2$SO$_4$ → the 3-nitro deriv. (99, 1261).
3. Rxn. with PhMgX → RR'$_2$PhAsX (97).
4. Rxn. with hot EtOH-HCl → MePh$_2$As(OH)Cl (146).
5. [Cu(OAsPh$_2$Me)$_4$][CuCl$_2$]$_2$, blue, forms Cu-O not Cu-As bonds (936).
6. [Cu(OAsPh$_2$Me)$_4$][CuCl$_3$]$_2$, brown, forms Cu-O not Cu-As bonds (936).
7. [Cu(OAsPh$_2$Me)$_4$][ClO$_4$]$_2$, blue, forms Cu-O not Cu-As bonds (936).
8. Rxn. with HCl → RR'R''As(OH)Cl (623, 1261).

TABLE L (cont'd.)
TRIORGANOARSINE OXIDES

R_3AsO	Prepd. from	Props. and Remarks	Ref.
C$_{18}$			
$[4,3\text{-}Br(O_2N)C_6H_3]_3AsO$	$(4\text{-}BrC_6H_4)_3AsO + HNO_3\text{-}H_2SO_4$	m. 252-254° (dec.)	99, 1261
$(4\text{-}BrC_6H_4)_3AsO$	$R_3As + KMnO_4$	m. 190-193°	99
	$R_3As + H_2O_2$	m. 204-205°, rxn.[2]	1261
$(3\text{-}O_2NC_6H_4)_3AsO$	$R_3As + H_2O_2$	m. 255-256°, rxns.[10, 12]	96
$Ph(4\text{-}BrC_6H_4)(4\text{-}ClC_6H_4)AsO$	RR'R'' As + Br, followed by NaOH	m. 154-156°	805
$(3\text{-}HOC_6H_4)_3AsO$	$[(3\text{-}H_2NC_6H_4)_3AsOH]_2SO_4$ by diazo-tization	White needles, m. >300°, rxn.[12]	96
$(4\text{-}HOC_6H_4)_3AsO$	$(3\text{-}H_2NC_6H_4)_3AsO$ by diazotization	m. 276-278° (dec.), rxn.[15]	96
$(3\text{-}H_2NC_6H_4)_3AsO$[14]		Hydrate, m. 262-263°	695
	$(3\text{-}AcNHC_6H_4)_3As(OH)_2$ by hydro-lysis	m. 272°, rxn.[16]	96
$(4\text{-}H_2NC_6H_4)_3AsO$	$R_3As + I$ followed by hydrolysis	Colorless crystals	64
$(cyclo\text{-}C_6H_{11})Ph_2AsO$	$RR_2'As + KMnO_4$	m. 206-207°, rxn.[3]	96
C$_{19}$			
$(2\text{-}HO_2CC_6H_4)Ph_2AsO$	$(2\text{-}MeC_6H_4)Ph_2As + KMnO_4$	m. 246-247°, rxn.[10]	613
$(3\text{-}HO_2CC_6H_4)Ph_2AsO$	$Ph_3As + BuLi$ in $Et_2O + CO_2$ and hydrolysis, followed by $KMnO_4$	m. 215°	473
$Ph(4\text{-}ClC_6H_4)(4\text{-}MeC_6H_4)AsO$	RR R As + Br + alkali	Deliquescent, m. 130°	805
C$_{24}$			
$Ph(4\text{-}PhC_6H_4)(4\text{-}ClC_6H_4)AsO$	RR'R'' As + Br + alkali	m. 163°	805
$[4,3\text{-}EtO(O_2N)C_6H_3]_3AsO$	$[4,3\text{-}Br(O_2N)C_6H_3]_3AsO$ + 10% EtOH-KOH	Pale yellow, m. 215-126° Rxns. 8, 17, 18	1261
$[3,4\text{-}H_2N(EtO)C_6H_3]_3AsO$ ( )	$[3,4\text{-}O_2N(EtO)C_6H_3]_3AsO + H_2/N1$ in EtOH	Trihydrate, m. 106-108°, at 140°/0.2 mm. → anhydr., 20-1 211-212°, salts[19], rxns.	1261
C$_{25}$			
$CH_2[As(O)Ph_2]_2$	$Ph_2AsCH_2As(O)Ph_2 + H_2O_2$	Deliquescent glass, rxn.[22]	245
C$_{30}$			
$[3,4\text{-}AcNH(EtO)C_6H_3]_3AsO$	$[3,4\text{-}H_2N(EtO)C_6H_3]_3AsO + Ac_2O$	m. 179-180°	1261

| C_{42} [:CAs(O)(C$_{10}$H$_7$)$_2$]$_2$ | [(1-C$_{10}$H$_7$)$_2$AsC:]$_2$ + BzO$_2$H | m. 218.5° (dec.), rxn.[23] | 516 |

9. Rxn. with H$_2$S → RR'R''AsS (623, 624).
10. Rxn. with SO$_2$ in H$_2$O contg. I → tertiary arsine (96, 242, 613, 623, 624).
11. Rxn. with concd. HCl → RR$_2'$AsCl$_2$ (242).
12. Rxn. with H$_3$PO$_2$ in AcOH contg. HI → tertiary arsine (96, 99).
13. Ethanolic KOH → [3,4-O$_2$N(EtO)C$_6$H$_3$]$_3$AsO (1261).
14. The compound is also formed as by-product in the preparation of 4-HOC$_6$H$_4$AsO$_3$H$_2$ by the Bart method (695).

15. Rxn. with CH$_2$N$_2$ → (4-MeOC$_6$H$_4$)$_3$AsO (96).
16. Diazotization followed by hydrolysis → (3-HOC$_6$H$_4$)$_3$AsO (96).
17. Rxn. with Ac$_2$O + H$_2$SO$_4$ → R$_3$As(OH)$_2$ (1261).
18. Catalytic redn. over Ni in EtOH → [4,3-EtO(H$_2$N)C$_6$H$_3$]AsO (1261).
19. Monopicrate, m. 200-201° (dec.); trihydrochloride, m. 221-222° (dec.) (1261).
20. Rxn. with Ac$_2$O → the triacetamido deriv. (1261).
21. Catalytic redn. over Raney Ni in EtOH at 80-100°/1000 psi → [3,4-H$_2$N(EtO)C$_6$H$_3$]$_3$As (1261).
22. Rxns. with aq. acids (HX) → Ph$_2$As(O)CH$_2$As(OH)(X)Ph$_2$ and CH$_2$[As(OH)(X)Ph$_2$]$_3$ (245).
23. Rxn. with alkali → hydrolysis (516).

TABLE LI
TRIORGANOARSINE SULFIDES

R_3AsS	Prepd. from	Props.	Ref.
C_{14}			
$MePh(2\text{-}HO_2CC_6H_4)AsS$	$RR'R''As(OH)Cl + H_2S$	m. 164-165°, rxn.[1]	623
$MePh(4\text{-}HO_2CC_6H_4)AsS$	$RR'R''As(OH)Cl + H_2S$	m. 159-160°	623
$MePh(3\text{-}HO_2CC_6H_4)AsS$	$RR'R''As(OH)Cl + H_2S$	m. 134-135°	623
C_{15}			
$(HO_2CCH_2CH_2)Ph_2AsS$	$RR'R''As + Br_2 + H_2S$	Colorless crystals, m. 131-134°	243
C_{18}			
Ph_3AsS	$R_3As(OPh)OH + H_2S$ in MeOH $PhN_2Cl + As$ in Me_2CO contg. $CaCO_3$ followed by H_2S		1274 1274
C_{19}			
$Et(1\text{-}C_{10}H_7)(4\text{-}HO_2CC_6H_4)AsS$	$RR'R''As + KMnO_4 + H_2S$	m. 123-124°	627
$Et(2\text{-}C_{10}H_7)(4\text{-}HO_2CC_6H_4)AsS$	$RR'R''As + KMnO_4 + H_2S$	m. 104-105°	627
C_{22}			
$[\text{-}CH_2As(S)Ph]_2$	$(\text{-}CH_2AsBuPh)_2 + Br + H_2S$ or Na_2S	Props.[2]	199
C_{25}			
$Ph_2AsCH_2As(S)Ph_2$	$Ph_2AsCH_2As(O)Ph_2 + HCl$ in $CHCl_3$, followed by H_2S	Colorless crystals, m. 138-139°	245

1. Rxn. with SO_2 in H_2O contg. I → tertiary arsine (623).
2. The sulfide occurs in two modifications: α-form, obtd. by using N_2S, m. 113-116°, can be converted into the β-form by heating; β-form obtd. by using H_2S for the rxn. with the arsine dibromide, m. 121°; mixed m. 95-103° (199).

TRIPHENYLARSINE SELENIDE $C_{18}H_{15}AsSe$ Ph_3AsSe

Prepn.: by treating Ph_3As with SeO_2 in C_6H_6 at r. t.; a mixc. of Ph_3AsO and Ph_3-AsSe is formed (842).

QUATERNARY ARSONIUM COMPOUNDS

Quaternary arsonium salts and hydroxides are prepared by:
1. Heating tertiary arsines with alkyl halides, especially with metoyl odide.
2. Reacting trialkylarsine dihalides with alkyllithium derivatives at -60 to -50° and treating the resulting pentaalkylarsenic compounds with acids, such as hydrochloric or picric acid, to form tetraalkylarsonium chlorides and picrates, respectively, or complexing the pentaorganoarsenic derivatives with triphenylborine to form tetraorganoarsonium organotriphenylborates.
3. Dissolving tertiary arsines in nitric acid to form triorganohydroxyarsonium nitrates, which can be converted to the picrates by means of picric acid.
4. Heating trialkylarsines with ethylene oxide in the presence of water to from trialkyl(2-hydroxyethyl)arsonium hydroxides.
5. Reacting tertiary arsines with diaryliodonium salts.
6. Treating tertiary arsines with aryl halides at elevated temperatures (180-190°) in the presence of aluminum chloride.
7. Refluxing arsenic trichloride in benzene in the presence of aluminum chloride to form tetraphenylarsonium chloride.
8. Disporportionation of dichlorophenyl-, chlorodiphenyl-, or triphenyl-arsine under reflux in the presence of aluminum chloride.
9. Complexing pentaphenylarsenic with triphenylborane to form tetraphenylarsonium tetraphenylborate.
10. Treating pentaphenylarsenic with halogens to form tetraphenylarsonium perhalides.
11. Reacting alkylidenetrialkylarsenic with alkyl halides, hydrohalic acids, or trialkylsilyl halides (addition of the R_3Si group to the C atom of the alkyldiene group and reduction of the double bond.
12. Reacting tertiary arsine oxides with Grignard compounds and hydrolyzing the product with hydrochloric acid.

Quaternary arsonium halides undergo metathetic reactions with moist silver oxide or with aqueous alkali to form the corresponding arsonium hydroxides, with silver nitrate or sulfate to form the nitrates and sulfates, respectively, and with sodium picrate to form the arsonium picrates. Moreover, arsonium iodides are converted into the chlorides directly by silver chloride or via arsonium hydroxide followed by acidification with hydrochloric acid. Arsonium bromides are readily converted into the iodides by a simple treatment with sodium iodide, Arsonium thiocyanates are obtained by reacting the chlorides with potassium thiocyanate.

Dibenzyldimethylarsonium bromide treated with phenyllithium in ether followed by water gave stilbene and an oily fraction which, after treatment with methyl iodide under nitrogen, gave trimethyl(1,2-diphenylethyl)arsonium iodide (Stevens rearrangement).

TABLE LII

QUATERNARY ARSONIUM COMPOUNDS

Quaternary Arsonium R₄As⁺	Salt X⁻	Prepd. from	Props. and Remarks	Ref.
C₄				
Me₃(CF₃)As	I	Me₂AsCF₃ + MeI	White cryst. solid	520
Me₄As	Cl	Me₃AsBr₂ + MeLi, followed by HCl	Isolated as M₄AsCl·2HgCl₂	1314
		Me₄AsI + Ag₂O +HCl	White crystals, m. 177-178°	1314
			Raman spectrum	1138
	OH	Me₄AsI + Ag₂O		1053
	BMePh₃	Me₃AsBr₂ + MeLi, followed by Ph₃B		1314
	I	Me₃As + MeI	m. 315-318°, rxns.¹⁻⁴	1314
	NO₃	Me₄AsI + AgNO₃	m. 328°	1053
	Picrate	Me₄AsI + PhLi, followed by picric acid	m. 268-270°	1053
			Yellow needles, decomp. 292-294°	1314
	OSO₃H	Me₄AsI + Ag₂SO₄	Rxn.⁵	993
C₅				
Me₃(HOCH₂CH₂)As	Picrate		Conductance in C₂H₄Cl₂	836
[Me₃(HOCH₂CH₂)As]₂	SO₄	Me₃(HOCH₂CH₂)AsI + Ag₂SO₄	Rxn.⁵	993
Me₃(HOCH₂CH₂)As	OH	Me₃As + CH₂CH₂O	Catalyst for prepn. of phenolic resins	118
C₆				
Me₃(HO₂CCH₂CH₂)As	OH	Me₄AsI + MeLi or PhLi, followed by MeI⁹	Rxn.⁵	990
Et₂Me₂As	I		Platelets, decomp. 305°	1314
		Me₂AsO₂H + H₃PO₂ in the presence of EtI		993
Et₂Me(MeO)As	OH	Et₂Me₂AsI + NaOH	Rxn.⁵	993
	I	Et₂AsI + NaOMe in Et₂O	Decompn. 283-284°	648
C₇				
Me₃(AcOCH₂CH₂)As	Picrate		Conductance in C₂H₄Cl₂	836

C9				
Me3[4,3-Br(O2N)C6H3]As	Picrate		Orange yellow, m. 91-93°	243
	Br	Me3[4,3-Br(O2N)C6H3]AsNO3 + NaBr	m. 255-275° (dec.)	100
	NO3	Me3(4-BrC6H4)AsNO3 + HNO3- H2SO4	m. 176-181° (dec.)	100
Me3(4-BrC6H4)As	I	Me2(4-BrC6H4)As + MeI	m. 253-255° (dec.)	100
			m. 240° (dec.)	612
	NO3	Me3(4-BrC6H4)AsI. + Ag2O + HNO3	m. 163-165°	100
Me3(3-O2NC6H4)As	Cl	Me3(3-O2NC6H4)AsI + Ag2O + HCl	m. 263-270° (dec.), rxn.[6]	100
	I	Me2(3-O2NC6H4)As + MeI	m. 285° (dec.)	998
	NO3	Me3(3-O2NC6H4)AsNO3 + NaI	m. 286-290° (dec.)	100
		Me3PhAsNO3 + HNO3-H2SO4	m. 278-279° (dec.,), rxn.[7]	100, 998
Me3[4,3-HO(O2N)C6H3]As	Br	Me3[4,3-Br(O2N)C6H3]AsBr + aq. KOH, followed by HBr	m. 269-271° (dec.)	100
	NO3	Me3[4,3-Br(O2N)C6H3]AsNO3 + aq. KOH + NaNO3	m. 225°	100
Me3PhAs	Br		m. 283-284°	1056
	OH		m. 106-116°	1053
	I	Me2PhAs + MeI	m. 224-245°	612
	NO3	Me3PhAsI + AgNO3	m. 195-196°	1053
			m. 194-196°	100
(Me3PhAs)2	BiI4		Deep red solid, m. 171°	312
	CdI4		m. 189°	313

1. Rxn. with deuterium → exchange of D for H (297).
2. Rxn. with Ag2SO4 → the corresponding arsonium sulfate (993).
3. Rxn. with AgNO3 → the corresponding arsonium nitrate (1053).
4. Rxn. with Ag2O → the corresponding arsonium hydroxide (1053, 1262).
5. Rxn. with H2SO4 at 320° → As(OH)3 (990, 991, 993).
6. Rxn. with SnCl2 in AcOH-HCl, followed by NaOH and NaI → redn. of the NO2 group (100, 998).
7. Rxn. with NaI → Me3(3-O2NC6H4)AsI (100, 998).

TABLE LIII (cont'd.)
QUATERNARY ARSONIUM COMPOUNDS

Quaternary Arsonium R_4As^+	Salt X^-	Prepd. from	Props. and Remarks	Ref.
$(Me_3PhAs)_2$	$Co(SCN)_4$		m. 89°	315
Me_3PhAs	Picrate		m. 142.5-143.2°	1056
$Me_3[3,4-H_2N(Br)C_6H_3]As$	I	$Me_3[4,3-Br(O_2N)C_6H_3]AsBr$ + $SnCl_2$ in AcOH-HCl, followed by NaOH + NaI	m. 235-237°	100
$Me_3(2-HOC_6H_4)As$	I	$Me_2(2-HOC_6H_4)AsI$ + MeI	m. 238-239° (dec.)	481
$Me_3(3-HOC_6H_4)As$	I	$Me_3(3-H_2NC_6H_4)AsCl$ by azotization	m. 208-211° (dec.)	100, 998
$Me_3(4-HOC_6H_4)As$	I	$Me_2(3-HOC_6H_4)As$ + MeI	m. 209-211°	481
$Me_2(MeO)PhAs$	I	$Me_2(4-HOC_6H_4)As$ + MeI	m. 224-246°	481
	NO_3	MePhAsOR + MeI	m. 109-110°	1160
			m. 143-144°	642
$Me_3(3-H_2NC_6H_4)As$	Cl	$Me(3-H_2NC_6H_4)AsI$ + Ag_2O + HCl	m. 243-244° (dec.)	100
	I	$Me_3(3-O_2NC_6H_4)AsCl$ by redn. with $SnCl_2$	m. 175-176°	100, 998
		$Me_2(3-O_2NC_6H_4)As$ + MeI, followed by redn.		998
$Me_3[3,4-H_2N(HO)C_6H_3]As$	Cl	$Me_3[3,4-O_2N(HO)C_6H_3]AsBr$ by redn. with $SnCl_2$	HCl salt, m. 211-215°	100
C_{10}				
$Me_3(PhCH_2)As$	I	Me_2AsI + $PhCH_2MgCl$ + MeI	Crystals, m. 202°	605
	BiI_4		Red, m. 124°	312
	CdI_4		m. 188°	313
	$Co(SCN)_4$		m. 129°	315
$Me_3(4-MeC_6H_4)As$	I			811
$[Me_3(4-MeC_6H_4)As]_2$				
$Me_3(3-MeOC_6H_4)As$	I		Prisms, m. 248-250°	811
$MeEtPh(MeO)As$	I	EtPhAsOPr + MeI	m. 112-114°	1160
C_{11}				
$Me(ClCH{:}CH)_2PhAs$	I	$(ClCH{:}CH)_2PhAs$ + MeI	m. 232°	263
	Picrate		m. 164.5-165.5°	810
$Me_2(NCCH_2CH_2)PhAs$	Toluene-sulfonate		m. 117-118°	810
			m. 144-145°	810

Compound	Salt	Preparation	Properties	Ref.
Me₃(PhAs:CH)As	I		m. 155	265
Me₂(HO₂CCH₂CH₂)PhAs	I		m. 110°, remelts at 151-153°	811
	S-benzyl-lthiuronium salt		m. 137°	811
MeEt₂(2-BrC₆H₄)As	Cl		m. 114-115°	612
Me₃(3-AcNHC₆H₄)As	I		m. 256-258° (dec.)	100
	I		m. 242-246° (dec.)	100
MeEt₂PhAs	Picrate	Et₂PhAs + MeI, followed by picrate	m. 85-86°	808
MeEt₂(3-HOC₆H₄)As	I	Et₂(3-HOC₆H₄)As + MeI	m. 87-89°	481
Me₃(2-Me₂NC₆H₄)As	Br		m. 260-261°	813
	Picrate	Me₂(2-Me₂NC₆H₄)As + MeI	m. 178-181°	810
2-(Me₂As)C₆H₄AsMe₃	Br		m. 221-222°	810
Et₂AsC:CAsMeEt₂	I		m. 225-226°	807
Et₃[MeCH(CO₂Et)]As	Br	Et₃As + MeCHBrCO₂Et	m. 133°, instable	516
Me₃(n-C₈H₁₇)As	I	n-C₈H₁₇AsMe₂ + MeI	m. 69-70°, racemic compd.	81
Me₃[3-MeNHCO₂)C₆H₄]As	I	Me₂(3-HOC₆H₄)As + MeNCO + MeI	m. 82°	605
	I		m. 187-189°	481
C₁₂				
Me₂[NC(CH₂)₃]PhAs	Br	Me₂PhAs + Br(CH₂)₃CN	m. 170°, dissoc. on heating into Me₂PhAs + Br(CH₂)₂CN	242
	Picrate		Orange-yellow plates, m. 119° rxn.8	242
Me₂(NCCH₂CH₂)(3-MeOC₆H₄)As	I	m.	m. 148°	811
Me₂(NCCH₂CH₂)(3-MeOC₆H₄)As	Picrate		m. 117-118°	811
Me₂(EtO₂CCH₂)PhAs	I	Me(EtO₂CCH₂)PhAs + MeI	m. 71-72°	630
Me₂(HO₂CCH₂CH₂)(3-MeOC₆H₄)As	I		Crystals, m. 138°	811
	S-benzylthiuronium salt		m. 138°	811
Et₂(BrCH₂CH₂)PhAs	Br	Et₂PhAs + (CH₂Br)₂	m. 106°	609
Me₃[4-(Me₂NCO₂)C₆H₄]As	I	Me₃(4-HOC₆H₄)AsI + MeNCOCl	m. 226.5°	481
(CH₂:CHCH₂)₄As	Br	(CH₂:CHCH₂)₃As + CH₂:CHCH₂Br	Crystals, m. 40-50°, HgCl₂ adduct, m. 66°	1262
	Picrate		Yellow crystals, m. 82.5°, rxn.⁴	1262
	OH	(CH₂:CHCH₂)₄AsBr + Ag₂O		1262

8. Rxn. with alkali → the corresponding arsonium hydroxide (242).

TABLE LII (cont'd.)

QUATERNARY ARSONIUM COMPOUNDS

Quaternary arsonium R₄As⁺	Salt X⁻	Prepd. from	Props. and Remarks	Ref.
Et₂(EtO)PhAs	I	Et(EtO)PhAs + EtI	m. 111-113°	1160
o-C₆H₄(AsMe₃)₂	Br₂		m. 222-223°	807
	I₂		m. 220-222°	612
p-C₆H₄(AsMe₃)₂	Br₂		m. 222-224°	807
			m. 253° (effervesc.)	610
C₁₃				
Me(NCCH₂CH₂)₂PhAs	Picrate	MeEtPrAs + PhCH₂Cl followed	Yellow needles, m. 115-118.5	242
MeEtPr(PhCH₂)As	Picrate	by Na picrate	m. 65-66°	193
Me(CH₂:CMeCH₂)₃As	I	(Me₃SiCH₂)₃As + MeI	Long rectangles, m. 129°	616
Me(Me₃SiCH₂)₃As	I		m. 193-195°, HgI₂ adduct, m. 134-135°	1121
C₁₄				
Me₂[4,3-Br(O₂N)C₆H₃]₂As	Br	Me₂[4,3-Br(O₂N)C₆H₃]₂AsNO₃ + NaBr	m. 183-185°	100
	I	Me[4,3-Br(O₂N)C₆H₃]₂As + MeI	m. 169-170°	100
	NO₃	Me₂(4-BrC₆H₄)₂AsNO₃ + HNO₃-H₂SO₄	m. 206-207° (dec.)	100
Me₂(4-BrC₆H₄)₂As	I	Me(4-BrC₆H₄)₂As + MeI	m. 221-224°, rxn.³	100
	NO₃	Me₂(4-BrC₆H₄)₂AsI + Ag₂O + HNO₃	m. 195-196°	100
Me₂Ph₂As	OH		m. 120-135°	1053
	I		m. 211-213°, rxn.³	1053, 1215
	NO₃	Me₂Ph₂AsI + AgNO₃	m. 190°	1053
(Me₂Ph₂As)₂	BiI₄		m. 149-151°	312
	BiOI₂		Reddish orange, m. 161°	312
	Bi₂O₃I₂		Chocolate colored	312
	CdI₄		Pale red-brown	313
	Co(SCN)₄		m. 173°	315
Me(HO₂CCH₂CH₂)₂(3-MeOC₆H₄)As	I		m. 120°	811
Me₂[NC(CH₂)₅]PhAs	Br	Me₂PhAs + Br(CH₂)₂CN	m. 162° Deliquesc. cryst. mass, dissoc. into Me₂PhAs + Br-(CH₂)₅CN	242

$Me_2[NC(CH_2)_5]PhAs$	Picrate		Yellow needles, m. 58-59°, rxn.[8]	242
$Me_2(C_6H_{11})_2As$	I		m. 186-187°	1053
	NO_3		m. 147°	1053
$Et_3(n\text{-}C_8H_{17})As$	I	$Et_3As + n\text{-}C_8H_{17}I$	Hydroscopic crystals, m. 68-70°	605
$Et(Me_3SiCH_2)_3As$	I	$(Me_3SiCH_2)_3As + EtI$	m. 112-115°	1121
C_{15}				
$Me_2Ph(PhCH_2)As$	Picrate	$Me_2PhAs + PhCH_2Cl$ + picric acid	m. 125-126°	193
$MeEtPh_2As$	I		m. 170°	1215
$Me_3(4\text{-}PhOC_6H_4)As$	I	$Me_2(4\text{-}PhOC_6H_4)As + MeI$	Glistening leaflets, m. 240°	271
$Me(HOCH_2CH_2)Ph_2As$	I	$Ph_2(HOCH_2CH_2)As + MeI$	m. 170-173° (dec.)	122
$Me(CH_2{:}CMeCH_2)_2PhAs$	I	$(CH_2{:}CMeCH_2)_2PhAs + MeI$	Needles, m. 163°	616
$Me_3(2\text{-}PhC_6H_4)As$	I		m. 259-260°, dissoc. at 280°/0.15 mm.	522, 523, 253
$Pr_2(PrO)PhAs$	Picrate	$Pr(PrO)PhAs + PrI$	m. 138-139°	1160
	I		m. 162-163°	642
	NO_3	$Pr_2(PrO)PhAsI + HNO_3$	m. 112-113°	1160
$MeBu(BuO)PhAs$	I	$Bu(BuO)PhAs + MeI$	m. 139-141°	605
$Me_3(n\text{-}C_{12}H_{25})As$	I	$Me_2(n\text{-}C_{12}H_{25})As + MeI$	Shiny needles, m. 121-122°	1160
C_{16}				
9-Fluorenyltrimethylarsonium	Br	$MeAs$ + 9-bromofluorene	m. 205-206°, rxn.[10-12]	1313
	I	From the preceding arsonium bromide or 9-fluorenylene-trimethylarsenic + KI	m. 182-184°	1313
$Me_3(2\text{-}PhC_6H_4)As$	I		m. 232-235°, UV spectrum rxn.[13]	244
$Me_2(PhCH_2)_2As$	Br	$Me_2(PhCH_2)As + PhCH_2Br$	m. 147-148°, rxn.[13]	1313

9. The ethyl groups are formed by metallation of the methyl groups, followed by reaction with MeI (1314).

10. Rxn. with KI → the arsonium iodide (1313).

11. Rxn. with PhLi → (9-fluorenylidine)trimethylarsenic (1313).

12. Rxn. with PhLi, followed by B_8Ph and by H_2O → 9-fluorenyldiphenylcarbinol (1313).

13. Rxn. with PhLi, followed by H_2O → stilbene and $Me_2(PhCH_2CHPh)As$ (1313).

TABLE LII (cont'd.)
QUATERNARY ARSONIUM COMPOUNDS

Quaternary Arsonium R_4As^+	Salt X^-	Prepd. from	Props. and Remarks	Ref.
$[Me_2(4\text{-}MeC_6H_4)_2As]$	BiI_4		Orange, m. 77°	312
$[Me_2(4\text{-}MeC_6H_4)_2As]_2$	CdI_4		m. 128°	313
	$Co(SC\text{-}N_4)$		m. 110°	315
$Me_2Et(4\text{-}PhOC_6H_4)As$	I	$Me(4\text{-}PhOC_6H_4)As + EtI$	Leaflets, m. 237°	271
$2\text{-}(MeEt_2P)C_6H_4AsEt_2Me$	I_2		m. 165°, P-monomethiodide, m. 148-149°	612
Bu_4As	Picrate		Electrolytic. cond. in C_2H_4-Cl_2	673
$(Me_3SiCH_2)_4As$	I	$(Me_3SiCH_2)_3As + Me_3SiCH_2I$	m. 143.5-145°	1121
C_{17}				
$Me(NCCHMeCH_2)Ph_2As$	Picrate	$Ph_2AsCH_2CHMeCN + MeI$	Yellow needles, m. 125-129°	242
$Me_3(PhCH_2CHPh)As$	I	$Me_2(PhCH_2)_2AsBr + PhLi + MeI$	Crystals, m. 226-228°	1313
$Me_2(PhCH_2)(n\text{-}C_8H_{17})As$	Cl	$Me_2(n\text{-}C_8H_{17})As + PhCH_2Cl$	White hydroscop. cryst. mass m. 144-146°	605
	I	$Me_2(PhCH_2)As + n\text{-}C_8H_{17}I$	Hard crystals, m. 71°	605
C_{18}				
$Me_3(Ph_2CHOCH_2CH_2)As$	I	$Me_3As + Ph_2CHOCH_2CH_2Cl$	m. 174-176°	1042
$(Me_2PhAsCH_2\text{-})_2$	I_2		Decomp. 250°	609
	Dipi-crate		m. 217-218° (dec.)	609
$(2\text{-}C_6H_4AsMe_3)_2$	Br_2		Dihydrate, m. 292-294°	523
	I_2		m. 286-288°, at 260°/0.05 mm. dissoc.	523
$Bu_2(BuO)PhAs$	I	$Bu(BuO)PhAs + BuI$	m. 135-136°	1160
	NO_3	$Bu_2(BuO)PhAsI + HNO_3$	m. 84-85°	642
$Et_3(n\text{-}C_{12}H_{25})As$	I	$Et_3As + n\text{-}C_{12}H_{25}I$	White crystals, m. 98°	605
C_{19}				
$Me[4,3\text{-}Br(O_2NC_6H_3)_3As]$	NO_3	$Me(4\text{-}BrC_6H_4)_3AsNO_3 + HNO_3\text{-}H_2SO_4$	m. 175-177° (dec.)	101
$Me(4\text{-}BrC_6H_4)_3As$	I	$(4\text{-}BrC_6H_4)_3As + MeI$	m. 178-180°	101
	NO_3	$Me(4\text{-}BrC_6H_4)_3AsI + Ag_2O + HNO_3$	m. 187-195°	101

Compound		Preparation	Properties	Ref.
Me(3-O$_2$NC$_6$H$_4$)$_3$As	Cl	Me(3-O$_2$NC$_6$H$_4$)$_3$AsNO$_3$ + NaCl	Decomp. 130°	101
	NO$_3$	MePh$_3$AsNO$_3$ + HNO$_3$	Colorless crystals, m. 197-198°	101
Ph(4-ClC$_6$H$_4$)(4-MeC$_6$H$_4$)As(OH)	Picrate		Greenish yellow, m. 174°	805
(ICH$_2$)Ph$_3$As	NO$_3$	ICH$_2$Ph$_3$AsI + HNO$_3$	m. 189-190°	98
Me(3-H$_2$NC$_6$H$_4$)$_3$As	Cl	Me(3-O$_2$NC$_6$H$_4$)$_3$AsCl + SnCl$_2$ in AcOH-HCl	m. 198-200°	101
MePh$_3$As	I	Me(3-H$_2$NC$_6$H$_4$)$_3$AsCl + NaI	m. 167-169°	101
	Br	Ph$_3$As + MeBr	Rxn.[14,15]	531, 1053
	OH		m. 124-126°	531
	I	Ph$_3$As + MeI; MePh$_3$AsO + PhMgI	m. 175-176°, dissoc. at elevated temp., rxns[3,15-6] use[17]	97
	NO$_3$	MePh$_3$AsI + HNO$_3$	m. 131-133°	98
	SCN		m.~100°. Analyt. reagent for Co and Cu	1053, 330, 331
(MePh$_3$As)$_2$	BiI$_4$		Light orange, m. 80°	312
	CdI$_4$	MePh$_3$AsI + CdI$_2$	m. 179°	313
	Cr$_2$O$_7$		Yellow salt, sol. in (CH$_2$Cl)$_2$	463
	CrO$_6$	MePh$_3$AsCl + K$_2$Cr$_2$O$_7$, followed by H$_2$O$_2$	mol. wt. and magnetic data	320
	Co(SC-N)$_4$		m. 47°	315
	CoX$_4$		X = Br, Cl, and I; blue, paramagentic crystals	481a
	CuX$_4$		X = Br, Cl, and I; yellow, paramagentic crystals	481a
	Cu-(SCN)$_4$			331
	FeX$_4$		X = Br, Cl, and I; cream-colored, paramagentic crystals	481a

14. Rxn. with PhLi in Et$_2$O → Ph$_3$As:CH$_2$ (495a).
15. Rxn. with PhLi, followed by BzPh and hydrolysis → Ph$_3$As, Ph$_2$C:CH$_2$, Ph$_2$CHCHO, and Ph$_3$AsO (531).
16. Rxn. with Os and Ru in HCl → characteristic crystals (890).
17. Useful reagent for detn. of Sb(381) and Cd (1284) in the presence of Zn (317).

TABLE III (cont'd.)

QUATERNARY ARSONIUM COMPOUNDS

Quaternary Arsonium R$_4$As$^+$	Salt X$^-$	Prepd. from	Props. and Remarks	Ref.
(MePh$_3$As)$_2$	MnX$_4$		X = Br, Cl, and I; green, paramagnetic crystals	481a
	NiX$_4$		X = Br, Cl, and I; green-blue paramagnetic crystals	481a
	ZnX$_4$		X = Br, Cl, and I; white, diamagnetic crystals	481a
(MePh$_3$As)$_3$	Fe-(SCN)$_6$	MePh$_3$AsCl	Dark red, cryst. powder, dissocn. data	318
MePh$_3$As	PdCl$_4$		Pink crystals	512a
[Pd(MePh$_3$As)Cl]$_2$	Cl$_2$		Brownish pink crystals	512a
MePh$_3$As	PtCl$_4$		Pale pink ppt.	512a
Me(3-H$_2$NC$_6$H$_4$)$_3$As	Cl	Me(3-O$_2$NC$_6$H$_4$)$_3$AsCl + SnCl$_2$ in AcOH-HCl	m. 198-200°; HgCl$_2$ complex, m. 191-192°	101
Me Ph$_2$(cyclo-C$_6$H$_{11}$)As	I	Ph$_2$(C$_6$H$_{11}$)As + MeI	m. 220-221°	97
Me$_2$(2-MeNC$_6$H$_4$)(2-MeOCH$_2$C$_6$-H$_4$CH$_2$)As	Br		m. 162-163°	813
Me$_2$(2-Me$_2$NC$_6$H$_4$)(2-MeOCH$_2$C$_6$-H$_4$CH$_2$)As	Cl		Hydroscopic, m. 140-142°	813
	Picrate		m. 104°	813
	ClO$_4$		Hemihydrate, m. 159-160°	813
	PtCl$_4$		Hydrate, m. 192-193°	813
Me$_3$(n-C$_{16}$H$_{33}$)As	I	Me$_2$(n-C$_{16}$H$_{33}$)As + MeI	Crystals, m. 132°	605
C$_{20}$				
MePh$_2$(2-HO$_2$CC$_6$H$_4$)As	I		m. 102-104°	613
EtPh$_3$As	NO$_3$		Cond. in (CH$_2$Cl)$_2$	490
	Picrate		Cond. in (CH$_2$Cl)$_2$	490
	ClO$_4$		Cond. in (CH$_2$Cl)$_2$	490
Ph$_3$(HOCH$_2$CH$_2$)As	NO$_3$	Ph$_3$(HOCH$_2$CH$_2$)AsCl + AgNO$_3$	m. 138-140°	98
C$_{21}$				
Ph$_3$(CH$_2$:CHCH$_2$)As	Br	Ph$_3$As + CH$_2$:CHCH$_2$Br	m. 180-181°	98
	I	Ph$_3$(CH$_2$:CHCH$_2$)AsBr + KI	m. 163-164°	98
Me$_2$Ph(2-PhC$_6$H$_4$CH$_2$)As	Picrate	Me$_2$PhAs + 2-PhC$_6$H$_4$CH$_2$Br, followed by Na picrate	Yellow crystals, m. 102°	244
Me$_2$Ph(4-PhC$_6$H$_4$CH$_2$)As	Br	Me$_2$PhAs + 4-PhC$_6$H$_4$CH$_2$Br	m. 192-193°, UV spectrum	244

Compound	Anion	Preparation	Properties	Ref.
Ph₃(CH₂:CHCH₂)AS	NO₃	Ph₃(CH₂:CHCH₂)AsBr + AgNO₃	m. 146-148	98
Me₂(PhCH₂)(n-C₁₂H₂₅)As	Cl	Me₂(n-C₁₂H₂₅)As + PhCH₂Cl	Hydroscopic needles, m. 159-161°	605
	I	Me₂(PhCH₂)As + n-C₁₂H₂₅I	Needles, m. 72°	605
C₂₂				
Me₂(PhCH₂)(C₁₃H₉)As	Br	Me₂(PhCH₂)As + 9-bromofluor-ene	m. 182-184° (dec.), rxn[12,18]	1313
	I	Me₂(PhCH₂)(C₁₃H₉)AsBr + KI	m. 174-176°	1313
[Me(4-MeC₆H₄)₃As	BiI₄		Orange yellow, m. 103°	312
[Me(4-M₃C₆H₄)₃As]₂	CdI₄		m. 203°	313
	Co(SCN)₄		m. 60°	315
o-C₆H₄(2-C₆H₄AsMe₃)₂	I₂		Dihydrate, m. 283-285°, an-hyd., m. 308-310°	523
Me(3-MeOC₆H₄)₃As	I	(3-MeOC₆H₄)₃As + MeI	m. 120-121°	101
Ph₃(Me₃SiCH₂)As	Br	Ph₃As:CH₂ + Me₃SiBr		495a
(Et₂PhAsCH₂-)₂	Br₂	Et₂PhAs + (CH₂Br)₂		609
	Dipi-crate		m. 166-167°	609
[Me₂(2-Me₂NC₆H₄)AsCH₂-]₂	Br₂	2-Me₂NC₆H₄AsMe₂ + (CH₂Br)₂	m. 191-192°	813
	Dipi-crate		m. 150°	813
Et₃(n-C₁₆H₃₃)As	I	Et₃As + n-C₁₆H₃₃I	Crystals, m. 114°	605
MePh₂(1-C₁₀H₇)As	I	Ph₂(1-C₁₀H₇)As + MeI	m. 190-191°	97
[Me₂(2-Me₂NC₆H₄)AsCH₂]₂CH₂	Br₂	2-Me₂NC₆H₄AsMe₂ + CH₂(CH₂Br)₂	m. 209-210°	813
C₂₄				
(3-O₂NC₆H₄)₄As	Br	(3-O₂NC₆H₄)₄AsNO₃ + NaBr	m. 252-258° (dec.)	101
	Cl	(3-O₂NC₆H₄)₄AsNO₃ + NaCl	m. 235-239° (dec.)	101
	I	(3-O₂NC₆H₄)₄AsNO₃ + NaI	m. 235-237° (dec.)	101
	NO₃	Ph₄AsNO₃ + HNO₃-H₂SO₄	m. 248-256° (dec.)	101
Ph₄As	Br	Ph₄AsI + KBr	Hydrate, white needles[19]	201
		Ph₅As·0.5C₆H₁₂ + HBr	m. 279-281	1311
	Br₃	Ph₅As·0.5C₆H₁₂ + Br	Orange leaflets, m. 215-216°	1311
	Cl	Ph₃AsO + PhMgBr + HCl	m. 255-257°, other data[20]	98, 1136

18. Rxn. with PhLi in Et₂O → benzyldimethylfluorenylidenearsenic (1313)
19. Hydrate loses H₂O at 200°, m. 277-281° (201). At elevat. temp. dissoc. (97); inhibits corrosion of steel (30); reacts with PtBr₃ to form (Ph₄As)₂PtBr (103).
20. M. 256° (1131a). UV spectrum (1023a). Cond. (1023a). Polarographic wave (1131a). Adducts: Ph₄AsCl.HCl, m. 204-208° (1136). (Ph₄As)₂HgCl₄, insol. (1300, 1303). Forms also adduct with Sn⁴⁺ Cd²⁺, Tl³⁺, and Tc⁴⁺ (1303).

TABLE LII (cont'd.)

QUATERNARY ARSONIUM COMPOUNDS

Quaternary Arsonium R_4As^+	Salt X^-	Prepd. from	Props. and Remarks	Ref.
Ph_4As	Cl	$Ph_2AsCl + AlCl_3$ at 200° $PhAsCl_2 + AlCl_3$	Rxns.[21] Use[22]	755 755
	Cl_3	$Ph_5As:0.5C_6H_{12} + Cl$	Unstable green leafltes, m. 169-180° (dec.).	1311
	I	$AsCl_3$ in C_6H_6 or $PhAsCl_2$, Ph_2AsCl, or Ph_3As under reflux[30]	White needles, m. 314-319° Rxn. with KBr → Ph_4AsBr Diffraction data Crystal structure	201 201 405 859
	I_3	$Ph_5As0.5C_6H_{12} + I$	Brown crystals, m. 228-230° (dec.), other data[23]	1311
$(Ph_4As)_2$	NO_3	$Ph_4AsBr + HNO_3$ $Ph_4AsOH + HNO_3$	m. 259-261° m. 260-262°	98 97
Ph_4As	SO_4	$Ph_4AsOH + H_2SO_4$	m. 257-258° (dec.)	97
	OAc	$Ph_4AsOH + AcOH$	m. 215-127°	97
	OCO_2H		m. 173-174° (dec.)	97
	Picrate		Yellow, m. 203-204°	97
	SCN		m. 201-202°	97
	BPh_4	$Ph_3As + Ph_2IBPh_4$	m. 268-270°	755
		$Ph_5As + BPh_3$	m. 302-308°	755 779, 905
		$Ph_4AsBr + LiBPh_4$	UV spectrum; cond.	1023a
	$AlBrCl_3$	$Ph_3As:AlCl_3 + PhBr$ at 200°	m. 309-311°	1311
	MnO_4		Insol. salt	1311
	ClO_4		Insol. salt	755
	RhO_4		Crystal nucleation White cryst. ppt.	1299 1299 304 1299, 1301, 1325a 1226
$(Ph_4As)_2$	$FeCl_4$	$Ph_4AsBr + PtBr_3$	Soly.	1335
	$PtBr_6$	$Ph_4AsCl + H_2TeO_3$	Crystal structure	103
	$TeCl_6$			104
Ph_4As	$Co-(SCN)_2$		Soly.	5

Compound	Anion	Rxn. with	Properties	Ref.
			Ph₃AsS m. > 325°	
(3-H₂NC₆H₄)₄As	Br	(3-H₂NC₆H₄)₄AsCl + NaBr		101
	Cl	(3-O₂NC₆H₄)₄AsCl + SnCl₂ + AcOH-HCl		101
(PhCH₂)₃(CH₂:CHCH₂)As	Br	(PhCH₂)₃As + CH₂:CHCH₂Br	m. 180-182°	98
Ph₃(cyclo-C₆H₁₁)As	Br	Ph₂(C₆H₁₁)As + PhMgBr	m. 183-184°, dissoc. >315°	97
(MeBuPhAsCH₂-)₂	Dipicrate	(BuPhAsCH₂-)₂ + MeI, followed by Na picrate	Occurs in two forms[24]	199
C₂₅				
Ph(4-BrC₆H₄)(4-ClC₆H₄)(4-MeC₆H₄)As	I	Ph(4-BrC₆H₄)(4-ClC₆H₄)As + 4-MeC₆H₄Br[30]	m. 206-208°, other salts[25]	805
Ph₃(2-O₂NC₆H₄CH₂)As	Br	Ph₃As + 2-O₂NC₆H₄CH₂Br	Prisms or plates, m. 135°	710
Ph₃(3-O₂NC₆H₄CH₂)As	Br	Ph₃As + 3-O₂NC₆H₄CH₂Br	Prismatic needles, m. 167-168°	710
Ph₃(4-O₂NC₆H₄CH₂)As	Br	Ph₃As + 4-O₂NC₆H₄CH₂Br	Monohydrate, m. 93-94°, anhyd. 150-152°	710
			m. 160-162°	710
			m. 185-187°	98
Ph₃(3-MeC₆H₄)As.	I	Ph₃As + 3-MeC₆H₄Br[30]		805
Ph₃(4-MeC₆H₄)As	I	Ph₃As or Ph₂(4-MeC₆H₄)As[26]	Monohydrate, m. 186-187°	755
	SCN	Ph₃As or Ph₂(4-MeC₆H₄)As[26]	Monohydrate, m. 147-148°	755
Ph₃(PhCH₂)As	Cl	Ph₃(PhCH₂)AsI + AgCl	m. 180-181	98
	I	Ph₃(PhCH₂)AsI	m. 155-157°	98
	NO₂	Ph₃(PhCH₂)AsI + HNO₃	m. 178-180°	98
		Ph₃(PhCH₂)AsCl + NH₄NO₃	Complex compds.[27]	838
MePh(BzCH₂CH₂)₂As	I		m. 145-146°	243
Ph₃[(EtO₂C)₂CH]As	Br	Ph₃As + BrCH(CO₂Et)₂	m. 169-171°	98
Me(3-AcNHC₆H₄)₃As	Cl		m. 181-190° (dec.)	101
Me₂(PhCH)(n-C₁₆H₃₃)As	Cl	Me₂(n-C₁₆H₃₃)As + PhCH₂I	Hydroscop., m. 181-183°	605
	I	Me₂(PhCH₂)As + n-C₁₆H₃₃I	Needles[2], m. 69°	605

21. Rxn. with: LiAlH₄ or LiBH₄ → Ph₃As (1289); PhLi → Ph₅As(1311); OsCl₆²⁻ in aq. soln. → yellow ppt. (spectroscopic detn. of Os) (890, 891); Bi³⁺, Cd²⁺, and Hg²⁺ → pptn. (1131a); Ti⁴⁺ + chelating agent → orang-yellow color (1152); Fe(III) salt of 7-iodo-8-quinolinol-5-sulfonic acid → HCCl₃-sol. salt (1340).

22. Quant. detn. of MnO₄⁻, ReO₄⁻, TcO₄⁻, ClO₄⁻, IO₄⁻, and BF₄⁻ (1303); Polarographic detn. of Sn (587, 676); Extraction of Re (1140); pptn. of Te (1284).

23. Very stable, extremely insol. (860); Crystal structure (860, 1147).

24. Meso- and racemic forms were isolated; α-form, m. 113-115°, sol. in EtOH; β-form, m. 139.5-140.5°, less sol. in EtOH; mixed m. 114-123° (199).

TABLE LII (cont'd.)
QUATERNARY ARSONIUM COMPOUNDS

Quaternary Arsonium R4As+	Salt X-	Prepd. from	Props. and Remarks	Ref.
C26				
Ph3(4-BrC6H4CH2CO)As	Br	Ph3As + 4-BrC6H4CH2COBr	m. 170-171°	98
Ph3(BzCH2)As	Br	Ph3As + BzCH2Cl	Prisms or leaflets, m. 182-183°	710
Ph(4-BrC6H4)(4-ClC6H4)(3,4-Me2C6H3)As	Cl	Ph3(BzCH2)AsCl + NaOH	Hexagonal leaflets, m. 166°	710
	OH	Ph(4-BrC6H4)(4-ClC6H4)As + 3,4-Me2C6H3Br[30]	m. 176° Other salts[28]	710 805
Ph(4-ClC6H4)(3-MeC6H4)(4-MeC6H4)As	I	Ph(4-ClC6H4)(4-MeC6H4)As + 3-MeC6H4Br[30]	m. 148-141°; other salts[29]	805
Ph2(4-MeC6H4)2As	I	Ph2(4-MeC6H4)As + 4-MeC6H4Br[30]	Monohydrate	755
Ph3(4-EtC6H4)As	I	Ph3As + 4-EtC6H4Br[30]		805
Et(PhCH2)(4-MeC6H4)(1-C10H7)As	Br	Et(4-MeC6H4)(1-C10H7)As + PhCH2Br at 145°	m. 164-165°, rxn.[31]	627
			m. 186.5-189°	627
C27				
(4-ClC6H4)(3-MeC6H4)(4-MeC6H4)2As	I	(4-ClC6H4)(4-MeC6H4)2As + 3-MeC6H4Br[30]		805
Ph(4-MeC6H4)3As	I	(4-MeC6H4)3As + PhBr[30]	m. 205-206°	755
Ph(4-MeC6H4)3As	SCN	(4-MeC6H4)3As + PhBr + AlCl3 under reflux, followed by aq. KSCN	m. 143-144°	755
(MePh2AsCH2-)2	I2	(Ph2AsCH2-)2 + MeI	m. 172°	1215
MePh[PhCH2O(CH2)3]2As	I	Ph[PhCH2O(CH2)3]2As + MeI	m. 67-69.5°	242
(C6H11)2AsC:CAsMe(C6H11)2	I	(C6H11)2AsC:CAsMe(C6H11)2	m. 133°	516
C28				
(4-MeC6H4)4As	I	(4-MeC6H4)3As + 4-MeC6H4Br[30]	m. 253-255°	755
	SCN	(4-MeC6H4)3As + 4-MeC6H4Br[30]	Hydrate, m. 207-209°, loses H2O -t 130°/15 mm	755
(3-MeC6H4)4As	I	(3-MeC6H4)3As + 3-MeC6H4Br[30]	m. 155-156°	755
o-C6H4[CH2AsMe2(C6H4NMe2)]2	Br2	2-MeNC6H4AsMe2 + o-C6H4(CH2-Br)2		813
C29				
Ph2(PhCH2)(1-C10H7)As	I	Ph2(1-C10H7)As + PhCH2I	m. 171-172°	98

Compound	Anion	Preparation	Properties	Ref
			m. 1/8-1/9.5 rxn.	805
Ph(4-ClC$_6$H$_4$)(3-MeC$_6$H$_4$)(4-Ph-C$_6$H$_4$)As	I	Ph(4-ClC$_6$H$_4$)(3-MeC$_6$H$_4$)(4-Ph-3-MeC$_6$H$_4$Br30 +	Monohydrate, m. 75-135°	805
Ph(4-ClC$_6$H$_4$)(4-MeC$_6$H$_4$)(4-Ph-C$_6$H$_4$)As	I	Ph(4-ClC$_6$H$_4$)(4-MeC$_6$H$_4$)(4-Ph-4-MeC$_6$H$_4$Br30 +	Monohydrate, m. 90-125° (dec.) other salts 33	805
C$_{32}$				
Ph(4-ClC$_6$H$_4$)(4-EtC$_6$H$_4$)(4-PhC$_6$-H$_4$)As	I	Ph(4-ClC$_6$H$_4$)(4-PhC$_6$H$_4$)As + 4-EtC$_6$H$_4$Br30	Monohydrate	805
(3-AcNHC$_6$H$_4$)$_4$As	Cl	(3-H$_2$NC$_6$H$_4$)$_4$AsCl + Ac$_2$O	m. 172-220° (dec.)	101
C$_{33}$				
Ph(PhCH$_2$)[PhCH$_2$O(CH$_2$)$_3$]As	Br	Ph[PhCH$_2$O(CH$_2$)$_3$]$_2$As + PhCH$_2$Br	Viscous liquid	242
{Ph(PhCH$_2$)[PhCH$_2$O(CH$_2$)$_3$]As}$_2$	PtCl$_6$	Ph(PhCH$_2$)[PhCH$_2$O(CH$_2$)$_3$]AsBr + H$_2$PtCl$_6$	Pinkish-yellow crystals, m. 119.5-121°	242
C$_{36}$				
4,1,2-ClC$_6$H$_3$(CH$_2$AsMe$_2$Ph)$_2$	Dipicrate	Me$_2$PhAs + 4,1,2-ClC$_6$H$_3$(CH$_2$-Br)$_2$, followed by Na picrate	Yellow salt, m. 151-152°	757
o-C$_6$H$_4$(CH$_2$AsMe$_2$Ph)$_2$	Dipicrate	Me$_2$PhAs + o-C$_6$H$_4$(CH$_2$Br)$_2$ followed by Na picrate	Orange, m. 170-173°	757
C$_{37}$				
Me(2-PhC$_6$H$_4$)$_3$As	I	(2-PhC$_6$H$_4$)$_3$As + MeI	Dissoc. at 154° into (2-PhC$_6$H$_4$)$_3$As + MeI	1318
C$_{39}$				
(CH$_2$:CHCH$_2$)(PhC$_6$H$_4$)$_3$As	Br	(PhC$_6$H$_4$)$_3$As + CH$_2$:CHCH$_2$Br	m. 264-266°	98

25. d-Bromocamphorsulfonate.H$_2$O, friable glass; d-camphornitronate, friable glass; 1-N-(1-phenylethyl)phthalamate, brittle glass; 1-menthoxyacetate, brown sirup; antimonyl-d-tartrate (805).

26. The arsonium salts prepd. by heating Ph$_3$As with 4-MeC$_6$H$_4$Br or Ph$_2$(4-MeC$_6$H$_4$)As with PhBr in the presence of AlCl$_3$ and treating bromide with KI and KSCN, resp. Monohydrates lose H$_2$O at 130°/15 mm. (755).

27. The nitrate forms complex salts: La(NO$_3$)$_3$.2[Ph$_3$(PhCH$_2$)As]NO$_3$, yellow, m. 163.75-164°; Pr(NO$_3$)$_3$.-2[Ph$_3$(PhCH$_2$)As]NO$_2$, yellow green, m. 165-165.25°; Nd(NO$_3$)$_3$.2[Ph$_3$(PhCH$_2$)As]NO$_3$, lilac, m. 166.25° (839).

28. 1-Menthoxyacetate, sirup; Antimonyl-d-tartrate, solid; Sulfate, hydroscopic; d-H-tartrate.2H$_2$O, brittle glass; 1-Tricatechylarsanate(805).

29. d-Bromocamphorsulfonate monohydrate, hydroscopic; d-Campornitronate, brittle hydroscopic glass (805).

30. The quaternization carried out at 180-190° in the presence of AlCl$_3$ and the arsonium bromide is treated with KI and KSCN, resp. (755, 805).

31. Rxn. with Ag d-π-bromocamphorsulfonate → solid, m. 162-164°, m. 170-172° (627).

32. Rxn. with alc. NaOH → (9-fluorenylidene)triphenylarsenic (606).

33. d-Camphorsulfonate, friable glass; d-Bromocamphorsulfonate, friable glass (805).

PENTAPHENYLARSENIC $C_{30}H_{25}As$ Ph_5As

Prepn.: By reacting Ph_4AsBr with PhLi· in Et_2O in a sealed tube at 40° (1311).
 By adding dropwise PhLi in Et_2O to Ph_3AsCl_2 in Et_2O (1311).
Props.: $Ph_5As.0.5C_6H_{12}$ (from cyclohexane) prisms, m. 149-150°, loses C_6H_{12} in
high vacuum at 65°; at 150° under N → Ph_3As, C_6H_{12}, and Ph_2 (1311).
Rxns. with: Halogens in CCl_4 → Ph_4As perhalides (1311).
HBr → Ph_4AsBr (1311).
AcOH, followed by H_2O and KI → Ph_4AsI (1311).
Ph_3B in Et_2O under N → $(Ph_4As)(BPh_4)$ (1311).

DERIVATIVES WITH THE As-N BONDS

The group of organoarsenic compounds includes organoarsenic derivatives containing
single and double arsenic-nitrogen bonds. The compounds with a single As-N bond
are formed by reacting triarylarsine dihydroxides with sulfanilamide in dioxane,
while those with a double As=N bond are obtained by reacting tertiary arsines with
chloramine-T in absolute ethanol or with the sodium derivative of methyl N-chloro-
carbamate in benzene. The constitution of the derivative obtained from the carbamate is, however, questionable. The compounds with the single As-N bond are
named (4-Sulfanilamido)triarylarsonium hydroxides, while those with the double
As=N bond are named As,As,As-triaryl-N-p-tolylsulfonylarsinimines and Triarylarsin
carbomethoxyimides, respectively.

TABLE LIII
SULFAMIDOTRIARYLARSONIUM HYDROXIDES

$R_3As(OH)NHO_2SC_6H_4NH_4$	Prepd. from	Yield %	Props.	Ref.
$Ph_3As(OH)NHO_2SC_6H_4NH_4$	$Ph_3As(OH)_2 + 4\text{-}H_2NC_6H_4\text{-}SO_2NH_2$	79	m. 163-164°	1198
$Ph_3As(OH)NHO_2SC_6H_4NHAc$	$Ph_3As(OH)_2 + 4\text{-}AcNHC_6H_4\text{-}SO_2NH_2$	86	m. 158-159° (dec.)	1198
$(2\text{-}MeC_6H_4)_3As(OH)NHO_2SC_6H_4NH_2$	$(2\text{-}MeC_6H_4)_3As(OH)_2 + 4\text{-}H_2NC_6H_4SO_2NH_2$		m. 186-187°	1198
$(4\text{-}MeC_6H_4)_3As(OH)NHO_2S\text{-}C_6H_4NH_2$	$(4\text{-}MeC_6H_4)_3As(OH)_2 + 4\text{-}H_2NC_6H_4SO_2NH_2$	63	m. 149-150°	1198
$(2\text{-}MeC_6H_4)_3As(OH)NHO_2SC_6H_4NHAc$	$(2\text{-}MeC_6H_4)_3As(OH)_2 + 4\text{-}AcNHC_cH\ SO_2NH_2$	63	m. 215-217° (dec.)	1198

TABLE LIV
TRIARYLARSINIMINES

$R_3As:NO_2XR'$	Prepd. from	Props.	Ref.
$Ph_3As:NOCOMe$	$Ph_3As + MeOCONClNa$	m. 84°, questionable structure	191, 192
$Ph_3As:NO_2SC_6H_4Me$	$Ph_3As + 4-MeC_6H_4SO_2NClNa$	m. 192-193°	799
$(2-MeC_6H_4)_3As:NO_2SC_6H_4Me$	$(2-MeC_6H_4)_3As + 4-MeC_6H_4-SO_2NClNa$	m. 201-202°	799
$(3-MeC_6H_4)_3As:NO_2SC_6H_4Me$	$(3-MeC_6H_4)_3As + 4-MeC_6H_4-SO_2NClNa$	Waxy solid	799
$(4-MeC_6H_4)_3As:NO_2SC_6H_4Me$	$(4-MeC_6H_4)_3As + 4-MeC_6H_4SO_2-NClK$	m. 179-180° (dec.)	1198

YLIDENE-TRIORGANOARSENIC COMPOUNDS

The derivatives of pentavalent arsenic containing four organic substituents, one
of which is linked to the arsenic atom by a double C=As bond, are prepared by reacting quaternary arsonium halides which contain one organic substituent linked to
the arsenic atom by a primary or secondary carbon atom with organolithium compounds
or with aqueous alkali to eliminate one molecule of hydrogen halide. The compounds
are also prepared by reacting tertiary arsine dihalides with organic compounds
containing a carbon atom carrying two reactive hydrogen atoms.

(9-FLUORENYLIDENE)TRIMETHYLARSENIC $C_{16}H_{17}As$

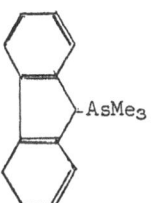

Prepn.: by treating 9-fluorenyltrimethylarsonium bromide with PhLi in Et_2O (1313).
Props.: Yellow compd. (1313).
Rxns. with: KI → 9-fluorenyltrimethylarsonium iodide (1313).
H_2O, NaOH, or MeONa → Me_3AsO (cleavage of the C=As bond) (1313).

METHYLENETRIPHENYLARSENIC $C_{20}H_{17}As$ $Ph_3As:CH_2$
Prepn.: by reacting $Ph_3MeAsBr$ with PhLi in Et_2O (495a).
Rxn. with Me_3SiBr → $Ph_3(Me_3SiCH_2)AsBr$ (495a).

TRIPHENYLARSENANYLIDENEMALONONITRILE $C_{21}H_{15}AsN_2$ $Ph_3As:C(CN)_2$
Prepn.: by reacting Ph_3AsCl_2 with $H_2C(CN)_2$ in C_6H_6 at 80° in the presence of Et_3N;
54% yield (563).
Props.: m. 190-191° (563).

CYANO(TRIPHENYLARSENANYLENE)ACETIC ACID METHYL ESTER $C_{22}H_{18}AsNO_2$ $Ph_3As:C(CN)CO_2$
Prepn.: by reacting Ph_3AsCl_2 with $NCCH_2CO_2Me$ in C_6H_6 at 80° in the presence of Et_3N;
62% yield (563).
Props.: m. 193-194° (563).

BENZYL(9-FLUORENYLIDENE)DIMETHYLARSENIC $C_{22}H_{21}As$

Prepn.: by treating benzyl-9-fluorenyldimethylarsonium bromide with PhLi in Et_2O
(not isolated) (1313).
Props.: Yellow compd..(1313).
Rxns. with: Dild. HCl, followed by aq. KI → benzylfluorenyldimethylarsonium
 iodide (1313).
H_2O → fluorene (1313).
Ph_2CO, followed by hydrolysis → 9-fluorenyldiphenylcarbinol (1313).

(PHENYLNITROMETHYLENE)TRIPHENYLARSENIC $C_{25}H_{20}AsNO_2$ $Ph_3As:C(Ph)NO_2$
Prepn.: by reacting Ph_3AsCl_2 with $PhCH_2NO_2$ in C_6H_6 at 80° in the presence of
Et_3N; 23% yield (563).
Props.: m. 128-129° (563).

TRIPHENYL[BIS(PHENYLSULFONYL)METHYLENE]ARSENIC $C_{31}H_{25}AsO_4S_2$ $Ph_3As:C(O_2SPh)_2$
Prepn.: by reacting Ph_3AsCl_2 with $H_2C(O_2SPh)_2$ in C_6H_4 at 80° in the presence of
Et_3N; 49% yield (563).
Props.: m. 252-254° (563).

(9-FLUORENYLIDENE)TRIPHENYLARSENIC $C_{21}H_{23}As$

Prepn.: by treating 9-fluorenyltriphenylarsonium bromide with 2.5 N NaOH in alc.
at r. t.; 78% yield (606).
Props.: Yellow plates, m. 188-190° (decompn.); UV spectrum (606).
Rxns. with: Alc. NaOH under reflux → Ph_3AsO(606).
RCHO → Ph_3AsO and $RCH:CC_6H_4C_6H_4$ (606).

MISCELLANEOUS DERIVATIVES OF PENTAVALENT ARSENIC

DIHYDROXYMETHYL-p-TOLYLARSONIUM CHLORIDE $C_8H_{12}AsClO_2$ $Me(p-MeC_6H_4)As(OH)_2Cl$
Prepn.: by reacting $[Me(p-MeC_6H_4)As]_2O$ or $Me(p-MeC_6H_4)AsCl$ with $PbCl_4$ in $CHCl_3$ at -50° (144).
Props.: Shiny leaflets, m. 133° (144).
Rxn.. with $H_2O \rightarrow Me(p-MeC_6H_4)AsOOH$ (144).

DIHYDROXYDIPHENYLARSONIUM CHLORIDE $C_{12}H_{12}AsCl$ $Ph_2As(OH)_2Cl$
Prepn.: by treating $(Ph_2As)_2O$ with dry HCl in $CHCl_3$ (145).
Props.: m. 128°, previously (Ann. <u>201</u>, 230(1880)) erroneously identified as $(Ph_2-AsCl_2)_2O$ (145).

TRI-o-TOLYLARSINE OXYDIBROMIDE $C_{42}H_{42}As_2Br_2O$ $[(2-MeC_6H_4)_3AsBr]_2O$
Prepn.: by recrystallizing $(2-MeC_6H_4)_3As(OH)OAsBr(C_6H_4Me-2)_3$ from dild. HBr (755).
Props.: m. 232° (755).

(4-ETHYLPHENYL)HYDROXYPHENYL-p-TOLYLARSONIUM OXIDE $C_{42}H_{44}AsO_3$ $[Ph(4-EtC_6H_4)-(4-MeC_6H_4)AsOH]_2O$
Prepn.: by treating $Ph(4-EtC_6H_4)(4-MeC_6H_4)As$ with Br_2, followed by NaOH (805).
Props.: Hydroscopic solid (805).

TRI-o-TOLYLARSINE OXYDIPICRATE $C_{54}H_{46}As_2N_6O_{15}$ $[(2-MeC_6H_4)_3AsOC_6H_2(NO_2)_3]_2O$
Props.: m. 169-170° (755).

TRI-o-TOLYLARSINE HYDROXY-OXY-BROMIDE $C_{42}H_{43}As_2BrO_2$ $(2-MeC_6H_4)_3As(OH)OAsBr-(C_6H_4Me-2)_3$
Prepn.: by heating $(2-MeC_6H_4)_3As$ with $2-MeC_6H_4Br$ at 100° in the presence of $AlCl_3$ and treating the rxn. prod. with aq. NaBr (755).
Props.: m. 148-152° (755).
Rxn. with dild. HBr → $[(2-MeC_6H_4)_3AsBr]_2O$ (755).

TRI-o-TOLYLARSINE HYDROXY-OXY-IODIDE $C_{42}H_{43}As_2IO_2$ $(2-MeC_6H_4)_3As(OH)OAsI-(C_6H_4Me-2)_3$
Prepn.: by heating $(2-MeC_6H_4)_3As$ with $2-MeC_6H_4Br$ at 100° in the presence of $AlCl_3$ and treating the rxn. prod. with aq. NaI (755).
Props.: Decomp. at elevated temp. (755).

DICHLORO(2-CHLOROVINYL)ARSINE OXIDE $C_2H_2AsCl_3O$ $ClCH:CHAsCl_2O$
Rxn. with iodo-Cu(I) reagent (from NaI + $CuSO_4$) → ppt. (tetection of traces of arsines) (301).

4-CHLORO-3-NITROPHENYLARSENIC DISULFIDE $C_6H_3AsClNO_2S_2$ $4,3-Cl(O_2N)C_6H_3AsS_2$
Prepn.: by treating $4,3-Cl(O_2N)C_6H_3AsO(OH)_2$ with H_2S in AcOH (147).
Props.: Discolors at 60°, decomp. 220° (147).
Rxn. with Na_2S in aq. alc. → $4,3-H_2N(HS)C_6H_4AsS_2$ (147).

3-DIAZONIUM-4-HYDROXYBENZENEDICHLOROARSONITE $C_6H_5AsCl_2N_2O$

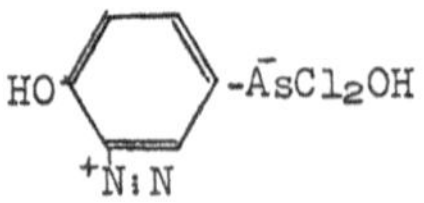

HO⟨⟩-ĀsCl₂OH
    ⁺N≀N

Prepn.: by treating 3-diazonium-4-hydroxybenzenetrichloroarsonite with abs. MeOH (48).
Props.: Bright yellow solid (48).

3-DIAZONIUM-4-HYDROXYBENZENETRICHLOROARSONITE $C_6H_4AsCl_3N_2O$

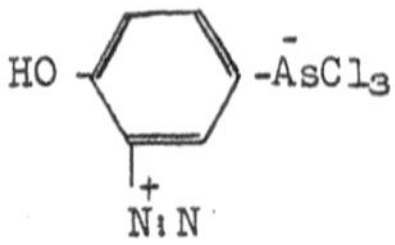

HO-⟨⟩-ĀsCl₃
      ⁺N≀N

Prepn.: by diazotization of 3-amino-4-hydroxybenzenearsonous acid, followed by a treatment with cold concd. HCl (48).
Props.: Pale yellow (48).
Rxns. with: β-Naphthol or α-naphthylamine in aq. soln. → deep colored dyes (48).
MeOH → replacement of one Cl by OH (48).

ARSENO DISULFIDES

The following compounds of the general formula RAs(S):As(S)R (?) were prepared from the corresponding arsonic acids by treating neutralized solutions of the arsonic acids with acetic acid, followed by anhydrous Na_2SO_3 at 30° and then by 50% H_3PO_2 at temperatures up to 70°, and reacting the arseno compounds with carbon disulfide.

TABLE LV
ARSENO DISULFIDES

Arseno Disulfide	Yield %	Props. and Remarks	Ref.
[MeAs(S):]₂	95	m. 93°, b. 250-260° (dec.)	655
[EtAs(S):]₂	64		655
[PrAs(S):]₂			655
[i-PrAs(S):]₂			655
[PhAs(S):]₂	92.3		655
[PhCH₂As(S):]₂	85.7		655
[4-HOC₆H₄As(S):]₂	82.5		655
[2-MeC₆H₄As(S):]₂			655
[3-ClC₆H₄As(S):]₂			655
[AcCH₂As(S):]₂			655
[5,8,7-Cl(HO)C₁₀H₄N-As(S):]₂*			655
MeAs(S):As(S)Et	60	Mixt. of MeAsO₃H₂ with EtAs-O₃H₂ used as the starting matl.	655

* 5-Chloro-8-hydroxy-7-quinolyl deriv.

3. HETEROCYCLIC DERIVATIVES OF ARSENIC

FIVE-MEMBERED RING SYSTEMS

ARSENOLE DERIVATIVES

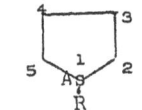

1-R-Arsenolidine

ARSENOLIDINE C_4H_9As $\overline{CH_2(CH_2)_3As}H$
Prepn.: by reducing $\overline{CH_2(CH_2)_3As}Cl$ with $LiBH_4$ at -60° in Et_2O in high vacuum; 55% yield (1279).
Props.: Colorless, water-clear liquid, sol. in org. solents, insol. in H_2O, m. -29 to -28°; extremely sensitive to oxygen (1297).

1-PHENYLARSENOLIDINE-2,5-DIONE $C_{10}H_9AsO_2$ $(CH_2CO)_2AsPh$
Prepn.: by adding a mixt. of $(CH_2COCl)_2$ and $PhAsCl_2$ to a mixt. of dry C_6H_6, Na wise, and AcOEt, and heating at 100° (267).
Props.: b. 119-120°/10 mm. (267).
Rxn. with Na + EtOH in MePh → $\overline{CH_2(CH_2)_3As}Ph(267)$.
Derivs.: Picrate, m. 117°; $HgCl_2$ adduct, m. 245°; Methiodide, m. 176°; Ethiodide, m. 165-167° (267).

1-PHENYLARSENOLIDINE $C_{10}H_{13}As$ $\overline{CH_2(CH_2)_3As}Ph$
Prepn.: by adding $(CH_2CO)_2AsPh$ in EtOH to hot MePh contg. Na (267).
Derivs.: Methiodide, m. 135-136°; $HgCl_2$ adduct, m. 160-162° (267).

2-(2-METHYLARSENOL-2-YL)SUCCINIC ACID DIMETHYL ESTER $C_{11}H_{19}AsO_4$ $\overline{HAs(CH_2)_3C}-$
(Me)CH(CO_2Me)CH_2CO_2Me
Prepn.: by passing AsH_3 into a soln. obtd. by reacting 5-chloro-2-pentanone with dimethyl bromosuccinate in the presence of activated Zn foil (961).

ARSINDOLE DERIVATIVES

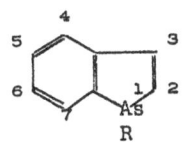

1-R-Arsindole

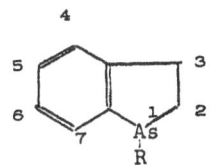

1-R-Arsindoline

1-CHLOROARSINDOLE C_8H_6AsCl
Prepn.: By heating under reflux $Ph(ClCH:CH)AsCl$ with $AlCl_3$ in CS_2 (260, 262, 263).
By heating under reflux $2\text{-}CH_2:CHC_6H_4AsCl_2$ with $AlCl_3$ in CS_2 (264).

By heating (PhCH:CH)$_3$As with excess AsCl$_3$ at 180° (intermidiate formation of PhCH:CHAsCl$_2$) (265).
Props.: Light yellow, skin-irritating oil of sharp odor, b. 128-135°/3 mm. (260, 262); b. 135-140°/5 mm (263); b. 125-130°/5 mm (264).
Rxns. with: RMgI → 1-R-arsindole (260, 262, 263).
HNO$_3$ → 2-HO$_2$CC$_6$H$_4$AsO$_3$H$_2$ (260, 262).
Complex compd. with HgCl$_2$, m. 151° (264).

1-METHYLARSINDOLE C$_9$H$_9$As*
Prepn.: By reacting 1-chloroarsindole with MeMgBr (260, 262, 263).
By heating (PhCH:CH)Me$_2$AsCl$_2$ (via unstable Me(PhCH:CH)AsCl) (265).
Props.: Malodorous, vesicant oil, b. 142-145°/6 mm (260, 262); b. 140-145°/5 mm. (265).
Rxn. with MeI → the methiodide (260, 262).
Derivs.: Methiodide, yellow needles, decomp. 216-218° (260, 262); Picrate, m. 106-107° (263); HgCl$_2$ complex, silky needles, m. 150-151° (260, 262, 263).

1-ETHYLARSINDOLE C$_{10}$H$_{11}$As*
Prepn.: by reacting 1-chloroarsindole with EtMgI (263).
Props.: b. 138-145°/6 mm., vesicant action (263).
Deriv.: Picrate, m. 100-102° (263).

1-PHENYLARSINDOLE C$_{14}$H$_{11}$As*
Prepn.: by reacting PhAsCl$_2$ with PhCH:CHBr in aq. NaOH at r. t., heating the mixture, isolating NaCl and NaBr, acidifying the org. prod. with HCl, and treating with SO$_2$ (265).
Props.: b. 165-170°/3 mm (265).

3-CHLORO-1-PHENYLARSINDOLE C$_{14}$H$_{10}$AsCl*
Prepn.: by reacting PhAsCl$_2$ with PhC:CH under reflux at 140-150° (266).
Props.: b. 168-175°/10 mm. (266).
Rxn. with H$_2$O$_2$ → Ph(2-HO$_2$CC$_6$H$_4$)AsO$_2$H (266).
Derivs.: Methiodide, m. 152-153° (266); Ethiodide, m. 161° (266); Picrate, m. 115-116° (266); HgCl$_2$ adduct, m. 232-233° (266),

1-PHENYLARSINDOLINE C$_{14}$H$_{13}$As*
Prepn.: By heating Ph(PhCH$_2$CH$_2$)AsO$_2$H with conc. H$_2$SO$_4$ at 100° and reducing the prod. with SO$_2$ in HCl contg. KI (614).
By cyclizing Ph(PhCH$_2$CH$_2$)AsCl at 100° in the presence of AlCl$_3$; 17% yield (614)
Props.: b. 126-128°/0.6 mm (614).
Derivs.: Methiodide.H$_2$O, m. 174-175° (effervesc.) (614).
[Cd(C$_{14}$H$_{13}$As)$_2$Br$_2$], m. 229-230° (614).

---

*For the structural formula see page 374.

ISOARSINDOLINE DERIVATIVES

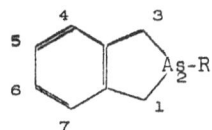

5-CHLORO-2-IODOISOARSINDOLINE C_8H_7AsClI
Prepn.: by heating 5-chloro-2-phenylisoarsindoline with HI at 120-130° in CO_2 (757).
Props.: m. 121° (757).
Rxn. with MeMgI → 5-chloro-2-methylisoarsindoline (757).

2-CHLOROISOARSINDOLINE C_8H_8AsCl
Prepn.: by hydrolyzing 2-iodoisoarsindoline with aq. Na_2CO_3 and treating the noncrystallizing hydroxy deriv. with concd. HCl (757).
Props.: m. 73-74° (757).

2-IODOISOARSINDOLINE C_8H_8AsI
Prepn.: by heating 2-phenylisoarsindoline or 2-(4-anisyl)isoarsindoline with HI at 120-130° in CO_2 (757).
Props.: Pale yellow, m. 107-108° (757).
Rxns. with: Aq. Na_2CO_3 → the 2-hydroxy deriv. which was converted to the 2-chloro deriv. by concd. HCl and to bis(2-isoarsindolinyl) sulfide by H_2S (757).
HNO_3 → $CH_2C_6H_4CH_2As(O)NO_3$ (757).
RMgX → 2-R-isoarsindole deriv. (813).

2,2-DIHYDROXYISOARSINDOLINIUM NITRATE $C_8H_{10}AsNO_5$
Prepn.: by reacting 2-iodoisoarsinoline with HNO_3 (757).
Props.: m. 144° (decompn.) (757).

5-CHLORO-2-METHYLISOARSINDOLINE $C_9H_{10}AsCl$
Prepn.: by treating 5-chloro-2-iodoisoarsindoline with MeMgI in Et_2O (757).
Props.: b. 81° /0.1 mm (757).
Rxn. with 5,1,2-$ClC_6H_3(CH_2Br)_2$ → 2,2'-spirobis(5-chloroisoarsindolinium bromide) (757).
Derivs.: Methiodide, m. 202-204° (257); Methopicrate, m. 270° (757); 5-Chloro-2-[4(or 5)-chloro-2-(bromomethyl)benzyl]-2-methylisoarsindolinium bromide, hard glass (757).

2-METHYLISOARSINDOLINE $C_9H_{11}As$
Prepn.: by reacting o-$C_6H_4(CH_2Br)_2$ with $MeAsCl_2$ and Na in Et_2O under reflux; 3.9% yield (756).
Props.: b. 115°/17 mm. (756).
Derivs.: Methiodide, m. 210-211° (756); Methopicrate, m. 209° (756); 2-[2-(Bromomethyl)benzyl]-2-methylisoarsindolinium bromide, a hydroscopic substance , which on heating at 14 mm was converted into 2, 2'-spirobis(isoarsindolinium bromide) (756).

5-CHLORO-2-PHENYLISOARSINDOLINE $C_{14}H_{12}AsCl$ *

Prepn.: By reacting 4-chloro-o-xylylene dibromide and $PhAsCl_2$ with Na in Et_2O under N with periodic addition of EtOAc (757).

By adding slowly $PhMe_2As$ to 4-chloro-o-xylylene dibromide in C_6H_6 under CO_2, removing the solvent, and heating the residue at 18 mm pressure until evolution of MeBr ceases; 63% yield (757).

Props.: b. 155-160°/0.2 mm (757).

Rxn. with HI → 5-chloro-2-iodoisoarsindoline (757).

Derivs.: 5-Chloro-2-hydroxy-2-phenylisoarsindolinium nitrate, m. 143.5° (decompn (757).

5-Chloro-2-hydroxy-2-phenylisoarsindolinium picrate, pale yellow, m. 154-155° (757).

2-(4-CHLOROPHENYL)ISOARSINDOLINE $C_{14}H_{12}AsCl$ *

Prepn.: by treating a mixt. of $4-ClC_6H_4AsCl_2$ and $o-C_6H_4(CH_2Br)_2$ with Na in Et_2O under reflux; the prod. was purified by conversion to the hydroxy nitrate, then t dihydroxide, followed by redn. with SO_2 (757).

Props.: b. 153-154°/0.05 mm (757).

Derivs.: 2-Hydroxy-nitrate, m. 144-145°; 2-Hydroxy-picrate, yellow, m. 147-148°; Dihydroxide.$1/2$ H_2O, m. 77-78°, loses H_2O in vacuo to form the oxide; Oxide, m. 145-146°; Methiodide, m. 204°; Methopicrate, yellow, m. 162° (757).

2-PHENYLISOARSINDOLINE $C_{14}H_{13}As$ *

Prepn.: By reacting $PhAs(MgBr)_2$ with $o-C_6H_4(CH_2Br)_2$ in C_6H_6 under reflux; 82% yield. $o-C_6H_4(CH_2Cl)_2$ instead of the dibromide gave a yield of 40% along with some $PhAsCl_2$ and s-tribenzocyclododecatriene (68).

By treating a mixt. of $PhAsCl_2$ and $o-C_6H_4(CH_2Br)_2$ with Na in Et_2O under N; 18.5% yield (756).

By reducing 2-phenylisoarsindoline 2,2-dihydroxide with SO_2 in aq. HCl contg. KI (756).

By treating $PhAsMe_2$ in $CHCl_3$ with $o-C_6H_4(CH_2Br)_2$ in CO_2, removing the solvent, and heating at 180°/17 mm; 59% yield (757).

Props.: b. 136-138°/0.3 mm (756).

Rxns. with: Concd. HNO_3 → the 2-hydroxide-2-nitrate (756).

Br_2 → the 2,2-dibromide (756).

H_2O_2 → the 2,2-dihydroxide (756).

$KMnO_4-Na_2CO_3$ → the 2,2-dihydroxide (756).

$K_2S_2O_8 + K_2Cr_2O_7$ → the 2,2-dihydroxide (765).

MeI → 2-methyl-2-phenylisoarsindolinium iodide (756).

HI at 120-130° → 2-iodoisoarsindoline (757).

Hydrated chloramine-T → 2-hydroxy-2-phenylisoarsindolinium p-toluenesulfonate (75

Derivs.: Methiodide, m. 189-191°; Picrate, m. 122-123° (756).

2-Oxide, prepd. by dehydration of 2,2-dihydroxide over P_2O_5 in vacuo; m. 130-131 on exposure air forms the dihydroxide (756).

2-Sulfide, prepd. by treating the 2,2-dihydroxide with H_2S in $CHCl_3$; m. 108° (756

2-Bromide-2-hydroxide, prepd. by treating the 2,2-dihydroxide with conc. HBr; m. 132-133° (756).

2-Hydroxide-2-nitrate, prepd. by treating 2-phenylisoarsindoline with concd. HNO_3 m. 149-150°; rxn. with 10% NaOH → the 2,2-dihydroxide (756).

---

* For the structural formula see page 375.

2,2-Dihydroxide, prepd. by: (1) treating 2-phenylisoarsindoline 2-hydroxide-2-nitrate with 10% NaOH; (2) treating 2-phenylisoarsindoline with H_2O_2, $KMnO_4$-Na_2CO_2, $K_2S_2O_8$-$K_2Cr_2O_7$, or Br_2 in $CHCl_3$, followed by aq. NaOH; m. 78-79°, loses H_2O in vacuo over P_2O_5 to form the 2-oxide. <u>Rxn. with:</u> SO_2 in aq. HCl contg. KI → 2-phenylisoarsindoline; with HBr → 2-hydroxide-2-bromide; Picric acid → 2-hydroxide-picrate; H_2S → 2-sulfide (756).

2-Hydroxide-2-picrate, prepd. by treating the 2,2-dihydroxide with picric acid; yellow, m. 164-166° (dec.) (756).

2-Hydroxide-p-toluenesulfonate, prepd. by treating 2-phenylisoarsindoline with hydrated chloramine-T in EtOH under reflux; m. 122°; rxn. with NaOH → 2-hydroxide-2-picrate (756).

2-p-TOLYLISOARSINDOLINE $C_{15}H_{15}As$ *

<u>Prepn.</u>: by treating a mixt. of 4-$MeC_6H_4AsCl_2$ and o-$C_6H_4(CH_2Br)_2$ with Na in Et_2O in the presence of AcOEt; 19% yield; purified by converting it to the 2-hydroxide-2-nitrate, treating the latter with dild. NaOH and reducing with SO_2 in dild. HCl contg. KI (757).
<u>Props.</u>: b. 138-139°/0.3 mm (747).
<u>Rxn.</u> with conc. HNO_3 → the 2-hydroxide-2-nitrate (757).
<u>Derivs</u>.: 2-Hydroxide-2-nitrate, m. 146-146.5°, prepd. in 13% yield from the arsine + conc. HNO_3 (757).
2-Hydroxy-2-picrate, light feathery crystals, m. 147°, prepd. from the hydroxide-nitrate (757).
5-Oxide, m. 134-136°, prepd. by shaking the hydroxide-nitrate with 5% NaOH, is highly hydroscopic and forms dihydroxide hemihydrate, m. 53-54° (757).
Methiodide, m. 156° (757).
Methopicrate, yellow, m. 134-135° (757).

2-(4-METHOXYPHENYL)ISOARSINDOLINE $C_{15}H_{15}AsO$ *

<u>Prepn.</u>: by treating a mixt. of 4-$MeOC_6H_4AsCl_2$ and o-$C_6H_4(CH_2Br)_2$ with Na in Et_2O contg. EtOAc; 12% yield (757).
<u>Props.</u>: m. 91-92°; b. 167-168°/0.03 mm (757).
<u>Derivs.</u>: Methiodide, m. 148° (757).
Methopicrate, yellow, m. 134° (757).
2-Hydroxide-picrate, yellow, m. 159° (759).
2-Hydroxide-2-nitrate, m. 414° (757).
2-Oxide, m. 150-151° (757).
2, 2-Dihydroxide, sirupy liquid (757).

DI(2-ISOARSINDOLINYL) SULFIDE $C_{16}H_{16}As_2S$ *

<u>Prepn.</u>: by hydrolyzing 2-iodoisoarsindoline with aq. NaOH and treating the hydroxy deriv. with H_2S (757).
<u>Props.</u>: m. 146° (757).

2-(2-DIMETHYLAMINOPHENYL)ISOARSINDOLINE $C_{16}H_{18}AsN$ *

<u>Prepn.</u>: by reacting 2-iodoisoarsindoline with 2-$Me_2NC_3H_4MgBr$ in C_6H_4; 68% yield (813).
<u>Props.</u>: b. 151-154°/0.5 mm; m. 75° (813).
<u>Deriv.</u>: Methobromide.hemihydrate, m.189-190°, Perchlorate, m. 156° (813).

---

* For the structural formula see page 375.

2,2'-o-PHENYLENEBIS(5-CHLOROISOARSINDOLINE) $C_{22}H_{18}As_2Cl_2$

<u>Prepn.</u>: by heat decompn. of 8-chloro-5,6,11,12-tetrahydro-5,5,12,12-tetramethyl-dibenzo[b,f][1,4]diarsocinium dibromide or 8-chloro-6,1-dihydro-5,12-dimethyl-5,12-ethanodibenzo[b,f][1,4]diarsocinium dibromide (610).
<u>Props.</u>: m. 191-195° (610).

2,2'-o-PHENYLENEDI(ISOARSINDOLINE) $C_{22}H_{20}As_2$

<u>Prepn.</u>: by heat decompn. of [Me$_2$AsC$_6$H$_4$As(Me$_2$)CH$_2$C$_6$H$_4$CH$_2$]Br$_2$ with or without o-C$_6$H$_4$(CH$_2$Br)$_2$ or [CH$_2$CH$_2$As(Me)C$_6$H$_4$As(Me)CH$_2$C$_6$H$_4$CH$_2$]Br$_2$ (610).
<u>Props.</u>: m. 129-131° (610).
<u>Rxns. with</u>: MeI → monomethiodide (610).
K$_2$PdBr$_4$ → mono-PdBr$_2$ complex (610).
(CH$_2$Br)$_2$ → 1,1'',2',3,3',3'' -hexahydrodispiro[isoarsindolinium-2,1'-[1,4]benzo-diasenin-4',2'' -isoarsindolinium]dibromide (810).
<u>Derivs.</u>: Monomethiodide, m. 180-181°; mono-PdBr$_2$, yellow crystals, m. 336-337°
(610).

1,1'',2',3,3',3'' -HEXAHYDRODISPIRO[ISOARSINDOLINIUM-2,1'-[1,4]BENZODIARSENIN-4',2'' -ISOARSINDOLINIUM] DIBROMIDE $C_{24}H_{24}As_2Br_2$

<u>Prepn.</u>: by heating at 125-130° (CH$_2$Br)$_2$ with 2,2 -o-phenylenedi(isoarsindoline) (

Props.: Monohydrate, m. 223° (610).
Derivs.: Dipicrate, yellow crystals, m. 216-217° (decomp.) (610)

OXADIARSENOLE AND BENZOXADIARSENOLE DERIVATIVES

1,3-R,R-2,1,3-Oxadiarsenolane   1,3-R,R-1,3-Dihydro-2,1,3-benzoxadiarsenole

1,3-DIPHENYL-2,1,3-OXADIARSENOLANE $C_{14}H_{14}AsO$
repn.: by reacting $(CH_2AsPhCl)_2$ with NaOH (609).
Props.: m. 94° (609).
Rxn. with: H_2O_2 → the 1,3-dioxide (609).

ETHYLENEBIS(PHENYLARSINIC ACID) CYCLIC ANHYDRIDE $C_{14}H_{14}As_2O_3$

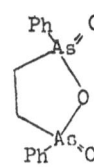

repn.: by treating 1,3-diphenyl-2,1,3-oxadiarsenolane with H_2O_2 (609).
Props.: Monohydrate, m. 191° (effervesc.) (609).
Rxn. with warm acids → $(-CH_2AsPhOOH)_2$ (609).

1,3-DICHLORO-1,3-DIHYDRO-2,1,3-BENZOXADIARSENOLE $C_6H_4As_2Cl_2O$
repn.: by reducing $4-MeOC_6H_4As(O)C_6H_4-2-As(O)(OH)O$ with SO_2 in hot HCl; 84%
yield (611).
Props.: Yellow crystals, m. 148-150° (611).
Rxn. with $SOCl_2$ → o-phenylenebis(dichloroarsine) (199).

1,3-DIHYDRO-1,1,3,3-TETRAMETHYL-2,1,3-BENZOXADIARSENOLINIUM DINITRATE
   $C_{10}H_{16}As_2N_2O_7$
repn.: by oxidation of $o-C_6H_4(AsMe_2)_2$ with cold HNO_3 (610).
Props.: Monohydrate, m. 223° (610).
Deriv.: Dipicrate, yellow needles, m. 138-140° (610).

2-(p-TOLYLARSINICO)BENZENEARSONIC ACID CYCLIC ANHYDRIDE $C_{13}H_{12}As_2O_5$

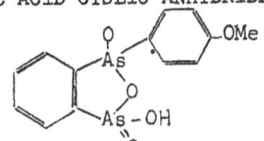

repn.: by diazotization of o-arsanilic acid, coupling with $4-MeOC_6H_4AsO$ in alc.

soln., and neutralizing the rxn. mixt.; 51% yield (611).
Props.: m. 373° (decompn.) (611).
Rxn. with SO_2 in hot HCl → $ClAsC_6H_4AsClO$ (611).

DIBENZARSENOLE DERIVATIVES

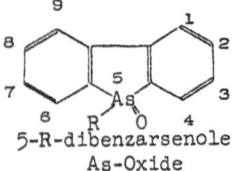

Dibenzoarsenole   Dibenzarsenolic Acid   5-R-dibenzarsenole
                                                          As-Oxide

3-BROMO-5-CHLORODIBENZARSENOLE $C_{12}H_7AsBrCl$
Prepn.: by adding concd. HCl and amalgamated Zn to 3-bromodibenzarsenolic acid in C_6H_6; 40% yield (451).
Props.: m. 176° (451).

5-BROMO-3-CHLORODIBENZARSENOLE $C_{12}H_7AsBrCl$
Prepn.: by reducing 3-chlorodibenzarsenolic acid with PBr_3 in $CHCl_3$; 43% yield (451).
Props.: m. 153° (451).

5-BROMO-3-NITRODIBENZARSENOLE $C_{12}H_7AsBrNO_2$
Prepn.: by treating 3-nitrodibenzarsenolic acid with PBr_3 in AcOH at 80°; 80.1% yield (449).
Props.: m. 199-200° (449).

5-CHLORO-3-NITRODIBENZARSENOLE $C_{12}H_7AsClNO_2$
Prepn.: by treating 3-nitrodibenzarsenolic acid with PCl_3 in AcOH at 80°; 79.3% yield (449).
Props.: m. 186-187° (449).

2,5-DICHLORODIBENZARSENOLE $C_{12}H_7AsCl_2$
Prepn.: by treating 2-chlorodibenzarsenolic acid with PCl_3 at 0°; 26.8% yield (451).
Props.: m. 135-136° (451).

3,5-DICHLORODIBENZARSENOLE $C_{12}H_7AsCl_2$
Prepn.: by reducing 3-chlorodibenzarsenolic acid with PCl_3 in AcOH at 80°; 50% yield (451).
Props.: m. 154-155° (451).

3,7-DINITRODIBENZARSENOLIC ACID $C_{12}H_7AsN_2O_6$.
Prepn.: by treating dibenzarsenolic acid with HNO_3 in concd. H_2SO_4 at 2°; 87% yield (377).
Props.: Pale yellow powder, m. >300° (377).

5-BROMODIBENZARSENOLE $C_{12}H_8AsBr$ *
Prepn.: by treating dibenzarsenolic acid with PBr_3 in $HCCl_3$; 65.1% yield (449).
Props.: m. 174-175° (449).

3-BROMODIBENZARSENOLIC ACID $C_{12}H_8AsBrO_2$ *
Prepn.: by diazotizing 3-aminodibenzarsenolic acid and adding the diazonium salt to Cu_2Br_2 in HBr; 54% yield (451).
Props.: m. 314-315° (451).
Rxn. with Hg-Zn in C_6H_6-HCl → 3-bromo-5-chlorodibenzarsenole (451).

5-CHLORODIBENZARSENOLE $C_{12}H_8AsCl$ *
Prepn.: by reducing dibenzarsenolic acid with PCl_3 in AcOH at 80°; 43.2% yield (449) or with SO_2 in HCl; 70% yield (179).
Props.: m. 160-161° (449); m. 161-162° (179).
Rxns. with: RMgX → 5-R-dibenzarsenole (179).
RCOSK → the 5-RCOS- deriv. (1249).

2-CHLORODIBENZARSENOLIC ACID $C_{12}H_8AsClO_2$ *
Prepn.: by diazotizing 2-aminodibenzarsenolic acid and adding the diazonium chloride soln. to an equimolar amt. of Cu_2Cl_2 in concd. HCl; 27% yield (451).
Props.: Decomp. 300° (451).
Rxn. with PCl_3 in $CHCl_3$ → 2,5-dichlorodibenzarsenole (451).

3-CHLORODIBENZARSENOLIC ACID $C_{12}H_8AsClO_2$ *
Prepn.: by diazotizing 3-aminodibenzarsenolic acid and adding the diazonium compd. to Cu_2Cl_2 in HCl; 44% yield (451).
Props.: m. 313-314° (451).
Rxn. with: PCl_3 in AcOH → 3,5-dichlorodibenzarsenole (451).
PBr_3 in $CHCl_3$ → 5-bromo-3-chlorodibenzarsenole (451).

2-NITRODIBENZARSENOLIC ACID $C_{12}H_8AsNO_4$ *
Prepn.: by heating 4-nitro-2-biphenylarsonic acid with concd. H_2SO_4 at 60°; 75% yield (377).
Props.: Cryst. powder, m. > 300° (377).
Rxn. with Fe (reduced in NaCl soln.) → the 2-amino deriv. (377).

3-NITRODIBENZARSENOLIC ACID $C_{12}H_8AsNO_4$ *
Prepn.: by treating dibenzarsenolic acid with HNO_3 in H_2SO_4 at 0°; 90% yield (377); at 5° (449).
Props.: Pale yellow crystals, m.>345° (377).
Rxns. with: $FeSO_4$ in NaOH → the 3-amino deriv. (377, 449).
PCl_3 in AcOH → the 5-chloro deriv. (449).
PBr_3 in AcOH → the 5-bromo deriv. (449).

DIBENZARSENOLIC ACID $C_{12}H_8AsO_2$ *
Prepn.: by heating $2\text{-PhC}_6H_4AsO(OH)_2$ with concd. H_2SO_4 at 50-60°; 80% yield (377, 449) or at 90°, 92% yield (179).
Props.: Needles, m. > 300° (377); m. 327-328° (449); m. 318-322° (179). UV spectrum (395).

---

* For the structural formula see page 380.

Rxns. with: HNO$_3$ in concd. H$_2$SO$_4$ at 0° → the 3-nitro and 3,7-dinotro derivs. (377, 449).

PCl$_3$ in AcOH → the 5-chloroarsenole (449).

PBr$_3$ in CHCl$_3$ → the 5-bromoarsenole (449).

SO$_2$ in HCl → 5-chlorodibenzarsenole (179).

1-AMINODIBENZARSENOLIC ACID C$_{12}$H$_{10}$AsNO$_2$ *

Prepn.: by cyclizing 2-AcNHC$_6$H$_4$-2-C$_6$H$_4$AsC(OH)$_2$ with concd. H$_2$SO$_4$ (377).

Props.: Microcrystals, m. > 300° (377).

2-AMINODIBENZARSENOLIC ACID C$_{12}$H$_{10}$AsNO$_2$ *

Prepn.: By adding 2-nitrodibenzarsenolic acid suspended in NaOH soln. to reduced Fe in NaCl soln. at 90°; 87% yield (377).

By heating 2,4-Ph(H$_2$N)C$_6$H$_3$AsO(OH)$_2$ with concd. H$_2$SO$_4$ at 60°; 80% yield (377).

Props.: Orange needles, m. > 350° (377).

Rxns. with: Ac$_2$O at 75-80° → the 2-acetamido deriv. (451).

ClCH$_2$CONH$_2$ → the 2-carbamylmethylamino deriv. (451).

NaNO$_2$, followed by Cu$_2$Cl$_2$ in HCl → the 2-chloro deriv. (451).

3-AMINODIBENZARSENOLIC ACID C$_{12}$H$_{10}$AsNO$_2$ *

Prepn.: By adding an alk. soln. of 3-nitrodibenzarsenolic acid to aq. soln. of FeSO$_4$ neutralized with NH$_4$OH; 75% yield (377, 449).

By cyclizing 2-(4-H$_2$NC$_6$H$_4$)C$_6$H$_4$AsO$_3$H$_2$ or 2-(4-EtO$_2$CNHC$_6$H$_4$)C$_6$H$_4$AsO$_3$H$_2$ with concd. H$_2$SO$_4$ (377).

Props.: Pink ppt., m. 240° (377); yellow color, decomp. without m. (449).

Rxns. with: NaNO$_2$, followed by Cu$_2$Br$_2$ in HBr → the 3-bromo deriv. (451).

NaNO$_2$, followed by Cu$_2$Cl$_2$ in HCl → the 3-chloro deriv. (451).

NaNO$_2$, followed by CuCN and NaCN → the 3-cyano deriv. (451).

3-CYANODIBENZARSENOLIC ACID C$_{13}$H$_8$AsNO$_2$ *

Prepn.: by diazotizing 3-aminodibenzarsenolic acid and reacting the diazonium compd. with CuCN and NaCN in H$_2$O at 60-70°; 38.7% yield (451).

Props.: decomp. 360° (451).

5-CHLORO-3-METHOXYDIBENZARSENOLE C$_{13}$H$_{10}$AsClO *

Prepn.: by heating 2-(4-MeOC$_6$H$_4$)C$_6$H$_4$AsCl$_2$ at 200°; 60% yield (179).

Props.: m. 136-137° (179).

Rxn. with p-MeC$_6$H$_4$MgBr → 3-methoxy-5-p-tolyldibenzarsenole (179).

5-METHYLDIBENZARSENOLE C$_{13}$H$_{11}$As *

Prepn.: By thermal decompn of 2,3,4-trihydro-1,1,5,5-tetramethyldibenzo[f,h]-1,5-diarsonianonin dibromide or 2,3-dihydro-1,1,4,4-tetramethyldibenzo[e,g]-1,4-diarsoniaocin dibromide (524).

Props.: m. 40-41°, b. 102-104°/0.01 mm. (528).

Derivs.: Methiodide monohydrate, m. 206-207° (528).

Methopicrate, m. 215° (528).

5-MERCAPTODIBENZARSENOLE ANHYDROSULFIDE WITH DIMETHYLDITHIOCARBAMIC ACID
  C$_{15}$H$_{14}$AsNS$_2$ *

Prepn.: by reacting 5-hydroxydibenzarsenole with Me$_2$NC(S)SNa in dild. HCl (1248).

---

* For the structural formula see page 3380.

Props.:   m. 142° (1248).

3,7-DIMETHYL-2,8-DINITRODIBENZARSENOLIC ACID $C_{14}H_{11}AsN_2O_6$ *
Prepn.:   by treating 3,7-Me$_2$-dibenzarsenolic acid with HNO$_3$ in H$_2$SO$_4$ at 0-5°;
76% yield (377).
Props.:   Feathery needles, m. >300° (377).

2-ACETAMIDODIBENZARSENOLIC ACID $C_{14}H_{12}AsNO_3$ *
Prepn.:   by heating 2-aminodibenzarsenolic acid with Ac$_2$O at 75-80°; 80.6% yield
(451).
Props.:   m. 245° (451).

2-CARBAMYLMETHYLAMINODIBENZARSENOLIC ACID $C_{14}H_{13}AsN_2O_3$ *
Prepn.:   by reacting on a steambath 2-aminodibenzarsenolic acid with ClCH$_2$CONH$_2$
in aq. KOH contg.KI; 12.5% yield (451).

3,7-DIMETHYLDIBENZARSENOLIC ACID $C_{14}H_{13}AsO_2$ *
Prepn.:   by cyclizing 4,4'-dimethyl-2-biphenylarsonic acid with H$_2$SO$_4$; 92% yield
(377).
Props.:   White needles, m. 321-322° (377).
Rxn. with HNO$_2$ in H$_2$SO$_4$ → the 2,8-dinitro deriv. (377).

3-NITRO-5-PHENYLDIBENZARSENOLE As-OXIDE $C_{18}H_{12}AsNO_3$ *
Prepn.:   by cyclizing (4'-nitro-2-biphenylyl)phenylarsinic acid in H$_2$SO$_4$ at 140-
150° (179).
Props.:   m. 120°; remelts at 272-273° (179).
Rxn. with SnCl$_2$ in HCl-EtOH → 3-amino-5-phenyldibenzarsenole (179).

5-PHENYLDIBENZARSENOLE $C_{18}H_{13}As$ *
Prepn.:   by reacting 5-chlorodibenzarsenole with PhMgBr (96).
Props.:   m. 87-88° (96).
Rxns. with: KMnO$_4$ in Me$_2$CO → (C$_{12}$H$_8$)PhAs(OH)$_2$ (96).
MeI → (C$_{12}$H$_8$)PhMeAsI (96).
Derivs.:   5,5-Diphenyldibenzarsenolinium bromide, prepd. by reacting 5-phenyl-
    dibenzarsenole 5-oxide with PhMgBr(23% yield), m. 240-241° (97).
5-Methyl-5-phenyldibenzarsenolinium iodide, prepd. by reacting 5-phenyldibenzar-
    senole with MeI (82% yield), m. 117-118° (97).

5-PHENYLDIBENZARSENOLE OXIDE $C_{18}H_{13}AsO$ *
Prepn.:   by oxidation of 5-phenyldibenzarsenole with KMnO$_4$ (96).
Props.:   m. 178-179° (96).
Rxn. with PhMgBr → (C$_{12}$H$_8$)Ph$_2$AsBr (97).

3-AMINO-5-PHENYLDIBENZARSENOLE $C_{18}H_{14}AsN$ *
Prepn.:   by reducing 3-nitro-5-phenyldibenzarsenole As-oxide with SnCl$_2$ in HCl-
EtOH under reflux; recovered as hydrochloride by decompg. the stannic chloride
of the base by NaOH and conversion to hydrochloride (179).
Props.:   d-isomer, m. 38-48°, [α]$_D^{25}$ + 255°; l-isomer, m. 37-47°, [α]$_D^{25}$ -251° (179).
Derivs.:   Hydrochloride, colorless needles, m. 220-224° (decompn.) (179).

---

* For the structural formula see page 380.

N-acetyl deriv., m. 156-157°; (+)- and (-) were isolated (179).

5-PHENYLDIBENZARSENOLE 5,5-DIHYDROXIDE $C_{18}H_{15}AsO_2$ *
Prepn.: by oxidation of 5-phenylbenzarsenole with $KMnO_4$ in Me_2CO (96).
Props.: m. 107-108° (96).

p-(5-DIBENZARSENOLYL)BENZOIC ACID $C_{19}H_{13}AsO_2$

Prepn.: by oxidizing 5-p-tolyldibenzarsenole by $KMnO_4$ in an alk. medium and
reducing the prod., p-(5-dibenzarsenolyl)benzoic acid As-oxide, by SO_2 in $CHCl_3$-
HCl in the presence of I_2 (179).
Props.: Colorless needles, m. 268-271°; attempted resolution (179).
Quinine salt, m. 207-208° attempted resolution (179).

p-(5-DIBENZARSENOLYL)BENZOIC ACID   As-OXIDE $C_{19}H_{13}AsO_3$

Prepn.: by oxidizing 5-p-tolyldibenzarsenole by $KMnO_4$ in an alk. medium (179).
Props.: m. 310-321° (315-318° in a bath preheated to 300°) (179).
Rxn. with SO_2 in $CHCl_3$-HCl + I_2 → p-(5-dibenzarsenolyl)benzoic acid (179).

5-p-TOLYLDIBENZARSENOLE $C_{19}H_{15}As$ *
Prepn.: by treating 5-chlorodibenzarsenole with p-MeC₆H₄MgBr in Et_2O; 70% yield
(179).
Props.: m. 130.5-131.5° (179); UV spectrum (174).
Salts: $(R_3As)_2(HgCl_2)_2$, m. 219-222° (179).
Rxns. with: Cl_2 → dichloride which is rapidly hydrolyzed by atm. moisture to the
corresponding dihydroxide, and the latter loses H_2O at 114-130°, giving presum-
ably an oxide, m. 206-207° (179).
$KMnO_4$ in an alk. medium → 5-p-carboxyphenyldibenzarsenole 5-oxide (179).

5-(4-METHOXYPHENYL)DIBENZARSENOLE $C_{19}H_{15}AsO$ *
Prepn.: by reacting 5-chlorodibenzarsenole with 4-MeOC₆H₄MgI (96).
Props.: m. 115-116° (96).

p-(3-METHOXY-5-DIBENZARSENOLYL)BENZOIC ACID $C_{20}H_{15}AsO_3$*
Prepn.: By oxidizing 3-methoxy-5-p-tolyldibenzarsenole with alk. $KMnO_4$ and reducing the rxn. prod. with SO_2 (179).

* For the structural formula see page 380.

384

By heating $(p-HO_2CC_6H_4)[2-(4-MeOC_6H_4]AsCl$ at $250°/14$ mm (179).

By heating $(p-HO_2CC_6H_4)[2-(4-MeOC_6H_4)C_6H_4]AsOOH$ with polyphosphoric acid at $160°$; 76% yield (179).

Props.: m. 223° (179).

3-METHOXY-5-PHENYLDIBENZARSENOLE $C_{20}H_{17}AsO$ *

Prepn.: by treating 5-chloro-3-methoxydibenzarsenole with PhMgBr in $Et_2O-C_6H_6$ (179).

Props.: m. 72-85° (crude prod.), resolution into optical antipodes (179).

Rxn . with: $KMnO_4$ in an alk. medium, followed by red. with SO_2 → p-(3-methoxy-5-dibenzarsenolyl)benzoic acid (179).

5-(1-NAPHTHYL)DIBENZARSENOLE $C_{22}H_{15}As$ *

Prepn.: by reacting 5-chlorodibenzarsenole with $1-C_{10}H_7MgBr$ (96).

Props.: m. 150-151° (96).

Rxn . with: $KMnO_4$ in Me_2CO → $(C_{12}H_8)(C_{10}H_7)As(OH)_2$ (96).

5-(1-NAPHTHYL)DIBENZARSENOLE 5,5-DIHYDROXIDE $C_{22}H_{17}AsO_2$ *

Prepn.: by oxidation of 5-(1-naphthyl)dibenzarsenole with $KMnO_4$ in Me_2CO (96).

Props.: m. 212-213° (96).

DIOXARSENOLANE DERIVATIVES

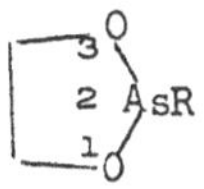

2-R-1,3,2-Dioxarsenolane

2-Organo-1,3,2-dioxarsenolane derivatives, which are cyclic esters of arsonous acids with α,β-glycols, are prepared by reacting organodihaloarsines, $RAsCl_2$, with glycols in ether or in a hydrocarbon solvent in the presence of amines such as pyridine or triethylamine, or by heating arsonous acids, $RAs(OH)_2$, or arsenoso compounds, RAsO, with glycols or pyrocatechol under atmospheric or reduced pressure or in ether in the presence of an amine. The compounds are compiled in Table LVI.

---

* For the structural formula see page 380.

TABLE LVI

2-ORGANO-1,3,2-DIOXARSENOLANE DERIVATIVES

$RAsOCH_2CHRO$	Prepd. from	Yield %	Props.	Ref.
C$_4$				
$EtAsOCH_2CH_2O$	RAsO	78.5	b. 62°/10 mm., d_{20} 1.5423, n_D^{20} 1.5230	641
C$_5$				
$EtAsOCH_2CH(Me)O$	RAsO	63.8	b. 52-53°/10 mm.; d_{20} 1.3859, n_D^{20} 1.4929, rxn.[1]	641
C$_6$				
$EtAsOCH_2CH(CH_2OMe)O$	RAsO	79.5	b. 90°/10 mm., d_{20} 1.3664, n_D^{20} 1.4880	641
C$_7$				
$EtAsOCH_2CH(CH_2OEt)O$	RAsO	55.3	b. 99°/10 mm., d_{20} 1.3167, n_D^{20} 1.4863, rxn.[2]	641
C$_8$				
$4-O_2NC_6H_4AsOCH_2CH_2O$	RAs(OH)$_2$	86.5	m. 119-121°	633, 641
$EtAsOC_6H_4O$-o	RAsO		b. 125°/13 mm., m. 49-50°	
$PhAsOCH_2CH_2O$	RAsCl$_2$	56	b. 122-123°/10 mm., d_{20} 1.5279, n_D^{20} 1.5961	632, 644
			b. 121-122°/9 mm., d_{20} 1.5365, n_D^{20} 1.6106, rxns.1,3	635
$EtAsOCH_2CH(CH_2OPr)O$	RAsO	74.3	b. 114°/11 mm., d_{20} 1.2602, n_D^{20} 1.4818, rxn.[4]	641
C$_9$				
$PhAsOCH_2CH(Me)O$	RAsCl$_2$	65.3	b. 156-158°/11 mm, d_{20} 1.3584, n_D^{20} 1.5540	635
$4-MeC_6H_4AsOCH_2CH_2O$	RAsCl$_2$	45.6	m. 138-140°	632
	RAsO	66.4	m. 134-135°, b. 138-139°/10 mm	635
$EtAsOCH_2CH(CH_2OBu)O$	RAsO		b. 124-125°/10 mm, d_{20} 1.2402, n_D^{20} 1.4805, rxn.[4]	641

		Yield %		Ref.
C_{10}				
4-$O_2NC_6H_4$AsOCH$_2$CH(CH$_2$OMe)O	RAs(OH)$_2$		b. 188-190°/2 mm (dec.), m. 54-55°	633
PhAsOCH$_2$CH(CH$_2$OMe)O	RAsCl$_2$	62	b. 155°/2 mm, d$_0^{20}$ 1.4373, n$_D^{20}$ 1.5720	632, 644
C_{11}				
4-$O_2NC_6H_4$AsOCH$_2$CH(CH$_2$OEt)O	RAs(OH)$_2$		b. 184-185°/3 mm, d$_{20}$ 1.4793	633
PhAsOCH$_2$CH(CH$_2$OEt)O	RAsCl$_2$	65.3	b. 158-159°/11 mm, d$_0^{20}$ 1.3584, n$_D^{20}$ 1.5700	632, 644
4-MeC$_6$H$_4$AsOCH$_2$CH(CH$_2$OMe)O	RAsCl$_2$	48.3	b. 164-165°/10 mm., d$_0^{20}$ 1.3717, n$_D^{20}$ 1.5621	632, 644
C_{12}				
PhAsOC$_6$H$_4$O-o	RAsO	86	b. 178-179°/9 mm., m. 85-87°	635
	RAsCl$_2$	20.1	b. 178-179°/9 mm., m. 85-86°	632, 644
1-$C_{10}H_7$AsOCH$_2$CH$_2$O	RAsCl$_2$		b. 169-170°/4 mm., d$_{30}$ 1.5043, n$_D^{3D}$ 1.6677	635
4-MeC$_6$H$_4$AsOCH$_2$CH(CH$_2$OEt)O	RAsCl$_2$	51.8	b. 169-170°/9 mm., d$_0^{20}$ 1.3237, n$_D^{20}$ 1.5505	632, 644
C_{13}				
4-MeC$_6$H$_4$AsOC$_6$H$_4$O-o	RAsCl$_2$		m. 106-107°$_1$, b. 186-188°/9 mm., rxn.	632, 634
PhAsOCH$_2$CH(CH$_2$OBu)O	RAsCl$_2$	52.8	b. 175-176°/10 mm., d$_0^{20}$ 1.2795, n$_D^{20}$ 1.5369	632, 634
C_{14}				
1-$C_{10}H_7$AsOCH$_2$CH(CH$_2$OMe)O	RAsCl$_2$		b. 185-186°/3 mm., d$_{20}$ 1.4209, n$_D^{20}$ 1.6282	633
4-MeC$_6$H$_4$AsOCH$_2$CH(CH$_2$OBu)O	RAsCl$_2$	63.1	b. 190-191°/10 mm., d$_0^{20}$ 1.2483, n$_D^{20}$ 1.5257	632, 634

1. Rxn. with Ac$_2$O → RAs(OAc)$_2$ (639, 641).
2. Rxn. with MeI → Me$_3$EtAsI, colorless crystals (641).
3. Rxn. with AcCl or AcBr → PhAsCl$_2$ or PhAsBr$_2$ and (-CH$_2$OAc)$_2$ (639).
4. Rxn. with AcCl → EtAsCl$_3$ + AcOCH$_2$CHRCH$_2$OAc (641).

DITHIARSENOLANE AND DITHIARSINDAN DERIVATIVES

2-R-1,3,2-Dithiarsenolane   2-R-1,3,2-Dithiarsindan

2-Organo-1,3,2-dithiarsenolane and 2-organo-1,3,2-dithiarsindan derivatives are cyclic esters of dithioarsonous acids with α,β-dimercaptans or with o-dithiophen.
They are obtained by heating organodihaloarsines, $RAsCl_2$, or arsenoso compounds, $(RAsO)n$, with dimercaptans, $HSCH_2CH(SH)R$, or with o-dithiophenols in aqueous alkaline solution, or in aqueous solution in the presence of an organic base, or in alcohol when water-insoluble mercaptans are employed. The cyclic esters were also prepared by reducing and esterifying arsonic acids, using excess of ammonium or alkali metal dithioglycolates in aqueous solution at elevated temperatures. The reduction and esterification reactions can be carried out in one step, using excess of a dimercaptide, or in two steps, first reducing the arsonic acid with a mercaptane or dimercaptane and then esterifying the arsonous derivative with the same or another dimercaptan.

TABLE LVII

1,3,2-DITHIARSENOLANE AND DITHIARSINDAN DERIVATIVES

RAsSCH₂CHRS	Prepd. from	Yield %	Props.	Ref.
C₄				
ClCH:CHAsSCH₂CH₂S	RAsCl₂ + (CH₂SNa)₂	56	b. 120°/0.4 mm	1167
MeAsSCH₂CH(CH₂OH)S	RAsCl₂ + HOCH₂CH(SNa)CH₂SNa	70	b. 158°/1.5 mm., rxn.[1]	1167
C₅				
ClCH:CHAsSCH₂CH(CH₂OH)S	RAsCl₂ + HOCH₂CH(SNa)CH₂SNa	65	b. 165°/0.2 mm.	1167
ClCH:CHAsSCH₂CH(CH₂SO₃H)S	RAsCl₂ + HSCH₂CH(SH)CH₂SO₃Na		Salts[2]	997
C₆				
ClCH:CHAsSCH₂CH(CH₂OMe)S	RAsCl₂ + MeOCH₂CH(SNa)CH₂SNa	78	b. 145°/0.6 mm.	1167
C₉				
PhAsSCH₂CH(CH₂OH)S	RAsCl₂ + HSCH₂CH(SH)CH₂OH + C₅H₅N	65	m. 97-98°	1167
4-HOC₆H₄AsSCH₂CH(Me)S	RAsCl₂ + MeCH(SH)CH₂SH			443
4-HOC₆H₄AsSCH₂CH(CH₂OH)S	RAsCl₂ + HOCH₂CH(SH)CH₂SH			442-4
PhAsSCH₂CH(CH₂SO₃H)S	RAsCl₂ + HSCH₂CH(SH)CH₂SO₃Na		Salts[3]	997
4-H₂NC₆H₄AsSCH₂CH(CH₂OH)S	RAsCl₂ + HSCH₂CH(SH)CH₂OH			441-4
	RAsO₃HNa + HSCH₂CH(SH)CH₂OH			444
3-H₂NC₆H₄AsSCH₂CH(Me)S	RAsCl₂ + MeCH(SH)CH₂SH			443
4-H₂NC₆H₄AsSCH₂CH(Me)S	RAsCl₂ + MeCH(SH)CH₂SH			443-4
3,4-H₂N(HO)C₆H₃AsSCH₂CH(Me)S	RAsO + MeCH(SH)CH₂SH			444
3,4-H₂N(HO)C₆H₃AsSCH₂CH(CH₂-OH)S	RAsCl₂ + HSCH₂CHSHCH₂OH			441-4
3,4-H₂N(HO)C₆H₃AsSCH₂CH(CH₂-SO₃H)S	RAsCl₂ + HSCH₂CHSHCH₂SO₃Na		m. 224-227°; Na salt, m. 185°	1105 997
C₁₀				
4-HO₂CC₆H₄AsSCH₂CH(CH₂OH)S	RAsCl₂ + HSCH₂CH(SH)CH₂OH + C₅H₅N		m. 178-180°	1167

1. Rxn. with Na in liq. Na and EtOH → Me₂AsH + HSCH₂CH(SH)CH₂OH (1167).
2. Na salt, m. 218-225°; S-benzylthiuronium salt, m. 110-112° (997).
3. Na salt, m. 215-128°; S-benzylthruronium salt, m. 166-167° (997).

TABLE LVII (cont'd.)

1,3,2-DITHIARSENOLANE AND DITHIARSINDAN DERIVATIVES

$RAsSCH_2CHRS$	Prepd. from	Yield %	Props.	Ref.
C_{11}				
4-$H_2NCONHC_6H_4AsSCH_2CH(CH_2OH)S$	$RAsCl_2 + HSCH_2CH_2(SH)CH_2OH$ $RAsO_3HNa + HSCH_2CO_2NH_4 + HS-CH_2CH(SH)CH_2OH$			441-3 442, 444
2,4-$HO(AcNH)C_6H_3AsSCH_2CH(CH_2-OH)S$	$RAsCl_2 + HSCH_2CH(SH)CH_2SO_3Na$ $RAsCl_2 + HSCH_2CH(SH)CH_2OH$		Na salt, m. 224°	997 441-4
3,4-$AcNH(HO)C_6H_3AsSCH_2CH-(CH_2OH)S$	$RAsCl_2 + HSCH_2CH(SH)CH_2OH$			441-4
4-$H_2NCOCH_2NHC_6H_4AsSCH_2CH-(CH_2OH)S$	$RAsCl_2 + HSCH_2CH(SH)CH_2OH$ $RAsO_3HNa + Na$ thiomalate + $HSCH_2CH(SH)CH_2OH$			441-4, 443-4
4-$HOCH_2CH_2NHC_6H_4AsSCH_2CH-(CH_2OH)S$	$RAsCl_2 + HSCH_2CH(SH)CH_2OH$ $RAsO_3HNa + HSCH_2CO_2NH_4 + HS-CH_2CH(SH)CH_2OH$			443 444
C_{12}				
3,4-$H_2N(HO)C_6H_3AsSC_6H_4S$-o	$RAsCl_2 + o-C_6H_4(SH)_2$			442-4 443 444
4-$[N:C(NH_2)N:C(NH_2)N:CNH]$-$C_6H_4AsSCH_2CH(CH_2OH)S$	$RAsCl_2 + HSCH_2CH(SH)CH_2OH$ $RAsO_3HNa + HSCH_2CH(SH)CH_2-OH$ in $MeCHOHCH_2OH$ $RAs(SCH_2CO_2K)_2 + HSCH_2CH(SH)-CH_2OH$			441-2 444 444
4,1,3-$HO[N:C(NH_2)N:C(NH_2)N:CNH]$-$C_6H_3AsSCH_2CH(CH_2OH)S$	$N:C(NH_2)N:C(NH_2)N:CCl + 4-H_2-NC_6H_4AsSCH_2CH(CH_2OH)S$ $RAsCl_2 + HSCH_2CH(SH)CH_2OH$			442, 444
4-$[HO(CH_2)_3NH]C_6H_4AsSCH_2CH-(CH_2OH)S$	$RAsO_3HNa + HSCH_2CONH_4 + HS-CH_2CH(SH)CH_2OH$			442, 444
C_{13}				
4-$[N:C(NH_2)N:C(Cl)CH:CNH]$-$AsSCH_2CH(CH_2OH)S$	$RAsO + HSCH_2CH(SH)CH_2OH$		m. 193-194°	8a
$PhAsSC_6H_3(Me)S$	$RAsCl_2 + 1,3,4-MeC_6H_3(SH)_2 +$ $NaOAc$ in $MeOH$		m. 53-55°.	180

C$_{14}$ 4-HO$_2$CC$_6$H$_4$AsSC$_6$H$_3$(Me)S	RAsCl$_2$ + 1,3,4-MeC$_6$H$_3$(SH)$_2$ + NaOAc in MeOH	m. 200-202°, d-isomer m. 201-202°, [α]$_D$ 6.8° (in C$_5$H$_5$N), 8.9° (in CHCl$_3$), mol. structure; salts[4]	180
4-[N:C(NH$_2$)N:C(Me)CH:CNH]-C$_6$H$_4$AsSCH$_2$CH(CH$_2$OH)S	RAsO + HSCH$_2$CH(SH)CH$_2$OH	HCl salt, m. 129-130°	8a
4-[N:C(NH$_2$)N:C(NH$_2$)N:CNH]-C$_6$H$_4$AsSCH$_2$CH(CH$_2$OEt)S	RAsCl$_2$ + HSCH$_2$CH(SH)CH$_2$OEt RAs(SCH$_2$CO$_2$K)$_2$ + HSCH$_2$CH(SH)-CH$_2$OEt		443 442
C$_{15}$ 4-H$_2$NC$_6$H$_4$AsSCH$_2$CH(CH$_2$OC$_6$H$_{11}$-O$_5$)S	RAsCl$_2$ + HSCH$_2$CH(SH)CH$_2$OC$_6$-H$_{11}$O$_5$ (glucoside) RAsO$_3$HNa + HSCH$_2$CH(SH)CH$_2$OC$_6$-H$_{11}$O$_5$		442-3 444

4. The (-)-acid (+)-PhCH(Me)NH$_2$ salt, m. 199-200°; the (+)-acid quinine salt, m. 206-207° (180)

SIX-MEMBERED RING SYSTEMS

ARSENANE DERIVATIVES

Arsenane   1-Hydroxyarsenane 1-Oxide

1-IODOARSENANE $C_5H_{10}AsI$
Prepn.: by treating $\overline{CH_2(CH_2)_4}AsH$ with I in Et_2O (1297).
Props.: Dark red crystals, m. 27° to a red-brown liquid (1297).

ARSENANE $C_5H_{11}As$
Prepn.: by reducing $\overline{CH_2(CH_2)_4}AsCl$ in Et_2O with $LiBH_4$ at -50° or with $LiAlH_4$ at
-60° in high vacuum; 77% yield (1297).
Props.: Colorless, skin-irritating liquid of phosphine-like odor, sol. in org.
solvents, insol. in H_2O; m. -11°, very sensitive to O (1297).
Rxns. with: $O_2 \rightarrow \overline{CH_2(CH_2)_4}AsO_2H$ (1297).
$I_2 \rightarrow \overline{CH_2(CH_2)_4}AsI$ (1297).

1-HYDROXYARSENANE 1-OXIDE $C_5H_{11}AsO_2$
Prepn.: by oxidation of $\overline{CH_2(CH_2)_4}AsH$ with atm. O (1297).
Props.: White solid, insol. in Et_2O (1297).

1,2-DIMETHYLARSENANE $C_7H_{16}As$
Prepn.: By reacting $MeAsCl_2$ with $MeCHMgCl(CH_2)_3CH_2MgCl$ in a H atm. (1334).
By pyrolysis of 1,1,2-trimethylarsenan ium chloride (1334).
Props.: B. 169°, b. 85°/22 mm. (1334).
Rxns. with: O → the 1-oxide (1334).
Alkyl halide → the arsenonium salts (1334).
Derivs.: Methiodide, sublimes at 340° (1334).
Picrate, m. 231° (1334).
Methochloride, deliquescent solid, decomp. at 250-270° to give 1,2-dimethylars-
enidine (1334).

1,2-DIMETHYLARSENANE 1-OXIDE $C_7H_{16}AsO$
Prepn.: by oxidation of 1,2-dimethylarsen ane with air (1334).

1-R-1,2,3,4-Tetrahydroarsinoline

5,6-Dihydro-5-R-11-R′-11H-arsinoline[4,3-b]-
indole

5,6-Dihydro-5-R-arsinoline[4,3-b]quinoline

1-BROMO-1,2,3,4-TETRAHYDROARSINOLINE $C_9H_{10}AsBr$
Prepn.: by treating 1,2,3,4-tetrahydro-1-methylarsinoline with 1 mole Br in
CCl$_4$ (560).
Rxns. with: Piperidine N,N-pentamethyleneditiocarbamate → S-(1,2,3,4-tetrahydro-
    arsinolin-1-yl)-N,N-pentamethyleneurethan (560).
Ph$_3$COCH$_2$CH$_2$CH$_2$C$_6$H$_4$-o-MgBr → 1,2,3,4-tetrahydro-1-[2(3-triphenylmethoxypropyl)-
    phenyl]arsinoline (560).

1,2,3,4-TETRAHYDRO-1-METHYLARSINOLINE $C_{10}H_{13}As$
Rxns. with: Br$_2$ in CCl$_4$ → 1-bromo-1,2,3,4-tetrahydroarsinoline (560).

1,2,3,4-TETRAHYDRO-7-METHOXY-1-METHYL-4-OXOARSINOLINE $C_{11}H_{13}AsO$
Prepn.: by heating under reflux Me(3-MeOC$_6$H$_4$)AsCH$_2$CH$_2$CO$_2$H with P$_2$O$_5$ in xylene
(38% yield) or with P$_2$O$_3$-H$_3$PO$_4$ at 100° under N (26% yield) (811).
Props.: m. 52° (811).
Rxns. with: 4-Me$_2$NC$_6$H$_4$CHO → the 3-(4-dimethylaminobenzylidine) deriv. (811).
4-Me$_2$NC$_6$H$_4$NO → the 3-(4-dimethylaminophenylimino) deriv. (811).
Derivs.: Methiodide, prisms, m. 239°; Methopicrate, yellow crystals, m. 226°
Metho-p-toluenesulfonate, prisms, m. 180-205°; Semicarbazone, needles, m. 208°;
Phenylhydrazone, m. 121-122°; 2,4-Dinitrophenylhydrazone, m. 183° (811).

1-[2-(3-CHLOROPROPYL)PHENYL]-1,2,3,4-TETRAHYDRO-1-HYDROXYARSINOLINIUM SALTS
$C_{18}H_{21}AsClOX$ (X = Cl, NO_3)

Prepn.: by reacting 1-bromo-1,2,3,4-tetrahydroarsinoline with $Ph_3COCH_2CH_2CH_2$-C_6H_4-o-MgBr in Et_2O and treating the product with $SOCl_2$ under reflux (560).
Props.: Chloride, m. 158-161°; nitrate, m. 128.5-129° (560).

1,2,3,4-TETRAHYDRO-1-[2-(3-HYDROXYPROPYL)PHEHYL]ARSINOLINE $C_{18}H_{21}AsO$ *
Prepn.: by reacting 1-bromo-1,2,3,4-tetrahydroarsinoline with $Ph_3COCH_2CH_2CH_2$-C_6H_4-o-MgBr and refluxing the product with AcOH (560).
Rxns. with: PBr_3 → 3,3´,4,4´-Tetrahydro-1,1 -spirobi(arsinolinium)bromide (560).

3-(4-DIMETHYLAMINOPHENYLIMINO)-1,2,3,4-TETRAHYDRO-7-METHOXY-1-METHYL-4-
OXOARSINOLINE $C_{19}H_{21}AsN_2O_2$

Prepn.: by reacting 1,2,3,4-tetrahydro-7-methoxy-1-methyl-4-oxoarsinoline with
4-nitrosodimethylaniline in alc. NaOH (811).
Props.: Orange brown crystals, m. 157° (811).

1,2,3,4-TETRAHYDRO-1-[2-(3-METHOXYPROPYL)PHENYL]ARSINOLINE $C_{19}H_{23}AsO$

Prepn.: by reacting 1-chloro-1,2,3,4-tetrahydroarsinoline with 2-($MeOCH_2CH_2CH_2$)-C_6H_4Li or with 2-($MeOCH_2CH_2CH_2$)C_6H_4MgBr in Et_2O under reflux and hydrolyzing the rxn. prod. (560).
Props.: b. 192-194°/1 mm (560).
Methiodide, needles, m. 122.5-123.5°(560).

---

* For the structural formula see page 393.

Rxns. with: 48% HBr in AcOH at 120° → cleavage of the methoxy group and of the exocyclic As-C bond (560).
50% HBr in AcOH at 120° and completed in a stream of HBr at 150° → 2-(3-bromo-propyl)benzenearsonic acid (560).

S-(1,2,3,4-TETRAHYDROARSINOLIN-1-YL)-N,N-PENTAMETHYLENEDITHIOURETHAN
$C_{20}H_{20}AsNS_2$

$$SC(S)\overline{N(CH_2)_4}CH_2$$

Prepn.: by treating 1-bromo-1,2,3,4-tetrahydroarsinoline with piperidine N,N-pentamethylenedithiocarbamate in C_6H_6 (560).
Props.: m. 159-160° (560).

3-(4-DIMETHYLAMINOBENZYLIDINE)-1,2,3,4-TETRAHYDRO-7-METHOXY-1-METHYL-4-OXO-ARSINOLINE $C_{20}H_{22}AsNO_2$ *
Prepn.: by boiling a soln. of 1,2,3,4-tetrahydro-7-methoxy-1-methyl-4-oxoarsin-oline and $4-Me_2NC_6H_4CHO$ in EtOH and 10% NaOH in N (811).
Props.: Yellow needles, m. 159° (811).

5,6-DIHYDRO-3-METHOXY-5-METHYL-11-PHENYL-11H-ARSINOLINE[4,3-b]INDOLE $C_{23}H_{20}AsNO$ *
Prepn.: by boiling under N a soln. of 1,2,3,4-tetrahydro-7-methoxy-1-methyl-4-oxoarsenoline and Ph_2NH_2 in EtOH contg. HOAc; 30% yield (821).
Props.: Colorless plates, m. 157° (812).

5,6-DIHYDRO-3-METHOXY-5-METHYL-11H-ARSINOLINO[4,3-b]INDOLE $C_{17}H_{16}AsNO$ *
Prepn.: by boiling under N a soln. of the phenylhydrazone deriv. of 1,2,3,4-tetrahydro-7-methoxy-1-methyl-4-oxoarsinoline in HOAc contg. $ZnCl_2$ (812).
Props.: m. 137-139° (812).

5,6-DIHYDRO-3-METHOXY-5,11-DIMETHYL-11H-ARSINOLINE[4,3-b]INDOLE $C_{18}H_{18}AsNO$ *
Prepn.: by boiling under N a soln. of 1,2,3,4-tetrahydro-7-methoxy-1-methyl-4-oxoarsenoline and $PhMeNNH_2$ in EtOH contg. HOAc; 48% yield (812).
Props.: Colorless crystals, m. 115° (812).

5,6-DIHYDRO-3-METHOXY-5-METHYLARSINOLINO[4,3-b]QUINOLINE $C_{18}H_{16}AsNO$ *
Prepn.: by treating an alc. soln. of 1,2,3,4-tetrahydro-7-methoxy-1-methyl-4-oxoarsinoline and $2-H_2NC_6H_4CHO$ with 10% aq. KOH at r. t.; 39% yield (812).
Props.: m. 136° (812).
Rxn. with $KMnO_4$ → the As-oxide, m. 232° (812).
Derivs.: Picrate, m. 192°; Methiodide, m. 225°; Methiodide hydriodide hydrate, indefinite m., completely molten at 225° (812).

5,6-DIHYDRO-3-METHOXY-5-METHYLARSINOLINO[4,3-b]QUINOLINE 5-OXIDE $C_{18}H_{16}AsNO_2$ *
Prepn.: by oxidation of 5,6-dihydro-3-methoxy-5-methylarsinolino[4,3-b]quinoline with $KMnO_4$ (812).

---

* For the structural formula see page 395.

Props.:   m. 232° (812).
Derivs.:   Picrate, m. 192° (812).

5,6-DIHYDRO-3-METHOXY-5-METHYLARSINOLINO[4,3-b]QUINOLINE-7-CARBOXYLIC ACID
$C_{16}H_{16}AsNO$ *

Prepn.:   by boiling under N a soln. of 1,2,3,4-tetrahydro-7-methoxy-1-methyl-4-
oxoarsinoline, isatin (1:1), and KOH in aq. EtOH; 53% yield (812).
Props.:   Yellow, m. 220-250° (decompn.) (812).

ISOARSINOLINE DERIVATIVES

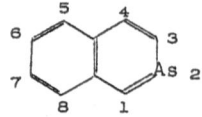

Isoarsinoline

2-CHLORO-1,2,3,4-TETRAHYDROISOARSINOLINE $C_9H_{10}AsCl$
Prepn.:   by heating 1,2,3,4-tetrahydro-2-methylisoarsinoline dichloride at 15 mm.
Hg pressure (557).
Props.:   b. 157°/14 mm (557).

1,2,3,4-TETRAHYDRO-2-IODOISOARSINOLINE $C_9H_{10}AsI$
Prepn.:   by refluxing the 2-phenyl derivative in const. boiling HI in CO_2 (68).
Props.:   Red oil (68).
Rxns. with: Piperidine N-pentamethylenedithiocarbamate → N-pentamethylene-1,
2,3,4-tetrahydro-2-isoarsinolyl-1-piperidinecarbodithioate, m.   121.5-2° (68).

1,2,3,4-TETRAHYDRO-2-METHYLISOARSINOLINE $C_{10}H_{13}As$
Prepn.:   by reacting $MeAsCl_2$ with $2\text{-}BrCH_2C_6H_4CH_2CH_2Br$ in the presence of Na in
refluxing EtOAc; 16% yield (557).
Props.:   b. 131°/18 mm (557).
Methiodide, m. 179-181°; Methopicrate, m. 163-164°; hydroxy picrate, m. 164-
165.5° (557).
Rxns. with: $2\text{-}BrCH_2C_6H_4CH_2CH_2Br$ → 3,3', 4,4'-tetrahydro-2,2'-(1,1')spirobi(iso-
arsinolinium) bromide (559).

1,2,3,4-TETRAHYDRO-2-METHYLISOARSINOLINE   5,5-DICHLORIDE $C_{10}H_{13}AsCl_2$
Prepn.:   by treating the isoarsinoline with Cl in CCl_4 at 0° (557).
Props.:   On heating at 15 mm pressure, MeCl is cleaved (557).

1,2,3,4-TETRAHYDRO-2-PHENYLISOARSINOLINE $C_{15}H_{15}As$
Prepn.:   By refluxing $PhAs(MgBr)_2$ with $o\text{-}BrCH_2C_6H_4CH_2CH_2Br$ in C_6H_6; 54% yield
along with PhAs:AsPh when $PhAs(MgBr)_2$ was used in excess (68).
By heating $o\text{-}BrCH_2C_6H_4CH_2CH_2Br$ with $PhAsMe_2$ and pyrolyzing the resulting
arsonium bromide at 180-200°/17 mm; 8% yield along with $PhMe_2As(OH)Br$ and MePh-
AsBr (68).
By reacting $PhAsCl_2$ with $o\text{-}BrCH_2C_6H_4CH_2CH_2Br$ in the presence of Na in EtOAc

---

* For the structural formula see page 393.

under reflux; 31.5% yield (557).
Props.: b. 127-132°/0.1 mm (68); b. 110-112°/0.001 mm; b. 128-130°/0.05 mm (557).
Derivs.: Methiodide, m. 136-137°; Hydroxy nitrate, m. 149-150°; Hydroxy picrate, m. 116-118° (557).
Rxns. with: $NHO_3 \rightarrow$ the corresponding hydroxy nitrate (68).
HI under reflux → 1,2,3,4-tetrahydro-2-iodoisoarsinoline (68).
$Br_2 \rightarrow$ colorless soln. which with NH_4OH gave a hygroscopic glass (oxide or dihydroxide); the latter was converted into hydroxy nitrate by treating with HNO_3 and to dichloride by treating with HCl (557).
Chloramine-T → amorphous glass which was hydrolyzed to $4\text{-}MeC_6H_4SO_2NH_2$ (557).

1,2,3,4-TETRAHYDRO-2-PHENYLISOARSINOLINE 5,5-DICHLORIDE $C_{15}H_{15}AsCl_2$ *
Prepn.: by treating 1,2,3,4-tetrahydro-2-phenylisoarsinoline oxide or dihydroxide with HCl (557).
Props.: m. 147-149° (557).

1,2,3,4-TETRAHYDRO-2-PHENYLISOARSINOLINE 5-SULFIDE $C_{15}H_{15}AsS$
Prepn.: by treating 1,2,3,4-tetrahydro-2-phenylisoarsinoline dibromide with H_2S (557).
Props.: m. 124° (557).

2-(4-CHLOROPHENACYL)-1,2,3,4-TETRAHYDRO-2-PHENYLISOARSINOLINIUM SALTS
$C_{23}H_{21}AsClOX$

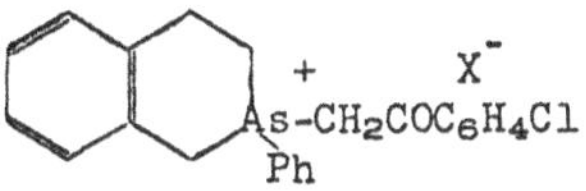

Prepn.: by treating 1,2,3,4-tetrahydro-2-phenylisoarsinoline with $4\text{-}ClC_6H_4COCH_2Br$ in C_6H_6 under reflux (558, 803).
Props.: The asymmetric 4-covalent As forms stable optically active salts.
Bromide, m. 190-191°, iodide, m. 190.5°; l-arsonium d -bromocamphorsulfonate, m. 256-238°; l-arsonium picrate; l-arsonium iodide, m. 178.5-179°; d-camphorsulfonate, m. 210-212°; chloroplatinate, m. 135-150° (decompn.); chloroaurate, m. 157-158° (558, 803).

ACRIDARSINE DERIVATIVES

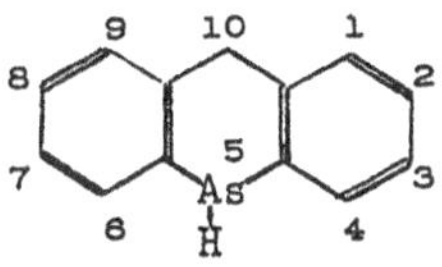

5,10-Dihydroacridarsine

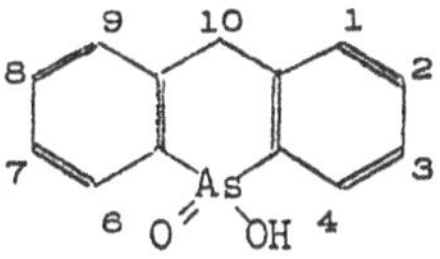

Acridarsinic Acid

---

* For the structural formula see page 393.

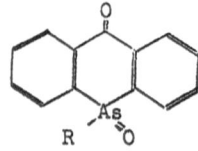
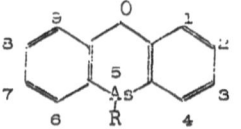

5-R-10(5H)-acridarsinone 5-Oxide   5-R-10(5H)-acridarsinone

2,5-DICHLORO-5,10-DIHYDROACRIDARSINE $C_{13}H_9AsCl_2$ *
Prepn.: by reducing 2-chloroacridarsinic acid with SO_2 in an HCl-HCCl$_3$ mixt. con
KI; 98.2% yield (270a).
Props.: m. 116° (270a).
Rxn . with: KCN in MeOH → the 2-chloro-5-cyano deriv. (270a).

5-CHLORO-5,10-DIHYDROACRIDARSINE $C_{13}H_{10}AsCl$ *
Prepn.: by reducing acridarsinic acid suspended in hydrochloric acid- HCCl$_3$ mixt
with SO_2-HCl gas mixt. in the presence of KI; 71% yield (537).
Props.: Pale yellow crystals, m. 110-111° (537).
Rxns. with: KCN → the 5-cyano deriv. (537).
RMgX → the 5-phenyl deriv. (613).

2-CHLOROACRIDARSINIC ACID $C_{13}H_{10}AsClO_2$ *
Prepn.: by cyclizing 4,2-Cl(PhCH$_2$)C$_6$H$_3$AsO$_3$H$_2$ in concd. H_2SO_4 at 100°; 88.5% yiel
(270a).
Props.: m. 210-212° (decompn.) (270a).
Rxn . with: SO_2 in HCl-HCCl$_3$ + KI → 2,5-dichloro-5,10-dihydroacridarsine (270a).

ACRIDARSINIC ACID $C_{13}H_{11}AsO_3$ *
Prepn.: by heating 2-(PhCH$_2$)C$_8$H$_4$AsO$_3$H$_2$ in concd. H_2SO_4 on a steam bath; 95% yiel
(537).
Rxn . with SO_2-HCl mixt. in hydrochloric acid-CHCl$_3$ in the presence of KI → 5-
chloro-5,10-dihydroacridarsine (537).

2-CHLORO-5(10H)-ACRIDARSINECARBONITRILE $C_{14}H_9NClN$ *
Prepn.: by refluxing 2,5-dichloro-5,10-dihydroacridarsine with KCN in MeOH; 49%
yield (270a).
Props.: m. 113-114° (270a)

5-CYANO-5,10-DIHYDROACRIDARSINE $C_{14}H_{10}AsN$ *
Prepn.: by reacting 5-chloro-5,10-dihydroacridarsine with KCN in MeOH under re-
flux; 56% yield (537).
Props.: Colorless crystals, m. 114-115° (537).

5-CHLORO-5,10-DIHYDRO-2-METHYLACRIDARSINE $C_{14}H_{12}AsCl$ *
Prepn.: by reducing the corresponding acridarsinic acid with SO_2 in an HCl-HCCl$_3$
mixt. contg. KI (270a).
Props.: m. 87° (270a).

---

* For the structural formual see page 397

5-CHLORO-5,10-DIHYDRO-3-METHYLACRIDARSINE $C_{14}H_{12}AsCl$ *
Prepn.: by converting 4-$MeC_6H_4CH_2C_6H_4$-2-NH_2 to the corresponding arsonic acid by the Bart reaction, cyclizing the arsonic acid in concd. H_2SO_4 on a steam bath, and reducing the acridarsinic acid with SO_2 in a mixt. of hydrochloric acid-$HCCl_3$ contg. KI (537).
Props.: Pale yellow prisms, m. 65.5-66.5° (537).

2-METHYLACRIDARSINIC ACID $C_{14}H_{13}AsO_2$ *
Prepn.: by cyclizing $2,4$-$Me(PhCH_2)C_6H_3AsO_3H_2$ in concd. H_2SO_4 at 100° (270a).
Props.: m. 184° (270a).
Rxn . with: SO_2 in an HCl-$HCCl_3$ mixt. → the 5-chloro deriv. (270a).

5,10-DIHYDRO-5-PHENYLACRIDARSINE $C_{19}H_{15}As$ *
Prepn.: by reacting 5-chloro-5,10-dihydroacridarsine with PhMgBr in Et_2O-C_6H_6; 98% yield (613).
Props.: m. 74-75° (613).
Derivs.: Methiodide, m. 151-152° (effervesc.); Methopicrate, m. 260° (decompn.) (613).
Rxn . with: $KMnO_4$ → the acridarsine oxide (613).

5-PHENYL-10(5H)-ACRIDARSINONE $C_{19}H_{13}AsO$ **
Prepn.: by reducing 5-phenyl-10(5H)-acridarsinone 5-oxide with SO_2 in HCl-H_2O contg. KI (613).
Props.: m. 138-139° (613).
Derivs.: Methotoluene-p-sulfonate, m. 200° (613).
$[Pd(C_{19}H_{13}AsO)_2Br_2]$, m. 298° (613).
Rxns. with: H_2O_2 → 5-phenyl-10(5H)-acridarsinone 5-oxide (613).
PhMgBr → 5,10-dihydro-5,10-diphenyl-10-acridarsinol (613)

5-PHENYL-10(5H)-ACRIDARSINONE 5-OXIDE $C_{19}H_{13}AsO_2$ **
Prepn.: by oxidation of 5,10-dihydro-5-phenylacridarsine with $KMnO_4$ or H_2O_2; theoret. yield (613).
Props.: m. 238° (613).
Rxns. with: SO_2 in HCl-H_2O contg. KI → 5-phenyl-10(5H)-acridarsinone (613).
PhNHNH_2 → 5-phenyl-10(5H)-acridarsinone (613).

5,10-DIHYDRO-5,10-DIPHENYL-10-ACRIDARSINOL $C_{25}H_{19}AsO$ *
Prepn.: by reacting 5-phenyl-10(5H)-acridarsinone with PhMgBr; 75% yield (613).
Props.: m. 194° (613).

---

* For the structural formula see page 397.
** For the structural formula see page 398.

ARSANTHRIDINE DERIVATIVES

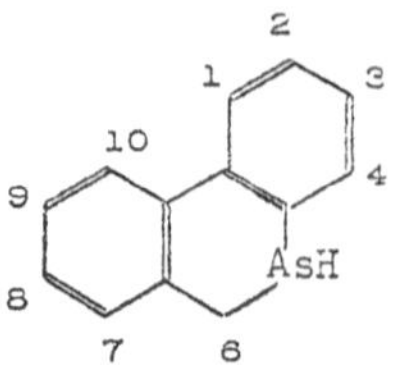

5,6-Dihydroarsanthridine

5,6-DIHYDRO-5-METHYLARSANTHRIDINE $C_{14}H_{13}As$
Prepn.: by treating o-PhC$_6$H$_4$CH$_2$AsMe$_2$ with Cl$_2$ in CCl$_4$ in a CO$_2$ atm. and heating
the prod. in CS$_2$ with AlCl$_3$ under reflux; isolated as the 5,6-dihydro-5,5-
dimethylarsanthridinium iodide (244).
Derivs.: Methiodide.0.5H$_2$O, m. 210-215°, anhydr. m. 212-215°; UV spectrum (244).

5,6-DIHYDRO-5-PHENYLARSANTHRIDINE $C_{19}H_{15}As$
Prepn.: by reacting PhAs(CH$_2$C$_6$H$_4$Ph-o)$_2$ with Cl$_2$ in CCl$_4$ (cleavage of one o-
PhC$_6$H$_4$CH$_2$ group) and treating the prod. with AlCl$_3$ in CS$_2$ under reflux; 15%
yield (244).
Derivs.: Methiodide, pale cream, m. 195° (decompn.); UV spectrum (244).
Methopicrate, yellow, m. 150-151° (244).
Complex compds.: (C$_{19}$H$_{15}$As)$_2$.PdCl$_2$, pale yellow, m. 244-245° (decompn.) (244).
(C$_{19}$H$_{15}$As)$_2$.PdBr$_2$, yellow, m. 244-245° (decompn.) (244).

BENZARSINOLIZINE DERIVATIVES

2,3,5,6-TETRAHYDRO-8-METHOXY-1H,7H-BENZ[ij]ARSINOLIZINE-1,7-DIONE $C_{13}H_{13}AsO_3$

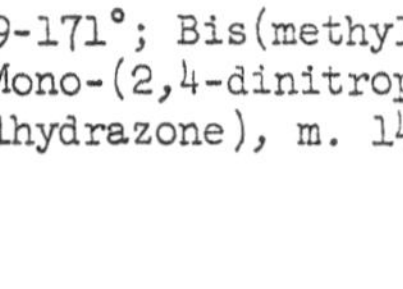

Prepn.: by heating 3-MeOC$_6$H$_4$As(CH$_2$CH$_2$CO$_2$H·)$_2$ with P$_2$O$_5$ in toluene under reflux
(11% yield) or with P$_2$O$_5$-H$_3$PO$_4$ at 100° (7% yield) (811).
Props.: Rods, m. 150° (811).
Derivs.: Bisphenylhydrazone, m. 169-171°; Bis(methylphenylhydrazone), m. 160°;
Bis(diphenylhydrazone), brown gum; Mono-(2,4-dinitrophenylhydrazone).-
EtOH, m. 125°; Bis(2,4-dinitrophenylhydrazone), m. 140°; Mono-oxime, prisms,
m. 181° (811).

5,6,8,13-TETRAHYDRO-3-METHOXY-4H-BENZ[1,9]ARSENOLIZINO[2,3-b]INDOL-4-ONE
C$_{19}$H$_{16}$AsNO$_2$

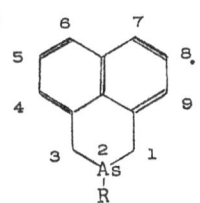

Prepn.: by boiling under N a soln. of bis(phenylhydrazone) of 2,3,5,6-tetrahydro-8-methoxy-1H,7H-benz[ij]arsinolizine-1,7-dione in HOAc contg. ZnCl$_2$ (812).
Props.: m. 232° (812).

5,6-DIHYDRO-3-METHOXY-4H,8H-BENZ[1,9]ARSINOLIZINO[2,3-b]QUINOLIN-4-ONE
C$_{20}$H$_{16}$AsNO$_2$

Prepn. by reacting 2,3,5,6-tetrahydro-8-methoxy-1H,7H-benz[ij]arsinolizine-1,7-dione with 2-H$_2$NC$_6$H$_4$CHO in EtOH contg. aq. KOH (812).
Props.: Glassy substance (821).
Derivs.: Picrate.EtOH, yellow plates, m. 164-166° (812).

ARSAPERINAPHTHANE DERIVATIVES

2-IODOARSAPERINAPHTHANE C$_{12}$H$_{10}$AsI
Prepn.: by heating 2-phenylarsaperinaphthane with HI at 125° (68).
Props.: Pale yellow crystals, m. 117-119° (68).

2-PHENYLARSAPERINAPHTHANE C$_{18}$H$_{15}$As
Prepn.: by refluxing PhAs(MgBr)$_2$ with 1,8-bis(bromomethyl)naphthalene; 60% yield (68).
Props.: brown gum (68).
Salts: Methiodide, monoalcoholate crystals, m. 196-199° (68).
Dichlorodi(2-phenylarsaperinaphthane)palladium, orange powder (68).
Rxn . with: HI → cleavage of the Ph group (68).

DIARSENANE AND BENZODIARSENIN DERIVATIVES

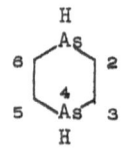

p-Diarsenane

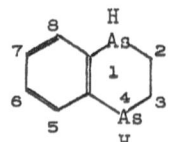

1,2,3,4-Tetrahydro-1,4-benzodiarsenin

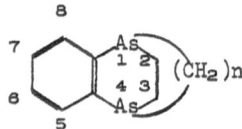

2,3-Dihydro-1,4-alkano-1,4-benzodiarsenin

1,4-DIBROMO-p-DIARSENANE $C_4H_8As_2Br_2$
Prepn.: by boiling PhAsCH$_2$CH$_2$As(Ph)CH$_2$CH$_2$ with HBr (609).
Props.: m. 168-170° (609).
Rxns. with: Piperidnium 1-piperidinecarbodithioate → C$_5$H$_{10}$NCSSAsCH$_2$CH$_2$As(SCSN-
C$_5$H$_{10}$)CH$_2$CH$_2$ (609).
RMgI → RAsCH$_2$CH$_2$As(R)CH$_2$CH$_2$ (609).

1,4-DIMETHYL-p-DIARSENANE $C_6H_{14}As_2$
Prepn.: by reacting BrAsCH$_2$CH$_2$As(Br)CH$_2$CH$_2$ with MeMgI (609).
Props.: b. 113-114°/24 mm (609).
Rxn . with: (CH$_2$Br)$_2$ → undetd. (609).

1-BROMO-1,2,3,4-TETRAHYDRO-4-METHYL-1,4-BENZODIARSENIN $C_9H_{11}As_2Br$
Prepn.: by-prod. formed during pyrolysis of 1,2,3,4-tetrahydro-1,1,4,4-tetra-
methyl-1,4-benzodiarsenium dibromide (807).
Props.: b. 120°/0.3 mm (807).
Derivs.: Monomethobromide, m. 240° (807).
Rxn . with: MeI → the 1-iodo methiodide deriv. (807).

1,2,3,4-TETRAHYDRO-1-IODO-4-METHYL-1,4-BENZODIARSENIN $C_9H_{11}As_2I$
Prepn.: by heating 1-bromo-1,2,3,4-tetrahydro-4-methyl-1,4-benzodiarsenin with
MeI; isolated as methiodide (807).
Props.: Methiodide, m. 199-200° (807).

2,3-DIHYDRO-1,4-BENZODIARSENIN $C_{10}H_{12}As_2$
Prepn.: by pyrolyzing 2,3-dihydro-1,4-dimethyl-1,4-ethano-1,4-benzodiarseniun
dibromide (807).
Props.: b. 95-96°/0.01 mm (807).
Derivs.: Monomethobromide, m. 244°; Monomethiodide, m. 240°; Dimethiodide, m. 260
Bis(hydroxy picrate), m. 184°.(807).
Rxn . with: H$_2$O$_2$ → the 1,4-dioxide (was converted to bis(hydroxy picrate))(807).

1,2,3,4-TETRAHYDRO-1,4-DIMETHYL-1,4-BENZODIARSENIN $C_{10}H_{14}As_2$ *

Prepn.: by pyrolyzing 1,2,3,4-tetrahydro-1,1,4,4-tramethyl-1,4-benzodiarseninium dibromide (807).
Props.: b. 94-97°/0.15 mm (807).
Derivs.: Methobromide, m. 228-230°; Dimethobromide, m. 254°; Methiodide, m. 250°; Dimethiodide, m. 245-246°; Dioxide, identified as bis(hydroxy picrate) m. 179-180° (807).
Rxn . with: (-CH_2Br)_2 → the 1,4-ethano deriv. (807).

2,3-DIHYDRO-1,4-DIMETHYL-1,4-ETHANO-1,4-BENZODIARSENINIUM DIBROMIDE $C_{12}H_{28}As_2Br_2$ *

Prepn.: by heating 1,2,3',4-tetrahydro-1,4-dimethyl-1,4-benzodiarsenin with (CH_2-Br)_2 at 120-125° in a N atm. (807).
Props.: m. 240° (807).
Derivs.: Dimethopicrate, m. 238° (807).

1,2,3,4-TETRAHYDRO-1,1,4,4-TETRAMETHYL-1,4-BENZODIARSENINIUM DIBROMIDE
$C_{12}H_{20}As_2Br_2$ *

Prepn.: by heating o-C_6H_4(AsMe_2)_2 with (-CH_2Br)_2 at 125-130° in N (489, 807).
Props.: m. 255° (489); Dipicrate, m. 236-238° (489).

1,4-DIPHENYL-p-DIARSENANE $C_{16}H_{18}As_2$ *

Prepn.: by heating in vacuo hexahydro-1,4-dimethyl-1,4-diphenyl-p-diarseninium dibromide (609).
Props.: m. 142-144° (609); Crystal structure and molecule configuration (927).
Derivs.: methiodide, m. 211° (decompn.); dimethiodide, m. 221° (effervesc.); bis(hydroxynitrate), m. 188° (effervesc.); bis(metho-p-toluenesulfonate), m. 266° (effervesc.) (609).
Rxns. with: HI → (CH_2AsI_2)_2 (609).
HBr → BrAsCH_2CH_2As(Br)CH_2CH_2 (609).

1,4-DIMERCAPTO-p-DIARSENANE DIESTER WITH 1-PIPERIDINECARBODITHIOIC ACID
$C_{16}H_{28}As_2N_2S_4$

Prepn.: by mixing BrAsCH_2CH_2As(Br)CH_2CH_2 with piperidinium 1-piperidinecarbo-dithioate (609).
Props.: m. 185-186° (609).

HEXAHYDRO-1,4-DIMETHYL-1,4-DIPHENYL-p-DIARSENIUM DIBROMIDE $C_{18}H_{24}As_2Br_2$ *

Prepn.: by heating equiv. amts. of (-CH_2AsMePh)_2 with (CH_2Br)_2 in MeOH at 100°; 21% yield. When the reaction was carreid out in MePh or PhCl, MePhAsOOH was obtained (609).
Props.: Sublimes at ~255°, loses MeBr when heated in vacuo and is converted to 1,4-diphenyl-p-diarsinane (609).
Derivs.: Dimethopicrate m. 244° (decompn.) (609).

---

* For the structural formula see page 402.

ARSANTHRENE DERIVATIVES

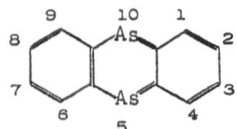

Arsanthrene

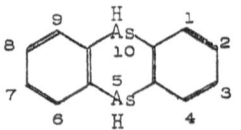

5,10-Dihydroarsanthrene

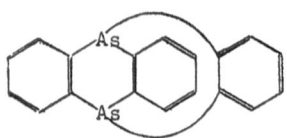

5,10-Dihydro-5,10-o-
benzenoarsanthrene

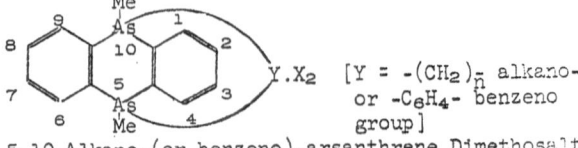

5,10-Alkano-(or benzeno)-arsanthrene Dimethosalts

5,10-DICHLORO-5,10-DIHYDROARSANTHRENE $C_{12}H_8As_2Cl_2$
Prepn.: by heating o-(PhAsCl)$C_6H_4AsCl_2$ under reflux at 12 mm pressure (200).
Props.: m. 179-184° (200).
Rxns. with: RMgX → 5,10-R$_2$ derivs. (200, 611).

5-BROMO-5,10-DIHYDRO-10-METHYLARSANTHRENE $C_{13}H_{11}As_2Br$
Prepn.: by heating arsanthrene dimethobromide with a gentle brush fire; 81%
yield (611).
Props.: Pale yellow crystals, m. 175-177° (611).
Rxns. with: NaI → the 5-iodo deriv. (611).
MeBr → the dimethobromide (611).
MeI → the dimethiodide (611).

5-IODO-5,10-DIHYDRO-10-METHYLARSANTHRENE $C_{13}H_{11}As_2I$
Prepn.: by boiling 5-bromo-5,10-dihydro-10-methylarsanthrene with NaI in Me$_2$CO-
EtOH (611).
Props.: Deep yellow crystals, m. 174-178° (611).

5,10-DICYANO-5,10-DIHYDROARSANTHRENE $C_{14}H_8As_2N_2$
Prepn.: by refluxing a mixt. of 5,10-dichloro-5,10-dihydroarsanthrene and AgCN
in MePh (494).
Props.: Colorless prisms from C_6H_6 contain 1 mol. C_6H_6, lose the solvent at 110-
120°, m. 165-167° (494).

ARSANTHRENE DIMETHOBROMIDE $C_{14}H_{14}As_2Br_2$
Prepn.: by heating 5,10-dihydro-5,10-dimethylarsanthrene with (CH$_2$Br)$_2$ at 125-
130° in a sealed tube (611).
Props.: m. 256° (effervesc.) (611).
Other salts: Dimethopicrate, m. 244° (dec.) (611).
Dimethiodide, orange crystals, m. 214° (dec.) (611).

5,10-DIHYDRO-5,10-DIMETHYLARSANTHRENE $C_{14}H_{14}As_2$ *

Prepn.: by reacting 5,10-dihydro-5,10-dichloroarsanthrene with MeMg-halides (611).
Props.: m. 191-192.5° (611).
Rxns. with: $(CH_2Br)_2$ → the dimethobromide, 5,10-ethanoarsanthrene dimethobromide, and 5,5'-ethylenebis(5,10-dihydro-10-methylarsantbrene)-5,5'-dimethobromide (611).
$CH_2(CH_2Br)_2$ → 5,10-propanoarsanthrene dimethobromide (611).
o-$C_6H_4(CH_2Br)_2$ (1:2) → coupling of two arsanthrene molecules (611).
o-$C_6H_4(CH_2Br)_2$ (1:2) → 5,10-benzenodibenzodiarsocinium deriv. (611).
Derivs.: Dibromide and diiodide, crystal structure (1179).
Monomethobromide, m. 219° (effervesc.) (611).
Dimethobromide, m. 302° (effervesc.) (611).
Dimethopicrate, m. 282° (611).
Monomethiodide, m. 219° (effervesc.) (611).
Dimethiodide, m. 263° (dec.) (611).
Bis(metho-p-toluenesulfonate), m. 316° (dec.) (611).
Bis(hydroxide nitrate), m. 224° (effervesc.) (611).
Bis(hydroxide picrate), m. 210° (611).

5,10-DIHYDRO-2,5,10-TRIMETHYLARSANTHRENE $C_{15}H_{16}As_2$ *

Prepn.: by diazotization of o-arsanilic acid, coupling with dichloro-p-tolylarsine in aq. alkali., isolating o-arsonotolylarsinic acid, reducing the arsonic and arsinic groups with SO_2 to the corresponding chloroarsines, distilling the crude prod., and methylating with MeMgI (611).
Props.: Foul-smelling gum (611).
Rxns. with: $(CH_2Br)_2$ or $CH_2(CH_2Br)_2$ → odorless hydroscopic gum (611).
Derivs.: Monomethiodide, m. 215-216°; Dimethobromide monohydrate, m. 265° (effervesc.); Dimethopicrate, yellow crystals, m. 258° (611).

5,10-ETHANOARSANTHRENE DIMETHOBROMIDE $C_{16}H_{18}As_2Br_2$ *

Prepn.: by heating 5,10-dihydro-5,10-dimethylarsanthrene with $(CH_2Br)_2$ in MeOH in a sealed tube at 100° (611).
Props.: m. 272° (effervesc.) and decomp. to 5,10-dihydro-5,10-dimethylarsanthrene (611).
Other salts: Dipicrate, yellow crystals, m. 273° (decompn.); Diiodide, m. 248° (effervesc.) (611).

5,10-PROPANOARSANTHRENE DIMETHOBROMIDE $C_{17}H_{20}As_2Br_2$ *

Prepn.: by heating 5,10-dihydro-5,10-dimethylarsanthrene with $CH_2(CH_2Br)_2$ at 155-160° 52% yield (611).
Props.: m. 267-268° (effervesc.); when heated in vacuo yields 5,10-dihydro-5,10-dimethylarsanthrene (611).
Other salts: Dipicrate, yellow crystals, m. 252° (decompn.).

5,10-o-BENZENOARSANTHRENE, TRI-o-PHENYLENEDIARSINE $C_{18}H_{12}As_2$ *

Prepn.: by-prod. formed on heating o-$(PhAsCl)C_6H_4AsCl_2$ at 12 mm pressure under reflux (200).

5,10-DIHYDRO-5,10-DI-p-TOLYLARSANTHRENE $C_{26}H_{22}As_2$ *

Prepn.: by reacting 5,10-dichloro-5,10-dihydroarsanthrene with 4-MeC_6H_4MgBr in C_6H_6-Et_2O; a by-prod., m. 216-217°, analyzing for tri-p-tolyldiarsine was also obtd. (200).

---

* For the structural formula see page 404.

Props.: the compd. occurs in two stereoisomeric forms; the α-isomer, forms leaflets, m. 178-179°; the β-isomer, bipyramidal crystals, m. 179-181°; mixed m. 144-
158° (200).
Rxns. with: Br$_2$ in CHCl$_3$ → the tetrabromide (200).
MeI → both isomers yield the monomethiodide, m. 176-179°; monohydrate of the β-isom.
m. 174-179°; mixed m. of the α- and β-monomethiodides 167-175° (200).

5,10-DI-p-TOLYLARSANTHRENE DIBROMIDE C$_{26}$H$_{22}$As$_2$Br$_2$ *
Prepn.: By passing damp H$_2$S through a soln. of 5,19-dihydro-5,10-di-p-tolylarsan-
threne tetrabromide in CHCl$_3$; the monosulfide is formed as by-prod. (200).
By boiling 5,10-di-p-tolylarsanthrene tetrabromide in CHCl$_3$ with gradual addn. o
Me$_2$CO (200).
By treating 5,10-dihydro-5,10-di-p-tolylarsanthrene with Br$_2$ in CHCl$_3$ (200).
Props.: White crystals, m. 298-300° (dec.); ionic character (200).
Rxn. with H$_2$S → the monosulfide (200).

5,10-DIHYDRO-5,10-DI-p-TOLYLARSANTHRENE TETRABROMIDE C$_{26}$H$_{22}$As$_2$Br$_4$ *
Prepn.: by treating 5,10-dihydro-5,10-di-p-tolylarsanthrene with Br$_2$ in CHCl$_3$
(200).
Props.: Pale yellow crystals, which readily evolve HBr on exposure to air (200).
Rxns. with: NH$_4$OH → the tetrahydroxide (200).
SO$_2$ in HCl contg. KI → 5,10-dihydro-5,10-di-p-tolylarsanthrene (200).
EtOH-Na$_2$S → the monosulfide (200).
H$_2$S (damp) in CHCl$_3$ → the dibromide and monosulfide (200).

5,10-DIHYDRO-5,10-DI-p-TOLYLARSANTHRENE DIOXIDE C$_{26}$H$_{22}$As$_2$O$_2$ *
Prepn.: by dehydration of 5,10-dihydro-5,10-di-p-tolylarsanthrene tetrahydroxide
at 250° (200).
Props.: Hydroscopic white powder (200).

5,10-DIHYDRO-5,10-p-TOLYLARSANTHRENE SULFIDE C$_{26}$H$_{22}$As$_2$S *
Prepn.: By treating 5,10-dihydro-5,10-di-p-tolylarsanthrene tetrabromide suspen-
ded in CHCl$_3$ with alc. Na$_2$S or with damp H$_2$S (200).
By reacting 5,10-di-p-tolylarsanthrene dibromide with H$_2$S in CHCl$_3$ (200).
Props.: White needles, m. 198-201°, stable in moist air; soly. data (200).
Rxns. with: MeI → the monomethiodide (200).
Br$_2$ in CHCl$_3$ → the dibromide (200).

5,10-DIHYDRO-5,10-DI-p-TOLYLARSANTHRENE TETRAHYDROXIDE C$_{26}$H$_{26}$As$_2$O$_4$ *
Prepn.: by hydrolizing 5,10-dihydro-5,10-di-p-tolylarsanthrene tetrabromide with
aq. NH$_3$ under reflux (200).
Props.: White crystals, which on heating lose H$_2$O and melt at 318-325° (decompn.
on heating at 250° is converted to the dioxide (200).

5,5'-ETHYLENEBIS(5,10-DIHYDRO-10-METHYLANTHRENE) 5,5'-DIMETHOBROMIDE
C$_{30}$H$_{32}$As$_4$Br$_2$ *
Prepn.: by heating a mixt. of 5,10-dihydro-5,10-dimethylarsanthrene, with (CH$_2$-
Br)$_2$ (1:1) and MeOH in a sealed tube at 100° for 9 hrs. (611).
Props.: m. 195° (effervesc.); at 210° decomp. to 5,10-dihydro-5,10-dimethylars-

---

* For the structural formula see page 404.

anthrene and 5-bromo-5,10-dihydro-10-methylarsanthrene (611).
Other salts: Dipicrate, yellow crystals, m. 237° (decompn.); diiodide dihydrate, m. 185° (effervesc.) (611).

5,12,17,24-TETRAHYDRO-6,11,18,23-DI-o-BENZENOTETRABENZO[b,f,j,n][1,4,9,12]-
 TETRARSACYCLOHEXADECITETRAINIUM TETRABROMIDE C$_{40}$H$_{32}$As$_4$Br$_4$

Prepn.: by heating 5,10-dihydro-5,10-dimethylarsanthrene with o-C$_6$H$_4$(CH$_2$Br)$_2$
(1:2) at 160°; 64% yield (611).
    By heating 6,11-dihydro-5-methyl-5,12-benzenodibenzo[b,f]diarsocinium bromide
with o-C$_6$H$_4$(CH$_2$Br)$_2$ (1:1) at 160°; 66% yield (611).
Props.: Cream-colored crystals, m. 273° (611).
Other salts: Tetraiodide, deep orange crystals, m. 244° (decompn.); Tetrapicrate
decahydrate, m. 232° (decompn.) (611).

5,12,17,24-TETRAHYDRO-6,8,11,18,21,23-HEXAMETHYL-6,11,18,23-DI-o-BENZENOTETRA-
 BENZO[b,f,j,n][1,4,9,12]TETRARSACYCLOHEXADECENIUM TETRAIODIDE C$_{46}$H$_{48}$As$_4$I$_4$

Prepn.: by heating at 100° a mixt. 5,10-dihydro-2,4,10-trimethylarsanthrene and
o-C$_6$H$_4$(CH$_2$Br)$_2$,and treating the prod. with alc. NaI soln.; 70% yield (611).
Props.: Monohydrate, m. 180-181° (decompn); at 170°/0.1 mm decomp. to a yellow
gummy prod. (611).
Other salts: Tetrapicrate, yellow crystals, m. 171-173°; Tetra-(+)-bromocamphor-
sulfonate dihydrate, m. 257-258°; Tetraantimony-tartrate dihydrate, m. 239° (611).

s-TETRAARSENIN DERIVATIVES

s-TETRAARSENIN-3,3,6,6-TETRACARBOXYLIC ACID $C_6H_4As_4O_8$ HO₂C   CO₂H ( )

Prepn.: by reducing $(HO_2C)_2C(AsO_3H_2)_2$ with excess H_3PO_2 (827).
Props.: Yellow-brown powder, decompg. 165° (827).
Rxn . with: $H_2O_2 \rightarrow (HO_2C)_2C(AsO_3H_2)_2$ (827).

1,4-OXARSINANE AND PHENOXARSINE DERIVATIVES

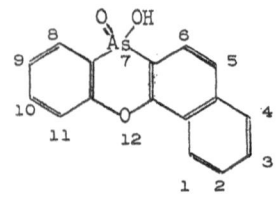

1,4-Oxarsinane   10-R-Phenoxarsine   Phenoxarsinic Acid

α-Benzophenoxarsinic Acid   γ-Benzophenoxarsinic Acid

4-PHENYL-1,4-OXARSINANE $C_{10}H_{13}AsO$
Prepn.: by refluxing $PhAs(MgBr)_2$ with $(BrCH_2CH_2)_2O$ in C_6H_6; isolated as oxar-
·sinane hydroxy picrate by oxidizing with H_2O_2 and pptg. with picric acid.
The free oxarsinane was obtd. by reducing the picrate with SO_2 in the presence
of KI (70).
Props.: Colorless liquid, b. 149-151°/18 mm (70).
Derivs.: Hydroxy picrate forms yellow crystals, m. 123° (70).
Hydroxy chloride, obtd.·by decompg. the picrate with HCl (70).
Sulfide, obtd. by alkalizing the hydroxy chloride with NH_4OH and treating the
prod. with H_2S; colorless crystals, m. 101.5-102° (70).
Methiodide, prepd. from the oxarsinane, colorless crystals, m. 162-162.5° (70).
$PdCl_2$ complex salt, orange crystals, m. 182° (70).

2-BROMO-10-CHLOROPHENOXARSINE $C_{12}H_7AsBrClO$ *
Prepn.: by heating $4\text{-}BrC_6H_4OC_6H_4\text{-}2\text{-}AsCl_2$ at 210° in CO_2 (858).
Props.: Prisms, m. 172-173° (858).

2,10-DICHLOROPHENOXARSINE $C_{12}H_7AsCl_2O$ *
Prepn.: by reducing 2-chlorophenoxarsinic acid with SO_2 in concd. HCl-$CHCl_3$
mixt. contg. I; $2\text{-}(4\text{-}ClC_6H_4O)C_6H_4AsCl_2$ is also formed and is converted into the
title compd. by distn. in vacuo (733).
Props.: m. 67-68° (733).
Rxns. with: RMgX → the corresponding 10-aryl deriv. (733).

10-CHLOROPHENOXARSINE $C_{12}H_8AsClO$ *
Prepn.: by heating o-$PhOC_6H_4AsCl_2$ at 200°/10 mm (271).
Props.: Crystals, m. 124° (271).
Rxns. with: N-substituted dithiocarbamic acid salts → the 10-RRNC(S)S- derivs.
(1250).

8,10-DICHLORO-2-METHYLPHENOXARSINE $C_{13}H_9AsCl_2O$ *
Prepn.: by reducing 8-chloro-2-methylphenoxarsinic acid with SO_2 in concd.
HCl contg. I (731).
Props.: Pale yellow, m. 171-172° (731).
Rxn. with: PhMgBr → the 10-phenyl deriv. (731).

10-CHLORO-3-METHYLPHENOXARSINE $C_{13}H_{10}AsClO$ *
Prepn.: by heating $3\text{-}MeC_6H_4OC_6H_4\text{-}2\text{-}AsCl_2$ at 160-180° (858).
Props.: Needles, m. 140-141° (858).

10-CHLORO-4-METHYLPHENOXARSINE $C_{13}H_{10}AsClO$ *
Prepn.: by heating $2\text{-}MeC_6H_4OC_6H_4\text{-}2\text{-}AsCl_2$ at 150-200° in CO_2 (858).
Props.: Prisms, m. 90-91° (858).

10-CHLORO-2-METHOXYPHENOXARSINE $C_{13}H_{10}AsClO_2$ *
Prepn.: by heating $4\text{-}MeOC_6H_4OC_6H_4\text{-}2\text{-}AsCl_2$ at 180-200° in CO_2 (858).
Props.: m. 108-109° (858).

8-CHLORO-2-METHYLPHENOXARSINIC ACID $C_{13}H_{10}AsClO_3$ *
Prepn.: by cyclizing $5,2\text{-}Cl(4\text{-}MeC_6H_4O)C_6H_3AsO_3H_2$ with concd. H_2O_4 (731).
Props.: m. 239-91 (731).
Rxn. with SO_2 in concd. HCl contg. I → 8,10-dichloro-2-methylphenoxarsine (731).

10-METHYLPHENOXARSINE $C_{13}H_{11}AsO$ *
Prepn.: by reacting 10-chlorophenoxarsine with MeMgI (96).
Props.: m. 192-195°/16 mm (96).
Rxn. with: $KMnO_4$ → methylphenoxarsine oxide (96).

10-METHYLPHENOXARSINE 10-OXIDE $C_{13}H_{11}AsO_2$ *
Prepn.: by oxidation of methylphenoxarsine by $KMnO_4$ (96).
Props.: m. 149-151° (96).
Rxn. with PhMgI → methylphenylphenoxarsonium iodide (97).

---

* For the structural formula see page 408.

10-CHLORO-1,3-DIMETHYLPHENOXARSINE $C_{14}H_{12}AsClO$ *
Prepn.: by heating $3,5\text{-}Me_2C_6H_3OC_6H_4\text{-}2\text{-}AsCl_2$ at 150-180° in CO_2 (858).
Props.: Prisms, m. 138-139° (858).

10-CHLORO-1,4-DIMETHYLPHENOXARSINE $C_{14}H_{12}AsClO$ *
Prepn.: by heating $2,4\text{-}Me_2C_6H_3OC_6H_4\text{-}2\text{-}AsCl_2$ at 150-180° in CO_2 (858).
Props.: Prisms, m. 146-147° (858).

10-CHLORO-2,4-DIMETHYLPHENOXARSINE $C_{14}H_{12}AsClO$ *
Prepn.: by heating $2,4\text{-}Me_2C_6H_3OC_6H_4\text{-}2\text{-}AsCl_2$ at 200° in CO_2 (858).
Props.: Prisms, m. 130-131° (858).

DIMETHYLTHIOCARBONYL 10-PHENOXARSINYL SULFIDE $C_{15}H_{14}AsNOS_2$

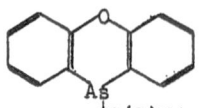

Prepn.: by reacting 10-chlorophenoxarsine or bis(phenoxarsine) oxide in Me_2CO
with aq. $Me_2NC(S)SNa$ or with Me_2NH, CS_2, and alkali under reflux (1250).
By treating $Me_2NC(S)SNa$ with MeEtCO and 10-chlorophenoxarsine (1250).
Props.: m. 163-164° (1250).

8-CHLORO-γ-BENZOPHENOXARSINIC ACID $C_{16}H_{10}AsClO_3$ *
Prepn.: by boiling 2-(2-chlorophenoxy)-1-naphthalenearsonic acid with AcOH (113).
Props.: m. 216-128° (113).

10-CHLORO-γ-BENZOPHENOXARSINIC ACID $C_{16}H_{10}AsClO_3$ *
Prepn.: by boiling 2-(2-chlorophenoxy)-1-naphthalenearsonic acid with AcOH (113).
Props.: m. 214° (113).

α-BENZOPHENOXARSINIC ACID $C_{16}H_{11}AsO_3$ *
Prepn.: by boiling 1-phenoxy-2-naphthalenearsonic acid with acetic acid (113).
Props.: m. 319° (113).

γ-BENZOPHENOXARSINIC ACID $C_{16}H_{11}AsNO_2S_2$ *
Prepn.: by boiling 2-phenoxy-1-naphthalenearsonic acid with acetic acid (113).
Props.: m. 278-280° (113).

MORPHOLINOTHIOCARBONYL 10-PHENOXARSINYL SULFIDE $C_{16}H_{16}AsNO_2S_2$
Prepn.: by reacting 10-chlorophenoxarsine in Me_2CO with aq. Na morpholinyl-
dithiocarbamate (1250).
Props.: m. 153° (1250).

11-METHYL-α-BENZOPHENOXARSINIC ACID $C_{17}H_{13}AsO_3$ *
Prepn.: by boiling 1-(2-anisyloxy)-2-naphthalenearsonic acid with AcOH (113).
Props.: m. 117° (113).

---

* For the structural formula see page 408.

10-METHYL-γ-BENZOPHENOXARSINIC ACID $C_{17}H_{13}AsO_3$ *

Prepn.: by boiling 2-(4-anisyloxy)-1-naphthalenearsonic acid with AcOH (113).
Props.: m. 215-127° (113).

10-PHENOXARSINYL 1-PYRROLIDINYLTHIOCARBONYL SULFIDE $C_{17}H_{16}AsNOS_2$

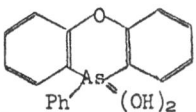

$\overline{SC(S)N(CH_2)_3CH_2}$

Prepn.: by reacting 10-chlorophenoxarsine with $\underline{CH_2(CH_2)_3N}C(S)SNa$ (1250).
Props.: m. 180° (1250).

10-PHENYLPHENOXARSINE $C_{18}H_{13}AsO$ *
Prepn.: by reacting 10-iodophenoxarsine with PhMgBr (96).
Props.: m. 108-109° (96).
Rxn. with KMnO$_4$ in acetone → the corresponding arsine oxide (96).
Derivs.: Methiodide, prepd. by treating phenylphenoxarsine oxide or methylphen-
oxarsine oxide with MeMgI and PhMgI, resp., m. 170-171° (97).
Phenyl bromide, prepd. by treating 10-phenylphenoxarsine oxide with PhMgBr (73%
yield). m. 229-230° (97).

10-PHENYLPHENOXARSINE 10-OXIDE $C_{18}H_{13}AsO_2$ *
Prepn.: by oxidation of the arsine by KMnO$_4$ in Me$_2$CO (96).
Props.: m. 188-189° (96).
Rxns. with: H$_2$O under reflux → the dihydroxide (96).
MeMgI → methylphenylphenoxarsonium iodide (97).
PhMgI → diphenylphenoxarsonium bromide (97).

10-PHENYLPHENOXARSINE 10,10-DIHYDROXIDE $C_{18}H_{15}AsO_3$

Prepn.: by heating phenylphenoxarsine oxide with H$_2$O (96).
Props.: m. 92-93° (96).

8-CHLORO-10-PHENYLPHENOXARSINE-2-CARBOXYLIC ACID $C_{19}H_{12}AsClO_3$ *
Prepn.: by oxidation of 8-chloro-2-methyl-10-phenylphenoxarsine with KMnO$_4$
(731).
Props.: dl. mixt., m. 220-221°, was resolved by means of 1-MePhCHNH$_2$, the d-
and l- isomer, both melt at 202-203° (731).

dl-4-(2-CHLORO-10-PHENOXARSINYL)BENZOIC ACID $C_{19}H_{12}AsClO_3$ *
Prepn.: by reducing the 10-oxide deriv. with SO$_2$ in HCl-CHCl$_3$ contg. I (733).
Props.: dl, m. 226-227°; resolved into d- and l- isomers by means of strychnine,
quinine, and cinchonine; both isomers melt at 219-220° (733).

---

* For the structural formula see page 408.

2-CHLORO-10-(4-CARBOXYPHENYL)PHENOXARSINE 10-OXIDE $C_{19}H_{12}AsClO_4$

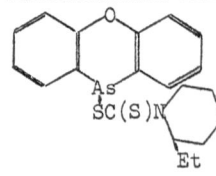

Prepn.: by oxidation of 2-chloro-10-p-tolylphenoxarsine with $KMnO_4$ (733).
Props.: m. 325-326° (733).
Rxn. with SO_2 in $HCl-CHCl_3$ contg I →(2-chloro-10-phenox arsine)benzoic acid (733).

dl-10-PHENYLPHENOXARSINE-2-CARBOXYLIC ACID $C_{19}H_{13}AsO_3$ *
Prepn.: by reducing the 10-oxide deriv. with SO_2 in $CHCl_3$-dild. HCl contg. I_2
(730).
Props.: Crystals, contg. crystn. solvent (EtOH), m. <100° (730).
Resolved into d- and l- isomers by means l-α-PhEtNH; the l-isomer, m. 199-200°,
the d-isomer, m. 189-190° (730).
Rxn . with l-α-PhEtNH → resolution (730).

dl-10-PHENYLPHENOXARSINE-2-CARBOXYLIC ACID 10-OXIDE $C_{19}H_{13}AsO_4$ *
Prepn.: by oxidation of 2-methyl-10-phenylphenoxarsine with $KMnO_4$ (730).
Props.: m. 320° (730); the l-isomer, m. 313-315° and the d-isomer, m. 312-134°;
the morphine salt of the l-isomer, m. 245-246° and of the d-isomer, m. 241-243°
(732).
Rxn . with: SO_2 in $CHCl_3$-dild. HCl contg. I_2 → 10-phenylphenoxarsine-2-carboxy-
lic acid (730).

2-CHLORO-10-p-TOLYLPHENOXARSINE $C_{19}H_{14}AsClO$ *
Prepn.: by reacting 2,10-dichlorphenoxarsine with $4-MeC_6H_4MgBr$ in Et_2O; 80%
yield (733).
Props.: b. 260-264°/10 mm; m. 70-71° (733).
Rxn . with: $KMnO_4$ → 2-chloro-10-(4-carboxyphenyl)phenoxarsine 10-oxide (733).

8-CHLORO-2-METHYL-10-PHENYLPHENOXARSINE $C_{19}H_{14}AsClO$ *
Prepn.: by reacting 8,10-dichloro-2-methylphenoxarsine with PhMgBr (731).
Props.: m. 75-76°, b. 258°/3 mm (731).
Rxn . with $KMnO_4$ → the 2-carboxylic acid (731).

2-METHYL-10-PHENYLPHENOXARSINE $C_{19}H_{15}AsO$
Rxn . with: $KMnO_4$ → 10-phenylphenoxarsine-2-carboxylic acid 10-oxide (730).

2-ETHYLPIPERIDINOTHIOCARBONYL 10-PHENOXARSINYL SULFIDE $C_{20}H_{21}AsNOS_2$

Prepn.: by reacting 10-chlorophenoxarsine in Me_2CO with aq. Na 2-Et-piperidyl-

* For the structural formula see page 408.

dithiocarbamate (1250).
<u>Props.</u>: m. 125-126° (1250).

DIBUTYLTHIOCARBAMOYL 10-PHENOXARSINYL SULFIDE $C_{20}H_{26}AsNOS_2$

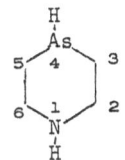

<u>Prepn.</u>: by reacting 10-chlorophenoxarsine in Me$_2$CO with aq. Bu$_2$NC(S)SNa (1250).
<u>Props.</u>: m. 91° (1250).

10-NAPHTHYLPHENOXARSINE $C_{22}H_{15}AsO$ *
<u>Prepn.</u>: by reacting 10-chlorophenoxarsine with α-$C_{10}H_7$MgBr (96).
<u>Props.</u>: m. 137-138° (96).
<u>Rxns. with</u>: KMnO$_4$ in acetone → the corresponding arsine dihydroxide (96).
MeI → methyl-α-naphthylphenoxarsonium iodide, m. 143-144° (97).

10-NAPHTHYLPHENOXARSINE 10,10-DIHYDROXIDE $C_{22}H_{17}AsO_3$ *
<u>Prepn.</u>: by oxidation of the arsine by KMnO$_4$ in acetone (96).
<u>Props.</u>: m. 191-192° (96).

HEXAHYDROAZARSINE DERIVATIVES

Hexahydro-1,4-azarsine

HEXAHYDRO-4-IODO-1-PHENYL-1,4-AZARSINE HYDRIODIDE $C_{10}H_{13}AsIN.HI$
<u>Prepn.</u>: by reacting hexahydro-1,4-diphenyl-1,4-azarsine with HI (70).
<u>Props.</u>: m. 173-174° (70).
<u>Rxn. with</u> NaHCO$_3$ → hexahydro-4-hydroxy-1-phenyl-1,4-azarsine (70).

HEXAHYDRO-4,4,4-TRIIODO-1-PHENYL-1,4-AZARSINE HYDRIODIDE
<u>Prepn.</u>: By reacting hexahydro-1,4-diphenyl-1,4-azarsine with HI contg. I (70).
<u>Props.</u>: m. 136-158° (70).
<u>Rxn. with</u> SO$_2$ → the corresp. 4-iodo deriv. (40).

HEXAHYDRO-4-HYDROXY-1-PHENYL-1,4-AZARSINE
<u>Prepn.</u>: by treating hexahydro-4-iodo-1-phenyl-1,4-azarsine hydriodide with
NaHCO$_3$ (70).
<u>Props.</u>: m. 116-116.5° (70).

---

* For the structural formula see page 408.

HEXAHYDRO-1,4-DIPHENYL-1,4-AZARSINE $C_{16}H_{18}AsN$ *

Prepn.: by reacting $PhAs(MgBr)_2$ with $PhN(CH_2CH_2Br)_2$ in C_6H_6 under reflux (70).
Props.: Colorless crystals, m. 96-97.5° (70).
Rxns. with: $(CH_2Br)_2$ → 4,4'-ethylenebis(hexahydro-1,4-diphenyl-1,4-azarsin-4-onium bromide)(70).
Dild. H_2O_2 in Me_2CO → azarsine 4-oxide (70).
Chloramine T → the azarsine 4-oxide (70).
Concd. H_2O_2 → the azarsine 1,4-dioxide (70).
HI → hexahydro-4-iodo-1-phenyl-1,4-azarsine hydriodide (70).
HI-I_2 → hexahydro-4,4,4-triodo-1-phenyl-1,4-azarsine (70).
Derivs.: Picrate, yellow crystals, m. 172-173° (70).
Monomethiodide, colorless crystlas, m. 181-182° (70).
Monoethobromide, colorless crystals, m. 179-179.5° (70).
Ethobromide hydrobromide, colorless crystals, m. 207-208° (70).
p-Chlorophenacyl azarsinium picrate, hydroscopic yellow crystals, m. 93-94° (70).
The 4-oxide dihydrate, m. 123-124° (70).
Hydroxy picrate, orange-red crystals, m. 188-189° (70).
Dihydrochloride, m. 156-156.5° (70).

HEXAHYDRO-1,4-DIPHENYL-1,4-AZARSINE 4-OXIDE $C_{16}H_{18}AsNO$ *

Prepn.: by oxidation of hexahydro-1,4-diphenyl-1,4-azarsine with H_2O_2 in Me_2CO
or with chloramine-T (70).
Props.: Dihydrate, m. 123-124°, loses some H_2O in vacuo over P_2O_5 (70).
Rxns. with: Picric acid → the hydroxy picrate, orange red crystals, m. 188-189° (70).
HCl → the dihydrochloride, m. 156-156.5° (70).
4-$MeC_6H_4SO_3Me$, followed by picric acid → hexahydro-4-hydroxy-1-methyl-1,4-diphenyl-1,4-azarsiniumdipicrate (70).

HEXAHYDRO-1,4-DIPHENYL-1,4-AZARSINE 1,4-DIOXIDE $C_{16}H_{18}AsNO_2$ *

Prepn.: by reacting hexahydro-1,4-diphenyl-1,4-azarsine with concd. H_2O_2 in
acetone (isolated as monopicrate) (70).
Deriv.: Monopicrate, m. 179° (70).

4,4'-ETHYLENEBIS(HEXAHYDRO-1,4-DIPHENYL-1,4-AZARSONIUM BROMIDE) $C_{34}H_{40}As_2Br_2N_2$
Prepn.: by heating hexahydro-1,4-diphenyl-1,4-azarsine with $(CH_2Br)_2$ at 115-120° (70).
Props.: Crystallizes with 2 mols. of EtOH, m. 227° (70).
Deriv.: Dipicrate, yellow, m. 166-167° (70).

PHENARSAZINE DERIVATIVES

5,10-Dihydrophenarsazine

Phenazarsinic Acid

* For the structural formula see page 413.

10-CHLORO-5,10-DIHYDROPHENARSAZINE, ADAMSITE, (RADIOACTIVE) $C_{12}H_9AsClN$
Prepn.: by reacting $AsCl_3$ with excess Ph_2NH; 70% yield (157).
Props.: Greenish yellow crystals, m. 194-195° (157); Identification (249, 530);
Light absorption data (853, 856); Diple moment (854).
Rxns. with: MeMgI → the 10-methyl deriv. (157).
HI (1:2) → Ph_2NH + $AsClI_2$ (530).
RSH → the 10-RS- deriv. (1093).
S_2Cl_2 → elimination of As and formation of 2,4,5,7-tetrachlorophenothiazine (1157).
CF_2:$CFMgI$ → the 10-trifluorovinyl deriv. (1165).

5,10-DIHYDROPHENARSAZINE $C_{12}H_{10}AsN$
Rxns. with HCO_2H → intensively colored prod., as a result of addn. of 1 H per
phenarsazine ring (894).

PHENAZARSINIC ACID $C_{12}H_{10}AsNO_2$
Rxn. with VO_3^- → ppt. (quant. analysis) (1014).

10-CYANO-5,10-DIHYDROPHENARSAZINE $C_{13}H_9AsN_2$
Props.: Identification (249).

5,10-DIHYDRO-1-METHYLPHENARSAZINE (RADIOACTIVE) $C_{13}H_{12}AsN$
Prepn.: by treating radiactive 10-chloro-5,10-dihydrophenarsazine with MeMgI (157).
Props.: Colorless needles, m. 108-109° (157).
Rxn. with S_2Cl_2 → elimination of As and formation of tetrachlorophenothiazine (157).

10-TRIFLUOROVINYL-5,10-DIHYDROPHENARSAZINE $C_{14}H_9AsF_3N$
Prepn.: by reacting 10-chloro-5,10-dihydrophenarsazine with CF_3:$CFMgI$ in Et_2O
at -20°; 75% yield (1165).
Props.: m. 122° (1165).

5,10-DIHYDRO-10-(2-HYDROXYETHYLTHIO)PHENARSAZINE $C_{14}H_{14}AsNOS$
Prepn.: by adding dropwise 5,10-dihydro-10-chlorophenarsazine dissolved in Me_2CO
to a soln. of $HOCH_2CH_2SH$ and KOH in EtOH (1093).
Props.: m. 164-166° (1093).

10-(BENZOYLTHIO)-5,10-DIHYDROPHENARSAZINE $C_{19}H_{14}AsNOS$
Prepn.: by reacting 10-chloro-5,10-dihydrophenarsazine with PhCOSK in aq. soln.
(1249).
Props.: m. 169° (1249).

BENZO[a]PHENARSAZINE DERIVATIVES

7,12-Dihydrobenzo[a]phenarsazine

12-Chloro-7,12-dihydrobenzo[a]phenarsazine are formed by reacting arsenic trichlor
with N-phenyl-2-naphthylamine in o-dichlorobenzene under reflux. Ring substituted
N-phenyl-2-naphthylamine derivatives yield the corresponding substituted 12-chloro
7,12-dihydrobenzo[a]phenarsazine derivatives. The chlorine atom at position 12 is
readily exchanged for an alkyl or aryl group in reactions with alkyl- or arylmagnesium halides.

The 12-chloro derivatives show hallochromic effect in H_2SO_4. They melt with
decomposition which starts below the melting temperature.

7,12-DIHYDROBENZO[a]PHENARSAZINE DERIVATIVES

7,12-Dihydrobenzo[a]phenarsazine substituted at... by	Prepd. from 2-(RNH)C₁₀H₆R' where R and	R'are:	Props. and Remarks	Ref.
C₁₆				
12-Chloro-10-fluoro-	4-FC₆H₄-	H	m. 297° (dec. 270°), rxn.[1]	163
8,12-Dichloro-	2-ClC₆H₄-	H	Yellow microcryst. powder, rxn.[2]	154
9,12-Dichloro-	3-ClC₆H₄	H	Lemon-yellow microcrystals, m. 265°, sol. in H₂SO₄, rxn[2]	154
10,12-Dichloro-	4-ClC₆H₄-	H	m. 262°, decomp. > 240°	158
C₁₇				
12-Chloro-10-fluoro-2-methyl-	4-FC₆H₄-	7-Me-	m. 250°, decomp. >212°	163
12-Chloro-10-fluoro-3-methyl-	4-FC₆H₄-	6-Me-	m. 285°, decomp. >270°	163
9,12-Dichloro-8-methyl-	2,3-Me(Cl)C₆H₃-	H	m. 290°, decomp. >270°, rxn.[1]	158
12-Chloro-3-methyl-	Ph-	6-Me-	m. 275-279° (dec.), forms yellow-orange sulfate	153
12-Chloro-8-methyl-	2-MeC₆H₄-	H	Lemon-yellow needles, m. 227°, sol. in H₂SO₄, rxn.[2]	154
12-Chloro-9-methyl-	3-MeC₆H₄-	H	Orange-yellow needles, m. 271°, sol. in H₂SO₄, rxn.[2]	154
12-Chloro-10-methoxy-	4-MeOC₆H₄-	H	m. 245°, decomp. >235°, rxn.[1]	158
10-Fluoro-12-methyl-[3]			m. 120°	163
C₁₈				
9-Chloro-8,12-dimethyl-[3]	2-MeC₆H₄-	6-Me-	m. 143°	158
12-Chloro-3,8-dimethyl-	3-MeC₆H₄-	6-Me-	m. ~246-248° (dec.), rxn.[1,2]	153
12-Chloro-3,9-dimethyl-	4-MeC₆H₄-	6-Me-	m. ~275° (dec.), rxns.[1,2]	153
12-Chloro-3,10-dimethyl-			Orange-yellow crystals, m. 258-261° (dec.).	153
12-Chloro-8,9-dimethyl-	2,3-Me₂C₆H₃-	H	m. 249°, decomp. > 235°, rxn[1]	158
12-Chloro-8,10-dimethyl-	2,4-Me₂C₆H₃-	H	Ocker yellow crystals, m. 269°, sol. in H₂SO₄, rxn.[2]	154
12-Chloro-8,11-dimethyl-	2,5-Me₂C₆H₃-		m. 250°, decomp. >220°, rxn.[1]	158

1. Rxn. with RMgX → the corresponding 12-R- deriv. (158, 163).
2. Rxn. with H₂O₂ → the corresponding benzo[a]phenazarsinic acid (154).
3. Prepd. by treating the corresponding 12-chloro deriv. with appropriate alkylmagnesium halide (153, 158, 163).

TABLE LVIII (cont'd.)

7,12-DIHYDROBENZO[a]PHENARSAZINE DERIVATIVES

7,12-Dihydrobenzo[a]phenarsazine substituted at... by	Prepd. from 2-(RNH)$C_{10}H_6R'$ where R and	$C_{10}H_6R'$ R' are:	Props. and Remarks	Ref.
12-Chloro-9,10-dimethyl-	3,4-Me$_2$C$_6$H$_3$-	H	m. 285°, decomp. >250°, rxn.[1]	158
12-Chloro-9,11-dimethyl-	3,5-Me$_2$C$_6$H$_3$-	H	m. 314°, decomp. >275°, rxn.[1]	158
12-Chloro-3-ethyl-	Ph-	6-Et-	Shiny, orange-yellow leaflets, m. 220°, decomp. >212°	160
12-Chloro-10-ethyl-	4-EtC$_6$H$_4$-	H	Silky, orange-yellow needles, m. 241°, decomp. >225°	160
12-Chloro-8-ethoxy-	2-EtOC$_6$H$_4$-	H	m. 192°, decomp. >170°	158
10-Methoxy-12-methyl-[3]			m. 158°	158
C$_{19}$				
9-Chloro-12-ethyl-8-methyl-[3]	2-MeC$_6$H$_4$-	6-Et-	m. 110°	158
12-Chloro-3-ethyl-8-methyl-			Silky, golden-yellow prisms, m. 275°, decomp. >270°	160
12-Chloro-11-ethyl-8-methyl-	2,5-Me(Et)C$_6$H$_3$-	H	Silky, yellow needles, decomp. >245°	160
12-Chloro-2,8,10-trimethyl-	2,4-Me$_2$C$_6$H$_3$-	7-Me-	m. 246°, decomp. >233°	163
12-Chloro-8,10,11-trimethyl-	2,4,5-Me$_3$C$_6$H$_2$-	H	Orange-yellow needles, m. 260°, sol. in H$_2$SO$_4$, rxns.[1,2]	154
			Radioactive AsCl$_3$ gave radio-active prod.	77
3,8,12-Trimethyl-[3]			m. 155°	153
3,9,12-Trimethyl-[3]			m. 140°	153
8,9,12-Trimethyl-[3]			m. 137°	158
8,11,12-Trimethyl-[3]			m. 158°	158
9,10,12-Trimethyl-[3]			m. 142°	158
C$_{20}$				
12-Chloro-9-ethyl-8,11-dimethyl-	3,2,5-Et(Me)$_2$C$_6$H$_2$-	H	Shiny, orange-yellow needles, m. 253°, decomp. >220°	161
12-Chloro-11-ethyl-8,9-dimethyl-	5,2,3-Et(Me)$_2$C$_6$H$_2$-	H	m. 245°, decomp. >213°	328
12-Chloro-11-isopropyl-8-methyl-	2,5-Me(Me$_2$CH)C$_6$H$_3$-	H	m. 265°, decomp. >220°	158
12-Ethyl-8,9-dimethyl-[3]			m. 128°	158
12-Ethyl-9,11-dimethyl-[3]			m. 104°	158
8,10,11,12-Tetramethyl-[3]			Radioactive compd.	77

C$_{21}$				
12-Butyl-9-chloro-8-methyl-[3]	4-MeC$_6$H$_4$-		m. 105°	153
3-tert-Butyl-12-chloro-10-methyl		6-Me$_3$C-	m. 271°, decomp. > 268°	158
12-Chloro-10-(1-methylbutyl)-	4-(MePrCH)C$_6$H$_4$-	H	m. 174°, decomp. >165°, rxn.[1]	158
12-Chloro-10-tert-pentyl-	4-(EtMe$_2$C)C$_6$H$_4$-	H	m. 228°, decomp. > 210°, rxn.[1]	158
12-Isopropyl-9,10-dimethyl-[3]			Decomp. > 180°	158
C$_{22}$				
12-Chloro-8-phenyl-	2-PhC$_6$H$_4$-	H	m. 242°, decomp. > 215°, rxn.[1]	158
12-Chloro-10-phenyl-	4-PhC$_6$H$_4$-	H	m. 260-265°, decomp. > 245° rxn.[1]	158
12-Chloro-10-cyclohexyl-	4-C$_6$H$_{11}$C$_6$H$_4$-	H	m. 220° (dec.)	168
10-tert-Amyl-12-methyl-[3]			m. 151°	158
12-Methyl-10-(1-methylbutyl)-[3]			m. 101°	158
C$_{23}$				
12-Methyl-8-phenyl-[3]			m. 135°	158
12-Methyl-10-phenyl-[3]			m. 189°	158
12-Chloro-3-(1-methylcyclo-hexyl)-	Ph-	6-(1-Me-C$_6$H$_{10}$)-	Shiny yellow prisms, m. > 270° (dec.), rxn.[1]	159
C$_{24}$				
12-Ethyl-8-phenyl-[3]			m. 116°	158
12-Chloro-10-methyl-3-(1-me-thylcyclohexyl)-	4-MeC$_6$H$_4$-	6-(1-Me-C$_6$H$_{10}$)-	Shiny, bright yellow needles, m. 275° (dec.)	159
12-Methyl-3-(1-methylcyclohexyl]-[3]				159
C$_{28}$				
8,12-Diphenyl-[3]			m. 162°	158

BENZO[a]PHENAZARSINIC ACIDS

7,12-Dihydrobenzo[a]phenazarsinic Acid

7,12-Dihydrobenzo[a]phenazarsinic acid and its derivatives, compiled in Table LIX were prepared by oxidation of 12-chloro-7,12-dihydrobenzo[a]phenarsazine or its derivatives by hydrogen peroxide.

TABLE LIX
SUBSTITUTED 7,12-DIHYDROBENZO[a]PHENAZARSINIC ACIDS

7,12-Dihydrobenzo[a]phenazarsinic Acid containing substituent(s):	Props.	Ref.
8-Chloro-	Yellowish white, microcryst. powder	154
9-Chloro-	Yellowish white, microcryst. powder	154
8-Methyl-	Yellowish white, microcryst. powder	154
8-Methyl-	Yellowish white, microcryst. powder	154
8,9-Dimethyl-	Yellowish white, microcryst. powder	154
8,10,11-Trimethyl-	Yellowish white, microcryst. powder	154

BENZO[b]PHENARSAZINE AND BENZO[b]PHENAZARSINIC ACID DERIVATIVES

7,12-Dihydrobenzo[b]phenarsazine

7,12-Dihydrobenzo[b]phenazarsinic Acid

12-Chloro-7,12-dihydrobenzo[b]phenarsazine derivatives are formed by reacting arsenic trichloride with 1-aklyl-N-phenyl-2-naphthyl amines in o-dichlorobenzene under reflux. The 12-chloro compounds are readily converted into the corresponding 12-alkyl- or 12-aryl derivatives by reacting with alkyl- or arylmagnesium halides. Substituted 7,12-dihydrobenzo[b]phenazarsinic acids are obtained from substituted 12-chloro-7,12-dihydrobenzo[b]phenarsazine by oxidation with hydrogen peroxide.

TABLE LX

SUBSTITUTED BENZO[b]PHENARSAZINES

7,12-Dihydrobenzo[b]phenarsazine substituted at... by	Prepd. from 1-RC$_{10}$H$_6$-2-NHR' R	R'	Props.	Ref.
C$_{17}$				
12-Chloro-6-methyl-	Me-	Ph-	m. 227-228°, rxns.[1,2]	155
C$_{18}$				
12-Chloro-6,9-dimethyl-	Me-	3-MeC$_6$H$_4$-	m. 237°, rxns.[1,2]	155
12-Chloro-6,10-dimethyl-	Me-	4-MeC$_6$H$_4$-	m. 233-235°, rxns.[1,2]	155
6,12-Dimethyl-[3]			Yellowish white crystals, m. 124°, sol. in H$_2$SO$_4$	155
C$_{19}$				
6,9,12-Trimethyl-[3]			Colorless platelets, m. 139-140°, sol. in H$_2$SO$_4$	155
6,10,12-Trimethyl-[3]			Pale yellow needles, m. 164-165°	155
C$_{21}$				
12-Chloro-11-isopropyl-10-methoxy-8-methyl-	H	2,4,5-Me(MeO)(Me$_2$CH)C$_6$H$_2$	Yellow-ocher needles, decomp. 232-253°	1087
C$_{22}$				
12-Butyl-6,9-dimethyl-[3]			Yellowish white plates, m. 72°	155
12-Butyl-6,10-dimethyl-[3]			Yellowish plates, m. 88-89°	155
C$_{23}$				
6-Methyl-12-phenyl-[3]			Colorless plates, m. 149°	155
C$_{24}$				
6,10-Dimethyl-12-phenyl-[3]			Yellowish white plates, m. 178°	155
C$_{28}$				
6,10-Dimethyl-12-α-naphthyl-[3]			Bright yellow crystals, m. 206°	155

1. Rxn. with RMgBr → the corresponding 12-R- derivative (155).
2. Rxn. with H$_2$O$_2$ → the corresponding 7,12-dihydrobenzo[b]phenarsazinic acid (155).
3. Prepd. by treating the 12-chloro deriv. with appropriate alkyl- or arylmagnesium halide (155).

7,12-DIHYDRO-6-METHYLBENZO[b]PHENAZARSINIC ACID $C_{17}H_{14}AsNO_2$ *
Prepn.: by oxidizing 12-chloro-7,12-dihydro-6-methylbenzo[b]phenarsazine with
H_2O_2 (155).
Props.: Yellowish, microcryst. powder (155).

7,12-DIHYDRO-6,9-DIMETHYLBENZO[b]PHENAZARSINIC ACID $C_{18}H_{16}AsNO_2$ *
Prepn.: by treating 12-chloro-7,12-dihydro-6,9-dimethylbenzo[b]phenarsazine with
H_2O_2 (155).
Props.: Yellowish, microcryst. powder (155).

7,12-DIHYDRO-6,10-DIMETHYLBENZO[b]PHENAZARSINIC ACID $C_{18}H_{16}AsNO_2$ *
Prepn.: by treating 12-chloro-7,12-dihydro-6,10-dimethylbenzo[b]phenarsazine
with H_2O_2 (155).
Props.: Yellowish, microcryst. powder (155).

BENZO[c]PHENARSAZINE AND BENZO[c]PHENAZARSINIC ACID DERIVATIVES

7,10-Dihydrobenzo[c]phenarsazine   7,12-Dihydrobenzo[c]phenazarsinic Acid

7-Chloro-7,12-dihydrobenzo[c]phenarsazine derivatives are formed by reacting arsen
trichloride with N-phenyl-1-naphthylamine or with substituted N-phenyl-1-naphthyl-
amine derivatives in o-dichlorobenzene under reflux. The 7-chloro derivatives are
readily converted into the corresponding 7-alkyl or 7-aryl compounds by reacting
with alkyl- or arylmagnesium halides. Oxidation of 7-chloro-7,12-dihydrobenzo[c]-
phenarsazine compounds with hydrogen peroxide yields the corresponding 7,12-
dihydrobenzo[c]phenazarsinic acids.

---

* For the structural formual see page 420.

7,12-DIHYDROBENZO[c]PHENARSAZINE DERIVATIVES

7,12-Dihydrobenzo[c]phenarsazine substituted at... by	Prepd. from 1-(RNH)C₁₀H₆R' — R	R'	Props. and Remarks	Ref.
C₁₆				
7-Chloro-9-fluoro-	4-FC₆H₄-	H	m. 237°, decomp. >220°	163
7,9-Dichloro-	4-ClC₆H₄-	H	m. 221-222°, decomp. 206°, rxn.[1]	158
C₁₇				
7,10-Dichloro-11-methyl-	2,3-Me(Cl)C₆H₃-	H	m. 275-280°, decomp. 240°	158
7-Chloro-10-methyl-	3-MeC₆H₃-	H	Lemon-yellow crystals, m. 235°, sol. in H₂SO₄, rxns. 2, 3	154
7-Chloro-11-methyl-	2-MeC₆H₃-	H	Bright yellow needles, m. 208°, sol. in H₂SO₄, rxn.[2]	154
9-Chloro-7-methyl-[4]			m. 123°	158
7-Chloro-9-methoxy-	4-MeOC₆H₄-	H	m. 242°, decomp. >220°	158
7-Chloro-11-methoxy-	2-MeOC₆H₄-	H	m. 239°, decomp. >215°	158
9-Fluoro-7-methyl-[4]			m. 107°	163
C₁₈				
7-Chloro-9-ethyl-	4-EtC₆H₄-	H	Silky, golden-yellow prisms, m. 210°; decomp. >196°, rxn.[1]	160
9-Chloro-7-ethyl-[1]			m. 81°	158
7-Chloro-8,10-dimethyl-	3,5-Me₂C₆H₃-	H	m. 312-314°, decomp. 275°, rxn.[1]	158
7-Chloro-8,11-dimethyl-	2,5-Me₂C₆H₃-	H	m. 217°, decomp. 206°, rxn.[1]	158
7-Chloro-9,10-dimethyl-	3,4-Me₂C₆H₃-	H	m. 280°, decomp. 253°; rxn.[1]	158
7-Chloro-9,11-dimethyl-	2,4-Me₂C₆H₃-	H	m. 244°, decomp. 215°, rxn.[1]	158
7-Chloro-10,11-dimethyl-	2,3-Me₂C₆H₃-	H	m. 260°, decomp. 230°, rxn.[1]	158
7-Chloro-11-ethoxy-	2-EtOC₆H₄-	H	m. 204°, decomp. >190°	158
C₁₉				
7-Chloro-8-ethyl-11-methyl-	2,5-Me(Et)C₆H₃-	H	Silky, lemon-yellow needles, m. 205°, decomp. >188°	160

1. Rxn. with RMg halides → the corresponding 7-R derivs. (158).
2. Rxn. with H₂O₂ → the corresponding 7,12-dihydrobenzo[c]phenazarsinic acids (154).
3. Rxn. with Ag₂O in xylene → 7,7'-oxybis(7,12-dihydro-10-methylbenzo[c]phenarsazine) (154).
4. Prepd. by reacting the corresponding 7-chloro deriv. with appropriate alkyl- or arylmagnesium halides (158).

TABLE LXI (cont'd.)

7,12-DIHYDROBENZO[c]PHENARSAZINE DERIVATIVES

7,12-Dihydrobenzo[c]phenarsazine substituted at ... by	Prepd. from 1-(RNH)$C_{10}H_6R'$		Props. and Remarks	Ref.
	R	R'		
7-Chloro-5,8,9-trimethyl-	3,4-$Me_2C_6H_3$-	4-Me-	m. 250° (dec.)	167
7-Chloro-5,9,10-trimethyl-	3,4-$Me_2C_6H_3$-	4-Me-	m. 284° (dec.)	167
7-Chloro-5,9,11-trimethyl-	2,4-$Me_2C_6H_3$-	4-Me-	m. 261° (dec.)	167
7-Chloro-5,10,11-trimethyl-	2,3-$Me_2C_6H_3$-	4-Me-	m. 229° (dec.)	167
7-Chloro-8,9,11-trimethyl-	2,4,5-$Me_3C_6H_2$-	H	Deep yellow crystals, m. 222°, sol. in H_2SO_4, rxn.2	154
9-Ethyl-7-methyl-[4]			Shiny needles, m. 109-110°	160
7,8,10-Trimethyl-[4]			m. 121°	158
7,8,11-Trimethyl-[4]			m. 110°	158
7,9,10-Trimethy-[4]			m. 171°	158
7,9,11-Trimethyl-[4]			m. 92°	158
7,10,11-Trimethyl-[4]			m. 130°	158
C_{20}				
7-Chloro-8-isopropyl-11-methyl-	2,5-Me(Me_2CH)C_6H_3-	H	m. 230°, dec. 208°	158
7-Ethyl-8,11-dimethyl-[4]			m. 109°	158
C_{21}				
7-Chloro-9-(1-methylbutyl)-	4-(MePrCH)C_6H_4-	H•	m. 165°, decomp. >160°	158
7-Chloro-9-tert-pentyl-	4-(EtMe_2C)C_6H_4-	H	m. 199°, decomp. >190°	158
7-Chloro-8-isopropyl-11-methyl-	2,4,5-Me(MeO)-(Me_2CH)C_6H_2-	H	Orange needles, decomp. 203-217°	1087
C_{22}				
7-Chloro-9-phenyl-	4-PhC_6H_4-	H	m. 275°, decomp. > 255°	158
7-Chloro-11-phenyl-	2-PhC_6H_4-	H	m. 195°, decomp. > 181°	158
7-Chloro-9-cyclohexyl-	4-$C_6H_{11}C_6H_4$-	H	m. 234° (dec.)	168
C_{25}				
7-Benzyl-8,11-dimethyl-[4]			m. 174°	158
C_{34}				
7,7'-Oxybis(10-methyl-[5]			Graysih white powder, sol. in H_2SO_4	154

5. Prepd. by treating 7-chloro-7,12-dihydro-10-methylbenzo[c]phenarsazine with As_2O in xylene (154).

7,12-DIHYDRO-10-METHYLBENZO[c]PHENAZARSINIC ACID $C_{17}H_{14}AsNO_2$ *
Prepn.: by oxidizing 7-chloro-7,12-dihydro-10-methylbenzo[c]phenarsazine by H_2O_2
(154).
Props.: Yellowish white microcryst. powder, m. >350°, gives a green-yellow soln.
turning violet-black (154).

7,12-DIHYDRO-11-METHYLBENZO[c]PHENAZARSINIC ACID $C_{17}H_{14}AsNO_2$ *
Prepn.: by treating the 7-chloro deriv. with H_2O_2 (154).
Props.: Yellowish white, microcryst. powder (154).

7,12-DIHYDRO-8,9,11-TRIMETHYLBENZO[c]PHENAZARSINIC ACID $C_{19}H_{18}AsNO_2$ *
Prepn.: by treating the 7-chlorobenzo[c]phenarsazine deriv. with H_2O_2 (154).
Props.: Yellowish white, microcryst. powder (154).

PYRIDO[3,2-a]PHENARSAZINE DERIVATIVES

7,12-Dihydropyrido[3,2-a]phenarsazine

12-CHLORO-7,12-DIHYDROPYRIDO[3,2-a]PHENARSAZINE $C_{15}H_{10}AsClN_2$
Prepn.: by heating 6-anilinoquinoline with $AsCl_3$ in o-$C_6H_4Cl_2$ (165).
Props.: Hydrochloride, m. 265-26 ° (decompn.) (165).

12-CHLORO-7,12-DIHYDRO-10-METHYLPYRIDO[3,2-a]PHENARSAZINE $C_{16}H_{12}AsClN_2$
Prepn.: by heating 6-p-toluinoquinoline with $AsCl_3$ in o-$C_6H_4Cl_2$ (165).
Props.: Hydrochloride, m. 290-291° (dec.) (165).

BENZOPENTENO- AND BENZINDENOPHENARSAZINE DERIVATIVES

7,13-Dihydrobenzopenteno[a,i]-
phenarsazine

7,13-Dihydrobenzopenteno[c,i]-
phenarsazine

---

* For the structural formula see page 422.

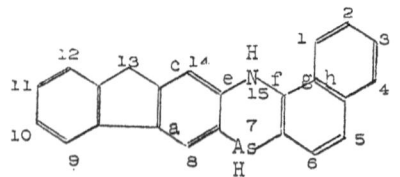

7,15-Dihydro-13H-benzo[h]indeno-[1,2-b]phenarsazine

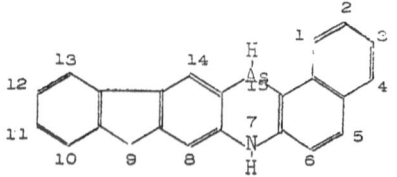

7,15-Dihydro-13H-benz[h]indeno[1,2-b]phenarsazine

13-CHLORO-7,13-DIHYDROBENZOPENTENO[a,i]PHENARSAZINE $C_{19}H_{15}AsClN$ *
Prepn.: by heating N-β-naphthyl-5-indanylamine with $AlCl_3$ (166).
Props.: Orange yellow leaflets, m. 290°, decomp. > 270° (166).

7-CHLORO-7,13-DIHYDROBENZOPENTENO[c,i]PHENARSAZINE $C_{19}H_{15}AsClN$ *
Prepn.: by heating N-α-naphthyl-5-indanylamine with $AsCl_3$ (166)..
Props.: Yellow leaflets, m. 278° (166).

7,13-DIHYDRO-13-METHYLBENZOCYCLOPENTENO[a,i]PHENARSAZINE $C_{20}H_{18}AsN$ *
Prepn.: by treating 13-chloro-7,13-dihydrobenzopenteno[a,i]phenarsazine with
MeMgI (166).
Props.: Colorless prisms, m. 207° (166).

7-CHLORO-7,15-DIHYDRO-13H-BENZ[h]INDENO[1,2-b]PHENARSAZINE $C_{23}H_{15}AsClN$
Prepn.: by heating 2-(1-naphthylamino)fluorene with $AsCl_3$ in $1-C_{10}H_7Cl$ (156).
Props.: Chars > 300° (156).
Rxn. with MeMgI → the 7-methyl deriv. (156).

15-CHLORO-7,15-DIHYDRO-9H-BENZ[a]INDENO[2,1-i]PHENARSAZINE $C_{23}H_{15}AsClN$
Prepn.: by heating 2-(2-naphthylamino)fluorene with $AsCl_3$ in $1-C_{10}H_7Cl$ (156).
Props.: Chars above 300° (156).

7,15-DIHYDRO-7-METHYL-13H-BENZ[h]INDENO[1,2-b]PHENARSAZINE $C_{24}H_{18}AsN$
Prepn.: by treating 7-chloro-7,15-dihydro-13H-benz[h]indeno[1,2-b]phenarsazine
with MeMgI (156).
Props.: m. 205° (156).

DIBENZOPHENARSAZINE DERIVATIVES

7,14-Dihydrodibenzo[a,h]phenarsazine

1,2,3,4,7,14-Hexahydrodibenzo[c,h]phenarsazine

---

* For the structural formula see page 425.

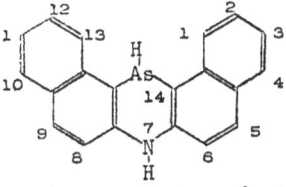

7,14-Dihydrodibenzo[a,j]phenarsazine

14-CHLORO-7,14-DIHYDRODIBENZO[a,h]PHENARSAZINE $C_{20}H_{13}AsClN$ *
Prepn.: by heating N-1-naphthyl-2-naphthylamine with $AsCl_3$ in o-$C_6H_4Cl_2$ (154).
Props.: Dark yellow crystals, m. 309°, sol. in H_2SO_4 with raspberry color (154).

14-CHLORO-7,8,9,10,11,14-HEXAHYDRODIBENZO[a,h]PHENARSAZINE $C_{20}H_{17}AsClN$ *
Prepn.: by heating N-2-naphthyl-5,6,7,8-tetrahydro-1-naphthylamine with $AsCl_3$
in o-$C_6H_4Cl_2$ (166).
Props.: Deep yellow needles, m. 290°, decomp. > 266° (166).
Rxn. with MeMgI → the 14-Me deriv. (166).

14-CHLORO-7,14-DIHYDRO-3-METHYLDIBENZO[a,h]PHENARSAZINE $C_{21}H_{15}AsClN$ *
Prepn.: by reacting N-(6-methyl-2-naphthyl)-1-naphthylamine with $AsCl_3$ in o-
$C_6H_4Cl_2$ (1085, 1086).
Props.: m. 283° (1085, 1086).
Rxn. with RMgX → the 14-R deriv. (1085).

7,8,9,10,11,14-HEXAHYDRO-14-METHYLDIBENZO[a,h]PHENARSAZINE $C_{21}H_{20}AsN$ *
Prepn.: by treating 14-chloro-7,8,9,10,11,14-hexahydrodibenzo[a,h]phenarsazine
with MeMgI (166).
Props.: Colorless prisms, m. 179° (166).

7,14-DIHYDRO-3,14-DIMETHYLDIBENZO[a,h]PHENARSAZINE $C_{22}H_{18}AsN$ *
Prepn.: by treating 14-chloro-7,14-dihydro-3-methyldibenzo[a,h]phenarsazine with
MeMgI (1085, 1086).
Props.: Yellowish prisms, m. 171° (1085, 1086).

14-ETHYL-7,14-DIHYDRO-3-METHYLDIBENZO[a,h]PHENARSAZINE $C_{23}H_{20}AsN$ *
Prepn.: by reacting 14-chloro-7,14-dihydro-3-methyldibenzo[a,h]phenarsazine with
EtMgI (1085, 1086).
Props.: M. 158° (1085, 1086).

7,14-DIHYDRO-3-METHYL-14-PHENYLDIBENZO[a,h]PHENARSAZINE $C_{27}H_{20}AsN$ *
Prepn.: by reacting 14-chloro-7,14-dihydro-3-methyldibenzo[a,h]phenarsazine with
PhMgI (1085, 1086).
Props.: M. 180° (1085, 1086).

7-CHLORO-1,2,3,4,7,14-HEXAHYDRODIBENZO[c,h]PHENARSAZINE $C_{20}H_{17}AsClN$ *
Prepn.: by reacting N-α-naphthyl-5,6,7,8-tetrahydro-α-naphthylamine with $AsCl_3$
in o-$C_6H_4Cl_2$ under reflux (166).
Props.: m. 268°, decomp. 252° (166).

---

* For the structural formula see page 426.

1,2,3,4,7,14-HEXAHYDRO-7-METHYLDIBENZO[c,h]PHENARSAZINE $C_{21}H_{20}AsN$ *
Prepn.: by treating 7-chloro-1,2,3,4,7,14-hexahydrodibenzo[c,h]phenarsazine
with MeMgI in Et_2O (166).
Props.: Colorless prisms, m. 173° (166).

14-CHLORO-7,14-DIHYDRO-3-METHYLDIBENZO[a,j]PHENARSAZINE $C_{20}H_{13}AsClN$ **
Prepn.: by reacting 6-methyldi-2-naphthylamine with $AsCl_3$ in 2-$C_6H_4Cl_2$ (1085,
1086).
Props.: Orange needles, m. 292° (1085, 1086).
Rxn. with RMgX → the 14-R deriv. (1085).

7,14-DIHYDRO-3,14-DIMETHYLDIBENZO[a,j]PHENARSAZINE $C_{21}H_{16}AsN$ **
Prepn.: by reacting 14-chloro-7,14-dihydro-3-methyldibenzo[a,j]phenarsazine with
MeMgI (1085, 1086).
Props.: M. 200° (1085, 1086).

14-ETHYL-7,14-DIHYDRO-3-METHYLDIBENZO[a,j]PHENARSAZINE $C_{22}H_{18}AsN$ **
Prepn.: by treating 14-chloro-7,14-dihydro-3-methyldibenzo[a,j]phenarsazine
with EtMg halides (1085, 1086).
Props.: M. 115° (1085, 1086).

14-CHLORO-1,13-DIETHYL-7,14-DIHYDRODIBENZO[a,j]PHENARSAZINE $C_{24}H_{21}AsClN$ **
Prepn.: by reacting bis(1-ethyl-7-naphthyl)amine with $AsCl_3$ (169).
Props.: m.∼262° (dec.) (169).

14-CHLORO-1,13-DIPROPYL-7,14-DIHYDRODIBENZO[a,j]PHENARSAZINE $C_{26}H_{25}AsClN$ **
Prepn.: by reacting bis(1-propyl-7-naphthyl)amine with $AsCl_3$ (169).
Props.: m.∼ 255° (169).

7,14-DIHYDRO-3-METHYL-14-PHENYLDIBENZO[a,j]PHENARSAZINE $C_{27}H_{20}AsN$ **
Prepn.: by treating 14-chloro-7,14-dihydro-3-methyldibenzo[a,j]phenarsazine
with PhMgI (1085, 1086).
Props.: M. 203° (1085, 1086).

PHOSPHARSENINE DERIVATIVES

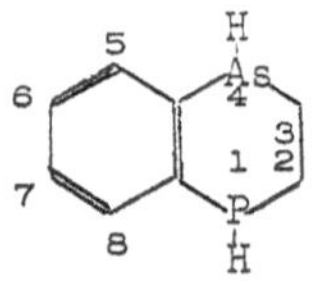

1,2,3,4-Tetrahydro-1,4-benzophospharsenine

1,4-DIETHYL-1,2,3,4-TETRAHYDRO-1,4-BENZOPHOSPHARSENIN 1-ETHOBROMIDE $C_{14}H_{23}As$.
Prepn.: by heating $(Et_2AsC_6H_4PEt_2CH_2CH_2)Br$ at 270°/15 mm for 15 min. (612).
Props.: m. 167-168° (612).

---

* For the structural formula see apge 426.

1-ETHYL-1,2,3,4-TETRAHYDRO-4-METHYL-1,4-BENZOPHOSPHARSENIN 1-ETHOBROMIDE
    4-METHOBROMIDE $C_{14}H_{24}AsBr_2P$
Prepn.: by heating at 90° 2-$Et_2PC_6H_4AsMe_2$ with $(CH_2Br)_2$ under N (612).
Props.: Dihydrate, m. 235-245° (612).
Derivs.: Dipicrate, m. 196-197° (612).

1,4-DIETHYL-1,2,3,4-TETRAHYDRO-1,4-BENZOPHOSPHARSENIN DIETHOBROMIDE $C_{16}H_{28}AsBr_2P$
Prepn.: by heating 2-$Et_2PC_6H_4AsEt_2$ with $(CH_2Br)_2$ at 90-120° under N (612).
Props.: m. 223°, at 270°/15 mm loses one EtBr (612).
Deriv.: Dipicrate, m. 215° (decompn.) (612).

SEVEN-MEMBERED RING SYSTEMS

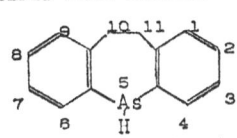

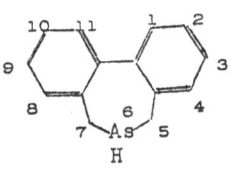

10 11-Dihydro-5H-dibenz[b,f]arsepin   5,7-Dihydro-6H-dibenz[c,e]arsepin

6H-5,7-DIHYDRO-6-IODODIBENZ[c,e]ARSEPIN $C_{14}H_{12}AsI$
Prepn.: by treating the corresponding 6-phenyl derivative with HI (69).
Props.: Yellow crystals, m. 117-117.5° (69).

10,11-DIHYDRO-5-PHENYL-5H-DIBENZ[b,f]ARSEPIN $C_{20}H_{17}As$
Prepn.: by treating (2-$BrC_6H_4CH_2$-)$_2$ with BuLi in petr. ether and reacting the
prod. with $PhAsCl_2$ in C_6H_6 (809).
Props.: m. 59-59.5°; b. 180-183°/0.15 mm; UV spectrum (809).
Derivs.: Methiodide, m. 215-227° (in vacuo); Oxide, m. 188-190°; Dichloro(bis-
arsine)palladium, m. 275-277° (809).
Rxns. with: H_2O_2 → the arsine oxide (809).
HI → AsI_3 (809).

5,7-DIHYDRO-6-PHENYL-6H-DIBENZ[c,e]ARSEPIN $C_{20}H_{17}As$
Prepn.: by refluxing $PhAs(MgBr)_2$ with (o-$BrCH_2C_6H_4$-)$_2$ in C_6H_6; 45% yield (69).
Props.: m. 118-118.5°
Salts: Methiodide crystallizes with 1 mol. MeOH, m. 223-224° (69).
Rxns. with: HI.→ cleavage of the Ph group with the formation of 6-iodo deriv.
(69).
p-$ClC_6H_4CH_2COBr$ → 6-(p-chlorophenacyl)-5,7-dihydro-6-phenyl-6H-benz[c,e]arsepin-
ium bromide (69).

6-(p-CHLOROPHENACYL)-5,7-DIHYDRO-6-PHENYL-6H-DIBENZ[c,e]ARSEPINIUM BROMIDE
    $C_{28}H_{23}AsBrClO$
Prepn.: by heating under reflux the arsepin deriv. with p-$ClC_6H_4CH_2COBr$ (69).
Props.: Colorless crystals, m. 110-115° (69).

DIARSEPIN DERIVATIVES

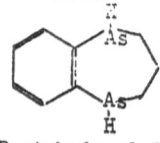

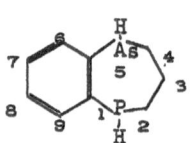

1,2,3,4,5-Pentahydro-1,5-benzodiarsepin   2-Hydrodibenzo[d,f]-1,3-diarsepin

1,2,3,4,5-PENTAHYDRO-1,1,5,5-TETRAMETHYL-1,5-BENZODIARSEPINIUM DIBROMIDE
    $C_{13}H_{22}As_2Br_2$

Prepn.: by reacting o-$C_6H_4(AsMe_2)_2$ with $(CH_2)_3Br_2$ at 156-160° (489).
Props.: Hydroscopic crystals, m. 258° (decompn.) (489).
Dipicrate, yellow, m. 260° (489).

2-HYDRO-1,3-DIMETHYLDIBENZ[d,f]-1,3-DIARSEPIN $C_{15}H_{16}As_2$
Prepn.: by pyrolysis of the 1,1,3,3-tetramethyldiarsonium dibromide deriv. (524)
Props.: m. 95.5-96° (524).
Derivs.: PdBr₂ complex, m. 318-319° (524).

1,1,3,3-TETRAMETHYLDIBENZO[d,f]-1,3-DIARSONIA.EPIN DIBROMIDE $C_{17}H_{22}As_2Br_2$
Prepn.: by heating (2-$C_6H_4AsMe_2$)$_2$ with CH_2Br_2 (524).
Props.: Tetrahydrate, m. 224-225° (decompn.); decomp. at elevated temp. to
2-hydro-1,3-dimethyldibenz[d,f]-1,3-diarsepin (524).
Derivs.: Dipicrate, m. 283-284° (524).

PHOSPHARSEPIN DERIVATIVES

2,3,4,5-Tetrahydro-1H-1,5-benzophospharsepin

1,5-DIETHYL-2,3,4,5-TETRAHYDRO-1H-1,5-BENZOPHOSPHARSEPIN 1-ETHIODIDE
    $C_{15}H_{25}AsIP$
Prepn.: by heating [$Et_2AsC_6H_4PEt_2(CH_2)_3$]$Br_2$ at 270°/15 mm and treating the oily
P-ethobromide with NaI (612).
Props.: m. 222-224° (612).

1-ETHYL-2,3,4,5-TETRAHYDRO-5-METHYL-1H-1,5-BENZOPHOSPHARSEPIN 1-ETHOBROMIDE
    5-METHOBROMIDE $C_{15}H_{26}AsBr_2P$
Prepn.: by heating at 90-120° 2-$Et_2PC_6H_4AsMe_2$ with $CH_2(CH_2Br)_2$ under N (612).
Props.: Monohydrate, m. 245-257° (612).
Derivs.: Dipicrate, m. 201-202° (612).

1,5-DIETHYL-2,3,4,5-TETRAHYDRO-1H-1,5-BENZOPHOSPHARSEPIN DIETHOBROMIDE
    $C_{17}H_{30}AsBr_2P$
Prepn.: by heating 2-$Et_2PC_6H_4AsEt_2$ with $CH_2(CH_2Br)_2$ under N (612).
Props.: m. 240°, at 270°/15 mm loses one EtBr at the As atom (612).
Derivs.: Dipicrate, m. 221° (decompn.) (612).

DIARSOCIN DERIVATIVES

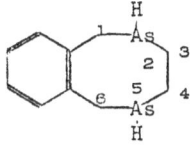

6,11-Dihydro-5,12-alkano(or benzeno)-
dibenzo[b,f][1,4]diarsocin

(Y = alkano or benzeno group)

1,2,3,4-Tetrahydrodibenzo[e,g][1,4]diarsocin

5,6,11,12-Tetrahydrodibenzo[b,f][1,4]-
diarsocin

1,2,3,4,5,6-Hexahydro-2,5-benzodiarsocin

8-CHLORO-6,11-DIHYDRO-5,12-DIMETHYL-5,12-ETHANODIBENZO[b,f][1,4]DIARSOCINIUM
DIBROMIDE $C_{18}H_{21}AsBr_2Cl$

Prepn.: by heating o-C_6H_4(AsMe$_2$)$_2$ with 4,2-Cl(BrCH$_2$)$C_6H_3CH_2$Br (610).
Props.: Monohydrate, m. 263°; decomp. at elevated temp. to o-phenylenebis-
[2-(5-chloroisoarsindoline)] (610).
Derivs.: Dimethopicrate, m. 213-214°; dimetho(+)-camphorsulfonate, m. 298-299°;
dimetho(+)-bromocamphorsulfonate dihydrate, m. 271° (decompn.) (610).

6,11-DIHYDRO-5,12-DIMETHYL-5,12-ETHANODIBENZO[b,f][1,4]DIARSOCINIUM DIBROMIDE
$C_{18}H_{22}As_2Br_2$

Prepn.: by heating o-C_6H_4(CH$_2$Br)$_2$ with MeAsC$_6H_4$As(Me)CH$_2$CH$_2$; 47% yield (610).
Props.: Monohydrate, m. 214-216°; IR spectrum; decomp. at elevated temp. to o-
phenylene[2-(5-chloroisoarsindoline)] (610).
Derivs.: Dimethopicrate, yellow crystals, m. 221-222° (610).

8-CHLORO-5,6,11,12-TETRAHYDRO-5,5,12,12-TETRAMETHYLDIBENZO[b,f][1,4]DIARSOCINIUM
DIBROMIDE $C_{18}H_{23}As_2Br_2Cl$

Prepn.: by heating 42-Cl(BrCH$_2$)$C_6H_3CH_2$Br with o-C_6H_4(AsMe$_2$)$_2$ (610).
Props.: m. 241-242°, decomp. at elevated temp. to o-phenylenebis[2-(5-chloroiso-
arsindoline)] (610).
Derivs.: Dimethopicrate, yellow crystals, m. 213° (610).

2,3-DIHYDRO-1,1,4,4-TETRAMETHYLDIBENZO[e,g][1,4]DIARSOCINIUM DIBROMIDE
$C_{18}H_{24}As_2Br_2$

Prepn.: by heating $(2-C_6H_4AsMe_2)_2$ with $(CH_2Br)_2$ (523, 524).
Props.: Decomp. at the m. temp. to 5-methyldibenzarsenol (524).

5,6,11,12-TETRAHYDRO-5,5,12,12-TETRAMETHYLDIBENZO[b,f][1,4]DIARSOCINIUM DIBROMI
$C_{18}H_{24}As_2Br_2$

Prepn.: by mixing $o-C_6H_4(AsMe_2)_2$ with $o-C_6H_4(CH_2Br)_2$ and heating the mixt. at 10C
75% yield (610).
Props.: Monohydrate, m. 214-216°, crystal structure and lattice data; anhydrous,
m. 213-215°; on heating yields o-phenylenebis(2-isoarsindoline) (610).
Derivs.: Dimethopicrate, yellow crystals, m. 219-220° (decomp.) (610).

6,11-DIHYDRO-5-METHYL-5,12-o-BENZENO [b,f][1,4]DIARSOCINIUM BROMIDE $C_{21}H_{19}As_2Br$
Prepn.: by heating 5,10-dihydro-5,10-dimethylarsanthrene with $o-C_6H_4(CH_2Br)_2$
(1:1) at 130° for 1 hr., 65% yield (611).
Props.: m. 200°; decomp. when heated in vacuo (611).
Derivs.: Picrate, yellow crystals, m. 211°; Iodide, m. 197° (611).
Rxn. with $o-C_6H_4(CH_2Br)_2 \rightarrow$ the tetraarsacyclohexadecitetrainium deriv. (611).

PHOSPHARSOCIN DERIVATIVES

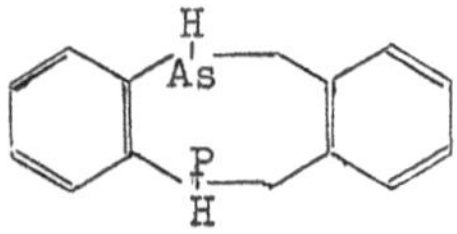

5-ETHYL-5,6,11,12-TETRAHYDRO-12-METHYLDIBENZO[b,f][1,4]PHOSPHARSOCIN 5-ETHO-
BROMIDE 12-METHOBROMIDE $C_{20}H_{28}AsBr_2P$

Prepn.: by reacting $2-Et_2PC_6H_4AsMe_2$ with $o-C_6H_4(CH_2Br)_2$ in warm MeOH under N (61:
Props.: Hemihydrate, m. 208-209° (212).
Derivs.: Dipicrate, m. 188° (212).

5,12-DIETHYL-5,6,11,12-TETRAHYDRODIBENZO[b,f]PHOSPHARSOCIN DIETHOBROMIDE
$C_{22}H_{32}AsBr_2P$

Prepn.: by reacting $2-Et_2PC_6H_4AsEt_2$ with $o-C_6H_4(CH_2Br)$ in warm MeOH under N (612
Props.: m. 234° (612).
Deriv.: Dipicrate, m. 208°; DI-(+)-camphorsulfonate monohydrate, m. 252° (decomp
was not resolved (612).

NINE-MEMBERED RING SYSTEM

DIBENZO[f,h][1,5]DIARSONIN DERIVATIVES

6,7,8,9-TETRAHYDRO-5,5,9,9-TETRAMETHYL-5H-DIBENZO[f,h]DIARSONINIUM DIBROMIDE
$C_{19}H_{26}As_2Br_2$

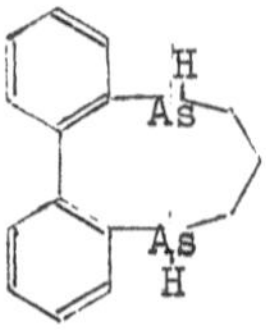

Prepn.: by heating $(2\text{-}C_6H_4AsMe_2)_2$ with $(CH_2)_aBr_2$ (524).
Props.: Monohydrate, m. 261-262° (decompn.), on heating decomp. with contraction
of the ring to 5-methyldibenzarsenol and $Me_2AsCH:CH_2$ (524).
Derivs.: Dipicrate, m. 201-203° (524).

TEN-MEMBERED RING SYSTEM

TRIBENZO[b,d,h][1,6]DIARSECIN DERIVATIVES

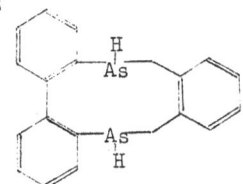

   9,10,15,16-TETRAHYDRO-9,16-DIMETHYLTRIBENZO[b,d,h][1,6]DIARSECIN
Prepn.: by pyrolysis of 9,10,15,16-tetrahydro-9,9,16,16-tetramethyltribenzo[b,d,h]-
[1,6]diarsecinium dibromide (524).
Props.: m. 110-111° (524).

   9,10,15,16-TETRAHYDRO-9,9,16,16-TETRAMETHYLTRIBENZO[b,d,h][1,6]DIARSECINIUM
      DIBROMIDE $C_{24}H_{28}As_2Br_2$
Prepn.: by heating $(2\text{-}C_6H_4AsMe_2)_2$ with $o\text{-}C_6H_4(CH_2Br)_2$ (524).
Props.: Monohydrate, m. 209-210° (dec.); on heating → cleavage of two MeBr
groups and formation of the tribenzo[b,d,h][1,6]diarsecin deriv. (preceding compd.)
(524).
Deriv.: Dipicrate, m. 225-227° (524).

   1,2,3,4,5,6-HEXAHYDRO-2,5-DIMETHYL-2,5-DIPHENYL-2,5-BENZODIARSECINIUM DIBROMIDE
      $C_{24}H_{28}As_2Br_2$
Prepn.: by heating at 100° a mixt. of $(MePhAsCH_2\text{-})_2$ with $o\text{-}C_6H_4(CH_2Br)_2$ (609).
Derivs.: Dimethiodide.2EtOH, m. 160-161° (effervesc.) (609).

SPIRO HETEROCYCLIC ARSENIC COMPOUNDS

SPIROBISISOARSINDOLINIUM SALTS

2,2´-Spirobisisoarsindolinium Salt

2,2'-SPIROBISISOARSINDOLINIUM BROMIDE $C_{16}H_{16}AsBr$ *
Prepn.: by reacting 2-methylisoarsindoline with o-$C_6H_4(CH_2Br)_2$ until the mixt.
forms a homogeneous liquid containing 2-[2-(bromomethyl)benzyl]-2-methylisoarsin-
dolinium bromide, and heating the product at 200°/15 mm (756).
Props.: Decomp. 235-239° (756).
Other salts: Picrate, yellow, m. 188-189° (756).
Iodide, decomp. ~200° (756).

2,2'-SPIROBIS(5-CHLOROISOARSINDOLINIUM BROMIDE) $C_{16}H_{14}AsBr^{Cl}_2$ *
Prepn.: by warming gently 5-chloro-o-xylylene dibromide with 5-chloro-2-methyl-
isoarsindoline and heating the resulting glass at 160°/15 mm for 30 min. and then
at 175°/15 mm. for 15 min.; 63% yield (756).
Props.: m. 244-246° (dec.) (757).
Other salts: Iodide, m. 258-262°; Thiocyanate, m. 223° (dec.); Picrate, m. 108-
112°; d-Bromocamphorsulfonate monohydrate, m. 143-147°, optically active; d-
Camphorsulfonate monohydrate, m. 302-303°; 1-N-1-Phenylethylphthalamate monohydrat
m. 145-146° (757).

1,1'',2',3,3',3''-HEXAHYDRODISPIRO[ISOARSINDOLINIUM-2,1'-[1,4]BENZODIARSENIN-
  4',2''-ISOARSINDOLINIUM] DIBROMIDE $C_{24}H_{24}As_2Br_2$

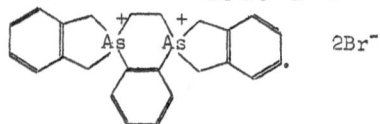

2Br⁻

Prepn.: by heating at 125-130° $(CH_2Br)_2$ with 2,2'-o-phenylenedi(isoarsindoline)
(610).
Props.: Monohydrate, m. 223° (610).
Derivs.: Dipicrate, yellow crystals, m. 216-217° (dec.)

TETROX-5-ARSENASPIRO[4,4]NONANE DERIVATIVES

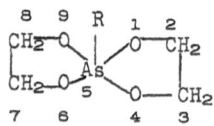

5-R-1,4,6,9-Tetrox-5-arsenaspiro[4,4]nonane

5-METHYL-1,4,6,9-TETROX-5-ARSENASPIRO[4,4]NONANE $C_5H_{11}AsO_4$
Prepn.: by azeotrop. distn. of $MeAsO_3H_2$ with ethylene glycol in benzene (1097).
Props.: b. 110-111°/4 mm (1097).

2,5,7-TRIMETHYL-1,4,6,9-TETROX-5-ARSENASPIRO[4,4]NONANE $C_7H_{15}AsO_4$
Prepn.: by azeotrop. distn. of $MeAsO_3H_2$ with propylene glycol in C_6H_6 (1097).
Props.: b. 115.5-116°/3 mm, d_4^{20} 1.41819, n_α^{20} 1.47992, n_D^{20} 1.48300, n_β^{20}
1.4810, n_J^{20} 1.49445 (1097).

---

* For the structural formula see page 433.

5,BUTYL-1,4,6,9-TETROX-5-ARSENASPIRO[4,4]NONANE $C_8H_{17}AsO_4$
Prepn.: by azeotropic dist. of $BuAsO_3H_2$ with ethylene glycol in C_6H_6 (1097).
Props.: M. 20°, b. 140.5-141.5°/3 mm, d_4^{20} 1.4023, n_α^{20} 1.49094, n_D^{20} 1.49386, n_β^{20} 1.50040, n^{26} 1.50590 (1097).

2,3,5,7,8-PENTAMETHYL-1,4,6,9-TETROX-5-ARSENASPIRO[4,4]NONANE $C_9H_{19}AsO_4$
Prepn.: by azeotrop. dist. of $MeAsO_3H_2$ with pinacol in C_6H_6 (1097).
Props.: b. 131-132°/3 mm (1097).

5-PHENYL-1,4,6,9-TETROX-5-ARSENASPIRO[4,4]NONANE $C_{10}H_{13}AsO_4$
Prepn.: by heating $PhAsO_3H_2$ with ethylene glycol in Ac_2O (1097).
Props.: M. 105.5° (1097).

5-BUTYL-2,7-DIMETHYL-1,4,6,9-TETROX-5-ARSENASPIRO[4,4]NONANE $C_{10}H_{13}AsO_4$
Prepn.: by azeotrop. distn. of $BuAsO_3H_2$ with propylene glycol in C_6H_6 (1097).
Props.: B. 142.6-143.4°, d_4^{20} 1.28156, n_D^{20} 1.47849 (1097).

5-BUTYL-2,3,7,8-TETRAMETHYL-1,4,6,9-TETROX-5-ARSENASPIRO[4,4]NONANE $C_{12}H_{25}AsO_4$
Prepn.: by azeotrop. dist. of $BuAsO_3H_2$ with pinacol in C_6H_6 (1097).
Props.: B. 169.0-170.0°/4 mm (1097).

2,2,7,7-TETRAMETHYL-5-PHENYL-1,4,6,9-TETROX-5-ARSENASPIRO[4,4]NONANE-3,8-DIONE
    $C_{14}H_{17}AsO_6$
Prepn.: by heating $PhAsO_3H_2$ with α-hydroxybutyric acid in Ac_2O (1097).
Props.: M. 138.2° (1097).

2,3,7,8-TETRAMETHYL-5-PHENYL-1,4,6,9-TETROX-5-ARSENASPIRO[4,4]NONANE $C_{14}H_{21}AsO_4$
Prepn.: by heating $PhAsO_3H_2$ with pinacol in Ac_2O (1097).
Props.: Canary-yellow crystals, m.176° (1097).

2,2,7,7-TETRAMETHYL-5-BENZYL-1,4,6,9-TETROX-5-ARSENASPIRO[4,4]NONANE-3,8-DIONE
    $C_{15}H_{19}AsO_6$
Prepn.: by heating $PhCH_2AsO_3H_2$ with $Me_2C(OH)CO_2H$ in Ac_2O (1097).
Props.: M. 169° (1097).

2,7-DIETHYL-2,7-DIMETHYL-5-PHENYL-1,4,6,9-TETROX-5-ARSENASPIRO[4,4]NONANE-
    3,8-DIONE $C_{16}H_{21}AsO_6$
Prepn.: by heating $PhAsO_3H_2$ with $EtMeC(OH)CO_2H$ in Ac_2O (1097).
Props.: m. 91.4° (1097).

2,7-DIETHYL-2,7-DIMETHYL-5-BENZYL-1,4,6,9-TETROX-5-ARSENASPIRO[4,4]NONANE-3,8-
    DIONE $C_{17}H_{23}AsO_6$
Prepn.: By heating $PhCH_2AsO_3H_2$ with $EtMeC(OH)CO_2H$ in Ac_2O (1097).
Props.: m. 161° (1097).

2,2'-SPIROBI-1,3,2-BENZODIOXARSENOLE DERIVATIVES

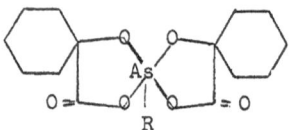

2-R-2,2'-Spiro-1,3,2-benzodioxarsenole

2-METHYL-2,2'-SPIROBI-1,3,2-BENZODIOXARSENOLE $C_{13}H_{11}AsO_4$
Prepn.: by azeotrop. distn. of $MeAsO_3H_2$ with pyrocatechol in C_6H_6 (1097).
Props.: Canary-yellow crystals (1097).

2-BUTYL-2,2'-SPIROBI-1,3,2-BENZODIOXARSENOLE $C_{16}H_{17}AsO_4$
Prepn.: by azeotrop. distn. of $BuAsO_3H_2$ with pyrocatechol in C_6H_6 (1097).
Props.: Canary-yellow crystals, m. 83° (1097).

2-α-TOLYL-2,2'-SPIROBI-1,3,2-BENZODIOXARSENOLE $C_{19}H_{15}AsO_4$
Prepn.: by heating $PhCH_2AsO_3H_2$ with pyrocatechol in Ac_2O (1097).
Props.: Canary-yellow crystals, m. 147° (1097).

3,3'-(ARYLARSYLENE)DIPROPIONIC ACID DIHYDROXIDE SPIROCYCLIC SALTS

3,3'-(p-CHLOROPHENYLARSYLENE)DIPROPIONIC ACID.DIHYDROXY SPIROCYCLIC INNER SALT
 $C_{12}H_{12}AsClO_4$
Prepn.: by oxidation of $p\text{-}ClC_6H_4As(CH_2CH_2CO_2H)_2$ by H_2O_2 in Me_2O (119).
Props.: 223° (119).

3,3'-(PHENYLARSYLENE)DIPROPIONIC ACID DIHYDROXIDE SPIROCYCLIC INNER SALT
 $C_{12}H_{13}AsO_4$
Prepn.: by oxidation of $PhAs(CH_2CH_2CO_2H)_2$ by H_2O_2 in Me_2CO (119).
Props.: m. 235° (119).

ARSENATRISPIRO[5.1.1.5.2.2]NONADECANE DERIVATIVES

8-R-7,9,17,18-Tetrox-8-arsenatrispiro[5.1.1.5.2.2]nonane-16,1
dione

8-METHYL-7,9,17,18-TETROX-8-ARSENATRISPIRO[5.1.1.5.2.2]NONADECANE-16,19-DIONE
$C_{15}H_{23}AsO_6$
Prepn.: by azeotrop. dist. of $MeAsO_3H_2$ with cyclohexanol-1-carboxylic acid in
C_6H_6 (1097).
Props.: m. 213.2° (1097).

8-BUTYL-7,9,17,18-TETROX-8-ARSENATRISPIRO[5.1.1.5.2.2]NONADECANE-16,19-DIONE
$C_{18}H_{29}AsO_6$
Prepn.: by azeotrop. distn. of $BuAsO_3H_2$ with cyclohexanol-1-carboxylic acid in
C_6H_6 (1097).
Props.: m. 194° (1097).

8-BENZYL-7,9,17,18-TETROX-8-ARSENATRISPIRO[5.1.5.2.2]NONADECANE-16,19-DIONE
$C_{21}H_{27}AsO_6$
Prepn.: by heating $PhCH_2AsO_3H_2$ with cyclohexanol-1-carboxylic acid in C_6H_6 (1097).
Props.: m. 175° (1097).

SPRIOBISARSINOLINIUM AND SPIROBISISOARSINOLINIUM SALTS

3,3′,4,4′-TETRAHYDRO-1,1′-SPIROBISARSINOLINIUM BROMIDE

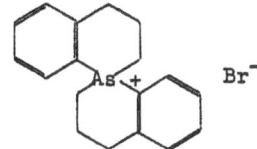

Prepn.: by treating 1,2,3,4-tetrahydro-1-[2-(3-hydroxypropyl)phenyl]arsinoline with
PBr_3 (560).
Props.: Dihydrate, m. 253-254°; anhydrous salt, m. 270-271° (560).
Other salts: Iodide, plates, m. 277-278°; Picrate, m. 102-103°; the l-arsinolinium l-menthyloxyacetate dihydrate, m. 87-89°; anhydrous salt, m. 88.5-90°; the l-
arsinolinium picrate, yellow crystals, m. 95-97°; the d-arsinolinium iodide,
needles, m. 223-223.5°; the dl-arsinolinium d-bromocamphorsulfonate, m. 191-192°
(560).

3,3′,4,4′-TETRAHYDRO-2,2′-(1,1′)SPIROBISISOARSINOLINIUM BROMIDE

Prepn.: by treating 1,2,3,4-tetrahydro-2-methylisoarsinoline with $2-BrCH_2C_6H_4-$
CH_2CH_2Br, heating the resulting glass at 14 mm pressure, and completing the rxn.
at 200°; 38% yield (559).
Props.: dl-salt, m. 228-290° (559).
Other salts: Iodide, m. 231-233°; the l-arsinolinium d-bromocamphorsulfonate hydrate, m. 70-74°; the l-arsinolinium iodide, m. 225-228°; the d-arsinolinium l-
bromocamphorsulfonate monohydrate, m. 185-187°; the d-arsinolinium iodide, m. 226-
228° (559).

B. ANTIMONY

1. COMPOUNDS OF TRIVALENT ANTIMONY

PRIMARY STIBINES

Primary stibines, $RSbH_2$, are prepared by reducing organoantimony tetrahalides, $RSbX_4$, or organodihalostibines, $RSbX_2$, with lithiumaluminum hydride or lithium tetrahydroborate in ether.

METHYLSTIBINE CH_5Sb $MeSbH_2$
Formation: by-prod. in the rxn. of Me_2SbBr with $NaBH_4$ at -78° in $(-CH_2OMe)_2$ (139a).
Props.: Fairly stable at -78°; b. (calcd.) 41°; thermodynamic data; decompn. above -78° to form a black solid, $(MeSb)x$ (?) (139a).

PHENYLSTIBINE C_6H_7Sb $PhSbH_2$
Prepn.: By reducing $PhSbCl_4$ with $LiAlH_4$ or $LiBH_4$ in Et_2O (1290).
By reducing $PhSbI_2$ with $LiBH_4$ or $LiAlH_4$ in Et_2O at -50°; 42% yeild (lower yield with $LiAlH_4$) (1294).
Props.: Colorless oil with phosphine-like odor, m. -38° (1290, 1294).
Rxns. with: O in $Et_2O \rightarrow PhSb(OH)_2$ and $PhSbO_3H_2$ (1294).
I in $Et_2O \rightarrow PhSbI_2$ (1294).
$PhSbI_2 \rightarrow (PhSb)x$ (?) (1294).

SECONDARY STIBINES

Secondary stibines, R_2SbH, are prepared by reducing diorganohalostibines, R_2SbX, or diorganoantimony trihalides, R_2SbCl_3, with $LiAlH_4$, or $LiBH_4$.

DIMETHYLSTIBINE C_2H_7Sb Me_2SbH
Prepn.: by reducing Me_2SbBr with $NaBH_4$ or $LiHB(OMe)_3$ in $(-CH_2OMe)_2$ below -20°; 35% yield (139a).
Props.: Apparently stable at -78°, loses H at r. t. to form $(Me_2Sb)_2$, b. (calcd. 60.7; thermodynamic data (139a).
Rxns. with: $HCl \rightarrow Me_2SbCl + H_2$ (139a).
$B_2H_5Cl \rightarrow B_2H_6 + Me_2SbCl$ (139a).
$B_2H_5Br \rightarrow B_2H_6 + Me_2SbBr$ (139a).
$Me_3NBH_3 \rightarrow (Me_2Sb)_2$ (139a).

DIPHENYLSTIBINE $C_{12}H_{11}Sb$ Ph_2SbH
Prepn.: By reducing Ph_2SbCl_3 with $LiAlH_4$ or $LiBH_4$ in Et_2O (1290).

By reducing Ph_2SbCl with $LiBH_4$ or $LiAlH_4$ in Et_2O at -60°; 65 and 22% yield, respectively (1295).

rops.: Water-clear, colorless, oxidizable oil, m. -2° (1290, 1295).

isproportionates, in Et_2O in the presence of $LiBH_4$ or B_2H_6 at 0° (1295).

xns. with: O in $Et_2O \rightarrow Ph_2SbO_2H$ (1295).

I_2 in $Et_2O \rightarrow Ph_2SbCl_3$ (1295).

2 in $Et_2O \rightarrow Ph_2SbI_3$ (1295).

ERTIARY STIBINES

YMMETRICAL TERTIARY STIBINES

ymmetrical tertiary stibines are prepared or formed by:

1. Heating antimony powder with trifluoroiodomethane in a sealed steel cylinder at 165-175° to form tri(trifluoromethyl)stibine.
2. Reacting dichloro(chloromethyl)stibine with diazomethane in benzene at 5-6° to form tris(chloromethyl)stibine.
3. Reacting antimony trihalides with organometallic compounds, such as trialkylaluminum or trialkylaluminum etherates, Grignard compounds, alkenyl- and aryllithium, and cyclopentadienyl- and alkynylsodium compounds, using ether, tetrahydrofuran, or benzene as solvents.
4. Treating antimony trihalides with aryl halides and metallic sodium in benzene or ether.
5. Reacting anitmony powder with organolithium compounds in benzene or ether under reflux or with dialkylmercury in the presence of mercury(II) chloride in a sealed tube at 150°.
6. Reacting antimony powder with aryldiazonium halides, RN_2X, or aryldiazonium-zinc chloride double salts, $(RN_2)_2ZnCl_4$, in the presence or absence of calcium carbonate, using acetone, ethyl acetate, or acetic acid as solvents.
7. Reacting antimony trichloride with aryldiazonium fluoroborates in acetone in the presence of zinc dust to form tertiary stibines along with stiboso derivatives and bis(diarylantimony) oxides.
8. Diazotization of aromatic amines in ethyl acetate in the presence of antimony trichloride and concentrated hydrochloric acid and decomposing the reaction product with zinc dust.
9. Metathesis of antimony trichloride with cyanotrimethylsilane in xylene under reflux to form tricyanostibine.
10. Reduction of tertiary stibine dichlorides with sodium bisulfite or with hydrazine hydrate in aqueous alcohol.
11. Electrolysis of silver cyanide or triethylaluminum in an electrolyte using an electrode of antimony to form tricyano- and triethylstibine, respectively.
12. Decomposition of aryldiazonium tetrachloroantimonites with zinc or iron powder in ethyl acetate at 70-90° to yield tertiary stibines along with stiboso and aryldichlorostibine derivatives.
13. Treating antimony pentachloride with aryllithium compounds in ether at -70° in an inert atmosphere; tertiary stibines are obtained as by-products.
14. Neutron irradiation of antimony trichloride in benzene.

TRIS(TRIFLUOROMETHYL)STIBINE C_3F_9Sb $(CF_3)_3Sb$

Prepn.: by heating Sb powder with CF_3I in the presence or absence of SbI_3 in a sealed steel cylinder at 165-175°; the presence of SbI_3 favors formation of the product (258).

Props.: b. 72° (334); m. -58°, b. 71.7°, thermodynamic data, is oxidized by atmospheric oxygen (258). Electron diffraction data (112).

Rxns. with: NH_3 in a sealed tube → CHF_3 and SbN (?) (253).
SiO_2 at 180-240° → SiF_4, CO and fluorocarbons (23).
Me_2NH at 20° → CHF_3 (253).
Aq. KOH → hydrolysis (253, 521).
Cl_2 → $(CF_3)_3SbCl_2$ (258).
Br_2 → $(CF_3)_3SbBr_2$ (258).
I_2 → CF_3SbI_2, $(CF_3)_2SbI$, CF_3I (258).
SbI_3 → $(CF_3)_2SbI$ (258).

Complex compd.: $(CF_3)_3Sb \cdot C_5H_5N$, m. 39°, b. 127°, dissoc. in the vapor phase (258).

TRIMETHYLSTIBINE C_3H_9Sb Me_3Sb

Prepn.: By heating metallic Sb with Me_2Hg containing a trace of $HgCl_2$ in a sealed tube at 150° for 160 hrs. (747).
By treating $SbCl_3$ with MeMgI in Et_2O under N at -20° (1313).

Props.: b. 80.6°, freezes at -62.0°, $\log_{10}P = 7.496-1627/T$ (31); b. 79.4° (extrapolated), m. -87.6° (extrapolated), d. 1.528 (747). Thermodynamic data (746, 747, 1081). Raman spectrum (359, 1079). Intramol. potential (1131). Force const. (1138). Azeotropic mixt. with Bu_2O, b. 72-74° (1115). Pyrolysis (1035a). Relation between polarity and reactivity with O, CO, and H_2O (1098). Vibrational ener level splitting and optical isomerism (1285).

Rxns. with: O → rapid oxidation (32) sometimes explosive (1116).
CF_3I → Me_2SbCF_3 (520).
PCl_3 or $SbCl_5$ → Me_3SbCl_2 (561).
Br_2 → Me_3SbBr_2 (1313).
9-Bromofluorene in Et_2O → difluorenyl (1313).
MeI → Me_4SbI (1314).
MeBr → Me_4SbBr (1314).
Me_2BBr → Me_2SbBr (139a).
HCl → Me_3SbCl_2 + H_2 (139a).

Complex compds.: $Me_3Sb \cdot BH_3$, m. -35° (dec.) (540).
$[Pd(Me_3Sb)_2Cl_2]$, cis- trans- equilibrium (205).
trans-$[Pt(Me_3Sb)Cl_2]_2$, reddish orange, m. 150-170° (dec.), prepd. by reacting Me_3Sb with $[Pt(CH_2:CH_2)_2Cl_2]_2$ (207).
$Me_3Sb \cdot GaMe_3$, dissocn. data (235).

Use: Catalyst in the polymerization of vinyl monomers (247). Fuel additive for internal-combustion engines (496).

TRICYANOSTIBINE C_3N_3Sb $Sb(CN)_3$

Prepn.: By electrolysis of AgCN in pyridine using Sb anode (1108).
By refluxing $SbCl_3$ with Me_3SiCN (1:3) in xylene under reflux; 64% yield (87a).

Props.: Colorless hydroscopic crystals decomp. at elevated temps. (1108). White powder, decomp. > 200° (87a).

TRI(2-CHLOROVINYL)STIBINE $C_6H_6Cl_3Sb$ (ClCH:CH)$_3$Sb

Prepn.: by reducing (ClCH:CH)$_3$SbCl$_2$ isomers with NaHSO$_3$ in aq. alc. (898, 899).
Props.: The compd. occurs in two stereoisomeric forms. The trans isomer m. 48-49°
(898, 899); b. 121°/2 mm., d_4^{100} 1.7021, γ^{100} 35.10(899); n_D^{50} 1.6298, d$_{50}$ 1.7617;
R.Sb 17.54 (915). The cis isomer, b. 121-122°/3 mm. (898, 899); d_4^{15} 1.8070, d_4^{100}
1.7181, γ^{100} 35.93 (899); n_D^{50} 1.6180, d$_{50}$ 1.7580, P_{Sb} 16.71 (915). Raman spectra
of both isomers (902). Both isomers have properties of organometallic compounds
and of complex compounds since they do not dissociate to give off C_2H_2 (898).
Rxns. with: BCl$_3$ in $C_6H_6 \rightarrow$ SbCl$_3$ + ClCH:CHB(OH)$_2$ (107).
HgCl$_2$ (with both isomers) $\rightarrow$ ClCH:CHHgCl + SbCl$_3$ (898, 901).

TRIVINYLSTIBINE C_6H_9Sb (CH$_2$:CH)$_3$Sb

Prepn.: by reacting CH$_2$:CHMgBr with SbCl$_3$ in C_4H_8O under reflux, 66% yield (773,
774, 914).
Props.: b. 149.9° (calcd.), thermodynamic data, inflames spontaneously on exposure
to air (773, 774); b. 46°/15 mm., n_D^{20} 1.5614, d$_{20}$ 1.4341 (914, 915).
Rxns. with: I$_2 \rightarrow$ (CH$_2$CH)$_3$SbI$_2$ (773).
SbBr$_3 \rightarrow$ (CH$_2$:CH)$_2$SbBr + CH$_2$:CHSbBr$_2$ (773, 774).
Br$_2 \rightarrow$ (CH$_2$:CH)$_3$SbBr$_2$ (914).
Deriv.: Methiodide, m. 119-120° (dec), dissoc. in vacuo into (CH$_2$:CH)$_3$Sb + MeI
(773).
Complex compd.: cis-{Pt[(CH$_2$:CH)$_3$Sb]$_2$}Cl$_2$, m. 113-114° (773, 774).

TRIETHYLSTIBINE $C_6H_{15}Sb$ Et$_3$Sb

Prepn.: By reacting SbF$_3$ with Et$_3$Al in C_6H_{14} (622) or in Et$_2$O, 70.5% yield (1331).
   By electrolysis of Et$_3$Al in an electrolyte in an inert atmosphere using an
antimony electrode (1341).
Props.: b. 158-160°, d$_{20}$ 1.322 (1331); b. 161.4°, freezing point -98.0°, $\log_{10}P =$
7.094-2183/T (31); b. 160° (622). Heat of combustion (725b).
Use: Radical-type catalyst for polymerization of vinyl compds. (445a).
Complex compds.: [Pt(Et$_3$Sb)(C$_5H_{11}$N)Cl$_2$], yellowish orange prisms, m. 76-77°, prepd. by
reacting trans-[Pt(EtSb)Cl$_2$]$_2$ with piperidine in Me$_2$CO (208).
cis- and trans-[Pt(Et$_3$Sb)$_2$Cl$_2$] prepd. by reacting Et$_3$Sb with K$_2$PtCl$_4$, m. 102.5-
103° (204); cis-isomer, light greensih yellow, m. 104-104.5° (600); dipole moment
(204, 600); conductivity (601).
[Pt(Et$_3$Sb)Cl$_2$]$_2$, reddish orange, m. 130-135° (dec.), prepd. by reacting Et$_3$Sb with
[Pt(C$_2H_4$)Cl$_2$]$_2$ (207).
[Pt(Et$_3$Sb)$_2$I$_2$], cis-trans equilibrium (210).
[Pt(Et$_3$Sb)$_2$Br$_2$], light greenish yellow crystals, m. 132-132.5°; dipole moment (600).
[Pt(Et$_3$Sb)$_2$I$_2$], chrome yellow crystals, m. 68-69°; dipole moment (600).
[Pt(Et$_3$Sb)$_2$(NO$_2$)$_2$], almost white, m. 200-205°; dipole moment (600).
[Pd(Et$_3$Sb)Cl$_2$]$_2$, dark red powder, m. 87-88° (dec.), prepd. by reacting Et$_3$Sb with
Na$_2$PdCl$_4$ in Me$_2$CO at -70° (212).
cis-[Pd(Et$_3$Sb)$_2$Cl$_2$], golden yellow, decomp. 90°, dipole moment (600); cis- trans
equilibrium (205, 210).

TRIPROPENYLSTIBINE $C_9H_{15}Sb$ (CH$_3$:CH:CH)$_3$Sb

Prepn.: by reacting SbCl$_3$ with cis- or trans-allyllithium (913, 916).
Props.: Occurs in cis and trans form. Cis isomer: n_D^{20} 1.5590, d$_{20}$ 1.3092,

^{R}Sb 1696 (915). Trans isomer: n_D^{20} 1.5511, d_{20} 1.2870, ^{R}Sb 17.28 (915). IR
spectra (916).
Rxns. with: Halogens → the stibine dihalides (913, 916).
TlCl$_3$ → the stibine dichloride· (916).

TRIPROPYLSTIBINE C$_9$H$_{21}$Sb Pr$_3$Sb
Prepn.: by reacting Pr$_3$Al with powdered SbF$_3$ in Et$_2$O; 66% yield (1331).
Props.: b. 82-83°/14 mm., d_{21} 1.242 (1331).
Use: Catalyst for polymerization of CH$_2$:CHCl (1151b).
Complex compds.: cis-[Pt(Pr$_3$Sb)$_2$Cl$_2$], greenish yellow crystals, m. 80-81° (600);
 prepd. by treating Pr$_3$Sb with K$_2$PtCl$_3$ (203). Dipole moment (600).
[Pt(Pr$_3$Sb)Cl$_2$]$_2$, red-orange solid, m. 133-134°, prepd. by reacting Pr$_3$Sb with
 [Pt(C$_2$H$_4$)Cl$_2$]$_2$ in Me$_2$CO (202, 207).
trans-[Pt(Pr$_3$Sb) (C$_7$H$_7$NH$_2$)Cl$_2$], orange, m. 69°, prepd. by reacting [Pt(Pr$_3$Sb)Cl$_2$]$_2$
with toluidene (202). IR spectrum (209, 211). Assocn. const. (211).
[Pd(Pr$_3$Sb)$_2$Cl$_2$], cis-trans equilibrium (205, 210).

TRIBUTYLSTIBINE C$_{12}$H$_{27}$Sb Bu$_3$Sb
Prepn.: by reacting SbCl$_3$ with BuMgBr (1:3.3) in Et$_2$O; 70% yield (1115).
Use: Catalyst for polymerization of vinyl monomers (247).
Complex compds.: [Pt(Bu$_3$Sb)$_2$Cl$_2$], cis-trans equilibrium (210).
cis-[Pt(Bu$_3$Sb)$_2$Cl$_2$], greenish yellow crystals, m. 62-63°; dipole moment (600).

TRICYCLOPENTADIENYLSTIBINE C$_{15}$H$_{15}$Sb (ĊH:CHCH:CHĊH)$_3$Sb
Prepn.: by reacting SbCl$_3$ with cyclopentadienylsodium in C$_4$H$_8$O at 0°; 50-60%
yield (382a, 383).
Props.: Carmine-red substance (382a). Purple-red substance, very sensitive to
air, stable below 0°, above 10° disproportionates into [(C$_5$H$_5$)$_2$Sb]$_2$ (383). Magnet
data (383b).

TRI(4-BROMOPHENYL)STIBINE C$_{18}$H$_{12}$Br$_3$Sb (4-BrC$_6$H$_4$)$_3$Sb
Prepn.: by reacting SbCl$_3$ with 4-BrC$_6$H$_4$MgBr in Et$_2$O; 30% yield (491, 513); 73%
yield (949).
Props.: Glystening crystals, m. 109.8-110.4° (513); m. 136-136.5° (491); m. 134-
135° (949).
Rxns. with: Cl$_2$ → (4-BrC$_6$H$_4$)$_3$SbCl$_2$ (491, 949).
RLi → cleavage of the C-Sb bonds (949, 1195a).

TRI-4-CHLOROPHENYLSTIBINE C$_{18}$H$_{12}$Cl$_3$Sb (4-ClC$_6$H$_4$)$_3$Sb
Prepn.: by reacting SbCl$_3$ with 4-ClC$_6$H$_4$MgBr; 77.5% yield (491, 513, 1316).
Props.: Feathery crystals, m. 99.5-100.5° (513); m. 110.5-111°, crystal structure
(491); m. 101-101.5° (1316).
Rxns. with: Cl$_2$ → (4-ClC$_6$H$_4$)$_3$SbCl$_2$ (491).
RLi → (4-ClC$_6$H$_4$)$_2$SbR + 4-ClC$_6$H$_4$Li (1195a, 1316).
Use: Stabilizer for poly(vinyl chloride) (970a).
Complex compds.: {Co[(ClC$_6$H$_4$)$_3$Sb](NO)(CO)$_2$}, red crystals, m. 150° (781, 783).
[Rh[(ClC$_6$H$_4$)$_3$Sb]$_2$(CO)Cl], orange-red plates, decomp. 160-162° (1252).

TRIPHENYLSTIBINE $C_{18}H_{15}Sb$ Ph_3Sb

Prepn.: By treating $SbCl_3$ with PhN_2BF_4 (1:3) in Ac_2O, followed by Zn dust below 20°; 12% yield along with $PhSbO$ (8%) and $(Ph_2Sb)_2O$ (30%) (886).

By treating a mixt. of $PhNH_2$, $SbCl_3$, and concd. HCl in EtOAc with $NaNO_2$ at 5-10°, decompg. the double salt with Zn dust, and working up the rxn. mixt.; 50% yield. If $CaCl_2$ is added at the end of diazotization, Ph_3SbCl_2 is formed (903).

By reacting Sb powder with PhLi in Et_2O or xylene under reflux; 8.9% isolated as dibromide (1192).

By reacting $SbCl_3$ with PhLi in Et_2O under reflux; 96-97% yield (1193).

By-prod. from a rxn. of $SbCl_5$ with PhLi in Et_2O under N at -70° (1311).

By-prod. from neutron irradiation of $SbCl_3$ in C_6H_6 (907).

By reducing Ph_3SbCl_2 with $N_2H_4.H_2O$ in 95% EtOH (1316).

Props.: m. 52° (903, 1193); m. 49-50°, b. 220°/12 mm (996); Vapor pressure (386). Magnetic susceptibility (968, 969); Diamagnetic anomaly (975). Dipole moment (1147b). UV spectrum (178, 598); Near UV spectrum (1050); IR spectrum (711, 821, 1050); NMR data (60); Crystal structure (593, 1286). Decomp. at 325° (513). Corrosion inhibiting effect on steel (30).

Rxns. with: $CHCl_3$ + $AlCl_3$ → a small amt. of Ph_2 and $SbBr_3$ (791).

EtBr + $AlCl_3$ → EtPh + $Ph_2EtSbBr_2$ (791).

BuBr + $AlCl_3$ in CS_2 → BuPh + Sb (791).

PhBr + $AlCl_3$ in CS_2 → Ph_2, Ph_2SbBr, and $SbBr_3$ (791).

$AlCl_3$ with or without PhBr at 230°, followed by KBr → Ph_4SbBr (201).

$AlCl_3$ in CS_2 → C_6H_6, $PhSbCl_2$ and Sb_2S_3 after H_2S treatment (790).

$AlCl_3$ at an elevated temp. → Ph_2SbCl, $PhSbCl_2$ and $SbCl_3$ (815).

$AlCl_3$ in the cold → Ph_3SbCl_2 + $SbOCl$ (815).

$FeCl_3$ in $CHCl_3$ → Ph_3SbCl_2 + $FeCl_2$ (815).

$BiCl_3$ in $CHCl_3$ → Ph_3SbCl_2 + Ph_2SbCl (in CO_2) (817).

RCOCl + $AlCl_3$ → RCOPh, $PhSbCl_2$, and Sb_2S_3 (after H_2S treatment) (790).

Ph_2IBF_4 → Ph_4SbBF_4 (779, 904).

PhN_2Cl in AcOH → Ph_3SbCl_2 (130).

Diallyl disulfide (195).

AcNNaBr → Ph_3SbNAc (996).

PhI in UV light → Ph_3SbI_2 (1054).

$-IC_6H_4OMe$ in MeOH in UV light → $Ph_3Sb(OH)_2$ + MeOPh (1137).

Li in C_4H_8O, followed by H_2O → Ph_2SbO_2H (1309).

O_2 → Ph_2SbO_2H (1316).

RLi → Ph_2SbR + $PhLi$ (1195a, 1316).

Neutron bombardment → radioactive prod. (500, 620, 772, 875, 1304).

Use: Wear inhibitor for Fe carbonyl (63). Fire-retardant for plastics (246). Vulcanization agent for rubber (847). Catalyst for tetramerization of acetylene to cyclooctatetraene (1323b).

Complex compds.: $[Cr(Ph_3Sb)(CO)_5]$, m. 147-149°, prepd. by treating Ph_3Sb with $Cr(CO)_6$ in $(-CH_2OMe)_2$ (833a).

$Co(Ph_3Sb)(NO)(CO)_2]$, red crystals, m. 100° (781, 783).

$[Co(Ph_3Sb)(CO)_3]_2$, yellow microcryst. solid (544).

$Fe(Ph_3Sb)_2(CO)_2Br_2]$ (542).

$Fe(Ph_3Sb)_2(CO)_2Cl_2]$ (542).

$Fe(Ph_3Sb)(CO)_3I_2]$ (542).

$Fe(Ph_3Sb)(Ph_3P)(CO)_3]$, m. 242° (1058).

[Fe(Ph$_3$Sb)(NO)$_2$CO], orange-red, decomp. 105-115° (783).
[Fe(Ph$_3$Sb)(C$_8$H$_8$)(CO)$_2$] (814).
[Fe(Ph$_3$Sb)(CO$_4$], m. 137-138° (1111).
[Me(Ph$_3$Sb)(CO)$_3$], yellow crystals, decomp. 220°; soly. data; prepd. by refluxing
  cycloheptatrienemolybdenum tricarbonyl with Ph$_3$Sb in C$_6$H$_6$ (2).
[Ni(Ph$_3$Sb)(CO)$_3$], m. 96-100° (1058); m. 105-109° (dec.) (1111).
[Pd(Ph$_3$Sb)$_2$Cl$_2$], cis-trans equilibrium (205).
cis-[Pt(Ph$_3$Sb)$_2$Cl$_2$], yellow crystals, decomp. 140°, dipole moment (600).
[Rh(Ph$_3$Sb)$_2$(CO)Cl], purple-red prisms, decomp. 151-153°, diamagnetic non-electroly
  (1252); red-orange, m. 160-170° (1254).
[Rh(Ph$_3$Sb)$_2$(CO)$_2$Cl], purple (541) or red needles (545).
[Rh(Ph$_3$Sb)$_2$(CO)Cl$_3$], decomp. 150-155° (1253).
[Rh(Ph$_3$Sb)$_2$(CO)ClI$_2$], decomp. 220-222° (1253).
[Rh(Ph$_3$Sb)$_2$(CO)I$_3$], decomp. 198-200° (1253).
[Ru(Ph$_3$Sb)$_2$(CO)$_2$I$_2$] (541, 543).

TRIBENZYLSTIBINE C$_{21}$H$_{21}$Sb (PhCH$_2$)$_3$Sb
Prepn.: by reacting SbCl$_3$ with PhCH$_2$MgCl in Et$_2$O in an inert atm.; 50-55% yield
(1195, 1230).
Props.: m. 87° (1195); m. 90° (1230).
Rxns. with: EtLi → PhCH$_2$Li + Et$_3$Sb (1194, 1195).
Atm. O → (PhCH$_2$)$_3$SbO (1230).
Cl$_2$ → (PhCH$_2$)$_3$SbCl$_2$ (1230).
Br$_2$ → (PhCH$_2$)$_3$SbBr$_2$ (1230).
SbCl$_3$ → (PhCH$_2$)$_2$SbCl$_3$ (1230).

TRI-p-TOLYLSTIBINE C$_{21}$H$_{21}$Sb (4-MeC$_6$H$_4$)$_3$Sb
Prepn.: By reacting SbCl$_3$ with 4-MeC$_6$H$_4$Br(1:3) and excess Na in Et$_2$O in the presence of catalytic amts. of AcOEt; 70% yield (793).
    By treating SbCl$_3$ with 4-MeC$_6$H$_4$N$_2$BF$_4$ (1:3) in Me$_2$CO, followed by Zn dust below
20°; 42% yield along with 4-MeC$_6$H$_4$SbO and [(4-MeC$_6$H$_4$)$_2$Sb]$_2$O (886).
    By reacting SbCl$_3$ with 4-MeC$_6$H$_4$Li in Et$_2$O under reflux; 95.3% yield (1193).
    By reducing (4-MeC$_6$H$_4$)$_3$SbCl$_2$ with N$_2$H$_4$.H$_2$O in 95% EtOH (1316).
Props.: m. 122° (793); m. 127° (1193). Crystal structure (491). Magnetic data
(968, 969).
Rxns. with: RCOCl + AlCl$_3$ in CS$_2$ → 4-MeC$_6$H$_4$COR (793).
RCl + AlCl$_3$ in CS$_2$ → 4-MeC$_6$H$_4$R (793).
BuLi → (4-MeC$_6$H$_4$)$_2$SbBu + 4-MeC$_6$H$_4$Li (1316).
Complex compds.: {Co[(MeC$_6$H$_4$)$_3$Sb](NO)(CO)$_2$}, red crystals, m.~135° (781, 783);
IR spectrum (733a).
{Fe[(MeC$_6$H$_4$)$_3$Sb](NO)$_2$(CO)}, red crystals, decomp.~110° (783).
{Rh[(MeC$_6$H$_4$)$_3$Sb]$_2$(CO)Cl}, red needles, decomp. 156-158° (1252); red-orange, m. 16,
170° (dec.) (1254).

TRIS[4-(2,5-DIMETHYL-1-PYRRYL)PHENYL]STIBINE C$_{24}$H$_{36}$N$_3$Sb [4-(MeC:CHCH:C(Me)N]
  C$_6$H$_4$]$_3$Sb
Prepn.: by reacting SbCl$_3$ with 4-[2,5-dimethyl-1-pyrryl)phenylmagnesium bromide;
70% yield (949).
Props.: Light brown solid sublimes at 255° (494).

TRI-1-NAPHTHYLSTIBINE $C_{30}H_{21}Sb$ $(1-C_{10}H_7)_3Sb$
Prepn.: By reacting $1-C_{10}H_7MgBr$ with $SbCl_3$; 92.6% yield (491).
Props.: m. 223-223.5° (491); m. 218-219° (1193).
Rxns. with: $RCOCl + AlCl_3$ in $CS_2 \rightarrow 1-C_{10}H_7COR$ (793).
$RCl + AlCl_3$ in $CS_2 \rightarrow 1-C_{10}H_7R$ (793).
$RLi \rightarrow (1-C_{10}H_7)_2SbR + 1-C_{10}H_7Li$ (1195a, 1316).

TRI-o-BIPHENYLYLSTIBINE $C_{36}H_{27}Sb$ $(2-PhC_6H_4)_3Sb$
Prepn.: by heating $SbCl_3$ with $2-PhC_6H_4Cl$ and Na in C_6H_6 (1318).
Props.: m. 208-209° (1313).
Rxns. with: Br_2 or $Cl_2 \rightarrow$ the dibromide and dichloride, resp. (1318).
$SbCl_3$ in xylene in a sealed tube $\rightarrow 2-PhC_6H_4Sb(OH)Cl$ (1318).

TABLE LXII
SYMMETRICAL TERTIARY STIBINES

R_3Sb	Prepd. from	Yield %	Props. and Remarks	Ref.
C_3				
$(CCH_2)_3Sb$	$ClCH_2SbCl_2 + CH_2N_2$		b. 105°/3 mm., d^{20}_{20} 2.038, rxns.1, 2	1323, 1321
C_6				
$(F_2C:CF)_3Sb$	$SbCl_3 + F_2C:CFMgI$	41	b. 75-75.5°/74 mm., n^{24}_D 1.4190, d_{24} 2.06	1165
$(MeCO)_3Sb$		75.5	White solid, m. 77-78°	949
C_{12}				
$(\overline{OCH:CHCH:C})_3Sb$	$SbCl_3 + \overline{OCH:CHCH:C}Cl + Na$	70	m. 46-47°, b. 181-182°/17 mm rxns.3-5	345, 348
$(\overline{SCH:CHCH:C})_3Sb$			Used as corrosion inhibitor for dielec. compns.	599
$(Me_2CHCH_2)_3Sb$	$SbF_3 + (Me_2CHCH_2)_3Al.Et_2O$	77	b. 101.5°/8 mm., $n^{19.5}_D$ 1.4955, d_{20} 1.124, rxn.6	1330
	$SbCl_3 + (Me_2CHCH_2)_3Al.Et_2O$	63		1331
$(Me_3SiCH_2)_3Sb$	$SbCl_3 + Me_3SiCH_2MgCl$	74	m. 64-65°, ingnites spontaneously on contact with air	1121
C_{18}				
$(2\text{-}ClC_6H_4)_3Sb$	$SbCl_3 + 2\text{-}ClC_6H_4MgBr$		m. 136-137°	513
	$2\text{-}ClC_6H_4N_2.SbCl_4 + Zn$ dust		m. 140°	897
$(4\text{-}O_2NC_6H_4)_3Sb$	$SbCl_3 + 4\text{-}O_2NC_6H_4N_2BF_4 + Zn$ in Me_2CO	17		836
$(4\text{-}HOC_6H_4)_3Sb$	$4\text{-}HOC_6H_4N_2SbCl_4 + Zn$ or Fe in AcOEt			897
$(cyclo\text{-}C_6H_{11})_3Sb$	$SbCl_3 + C_6H_{11}MgBr$		b. 206-209°/11-12 mm., rxn?	816
C_{21}				
$(3\text{-}CF_3C_6H_4)_3Sb$	$SbCl_3 + 3\text{-}CF_3C_6H_4MgCl$	60	Colorless liq., b. 174-175°, d^{20}_{20} 1.599, nD 1.5413	949
$[5,2\text{-}Cl(MeO)C_6H_3N_2]_3Sb$	$Sb + 5,2\text{-}Cl(MeO)C_6H_3N_2Cl$		m. 188°	780
$[4,2\text{-}Cl(Me)C_6H_3]_3Sb$	$Sb + 4,2\text{-}Cl(Me)C_6H_3N_2Cl + CaCO_3$		m. 226°	780
	$Sb + 2,4\text{-}Cl(Me)C_6H_3N_2Cl.ZnCl_2$			780

Compound	Preparation	Properties	Yield (%)	Ref.
(5,2-Cl(Me)C₆H₃)₃Sb		m. 176		780
(2-MeOC₆H₄)₃Sb	Sb + 5,2-Cl(Me)C₆H₃N₂Cl; SbCl₃ + 2-MeOC₆H₄MgBr	Lustrous plates, m. 189°, complex[8]		513
(3-MeOC₆H₄)₃Sb	SbCl₃ + 3-MeOC₆H₄MgI	Plates, m. 88.5–89°, complex[9]	50	513, 1193, 969
(4-MeOC₆H₄)₃Sb	SbCl₃ + 4-MeOC₆H₄Li	m. 180°	31	491
(2-MeC₆H₄)₃Sb	SbCl₃ + 2-MeC₆H₄Li	Magnetic data; m. 102°, crystal structure, rxn.10	84	491
(3-MeC₆H₄)₃Sb	SbCl₃ + 3-MeC₆H₄Li	m. 72°, crystal structure; Magnetic data, rxn.10	71.8	968-9
(n-C₇H₁₅)₃Sb	SbCl₃ + n-C₇H₁₅MgBr	b. 229–231°/50 mm., rxns.11-2	18	1227
C₂₄				
(PhC:C)₃Sb	SbCl₃ + PhC:CMgBr or PhC:CNa	Crystals, m. 159° (dec.), rxn.13		517
(2-EtOC₆H₄)₃Sb	SbCl₃ + 2-EtOC₆H₄MgBr	Needles, m. 123.5–123.8°, complex[14]	56	513
(4-EtOC₆H₄)₃Sb	SbCl₃ + 4-EtOC₆H₄Li	m. 82°; Magnetic data	85.4	1193, 969
(4-Me₂NC₆H₄)₃Sb	SbCl₃ + 4-Me₂NC₆H₄Li	m. 229°; m. 239–241°	78.1, 35	1193, 1316
[Et(n-C₅H₁₁)CH]₃Sb		Curing agent for polyepoxides		1132
C₂₇				
(3,3,5-Me₃C₆H₈)₃Sb		Lead-scavenger for gasoline		1328

1. Rxn. with H₂O → white ppt. (1321).
2. Rxn. with Br in CCl₄ at 5° → (ClCH₂)₃SbBr₂ (1320, 1321).
3. Rxn. with Cl₂ → tri-2-furylstibine dichloride (345, 348).
4. Rxn. with HCl → furan + SbCl₃ (345, 348).
5. Rxn. with H₂O₂ → furan + Sb₂O₃ (348).
6. Rxn. with Br → (Me₂CHCH₂)₃SbBr₂ (1330, 1331).
7. Rxn. with atm. O → (C₆H₁₁)₃SbO.Sb₂O₃ (816).
8. Complex compd. with HgCl₂ (1:1), white solid, m. 160–161° (513).
9. Complex compd. with HgCl₂ (1:1), white solid, m. 180° (dec.) (513).
10. Rxn. with EtLi → MeC₆H₄Li (1195a).
11. Rxn. with I₂ in Et₂O → (n-C₇H₁₅)₃SbI₂ (1227).
12. Rxn. with HgO → (n-C₇H₁₅)₃SbO (1227).
13. Rxn. with aq. alkali → evolution of PhC:CH (517).
14. Complex compd. with HgCl₂ (1:1), white solid, m. 170.9° (513).

TABLE LXII (cont'd.)

SYMMETRICAL TERTIARY STIBINES

R_3Sb	Prepd. from	Yield %	Props. and Remarks	Ref.
C_{30}				
$(2\text{-}C_{10}H_7)_3Sb$	$SbCl_3 + 2\text{-}C_{10}H_7MgI$		Yellow solid, m. 195°	513
$(4\text{-}Et_2NC_6H_4)_3Sb$	$SbCl_3 + 4\text{-}Et_2NC_6H_4Li$	62	m. 225-226°	949
C_{36}				
$(OC_6H_4C_6H_3)_3Sb$	$SbCl_3$ + 4-dibenzofuryllithium	80.4	White solid, m. 243-244°	949
$(4\text{-}PhOC_6H_4)_3Sb$	$SbCl_3 + 4\text{-}PhOC_6H_4MgBr$		White solid, m. 157°, complex[15]	513
$(4\text{-}PhC_6H_4)_3Sb$	$SbCl_3 + 4\text{-}PhC_6H_4Li$	57	m. 176°	1193
$\{4\text{-}[MeC:CHCH:C(Me)N]C_6H_4\}_3Sb$	$SbCl_3 + MeC:CHCH:C(Me)NC_6H_4MgCl$	70	Light brown solid, sublimes at 235°	949
C_{48}				
$(n\text{-}C_{16}H_{33})_3Sb$	$SbCl_3 + n\text{-}C_{16}H_{33}MgCl$	75.5	White solid, m. 77-78°	949

15. Complex compd. with $HgCl_2$ (1:1), white crystals, m. 175-176° (dec.) (513).

ASYMMETRICAL TERTIARY STIBINES

Asymmetrical tertiary stibines of the general formulae $R_2R'Sb$ and $RR'R''Sb$ are prepared by:

1. Treating symmetrical tertiary stibines, R_3Sb, with trifluoroiodomethane in a sealed tube at room temperature. Thus trimethylstibine yielded trifluoromethyldimethylstibine (Method I).
2. Treating organohalostibines, $\overline{RSbX_2}$ or R_2SbX, with organometallic compounds, such as alkyl- or aryllithium, alkynylsodium,, or Grignard compounds. The reactions of diorganohalostibines with sodium acetylide in liquid ammonia or in tetrahydrofuran yield mixtures of the corresponding ethynyldiorgano stibines with ethynylenebis(diorganostibines), whereas arylethynylsodium yields diorgano(arylethynyl)stibines. Ethynylenebis(magnesium bromide) yields ethynylenebis(diorganostibines) (Method II).
3. Reacting organodihalostibines, $RSbX_2$, with arylethynylsodium compounds, $RC{:}CNa$, in tetrahydrofuran to form organodi(arylethynyl)stibines, $RSb(C R)_2$ (Method III).
4. Treating diorganoiodostibines, R_2SbI, with metallic sodium followed by alkyl or aryl iodides in liquid ammonia (Method IV).
5. Reducing asymmetrical tertiary stibine dihalides, $R_2R'SbX_2$ or $RR'R''SbX_2$, with hydrogen sulfide in hot ethanol containing ammonia (Method V).
6. Reacting diorganoantimony trihalides, R_2SbX_3, with diaryliodonium chlorides in acetone and decomposing the reaction product with zinc dust. The reaction yields mixtures of asymmetrical tertiary arsines with their dichlorides (Method VI).
7. Treating antimony powder with methyl chloride at $372°$ in the presence of copper and reacting the methylation products with alkylmagnesium halides to form mixtures of dimethylalkyl-, methyldialkyl-, and trialkylstibines (Method VII).

TABLE LXIII

ASYMMETRICAL TERTIARY STIBINES

$R_2R'Sb$ or $RR'R''Sb$	Prepd. from	Method	Yield %	Props. and Remarks	Ref.
C3 $Me_2(CF_3)Sb$	$Me_3Sb + CF_3I$	I	50	Colorless liq., b. (calcd.) 85.8°, rxn.1	520
C6 $Me_2SbC:CSbMe_2$	$R_2SbBr + CH:CNa$	II		b. 116°/17 mm.	515
Me_2BuSb	$MeCl + Sb$, followed by $BuMgBr$	VII			774a
C8 Me_2PhSb				UV spectrum	111
$(Me_2CH)_2(CH:C)Sb$	$R_2SbBr + CH:CNa$ in C_4H_8O	II	~70	Colorless oil, b. 68°/15 mm., sensitive to O, rxns.2, 3	515
C9 $Me_2(PhCH_2)Sb$	$R_2SbBr + R$ $MgCl$	II		b. 110-112°/0.4 mm., rxn4.	1313
Bu_2MeSb	$MeCl + Sb$, followed by $BuMgBr$	VII			774a
C14 $[(Me_2CH)_2SbC:]_2$	$R_2SbBr + CH:CNa$ in C_4H_8O	II		Colorless oil, b. 118°/1 mm., sensitive to O, rxn.5	515
C15 Ph_2PrSb	$R_2SbCl + R'MgBr$	II	65	b. 180-182°/12-13 mm., $n^{19.5}_D$ 1.6428, $d_{19.5}$ 1.4160	1200
$Ph_2(Me_2CH)Sb$	$R_2SbCl + R'MgBr$	II	65	b. 167-169°/6-7 mm., $n^{19.5}_D$ 1.6340, $d_{19.5}$ 1.3990	1200
C16 Ph_2BuSb	$R_2SbCl + R'MgBr$	II	75	b. 188-192°/6-7 mm., d 1.3810, n^{28}_D 1.6180	816

Compound	Reaction	Method	Yield (%)	Properties	Ref.
Ph$_2$(Me$_3$C)Sb	R$_2$SbCl + R'MgBr	II	65	1.3690, n$_D$ 1.6278 b. 176–178°/6–7 mm., n$_D^{18}$ 1.6260, d$_{18}$ 1.3750	1200
C$_{17}$ Ph$_2$(n-C$_5$H$_{11}$)Sb	R$_2$SbCl + R'MgBr	II	65	b. 203–205°/6–8 mm., n$_D^{17.5}$ 1.6370, d$_{17.5}$ 1.3610	1200
Ph$_2$(Me$_2$CHCH$_2$CH$_2$)Sb	R$_2$SbCl + R'MgBr	II	75	b. 195–200°/8–9 mm., d$_{29}$ 1.3770, n$_D^{29}$ 1.6319	816
C$_{18}$ Ph$_2$(4-BrC$_6$H$_4$)Sb	R$_2$SbCl + R'MgBr	II	60	b. 220–225°/1–2 mm., d$_{50}$ 1.5227, n$_D^{56}$ 1.6706	1329
Ph$_2$(4-ClC$_6$H$_4$)Sb	R$_2$SbCl + R'MgBr	II		Rxn. 6	1315
	R$_2$SbCl + Na + R'I	IV			1315
Ph$_2$(cyclo-C$_6$H$_{11}$)Sb	R$_2$SbCl + R'MgBr	II	75	b. 213–216°/7–8 mm.	816
		II	55	b. 223–225°/2 mm., d$_{50}$ 1.3313, n$_D^{56}$ 1.6432	1329
Ph$_2$(n-C$_6$H$_{13}$)Sb	R$_2$SbCl + R'MgBr	II	65	b. 212–214°/6–7 mm., n$_D^{26}$ 1.6221, d$_{26}$ 1.3440, rxn. 7	1200
C$_{19}$ Ph$_2$(3-F$_3$CC$_6$H$_4$)Sb	R$_2$SbCl + R'MgBr	II	47.7	b. 181–182°, d$_4^{20}$ 1.491, n$_D^{20}$ 1.6212	949
Ph$_2$(2-MeOC$_6$H$_4$)Sb	R$_2$SbCl + R'MgBr	II	55	b. 178–180°/2 mm., d$_{50}$ 1.3983, n$_D^{50}$ 1.6455	1329
Ph$_2$(4-MeOC$_6$H$_4$)Sb	R$_2$SbCl + R'MgBr	II	55	b. 225–230°/12–13 mm., d$_{50}$ 1.3850, n$_D^{50}$ 1.6420	1329
Ph$_2$(PhCH$_2$)Sb	R$_2$SbCl + R'MgBr	II	75	b. 224–225°/15–17 mm., d$_{31}$ 1.3440, n$_D^{31}$ 1.6510	816

1. Rxn. with aq. alkali → complete hydrolysis (521).
2. Rxn. with H$_2$O$_2$ → crystalline compd., mol. wt. 810 (515).
3. Rxn. with MeI → evolution of C$_2$H$_2$ (515).
4. Rxn. with PhCH$_2$Br → Me$_2$(PhCH$_2$)$_2$SbBr (1313).
5. Rxn. with MeI → [Me(Me$_2$CH)$_2$SbC:]$_2$I$_2$, cryst. salt (515).
6. Rxn. with BuLi → Ph$_2$BuSb + 4-ClC$_6$H$_4$Li (1315).

TABLE LXIII (cont'd.)
ASYMMETRICAL TERTIARY STIBINES

$R_2R'Sb$ or $RR'R''Sb$	Prepd. from	Method	Yield %	Props. and Remarks	Ref.
C_{20}					
$Ph(4\text{-}MeC_6H_4)(4\text{-}HO_2CC_6H_4)Sb$	$(4\text{-}MeC_6H_4)(4\text{-}EtO_2CC_6H_4)SbCl + PhMgBr$, followed by alk. hydrolysis		10	m. 161-163°	181
$(4\text{-}MeC_6H_4)_2(4\text{-}ClC_6H_4)Sb$	$R_2SbCl_3 + R'_2ICl$	VI	24	m. 113-113.5°, rxn.8	1074a
$Ph_2(4\text{-}EtOC_6H_4)Sb$	$R_2SbCl + R'MgBr$	II	55	b. 235-238°/7-8 mm., d_{50} 1.3442, n_D^{26} 1.6215	1329
$Ph_2(3,5\text{-}Me_2C_6H_3)Sb$	$R_2SbCl + R'MgBr$	II	55	b. 190-195°/2-3 mm., d_{35} 1.3790, n^{35} 1.6420	1329
$Ph_2(n\text{-}C_8H_{17})Sb$	$R_2SbCl + R'MgBr$	II	65	b. 227-228°/3-4 mm., $n_D^{27.6}$ 1.5820, $d_{27.6}$ 1.224	1200
C_{21}					
$Ph_2(2,4,6\text{-}Me_3C_6H_2)Sb$	$R_2SbCl + R'MgCl$	II		m. 92-93°, rxn.9	1315
$Ph_2(n\text{-}C_9H_{19})Sb$	$R_2SbI + R'MgBr$	II	65	b. 236-238°/5-7 mm., n_D^{26} 1.5600, d_{26} 1.191, rxn.10	1200
C_{22}					
$Ph_2(1\text{-}C_{10}H_7)Sb$	$R_2SbCl + R'MgBr$	II		m. 99-100°, rxn.11	1315
	$R_2SbI + Na + R'I$	IV			
C_{23}					
$Ph(4\text{-}HO_2CC_6H_4)(1\text{-}C_{10}H_7)Sb$	$R'R''SbCl + RMgBr$	II	38	m. 195-196°, d- and l-forms resolved by d- and l-MeCHPhNH₂	181
$Ph(4\text{-}HO_2CC_6H_4)(2\text{-}C_{10}H_7)Sb$	$R'R''SbCl + RMgBr$	II	18	m. 170°	181
$(PhC\!:\!C)_2(4\text{-}MeC_6H_4)Sb$	$R'SbCl_2 + PhC\!:\!CNa$ in C_4H_8O	II		m. 159°	518
C_{24}					
$Ph_2(o\text{-}\overline{C_6H_4C_6H_3O})Sb$	$R_2SbCl + 2$-dibenzofuryl-lithium	II	51.5	m. 125-128°	949
$Ph_2(4\text{-}PhOC_6H_4)Sb$	$R_2SbCl + R'MgBr$	II	~60	b. 260-265°/2-3 mm., d_{50} 1.3344, n^{50} $_D$ 1.6502	1329
$Ph_2[4\text{-}(Me\overline{C\!:\!CHCH\!:\!CMeN})C_6H_4]Sb$	$R_2SbCl + R'MgBr$	II	49.8	m. 63-84° D	949

Compound	Reaction	Method		Properties	Ref.
Ph(2-ClC6H4)(4-HO2CC6H4)Sb	Ph(2-PhC6H4)(4-EtO2CC6H4)Sb+5% alc. KOH			Racemate, m. 199-200°, d-form, m. 76-80°; l-form, m. 76-79, resolution[12], rxn.[13]	178
Ph(2-PhC6H4)(4-MeC6H4)Sb	RR'R''SbCl2 + H2S	V		m. 81-83°, UV spectrum	178
	R'R''SbCl + RMgBr	II			178
Ph(4-HO2CC6H4)(cyclo-C6H11)Sb	R'R''SbCl + RMgBr	II		m. 165-166°	181
C26 [(4-ClC6H4)2SbC:]2	(4-ClC6H4)2SbCl + CH:CNa in liq. NH3		25	m. 149°, stable in light and moisture	518
(PhC:C)2(1-C10H7)Sb	R'SbCl2 + PhC:CNa in Et2O			m. 157°	518
(Ph2SbC:)2	Ph2SbCl + CH:CNa in liq. NH3	II		m.110.5°	514
	Ph2SbCl + (:CMgBr)2 in CHCl3-Et2O	II		Rxn. with aq. KOH → cleavage of C2H2	514
Ph(4-HO2CC6H4)[3,4-Ph-(MeO)C6H3]Sb	R'R''SbCl + RMgBr	II	20	m. 196-197°	181
C27 Ph(2-PhC6H4)(4-EtO2CC6H4)-Sb	R'R''SbCl + RMgBr	II		m. 105-107°	178
	RR'R''SbX2 (X = Cl or Br) + H2S	V			
C30 [(4-MeC6H4)2SbC:]2	(4-MeC6H4)2SbCl + CH:CNa in liq. NH3			m. 121°, stable in light and moist air	518
C31 (2-PhC6H4)2(4-MeC6H4)Sb	R2R'SbCl2 + H2S	V		m. 130-131°	173

7. Rxn. with Br in CHCl3 → Ph(n-C6H13)SbBr (1200).
8. Rxn. with alc. CuCl2 → R2R'SbCl2 (1074a).
9. Rxn. with BuLi → Ph2BuSb + Me3C6H2Li (1315).
10. Rxn. with halogens (Br or I) in CHCl3 → Ph(n-C9H19)As halide (1200).
11. Rxn. with BuLi → Ph2BuSb + C10H7Li (1315).
12. The racemic mixtures was resolved via d-MeCH(Ph)NH2; the l- salt of d-MeCH(Ph)NH2, m. 174-176°, and the d-salt of d-MeCH(Ph)NH2, m. 168-170° (178).
13. Rxn. with atm. O in C10H18 → the stibine oxide (178).

DIORGANOHALOSTIBINES

Dialkyl-, diaryl-, and alkylarylhalostibines, R_2SbX and $RR'SbX$, are prepared or formed by:

1. Thermal decomposition of tertiary stibine dihalides, R_3SbX_2 or $R_2R'SbX_2$.
2. Reduction of diorganoantimony trihalides, R_2SbX_3, with sulfur dioxide in hydrochloric acid in the presence of iodide ions, or with tin(II) chloride in hydrochloric acid or in acetone, or with antimony powder in ether.
3. Reducing stibinic acids, R_2SbO_2H, with tin(II) chloride.
4. Alkylation of antimony trichloride with diazoalkanes, $RCHN_2$, using a proper ratio of the reactants, to yield chlorobis(1-chloroalkyl)stibines, $(RCHCl)_2SbCl$.
5. Treating teriary stibines, R_3Sb or $R_2R'Sb$, with halogens or with antimony trihalides or with bismuth trichloride at elevated temperatures to yield mixtures of diorganohalo- and organodihalostibines.
6. Treating tertiary stibines, R_3Sb, with bromodimethylborine.
7. Reacting antimony trichloride with tetraphenyltin, Ph_4Sn.
8. Decomposition of aryldiazonium chloroantimonites, RN_2SbCl_4 or $(RN_2)_2SbCl_5$, with zinc dust or iron powder in ethyl acetate at 70-90° or with copper bronze in acetic acid containing hydrochloric acid to yield diarylchlorostibines along with tertiary stibine dichlorides, R_3SbCl_2. Aryldiazonium chloroantimonites can be prepared by reacting aryldiazonium chlorides with antimony trioxide in hydrochloric acid.
9. Reacting aryldiazonium chlorides with antimony powder in acetone under reflux.
10. Reacting aryldiazonium chloroantimonites, RN_2SbCl_4, with aryldihalostibines in ethanol at 50° in the presence of copper and reducing the reaction product with tin(II) chloride in acetone.
11. Decomposition of aryldiazonium chlorozincates, RN_2ZnCl_3, with antimony powder in acetone in the presence of calcium carbonate.
12. Disproportionation of stiboso compounds, $RSbO$, in concentrated hydrochloric acid.
13. Decomposition of diaryliodonium chloroantimonites, $R_2I.SbCl_4$ with antimony powder in acetone to yield diarylchlorostibines along with triarylstibine dichlorides, R_3SbCl_2, diarylantimony trichlorides, R_2SbCl_3, and aryldichlorostibines, $RSbCl_2$.
14. Reacting antimony trichloride with Grignard compounds.
15. Reacting antimony powder with trifluoroiodomethane in a sealed steel tube at 165-175° to yield bis(trifluoromethyl)iodo- and trifluoromethyldiiodostibine.
16. Neutron irradiation of antimony trichloride in benzene to yield a mixture of products containing radioactive chlorodiphenylstibine.

IODOBIS(TRIFLUOROMETHYL)STIBINE C_2F_6ISb $(CF_3)_2SbI$

Prepn.: By reacting $(CF_3)_3Sb$ with I_2 in a sealed tube, a mixt. of $(CF_3)_2SbI$ and CF_3SbI_2 was obtd. (258).

By heating $(CF_3)_3Sb$ with SbI_3 at 120° (258).

By-prod. in the rxn. of Sb powder with CF_3I in a sealed steel tube at 165-175° (258, 334).

Props.: m. -42°, b. 16°/8 mm; disproportionates on exposure to air or upon heating to $(CF_3)_3SbI_2$ (258).
Rxns. with: $AgCl \rightarrow (CF_3)_2SbCl$ (258).
Hg or Zn $\rightarrow [(CF_3)_2Sb]_2$ (258, 334).

BROMODIMETHYLSTIBINE C_2H_6BrSb Me_2SbBr
Prepn.: By treating Me_3Sb with Me_2BBr (139a).
By disproportionation of Me_3SbBr_2 (139a).
Props.: b. 173-177°/735 mm. (1313).
Rxns. with: $CH:CNa$ in $NH_3 \rightarrow Me_2SbC:CSbMe_2$ (515).
$CH:CNa$ in $C_4H_8O \rightarrow Me_2SbC:CH$ (515).
$RMgCl$ in Et_2O in a N atm. $\rightarrow Me_2RSb$ (1313).
$NaBH_4 \rightarrow Me_2SbBH_2$ (?), Me_2SbH and BH_3 complex with $(-CH_2OMe)_2$ (139a).
Na in liq. $NH_3 \rightarrow (Me_2Sb)_2$ (139a).

CHLORODIPHENYLSTIBINE $C_{12}H_{10}ClSb$ Ph_2SbCl
Prepn.: By reducing Ph_2SbCl_3 with SO_2 in HCl contg. HI; 77% yield (134, 462, 1119) or with $SnCl_2$ (1316).
By heating $SbCl_3$ with Ph_4Sn (816).
By heating Ph_3Sb with $BiCl_3$ in $CHCl_3$ under CO_2 (817).
By diazotization of $PhNH_2.HCl$ in MeOH with $n-C_5H_{11}ONO$ at -5°, followed by addition of $SbCl_3$ in MeOH, decompn. of the resulting double salt $(PhN_2)_2SbCl_5$ with Zn dust in EtOAc, hydrolysis with 5% aq. NH_4OH, extraction of $(Ph_2Sb)_2O$ with $CHCl$, and treatment with EtOH-HCl; 42-64% yield (903).
By decompn. of $Ph_2I.SbCl_4$ with Sb powder in Me_2CO to yield a mixt. of Ph_2SbCl, Ph_2SbCl_3, and Ph_3SbCl_2 (1072). From $Ph_2Cl.SbCl_4$ and $Ph_2Br.SbCl_4$ with Sb, 24 and 17% yields, respectively, were obtd. (912).
By reducing Ph_2SbO_2H with $SnCl_2$ (1315).
Props.: m. 69-70° (134, 1119); m. 68° (816, 1295); m. 67-68° (903).
Rxns. with: PhN_2Cl (1:1) in AcOH, followed by decompn. with dild. HCl $\rightarrow Ph_2SbCl_3$, while decompn. in $CHCl_3 \rightarrow Ph_3SbCl_2$ (130).
$(PhN_2)_2ZnCl_4 \rightarrow Ph_3SbCl_2$ (462).
$CH:CNa \rightarrow Ph_2SbC:CSbPh_2$ (514).
$(:CMgBr)_2 \rightarrow (Ph_2SbC:)_2$ (514).
$RCOCl + AlCl_3$ in $CS_2 \rightarrow PhCOR$ (792).
$RLi \rightarrow RPh_2Sb$ (949).
$RCl + AlCl_3$ in $CS_2 \rightarrow PhR$ (792).
$RMgX \rightarrow RPh_2Sb$ (816, 949, 1200, 1315, 1329).
$H_2O \rightarrow (Ph_2Sb)_2O$ (912).
NO_2^- in AcOH, followed by $NH_4OH \rightarrow Ph_2SbO_2H$ (1074).
$LiBH_4$ or $LiAlH_4 \rightarrow Ph_2SbH$ (1295).
RN_2Cl (1:3) in EtOH $\rightarrow RN_2.(Ph_2SbCl_4)$ (1074).
$RN_2Cl.MCl_n$ (M = Fe, Zn, Sb; n = valency of M) $\rightarrow RN_2.(Ph_2SbCl_4)$ (1074).
Complex. compds.: $PhN_2.(Ph_2SbCl_2)$, white powder, m. 76-77°, which decomp. in HCl to Ph_2SbCl_3 and in $CHCl_3$ to Ph_3SbCl_2 (130). The complex was identified to be $PhN_2.(Ph_2SbCl_4)$ (1074).

o-BIPHENYLYLCHLORO-p-TOLYLSTIBINE $C_{19}H_{16}ClSb$ $(4-MeC_6H_4)(2-PhC_6H_4)SbCl$

<u>Prepn</u>: by reducing $(4-MeC_6H_4)(2-PhC_6H_4)SbCl_3$ with $SnCl_2$ in HCl (173).

<u>Props</u>.:   m. 94-95° (173).

<u>Rxns. with</u>: $AlCl_3$ in C_6H_6 at 100°/10⁻⁴ mm → crystals, m. 174-190° (173).

$RMgX$ → R·$(4-MeC_6H_4)(2-PhC_6H_4)Sb$ (178).

EtOH in Me_2CO + NaOAc in H_2O → RR'SbOH, which on heating at 90° in Ac_2O contg. H_2SO_4, followed by redn. with $SnCl_2$ in Me_2CO-HCl → 5-p-tolyldibenzostibiole (173.

R_2SbX or $RR'SbX$	Prepd. from	Props. and Remarks	Ref.
C_2			
$(CF_3)_2SbBr$	$(CF_3)_3SbBr_2$ by pyrolysis	b. 113°	258
$(CF_3)_2SbCl$	$R_2SbI + AgCl$	Colorless liquid, b. ~88°, disproportionates → $(CF_3)_3Sb$ + $SbCl_3$, rxn.[1]	258
Me_2SbBr		Polagraphic redn.	1094
Me_2SbCl		Rxn.[2]	
C_4			
$(ClCH:CH)_2SbCl$	$R_2SbCl_3 + Sb$ in Et_2O	Colorless liquid, b. 113-115°/2 mm.	899
$(CH_2:CH)_2SbBr$ [3]	$R_3Sb + SbBr_3$ (by-prod.)	b. 42.5°/0.7 mm.	773-4
$(CH_2:CH)_2SbI$ [3]	R_3SbI_2 by pyrolysis	b. 69°/1.3 mm.	773
$(MeCHCl)_2SbCl$	$SbCl_3 + MeCHN_2$	b. 88-88.5°/3 mm., b. 69.5°/1 mm., d20 1.831, rxns.4, 5	1320-1
C_6			
$(Me_2CH)_2SbBr$		Rxn.[6]	515
C_{12}			
$(4-BrC_6H_4)_2SbCl$ [3]	$R_2I.SbCl_4 + Sb$ in Me_2CO	m. 55°, rxn.[7]	1072
$(4-FC_6H_4)_2SbCl$	$R_2SbCl_3 + SO_2$ in HCl		132
$(4-IC_6H_4)_2SbCl$ [3]	$R_2I.SbCl_4 + Sb$ in Me_2CO		1072
$(2-ClC_6H_4)_2SbCl$ [3]	$RN_2.SbCl_4 + Fe$ in Me_2CO	m. 94-95°	904
	$RSbO$ + conc. HCl	Rxn.8	910
$(3-ClC_6H_4)_2SbCl$ [3]	$RN_2.SbCl_4 + Zn$ or Fe	77% yield	897
$(4-ClC_6H_4)_2SbCl$ [3]	$R_2I.SbCl_4 + Sb$ in Me_2CO	Rxn.8	1072
Ph_2SbI		Rxn.[9]	

1. Rxn. with H_2O or aq. alkali → complete hydrolysis and CHF_3 formation (258).
2. Rxn. with NH_4F under reflux → disproportionation (1298).
3. The compound was obtained as by-prod.
4. Rxn. with H_2O → cloudy soln. but without noticeable decompn. (1321).
5. Rxn. with H_2O_2 → $(MeCHCl)_2SbO_2H$ (1320, 1321).
6. Rxn. with CH CNa in C_4H_8O → $(Me_2CH)_2SbC:CH$ + $[(Me_2CH)_2SbC:]_2$ (515).
7. Rxn. with 5% NH_4OH → $[(4-FC_6H_4)_2Sb]_2O$ (132).
8. Rxn. with $RN_2.SbCl_4$ → $R(ClC_6H_4)_2SbCl_2$ (910, 1063).

TABLE LXIV (cont'd.)
DIORGANOHALOSTIBINES

R$_2$SbX or RR'SbX	Prepd. from	Props. and Remarks	Ref.
Ph(n-C$_6$H$_{13}$)SbBr	R$_2$R'Sb + Br$_2$ in CHCl$_3$	m. 195°	1200
C$_{14}$			
[5,2-Cl(MeO)C$_6$H$_3$]$_2$SbCl$_3$	RN$_2$Cl + Sb in Me$_2$CO	m. 144°	780
[4,2-Cl(Me)C$_6$H$_3$]$_2$SbCl$_3$	RN$_2$Cl.ZnCl$_2$ + Sb + CaCO$_3$ in Me$_2$CO	m. 131°, rxn.10	780
(2-MeOC$_6$H$_4$)$_2$SbCl$_3$	RN$_2$.SbCl$_4$ + Fe in Me$_2$CO	m. 111-112°	904
(4-MeOC$_6$H$_4$)$_2$SbCl$_3$	SbCl$_3$ + RMgBr (1:4)	Plates, m. 116-117°	513
(4-MeC$_6$H$_4$)$_2$SbCl	R$_2$I.SbCl$_4$ + Sb in Me$_2$CO		1072
C$_{15}$			
Ph(n-C$_9$H$_{19}$)SbBr	R$_2$R'Sb + Br$_2$ in CHCl$_3$	m. 208°	1200
Ph(n-C$_9$H$_{19}$)SbI	R$_2$R'Sb + I in CHCl$_3$	m. 227°	1200
C$_{16}$			
4-MeC$_6$H$_4$(4-EtO$_2$CC$_6$H$_4$)SbCl	RR'SbCl$_3$ + SnCl$_2$ in HCl	m. 63-64°, rxn.11	181
(2-BtOC$_6$H$_4$)$_2$SbCl	RSbO by disproportionation in conc. HCl	m. 107-108°	910
	RN$_2$.SbCl$_4$ + Zn	40% yield, rxn.12	897
C$_{19}$			
(4-HO$_2$CC$_6$H$_4$)[4,2-Br(Ph)C$_6$H$_3$]-SbCl	RR'SbCl$_3$ + SnCl$_2$ in Me$_2$CO-HCl	m. 170°	173
(4-MeC$_6$H$_4$)[4,3-Br(Ph)C$_6$H$_3$]SbCl	RR'SbCl$_3$ + SnCl$_2$ in Me$_2$CO	m. 140-141° 11	173
(4-EtO$_2$CC$_6$H$_4$)(1-C$_{10}$H$_7$)SbCl	RR'SbCl$_3$ + SnCl$_2$ in HCl	m. 90°, rxn.11	181
(4-EtO$_2$CC$_6$H$_4$)(2-C$_{10}$H$_7$)SbCl	RR'SbCl$_3$ + SnCl$_2$ in HCl	m. 93.5-95°, rxn.11	181
C$_{21}$			
(4-EtO$_2$CC$_6$H$_4$)[4,2-Br(Ph)C$_6$H$_3$]-SbCl	RR'SbCl$_3$ + SnCl$_2$ in Me$_2$CO	m. 108°	173
(4-EtO$_2$CC$_6$H$_4$)[2-(4-O$_2$NC$_6$H$_4$)C$_6$-H$_4$]SbCl	RR'SbCl$_3$ + SnCl$_2$ in HCl	m. 156-157°	175
(4-EtO$_2$CC$_6$H$_4$)(2-PhC$_6$H$_4$)SbCl	RR'SbCl$_3$ + SnCl$_2$ in HCl	m. 84-85°, rxn.11	175
(4-EtO$_2$CC$_6$H$_4$)[4-(C$_6$H$_{11}$)C$_6$H$_4$]-SbCl	RR'SbCl$_3$ + SnCl$_2$ in HCl	m. 110-111°; R' = 4-cyclohexyl-phenyl, rxn.11	175 / 181

C22 (4-EtO2CC6H4)[5,2-Me(Ph)C6H3]-SbCl	RR'SbCl3 + SnCl2 in EtOH	m. 121-122°	177
(4-MeC6H4)[2-(4-EtO2CC6H4)C6-H4]SbCl	RR'SbCl3 + SnCl2 in Me2CO	Needles, m. 90-91°	173
(4-EtO2CC6H4)[2-(4-MeOC6H4)C6-H4]SbCl	RR'SbO2H + SnCl2 in HCl	m. 96°	182
(4-EtO2CC6H4)[3,4-Ph(MeO)C6H3]-SbCl	RR'SbCl3 + SnCl2 in HCl	m. 154-155°, rxn.[11]	181
C24 (4-PhC6H4)2SbCl	RN2SbCl4 + 4-EtO2CC6H4SbCl2 in EtOH + Cu, followed by SnCl2	m. 200°	178
C26 [2-(4-HO2CC6H4)C6H4]2SbCl	RN2Cl + Sb2O3 in EtOH + Cu, followed by SnCl2	m. 169-170°	173

9. Rxn. with Na in liq. NH3 → green color (Ph2Sb formation) (1316).
10. Rxn. with Cl2 in CCl4 → [4,2-Cl(Me)C6H3]2SbCl3 (780).
11. Rxn. with R''MgX → the corresponding tert. stibine RR'R''Sb (178, 181).
12. Rxn. with RN2·SbCl4 or (RN2)2ZnCl4 → R(2-EtOC6H4)2SbCl2 (910, 1063).

459

Organodihalostibines, $RSbX_2$, are prepared by:

1. Treating stiboso compounds, RSbO, with hydrohalic acids of hydrogen halides in alcohol. In the preparation of organodiiodostibines stiboso compounds are dissolved in hydrochloric or acetic acid and treated with potassium iodide. When acetic acid is used, the product is precipitated with hydrochloric acid.
2. Reducing organoantimony tetrahlides, $RSbX_4$, with sulfur dioxide in aqueous hydrochloric acid in the presence of potassium iodide or with tin(II) chloride in hydrochloric acid.
3. Reducing stibonic acids, $RSbO_3H_2$, with sulfur dioxide in hydrochloric acid or in ethanol in·the presence of potassium iodide or iodine, or with tin(II) chloride in hydrochloric acid. Arenestibonic acids, $RSbO_3H_2$, are obtained by complexing aryldiazonium chlorides with antimony trichloride in concentrated hydrochloric acid, decomposing the double salts, RN_2SbCl_4, in absolute ethanol in the presence of copper(I) chloride or bromide, and hydrolyzing the product, $RSbCl_4$, with water.
4. Decomposition and reduction of aryldiazonium tetrachloroantimonites, RN_2-$SbCl_4$, with zinc dust or iron powder in ethanol or pentanol at 70-90°. The reaction yields aryldichlorostibines, as by-products, along with stiboso compounds and triarylstibines or diarylstibonous acids.
5. Reacting tertiary stibines, R_3Sb, with iodine in a sealed tube to form organodiiodostibines and diorganoiodostibines.
6. Treating antimony trichloride with diazomethane in benzene at 2-5° to form dichloro(chloromethyl)stibine.
7. Reacting antimony trichloride with Grignard compounds in an inert atmosphere.
8. Reacting antimony powder with diarylchloronium or diarylbromonium tetrachloroantimonites.
9. Reacting diorganoantimony trihalides with antimony in ether.
10. Heating antimony trihalides with tetraethyllead in benzene in an inert atmosphere.
11. Neutron irradiation of antimony trichloride in benzene; dichlorophenylstibine is one of the reaction products.
12. Reacting tertiary stibines, R_3Sb, with antimony trihalides to form organodihalo- and diorganohalostibines.
13. Disproportionation of tertiary stibine dihalides. Thus disproportionation of tris(trifluoromethyl)stibine dibromide, $(CF_3)_3SbBr_2$, at 20° yielded dibromo(trifluoromethyl)stibine as by-product.
14. Heating antimony powder with trifluoroiodomethane in a sealed tube at 165-175°.

TRIFLUOROMETHYLDIIODOSTIBINE CF_3I_2Sb CF_3SbI_2
Prepn.: By reacting $(CF_3)_3Sb$ with I_2 in a sealed tube, mixt. of CF_3SbI_2 with $(CF_3)_2SbI$ is formed (258).
    By-prod. in the rxn. of Sb powder with trifluoroiodomethane in a sealed tube at 165-175° (258).
Props.: Viscous bright-yellow liquid which freezes to a pale yellow solid, m. (crude) 4-8°, b. >200° with disproportionation (258).

Rxn. with aq. KOH → complete hydrolysis (258).

DICHLORO(3-CHLOROPHENYL)STIBINE $C_6H_4SbCl_3$ 3-$ClC_6H_4SbCl_2$
Salts: $C_5H_5N.H(3-ClC_6H_4SbCl_3)$, m. 117-118° (pyridinium salt); $C_9H_7N.H(3-ClC_6H_4-SbCl_3)$, m. 118-119° (quinolinium salt) (1001).

DICHLOROPHENYLSTIBINE $C_6H_5Cl_2Sb$ $PhSbCl_2$
Prepn.: By reacting $SbCl_3$ with PhMgBr in Et_2O under N (728).
By adding $PhN_2.SbCl_4$ to a suspension of Zn dust or Fe in EtOAc and decompg. the double salt at 70-90°; 18% yield along with PhSbO and Ph_2SbOH (897).
By reducing $PhSbCl_4$ with SO_2 in HCl contg. KI (462).
By reducing $PhSbO_3H_2$ with SO_2 in HCl contg. KI; 58% yield (294, 1119) or by $SnCl_2$ in HCl; 35% yield (294).
By decompg. $(Ph_2Cl)SbCl_4$ or $(Ph_2Br)SbCl_4$ with Sb powder in Me_2CO; 9 and 18% yield, respectively, along with 17% Ph_2SbCl (912, 1073).
By treating PhSbO with HCl in EtOH (1070).
By neutron irradiation of $SbCl_3$ in C_6H_6 (907).
Props.: m. 60° (294); m. 58.5°, d_{15} 2.33 (728). Polarographic redn. (1094).
Rxns. with: Aq. NaOH → PhSbO (294, 728, 912).
Cl_2 → $PhSbCl_4$ (728).
H_2O_2 → $PhSbO_3H_2$ (728).
$(RN_2)_2.ZnCl_4$ → $RPhSbCl_3$ (462, 910).
$o-C_6H_4(CH_2Br)_2$ + Na, followed by NHO_3 → bis(triphenylstibine) oxide dinitrate (757).
Complex compds.: $C_5H_5N.H(PhSbCl_3)$, m. 105° (dec.) (1001).
$PhN_2.(PhSbCl_3)$, grayish powder, m. 58-60°, prepd. by reacting PhN_2Cl with $PhSbCl_2$ in AcOH at 0-2°, or by reacting $SbCl_5$ and $FeCl_3$ with $PhNHNH_2.HCl$, followed by $CuCl_2$ in HCl, or by adding PhN_2Cl under the surface of a mixt. of $SbCl_5$, $PhNHNH_2.HCl$, and $CuCl_2$ at -5°. The complex is sol. in Me_2CO with evolution of N_2. On heating with dild. HCl yields Ph_2SbCl_3 (127, 134).
The following complex salts were prepared by treating PhSbO with HCl in EtOH and reacting the soln. with RN_2FeCl_4 in Me_2CO: 2-$EtOC_6H_4N_2.(PhSbCl_3)$, m. 82°; 2-$MeOC_6H_4N_2.(PhSbCl_3)$, m. 99-100°; $PhN_2.(PhSbCl_3)$, m. 71-72°; 4-$MeC_6H_4N_2.(PhSbCl_3)$; 4-$MeOC_6H_4N_2.(PhSbCl_3)$; 4-$Me_2NC_6H_4N_2.(PhSbCl_3)$; 4-$EtOC_6H_4N_2.(PhSbCl_3)$ (1070).

DIIODOPHENYLSTIBINE $C_6H_5I_2Sb$ $PhSbI_2$
Props.: m. 69° (1294).
Rxns. with: 5-Methyl-(2,2'-oxydiphenylene)bis(magnesium bromide) → 2-methyl-10-phenylphenoxastibine (172).
$(2-LiC_6H_4-)_2$ → 5-phenyldibenzostibiole (528).
Zn + HCl → $PhOI_{0.85}Sb_8$ but no $(Ph_4Sb)_2$ (672).
$RCOCl$ + $AlCl_3$ in CS_2 → PhCOR (792).
RCl + $AlCl_3$ in CS_2 → PhR (792).
$LiBH_4$ or $LiAlH_4$ in Et_2O → $PhSbH_2$ (1294).

DICHLORO(4-SULFAMYLPHENYL)STIBINE $C_6H_6Cl_2NO_2SSb$ 4-$H_2NO_2SC_6H_4SbCl_2$
Prepn.: by treating 4-$H_2NO_2SC_6H_4SbO$ with HCl in EtOH (1070).
Complex compds.: 4-$EtOC_6H_4N_2.(4-H_2NO_2SC_6H_4SbCl_3)$, m. 114° (1070).
4-$Me_2NC_6H_4N_2.(4-H_2NO_2SC_6H_4SbCl_3)$ (1070).
$PhN_2.(4-H_2NO_2SC_6H_4SbCl_3)$ (1070).

3-DICHLOROSTIBINOANILINE $C_6H_6Cl_2NSb$ $3-H_2NC_6H_4SbCl_2$

Rxns. with: $N:C(NH_2)N:C(NH_2)N:CCl \rightarrow 3-[\overline{N:C(NH_2)N:C(NH_2)N:CNH}]C_6H_4SbCl_2$ (430).
$(ClCN)_3$, followed by cold 10% $NH_4OH \rightarrow 3-(\overline{N:CClN:CClN:CNH})C_6H_4SbCl_2$ (435).

4-DICHLOROSTIBINOANILINE $C_6H_6Cl_2NSb$ $4-H_2NC_6H_4SbCl_2$

Rxns. with: $RSH \rightarrow 4-H_2NC_6H_4Sb(SR)_2$ (430, 435).
$\overline{N:C(NH_2)}N:C(NH_2)N:CCl \rightarrow 4-[\overline{N:C(NH_2)N:C(NH_2)N:CNH}]C_6H_4SbCl_2$ (430).

DICHLORO-p-TOLYLSTIBINE $C_7H_7Cl_2Sb$ $4-MeC_6H_4SbCl_2$

Prepn.:   By reducing $4-MeC_6H_4SbO_3H_2$ or $4-MeC_6H_4SbCl_4$ with SO_2 in HCl contg. HI (133 or with $SnCl_2$ in HCl (294).

By adding $4-MeC_6H_4N_2SbCl_4$ to a suspension of Zn dust or Fe in Me_2CO; 20% yield (397).

By treating $4-MeC_6H_4SbO$ with 5N HCl in AcOH; 58% yield (910).
Props.:   m. 92-93° (133); m. 91-92° (910); m. 94.5° (294).
Rxn. with: $RN_2SbCl_4 \rightarrow R(4-MeC_6H_4)SbCl_3 + R_2(4-MeC_6H_4)SbCl_2$ (173, 177, 910).
$PhC:CNa \rightarrow 4-MeC_6H_4Sb(C:CPh)_2$ (518).
RN_2OAc, followed by hydrolysis $\rightarrow R(4-MeC_6H_4)SbO_2H$ (910, 1063).
$4-MeC_6H_4N_2OAc(1:2) \rightarrow 4-MeC_6H_4Sb(OAc)_2 \cdot 4-MeC_6H_4N_2OAc$ (910).
$RN_2Cl \cdot MX_n$ (M = metal, X = halogen) $\rightarrow R(4-MeC_6H_4)SbCl_3$ (910).
$RN_2OAc \rightarrow 4-MeC_6H_4As(OAc)_2 \cdot RN_2OAc$ (1063).
Complex compds.: $4-MeC_6H_4N_2 \cdot (4-MeC_6H_4AsCl_3)$, decomp. 115°, prepd. by treating $4-MeC_6H_4N_2SbCl_4$ and $4-MeC_6H_4NHNH_2 \cdot HCl$ with $CuCl_2$ (133) or by adding $4-MeC_6H_4SbCl_2$ to $4-MeC_6H_4N_2Cl$ in AcOH at $0-2°$; m. 90-92° (dec). On boiling in dild. HCl yields $(4-MeC_6H_4)_2SbCl_3$ (128).

$(4-MeC_6H_4N_2Cl)_2 \cdot 4-MeC_6H_4SbCl_2$, m. 108-110°, prepd. by reacting $4-MeC_6H_4SbCl_2$ with excess of $4-MeC_6H_4N_2Cl$. On treating with HCl bitolyl and $4-MeC_6H_4SbCl_4$ are formed (128, 133).

The following complex salts were prepared by treating $4-MeC_6H_4SbO$ with HCl in EtOH, followed by aryldiazonium tetrachloroferrate in Me_2CO: $4-MeC_6H_4N_2 \cdot (4-MeC_6H_4SbCl_3)$; $4-MeOC_6H_4N_2 \cdot (4-MeC_6H_4SbCl_3)$; $4-Me_2NC_6H_4N_2 \cdot (4-MeC_6H_4SbCl_3)$ (1070).

2,4-DICHLORO-6-(3-DICHLOROSTIBINOANILINO)-s-TRIAZINE $C_9H_5N_4Sb$
     $3-(\overline{N:CClN:CClN:CNH})C_6H_4SbCl_2$

Prepn.:   by reacting $3-Cl_2SbC_6H_4NH_2 \cdot HCl$ with $(ClCN)_3$ in H_2O (430, 435).
Rxns. with: Aq. $NH_4OH \rightarrow$ the corresponding stiboso deriv. (430).
$HSCH_2CO_2K \rightarrow$ the corresponding $-Sb(SCH_2CO_2K)_2$ deriv. (430).

[3-(4,6-DIAMINO-s-TRIAZIN-2-YLAMINO)-4-CHLOROPHENYL]DICHLOROSTIBINE
     $C_9H_8Cl_3N_6Sb$ $4,3-Cl[\overline{N:C(NH_2)N:C(NH_2)N:CNH}]C_6H_3SbCl_2$
Prepn.:   by reacting $3,4-H_2N(Cl)C_6H_3SbCl_2$ with chlorocyanuric diamide in 1% HCl in an inert atm. (430, 435).

2-(4,6-DIAMINO-s-TRIAZIN-2-YLAMINO)-4-DICHLOROSTIBINOPHENOL $C_9H_9Cl_2N_6OSb$
     $4,3-HO[\overline{N:C(NH_2)N:C(NH_2)N:CNH}]C_6H_3SbCl_2$
Prepn.:   by refluxing 2-amino-4-dichlorostibinophenol with 2,4-diamino-6-chloro-s-triazine in 1% HCl in an inert atm. (430, 435).

By treating 3-(4,6-Cl_2-s-triazine-2-ylamino)4-hydroxybenzenestibonic acid with 25% aq. NH_3 at 110-130° and reducing the rxn. prod. with $SnCl_2$ in dil. HCl (431).

4-DICHLOROSTIBINOBENZOIC ACID ETHYL ESTER $C_9H_9Cl_2O_2Sb$ 4-$EtO_2CO.H_4SbCl_2$

Prepn.: by treating 4-$EtO_2CC_6H_4N_2SbCl_4$ with Cu_2Cl_2 in EtOH, hydrolyzing the rxn. mixt. to 4-$EtO_2CC_6H_4SbO_3H_2$, and reducing with $SnCl_2$ in HCl (172).

Props.: Flat elongated plates, m. 127°, sternutatory props. (172).

Rxns. with: KI in Me_2CO → 4-$EtO_2CC_6H_4SbI_2$ (172).

RN_2Cl → R(4-$EtO_2CC_6H_4$)$SbCl_3$ + R_2(4-$EtO_2CC_6H_4$)$SbCl_2$ (173, 175), and R_2SbCl_3 (178).

RN_2SbCl_4 → R(4-$EtO_2CC_6H_4$)$SbCl_3$; when R was 2-MeC_6H_4-, 2-ClC_6H_4-, and 4-BrC_6H_4-, (4-$EtO_2CC_6H_4$)$_2SbCl_3$ was obtd. (181).

5-Methyl(2,2′-oxydiphenylene)bis(magnesium bromide) → 10-(4-carbethoxyphenyl)-2-methylphenoxastibine (172).

DICHLORO-2-NAPHTHYLSTIBINE $C_{10}H_7Cl_2Sb$ 2-$C_{10}H_7SbCl_2$

Prepn.: by reducing 2-$C_{10}H_7SbCl_4.NH_4Cl$ with SO_2 in HCl contg. HI (133) or in EtOH contg. I, decompg. the rxn. prod. with H_2O, and treating the stiboso deriv. with concd. HCl (462).

Props.: m. 102° (462).

Rxns. with: H_2O → 2-$C_{10}H_7SbO_3H_2$ (133).

NH_4OH → 2-$C_{10}H_7SbO$ (133).

(2-$C_{10}H_7N_2Cl$)$_2.ZnCl_2$ → (2-$C_{10}H_7$)$_2SbCl_3$ (462).

Complex salt: (2-$C_{10}H_7N_2Cl$)$_2.2$-$C_{10}H_7SbCl_2$, prepd. from 2-$C_{10}H_7SbCl_2$ and excess of 2-$C_{10}H_7N_2Cl$. Decompn. in HCl under reflux yielded 2-$C_{10}H_7SbCl_4$ (131). Decompn. with NH_4Cl in HCl yielded 2-$C_{10}H_7SbCl_4.NH_4Cl$ (133).

TABLE LXV

ORGANODIHALOSTIBINES

RSbX₂	Prepd. from	Yield %	Props. and Remarks	Ref.
C₁				
CF₃SbBr₂	R₃SbBr₂ by disproportionation at 20°		b. 34°/2.5 mm., b. ~155°/760 mm.	258
ClCH₂SbCl₂	SbCl₃ + CH₂N₂		m. 36-38°, b. 86.5°/2 mm., d28 2.677	1320-1
MeSbCl₂			b. 96-99°/4 mm., rxns.[1-3] Rxn.[4]	1321
C₂				
ClCH:CHSbCl₂	R₂SbCl₃ + Sb		Colorless liq., b. 102-105°/1.5 mm., d20/4 2.1486	899
CH₂:CHSbBr₂	R₃Sb + SbBr₃	71		773-4
EtSbCl₂	SbCl₃ + Et₄Pb		b. 62-83°/1 mm., d 2.182	661-3
C₃				
SCH:CHN:CSbCl₂	RSbCl₄ + SnCl₂	80	HCl salt, m. 138°	844
	RSbCl₄ + SO₂ in MeOH + KI		Rxn. with aq. NaOH → RSbO	844
C₆				
4-BrC₆H₄SbCl₂	RSbO₃H₂ + SnCl₂	31	m. 90°, rxn.[5]	294
	RN₂.SbCl₄ + Zn or Fe in EtOAc	21	m. 198°	897
3-ClC₆H₄SbI₂	RSbO + HCl + KI		m. 67-68°	294
4-ClC₆H₄SbI₂	RSbO + AcOH + KI	51	m. 87-89°, rxn.[6]	910
4-FC₆H₄SbCl₂	RSbCl₄ + SO₂		m. 46-47°, rxn.[5]	132
4-IC₆H₄SbCl₂	RN₂SbCl₄ + Zn or Fe in EtOAc at 70-90° or in AmOH at 110-120°	23	m. 200°	897
4-O₂NC₆H₄SbCl₂	RSbO₃H₂ + SO₂	5.7	m. 95°, rxn.[5], complex compd.[7]	897
4-ClC₆H₄SbCl₂	RSbO₃H₂ + SnCl₂	42	m. 74°, rxn.[5]	294
	RN₂.SbCl₄ + Zn or Fe	28	m. 193°	294
2-ClC₆H₄SbCl₂	RN₂.SbCl₄ + Zn or Fe	10	m. 185°	897
3-ClC₆H₄SbCl₂	RN₂.SbCl₄ + Zn or Fe	25	Salts 8	897

C₇				
4-NCC₆H₄SbCl₂	RSbO₃H₂ + SnCl₂ in HCl		m. 74-75°, rxn.8	172
4-NCC₆H₄SbI₂	RSbCl₂ + KI		m. 98.5-99°, rxn.9	172
2-MeOC₆H₄SbCl₂	RN₂.SbCl₄ + Zn or Fe in EtOAc	20	m. 245°	897
	RSbO + HCl in EtOH		Complex compd. 10	1070
4-MeOC₆H₄SbCl₂	RN₂SbCl₄ + Zn or Fe	9.5	m. 108°, rxn.5	897
3-MeC₆H₄SbCl₂	RSbO₃H₂ + SnCl₂ in HCl	16.5	m. 51°	294
4-MeC₆H₄SbI₂			Rxn.11	910
C₈				
4-AcC₆H₄SbCl₂	RSbO₃H₂ + SnCl₂ in HCl	24	m. 141°, rxn.5	294
2-EtOC₆H₄SbCl₂ 12	RN₂SbCl₄ + Zn in EtOAc	40		897
4-EtC₆H₄SbCl₂	RSbO₃H₂ + SnCl₂ in HCl	6	m. 43°, rxn.5	294
4-Me₂NO₂SC₆H₄SbCl₂	RSbO₃H₂ + SO₂ in HCl-AcOH		m. 215°, rxn.13	1106
C₉				
3-[N̄:C(NH₂)N:C(NH₂)N:C̄NH.]C₆H₄ SbCl₂	3-H₂NC₆H₄SbO + N̄:C(NH₂)N:C(N-H₂)N:C̄Cl in HCl		Rxn.5	430, 435
4-[N̄:C(NH₂)N:C(NH₂)N:C̄NH]C₆H₄- SbCl₂	4-H₂NC₆H₄SbCl₂ + N̄:C(NH₂)N:C-(NH₂)N:C̄Cl		White ppt., rxn.5	431, 430
3,4-H₂N[N̄:C(NH₂)N:C(NH₂)N:C̄NH]C₆H₄SbCl₂	RSbO₃H₂ + SnCl₂ in HCl			431
C₁₀				
1-C₁₀H₇SbCl₂	RSbCl₄ + SO₂		Colorless prisms, m. 109° m. 105°, complex salt 14	462, 1001

1. Rxn. with H₂O → Sb oxides (1321).
2. Rxn. with alkali → dark ppt. (1321).
3. Rxn. with CH₂N₂ → (ClCH₂)₃Sb (1321).
4. Rxn. with NH₄F under reflux → disproportionation (1298).
5. Hydrolysis with dild. NaOH or NH₄OH → the corresponding stiboso deriv. (132, 294).
6. Rxn. with RN₂Cl.MCl_n (M = metal) → R₂(4-ClC₆H₄)SbCl₂ and R(4-ClC₆H₄)SbCl₃ (910).
7. Complex compds. pred. by treating 4-O₂NC₆H₄SbO (RSbO) with HCl in EtOH, followed by R'N₂Cl.FeCl₃ in Me₂CO: RSbCl₂.4-AcC₆H₄N₂Cl, m. 98-99°; RSbCl₂.PhN₂Cl, m. 74-75°; RSbCl₂.4-MeC₆H₄N₂Cl, m. 95°, RSbCl₂.4-MeOC₆H₄N₂Cl, m. 111-112°; RSbCl₂.4-Me₂NC₆H₄N₂Cl (1070).
8. Rxn. with KI → 4-NCC₆H₄SbI₂ (172).
9. Rxn. with 5-methyl(2,2'-oxydiphenylene)bis(magnesium bromide) → 10-p-cyanophenyl-2-methylphen- oxastibine (172).
10. PhN₂.(2-MeCC₆H₄SbCl₃), m. 77-78° (1070).

TABLE LXV (cont'd.)
ORGANODIHALOSTIBINES

RSbX$_2$	Prepd. from	Yield %	Props. and Remarks	Ref.
C$_{15}$ 2-(4-EtO$_2$CC$_6$H$_4$)C$_6$H$_4$SbCl$_2$	RN$_2$SbCl$_4$ + 4-MeC$_6$H$_4$SbCl$_2$		m. 155-156°	173
C$_{16}$ 2-(4-EtO$_2$CCH$_2$OC$_6$H$_4$)C$_6$H$_4$SbCl$_2$	RN$_2$SbCl$_4$ + 4-MeC$_6$H$_4$SbCl$_2$		m. 142-143°	177

11. Rxn. with RN$_2$Cl.MCl$_n$ (M = metal, n = valency of M) → R(4-MeC$_6$H$_4$)SbCl$_3$ (910).
12. The compound was indentified to be (2-EtOC$_6$H$_4$)$_2$SbCl (910).
13. Rxn. with sodium pyrogalloldisulfonate → complex salt (1106).
14. Pyrimidinium·C$_{10}$H$_7$SbCl$_3$ salt, m. ~90°, decomp. in air (1001).

STIBOSO COMPOUNDS

Organoantimony oxides, called stiboso compounds, occur in the form of amorphous powder . Their purification is very difficult, and therefore their molecular weights and structure have not been estabilished. Aromatic stiboso derivatives disproportionate in the solid state at moderately elevated temperatures and yield bis(diorganoantimony)oxides, (R₂Sb)₂0. The rates of disproportionation are influenced by the electronic character of substituents in the benzene ring.

The stiboso compounds, RSbO, are prepared by:
1. Hydrolyzing organodihalostibines, RSbX₂, dissolved in alcohol with aqueous alkali or ammonia.
2. Reducing stibonic acids, RSbO₃H₂, with sulfur dioxide in aqueous or alcoholic acid in the presence of potassium iodide and neutralizing the reduction product with aqueous ammonia.
3. Decomposing aryldiazonium tetrachloroantiminites, RN₂SbCl₄, with zinc dust or iron powder in ethyl acetate at 70-90°, or in pentyl alcohol at 110- 120°, or in acetone to form stiboso compounds, RSbO, along with aryldihalostibines, diarylstibonous acids, and tertiary stibines.
4. Reacting potassium aryldiazoniumcarboxylates, RN₂CO₂K, with antimony trichloride in ethyl acetate at 60-70° and treating the reaction product with aqueous ammonia.
5. Hydrolyzing organochlorohydroxystibines, RSb(OH)Cl, with alcoholic ammonia.
6. Reacting antimony trichloride with aryldiazonium tetrafluoroborates in acetone, followed by decomposition with zinc dust below 20°, to form stiboso compounds, as by-products, beside bis(diarylantimony) oxides.

STIBOSOBENZENE C₆H₅OSb PhSbO

repn.: By hydrolyzing PhSbCl₂ in aq. NH₃ or NaOH (294, 596, 728).
By adding PhN₂SbCl₄ to a suspension of Zn dust or Fe in EtOAc at ~70°; 46% yield along with PhAsCl₂ and Ph₂SbOH (897, 903); 42% along with (Ph₂Sb)₂O (1059).
By reacting PhN₂CO₂K with SbCl₃ in EtOAc at 60-70° and evapg. the soln. at r. t.; 7% yield along with PhN₂Cl.Sb₂O₃ (1060).
By-prod. in the rxn. of SbCl₃ with PhN₂BF₄ in Me₂CO, followed by Zn dust at 20°; 8% yield (886).
props.: Amorph. colorless solid, sinters at ~154°; probably polymeric structure; at 100° disproportionates to (Ph₂Sb)₂O and Sb₂O₃ (596, 597).
rxns. with: PhN₂Cl in AcOH at 0-2° → PhN₂(PhSbCl₃)(127).
aH₃PO₂ → Ph₁O₁.₈Sb₁.₂₅, but no (PhSb:)n (672).
Cl in EtOH, followed by RN₂FeCl₄ in Me₂CO → RN₂(PhSbCl₃) (1070).
g(OAc)₂ in AcOH → 4-AcHgC₆H₄SbO (552).

4-STIBOSOTOLUENE C₇H₇OSb 4-MeC₆H₄SbO

repn.: By adding alc. solution of 4-MeC₆H₄SbCl₂ to dild. aq. NaOH (294).
By reducing 4-MeC₆H₄SbO₃H₂ with SO₂ in MeOH-HCl contg. KI and treating the rod. with NH₄OH (552).
By adding 4-MeC₆H₄N₂SbCl₄ to a suspension of Zn dust or Fe in EtOAc at 70-90°; 1% yield. When Me₂CO is used instead of EtOAc, the title compd. is obtd. in 18% ield, along with 20% (4-MeC₆H₄)₂SbOH and 20% 4-MeC₆H₄SbCl₂ (897).

By reacting 4-MeC$_6$H$_4$N$_2$CO$_2$K with SbCl$_3$ in EtOAc at r.t., evapg. the soln. at r. t., and treating the residue with cold HCl, followed by 5% NH$_4$OH; 44% yield along with 11.5% 4-MeC$_6$H$_4$SbOCl$_2$ (1060).

By-prod. in the rxn. of SbCl$_3$ with 4-MeC$_6$H$_4$N$_2$BF$_4$ in Me$_2$CO, followed by treatment with Zn dust below 20°; 8% yield (886).

Props.: Disproportionation (597).

Rxns. with: 5 N HCl → 4-MeC$_6$H$_4$SbCl$_2$ (910).

HCl in EtOH, followed by RN$_2$FeCl$_4$ in Me$_2$CO → RN$_2$(4-MeC$_6$H$_4$SbCl$_3$) (1070).

Hg(OAc)$_2$ in AcOH → 4,2-Me(AcOHg)C$_6$H$_3$SbO (552).

2,4-DICHLORO-6-(3-STIBOSOANILINO)-s-TRIAZINE   C$_9$H$_5$Cl$_2$N$_4$OSb   3-[$\overline{\text{N}:\text{C}(\text{Cl})\text{N}:\text{C}(\text{Cl})\text{N}:\text{C}}$ NH]C$_6$H$_4$SbO

Prepn.: by treating 3-[$\overline{\text{N}:\text{C}(\text{Cl})\text{N}:\text{C}(\text{Cl})\text{N}:\text{CNH}}$]C$_6H_4$SbCl$_2$ with aq. NH$_3$ (430, 435).

Props.: White ppt. (430).

Rxn. with 10% aq. NH$_3$ at 45° → 2-amino-4-chloro-6-(3-stibosoanilino)-s-triazine (430,431).

2-AMINO-4-CHLORO-6-(3-STIBOSOANILINO)-s-TRIAZINE   C$_9$H$_7$ClN$_5$OSb   3-[$\overline{\text{N}:\text{C}(\text{Cl})\text{N}:\text{C}(\text{NH}_2)\text{N}:\text{CNH}}$]C$_6H_4$SbO

Prepn.: by treating 2,4-dichloro-6-(3-stibosoaniline)-s-triazine with 10% NH$_3$ at 45° (435).

Rxn. with aq. NH$_3$ at elevated temp. under pressure → the 2,4-diamino-s-triazine deriv. (435).

2,4-DIAMINO-6-(3-STIBOSOANILINO)-s-TRIAZINE   C$_9$H$_9$OSb   3-[$\overline{\text{N}:\text{C}(\text{NH}_2)\text{N}:\text{C}(\text{NH}_2)\text{N}:\text{CNH}}$ C$_6$H$_4$SbO

Prepn.: by treating 2-amino-4-chloro-6-(3-stibosoanilino)-s-triazine with aq. NH$_3$ at elevated temp. under pressure (435).

Rxn. with: RSH → the -Sb(SR)$_2$ derivs.(430).

TABLE LXVI
STIBOSO COMPOUNDS

RSbO	Prepd. from	Yield %	Props. and Remarks	Ref.
C_3				
SCH:CHN:CSbO	$RSbCl_2$ + aq. NaOH	95	m. 250°	844
C_6				
4-BrC_6H_4SbO	$RSbCl_2$[1] + aq. NaOH	56	m. 142–143°	294
	RN_2SbCl_4 + Zn or Fe in EtOAc		m. 131–132°	897
	RN_2CO_2K + $SbCl_3$ in EtOAc followed by NH_4OH		Disproportionation	1060
2-ClC_6H_4SbO	RN_2SbCl_4 + Zn or Fe	40	m. ~114°, rxn.[2]	597 / 897
3-ClC_6H_4SbO	$RSbCl_2$[1] + aq. NaOH		Floculent ppt., rxn. 3	294
	$RSbO_3H_2$ + SO_2		Disproportionation	1001
4-ClC_6H_4SbO	$RSbCl_2$[1] + aq. NaOH	35	m. 115–120°, rxn.[4]	597 / 294
	RN_2SbCl_4 + Zn or Fe		Disproportionation	897
4-FC_6H_4SbO	$RSbCl_2$ + aq. NH_4OH		m. >280°, rxn.[5]	132 / 897
4-IC_6H_4SbO	RN_2SbCl_4 + Zn or Fe in EtOAc in AmOH	40	m. ~140–145°	897
4-$O_2NC_6H_4SbO$	$RSbCl_2$[1] + aq. NaOH	38	Rxn.[6]	294 / 897
	RN_2SbCl_4 + Zn in EtOAc	58		886
	RN_2BF_4 + $SbCl_3$ in Me_2CO, followed by Zn			
4-$H_2NO_2SC_6H_4SbO$	$RSbO_3H_2$ + SO_2		Disproportionation Light yellow ppt., complex[7]	597 / 1106
C_7				
3-MeC_6H_4SbO	$RSbCl_2$[1] + aq. NaOH	24.5	m. 101°, rxn.[8]	294
2-$MeOC_6H_4SbO$	RN_2SbCl_4 + Zn or Fe		Decomp. 103°	897
4-$MeOC_6H_4SbO$	RN_2SbCl_4 + Zn or Fe	17		897

1. Alcoholic solution used for hydrolysis with aqueous alkali.
2. Rxn. with conc. HCl → (2-$ClC_6H_4)_2SbCl$ (910).
3. Rxn. with H_2S → yellow ppt. (1001).
4. Rxn. with KI in HOAc → 4-$ClC_6H_4SbI_2$ (910).
5. Rxn. with AcOH → acetate, m. 120–121° (132).
6. Rxn. with HCl in EtOH, followed by RN_2FeCl_4 in Me_2CO → $RN_2.$(4-$O_2NC_6H_4SbCl_3$) (1070).
7. Complex salt with pyrocatecholdisulfonic acid, faintly colored powder (1106).

TABLE LXVI (cont'd.)
STIBOSO COMPOUNDS

RSbO	Prepd. from	Yield %	Props. and Remarks	Ref.
C8				
4,2-Br(AcOHg)C6H3SbO	4-BrC6H4SbO + Hg(OAc)2		Very stable compd., rxn.9	552
4-AcOC6H4SbO	PhSbO + Hg(OAc)2		Very stable compd., rxn.9	552
4-AcC6H4SbO	RSbCl2 1 + aq. NaOH			294
4-EtC6H4SbO	RSbCl2 1 + aq. NaOH		Disproportionation	597 294
C9				
4,2-Me(AcOHg)C6H3SbO	4-MeC6H4SbO + Hg(OAc)2		Very stable, decomp. >300°	552
C10				
2-C10H7SbO	RSbCl2 + NH4OH	88	Decomp. 160°	133
	RN2CO2K + SbCl3	90	Disproportionation	1060
			m.~135-140°, rxn.10	1001
2-PhC6H4SbO	RSb(OH)Cl + alc. NH3		Tiny needles, m. 195-196°	1318
C12				
4,4'-SO2(C6H4SbO)2	4,4'-SO2(C6H4SbO3H2)2 + SO2 in MeOH-HNO3		m. 240°	220a

8. Rxn. with HCl in EtOH, followed by RN2FeCl4 in Me2CO → RN2(2-MeOC6H4SbCl3) (1070).
9. Rxn. with HCl → hydrolysis of the AcOHg-group and formation of ClHg- deriv. (552).
10. Rxn. with pyridine in conc. HCl → C5H5N.H(2-C10H7SbCl3), crystals (1001).

470

Diarylstibonous acids are formed by reducing arenestibonic acids, $RSbO_3H_2$, with sulfur dioxide in aqueous ethanol containing hydrochloric acid, precipitating the reduction product with water, dissolving the precipitate in aqueous sodium carbonate solution, and acidifying the solution with acetic acid. The acids are also formed as by-products during decomposition of aryldiazonium chlorcantimonites, RN_2SbCl_4, with zinc dust or iron powder in organic solvents, from which they are recovered as anhydrides with acetic acid.

BIS(2,5-DIMETHYL-4-SULFAMYLPHENYL)STIBINOUS ACID (?) $C_{16}H_{21}N_2O_5S_2Sb$ [2,5,4-Me$_2$(H$_2$NO$_2$S)C$_6$H$_2$]$_2$SbOH

Prepn.: by reducing 2,4,5-Me$_2$(H$_2$NO$_2$S)C$_6$H$_2$SbO$_3$H$_2$ with SO$_2$ in H$_2$O-EtOH-HCl, pptg. the prod. with H$_2$O, dissolving the solid in Na$_2$CO$_3$ soln. and acidifying with AcOH (1106).

DI(4-BROMOPHENYL)STIBINOUS ACID ANHYDRIDE WITH ACETIC ACID $C_{14}H_{11}Br_2O_2Sb$ (4-BrC$_6$H$_4$)$_2$SbOAc

Prepn.: by-prod. from décompn. of 4-BrC$_6$H$_4$N$_2$SbCl$_4$ with Zn dust or Fe in EtOAc at 70-90°, followed by a treatment with aq. NH$_3$ and AcOH; 15% yield (897).

Props.: m. 131-132° (897).

DI(4-CHLOROPHENYL)STIBINOUS ACID ANHYDRIDE WITH ACETIC ACID $C_{14}H_{11}Cl_2O_2Sb$ (4-ClC$_6$H$_4$)$_2$SbOAc

Prepn.: by-prod. from decompn. of 4-ClC$_6$H$_4$N$_2$SbCl$_4$ with Zn dust or Fe in EtOAc, followed by a treatment with aq. NH$_3$ and AcOH; 33% yield (897).

Props.: m. 126° (897).

DI(4-IODOPHENYL)STIBINOUS ACID ANHYDRIDE WITH ACETIC ACID $C_{14}H_{11}I_2O_2Sb$ (4-IC$_6$H$_4$)$_2$SbOAc

Prepn.: by-prod. from decompn. of 4-IC$_6$H$_4$N$_2$SbCl$_4$ with Zn dust or Fe in EtOAc at 70-90°, followed by a treatment with aq. NH$_3$ and AcOH; 9% yield along with -IC$_6$H$_4$SbO. When n-C$_5$H$_{11}$OH is used as solvent and the decompn. is carried out at 110-120°, 20% yield was obtd. (897).

Props.: m. 140° (897).

DIPHENYLSTIBINOUS ACID ANHYDRIDE WITH ACETIC ACID $C_{14}H_{13}O_2Sb$ Ph$_2$SbOAc

Prepn.: by-prod. in the rxn. of PhN$_2$.SbCl$_4$ with Zn dust or Fe in EtOAc at 70-90°, followed by decompn. with aq. NH$_3$ and conversion to the acetate; 16% yield (897, see also 903, 1059).

Props.: m. 131-132° (1059).

Rxns. with: PhN$_2$OAc → Ph$_3$As(OAc)$_2$ (1063).

DI-p-TOLYLSTIBINOUS ACID ANHYDRIDE WITH ACETIC ACID $C_{16}H_{17}O_2Sb$ (4-MeC$_6$H$_4$)$_2$SbOAc

Prepn.: by-prod. from decompn. of 4-MeC$_6$H$_4$N$_2$SbCl$_4$ in EtOAc, followed by a treatment with aq. NH$_3$ and AcOH; 0.9-4.5% yield. When Me$_2$CO was used as the solvent for the decompn., 20% yield was obtd. (897).

Rxns. with: RN$_2$OAc → R(4-MeC$_6$H$_4$)$_2$Sb(OAc)$_2$ (1063).

DI-p-ANISYLSTIBINOUS ACID ANHYDRIDE WITH ACETIC ACID $C_{16}H_{17}O_4Sb$ $(4-MeOC_6H_4)_2SbOA$
Prepn.: by-prod. from decompn. of $4-MeOC_6H_4N_2SbCl_4$ in EtOAc at 70-90°, followed by a treatment with aq. NH_3 and AcOH; 8.8% yield (897).

STIBONOUS ACID ESTERS WITH MERCAPTANS

Stibonous acid esters with mercaptans are prepared by reacting organodichlorostibines $RSbCl_2$, or stiboso compounds, RSbO, with mercaptans in aqueous alkali or in alcohol, or by reducing stibonic acid sodium salts, $RSbO(ONa)_2$ with ammonium mercaptoacetate or with mercaptans and esterifying the product with the same or with different mercaptans. Dimercaptans and arenedithiols yield cyclic esters, 1,3,2-dioxastibolane and 1,3,2-dioxastibinane derivatives and analogs, which are compiled on page 529. Functionally substituted benzenestibonous acids thioesters, such as aminobenzenestibonous acid thioesters, may be further converted into N-substituted derivatives. Thus reactions with chloro-s-triazine derivatives yield the corresponding s-triazinylaminobenzenestibonous acid thioesters.

TABLE LXVII
STIBONOUS ACID ESTERS WITH MERCAPTANS

$RSb(SR')_2$	Prepd. from	Ref.
$4-H_2NC_6H_4Sb(SCH_2CO_2H)_2$	$RSbCl_2 + HSCH_2CO_2H$	435
$3-[\overline{N:C(NH_2)N:C(NH_2)N:CNH}]C_6H_4Sb-$ $(SCH_2CO_2H)_2$	$RSbCl_2 + HSCH_2CO_2K$ in H_2O	430
$4-[\overline{N:C(NH_2)N:C(NH_2)N:CNH}]C_6H_4Sb-$ $(SCH_2CO_2H)_2$	$RSbO_3H_2 + HSCH_2CO_2H$ in aq. K_2CO_3	431
	$RSbCl_2 + HSCH_2CO_2H$ in MeOH	430
	$4-H_2NC_6H_4Sb(SCH_2CO_2H)_2 + \overline{N:C(NH_2)-}$ $\overline{N:C(NH_2)N:CCl}$	430, 435
$4-H_2NC_6H_4Sb(SC_6H_4CO_2H-o)_2$	$RSbCl_2 + 2-HSC_6H_4CO_2H$	430
$4-[\overline{N:C(NH_2)N:C(NH_2)N:CNH}]C_6H_4Sb-$ $(SC_6H_4SO_3H-p)_2$	$RSbO + 4-HSC_6H_4SO_3H$	430
$4-[\overline{N:C(NH_2)N:C(NH_2)N:CNH}]C_6H_4Sb-$ $(SC_6H_4CO_2H-o)_2$	$4-H_2NC_6H_4Sb(SC_6H_4CO_2H-o)_2 + \overline{N:C-}$ $\overline{(NH_2)N:C(NH_2)N:CCl}$	430, 435

BIS(DIORGANOANTIMONY) OXIDES AND SULFIDES

Bis(diarylantimony) oxides are formed by hydrolysis of diarylhalostibines, R_2SbX, and by disproportionation of stibosoaromatic derivatives at 100°. They are also obtained as by-products in the preparation of stiboso compounds and tertiary stibines by arylation of antimony trichloride with aryldiazonium fluoroborates or by decomposition of aryldiazonium chloroantimonites with zinc dust.

BIS(DI-4-CHLOROPHENYLANTIMONY) OXIDE $C_{24}H_{16}Cl_4OSb_2$ $[(4-ClC_6H_4)_2Sb]_2O$
Prepn.: by disproportionation of $4-ClC_6H_4SbO$ at 100° (597).
Props.: m. 147° (597).

BIS[DI(p-FLUOROPHENYL)ANTIMONY] OXIDE $C_{24}H_{16}F_4OSb_2$ $[(4-FC_6H_4)_2Sb]_2O$
Prepn.: by treating $(FC_6H_4)_2SbCl$ with 5% NH_4OH (132).
Props.: m. 85° (132).

BIS(DI-4-NITORPHENYLANTIMONY) OXIDE $C_{24}H_{16}N_4O_9Sb$ $[(4-O_2NC_6H_4)_2Sb]_2O$
Prepn.: by-prod. in the prepn. of $(4-O_2NC_6H_4)_3Sb$ by reacting $SbCl_3$ with $4-O_2NC_6$-$H_4N_2BF_4$ (1:3) in Me_2CO, followed by decompn. with Zn dust below 20° (886).

BIS(DIPHENYLANTIMONY) OXIDE $C_{24}H_{20}OSb_2$ $(Ph_2Sb)_2O$
Prepn.: by disproportionation of PhSbO at 100° (596, 597).
      By reacting $SbCl_3$ with PhN_2BF_4 (1:3) in Ac_2O, decomposing the rxn. prod. with Zn dust below 20°, and extracting the oxide with ether; 30% yield along with Ph_3Sb and PhSbO (886).
      By-prod. from the prepn. of PhSbO by decompn. of PhN_2SbCl_4 with Zn dust in EtOAc at 60° (1059).
Props.: m. 82° (597).

BIS(DI-p-TOLYLANTIMONY) OXIDE $C_{28}H_{28}OSb_2$ $[(4-MeC_6H_4)_2Sb]_2O$
Prepn.: By disproportionation of $4-MeC_6H_4SbO$ at 100° (597).
      By-prod. in the prepn. of $(4-MeC_6H_4)_3Sb$ by reacting $SbCl_3$ with $4-MeC_6H_4N_2BF_4$ (1:3) in Me_2CO and decompg. the prod. with Zn dust below 20°; 4% yield (886).
Props.: m. 107° (597).

BIS(DIPHENYLANTIMONY) SULFIDE $C_{24}H_{20}SSb_2$ $(Ph_2Sb)_2S$
Use: Stabilizer for synthetic rubber (10).

DISTIBINES

Tetraorganodistibines are prepared by reacting diorganohalostibines with metals, such as mercury, zinc, or sodium; by decomposing secondary stibines with mercury, sodium, dibonane, or trimethylamine-borine; and by reacting antimony trichloride with cyclopentadienylsodium in tetrahydrofuran under reflux to form tetracyclo-pentadienyldistibine.

TETRAKIS(TRIFLUOROMETHYL)DISTIBINE $C_4F_{12}Sb_2$ $[(CF_3)_2Sb]_2$
Prepn.: by reacting $(CF_3)_2SbI$ with excess Hg or Zn; theoret yield (258).
Props.: Pale yellow liq., b. 16.8°/4.1 mm., b. ~136°/760 mm.; freezes to a color-less solid (258).
Rxns. with: $Cl_2 \rightarrow (CF_3)_2SbCl_3$ (258).
$Br_2 \rightarrow CF_3Br$ and $SbBr_3$ (258).
$I_2 \rightarrow CF_3I$ and SbI_3 (258).

TETRAMETHYLDISTIBINE $C_4H_{12}Sb_2$ $(Me_2Sb)_2$
Prepn.: By decompn. of Me_2SbH at r. t., expecially in the presence of Hg (139a).
      By treating Me_2SbH with Me_3NBH_3 (139a).
      By treating Me_2SbH with Na in liq NH_3 (139a).
      By reacting Me_2SbH with B_2H_6 at -78 to-10° (139a).
      By reacting Me_2SbBr with Na in liq. NH_3; 75% yield (139a).

Props.: Yellow oil; b. (calc.) 224°; m. 17°; stable at 100°, decomp. at 160° (139a).
Rxns. with: B$_2$H$_6$ above 60° → Me$_2$SbBH$_2$ (139a).
B$_2$H$_5$Br → Me$_2$SbBr + B$_2$H$_6$ (139a).
HCl at r. t. → Me$_2$SbCl + H$_2$ (139a).
BCl$_3$ → Me$_2$SbCl (139a).

TETRACYCLOPENTADIENYLDISTIBINE C$_{20}$H$_{20}$Sb$_2$ [(C$_5$H$_5$)$_2$Sb]$_2$
Prepn.: by reacting C$_5$H$_5$Na with SbCl$_3$ in C$_4$H$_8$O under reflux; 83% yield (383).
Props.: Purple-red crystals (383).

MISCELLANEOUS DERIVATIVES OF TRIVALENT ANTIMONY

DIMETHYLSTIBINOBORINE C$_2$H$_8$BSb Me$_2$SbBH$_2$
Prepn.: By reacting Me$_2$SbBr with NaBH$_4$ in (CH$_2$OMe)$_2$ below -40° (139a).
   By heating rapidly a condensed mixt. of Me$_2$SbH with B$_2$H$_6$ from -196° to r. t.
(139a).
   By reacting (Me$_2$Sb)$_2$ with B$_2$H$_6$ above 60° (139a).
Props.: b. (calcd) 70°; relatively stable at 210° (139a).
Rxn. with Me$_3$N at 80° → Me$_3$NBH$_3$ and Me$_3$Sb (139a).

ANTIMONOBENZENE (C$_6$H$_5$Sb)x (PhSb)x
Prepn.: By reacting PhSbH$_2$ with PhSbI$_2$ (1294).
   By reducing PhSbO$_3$H$_2$ with Na-Hg + HCl or by H$_3$PO$_2$ (672).
   By reducing PhSbI$_2$ with Zn in alc. HCl; the prod. was not exactly indentified
(672).
Props.: Black-brown solid, insol. in Et$_2$O, readily oxidizable to (PhSbO)x (1294).

DIACETOXY-p-TOLYLSTIBINE C$_{11}$H$_{13}$O$_4$Sb 4-MeC$_6$H$_4$Sb(OAc)$_2$
Prepn.: by reacting 4-MeC$_6$H$_4$SbCl$_2$ with 4-MeC$_6$H$_4$N$_2$OAc (1:2) in Me$_2$CO at an elevated
temp.; 60% yield isolated in the form of 4-MeC$_6$H$_4$Sb(OAc)$_2$·4-MeC$_6$H$_4$N$_2$OAc (910).
Complex salt 4-MeC$_6$H$_4$Sb(OAc)$_2$·4-MeC$_6$H$_4$N$_2$OAc, m. 122-123° (910).

o-BIPHENYLYLCHLOROHYDROXYSTIBINE C$_{12}$H$_{10}$ClOSb 2-PhC$_6$H$_4$Sb(OH)Cl
Prepn.: by heating (2-PhC$_6$H$_4$)$_3$Sb with SbCl$_3$ in xylene in sealed tube at 220-250°
(1318).
Props.: m. 201-202° (1318).
Rxns. with: EtOH-NH$_3$ → 2-PhC$_6$H$_4$SbO (1318).
Cl$_2$ followed by H$_2$O → 2-PhC$_6$H$_4$SbO$_3$H$_2$ (1318).

2. COMPPOUND OF PENTAVALENT ANTIMONY

STIBONIC ACIDS

Arene- and heterocycle-stibonic acids are prepared similarly to arsonic acids as
follows:

1. According to Bart method, aryldiazonium chlorides are reacted with alkali metal antimonites or metaantimonites in water, containing glycerol, in the presence of copper or copper salts. Alkali metal antimonites can be prepared in situ by dissolving antimony trioxide in aqueous alkali solution (Method I).
2. According to Bart method modified by Schmidt (Bart-Schmidt method), aryldiazonium salts are coupled with alkali metal antimonites in a neutral solution in the absence of catalysts (Method II).
3. By the Bart method modified by Scheller (Bart-Scheller method), diazotization of aromatic amines and coupling of diazonium salts with antimony compounds are carried out in one step. Aromatic amines dissolved in alcohol containing sulfuric or hydrochloric acid are diazotized in the presence of antimony trichloride and catalytic amounts of copper(I) bromide. After nitrogen evolution is completed, the reaction product is hydrolyzed directly or after purification (Method III).
4. According to Doak modification of the Bart-Scheller method, the catalyst is added to the reaction mixture after completed diazotization of anitmony trichloride in alcohol (Method IV).
5. Aryldiazonium fluoroborates react with sodium antimonite in aqueous, alcoholic, or acetone solution in the presence of copper, or with antimony trichloride in alcoholic or acetone solution in the presence of copper(I) chloride to yield the corresponding stibonic acids (Method V).
6. Diazotization of aromatic amines in hydrochloric acid in the presence of antimony trichloride as well as reactions of performed aryldiazonium chlorides with antimony trichloride in hydrochloric acid yield aryldiazonium tetrachloroantimonites, RN_2SbCl_4, which are decomposed by aqueous alkali in the presence of copper bronze to yield stibonic acids. The decomposition of aryldiazonium tetrachloroantimonites can also be carried out in acetone in the presence of copper(I) chloride or in ethanol saturated with hydrogen chloride in the presence of copper bronze to yield arylantimony tetrachlorides, which are then hydrolyzed by alkali (Method VI).
7. Preformed arylantimony tetrahalides are hydrolyzed by aqueous alkali to stibonic acids (Method VII).
8. Reactions of arylhydrazines with antimony trichloride in diluted hydrochloric acid in the presence of copper(II) chloride yield mixtures of $RSbCl_4$ with R_2SbCl_3. The trichlorides are separated by filtration and the tetrachlorides are isolated as the ammonium complex salts $NH_4(RSbCl_5)$ and hydrolyzed by water (Method VIII).
9. Reactions of aryldiazonium chloroferrates, RN_2FeCl_4, with antimony trichloride in hydrochloric acid yield aryldiazonium chloroantimonates, RN_2SbCl_6, which are decomposed with excess copper(I) chloride in acetone at 25° and hydrolyzed with alkali to stibonic acids (Method IX).
10. Arylation of antimony trichloride by potassium aryldiazocarboxylates, RN_2-CO_2K, in ethyl acetate, followed by hydrolysis with aqueous ammonia, yield stibonic acids along with stiboso derivatives (Method X).
11. Aryldichlorostibines, $RSbCl_2$, are oxidized by hydrogen peroxide to the corresponding stibonic acids (Method XI).
12. Benzenestibonic acid is formed by treating a mixture of antimony pentachloride and dichlorodiphenylsilane, Ph_2SiCl_2, with cold concentrated hydrochloric acid. (Method XII).

Stibonic acids are purified by dissolving in hydrochloric acid, precipitating the arylantimony tetrachloride thus formed with pyridine hydrochloride or ammonium chloride as a complex salt $C_5H_5N.H(RSbCl_5)$ or $NH_4(RSbCl_5)$, and hydrolyzing the salt with aqueous alkali.

4-BROMOBENZENESTIBONIC ACID $C_6H_6BrO_3Sb$ 4-$BrC_6H_4SbO_3H_2$
Prepn.: By reacting 4-$BrC_6H_4N_2Cl$ with $SbCl_3$ in HCl at 5°, and decompg. the double salt with 2 N NaOH (324).
By decompg. 4-$BrC_6H_4N_2SbCl_6$ (prepd. from 4-$BrC_6H_4N_2FeCl_4$ + $SbCl_5$) with excess Cu_2Cl_2 in Me_2CO at 25°, followed by alk. hydrolysis, 1% yield (906).
By reacting 4-$BrC_6H_4N_2CO_2K$ with $SbCl_3$ in EtOAc and treating the soln. with NH_4OH; 53.8% yield along with 4-BrC_6H_4SbO (1060).
By diazotization of 4-$BrC_6H_4NH_2$ in ethanol contg. H_2SO_4 in the presence of $SbCl_3$, followed by decompn. with Cu_2Br_2 (Bart-Scheller-Doak method); 40% yield (293).
Props.: Pale buff powder (324). Mol. wt. detn. (292).
Rxns. with: NaOH → 4-$HOC_6H_4SbO_3H_2$ (324).
HNO_3 in H_2SO_4, followed by $Na_2S_2O_4$ → the 3-amino deriv. (324).
Complex compd.: $C_5H_5N.H$(4-$BrC_6H_4SbCl_5$), m. 154° (293).

3-CHLOROBENZENESTIBONIC ACID $C_6H_6ClO_3Sb$ 3-$ClC_6H_4SbO_3H_2$
Prepn.: by adding Sb_2O_3 dissolved in fuming HCl to 3-$ClC_6H_4N_2Cl$, decompg. 3-$ClC_6H_4N_2SbCl_4$ thus formed with excess 10% NaOH, and purifying by conversion to NH_4(3-$ClC_6H_4SbCl_5$) and hydrolysis (1001).
Props.: Colorless solid (1001).
Rxns. with: SO_2 in $EtOH-H_2O-HCl$ → 3-ClC_6H_4SbO (1001).
Pyridine in HCl-AcOH mixt. → $C_5H_5N.H$(3-$ClC_6H_4SbCl_3$) (1001).
NH_4Cl in concd. HCl → NH_4(3-$ClC_6H_4SbCl_5$) (1001).

4-CHLOROBENZENESTIBONIC ACID $C_6H_5ClO_3Sb$ 4-$ClC_6H_4SbO(OH)_2$
Prepn.: By decompn. of 4-$ClC_6H_4N_2Cl.SbCl_5$ (from 4-$ClC_6H_4N_2Cl.FeCl_3$ + $SbCl_5$) with excess Cu_2Cl_2 in Me_2CO at 25°, followed by alk. hydrolysis; 5% yield (906).
By diazotization of 4-$ClC_6H_4NH_2$ and coupling with $SbCl_3$ by the Bart-Scheller method; 78% yield (293).
Props.: Mol. wt. detn. (292).
Rxns. with: HCl in EtOH, followed by RN_2FeCl_4 in Me_2CO → RN_2(4-$ClC_6H_4SbCl_5$) (1067).
Complex compd.: $C_5H_5N.H$(4-$ClC_6H_4SbCl_5$), m. 160° (293).
Deriv.: Gluconic acid 4-chlorobenzenestibonate Na salt (1166).

4-NITROBENZENESTIBONIC ACID $C_6H_6NO_5Sb$ 4-$O_2NC_6H_4SbO_3H_2$
Prepn.: By coupling 4-$O_2NC_6H_4N_2Cl$ with Na_3SbO_3 in alk. solution in the presence of a copper catalyst; 35% yield (1222).
By reacting 4-$O_2NC_6H_4N_2BF_4$ with $SbCl_3$ in MeOH in the presence of Cu_2Cl_2; 4.1% yield (1222).
By reacting 4-$O_2NC_6H_4N_2Cl$ with $SbCl_3$ in Me_2CO in the presence of Cu_2Cl_2; 55% yield (1222).
By decompg. 4-$O_2NC_6H_4N_2SbCl_4$ with Cu_2Cl_2 in Me_2CO or with 5 N NaOH (1221).
By reacting 4-$O_2NC_6H_4N_2CO_2K$ with $SbCl_3$ in EtOAc and treating the prod. with HCl

followed by NH$_4$OH; 50.7% yield (1060).

By decompg. 4-O$_2$NC$_6$H$_4$N$_2$SbCl$_6$ (prepd. from 4-O$_2$NC$_6$H$_4$N$_2$FeCl$_4$ dissolved in Me$_2$CO + Sb-Cl$_5$ in CHCl$_3$) with excess Cu$_2$Cl$_2$ in Me$_2$CO at 25° and hydrolyzing the rxn. prod.; 12% yield (906).

By diazotization of 4-O$_2$NC$_6$H$_4$NH$_2$ in alc. contg. H$_2$SO$_4$ in the presence of SbCl$_3$ and treating the diazonium compd. with Cu$_2$Br$_2$; 53% yield (293).

Props.: Mol. wt. detn. (292).
Rxns. with: H$_2$/Raney Ni in alk. soln. → the 4-amino deriv. (291).
HCl → H[4-O$_2$NC$_6$H$_4$SbCl$_5$], which gives a color test with Rhodamine B (714).
HCl + C$_5$H$_5$N → C$_5$H$_5$N.H(4-O$_2$NC$_6$H$_4$SbCl$_5$) (906); m. 168.5° (293).
HCl in EtOH, followed by RN$_2$FeCl$_4$ in Me$_2$CO → RN$_2$(4-O$_2$NC$_6$H$_4$SbCl$_5$) (1067).
Deriv.: Gluconic acid 4-nitrobenzenestibonate Na salt (1166).

BENZENESTIBONIC ACID C$_6$H$_7$O$_3$Sb PhSbO$_3$H$_2$
Prepn.: By diazotizing PhNH$_2$ in the presence of SbCl$_3$ in EtOH contg. H$_2$SO$_4$ and decompg. the rxn. mixt. with Cu$_2$Br$_2$ (Bart-Scheller-Doak method); 39% yield (293).

By oxidation of PhSbCl$_2$ with H$_2$O$_2$ (728).

From PhNH$_2$ by the Bart method, 35-40% yield (884); 48% yield (1222).

By decompg. PhN$_2$SbCl$_4$ with 5 N NaOH (552) or with Cu-bronze in alcohol (949, 1219).

By decompg. PhN$_2$SbCl$_6$ (prepd. from PhN$_2$FeCl$_4$ + SbCl$_5$ in Me$_2$CO-CHCl$_3$) with excess Cu$_2$Cl$_2$ in Me$_2$CO at 25° and hydrolyzing the rxn. mixt.; 37% yield (906).

By adding PhNHNH$_2$ to a soln. of SbCl$_3$ and CuCl$_2$ in dild. HCl and stirring the mixt in contact with air; 20-22% yield along with Ph$_2$SbO$_2$H (1119).

By reacting PhN$_2$BF$_4$ with Na$_3$SbO$_3$ or with SbCl$_3$ in the presence of Cu in Me$_2$CO or in MeOH; up to 25% yield (1222).

By reacting SbCl$_3$ with PhN$_2$Cl in Me$_2$CO in the presence of Cu$_2$Cl$_2$; 60% yield (1222).

By treating a mixt. of Ph$_2$SiCl$_2$ and SbCl$_5$ with cold conc. HCl; 55% yield along with polysiloxane (1322).

Purification: by dissolving in hot conc. HCl, adding C$_5$H$_5$N.HCl, and decompg. the adduct C$_5$H$_5$N.H(PhSbCl$_5$) with Na$_2$CO$_3$ (293).

Props.: Monobasic acid for which the H[PhSb(OH)$_5$] formula was proposed in which the two H$_2$O molecules are considered to be water of constitution, not of crystallization (569). On drying under a reduced pressure at 20° and 100°, the acid gradually loses 2 1/2 mols. of H$_2$O and forms a monobasic dimeric acid (881). Mol. wt. measurements indicate a polymeric structure in the solid state, assocd. by the H-bond formation (292).

Rxns. with: Na-Hg + HCl → PhO$_{0.4}$Sb$_{1.25}$, but no (PhSb:)$_2$ (672).
H$_3$PO$_2$ → Ph$_{0.4}$Sb$_{1.25}$, but no (PhSb:)$_2$ (672).
AcOH → the dimer monoacetate (881).
Tartaric acid → phenylstibonyltartaric acid (882).
HCl + C$_5$H$_5$N.HCl → PhSbCl$_4$.C$_5$H$_5$.HCl (906).
HCl in EtOH, followed by RN$_2$Cl.FeCl$_3$ in Me$_2$CO → PhSbCl$_4$.RN$_2$Cl (1067).
SO$_2$ in HCl contg. KI → PhSbCl$_2$ (1119).
Derivs.: Gluconic acid phenylstibonate Na salt, prepd. from C$_5$H$_5$N.H(PhSbCl$_5$) and Na gluconate in the presence of Me$_2$NCH$_2$CH$_2$OH (1166).
Monoanhydride with AcOH (881).
Tartaric acid phenylstibonate Na salt (882).
Tl salt, shiny yellow ppt., m. 270° (949).

p-AMINOBENZENESTIBONIC ACID, STIBANILIC ACID, $C_6H_8NO_3Sb$ $p-H_2NC_6H_4SbO(OH)_2$
Prepn.: by catalytic redn. of $p-O_2NC_6H_4SbO_3H_2$ in alk. soln. over Raney Ni; 62%
yield (291).
Salts: with glucamine and N-methylglucamine (280, 492). Na salt (495). Diethyl-
amine salt "Neostibosan," analysis (568).
Rxns. with: Urea → $H_2NCONHC_6H_4SbO(OH)ONH_4$ (229, 268, 269).
Ac_2O → $AcNHC_6H_4SbO_3H_2$ (291).
RCOCl → $RCONHC_6H_4SbO_3H_2$ (291, 1124).
$NaNO_2$ in N HCl, followed by PhONa → $4-(4-HOC_6H_4N_2)C_6H_4SbO_3H_2$ (291).
$NaNO_2$, followed by $Cu_2(CN)_2$ → $p-NCC_6H_4SbO_3H_2$ (291, 460).
$CSCl_2$ in HCl → $SCNC_6H_4SbO_3H_2$ (324).
$NaNO_2$, followed by salicylic acid → $4-[4,3-HO(HO_2C)C_6H_3N_2]C_6H_4SbO_3H_2$ (324).
$NaNO_2$, followed by $o-C_6H_4(OH)_2$ → $4-(2,4-(HO)_2C_6H_3N_2)C_6H_4SbO_3H_2$ (324).
CS_2 → $CS[NHC_6H_4SbO_3H_2]_2$ (324).
Glucose → pale brown compd. (324).
$3-HO_2CC_6H_4NCS$ → $4-(3-HO_2CC_6H_4NHCSNH)C_6H_4SbO_3H_2$ (324).
HCl at 90° → solid, insol. in H_2O, sol. in alkali (324).
HOC_6H_4NCS → $4-(HOC_6H_4NHCSNH)C_6H_4SbO_3H_2$ (324).
$3,5-(SCN)_2C_6H_3CO_2H$ → $3,5-(4-H_2O_3SbC_6H_4NHCSNH)_2C_6H_3CO_2H$ (324).
$3-SCNC_6H_4SO_3Na$ → $4-(3-HO_3SC_6H_4NHCSNH)C_6H_4SbO_3H_2$ (324).
$1,3,5-C_6H_3(NCS)_3$ → $C_6H_3(NHCSNHC_6H_4SbO_3H_2)_3$ (324).
$4-SCNC_6H_4AsO_3H_2$ → $4-(4-H_2O_3AsC_6H_4NHCSNH)C_6H_4SbO_3H_2$ (324).
ClCHRCONR'R″ → $H_2O_3SbC_6H_4NHCHRCONR'R″$ (102).
$NaNO_2$, followed by thioammeline → p-(4,6-diamino-s-triazin-1-ylthio)benzenestibonic
   acid (423).
$(ClCN)_3$ → 4-(4,6-dichloro-s-triazin-2-ylamino)benzenestibonic acid (425, 427, 433,
   436).
Chlorocyanuric diamide → 4-(4,6-diamino-s-triazin-2-ylamino)benzenestibonic acid
   (431, 433).
Et_2NH, followed by Et_2NH antimonate → a white powder, $C_{18}H_{48}N_3O_{31}Sb$, which resembles
   Neostibosan (570).
H_3SbO_4 in $4-H_2NC_6H_4NHAc$, followed by Et_2NH and Et_2NH antimonate → a complex compd.,
   $C_{22}H_{46}N_4O_{17}Sb_4$, similar to Neostibosan (571).
Hexuronic acid—antimonic acid complex, followed by alkali → a complex salt of the
   formula $C_{42}H_{85}N_3Na_5O_{45}Sb_5$ ( ) (573).
HCl → $H[RSbCl_5]$ deriv. which gives a color test with Rhodamine B (714).
Pyridine in conc. HCl → $C_5H_5N.H[PhSbCl_5]$ (1001).

4-SULFAMYLBENZENESTIBONIC ACID $C_6H_8NO_5SSb$ $4-H_2NO_2SC_6H_4SbO_3H_2$
Prepn.: By reacting $4-H_2NO_2SC_6H_4N_2BF_4$ with $SbCl_3$ in 95% EtOH, treating the mixt.
with Cu_2Cl_2 at 60°, isolating the prod. with $C_5H_5N.HCl$ as $C_5H_5N.H(H_2NO_2SC_6H_4SbCl_5)$,
and hydrolyzing with aq. Na_2CO_3; 33% yield (339).
   By adding aq. $NaNO_2$ soln. to a mixt. of $4-H_2NO_2SC_6H_4NH_2$ and $SbCl_3$ in abs. EtOH
contg. H_2SO_4, decompg. the diazonium salt with Cu_2Cl_2, and purifying the prod. via
the pyridinium complex, as described in the preceding example; 45% yield (339).
   By adding $4-H_2NO_2SC_6H_4N_2Cl$ to a soln. of $SbCl_3$ in aq. NaOH (566, 574).
   By adding $4-H_2NO_2SC_6H_4N_2Cl$ to $SbCl_3$ in 10% KOH and decompg. the diazonium compd.
with Cu powder at an elevated temp. (848).
   By coupling $4-H_2NO_2SC_6H_4N_2Cl$ with $SbCl_3$ in HCl and adding a concd. NaOH soln.
(1106).

Salts.: Di-Na salt pptd. in Me₂CO (566). Pyridinium salt, m. 216° (339).
Na salt, C₁₈H₃₁N₃NaO₁₃S₃Sb₃(?), white powder (574). Et₂NH salt, C₂₂H₄₂N₄O₁₉S₃Sb₃, white powder (574).
Rxns. with: HCl in EtOH followed by RN₂FeCl₄ in Me₂CO → RN₂(4-H₂NO₂SC₆H₄SbCl₅) (1067).
SO₂ in MeOH-HCl cont. I, followed by hydrolysis → 4-H₂NO₂SC₆H₄SbO (1106).
Derivs.: Gluconic acid 4-sulfamylbenzenestibonate Na salt, prepd. from C₅H₅N.H-(H₂NO₂SC₆H₄SbCl₅) + Na gluconate in the presence of Me₂N(CH₂)₂OH (1166).
Gluconic acid 4-sulfamylbenzenestibonate dimethylaminoethanol salt (1166).

p-BENZENEDISTIBONIC ACID C₆H₈O₆Sb₂ p-C₆H₄(SbO₃H₂)₂
Prepn.: by reacting p-C₆H₄(N₂Cl)₂ with SbCl₃ in HCl at 0° and decompg. the double satl with 2 N NaOH (324).
Props.: Grayish powder (342).

4-CARBAMYLBENZENESTIBONIC ACID C₇H₈NO₄Sb 4-H₂NOCC₆H₄SbO₃H₂
Prepn.: By ammonolysis of 4-MeO₂CC₆H₄SbO₃H₂ at r. t.; 85% yield (291).
By diazotization of 4-H₂NOCC₆H₄NH₂ in EtOH contg. H₂SO₄ in the presence of SbCl₃ and decompg. the diazonium compd. with Cu₂Br₂; purified by conversion to C₅H₅N.H(4-H₂NOCC₆H₄SbCl₅) and alkaline hydrolysis; 49% yield (293).
Deriv.: Gluconic acid 4-carbamylbenzinestibonate Na salt (1166).

5-STIBONO-2-PYRIDYLMERCAPTOACETIC ACID C₇H₈NO₅SSb

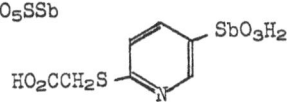

Prepn.: by coupling (2-carboxymethylmercapto)-5-pyridyldiazonium chloride with Na₃SbO₃ in alk. soln. contg. glycerol and Cu dust (1162).
Props.: Tan powder (1162).

3-AMINO-5-STIBONOBENZOIC ACID C₇H₈NO₅Sb 3,5-H₂N(HO₂C)C₆H₃SbO₃H₂
Prepn.: by treating 3-HO₂CC₆H₄SbO₃H₂ with HNO₃ in H₂SO₄ and reducing the 5-nitro deriv. with Na₂S₂O₄ (324).
Rxn. with urea → urea adduct, a buff colored ppt. (324).

p-GUANYLBENZENESTIBONIC ACID C₇H₉N₂O₃Sb H₂NC(:NH)C₆H₄SbO₃H₂
Prepn.: by treating p-NCC₆H₄SbO₃H₂ with dry HCl in alc. and heating the imido ester-HCl with alc. NH₃ at 60-65° (460).
Props.: m. >300° (460).

p-TOLUENESTIBONIC ACID C₇H₉O₃Sb 4-MeC₆H₄SbO₃H₂
Prepn.: By hydrolyzing 4-MeC₆H₄SbCl₄ (133).
By diazotization of 4-MeC₆H₄NH₂ in alcoholic soln. contg. H₂SO₄ in the presence of Cu₂Br₂ (Bart-Scheller method); 47% yield (293).
By diazotization of 4-MeC₆H₄NH₂ and coupling the diazonium chloride with Na₃SbO₃ in alk. solution in the presence of a Cu catalyst (Bart method); 55% yield (1222).
By decompg. 4-MeC₆H₄N₂SbCl₄ in 5 N NaOH (552) or in Me₂CO at 50° in the presence of Cu₂Cl₂ (885).

By decompg. 4-MeC₆H₄N₂SbCl₂ (prepd. from 4-MeC₆H₄N₂FeCl₄ in Me₂CO + SbCl₅ in CHCl₃) with excess Cu₂Cl₂ in Me₂CO at 25° and hydrolyzing the rxn. prod.; 26% yield (906).

By reacting 4-MeC₆H₄N₂Cl with SbCl₃ in Me₂CO in the presence of Cu₂Cl₂; 87% yield. Rxn. of 4-MeC₆H₄N₂BF₄ with Na₃SbO₃ in aq. soln. in the presence of Cu gave 41.1% yield. SbCl₃ treated with 4-MeC₆H₄N₂BF₄ in MeOH or in Me₂CO in the presence of Cu₂Cl₂ gave 31.9 and 25.1% yield, resp. (1222).
Purification: 4-MeC₆H₄SbO₃H₂ is purified by dissolving in hydrochloric acid, treating with pyridine to form C₅H₅N.H(4-MeC₆H₄SbCl₅), isolating the complex salt, and hydrolyzing with aq. alkali (293).
Props.: Mol. wt. detn. (292).
Rxns. with: SO₂ in HCl contg. HI → 4-MeC₆H₄SbCl₂ (133).
HNO₃-H₂SO₄ → the 3-nitro deriv. (291).
KMnO₄ → 4-HO₂CC₆H₄SbO₃H₂ (291).
HNO₃-H₂SO₄, followed by Na₂S₂O₄ → the 3-amino deriv. (324).
Concd. HCl + pyridine → C₅H₅N.H(4-MeC₆H₄SbCl₅) (906).
HCl in EtOH, followed by RN₂FeCl₄ in Me₂CO → RN₂(4-MeC₆H₄SbCl₅) (1067).
SO₂ in MeOH-HCl contg. KI, followed by NH₄OH → 4-MeC₆H₄SbO (552).
Deriv.: Gluconic acid p-toluenestibonate Na salt (1166).

$$4\text{-METHOXYBENZENESTIBONIC ACID} \quad C_7H_9O_4Sb \quad 4\text{-MeOC}_6H_4SbO_3H_2$$
Prepn.: By reacting 4-MeOC₆H₄N₂Cl with SbCl₃ in HCl at 0° and treating the double salt with 2 N NaOH (324, 1000).
By decompg. 4-MeOC₆H₄N₂SbCl₄ at 50° in Me₂CO in the presence of Cu₂Cl₂; 52% yield (without Cu₂Cl₂, 4-ClC₆H₄OMe and PhOMe are formed) (885).
By decompn. of 4-MeOC₆H₄N₂SbCl₈ (from 4-MeOC₆H₄N₂FeCl₄ + SbCl₅) with excess Cu₂Cl₂ in Me₂CO at 25°, followed by alk. hydrolysis; 4% yield (906).
Purification: by converting the acid to 4-anisylantimony tetrachloride, pptg. with pyridine hydrochloride as C₅H₅N.H(4-MeOC₆H₄SbCl₅), and hydrolyzing the prod. with aq. ammonia (1000).
Props.: Bulky, pale brown powder (324). White flakes (1000).
Rxns. with: HNO₃ in H₂SO₄, followed by H₂S₂O₄ → the 3-amino deriv. (324).
H₂S → yellow ppt. (1000).
Derivs.: Gluconic acid 4-anisolestibonate Na salt (1166).

$$4\text{-UREIDOBENZENESTIBONIC ACID, STIBAMINUREA,} \quad C_7H_{12}N_3O_4Sb \quad H_2NCONHC_6H_4SbO_3H_2$$
$$\text{or} \quad H_2NCONHC_6H_4SbO(OH)(ONH_4)$$
Prepn.: By heating 4-H₂NC₆H₄SbO₃H₂ with urea in H₂O (229).
By heating 4-H₂NC₆H₄SbO₃H₂ with urea in the presence of 4-AcNHC₆H₄SbO₃H₂ at 80° (268, 269).
Props.: Pptn. by Me₂CO from aq. soln. yielded a buff powder useful against tropical diseases (229). Commercial stibamineurea was sepd. into urea, 4-AcNHC₆H₄SbO₃H₂, CO(NHC₆H₄SbO₃H₂)₂, and an unknown Sb org. compd. (270).

$$3,4\text{-DIHYDRO-3-OXO-1,4,2H-BENZOTHIAZINE-6-STIBONIC ACID} \quad C_8H_8AsNO_4S$$

Prepn.: by reacting 3,4-dihydro-3-oxo-1,4,2H-benzothiazin-6-yldiazonium chloride

with $NaSbO_2$ in aq. alk.· (770).
Props.: Reddish brown, amorph. prod., decomp. $>300°$ (770).

4-ACETAMIDO-3-CHLOROBENZENESTIBONIC ACID $C_8H_9ClNO_4Sb$ $3,4-Cl(AcNH)C_6H_3SbO_3H_2$
Rxn. with alkali → $2-ClC_6H_4NH_2$ (567).
Deriv.: Gluconic acid 4-acetamido-3-chlorobenzenestibonate Na salt, prepd. from
$C_5N_5N.H[3,4-Cl(AcNH)C_6H_3SbCl_5]$ + Na gluconate in $Me_2NCH_2CH_2OH$ (1166).

2-STIBONOPHENYLMERCAPTOACETIC ACID $C_8H_9O_5SSb$ $2-(HO_2CCH_2S)C_6H_4SbO_3H_2$
Prepn.: by coupling $2-(HO_2CCH_2S)C_6H_4N_2Cl$ with Na_3SbO_3 in alk. soln. contg.
glycerol and Cu dust (1162).
Props.: Amorphous substance (1162).
Salts with: $HN(CH_2CH_2OH)_2$, Et_2NH, $N(CH_2CH_2OH)_2$, Ca, Mg, and Ba (1162).

4-ACETYL-3-AMINOBENZENESTIBONIC ACID $C_8H_{10}NO_3Sb$ $3,4-H_2N(Ac)C_6H_3SbO_3H_2$
Prepn.: by nitration of $4-AcC_6H_4SbO_3H_2$, followed by conversion to the semicarb-
azone, reduction with Al-Hg in H_2O, and acid hydrolysis (25% HCl) (923).
Rxn. with $NaNO_2$ followed by decompn. → $3,4-HO(Ac)C_6H_3SbO_3H_2$ (923).

p-ACETAMIDOBENZENESTIBONIC ACID $C_8H_{10}NO_4Sb$ $p-AcNHC_6H_4SbO_3H_2$
Prepn.: by reacting $p-H_2NC_6H_4SbO_3H_2$ with Ac_2O; 41% yield (291).
From $4-AcNHC_6H_4NH_2$ by the Bart method; 26% yield (884).
Rxns. with: Alkali → $PhNH_2$ (567).
HCl → $H[RSbCl_5]$ deriv. which gives color-reaction with Rhodamine B or methyl vio-
let (714).
Derivs.: Gluconic acid 4-acetamidobenzenestibonate Na salt (1166).

2,5-DIMETHYL-4-SULFAMYLBENZENESTIBONIC ACID $C_8H_{12}NO_5SSb$ $2,5,4-Me_2(H_2NO_2S)-$
$C_6H_2SbO_3H_2$
Prepn.: by reacting $2,5-Me_2(H_2NO_2S)C_6H_2N_2Cl$ with Sb_2O_3 in dild. HCl and alkalizing the rxn. mixt. with Na_2CO_3 (1106).
Rxn. with SO_2 in aq. EtOH contg. HCl, followed by hydrolysis and neutralization →
$[2,5,4-Me_2(H_2NO_2S)C_6H_2]_2SbOH$ (1106).

4-(4,6-DICHLORO-s-TRIAZIN-2-YLAMINO)BENZENESTIBONIC ACID $C_9H_7Cl_2N_4O_3Sb$
$4-(\overline{N:CClN:CClN:CNH})C_6H_4SbO_3H_2$
Prepn.: by adding $p-H_2NC_6H_4SbO(ONa)_2$ soln. to a suspension of $(ClCN)_3$ in H_2O
(425, 427).
Props.: White powder (425).
Rxns. with: 10% Aq. NH_3 at 45° → the p-(4-amino-6-chloro-s-triazinyl) deriv. (425,
427).
25% Aq. NH_3 at 110-130° → the 4,6-diaminotriazinyl deriv. (425, 427).
17% Aq. $MeNH_2$ → 4,6-di(methylamino)triazinyl deriv. (425, 427, 433).

3-(4,6-DICHLORO-s-TRIAZIN-2-YLAMINO)-4-HYDROXYBENZENESTIBONIC ACID
$C_9H_7Cl_2N_4O_4Sb$ $4,3-HO(\overline{N:CClN:CClN:CNH})C_6H_3SbO_3H_2$
Prepn.: by reacting $3,4-H_2N(HO)C_6H_3SbO_3H_2$ with $(ClCN)_3$ in dild. NaOH (431).
Props.: White ppt. (431).
Rxn. with 25% NH_4OH at 110-130°, followed by $SnCl_2$ in HCl → $4,3-HO[\overline{N:C(NH_2)N:C-}$
$\overline{(NH_2)N:CNH}]C_6H_3SbCl_2$ (431).

8-HYDROXY-7-IODO-5-QUINOLINESTIBONIC ACID $C_9H_7INO_4Sb$

Prepn.: by treating 8-hydroxy-5-quinolinestibonic acid with a calc. amt. of KI + KBrO$_3$ in H$_2$SO$_4$ (197).

8-HYDROXY-7-NITRO-5-QUINOLINESTIBONIC ACID $C_9H_7N_2O_6Sb$

Prepn.: by treating 8-hydroxy-5-quinolinestibonic acid with HNO$_3$ in H$_2$SO$_4$ (197).
Rxn. with FeSO$_4$ in dild. H$_2$SO$_4$ → the 7-amino acid (197).

8-HYDROXY-5-QUINOLINESTIBONIC ACID $C_9H_8NO_4Sb$

Prepn.: by reacting 8-hydroxy-5-quinolinediazonium chloride with a soln. of Sb$_2$O$_3$ and equiv. amt. 5 N NaOH in 3 N Na$_2$CO$_3$ in the presence of CuSO$_4$ (197).
Rxns. with: KI + KBrO$_3$ in H$_2$SO$_4$ → 8-hydroxy-7-iodo-5-quinolinestibonic acid (197).
HNO$_3$ in H$_2$SO$_4$ → 8-hydroxy-7-nitro-5-quinolinestibonic acid (197).

p-(4-AMINO-6-CHLORO-s-TRIAZIN-2-YLAMINO)BENZENESTIBONIC ACID $C_9H_9ClN_5O_3Sb$
     [N:C(NH$_2$)N:C(Cl)N:CNH]C$_6$H$_4$SbO$_3$H$_2$
Prepn.: by treating the 4,6-dichloro deriv. with 10% aq. NH$_3$ at 45° (425, 427, 433).
Props.: Small white crystals (425).
Rxns. with: 10% NH$_4$OH at 45° → the p-(4-amino-6-chloro-s-triazinyl-) deriv. (433).
25% NH$_4$OH at 110-130° → the 4-(4,6-diamino-s-triazinyl-) deriv. (433).

7-AMINO-8-HYDROXY-5-QUINOLINESTIBONIC ACID $C_9H_9N_2O_4Sb$

Prepn.: by reducing 8-hydroxy-7-nitro-5-quinolinestibonic acid by FeSO$_4$ in dild. NaOH (197).

4-(2-THIAZOLYLSULFAMYL)BENZENESTIBONIC ACID $C_9H_9N_2O_5S_2Sb$ 4-($\overline{SCH:CHN:C}NHO_2S$)-
  $C_6H_4SbO_3H_2$
Prepn.: By reacting 2-thiazolylsulfamylphenyldiazonium fluoroborate with $SbCl_3$
in 95% EtOH, treating the double salt with Cu_2Cl_2 at 60°, isolating the prod. with
$C_5H_5N.HCl$ as the pyridinium arylpentachloroantimonate, and hydrolyzing with 1% Na_2-
CO_3; 35% yield (339).
  By adding 2-thiazolylsulfamylphenyldiazonium chloride to $SbCl_3$ in 10% KOH and
decompg. the rxn. with Cu powder (848).
Derivs.: Gluconic acid thiazolylsulfamylbenzenestibonate Na salt (1166).
Glucoheptonic acid thiazolylsulfamylbenzenestibonate Na salt (1166).
Tartaric acid thiazolylsulfamylbenzenestibonate Na salt (1166).

4-(4,6-DIAMINO-s-TRIAZIN-2-YLAMINO)-3-NITROBENZENESTIBONIC ACID $C_9H_{10}N_7O_5Sb$
  3,4-$O_2N[\overline{N:C(NH_2)N:C(NH_2)N:C}NH]C_6H_3SbO_3H_2$
Prepn.: by reacting 4,3-$Cl(O_2N)C_6H_3SbO_3H_2$ with melamine in aq. NaOH (431).
Rxn. with $SnCl_2$ in HCl → redn. of the NO_2 and SbO_3H_2 groups to NH_2 and $SbCl_2$,
resp. (431).

4-(4,6-DIAMINO-s-TRIAZIN-2-YLAMINO)BENZENESTIBONIC ACID $C_9H_{11}N_6O_3Sb$
  4-$[\overline{N:C(NH_2)N:C(NH_2)N:C}NH]C_6H_4SbO_3H_2$
Prepn.: By reacting under reflux 4-$H_2NC_6H_4SbO(ONa)_2$ with chlorocyanuric diamide
in H_2O (425, 427, 431, 433).
  By reacting 4-$H_2NC_6H_4SbO(ONa)_2$ with $(ClCN)_3$ in aq. K_2CO_3 and treating the rxn.
prod. with 28% aq. NH_3 at 95°; 40% yield (436, 440).
  By treating 4-$(\overline{N:CClN:CClN:C}NH)C_6H_4SbO_3H_2$ with 25% aq. NH_3 at 110-130° (425,
427, 433).
Props.: Small white crystals (425). Decomp. > 250° (436).
Rxn. of 4-$[\overline{N:C(NH_2)N:C(NH_2)N:C}NH]C_6H_4SbO(ONa)_2.8H_2O$ with urea at 56° → polymeric
Na salt, amorphous, sol. in H_2O (436, 440).
Rxns. with: SO_2 in HCl cont. HI → $RSbCl_2$ (431).
R'SH → the corresponding $RSb(SR')_2$ (431).

1-NAPHTHALENESTIBONIC ACID $C_{10}H_9O_3Sb$ 1-$C_{10}H_7SbO_3H_2$
Prepn.: By reacting 1-$C_{10}H_7N_2Cl$ with $SbCl_3$ in HCl at 0° and decompg. the double
salt with 2 N NaOH (324, 1001).
  By decompn. of 1-$C_{10}H_7N_2SbCl_6$ (from 1-$C_{10}H_7N_2FeCl_4$ + $SbCl_5$) with excess Cu_2Cl_2
in Me_2CO at 25°, followed by alk. hydrolysis; 4% yield (906).
Props.: Bulky, pink powder (324).
Rxns. with: HCl → 1-$C_{10}H_7SbCl_4$ (906, 1001).
SO_2 in MeOH-HCl contg. KI → $C_{10}H_7SbCl_2$ (1001).
Pyridine in HCl → $C_5H_5N.H(1-C_{10}H_7SbCl_5)$ (1001).

2-NAPHTHALENESTIBONIC ACID $C_{10}H_9O_3Sb$ 2-$C_{10}H_7SbO_3H_2$
Prepn.: By reacting 2-$C_{10}H_7N_2Cl$ with $SbCl_3$ in HCl at 0° and decomp. the double salt
with 2 N NaOH (324, 1001).
  By decompn. of 2-$C_{10}H_7N_2SbCl_6$ (from 2-$C_{10}H_7N_2FeCl_4$ + $SbCl_5$) with excess Cu_2Cl_2
in Me_2CO at 25°, followed by alk. hydrolysis; 8% yield (906).
  By hydrolyzing $NH_4(2-C_{10}H_7SbCl_5)$ with H_2O (131, 133).
Props.: White powder (324).

Rxns. with: SO$_2$ in HCl contg. HI → 2-C$_{10}$H$_7$SbCl$_2$ (133, 1001).
HCl → 2-C$_{10}$H$_7$SbCl$_4$ (906).
Org. bases in hydrohalic acids → the corresponding ammonium 2-naphthylpentahalo-stibonates (1001).

[(4,6-DISTIBONO-m-PHENYLENE)DITHIO]DIACETIC ACID C$_{10}$H$_{12}$O$_{10}$S$_2$Sb$_2$

$$H_2O_3Sb\text{-}\underset{HO_2CCH_2S}{\overset{}{\bigcirc}}\text{-}SbO_3H_2 \quad SCH_2CO_2H$$

Prepn.: by coupling 2,4,1,5-(HO$_2$CCH$_2$S)$_2$ C$_6$H$_2$(N$_2$Cl)$_2$ with Na$_3$SbO$_3$ in alk. soln. contg. glycerol and Cu powder (1162).
Props.: Ppt. (1162).

BIPHENYL-4,4′-DISTIBONIC ACID C$_{12}$H$_{12}$O$_6$Sb$_2$ (p-C$_6$H$_4$SbO$_3$H$_2$)$_2$
Prepn.: by reacting (p-C$_6$H$_4$N$_2$Cl)$_2$ with SbCl$_3$ in HCl at 5° and decompg. the double salt with 2 N NaOH (324).
Props.: Pale brown powder (324).

4,4′-THIOBIS(BENZENESTIBONIC ACID) C$_{12}$H$_{12}$O$_6$SSb$_2$ S(C$_6$H$_4$SbO$_3$H$_2$)$_2$
Prepn.: by reacting S(C$_6$H$_4$N$_2$Cl)$_2$ with SbCl$_3$ in 10% KOH and decompg. the prod. with Cu powder (849).

4,4′-DITHIOBIS(BENZENESTIBONIC ACID) C$_{12}$H$_{12}$O$_6$S$_2$Sb$_2$ (-SC$_6$H$_4$SbO$_3$H$_2$)$_2$
Prepn.: by reacting (-SC$_6$H$_4$N$_2$Cl)$_2$ with SbCl$_3$ in 10% KOH and decompg. the prod. with Cu powder (849).

4,4′-SULFONYLBIS(BENZENESTIBONIC ACID) C$_{12}$H$_{12}$O$_8$SSb$_2$ SO$_2$(C$_6$H$_4$SbO$_3$H$_2$)$_2$
Prepn.: By reacting SO$_2$(C$_6$H$_4$N$_2$Cl)$_2$ with SbCl$_3$ in 10% KOH and decompg. the prod. with Cu powder (849).
By diazotization of a mixt. of SO$_2$(C$_6$H$_4$NH$_2$)$_2$ + H$_2$SO$_4$ + SbCl$_3$ in EtOH and de-compg. the rxn. mixt. with Cu$_2$Br$_2$ (220a).
Rxn. with SO$_2$ in HNO$_3$-MeOH cont. NaI, followed by neutralization → the stiboso deriv. (220a).

2-PHENYL-6-BENZOTHIAZOLESTIBONIC ACID C$_{13}$H$_{10}$NO$_3$SSb

$$Ph\text{-}C\underset{S}{\overset{N}{\big\langle}}\bigcirc\text{-}SbO_3H$$

Prepn.: by treating a mixt. of 6-amino-2-phenylbenzothiazole and SbCl$_3$ (1:1) in HCl with NaNO$_2$ soln., isolating the diazonium double salt, and decompg. with NaOH soln. in the presence of Cu-bronze; 29% yield (949).

2-PHENYL-6-BENZOXAZOLESTIBONIC ACID C$_{13}$H$_{10}$NO$_4$Sb

$$Ph\text{-}\underset{O}{\overset{N}{\big\langle}}\bigcirc\text{-}SbO_3H_2$$

Prepn.: by treating a mixt. of 6-amino-2-phenylbenzoxazole and SbCl$_3$ (1:1) in HCl with NaNO$_2$ soln., isolating the diazonium double salt, and decomp.g with NaOH soln. in the presence of Cu-bronze; 29% yield (949).

3,6-ACRIDINEDISTIBONIC ACID $C_{13}H_{11}NO_6Sb_2$

H_2O_3Sb —[structure]— N —[structure]— SbO_3H_2

Prepn.: by reacting a mixt. of 3,6-diaminoacridine and $SbCl_3$ (1:1) in HCl with $NaNO_2$ soln., isolating the bisdiazonium double salt, and decompg. with NaOH soln. in the presence of Cu-bronze; 13% yield (949).

4,4'-(THIOURYLENE)DIBENZENESTIBONIC ACID $C_{13}H_{14}N_2O_6SSb_2$ $CS(NHC_6H_4SbO_3H_2)_2$
Prepn.: by reacting $4-H_2NC_6H_4SbO(OH)(ONa)$ with CS_2 in water at 35-40° (323, 324, 325).
Props.: Pale yellow powder (324).
Salts: Na salt, colorless solid, sol. in H_2O (324).

1-ANTHRAQUINONESTIBONIC ACID $C_{14}H_9O_5Sb$

[structure with SbO_3H_2]

Prepn.: by treating a mixture of 1-aminoanthraquinone and $SbCl_3$ (1:1) in HCl with $NaNO_2$ soln., isolating the 1-anthraquinonediazonium chloride.$SbCl_3$ complex, and decompg. with NaOH soln. the presence of Cu-bronze; 20% yield (949).

3-[3-(4-STIBONOPHENYL)THIOUREIDO]BENZOIC ACID $C_{14}H_{13}N_2O_5SSb$ $4-(3-HO_2CC_6H_4NH-$
   $CSNH)C_6H_4SbO_3H_2$
Prepn.: by reacting $4-H_2NC_6H_4SbO_3H_2$ with $3-SCNC_6H_4CO_2H$ or their Na salts in dild. HCl and H_2O, respectively, at 90° (323, 324, 325).
Props.: Flesh-colored ppt. (324).
Salts: Na salt, pale brown powder (324).

4-[3-(4-STIBONOPHENYL)THIOUREIDO]BENZOIC ACID $C_{14}H_{13}N_2O_5SSb$ $4-(4-HO_2CC_6H_4NH-$
   $CSNH)C_6H_4SbO_3H_2$
Prepn.: by reacting $4-H_2NC_6H_4SbO_3H_2$ with $4-SCNC_6H_4CO_2H$ in dild. HCl or their Na salts in H_2O at 90° (324).
Props.: White powder (324).

3,3'-DIMETHOXY-4,4'-BIPHENYLDISTIBONIC ACID $C_{14}H_{16}O_8Sb_2$

[structure: [benzene ring with SbO_3H_2 and OMe]$_2$]

Prepn.: by reacting $[3,4-Me(ClN_2)C_6H_3-]_2$ with $SbCl_3$ in HCl at 0° and treating the prod. with 2 N NaOH (324).
Props.: White powder (324).

4-[(4-STIBONOPHENYLCARBAMYL)METHYL]-1-PIPERAZINECARBOXYLIC ACID ETHYL ESTER
   $C_{15}H_{22}N_3O_6Sb$ $EtO_2CN(CH_2CH_2)_2NCH_2CONHC_6H_4SbO_3H_2$
Prepn.: From $EtO_2CN(CH_2CH_2)_2NCH_2CONHC_6H_4NH_2$ by the Bart-Schmidt method (983).
Props.: Brown solid, sol. in dild. acids (983).

p-[(N-GLUCOSYL-N-METHYLCARBAMYLMETHYL)AMINO]BENZENESTIBONIC ACID $C_{15}H_{25}N_2O_9Sb$
   $HOCH_2(CHOH)_4CH_2NMeCOCH_2NHC_6H_4SbO(OH)_2$
Prepn.: by heating p-$H_2NC_6H_4AsO(ONa)_2$ with $ClCH_2CONMeCH_2(CHOH)_4CH_2OH$ (102, 577).
Props.: Decomp.~275° (102, 577).

1,4-DI(4-STIBONOPHENYL)PIPERAZINE $C_{16}H_{20}N_2O_6Sb_2$ $H_2O_3SbC_6H_4N(CH_2CH_2)_2NC_6H_4SbO_3H_2$
Prepn.: From 1,4-bis(4-aminophenyl)piperazine by the Bart-Schmidt method (983).

4,4'-(ETHYLMALONYLDIAMIDO)DIBENZENESTIBONIC ACID $C_{17}H_{20}N_2O_8Sb_2$ $EtCH(CONH-$
   $C_6H_4SbO_3H_2)_2$
Prepn.: by reacting $EtCH(COCl)_2$ with $4-H_2NC_6H_4SbO_3H_2$ in aq. NaOH (1124).

4,4'-(DIETHYLMALONYLDIAMIDO)BENZENESTIBONIC ACID $C_{19}H_{24}N_2O_8Sb_2$ $Et_2C(CONH-$
   $C_6H_4SbO_3H_2)_2$
Prepn.: by reacting $Et_2C(COCl)_2$ with $4-H_2NC_6H_4SbO_3H_2$ in alk. soln. (1124).

1,4-BIS[(4-STIBONOPHENYLCARBAMYL)METHYL]PIPERAZINE $C_{20}H_{28}N_4O_8Sb_2$
   $H_2O_3SbC_6H_4NHCOCH_2N(CH_2CH_2)_2NCH_2CONHC_6H_4SbO_3H_2$
Prepd.: From 4,4''-diamino-1,4-piperazinediacetanilide by the Bart-Schmidt method
(983).
Props.: Light brown solid (983).

3,5-BIS(m-STIBONOPHENYL-2-THIOUREIDO)BENZOIC ACID $C_{21}H_{20}N_4O_8S_2Sb_2$
   $3,5-(m-H_2O_3SbC_6H_4NHCSNH)_2C_6H_3CO_2H$
Prepn.: by reacting $3,4-(SCN)_2C_6H_3CO_2Na$ with $3-H_2NC_6H_4SbO(OH)ONa$ in H_2O at 80°
(323, 325).

3,5-BIS(p-STIBONOPHENYL-2-THIOUREIDO)BENZOIC ACID $C_{21}H_{20}N_4O_8S_2Sb_2$
   $3,5-(p-H_2O_3SbC_6H_4NHCSNH)_2C_6H_3CO_2H$
Prepn.: by reacting $3,5-(SCN)_2C_6H_3CO_2Na$ with $4-H_2NC_6H_4SbO(OH)ONa$ in H_2O at 60°
(323, 324, 325).
Props.: White ppt. (324).
Salts: Na salt, pale brown powder (324).

4,4'-(DIALLYLMALONYLDIAMIDO)DIBENZENESTOBONIC ACID $C_{21}H_{24}N_2O_8Sb$
   $(CH_2:CHCH_2)_2C(CONHC_6H_4SbO_3H_2)_2$
Prepn.: by reacting $(CH_2:CHCH_2)_2C(COCl)_2$ with $4-H_2NC_6H_4SbO_3H_2$ in aq. NaOH (1124).

p,p',p''-[s-PHENENYLTRIS(THIOURYLENE)]TRIBENZENESTIBONIC ACID $C_{27}H_{27}N_6O_9S_3Sb_3$
   $1,3,5-C_6H_3(NHCSNH-p-C_6H_4SbO_3H_2)_3$
Prepn.: by reacting $1,3,5-C_6H_3(NCS)_3$ with $4-H_2NC_6H_4SbO(OH)Ona$ in H_2O at 80° (324).
Props.: Pale buff powder (324).

$RSbO_3H_2$ where R is:	Prepd. from	Method *	Yield %	Props. and Remarks	Ref.
C₃					
SCH:CHN:C-	$RSbCl_4$	VII			844
C₄					
N:CHCH:CHN:C-	$RSbCl_4$	VII		White powder	668
C₅					
CH:NC(NH₂):CHCH:C-	RNH_2	VI			86
C₆					
2-BrC₆H₄-	RNH_2	IV	2	$C_5H_5N.H(RSbCl_5)$, m. 164°	293
	RNH_2	I	37		293
2-ClC₆H₄-	RN_2SbCl_6	IX	1		906
4-FC₆H₄-	$RN_2Cl + SbCl_3$	VI		Pale yellow ppt.	324
	$RNHNH_2$	VIII		Rxn. with SO₂ in HCl → 4-FC₆H₄SbCl₂	132
4-IC₆H₄	$RN_2Cl + SbCl_3$	VI		Brownish powder	324
	RN_2SbCl_6	IX	9	Rxn.[1]	905
2-O₂NC₆H₄-	RNH_2	IV	16	Rxn.[2]	293
3-O₂NC₆H₄-	RN_2SbCl_6	IX	6		906
	RN_2SbCl_6	IV	46	$C_5H_5N.H(RSbCl_5)$, m. 187°	293
					324
3,4-H₂N(Br)C₆H₃-	4-BrC₆H₄SbO₃H₂ + HNO₃ followed by Na₂S₂O₄	VI		Rxn.[3]	1106
2,4-Cl(H₂NO₂S)C₆H₃-	RN_2Cl	VI			1106
4-HOC₆H₄-	$RSbCl_4$	VII		White gelatinous mass, rxn.[4]	324
3-H₂NO₂SC₆H₄-	RN_2Cl	VI			1106
C₇					
3-F₃CC₆H₄-	RNH_2	VI	19.1		946

* See pages 474-5.

1. On heating with concd. HCl disproportionates → $(2-O_2NC_6H_4)_2SbO_2H$ (293).
2. Rxn. with hydrochloric acid followed by pyridine → the $C_5H_5N.H(RSbCl_5)$ complex (906).
3. Redn. by SO₂ or SnCl₂ in HCl, followed by hydrolysis → RSbO (1106).
4. Rxn. with HCl in EtOH, followed by R'N₂F₆Cl₄ in Me₂CO → R'N₂(RSbCl₅) (1057).

TABLE LXVIII (cont'd.)
STIBONIC ACIDS

RSbO₃H₂ where R is:	Prepd. from	Method*	Yield %	Props. and Remarks	Ref.
4-SCNC₆H₄-	4-H₂NC₆H₄SbO₃H₂ + CSCl₂	VI		White powder	323-5
4-NCC₆H₄-	RN₂Cl		40	m. > 320°, rxns.5-7	172
	4-H₂O₃SbC₆H₄N₂Cl + CuCN				291, 459
3,4-O₂N(HO₂C)C₆H₃-	3,4-O₂N(Me)C₆H₃SbO₃H₂ + alk. KMnO₄				291
2-HO₂CC₆H₄-	RN₂Cl	II	83		293
3-HO₂CC₆H₄-	RNH₂	IV	72	C₅H₅N.H(RSbCl₅), m. 293°, rxn.8	293
4-HO₂CC₆H₄-	RNH₂	IV	72	C₅H₅N.H(RSbCl₅), m. 252°	293
	4-MeC₆H₄SbO₃H₂ + alk. KMnO₄			Rxn. with ROH + HCl → RO₂CC₆H₄SbO₃H₂	291
	4-NCC₆H₄SbO₃H₂ by hydrolysis				459
2,4-Cl(Me)C₆H₃-	RN₂Cl	VI	83	Pale brown powder	324
3,4-O₂N(Me)C₆H₃-	4-MeC₆H₄SbO₃H₂ + HNO₃ in H₂SO₄			Rxn. with KMnO₄ → 3,4-O₂N(HO₂C)C₆H₃SbO₃H₂	291
2,6-O₂N(Me)C₆H₃-	RNH₂	IV	27	C₅H₅N.H(RSbCl₅), m. 191°	293
2,4-O₂N(HOCH₂)C₆H₃-	RNH₂	IV	30		293
2-MeC₆H₄-	RN₂Cl	VI	2	Rxn.9	924
	RNH₂	IV			293
	RNH₂	I	37	m. 172.6°	293
3-MeC₆H₄-	RNH₂	IV	24	Mol. wt. detn.	292
				Mol. wt. detn.	293
2-HOCH₂C₆H₄-	RN₂Cl	VI			924
3-HOCH₂C₆H₄-	RN₂Cl	VI			924
4-HOCH₂C₆H₄-	RN₂Cl	VI			924
2,4-H₂N(HOCH₂)C₆H₃-	2,4-O₂N(HOCH₂)C₆H₃SbO₃H₂ + FeSO₄ in dild. NaOH			Rxn. with AcCl → 2,4-AcNH(HOCH₂)C₆H₃SbO₃H₂	924
3,4-H₂N(Me)C₆H₃-	4-MeC₆H₄SbO₃H₂ + HNO₃-H₂SO₄ followed by Na₂S₂O₄				324
3,4-H₂N(MeO)C₆H₃-	4-MeC₆H₄SbO₃H₂ + HNO₃-H₂SO₄ followed by Na₂S₂O₄			Forms adduct with urea	324
4-H₂NC(:NH)NHO₂SC₆H₄-	RN₂Cl	VI			848

	Method		Yield (%)	Remarks	Ref.
4,2-Cl(HO₂CCH₂S)C₆H₃-	RN_2Cl	I		Na salt, decomp. 220-240° rxn.10	1168
4-AcC₆H₄-	$RN_2Cl + SbCl_3$ in HCl	VI			923
2-HO₂CCH₂SC₆H₄-	RN_2Cl	I		Dipyridinium and diquinolinium salts	713
	RNH_2	IV			226
4-HO₂CCH₂SC₆H₄-	RNH_2	IV		Dipyridinium and diquinolinium salts	226
3-HO₂CCH₂SC₆H₄-	RN_2Cl	I		Tan powder	1168
3,4-HO(Ac)C₆H₃-	$3,4-H_2N(Ac)C_6H_3SbO_3H_2$ via diazotization				923
3-AcOC₆H₄-				Mol. wt. detn.	292
4-AcOC₆H₄-				Mol. wt. detn.	292
3-MeO₂CC₆H₄-	$3-HO_2CC_6H_4SbO_3H_2 + MeOH + H_2SO_4$		85	$C_5H_5N.H(RSbCl_5)$, m. 195°	291
	$3-HO_2CC_6H_4SbO_3H_2 + MeOH-HCl$		70		291
4-MeO₂CC₆H₄-	$4-HO_2CC_6H_4SbO_3H_2 + MeOH + HCl$		85	$C_5H_5N.H(RSbCl_5)$, m. 217°	291
	$4-HO_2CC_6H_4SbO_3H_2 + MeOH-HCl$		60	Ammonolysis → 4-H_2NOC-deriv.	291
4-HO₂CCH₂OC₆H₄-	RNH_2	IV	26	$C_5H_5N.H(RSbCl_5)$, m. 114° Decomp. 240°	293
3,4-HO(AcNH)C₆H₃-	RNH_2	I	13		924
2-AcNHC₆H₄-	RNH_2	I	8		884
3-AcNHC₆H₄-	RNH_2	I	90		884
4-H₂NOCCH₂OC₆H₄-	$4-MeO_2CCH_2OC_6H_4SbO_3H_2 + NH_4OH$				291
2,4-Me₂C₆H₃-	RN_2SbCl_6	IX	14		906
2-EtOC₆H₄-	RN_2SbCl_6	IX	4		906

5. Redn. by $SnCl_2$ in HCl at -5° → $RSbCl_2$ (172).
6. Hydrolysis by concd. alkali → $4-HO_2CC_6H_4SbO_3H_2$ (459).
7. Alc. HCl, followed by alc. NH_3 → $4-H_2NC(:NH)C_6H_4SbO_3H_2$ (460).
8. Nitration in H_2SO_4, followed by redn. by $Na_2S_2O_4$ → $3,5-H_2N(HO_2C)C_6H_3SbO_3H_2$ (324).
9. Rxn. with $FeSO_4$ in alk. soln. → redn. of the NO_2 group (924).
10. Rxn. with HNO_3-H_2SO_4 in HCO_2H → the 3-nitro deriv. (923).
* See pages 474-5.

TABLE LXVIII (cont'd.)
STIBONIC ACIDS

$RSbO_3H_2$ where R is:	Prepd. from	Method*	Yield %	Props. and Remarks	Ref.
4-EtOC_6H_4-	RN_2Cl	VI		Almost white powder	324, 1221
	RN_2SbCl_4	IX	3		906
	$RSbCl_4$	VII		Rxn.[11]	1096
$4\text{-Me}_2NC_6H_4-$	RN_2Cl	VI		White powder	324
$4\text{-Me}_2NO_2SC_6H_4-$	RN_2Cl	VI		Rxn.[12]	1106
C_9					
$4,2\text{-HO}_2C(HO_2CCH_2S)C_6H_3-$	RN_2Cl	I		Ppt.	1162
$4\text{-}[N\!:\!C(NH_2)N\!:\!C(NH_2)N\!:\!CS]\text{-}C_6H_3-$	$4\text{-H}_2O_3SbC_6H_4N_2Cl$ + $HSC\!:\!NC(NH_2)\!:\!NC(NH_2)\!:\!N$			Yellow ppt.	423
$2\text{-}[MeCH(CO_2H)S]C_6H_4-$	RN_2Cl	I		HCl salt, crystals	423
$4,2\text{-Me}(HO_2CCH_2S)C_6H_3-$	RN_2Cl	I			1162
$2\text{-HO}_2CCH_2CH_2SC_6H_4-$	RN_2Cl	I			1162
$3\text{-EtO}_2CC_6H_4-$	$3\text{-HO}_2CC_6H_4SbO_3H_2$ + EtOH-HCl	I	70		1162
$4\text{-EtO}_2CC_6H_4-$	RN_2Cl	VI		$C_5H_5N\cdot H(RSbCl_5)$, m. 148°	291
	$4\text{-HO}_2CC_6H_4SbO_3H_2$ + EtOH + HCl		90	Pale brown powder	324
	$4\text{-HO}_2CC_6H_4SbO_3H_2$ + EtOH-HCl			$C_5H_5N\cdot H(RSbCl_5)$, m. 187°	291
$4\text{-EtCO}_2C_6H_4-$	$4\text{-HO}_2CC_6H_4SbO_3H_2$ + EtOH-HCl		60	Mol. wt. detn.	291
$4,2\text{-MeO}(HO_2CCH_2S)C_6H_3-$	RN_2Cl	I			292
$4\text{-MeO}_2CCH_2OC_6H_4-$	$4\text{-HO}_2CCH_2OC_6H_4SbO_3H_2$ + MeOH + HCl		66	$C_5H_5N\cdot H(RSbCl_5)$, m. 199°	1162
	$4\text{-HO}_2CCH_2OC_6H_4SbO_3H_2$ + MeOH-HCl		90	Ammonolysis → $4\text{-H}_2NOCCH_2OC_6H_4SbO_3H_2$	291
$4\text{-Me}_2NOCC_6H_4-$	$4\text{-MeO}_2CC_6H_4SbO_3H_2$ + Me_2NH		97		291
$4\text{-HOCH}_2CH_2NHOCC_6H_4-$	$4\text{-MeO}_2CC_6H_4SbO_3H_2$ + $HOCH_2CH_2NH_2$		62		291
C_{10}					
$4\text{-O}_2NC_{10}H_6-$	RNH_2	VI	40		949
$4\text{-}(N\!:\!CHCH\!:\!CHN\!:\!CNHO_2S)C_6H_4-$	RN_2BF_4	V	38		339
$4\text{-}(N\!:\!CHCH\!:\!CHN\!:\!CNHO_2S)C_6H_4-$	RN_2Cl	VI		Deriv.[13]	843
	RN_2BF_4	V			339
	RN_2Cl	VI			848
$2,4\text{-HO}_2CCH_2S(AcNH)C_6H_3-$	RN_2Cl	I	26		1162

Compound	Reagent	Method	Yield	Appearance	Ref.
4-[(HO)C₆H₂)₃NHOC)C₆H₄-	RN₂Cl	I			1163
C₁₁					
4-[N:CMeN:CMeN:C)C₆H₄-	RN₂Cl	VI			1148
4-[N:CHCH:C(Me)N:CNHO₂S)-C₆H₄-	RN₂Cl	VI			848
4-[N:C(NHMe)N:C(NHMe)N:C(NHMe)N:C)C₆H₄-	4-(N:CC1N:CC1N:CNH)C₆H₄Sb O₃H₂ + MeNH₂			White cryst. needles	425, 427, 433
4-(HOCH₂CH₂NHCH₂CH₂NHOC)C₆-H₄-	RN₂Cl	I			1163
3-(HOCH₂CH₂NHCH₂CH₂NHOC)C₆-H₄-	RN₂Cl	I		White ppt.	1163
C₁₂					
2,3,5-Cl₂(2,3-Cl₂C₆H₃N₂)-C₆H₂-	2,3-Cl₂C₆H₃N₂Cl	VI		Brown-red	324
4-(4-O₂NC₆H₄S)C₆H₄-	RN₂Cl	VI			849
4-(4-O₂NC₆H₄SO₂)C₆H₄-	RN₂Cl	VI			849
4-(4-HOC₆H₄N₂C₆H₄-	4-H₂O₃SbC₆H₄N₂Cl + PhONa				291
4-[2,4-(HO)₂C₆H₃N₂]C₆H₃-	4-H₂O₃SbC₆H₄N₂Cl + m-C₆H₄(OH)₂		63	Red powder	324
2-PhC₆H₄	2-PhC₆H₄Sb(OH)Cl + Cl₂ + H₂O			White powder, m. > 300°	1318
2,1-(HO₂CCH₂S)C₁₀H₆-	RN₂Cl	I			1162
4-(4-H₂NC₆H₄S)C₆H₄-	4-(4-AcNHC₆H₄S)C₆H₄N₂Cl	VI			849
4-(4-H₂NC₆H₄SO₂)C₆H₄-	4-(4-AcNHC₆H₄SO₂)C₆H₄N₂Cl	VI			849
4-(4-H₂NC₆H₄SO₂NH)C₆H₄-	4-(4-AcNHC₆H₄SO₂NH)C₆H₄NH₂				666
C₁₃					
4-BzNHC₆H₄-	4-H₂NC₆H₄SbO₃H₂ + BzCl		73	Red powder	291
4-[4,3-HO(HO₂C)C₆H₃N₂]C₆H₄-	4-H₂O₃SbC₆H₄N₂Cl + 2-HOC₆-H₄CO₂H				324
4-(HOC₆H₄NHCSNH)C₆H₄-	4-H₂NC₆H₄SbO(OH)(ONa) + 4-SCNC₆H₄ONa			White ppt.; Na salt, white	324
4-(3-HO₃SC₆H₄NHCSNH)C₆H₄-	4-H₂NC₆H₄SbO(OH)(ONa) + 3-SCNC₆H₄SO₃Na			Pale brown powder	324

11. Rxn. with HCl in EtOH, followed by RN₂FeCl₄ in Me₂CO → RN₂(4-EtOC₆H₄SbCl₅) (1067, 1069).
12. Redn. by SO₂ in AcOH-HCl → 4-Me₂NO₂SC₆H₄SbCl₂ (1106).
13. Gluconic acid pyrididylsulfamylbenzenestibonate Na salt (1166).
* See pages 474-5.

TABLE LXVIII (cont'd).
STIBONIC ACIDS

RSbO₃H₂ where R is	Prepd. from	Method *	Yield %	Props. and Remarks	Ref.
4-(4-HO₃SC₆H₄NHCSNH)C₆H₄-	4-H₂NC₆H₄SbO(OH)(ONa) + 4-SCNC₆H₄SO₃Na	*		Na salt, pale brown powder	324
4-(4-H₂O₃AsC₆H₄NHCSNH)C₆H₄-	4-H₂NC₆H₄SbO(OH)(ONa) + 4-SCNC₆H₄AsO₃H₂			Brown ppt.	324
4-(4-H₂NO₂SC₆H₄NHCSNH)C₆H₄-	4-H₂NC₆H₄SbO₃H₂ + 4-H₂NO₂-SC₆H₄NCS				326
4,4'-CH₂(C₆H₄-)₂	4,4'-CH₂(C₆H₄N₂Cl)₂	VI		White powder	324
C₁₆ 2,4-Ac(4-AcC₆H₄N:N)C₆H₃-	4-AcC₆H₄N₂Cl	VI		Bright orange; redn. by Zn → 4-AcC₆H₄NH₂	324

* See pages 474-5.

STIBINIC ACIDS

Dialkyl- and diarylstibinic acids, R_2SbO_2H and $RR'SbO_2H$, are prepared or formed by:

1. Hydrolysis of diorganoantimony trihalides, R_2SbX_3, with aqueous alkali, ammonia, or aqueous sodium acetate in acetone.

2. Decompsoition of aryldiazonium chloroantimonites, RN_2SbCl_4, or complex compounds of aryldiazonium chlorides with organoantimony tetrachlorides, $RN_2(R'SbCl_5)$, in acetone in the presence of iron powder or zinc dust at a low temperature, followed by hydrolysis of the reaction product. Aryldiazonium chloroantimonites yield homogeneous stibinic acids, R_2SbO_2H, while the complex compounds, $RN_2(R'SbCl_5)$, yield mixed stibinic acids, $RR'SbO_2H$.

3. Decomposition of aryldiazonium tetrachloroantimonites, RN_2SbCl_4, with Zn dust in ethyl acetate at $60°$, followed by treatment of the reaction products with aryldiazonium tetrachloroantimonites and hydrolysis by exposure to air at $30-35°$ or by aqueous ammonia.

4. Decomposition of aryldiazonium chloroantimonates, RN_2SbCl_6, with iron powder in acetone, followed by treatment of the reaction product with hydrochloric acid and with ethanol and by hydrolysis with aqueous ammonia. The reaction yields also by-products such as tertiary stibine dichlorides, R_3SbCl_2.

5. Decomposition of complex compounds of aryldiazonium chlorides with diarylantimony trichlorides, $RN_2(R_2'SbCl_4)$, by boiling in chloroform, followed by hydrolysis.

6. Arylation of antimony trichloride by potassium arylazocarboxylates, RN_2CO_2K, in ethyl acetate at $60-70°$, followed by introduction of a second aryl group, using aryldiazonium chloroantimonites, RN_2SbCl_4, at $35-40°$, and hydrolysis of the product. The arylation of antimony trichloride with potassium arylazocarboxylates in acetone yields aryl-diazonium chloride-antimony trioxide complexes, $RN_2Cl.Sb_2O_3$, which decompose in the presence of copper in acetone to yield, after hydrolysis, stibinic acids along with bis(diarylantimony) oxides, $(R_2Sb)_2O$.

7. Arylation of antimony trichloride with aryldiazonium fluoroborates, RN_2BF_4, in isopropanol in the presence of copper, followed by isolation of diarylantimony trichlorides as complex compounds with pyridinium chloride, which are then hydrolyzed. The reaction yields stibinic acids, as by-products, along with the corresponding stibonic acids.

8. Reaction of antimony trichloride with phenylhydrazine in hydrochloric acid, followed by decomposition of the resulting complex in acetone in the presence of copper(II) chloride and by hydrolysis. The reaction product contains phenylstibonic and diphenylstibinic acids.

9. Arylation of organodihalostibines, $RSbX_2$, with aryldiazonium chloroantimonites, acetates, or chlorozincates, RN_2SbCl_4, RN_2OAc, or $(RN_2)_2ZnCl_4$, in acetone under reflux, followed by hydrolysis.

10. Reaction of tertiary stibines, R_3Sb, with lithium in tetrahydrofuran, followed by hydrolysis.

11. Oxidation of tertiary stibines, R_3Sb, with hydrogen peroxide.

12. Disproportionation of 2-nitrobenzenestibonic acid in concentrated hydrochloric acid 2,2'-dinitrodiphenylstibinic acid was obtained.

DIPHENYLSTIBINIC ACID $C_{12}H_{11}O_2Sb$ Ph_2SbO_2H

Prepn.: By hydrolysis of Ph_2SbCl_3 with aq. NaOH; 79.6% yield (462).

By decomposing the $PhN_2(Ph_2SbCl_4)$ complex in boiling $CHCl_3$ and hydrolyzing the prod. with NH_4OH; 73% yield (1074).

By treating PhN_2SbCl_4 with iron powder at 0° in acetone, isolating the prod., and hydrolyzing with ice-cold 5% NH_4OH; 85% yield (904); 67% yield (1065).

By decompg. PhN_2SbCl_4 with Zn dust in EtOAc at 60°, isolating the sol. prod. from inorganic matter, treating with PhN_2SbCl_4 at 30-35° or 60°, and hydrolyzing by exposure to atmospheric moisture at 30-35°; isolated as $Ph_2SbO_2H.Sb_2O_3$ (1059); 60.4% yield (1060).

By decompg. PhN_2SbCl_6 with Fe powder in Me_2CO and hydrolyzing the prod. with aq. NaOH; 67% yield (1061).

By reacting PhN_2CO_2K with $SbCl_3$ in EtOAc at 60-70°, evaptg. the soln. at 30-35°, and treating the residue with 5 N HCl followed by 5% NH_4OH; 26.3% yield was isolated in the form of $Ph_2SbO_2H.Sb_2O_3$. The yield was increased to 37.5% by treating the rxn. prod. of PhN_2CO_2K and $SbCl_3$ with PhN_2SbCl_4 at 35-40° and then hydrolyzing the prod. When Me_2CO was used instead of EtOAc, as solvent, $PhN_2Cl.Sb_2O_3$ was formed, which was decompd. with Cu powder to give 32% yield of $Ph_2SbO_2H.Sb_2O_3$, along with $(Ph_2Sb)_2O$ (1060).

By oxidation of Ph_2SbCl with a nitrite or atm. oxygen in AcOH, followed by hydrolysis with NH_4OH (1074).

By adding $PhNHNH_2$ to a soln. of $SbCl_3$ and $CuCl_2$ in aq. HCl and stirring the mixt. in contact with air; 15-18% yield (1119).

By oxidation of Ph_3Sb with H_2O_2 (1316).

By treating Ph_3Sb with Li in C_4H_8O and hydrolyzing the prod.; 60% yield (1309).

By reacting $SbCl_3$ dissolved in Me_2CHOH with PhN_2BF_4 in the presence of Cu bronze, isolating the product, Ph_2SbCl_3, as a complex with pyridinium hydrochloride, and hydrolyzing the complex; 5% yield along with $PhSbO_3H_2$ (296).

Props.: m. 283-286° (1309).

Rxns. with: Aq. HCl → Ph_2SbCl_3 (904, 1061, 1065, 1074, 1119, 1316).
HCl in MeOH, followed by RN_2FeCl_4 in Me_2CO → $RN_2(Ph_2SbCl_4)$ (1068).
$SnCl_2$ → Ph_2SbCl (1315).

DI-p-TOLYLSTIBINIC ACID $C_{14}H_{15}O_2Sb$ $(4-MeC_6H_4)_2SbO_2H$

Prepn.: By reacting $4-MeC_6H_4SbI_2$ with $4-MeC_6H_4N_2SbCl_4$ in cold Me_2CO and hydrolyzing the prod. with NH_4OH; 84% yield (910).

By decompg. $4-MeC_6H_4N_2SbCl_6$ with Fe in Me_2CO and hydrolyzing the rxn. prod.; 58% yield along with $(4-MeC_6H_4)_3SbCl_2$ (1061, 1065).

By decompg. $4-MeC_6H_4N_2SbCl_4$ with Fe in Me_2CO at 0° and hydrolyzing the prod.; 71% yield (904).

By reacting $4-MeC_6H_4SbCl_2$ with $4-MeC_6H_4N_2OAc$ in Me_2CO; 20% yield along with 60% of $4-MeC_6H_4Sb(OAc)_2.4-MeC_6H_4N_2OAc$ (910, 1063).

By reacting $SbCl_3$ dissolved in Me_2CHOH with $4-MeC_6H_4N_2BF_4$ in the presence of Cu-bronze, isolating the prod., $(4-MeC_6H_4)_2SbCl_3$, as a complex with pyridinium hydrochloride, and hydrolyzing the complex; 8% yield along with $4-MeC_6H_4SbO_3H_2$ (296).

Rxns. with: 5 N HCl → $(4-MeC_6H_4)_2SbCl_3$ (910).
Hot HCl → $(4-MeC_6H_4)_2Sb(OH)Cl_2$ (1061, 1065).

TABLE LXIX

STIBINIC ACIDS

Stibinic Acids R_2SbO_2H or $RR'SbO_2H$	Prepd. from	Yield %	Props. and Remarks	Ref.
C_4				
$(MeCHCl)_2SbO_2H$	$R_2SbCl + H_2O_2$		m.>190°	1320-1
C_{12}				
$(4-BrC_6H_4)_2SbO_2H$	$RN_2SbCl_4 + Fe$ in Me_2CO [3]	90	Rxn.[1]	904
$(4-ClC_6H_4)(4-O_2NC_6H_4)SbO_2H$	$RSbI_2 + R'N_2SbCl_4$ in Me_2CO [3]		Rxn.[1]	910
$(2-ClC_6H_4)_2SbO_2H$	$RN_2SbCl_4 + Fe$ in Me_2CO [3]	86	Rxn.[2]	904
$(3-ClC_6H_4)_2SbO_2H$	$SbCl_3 + RN_2BF_4$ in $EtOH + Cu$ [4]	6		296
$(4-ClC_6H_4)_2SbO_2H$	$SbCl_3 + RN_2BF_4$ in alc. $+ Cu$ [4]	12	Rxn.[1]	296
	$RN_2SbCl_4 + Fe$ [3]	85		904
	$RSbI_2 + RN_2SbCl_4$ in Me_2CO [3]	86		910
	$RN_2SbCl_6 + Fe$ in Me_2CO [3]	59		1065
$(4-FC_6H_4)_2SbO_2H$	$RNHNH_2 + SbCl_3 + CuCl_2$ in HCl and decompn. in Me_2CO	14	Rxn.[1]	132
$(4-IC_6H_4)_2SbO_2H$	$RN_2SbCl_4 + Fe$ in Me_2CO [3]	85		904
	$RN_2SbCl_6 + Fe$ in Me_2CO [3]	81	Rxn.[1]	1065
$(2-O_2NC_6H_4)_2SbO_2H$	$RSbO_3H_2 + $ concd. HCl			293
$(3-O_2NC_6H_4)_2SbO_2H$	$RN_2SbCl_6 + Fe$ in Me_2CO [3]	72	Rxn.[1]	1061, 1065
$(4-O_2NC_6H_4)_2SbO_2H$	$SbCl_3 + RN_2BF_4 + Cu$ in alc. [4]	20		296
	$RN_2SbCl_4 + Fe$ in Me_2CO [3]	97		904
	$RN_2SbCl_6 + Fe$ in Me_2CO [3]	88	Rxn.[1]	1061, 1065
$Ph(4-BrC_6H_4)SbO_2H$	$RSbCl_4.R'N_2Cl + Fe$ in Me_2CO [3]	86	Rxn.[5]	1069
$Ph(4-ClC_6H_4)SbO_2H$	$RSbCl_2 + R'N_2SbCl_4$ in Me_2CO [3]	60		910
	$R'SbI_2 + RN_2SbCl_4$ in Me_2CO [3]	95		910
	$R'SbCl_4.R'N_2Cl + Fe$ in Me_2CO [3]	80	Rxns.[1,6]	1069
$Ph(4-IC_6H_4)SbO_2H$	$RSbCl_4.R'N_2Cl + Fe$ in Me_2CO [3]	75	Rxn.[1]	1069
$Ph(2-O_2NC_6H_4)SbO_2H$	$RSbCl_4.R'N_2Cl + Fe$ in Me_2CO [3]	71		1069

1. Rxn. with hydrochloric acid → the corresponding diorganoantimony trichloride (132, 904, 910, 1061, 1065).

2. Rxn. with aq. AcOH → $(2-ClC_6H_4)_2SbO_2Ac$ (904).

3. The arylation is followed by hydrolysis.

4. The resulting diarylantimony trichloride is isolated as a complex with pyridinium hydrochloride and hydrolyzed to yield the corresponding stibinic and stibonic acids (296).

5. Rxn. with $PhN_2FeCl_4 \rightarrow PhN_2.[Ph(4-BrC_6H_4)SbCl_4]$ (1069).

TABLE LXIX (cont'd.)
STIBINIC ACIDS

Stibinic acids R_2SbO_2H or $RR'SbO_2H$	Prepd. from	Yield %	Props. and Remarks	Ref.
C_{13}				
$(4\text{-}ClC_6H_4)(4\text{-}MeC_6H_4)SbO_2H$	$R'SbCl_2 + RN_2SbCl_4$ in Me_2CO^3	45	Rxn.[1]	910
	$RSbI_2 + R'N_2SbCl_4$ in Me_2CO^3	85		910
$(4\text{-}ClC_6H_4)(4\text{-}MeOC_6H_4)SbO_2H$	$RSbI_2 + R'N_2SbCl_4$ in Me_2CO^3	69	Rxn.[1]	910
$(4\text{-}O_2NC_6H_4)(4\text{-}MeC_6H_4)SbO_2H$	$R'SbCl_2 + RN_2SbCl_4$ in Me_2CO^3	87	Rxn.[1]	910
$Ph(4\text{-}MeC_6H_4)SbO_2H$	$R'SbCl_2 + RN_2OAc$ in Me_2CO^3	71	m. 155-160°	1063
	$R'SbCl_2 + RN_2SbCl_4$ in Me_2CO^3	45		910
	$BSbCl_4.R'N_2Cl + Fe$ in Me_2CO^3	50		1069
	$R'SbCl_4.RN_2Cl + Fe$ in Me_2CO^3	95	Rxns. 6, 7	1069
C_{14}				
$(4\text{-}EtOC_6H_4)(4\text{-}O_2NC_6H_4)SbO_2H$	$RSbCl_4.R'N_2Cl + Fe$ in Me_2CO^3	25	Rxn[6]	1069
	$R'SbCl_4.RN_2Cl + Fe$ in Me_2CO^3	51		1069
$Ph(4\text{-}EtOC_6H_4)SbO_2H$	$RSbCl_4.R'N_2Cl + Fe$ in Me_2CO^3	55	Rxn.[6]	1069
$(4\text{-}MeOC_6H_4)(4\text{-}MeC_6H_4)SbO_2H$	$R'SbCl_2 + (RN_2)_2.ZnCl_4$ (1:2) in Me_2CO^3	32		910
	$R'SbI_2 + RN_2SbCl_4$ in Me_2CO^3	66	Rxn.[1]q	910
C_{16}				
$(2,4\text{-}Me_2C_6H_3)_2SbO_2H$	$RN_2SbCl_4 + Fe$ in Me_2CO^3	96	Rxn.[1]	904
$(2\text{-}EtOC_6H_4)_2SbO_2H$	$RN_2SbCl_4 + Fe$ in Me_2CO^3	42	Rxn.[1]	904
C_{21}				
$(4\text{-}EtO_2CC_6H_4)[2\text{-}(4\text{-}O_2NC_6H_4)\text{-}C_6H_4]SbO_2H$	$RR'SbCl_3 + $ aq. NaOAc in Me_2CO		Rxn.[8]	175
$(2\text{-}PhC_6H_4)(4\text{-}EtO_2CC_6H_4)SbO_2H$	$RR'SbCl_3 + $ Aq. NaOAc in Me_2CO		Rxn.[9]	175
C_{22}				
$(4\text{-}EtO_2CC_6H_4)[2\text{-}(4\text{-}MeOC_6H_4)\text{-}C_6H_4]SbO_2H$	$RR'SbCl_3 + $ Aq. NaOAc in Me_2CO		Rxns. 10, 11	177
C_{23}				
$[2\text{-}(4\text{-}EtO_2CCH_2OC_6H_4)C_6H_4]\text{-}(4\text{-}MeC_6H_4)SbO_2H$	$RR'SbCl_3 + $ aq. NaOAc in Me_2CO		Rxns. 12, 13	177

6. Rxn. with $4-MeC_6H_4N_2FeCl_4 \rightarrow 4-MeC_6H_4N_2 \cdot (RR'SbCl_4)$ (1069).

7. Rxn. with 5 N HCl $\rightarrow RR'SbCl$ and $RR'SbCl_3$ (?) (910, 1063, 1069).

8. Cyclization in Ac_2O contg. $H_2SO_4 \rightarrow$ p-(3-nitro-5-dibenzostibiolyl)benzoic acid ethyl ester (175).

9. Cyclization in Ac_2O contg. $H_2SO_4 \rightarrow$ p-(5-dibenzostiboilyl)benzoic acid ethyl ester (175).

10. Cyclization in Ac_2O contg. $H_2SO_4 \rightarrow$ 5-carbethoxyphenyl-3-methoxydibenzostiole Sb-dihydroxide (182).

11. Redn. with $SnCl_2$ in HCl $\rightarrow RR'SbCl$ (182).

12. Cyclization with Ac_2O contg. $H_2SO_4 \rightarrow$ 3-carbethoxymethoxy-5-p-tolyldibenzostibiole Sb-dihydroxide (182).

13. Rxn. with polyphosphoric acid $\rightarrow$ cleavage of the Sb-C bonds (182).

ORGANOANTIMONY TETRAHALIDES

Aryl- and heterocyclylantimony tetrahalides are prepared by:
1. Treating aryl- or heterocyclyl-dihalostibines, $RSbX_2$, with halogens.
2. Treating stibonic acids, $RSbO_3H_2$, with hydrohalic acids.
3. Reacting antimony trichloride with aryl- or heterocyclyldiazonium chlorides, RN_2Cl, or with aryldiazonium chlorozincates, $(RN_2)_2ZnCl_4$, and decomposing the organodiazonium chloroantimonites, RN_2SbCl_4, thus formed in ethanol or acetone in the presence of copper(I) chloride or copper-bronze. The diazotization of the corresponding amines can be carried out in the presence of antimony trichloride using alkyl nitrites in ethanol or sodium nitrite in hydrochloric or sulfuric acid.
4. Decomposition of complex compounds of aryldiazonium chlorides with aryldichlorostibines, $RN_2(RSbCl_3)$, in boiling hydrochloric acid. The complex compounds can be prepared by adding arylhydrazines to antimony trichloride dissolved in hydrochloric acid and containing copper(II) chloride or iron(III) chloride, or by adding aryldichlorostibines to aryldiazonium chlorides, used in excess, in acetic acid.

2-THIAZOLYLANTIMONY TETRACHLORIDE $C_3H_2Cl_4NSSb$ $\overline{SCH:CHN:C}SbCl_4$
Prepn.: by adding conc. H_2SO_4 to 2-aminothiazole in EtOH, followed by $SbCl_3$ in EtOH, cooling the mixt. to 0°, then adding $EtNO_2$ in small portions and completing the rxn. at 60°; 55% yield (844).
Props.: m. 133° (844).
Rxns. with: $SnCl_2$ + HCl gas in EtOH → $\overline{SCH:CHN:C}SbCl_2$ (844).
SO_2 in MeOH contg. KI → $\overline{SCH:CHN:C}SbCl_2$ (844).
Aq. NaOH → $\overline{SCH:CHN:C}SbO_3H_2$ (844).

2-OXO-5-PYRIDYLANTIMONY TETRACHLORIDE HYDROCHLORIDE $C_5H_4Cl_4NOSb.HCl$

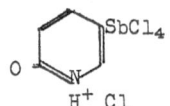

Prepn.: by treating a mixt. of 5-amino-2-hydroxypyridine, $SbCl_3$, and Cu_2Cl_2 in concd. HCl with aq. $NaNO_2$ (86).
Props.: m. 211° (86).

3-PYRIDYLSTIBINE TETRACHLORIDE HYDROCHLORIDE $C_5H_4Cl_4NSb.HCl$
$\overline{CH:CHCH:NCH:C}SbCl_4.HCl$
Prepn.: by treating a mixt. of 3-aminopyridine, $SbCl_3$, and Cu_2Cl_2 in concd. HCl with aq. $NaNO_2$; 50% yield (85,86).
Props.: Hydrochloride, m. 240° (85, 86).

4-BROMOPHENYLANTIMONY TETRACHLORIDE $C_6H_4BrCl_4Sb$ $4-BrC_6H_4SbCl_4$
Prepn.: by diazotizing $4-BrC_6H_4NH_2$ dissolved in MeOH contg. HCl, with n-pentyl nitrite in the presence of $SbCl_3$, and decompg. the rxn. in MeOH at 60°; 40% yield (668).

4-IODOPHENYLANTIMONY TETRACHLORIDE $C_6H_4Cl_4ISb$ $4-IC_6H_4SbCl_4$
Prepn.: by diazotizing $4-IC_6H_4NH_2$ dissolved in MeOH contg. HCl with n-pentyl
nitrite in the presence of $SbCl_3$, and decompg. the rxn. prod. in MeOH at 60°; 20%
yield (668).

3-NITROPHENYLANTIMONY TETRACHLORIDE $C_6H_4Cl_4NO_2Sb$ $3-O_2NC_6H_4SbCl_4$
Prepn.: by reacting $3-O_2NC_6H_4NH_2$ and $SbCl_3$ with $NaNO_2$ in H_2SO_4 and decompg. the
resulting m.-$O_2NC_6H_4N_2SbCl_4$ in EtOH in the presence of Cu_2Cl_2 (462).
Complex compd.: $C_5H_5N.H(3-O_2NC_6H_4SbCl_5)$, m. 187° (906).

4-NITROPHENYLANTIMONY TETRACHLORIDE $C_6H_4Cl_4NO_2Sb$ $4-O_2NC_6H_4SbCl_4$
Complex compds.: $PhN_2(4-O_2NC_6H_4SbCl_5)$, gray-yellow, decomp. 80° (1067). Kinetics
of decompn. (1071).
4-$EtOC_6H_4N_2(4-O_2NC_6H_4SbCl_5)$, colorless, decomp. 130° (1067).
$C_5H_5N.H(4-O_2NC_6H_4SbCl_5)$, m. 167-168° (906). Rxn. with Na gluconate gave gluconic
acid 4-nitrobenzenestibonate Na salt (1166).

3-CHLOROPEHNYLANTIMONY TETRACHLORIDE $C_6H_4Cl_5Sb$ $3-ClC_6H_4SbCl_4$
Prepn.: by treating $3-ClC_6H_4SbO_3H_2$ with concd. HCl, isolated as ammonium salt
(1001).
Complex compds.: $NH_4(3-ClC_6H_4SbCl_5)$, m. > 240° (1001).
$C_5H_5N.H(3-ClC_6H_4SbCl_5)$, crystals (1001).
Quinolinium. $(3-ClC_6H_4SbCl_5)$, crystals (1001).

4-CHLOROPHENYLANTIMONY TETRACHLORIDE $C_6H_4Cl_5Sb$ $4-ClC_6H_4SbCl_4$
Prepn.: By diazotizing $4-ClC_6H_4NH_2$ dissolved in MeOH contg. HCl with n-pentyl
nitrite in the presence of $SbCl_3$, and decompg. the rxn. prod. in MeOH at 60°;
50% yield (668).
By reacting $4-ClC_6H_4N_2Cl$ with $SbCl_3$, and decompg. the prod. in EtOH in the presence of Cu_2Cl_2 (462) or in Me_2CO at 0° in the presence of NaI, $FeCl_2$, Cu, or
Cu_2Cl_2 (1263).
Rxn. with: Na gluconate → gluconic acid 4-chlorobenzenestibonate Na salt (1166).
Aq. alkali → $4-ClC_6H_4SbO_3H_2$ (462).
Complex compd.: $PhN_2(4-ClC_6H_4SbCl_5)$, colorless, decomp. 106° (1067). Kinetics
of decompn. by Fe (1071).

3-HYDROXYPHENYLANTIMONY TETRACHLORIDE $C_6H_5Cl_4OSb$ $3-HOC_6H_4SbCl_4$
Prepn.: by diazotizing $3-HOC_6H_4NH_2$ dissolved in MeOH contg. HCl with n-pentyl
nitrite in the presence of $SbCl_3$, and decompg. the rxn. prod. in MeOH at 60°;
40% yield (668).

4-HYDROXYPHENYLANTIMONY TETRACHLORIDE $C_6H_5Sl_4OSb$ $4-HOC_6H_4SbCl_4$
Complex compd.: $4-O_2NC_6H_4N_2(4-HOC_6H_4SbCl_5)$, orange, decomp. 114° (1067).

PHENYLANTIMONY TETRACHLORIDE $C_6H_5Cl_4Sb$ $PhSbCl_4$
Prepn.: By adding a soln. of $SbCl_3$ in HCl to PhN_2Cl or to $(PhN_2)_2ZnCl_4$ and
decompg. the prod. in EtOH at 60° in the presence of $CuCl_2$ or Cu-bronze (462).
By treating $PhSbCl_2$ with chlorine (728).
By treating $PhSbO_3H_2$ with conc. HCl; the product was isolated as a complex with
pyridine hydrochloride (906).

By diazotizing aniline dissolved in MeOH contg. HCl with n-$C_5H_{11}NO_2$ in the presence of SbCl$_3$, and decompg. the prod. in MeOH at 60°; 55% yield (668).

Rxns. with: $(PhN_2)_2ZnCl_4 \rightarrow Ph_2SbCl_3$ (462).

Na gluconate → gluconic acid benzenestibonate, Na salt (1166).

LiAlH$_4$ or LiBH$_4$ → PhSbH$_2$ (1290).

H_2O or aq. alkali → PhSbO$_3$H$_2$ (462, 728).

SO_2 in HCl contg. HI → PhSbCl$_2$ (462).

Complex compds.: NH$_4$(PhSbCl$_5$), prepd. by decompg. PhN$_2$SbCl$_4$ in MeOH with Cu powder or with Cu$_2$Cl$_2$ at 33° in the presence of NH$_4$Cl (1219, 1220) or with Me$_2$CO-H$_2$O at 37° (1220).

$C_5H_5N.H$(PhSbCl$_5$), m. 133° (906).

PhN$_2$.(PhSbCl$_5$), kinetics of decompn. by Fe (1071).

The following complex compds. were prepd. by adding the corresponding aryldiazonium tetrachloroferrates, RN$_2$FeCl$_4$, dissolved in acetone to alc. soln. of PhAsO$_3$H$_2$ satd. with HCl; 4-MeC$_6$H$_4$N$_2$.(PhSbCl$_5$), colorless, m. 111° (dec.); 2,4-MeC$_6$H$_3$N$_2$.(PhSbCl$_5$), colorless, decomp. 83°; 4-ClC$_6$H$_4$N$_2$.(PhSbCl$_5$), colorless, decomp. 145°; 4-BrC$_6$H$_4$N$_2$.-(PhSbCl$_5$), decomp. 126°; 4-IC$_6$H$_4$N$_2$.(PhSbCl$_5$), decomp. 116-117°; 4-EtOC$_6$H$_4$N$_2$.(Ph-SbCl$_5$), decomp. 106°; 4-Me$_2$NC$_6$H$_4$N$_2$.(PhSbCl$_5$), orange, decomp. 136°; 2-O$_2$NC$_6$H$_4$N$_2$.-(PhSbCl$_5$), decomp. 126-127°; 3-O$_2$NC$_6$H$_4$N$_2$.(PhSbCl$_5$), decomp. 133.5°; 4-O$_2$NC$_6$H$_4$N$_2$.-(PhSbCl$_5$), decomp. 133°; 2-MeO$_2$CC$_6$H$_4$N$_2$.(PhSbCl$_5$), decomp. 107°; 4-EtO$_2$CC$_6$H$_4$N$_2$.-(PhSbCl$_5$), decomp. 94° (1067). The complex compds. decomp. in Me$_2$CO in the presence of Fe to the corresponding arylphenylstibinic acids, RPhSbO$_2$H (1069, 1071).

4-SULFAMYLPHENYLANTIMONY TETRACHLORIDE $C_6H_6Cl_4NO_2SSb$ 4-H$_2$NSO$_2$C$_6$H$_4$SbCl$_4$

Rxn. with Na gluconate → gluconic acid 4-sulfamylbenzenestibonate Na salt (1166).

Complex compd.: 4-EtOC$_6$H$_4$N$_2$(4-H$_2$NO$_2$SC$_6$H$_4$SbCl$_5$), colorless, decomp. 129° (1067).

4-AMINOPHENYLANTIMONY TETRACHLORIDE HYDROCHLORIDE $C_6H_6Cl_4NSb.HCl$
   4-(HCl.H$_2$N)C$_6$H$_4$SbCl$_4$

Rxn. with SbCl$_5$ in CHCl$_3$, followed by Et$_2$NH → product which could not be diazotized (572).

p-ANISYLANTIMONY TETRACHLORIDE $C_7H_7Cl_4OSb$ 4-MeOC$_6$H$_4$SbCl$_4$

Prepn.: by diazotizing 4-MeOHC$_6$H$_4$NH$_2$ dissolved in MeOH contg. HCl with n-C_5H_{11}-NO$_2$ in the presence of SbCl$_3$, and decompg. the prod. in MeOH at 60°; 50% yield (668).

p-TOLYLANTIMONY TETRACHLORIDE $C_7H_7Cl_4Sb$ 4-MeC$_6$H$_4$SbCl$_4$

Prepn.: By treating MeC$_6$H$_4$SbCl$_2$ with excess MeC$_6$H$_4$N$_2$Cl in AcOH and decompg. the complex by HCl; isolated as a double salt with NH$_4$Cl (128).

By adding p-MeC$_6$H$_4$NHNH$_2$HCl to a soln. of SbCl$_3$ and CuCl$_2$ in HCl under atm. oxygen, isolating the (p-MeC$_6$H$_4$N$_2$Cl)$_2$. p-MeC$_6$SbCl$_3$ ppt. and decompg. the complex by boiling with dild. HCl; isolated as p-MeC$_6$H$_4$SbCl$_4$.NH$_4$Cl (133).

By reacting p-MeC$_6$H$_4$N$_2$Cl with SbCl$_3$ in HCl and decompg. the resulting MeC$_6$H$_4$N$_2$-SbCl$_4$ with CuCl$_2$ in EtOH (462).

By diazotizing 4-MeC$_6$H$_4$NH$_2$ dissolved in MeOH contg. HCl with n-$C_5H_{11}NO_2$ in the presence of SbCl$_3$, and decompg. the rxn. prod. in MeOH at 60°; 50% yield (668).

Props.: Yellowish powder, does not m. 200° (133).

Rxns. with: H_2O → MeC$_6$H$_4$SbO$_3$H$_2$ (128).

SO_2 in alc. HCl contg. HI → MeC$_6$H$_4$SbCl$_2$ (133).

Na gluconate → gluconic acid 4-toluenestibonate Na salt (1166).
Complex compds.: $C_5H_5N.H(4-MeC_6H_4SbCl_5)$, m. 176-178° (906).
$PhN_2(4-MeC_6H_4SbCl_5)$, decomp. 104.5° (1066). Kinetics of decompn. by Fe (1071).
$NH_4(4-MeC_6H_4SbCl_5)$, prepd. by treating $4-MeC_6H_4N_2SbCl_4$ with 20% NaOH followed by HCl and NH_4Cl or with $CuCl_2$ (1221).

4-PHENETYLANTIMONY TETRACHLORIDE $C_8H_9Cl_4OSb$ $4-EtOC_6H_4SbCl_4$
Complex compds.: $PhN_2(4-EtOC_6H_4SbCl_5)$, yellow solid, decomp. 66° (1067). Kinetics of decompn. by Fe(1071).
$4-MeC_6H_4N_2(4-EtOC_6H_4SbCl_5)$ (1069).
$4-EtOC_6H_4N_2(4-EtOC_6H_4SbCl_5)$, yellow, decomp. 96.5-97° (1067).
$4-AcNHC_6H_4N_2(4-EtOC_6H_4SbCl_5)$, yellow, decomp. 84-85° (1067).

4-DIMETHYLAMINOPHENYLANTIMONY TETRACHLORIDE METHOCHLORIDE $C_9H_{13}Cl_5NSb$
$Me_3NC_6H_4SbCl_5$
Prepn.: by diazotizing $4-(Cl.Me_3N)C_6H_4NH_2.HCl$ with $NaNO_2$ at -5°, treating the soln. with Sb_2O_3 in fuming HCl, isolating the diazonium compd., and decompg. in HCl (d_{20} 1.075) on a steam bath; 10% yield (1000).
Props.: Colorless rosets; on heating decomp. without melting (1000).
Rxn. with H_2S → light yellow ppt. (1000).

2-NAPHTHYLANTIMONY TETRABROMIDE $C_{10}H_7Br_4Sb$ $2-C_{10}H_7SbBr_4$
Prepn.: by treating $2-C_{10}H_7SbO_3H_2$ with HBr (1001).
Complex compd.: $C_5H_5N.H(2-C_{10}H_7SbBr_4)$, orange-red crystals, m. 193-195 (1001).

1-NAPHTHYLANTIMONY TETRACHLORIDE $C_{10}H_7Cl_4Sb$ $1-C_{10}H_7SbCl_4$
Prepn.: By reacting $1-C_{10}H_7N_2Cl$ with $SbCl_3$ in HCl and decompg. the resulting $1-C_{10}H_7N_2SbCl_4$ in EtOH in the presence of Cu_2Cl_2 (462).
By treating $1-C_{10}H_7SbO_3H_2$ with HCl (906, 1001).
Rxn. with SO_2 in EtOH contg. I_2 → $1-C_{10}H_7SbCl_2$ (462).
Complex compds.: $C_5H_5N.H(1-C_{10}H_7SbCl_5)$, m. 186-188° (906); m. 187-189° (1001)
$NH_4(1-C_{10}H_7SbCl_5)$, orange-red ppt., m. >240° (1001).

2-NAPHTHYLANTIMONY TETRACHLORIDE $C_{10}H_7Cl_4Sb$ $2-C_{10}H_7SbCl_4$
Prepn.: By reacting $2-C_{10}H_7SbCl_2$ with excess $2-C_{10}H_7N_2Cl$ and decompg. the resulting $(2-C_{10}H_7N_2Cl)_2.2-C_{10}H_7SbCl_2$ complex in HCl under reflux; isolated as $NH_4(2-C_{10}H_7-SbCl_5)$ (131).
By reacting $SbCl_3$ and $CuCl_2$ or $FeCl_3$ with $2-C_{10}H_7NHNH_2.HCl$ in HCl, and decompg. the $(2-C_{10}H_7N_2Cl)_2.2-C_{10}H_7SbCl_2$ with NH_4Cl in HCl; isolated as $NH_4(2-C_{10}H_7SbCl_5)$ (133).
By reacting $2-C_{10}H_7N_2Cl$ with $SbCl_3$ in HCl, and decompg. the prod. with Cu_2Cl_2 in EtOH (462).
By treating $2-C_{10}H_7SbO_3H_2$ with HCl (906, 1001).
Props.: m. > 200° (131).
Rxns. with: H_2O → $2-C_{10}H_7SbO(OH)_2$ (131).
SO_2 in HCl contg. HI → $2-C_{10}H_7SbCl_2$ (133).
SO_2 in EtOH contg. I_2 → $2-C_{10}H_7SbCl_2$ (462).
Complex compds.: $C_5H_5N.H(2-C_{10}H_7SbCl_5)$, m. 199-200° (906); m. 200-202° (dec.) (1001).
$C_9H_7N.H(2-C_{10}H_7SbCl_5)$(quinolinium deriv.), decomp. 183° (906); m. 174-176° (dec.) (1001).

DIORGANOANTIMONY TRIHALIDES

Dialkyl- and diarylantimony trihalides, R_2SbX_3 and $RR'SbX_3$, are prepared or formed by:

1. Chlorination of diorganochlorostibines, R_2SbCl, or of secondary stibines, R_2SbH, in carbon tetrachloride and ether, respectively.
2. Treating tetraorganodistibines, $(R_2Sb)_2$, with chlorine in trichlorofluoromethane at $-78°$.
3. Recrystallization of stibinic acids, R_2SbO_2H or $RR'SbO_2H$, or of stibinic acid complex compounds $R_2SbO_2H.Sb_2O_3$ from hydrochloric acid.
4. Treating stibinic anhydrides $[R_2Sb(O)]_2O$, with hydrobromic acid to yield diorganoantimony tribromides.
5. Heating tertiary stibines, R_3Sb, with antimony trichloride.
6. Arylation of organodichlorostibines, $RSbCl_2$, with:
   (a) Aryldiazonium chlorides in ethanol in the presence of copper bronze or in acetic acid at $0-2°$ to form complex compounds $RN_2Cl.RSbCl_2$ and decomposing the latter in boiling diluted hydrochloric acid.
   (b) Aryldiazonium chloroantimonites, RN_2SbCl_4, or with aryldiazonium chlorozincates, $(RN_2)_2ZnCl_4$, in acetone, ethanol, or hydrochloric acid in the presence or absence of copper bronze.
   (c) Aryldiazonium chloroferrates, RN_2FeCl_4, in acetone.
7. Reacting aryldiazonium chloroantimonites, RN_2SbCl_4, with arylhydrazines in diluted hydrochloric acid in the presence of copper(II) chloride and decomposing the aryldiazonium complex compounds, $RN_2Cl.RSbCl_2$, with hydrochloric acid.
8. Reacting aryldiazonium chloroantimonates, RN_2SbCl_8, with phenylhydrazine in hydrochloric acid in the presence of iron and copper chlorides ($FeCl_3$ + $CuCl_2$).
9. Decomposing diaryliodonium chloroantimonites, $R_2I.SbCl_4$, with antimony powder in acetone to yield mixtures of diarylantimony trichlorides, triarylstibine dichlorides, and diarylchlorostibines.
10. Reacting antimony pentachloride with acetylene to yield bis(2-chlorovinyl)-antimony trichloride.
11. Reacting antimony trichloride with arylhydrazines in hydrochloric acid in the presence of copper(II) chloride and decomposing the product in boiling hydrochloric acid.

DI(2-CHLOROVINYL)ANTIMONY TRICHLORIDE $C_4H_4Cl_5Sb$ $(ClCH:CH)_2SbCl_3$
Prepn.: by reacting $SbCl_5$ with C_2H_2 at a temp. below $50°$; 75-80% yield (899).
Props.: Colorless, hydroscopic crystals, m. $51-52°$, sol. in dild. HCl (899).
Rxns. with: H_2O or aq. alkali $\rightarrow$ hydrolysis (899).
Alc. thiourea or benzyl sulfide $\rightarrow$ evolution of C_2H_2 (899).
$SO_2 \rightarrow$ di(2-chlorovinyl) sulfone (899).
Sb in $Et_2O \rightarrow (ClCH:CH)_2SbCl + ClCH:CHSbCl_2$ (899).

DIPHENYLANTIMONY TRICHLORIDE $C_{12}H_{10}Cl_3Sb$ Ph_2SbCl_3
Prepn.: By heating $PhN_2Cl.PhSbCl_2$ in dild. HCl; 26% yield (127, 130, 134).
By reacting PhN_2SbCl_8 with $PhNHNH_2$ in aq. HCl in the presence of $FeCl_3$ and $CuCl_2$; 16% yield (1061).

.By reacting PhSbCl$_2$ with (PhN$_2$)$_2$ZnCl$_4$ in EtOH (462).

By recrystallizing Ph$_2$SbO$_2$H from 5 N HCl (904, 1061, 1065, 1119, 1316).

By treating Ph$_2$SbO$_2$H.Sb$_2$O$_3$ with dild. HCl (1059, 1060).

By decompg. Ph$_2$I.SbCl$_4$ with Sb powder in Me$_2$CO, a mixt. of Ph$_2$SbCl$_3$, Ph$_3$SbCl$_2$, and Ph$_2$SbCl was obtd. (1072).

By reacting Ph$_2$SbH with Cl$_2$ in Et$_2$O (1295).

Props.: Needles, m. 171-172° (127, 904); m. 176-177° (1001); m. 175.5-176° (1059, 1060); m. 171-172° (1061, 1065); m. 172-175° (1074); m. 175° (1119); m. 176° (1295).
Monohydrate decomp. 173° (134). Inhibitory effect on corrosion of steel (30).
Rxns. with: SO$_2$ in HCl contg. HI → Ph$_2$SbI (134, 1119).
Aq. NaOH → Ph$_2$SbO$_2$H (462).
Slow neutrons → radioactive Ph$_2$SbCl$_3$ (875).
LiAlH$_4$ or LiBH$_4$ → Ph$_2$SbH (1290).
SnCl$_2$ → Ph$_2$SbCl (1316).
Ph$_2$ICl, followed by Zn dust in Me$_2$CO → Ph$_3$SbCl$_2$ (1074a).
Complex compds.: C$_5$H$_5$N.H(Ph$_2$SbCl$_4$), decomp. ~265° (1001); decomp. 263-264° (1060).
PhN$_2$(Ph$_2$SbCl$_4$), m. 148-162° (1074).
2-ClC$_6$H$_4$N$_2$(Ph$_2$SbCl$_4$), m. 146-148° (dec.) (1074).
4-ClC$_6$H$_4$N$_2$(Ph$_2$SbCl$_4$), decomp. 144-145.5° (1068); m. 146-147° (1074).
4-BrC$_6$H$_4$N$_2$(Ph$_2$SbCl$_4$), m. 139-140.5° (1068).
4-IC$_6$H$_4$N$_2$(Ph$_2$SbCl$_4$), yellow, decomp. 128-154° (1068).
2-O$_2$NC$_6$H$_4$N$_2$(Ph$_2$SbCl$_4$), yellow, decomp. 169.5-170° (1068).
3-O$_2$NC$_6$H$_4$N$_2$(Ph$_2$SbCl$_4$), pink, decomp. 138-140° (1068).
4-O$_2$NC$_6$H$_4$N$_2$(Ph$_2$SbCl$_4$), decomp. 134-135° (1068).
2-MeC$_6$H$_4$N$_2$(Ph$_2$SbCl$_4$), decomp. 132-164° (1068); m. 166-167° (1074).
4-MeC$_6$H$_4$N$_2$(Ph$_2$SbCl$_4$), decomp. 123° (1068); m. 128-130° (1074).
2-MeOC$_6$H$_4$N$_2$(Ph$_2$Cl$_4$), pink, m. 171-174° (dec.) (1074).
4-MeOC$_6$H$_4$N$_2$(Ph$_2$SbCl$_4$), colorless, m. 153° (1074).
2-EtOC$_6$H$_4$N$_2$(Ph$_2$SbCl$_4$), yellow, m. 163° (1074).
4-EtOC$_6$H$_4$N$_2$(Ph$_2$SbCl$_4$), decomp. 153-155.5° (1068).
4-Me$_2$NC$_6$H$_4$N$_2$(Ph$_2$SbCl$_4$), yellow, decomp. 147-147.5° (1068, 1074).
4-PhN$_2$C$_6$H$_4$N$_2$(Ph$_2$SbCl$_4$), brown, decompn. 136° (1068).
2-Antraquinonyldiazonium-(Ph$_2$SbCl$_4$), brown, decomp. 152-153.5° (1068).

DI-p-TOLYLANTIMONY TRICHLORIDE C$_{14}$H$_{14}$Cl$_3$Sb (4-MeC$_6$H$_4$)$_2$SbCl$_3$

Prepn.: By reacting 4-MeC$_6$H$_4$SbCl$_2$ with 4-MeC$_6$H$_4$N$_2$Cl in AcOH at 0-2° and boiling the adduct in dild. HCl (128).

By reacting 4-MeC$_6$H$_4$N$_2$SbCl$_4$ with 4-MeC$_6$H$_4$NHNH$_2$.HCl in the presence of CuCl$_2$ to form the 4-MeC$_6$H$_4$N$_2$Cl.4-MeC$_6$H$_4$SbCl$_4$ complex and treating the latter with HCl (133).

By decompg. 4-MeC$_6$H$_4$N$_2$SbCl$_4$ with Fe powder (2:1) in Me$_2$CO at 0°; 71% yield (904).

By crystallizing (4-MeC$_6$H$_4$)$_2$SbO$_2$H from 5 N HCl (910, 1065).

By decompn. of (4-MeC$_6$H$_4$)$_2$I.SbCl$_3$ with Sb powder in Me$_2$CO to form (4-MeC$_6$H$_4$)$_2$SbCl$_3$ along with (4-MeC$_6$H$_4$)$_3$SbCl$_2$ and (4-MeC$_6$H$_4$)$_2$SbCl (1072).

By-prod. in the rxn. of SbCl$_3$ with 4-MeC$_6$H$_4$NHNH$_2$ in hydrochloric acid in the presence of CuCl$_2$ and decompg. the prod. in boiling HCl (133).
Props.: White crystals, m. 155° (128, 133, 910, 1065); m. 155-156° (904).
Rxns. with: NH$_4$OH → (4-MeC$_6$H$_4$)$_2$SbO$_2$H (904).
R$_2$ICl in 5 N HCl, followed by decompn. with Zn dust in Me$_2$CO → R(4-MeC$_6$H$_4$)$_2$SbCl$_2$ → R(4-MeC$_6$H$_4$)$_2$Sb (1074a).
Complex compd.: 4-MeC$_6$H$_4$N$_2$[(4-MeC$_6$H$_4$)$_2$SbCl$_4$], m. 103-104° (dec.) (910).

TABLE LXX

DIORGANOANTIMONY TRIHALIDES

R_2SbX_3 or $RR'SbX_3$	Prepd. from	Props. and Remarks	Ref.
C_2			
$(F_3C)_2SbCl_3$	$(R_2Sb)_2 + Cl_2$	b. 13°/5 mm., m. 27°	258
C_{12}			
$(4-BrC_6H_4)_2SbCl_3$	$RSbO_2H + HCl$	m. 159°	904
	$R_2I.SbCl_4 + Sb$ in Me_2CO	Rxn. [1]	1072
$(4-FC_6H_4)_2SbCl_3$	$R_2SbO_2H + HCl$	m. 100°, remelts at 149-150°, rxn.[2]	132
$(4-IC_6H_4)_2SbCl_3$	$R_2SbO_2H + HCl$	m. 149°	1065
	$R_2I.SbCl_4 + Sb$ in Me_2CO		1072
$(3-O_2NC_6H_4)_2SbCl_3$	$R_2SbO_2H + HCl$	Decomp. 182-185° (sealed tube)	1061
$(4-O_2NC_6H_4)_2SbCl_3$	$R_2SbO_2H + HCl$	m. 183-185° Shrinks at 190°	1065, 904, 1055
$(4-ClC_6H_4)(4-O_2NC_6H_4)SbCl_3$	$RR'SbO_2H + HCl$	Decomp. 212-217°	1061
$(4-ClC_6H_4)_2SbCl_3$	$R_2SbO_2H + HCl$	Complex salts[3] m. 149-150°	910, 904, 910, 1065
$Ph(4-BrC_6H_4)SbCl_3$	$R_2I.SbCl_4 + Sb$ in Me_2CO	Rxn. [1], complex salt[4] $PhN_2.[Ph(4-BrC_6H_4)-SbCl_4]$, decomp. 150-165°	1072, 1059
$Ph(4-IC_6H_4)SbCl_3$	$RR'SbO_2H + HCl$	m. 137-138°	1059, 1069,
$Ph(4-O_2NC_6H_4)SbCl_3$	$RSbCl_2 + R'N_2FeCl_4$	$4-MeC_6H_4N_2.[Ph(4-O_2N-C_6H_4)SbCl_4]$, decomp. 98°	1070
$Ph(4-ClC_6H_4)SbCl_3$	$RR'SbO_2H + HCl$	m. 121°, adducts [5]	910, 1059
C_{13}			
$(4-O_2NC_6H_4)(4-MeC_6H_4)SbCl_3$	$RR'SbO_2H + HCl$	$PhN_2[(4-O_2NC_6H_4)(4-MeC_6H_4)SbCl_4]$, m. 74-76° (dec.)	910

$(4\text{-}ClC_6H_4)(4\text{-}MeOC_6H_4)SbCl_3$	$RR'SbO_2H + HCl$	$4\text{-}MeOC_6H_4N_2[(4\text{-}ClC_6H_4)(4\text{-}MeOC_6H_4)SbCl_4]$, m. $158°$ (dec.)	910
$(4\text{-}ClC_6H_4)(4\text{-}MeC_6H_4)SbCl_3$	$RN_2SbCl_4 + R'SbCl_2$ $RR'SbO_2H + HCl$	Complex salts [6], rxn. [7]	910
$Ph(4\text{-}MeC_6H_4)SbCl_3$	$RR'SbO_2H + HCl$	m. $150\text{-}151.5°$ m. $153.5\text{-}154°$, complex[8]	910, 1063 1069
C_{14}			
$[2,4\text{-}Me(Cl)C_6H_3]_2SbCl_3$	$R_2SbCl + Cl_2$ in CCl_4	m. $162°$	780
$(4\text{-}O_2NC_6H_4)(4\text{-}EtOC_6H_4)SbCl_3$		$4\text{-}MeC_6H_4N_2[(4\text{-}O_2NC_6H_4)(4\text{-}EtOC_6H_4)SbCl_4]$, decomp. $106°$	1069
$(PhCH_2)_2SbBr_3$	$[R_2Sb(O)]_2O + HBr$	m. $150\text{-}151°$ (dec.)	1230
$Ph(4\text{-}EtOC_6H_4)SbCl_3$		$4\text{-}MeC_6H_4N_2$. $[Ph(4\text{-}EtOC_6H_4)SbCl_4]$, decomp. $86°$	1059
$(4\text{-}MeOC_6H_4)(4\text{-}MeC_6H_4)SbCl_3$	$(RN_2)_2ZnCl_2 + R'SbCl_2$	$4\text{-}MeC_6H_4N_2[(4\text{-}MeOC_6H_4)SbCl_4]$, m. $99\text{-}100°$ (dec.)	910
$(PhCH_2)_2SbCl_3$	$R_3Sb + SbCl_3$	m. $157\text{-}158°$, rxn. [9]	1230
C_{16}			
$Ph(1\text{-}C_{10}H_7)SbCl_3$	$RSbCl_2 + (R'N_2)_2ZnCl_4$ in HCl + Cu	Colorless prisms, m.-155°	462
$Ph(2\text{-}C_{10}H_7)SbCl_3$	$RSbCl_2 + (R'N_2)_2ZnCl_2$ in HCl + Cu	m. $155°$	462
$(2\text{-}EtOC_6H_4)_2SbCl_3$	$R_2SbO_2H + HCl$	m. $149\text{-}150°$	904
$(4\text{-}EtO_2CC_6H_4)(4\text{-}MeC_6H_4)SbCl_3$	$RSbCl_2 + R'N_2SbCl_4$ in EtOH	m. $129°$; rxn.[10], salts [11]	131
$(2,4\text{-}Me_2C_6H_3)_2SbCl_3$	$R_2SbO_2H + HCl$	m. $169\text{-}170°$	904
$(3,5\text{-}Me_2C_6H_3)_2SbCl_3$		Rxn.[1]	1074a

1. Rxn. with R_2ICl in Me_2CO, followed by Zn dust → $R_2R'SbCl_2$ (1074a).
2. Rxn. with SO_2 in warm HCl contg. HI → R_2SbCl (132).
3. $4\text{-}MeC_6H_4N_2$. $[(4\text{-}ClC_6H_4)(4\text{-}O_2NC_6H_4)SbCl_4]$, decomp. $103\text{-}104°$ (910).
4. $4\text{-}MeC_6H_4N_2$. $[(4\text{-}ClC_6H_4)_2SbCl_4]$, decomp. $101°$ (910).
5. $4\text{-}ClC_6H_4N_2$. $[Ph(4\text{-}ClC_6H_4)SbCl_4]$, m. $109°$ (dec.) (910); $4\text{-}MeC_6H_4N_2$. $[Ph(4\text{-}ClC_6H_4)SbCl_4]$, m. $94\text{-}95°$ (dec.) (910), m. $111°$ (dec.) (1069).
6. $4\text{-}ClC_6H_4N_2[(4\text{-}ClC_6H_4)(4\text{-}MeC_6H_4)SbCl_4]$, m. $120\text{-}121°$; $4\text{-}MeC_6H_4N_2$. $[(4\text{-}ClC_6H_4)(4\text{-}MeC_6H_4)SbCl_4]$, m. $102°$ (dec.) (910).
7. Rxn. with NH_4OH → $RR'SbO_2H$ (910).
8. PhN_2. $[Ph(4\text{-}MeC_6H_4)SbCl_4]$, m. $63\text{-}64°$ (910); $4\text{-}MeC_6H_4N_2[Ph(4\text{-}MeC_6H_4)SbCl_4]$, decomp. $107°$ (1069).
9. Rxn. with dild. Na_2CO_3 soln.→ $[(PhCH_2)_2Sb(O)]_2O$ (1230).
10. Rxn. with $SnCl_2$ in HCl or in Me_2CO → $RR'SbCl$ (181, 173).

TABLE LXX (cont'd.)

DIORGANOANTIMONY TRIHALIDES

R_2SbX_3 or $RR'SbX_3$	Prepd. from	Props. and Remarks	Ref.
C18			
(4-EtO$_2$CC$_6$H$_4$)$_2$SbCl$_3$	RSbCl$_2$ + RN$_2$SbCl$_4$ in EtOH	m. 183-184°, salt[12]	181
C19			
(4-HO$_2$CC$_6$H$_4$)[4,2-Br(Ph)C$_6$H$_3$]SbCl$_3$	(4-EtO$_2$CC$_6$H$_4$)[4,2-Br(Ph)C$_6$H$_3$]SbCl$_3$ + AlCl$_3$ in C$_6$H$_6$	m. 190-192°, rxn.[10]	173
(4-MeC$_6$H$_4$)[4,2-Br(Ph)C$_6$H$_3$]SbCl$_3$	RSbCl$_2$ + R'N$_2$Cl in EtOH in Cu	m. 190-192°, rxn.[10]	173
(4-MeC$_6$H$_4$)[2(4-O$_2$NC$_6$H$_4$)C$_6$H$_4$]SbCl$_3$	RSbCl$_2$ + R'N$_2$SbCl$_4$ in EtOH	m. 160-162°, rxn.[13]	175
(4-MeC$_6$H$_4$)(2-PhC$_6$H$_4$)SbCl$_3$	RSbCl$_2$ + R'N$_2$SbCl$_4$ in EtOH + Cu	m. 153-154°, rxn.[10]	173
(4-EtO$_2$CC$_6$H$_4$)(1-C$_{10}$H$_7$)SbCl$_3$	RSbCl$_2$ + R'N$_2$SbCl$_4$ in EtOH	m. 135-136°, rxn.[10], salt[14]	181
(4-EtO$_2$CC$_6$H$_4$)(2-C$_{10}$H$_7$)SbCl$_3$	RSbCl$_2$ + R'N$_2$SbCl$_4$	m. 189-190°, rxn.[10]	181
C20			
(1-C$_{10}$H$_7$)$_2$SbCl$_3$	RSbCl$_2$ + (RN$_2$)$_2$ZnCl$_4$ in EtOH + Cu	Greenish crystals, m. 200°, rxn.[10]	462
(2-C$_{10}$H$_7$)$_2$SbCl$_3$	RSbCl$_2$ + (RN$_2$)$_2$ZnCl$_4$ in EtOH in Cu	m. 214°, rxn.[10]	462
(4-MeC$_6$H$_4$)[2-(4-MeOC$_6$H$_4$)C$_6$H$_4$]SbCl$_3$	RSbCl$_2$ + R'N$_2$SbCl$_4$ in EtOH + Cu	m. 153-154°, rxn.[13]	177
C21			
(EtO$_2$CC$_6$H$_4$)[4,2-Br(Ph)C$_6$H$_3$]SbCl$_3$	RSbCl$_2$ + R'N$_2$Cl in EtOH + Cu	m. 164-165°, rxns[10,13,14]	175
(4-EtO$_2$CC$_6$H$_4$)[2-(4-O$_2$NC$_6$H$_4$)C$_6$H$_4$]-SbCl$_3$	RSbCl$_2$ + R'N$_2$SbCl$_4$ in EtOH	m. 88-94°, rxns.[10,13]	175
(4-EtO$_2$CC$_6$H$_4$)(2-PhC$_6$H$_4$)SbCl$_3$	RSbCl$_2$ + R'N$_2$SbCl$_4$ in EtOH	m. 144-146°, rxn.[10,13]	175
(4-EtO$_2$CC$_6$H$_4$)(3-PhC$_6$H$_4$)SbCl$_3$	RSbCl$_2$ + R'N$_2$SbCl$_4$ in EtOH	m. 140°, salt[16]	181
(4-EtO$_2$CC$_6$H$_4$)(4-C$_6$H$_{11}$C$_6$H$_4$)SbCl$_3$	RSbCl$_2$ + 4-cyclo-C$_6$H$_{11}$-C$_6$H$_4$N$_2$Sb-Cl$_4$ in EtOH	m. 133-134°, salt[17], rxn.[10]	181
C22			
(4-MeC$_6$H$_4$)[2-(4-EtO$_2$CC$_6$H$_4$)C$_6$H$_4$]-SbCl$_3$	RSbCl$_2$ + R'N$_2$SbCl$_4$ in EtOH	Elongated prisms, m. 151-152°, rxn.[10]	173
(4-EtO$_2$CC$_6$H$_4$)[2,5-Ph(Me)C$_6$H$_3$]SbCl$_3$	RSbCl$_2$ + R'N$_2$SbCl$_4$ in EtOH + Cu	m. 164-166°, rxn.[10]	177
(4-EtO$_2$CC$_6$H$_4$)[2-(4-MeOC$_6$H$_4$)C$_6$H$_4$]-SbCl$_3$	RSbCl$_2$ + R'N$_2$SbCl$_4$ in EtOH + Cu	m. 169-170°, rxn.[13]	177
(4-EtO$_2$CC$_6$H$_4$)[3,4-Ph(MeO)C$_6$H$_3$]SbCl$_3$	RSbCl$_2$ + R'N$_2$SbCl$_4$ in EtOH	m. 162-163°, rxn.[10]	181
(4-EtO$_2$CC$_6$H$_4$)[2,5-Ph(MeO)C$_6$H$_3$]Sb-Cl$_3$	RSbCl$_2$ + R'N$_2$SbCl$_4$ in EtOH + Cu	m. 135-141°	177

C_{23}		
$(4\text{-MeC}_6\text{H}_4)[2\text{-}(4\text{-EtO}_2\text{CCH}_2\text{OC}_6\text{H}_4)\text{C}_6\text{H}_4]\text{RSbCl}_2 + \text{R}''_2\text{SbCl}_4$ in EtOH + Cu	m. 131-132° rxn.[13]	177
SbCl_3		

11. Complex salt with pyridine   hydrochloride, m. 230° (181).
12. Complex salt with pyridine   hydrochloride, m. 244° (dec.) (181).
13. Hydrolysis by NaOAc in Me$_2$CO → RR'SbO$_2$H (175, 177).
14. Complex salt with pyridine   hydrochloride, m. 216-218° (181)
15. Rxn. with AlCl$_3$ in Me$_2$CO → (4-HO$_2$CC$_6$H$_4$)[4,2-Br(Ph)C$_6$H$_3$]SbCl$_3$ (173).
16. Complex salt with pyridine   hydrochloride, m. 117-122° (dec.) (181).
17. Complex salt with pyridine   hydrochloride, m. 240° (dec.) (181).

DIORGANOANTIMONY HALIDE HYDROXIDES

Diorganoantimony dichloride hydroxides, R_2SbCl_2OH, were prepared by treating stibinic acids, R_2SbO_2H, with hydrochloric acid.

Diorganoantimony chloride dihydroxides, $R_2SbCl(OH)_2$, were prepared by decomposing aryldiazonium chloroantimonites, RN_2SbCl_4, with iron powder in acetone at $0°$ and diluting the reaction mixture with alcohol (904).

DI(4-IODOPHENYL)ANITMONY DICHLORIDE HYDROXIDE $C_{12}H_9Cl_2I_2OSb$ $(4-IC_6H_4)_2Sb(OH)Cl$
Prepn.: by treating $(4-IC_6H_4)_2SbO_2H$ with HCl (904).
Props.: m. 145-146° (904).

DI-p-TOLYLANTIMONY DICHLORIDE HYDROXIDE $C_{14}H_{15}Cl_2OSb$ $(4-MeC_6H_4)_2Sb(OH)Cl_2$
Prepn.: by treating $(4-MeC_6H_4)_2SbO_2H$ with hot HCl (1061).
Props.: Decomp. 205-207° (1061).

DI-(2-CHLOROPHENYL)ANTIMONY CHLORIDE DIHYDROXIDE $C_{12}H_{10}Cl_3O_2Sb$
   $(2-ClC_6H_4)_2Sb(OH)_2Cl$
Prepn.: by decompg. $2-ClC_6H_4N_2SbCl_4$ with Fe powder in Me_2CO at $0°$ and diluting
the rxn. mixt. with alc. (904).
Props.: m. 174-175° (904).

DI-2,4-XYLYLANTIMONY DICHLORIDE HYDROXIDE $C_{16}H_{19}Cl_2OSb$ $(2,4-Me_2C_6H_3)_2Sb(OH)Cl_2$
Prepn.: by treating $2,4-Me_2C_6H_3N_2SbCl_4$ with Fe powder in Me_2CO at $0°$ and diluting
the rxn. mixt. with alc.; 36% yield (904).
Rxn. with: $H_2O → (2,4-Me_2C_6H_3)_2SbO_2H$ (904).

TRIORGANOSTIBINE DIHALIDES

Trialkyl-, trialkenyl-, triaralkyl- and triarylstibine dihalides are prepared as
follows:
   1. From tertiary stibines, R_3Sb, by:
       (a) Reacting with halogens in halogenated hydrocarbons or with hydrohalic
            acids.
       (b) Reacting with alkyl halides in carbon disulfide.
       (c) Reacting with alkyl or aryl iodides in UV light in organic solvents.
       (d) Reacting with aryldiazonium chlorides in acetic acid.
       (e) Treating with aluminum chloride, iron(III) chloride, or bismuth(III)
            chloride.
   2. From halodiorganostibines, R_2SbX, by:
       (a) Reacting with Grignard compounds, isolating the tertiary stibines
            thus formed, and treating the latter with halogens in halohydrocarbons
       (b) Reacting with aryldiazonium chlorides, chloroantimonites, or chloro-
            zincates, $(RN_2Cl, RN_2SbCl_4,$ or $(RN_2)_2ZnCl_4)$ in acetic acid, acetone,
            or in ethanol in the presence of copper bronze.
   3. From dihaloörganostibines, $RSbX_2$, by:
       (a) Reacting with aryldiazonium chlorides or chloroantimonites $(RN_2Cl$ or

RN_2SbCl_4) in acetone or in ethanol in the presence of copper bronze to yield triorganostibine dichlorides along with diorganoantimony trichlorides.

4. From antimony trichloride by:
   (a) Forming complex salts with aryldiazonium chlorides, RN_2SbCl_4, and decomposing the salts with iron or antimony powder in acetone to yield triorganostibine dichlorides along with diorganoantimony trihalides and chlorodiorganostibines.
   (b) Forming complex salts with diaryliodonium chlorides, R_2ISbCl_4, and decomposing the salts with antimony powder in acetone to yield mixtures of triorganostibine dichlorides, diorganoantimony trichlorides, and chlorodiorganostibines.
   (c) Diazotization of aromatic amines in the presence of antimony trichloride in ethyl acetate containing hydrochloric acid, then treating the reaction mixture with calcium chloride and decomposing with zinc dust to yield triorganostibine dichlorides along with halodiorganostibines.
   (d) Treating with organolithium compounds in ether to form tertiary stibines along with small amount of tertiary stibine dichlorides.

5. From antimony pentachloride by:
   (a) Passing acetylene through a solution of antimony pentachloride in 50% alkali in the presence of mercury(II) chloride at 80-100° to form tris(2-chlorovinyl)stibine dichloride.
   (b) Complexing with aryldiazonium chlorides, decomposing the complex, RN_2SbCl_6, with iron powder in acetone, and treating the product with hydrochloric acid, followed by ethanol.

6. From antimony powder by reacting with aryldiazonium chlorides or aryldiazonium chlorozincates, $(RN_2)_2ZnCl_4$, in acetone under reflux in the presence of calcium carbonate.

7. From diorganoantimony trihalides by reacting with diaryldiodonium chlorides in acetone or in 5 N hydrochloric acid and treating the complex thus formed with zinc dust in acetone.

8. From quaternary stibonium perhalides by pyrolysis.

9. From pentaorganoantimony compounds by treating with concentrated hydrohalic acids or with halogens in halogenated hydrocarbons.

10. From tertiary stibine oxides, hydroxide halides, or diacetates, $(R_3SbO$, $R_3Sb(OH)Cl$, or $R_3Sb(OAc)_2.2AcOH)$ by treating with hydrohalic acids.

Tertiary stibine difluorides were prepared by treating the corresponding dichlorides with hot solutions of potassium fluoride in aqueous alcohol. Attempted reduction of tertiary stibine dibromides with tin(II) chloride in ethanol containing aqueous hydrochloric acid gave the corresponding tertiary stibine dichlorides.

TRIS(TRIFLUOROMETHYL)STIBINE DICHLORIDE $C_3Cl_2F_9Sb$ $(CF_3)_3SbCl_2$
Prepn.: by treating $(CF_3)_3Sb$ with Cl_2 (violent rxn.) (258, 336).
Props.: m. -34°, b. 110° (dec.); $log_{10}P = 7.290-2024/T$ between -30 and 50° (258); s. 101-102° (336). Monohydrate, white solid, m. 51° (336). Dihydrate, deliquescent crystals (336).
Rxns. with: $Hg \rightarrow (CF_3)_3Sb$ (258).
$H_2O \rightarrow$ mono- and dihydrate (336).

AgNO$_3$ → AgCl (336).
Ag$_2$O in H$_2$O, followed HCl → H[(CF$_3$)$_3$Sb(OH)$_3$] (336).
NOCl → NO[(CF$_3$)$_3$SbCl$_3$] (336).
Complex salt with pyridine, infusible substance (258).

TRIMETHYLSTIBINE DIBROMIDE C$_3$H$_9$Br$_2$Sb Me$_3$SbBr$_2$
Prepn.: by treating Me$_3$Sb with Br$_2$ in Et$_2$O (1313, 1314).
Props.: M. 192-193° (decompn.) (131). Dissoc. into Me$_2$SbBr + MeBr (139a).
Mol. structure and bonding (758).
Rxns. with: H$_2$O → hydrolysis to Me$_3$Sb(OH)Br (dissocn. const. detd.) (947).
Polarographic redn. (1094).
LiBH$_4$ or LiAlH$_4$ → Me$_3$Sb (1288, 1314).
MeLi in Et$_2$O under N → Me$_5$Sb (1314).

TRIS(2-CHLOROVINYL)STIBINE DICHLORIDE C$_6$H$_6$Cl$_5$Sb (ClCH:CH)$_3$SbCl$_2$
Prepn.: by passing C$_2$H$_2$ through a soln. of SbCl$_5$ in 50% alkali contg. HgCl$_2$ at
80-100°, completing the rxn. at 150-175°, and fractionating the prod. at 165-170°/
3 mm. The distillate contains the trans-isomer (80% yield), while the distn.
residue contains the cis-isomer (3-4% yield) (898, 899).
Props.: The compound occurs in two stereoisomeric forms. The cis-isomer, m. 61-
62°, is more sol. in org. solvents than the trans-isomer, m. 93-94° (898, 899).
The trans isomer decomp. at 200-250° into C$_2$H$_2$, ClCH:CHCl, (ClCH:CH-)$_2$, and SbCl$_3$
(899). Crystal structure and bond data (1177, 1178). Both isomers show some
characteristics of complex compds. (898).
Rxns. with: I and Br, both isomers yield ClCH:CHI and ClCH:CHBr with elimination
of SbCl$_2$I$_3$ and SbCl$_2$Br$_3$, resp. (899).
NaHSO$_3$ in H$_2$O-EtOH → cis- and trans-(ClCH:CH)$_3$Sb (898, 899).
HgCl$_2$, both isomers are converted into ClCH:CHHgCl + SbCl$_5$ (898).
AgNO$_3$ → [(ClCH:CH)$_2$Sb]$_2$AgNO$_3$ + AgC:CAg.AgNO$_3$ (899).
Use: Inhibitor of acid corrosion of steel (29).

TRIPROPENYLSTIBINE DIBROMIDE C$_9$H$_{15}$Br$_2$Sb (CH$_3$CH:CH)$_3$SbBr$_2$
Prepn.: by treating cis- and trans-(CH$_3$CH:CH)$_3$Sb with Cl$_2$ (913).
Props.: The compound forms two isomers. The cis isomer is crystalline substance,
while the trans form is liquid (913, 961). Refractive index, density, and mol. re
fraction data (915). IR data (916).
Rxn. with cis- and trans-CH$_3$CH:CHLi → cis- and trans-(CH$_3$CH:CH)$_5$Sb (913, 916).

TRIS(4-BROMOPHENYL)STIBINE DICHLORIDE C$_{18}$H$_{12}$Br$_3$Cl$_2$Sb (4-BrC$_6$H$_4$)$_3$SbCl$_2$
Prepn.: By treating (4-BrC$_6$H$_4$)$_3$Sb with dry Cl$_2$ (491, 513) or with Cl$_2$ in CHCl$_3$
(949).
By reacting 4-BrC$_6$H$_4$N$_2$Cl with Sb and CaCO$_3$ in Me$_2$CO under reflux (780).
By treating (4-BrC$_6$H$_4$)$_2$SbCl$_3$ with (4-BrC$_6$H$_4$)$_2$ICl in 5 N HCl and decompg. the
prod. with Zn dust in Me$_2$CO; 36% yield (1074a).
By decompg. (4-BrC$_6$HR)$_2$I.SbCl$_4$ with Sb powder in Me$_2$CO, a mixt. of (4-BrC$_6$H$_4$)$_3$
SbCl$_2$, (4-BrC$_6$H$_4$)$_2$SbCl$_3$, and (4-BrC$_6$H$_4$)$_2$SbCl is formed (1072).
By-prod. from decompn. of 4-BrC$_6$H$_4$N$_2$SbCl$_4$ with Fe in Me$_2$CO; 9% yield along wit
90% (4-BrC$_6$H$_4$)$_2$SbO$_2$H (904).
Props.: Needles, m. 200-201° (513); m. 185° (491); m. 200° (780); m. 198.5-199°
(904); m. 184-185° (949); m. 197° (1074a).

Rxns. with: KF in aq. EtCH → $(4\text{-}BrC_6H_4)_3SbF_2$ (491).
$H_2NNH_2 \cdot H_2O$ → $(4\text{-}BrC_6H_4)_3Sb$ (949).

TRI-4-CHLOROPHENYLSTIBINE DICHLORIDE $C_{18}H_{12}Cl_5Sb$ $(4\text{-}ClC_6H_4)_3SbCl_2$
Prepn.: By bubbling Cl through a soln. of $(4\text{-}ClC_6H_4)_3Sb$ in ice-cold $CHCl_3$ (1316).
By treating $(4\text{-}ClC_6H_4)_3Sb$ with dry Cl (491, 513).
By reacting $4\text{-}ClC_6H_4N_2Cl$ with Sb and $CaCO_3$ in Me_2CO under reflux (780).
By decomp.g $(4\text{-}ClC_6H_4)_2I.SbCl_4$ with Sb powder in Me_2CO, a mixt. of R_3SbCl_2,
R_2SbCl_3, and R_2SbCl was obtd. (1072).
By treating $(4\text{-}ClC_6H_4)_2SbCl_3$ with $(4\text{-}ClC_6H_4)_2ICl$ in 5 N HCl and decompg. the
prod. with Zn dust in Me_2CO; 46% yield (1074a).
By-prod. from decompn. of $4\text{-}ClC_6H_4N_2SbCl_4$ with Fe in Me_2CO; 30% yield along with
$(4\text{-}ClC_6H_4)_2SbO_2H$ (904, 1065).
By-prod. from a rxn. of $4\text{-}ClC_6H_4SbI_2$ with $4\text{-}ClC_6H_4N_2SbCl_4$ in Me_2CO (910).
Props.: Needles, m. 193-193.5° (513); m. 189.6° (491); m. 193° (780, 1065); m. 193-
194° (904); m. 190° (910); m. 190° (910); m. 186-186.5° (1074a); m. 189.5-190.5°
(1316).
Rxn. with KF in aq. EtOH → $(4\text{-}ClC_6H_4)_3SbF_2$ (491).

TRIPHENYLSTIBINE DICHLORIDE $C_{18}H_{15}Cl_2Sb$ Ph_3SbCl_2
Prepn.: By passing Cl gas into Ph_3Sb dissolved in petr. ether (736).
By reacting PhN_2Cl with Sb in Me_2CO in the presence of $CaCO_3$ (780, 1272, 1273).
By reacting PhN_2SbCl_4 or $(PhN_2)ZnCl_4$ with Sb in Me_2CO in the presence of $CaCO_3$
(780).
By reacting Ph_3Sb with $AlCl_3$ or $FeCl_3$ (815) or with $BiCl_3$ in $CHCl_3$ (817).
By reacting Ph_2SbCl with PhN_2Cl (1:1) in AcOH and decompg. the prod. in $CHCl_3$
(130).
By reacting Ph_2SbCl_3 with Ph_2ICl in 5 N HCl and decompg. the resulting complex
salt with Zn dust in Me_2CO; 50% yield (1074a).
By treating Ph_3Sb with PhN_2Cl in AcOH (130).
By reacting $PhSbCl_2$ with $(PhN_2)_2ZnCl_2$ in alc. soln. in the presence of a Cu
catalyst (462).
By decompg. $Ph_2I.SbCl_4$ with Sb powder in Me_2CO, a mixt. of Ph_3SbCl_2, Ph_2SbCl_3,
and Ph_2SbCl was obtd. (1072).
By hydrolyzing $Ph_3Sb(OAc)_2 \cdot 2AcOH$ with alc. HCl (910).
By treating $Ph_5Sb \cdot 0.5C_6H_{12}$ with excess Cl in CCl_4 (1311).
By diazotization of $PhNH_2$ in EtOAc contg. HCl in the presence of $SbCl_3$ at 5-
10°, followed by stirring with $CaCl_2$ for 1-2 hrs. and decompn. with Zn dust at 65-
75°; 46.2% yield along with Ph_2SbCl. When other solvents, e. g. EtOH, Me_2CO, C_6H_6,
$CHCl_3$, AmOAc, iso-Am_2O, MeOH, HCO_2Me, H_2O, or NH_3, were used instead of EtOAc, the
yield was lower, and PhSbO and Ph_2SbOAc were formed (903).
Props.: Colorless needles, m. 143° (736); m. 142-143° (130, 910); m. 141.5-142.5°
(757); m. 142-144° (1311); m. 139-142° (903); m. 142.5-143° (1074a). Dipole mc-
ment (603, 950, 1147b). Structure (950, 1147b). Magnetic data (968, 969); Polarographic redn. (1223). UV spectrum (598). Crystal structure (1176a).
Rxns. with: KF in aq. EtOH → Ph_3SbF_2 (491).
$LiAlH_4$ or $LiBH_4$ → Ph_3Sb (1290).
PhMgBr, followed by ice and 48% HBr → Ph_4SbBr (1302).
$H_2NNH_2 \cdot H_2O$ → Ph_3Sb (1316).
Slow neutrons → radioactive Sb deriv. (875).

TRI-p-TOLYLSTIBINE DICHLORIDE $C_{21}H_{21}Cl_2Sb$ $(4\text{-MeC}_6H_4)_3SbCl_2$

<u>Prepn.</u>: By treating $(4\text{-MeC}_6H_4)_2SbCl_3$ with $(4\text{-MeC}_6H_4)_2ICl$ in 5 N HCl and decompg. the prod. with Zn dust in Me_2CO at r. t. (1074a).

By decomp.g $(4\text{-MeC}_6H_4)_2ISbCl_4$ with Sb powder in Me_2O $(4\text{-MeC}_6H_4)_3SbCl_2$, $(4\text{-Me-}C_6H_4)_2SbCl_3$, and $(4\text{-MeC}_6H_4)_2SbCl$ are formed (1072).

By-prod. from decompn. of $4\text{-MeC}_6H_4N_2SbCl_6$ with Fe in Me_2CO, followed by HCl; 26.9% yield along with $(4\text{-MeC}_6H_4)_2SbO_2H$ (1061, 1065).

<u>Props.</u>: m. 156-156.5° (1061); m. 156° (1065); m. 155.5-156° (1074a). Crystal structure (491, 1337); Magnetic susceptibility (969).

<u>Rxn.</u> with $H_2NNH_2.H_2O \rightarrow (4\text{-MeC}_6H_4)_3Sb$ (1316).

TABLE LXXI
TRIORGANOSTIBINE DIHALIDES

R_3SbX_2	Prepd. from	Yield %	Props. and Remarks	Ref.
C_3				
$(CF_3)_3SbBr_2$	$R_3Sb + Br_2$ at $-30°$		m. $-16°$, decompn.[1]	258, 334
$(ClCH_2)_3SbBr_2$	$R_3Sb + Br_2$ in CCl_4		m. $90-90.5°$	1320-1 / 139a
Me_3SbCl_2	$R_3Sb + HCl$		Mol. structure and bonding / Polarographic redn.	758 / 1094, 1223
C_6				
$(CH_2{:}CH)_3SbBr_2$	$R_3Sb + Br_2$		b. $117°/2$ mm., n^{29} 1.6400, d_{20} 2.1152	914-5
$(CH_2{:}CH)_3SbI_2$	$R_3Sb + I_2$		RSb 15.05, rxn.[2] / Yellow oil, decompn.[3]	915 / 773
C_9				
$(CH_3CH{:}CH)_3SbI_2$	$R_3Sb + I_2$		cis- and trans-isomer, IR data	916
C_{12}				
$(OCH{:}CHCH{:}C{-})_3SbCl_2$	$R_3Sb + Cl_2$	60	Fine colorless crystal, m. 172°	345, 348
$(Me_2CHCH_2)_3SbBr_2$	$R_3Sb + Br_2$		m. 95°	1330-1
$(Me_3SiCH_2)_3SbBr_2$	$R_3Sb + Br_2$		m. 118-120°	1121
C_{14}				
$EtPh_2SbBr_2$	$Ph_3Sb + EtBr$ in CS_2		m. 158°	731
C_{18}				
$(2,4\text{-}Cl_2C_6H_3)_3SbCl_2$	$RN_2{.}SbCl_6 + Fe$ in Me_2CO, followed by $HCl + EtOH$	60	m. 235°	1065
$(4\text{-}ClC_6H_4)_3SbBr_2$	$R_3Sb + Br_2$		Lustrous plates, m. 189.5-190°	513

1. Decomposes at 20° → $(CF_3)_2SbBr$, CF_3SbBr_2, $(CF_3)_3Sb$, CF_3Br, and $SbBr_3$ (258).
2. Rxn. with $CH_2{:}CHMgBr$ → $(CH_2{:}CH)_5Sb$ (914).
3. On heating in vacuo decomp. → SbI_3 and $(CH_2{:}CH)_2SbI$ (773).

TABLE IXXI (cont'd.)
TRIORGANOSTIBINE DIHALIDES

R_3SbX_2	Prepd. from	Yield %	Props. and Remarks	Ref.
$(4\text{-BrC}_6\text{H}_4)_2(4\text{-ClC}_6\text{H}_4)\text{SbCl}_2$	$R_2SbCl_3 + R_2ICl$ in Me_2CO + Zn dust	35	m. 193.5-194.5°	1074a
$(4\text{-BrC}_6\text{H}_4)_3\text{SbF}_2$	R_3SbCl_2 + KF		m. 149-149.5°	491
$(4\text{-BrC}_6\text{H}_4)_3\text{SbI}_2$	$R_3Sb + I_2$ soln.		Pale yellow needles, m. 155-156° (dec.)	513
$(4\text{-BrC}_6\text{H}_4)_3\text{SbBr}_2$	$R_3Sb + Br_2$ in $CHCl_3$		Needles, m. 182°	513
$(4\text{-IC}_6\text{H}_4)_3\text{SbCl}_2$	$R_2I.SbCl_4$ + Sb powder in Me_2CO		R_2SbCl_3 and R_2SbCl are formed as by-prod.	1072
$(4\text{-ClC}_6\text{H}_4)_3\text{SbF}_2$	R_3SbCl_2 + KF		m. 115.5-116°	491
$(4\text{-ClC}_6\text{H}_4)_3\text{SbI}_2$	$R_3Sb + I_2$ soln.		Yellow plates, m. 137-183° (dec.).	513
$(4\text{-ClC}_6\text{H}_4)_2(4\text{-O}_2\text{NC}_6\text{H}_4)\text{SbCl}_2$	$R_2SbCl + R'N_2SbCl_4$ in Me_2CO	45	m. 202°	910, 1063
$(2\text{-ClC}_6\text{H}_4)_3\text{SbCl}_2$	$R_2SbCl + RN_2SbCl_4$ in Me_2CO	73	m. 204-206°	910, 1063
	RN_2SbCl_6 + Fe in Me_2CO	73	m. 205-206°	1061, 1063
Ph_3SbBr_2	$R_3Sb + Br_2$		m. 214-215°	736
	Ph_4SbBr_3 at 135°		m. 217-218°	1311
	$Ph_5Sb.0.5C_6H_{12} + 45\%$ HBr	90	rxns.4, 5	1311
			m. 216°	1192
			Magnetic suscept.	969
Ph_3SbF_2	Ph_3SbCl_2 + KF		m. 115°, rxn.6	491
Ph_3SbI_2	Ph_3Sb + PhI or MeI in UV light		Rxn.7	1054, 1055
			Magnetic data	969
C_{19}				
$(4\text{-BrC}_6\text{H}_4)_2(4\text{-MeC}_6\text{H}_4)\text{SbCl}_2$	$R_2SbCl_3 + R_2ICl$ in Me_2CO + Zn dust	28	m. 182-183°	1074a
$(4\text{-ClC}_6\text{H}_4)_2(4\text{-MeC}_6\text{H}_4)\text{SbCl}_2$	$R_2SbCl_3 + R_2ICl$ in Me_2CO + Zn dust	30	m. 185°	1074a
C_{20}				
$(4\text{-MeC}_6\text{H}_4)_2(4\text{-BrC}_6\text{H}_4)\text{SbCl}_2$	$R_2SbCl_2 + R_2ICl$ in Me_2CO + Zn dust	47	m. 164-165°	1074a
C_{21}				
$[5,2\text{-Cl(MeO)C}_6\text{H}_3]_3\text{SbCl}_2$ 8	RN_2Cl + Sb in AcOH under reflux		m. 281° (dec.)	730

Compound	Method of preparation	Yield	Properties	References
[4,2-Cl(Me)C₆H₃]₃SbCl₂	(RO₂/2.2H₂O₄ + Sb + CaCO₃ in Me₂CO		m. 204	780
[5,2-Cl(Me)C₆H₃]₃SbCl₂ [8]	RN₂Cl + Sb in Me₂CO		m. 238°	780
(2-MeOC₆H₄)₃SbBr₂	R₃Sb + Br₂		Glystening crystals, m. 225-226°	513
(3-MeOC₆H₄)₃SbBr₂	R₃Sb + Br₂		Pale yellow needles, m. 74.5-75.5°	513
(3-MeC₆H₄)₃SbBr₂			Magnetic data	969
(4-MeC₆H₄)₃SbBr₂			Crystal structure Magnetic data	491, 1337 969
(PhCH₂)₃SbBr₂	R₃Sb + Br₂ R₃SbO + HBr		m. 107-109°	1230 1230
(2-MeOC₆H₄)₃SbCl₂	R₃Sb + Cl₂ RN₂SbCl₄ + Fe in Me₂CO	67	White crystals, m. 237-238° m. 245°	513 904
(3-MeOC₆H₄)₃SbCl₂	R₃Sb + Cl₂		White crystals, m. 81.5-82.5°	513
(3-MeC₆H₄)₃SbCl₂	RLi + SbCl₃ in Et₂O		m. 137°, crystal structure, rxn.[o]	491
(PhCH₂)₃SbCl₂	R₃Sb + Cl₂ R₃SbO + HCl		Magnetic data m. 100-101°	1230 1230 969
(2-MeC₆H₄)₃SbF₂	R₃SbCl₂ + KF		m. 195.5°	491
(3-MeC₆H₄)₃SbF₂	R₃SbCl₂ + KF		m. 108.5-109°; crystal structure	491
(4-MeC₆H₄)₃SbF₂	R₃SbCl₂ + KF		m. 118-118.5°, crystal structure	491, 1337
(2-MeOC₆H₄)₃SbI₂	R₃Sb + I₂		Yellow solid, m. 141-142° (dec.)	513
(3-MeOC₆H₄)₂SbI₂	R₃Sb + I₂		Pale yellow plates, m. 99.5-100°	513
(4-MeC₆H₄)₃SbI₂			Crystal structure Magnetic data	491, 1337 959
(n-C₇H₁₅)₃SbI₂	R₃Sb + I₂			1227

4. Rxn. with H_2S → Ph_3SbS (736).
5. Rxn. with PhLi → Ph_5Sb (1311).
6. Bombardment by slow neutrons → the Sb^{122} deriv. (893).
7. Rxn. with alc. KOH → $Ph_3Sb(OH)_2$ (1055).
8. By-prod. along with the tertiary stibine (780).
9. By-prod. along with [4,2-Cl(Me)C₆H₃]₂SbCl (780).
10. Rxn. with KF soln. in aq. EtOH → the corresponding tertiary stibine difluoride (491).

TABLE LXXI (cont'd.)

TRIORGANOSTIBINE DIHALIDES

R_3SbX_2	Prepd. from	Yield %	Props. and Remarks	Ref.
C₂₂				
$(2\text{-EtOC}_6H_4)_2(4\text{-O}_2NC_6H_4)SbCl_2$	$R_2SbCl + R'N_2SbCl_4$ in Me_2CO	82	m. 227-228°	910, 1063
$(2\text{-EtOC}_6H_4)_2(2\text{-ClC}_6H_4)SbCl_2$	$R_2SbCl + (R'N_2)_2ZnCl_4$ in Me_2CO	74	m. 189-190°	910, 1063
$(2,4\text{-Me}_2C_6H_3)_2PhSbCl_2$	$R_2R'Sb(OH)Cl + HCl$		m. 162°	1069
C₂₄				
$(4\text{-AcNHC}_6H_4)_3SbCl_2$	$RN_2SbCl_6 + Fe$ in Me_2CO followed by $HCl + EtOH$	76	m. 170°	1065
$(2\text{-EtOC}_6H_4)_3SbBr_2$	$R_3Sb + Br_2$		Pale yellow needles, m. 237-238° (dec.)	513
$(2\text{-EtOC}_6H_4)_3SbCl_2$	$RN_2SbCl_6 + Fe$ in Me_2CO; $R_3Sb + Cl_2$	49	m. 230-231°; Colorless crystals, m. 231-232° (dec.)	904, 513
$(3,5\text{-Me}_2C_6H_3)_3SbCl_2$	$R_2SbCl_3 + R_2ICl$ in HCl followed by Zn in Me_2CO	33	m. 189-190°	1074a
$(2,4\text{-Me}_2C_6H_3)_3SbCl_2$	$RN_2SbCl_6 + Fe$ in Me_2CO	52	m. 190.5-191.5°	1061, 1062
$(2\text{-EtOC}_6H_4)_3SbI_2$	$R_3Sb + I_2$		Pale yellow plates, m. 143° (dec.)	513
C₂₅				
$Ph(4\text{-MeC}_6H_4)(2\text{-PhC}_6H_4)SbBr_2$	$R'R''SbCl + RMgBr$, followed by Br_2 in CCl_4		m. 169-170°, rxns.[11,12]	178
$Ph(4\text{-MeC}_6H_4)(2\text{-PhC}_6H_4)SbCl_2$	$RR'R''SbBr_2 + SnCl_2$ in $EtOH$ contg. HCl		m. 125-126°, rxn.[12]	178
C₂₇				
$Ph(4\text{-EtO}_2CC_6H_4)(2\text{-PhC}_6H_4)SbBr_2$	$R'R''SbCl + RMgBr$, followed by Br_2 in CCl_4		m. 148-150°, rxns.[11,12]	178
$Ph(4\text{-EtO}_2CC_6H_4)(2\text{-PhC}_6H_4)SbCl_2$	$RR'R''SbBr_2 + SnCl_2$ in $EtOH$ contg. HCl		m. 109-110°, rxn.[12]	178
$(4\text{-EtO}_2CC_6H_4)_3SbCl_2$	$RN_2SbCl_6 + Fe$ in Me_2CO, followed by $HCl + EtOH$	20	m. 133-134°	1065
$Ph(4\text{-EtO}_2CC_6H_4)(cyclo\text{-}C_6H_{11}\text{-}C_6H_4)SbCl_2$	$R'R''SbCl + RMgBr$ followed by Cl_2 in $CHCl_3$		m. 168-169°	181

Compound	Preparation		Properties	Ref.
C30		79		
$(2\text{-}C_{10}H_7)_3SbCl_2$	RN_2SbCl_6 + Fe in Me_2CO		m. 158-160°	1061, 1065
$(1\text{-}C_{10}H_7)_3SbF_2$	R_3SbCl_2 + KF		m. 279-280°	491
C31				
$4\text{-}MeC_6H_4[5,2\text{-}Br(Ph)C_6H_3]_2Sb\text{-}Cl_2$	$R'N_2Cl$ + $RSbCl_2$ in EtOH + Cu bronze		m. 240-241°	173
$4\text{-}MeC_6H_4(2\text{-}PhC_6H_4)_2SbCl_2$	$RSbCl_2$ + $R'N_2SbCl_4$ in EtOH + Cu		m. 228-230°, rxn.[13]	173
C33				
$(4\text{-}EtO_2CC_6H_4)[4,2\text{-}Br(Ph)C_6\text{-}H_3]_2SbCl_2$	$RSbCl_2$ + $R'N_2Cl$ in EtOH + Cu		m. 214-215°	173
C35				
$4\text{-}EtO_2CC_6H_4[4,2\text{-}Me(Ph)C_6H_3]_2\text{-}SbCl_2$	$RSbCl_2$ + RN_2SbCl_4 in EtOH + Cu		m. 252°	177
$4\text{-}EtO_2CC_6H_4[2\text{-}(4\text{-}MeOC_6H_4)C_6\text{-}H_4]_2SbCl_2$	$RSbCl_2$ + $R'N_2SbCl_4$ in EtOH + Cu		m. 247-248°	177
C36				
$(4\text{-}PhOC_6H_4)_3SbBr_2$	R_3Sb + Br_2		White needles, m. 151-152°	513
$(2\text{-}PhC_6H_4)_3SbBr_2$	R_3Sb + Br_2		Clusters of needles, m. 152-154°, rxn.[14]	1318
$(4\text{-}PhOC_6H_4)_3SbCl_2$	R_3Sb + Cl_2		Needles, m. 106-107°	513
$(2\text{-}PhC_6H_4)_3SbCl_2$	R_3Sb + Cl_2		m. 174-175°	1318
$(4\text{-}PhOC_6H_4)_3SbI_2$	R_3Sb + I_2		Pale yellow plates, m. 140° (dec.)	513
C37				
$4\text{-}MeC_6H_4[2\text{-}(4\text{-}EtO_2CC_6H_4)C_6\text{-}H_4]_2SbCl_2$	$RSbCl_2$ + $R'N_2SbCl_4$ in EtOH satd. with HCl		Squat prisms, m. 251-252.5°	173
C39				
$4\text{-}MeC_6H_4[2\text{-}(4\text{-}EtO_2CCH_2O)C_6\text{-}H_4)C_6H_4]SbCl_2$	$RSbCl_2$ + $R'N_2SbCl_4$ in EtOH + Cu		m. 196-197°	177

11. Rxn. with $SnCl_2$ in EtOH contg. HCl → $RR'R''SbCl_2$ (178).
12. Rxn. with H_2S in hot EtOH contg. NH_3 → $RR'R''Sb$ (178).
13. Rxn. with H_2S in hot EtOH contg. NH_3 → RR'_2Sb (173).
14. Rxn. with alc. NH_3 → $(2\text{-}PhC_6H_4)_3Sb(OH)_2$ (1318).

TRIORGANOSTIBINE DIHYDROXIDES, HYDROXIDE AND OXIDE SALTS, AND DICARBOXYLATES

Complete hydrolysis of triorganostibine dihalides yields the corresponding tri-
organoantimony dihydroxides. Hydrolysis of tris(trifluoromethyl)stibine dichloride
with silver oxide in aqueous solution yielded tris(trifluoromethyl)antimonic acid,
$H[(CF_3)_3Sb(OH)_3]$, a mobasic acid. Triorganostibine dihydroxides are also formed
from tertiary stibines by reacting with 4-iodoanisole in methanol in ultraviolet
light.

Partial hydrolysis of triorganostibine dihalides yields triorganostibine halide
hydroxides, $R_3Sb(X)OH$. The latter compounds are also formed from pentaorganoantimony compounds in reactions with hydrohalic acids, followed by extraction with
hot water.

Arylation of diarylantimony acetates, R_2SbOAc, with aryldiazonium acetates in
acetone under reflux yields triorganostibine acetate hydroxides and diacetates.

Attempted synthesis of 2-phenylisostibindoline by reacting a mixture of dichlorophenylstibine and o-phenylenedimethyl dibromide with metallic sodium gave a product
which after oxidation with nitric acid yielded oxybis(triphenylstibonium nitrate).

TRIS(TRIFLUOROMETHYL)ANTIMONIC ACID $C_3H_4F_9O_3Sb$ $H[(CF_3)_3Sb(OH)_3]$
Prepn.: by reacting $(CF_3)_3SbCl_2$ with Ag_2O in aq. soln. and treating the salt with
one equiv. of pyridine hydrochloride (336).
Props.: Monobasic acid, forms pyridinium salt, white solid, which reacts with HBr
or HCl to yield the corresponding pyridinium trihalotris(trifluoromethyl)antimonate,
$C_5H_5N.H[(CF_3)_3SbX_3]$ (336).
Rxn. with 2 N KOH → CHF_3 (336).

TRIPHENYLSTIBINE DIHYDROXIDE $C_{18}H_{17}SbO_2$ $Ph_3Sb(CH)_2$
Prepn.: By treating Ph_3SbI_2 with alc. KOH (1055).
By reacting Ph_3Sb with $4-IC_6H_4OMe$ in MeOH in UV light· 43% yield (1137).
Props.: m. 206-208° (1055); m. 205-206° (1137); dipole moment (603).

TRI-o-BIPHENYLSTIBINE DIHYDROXIDE $C_{36}H_{29}O_2Sb$ $(2-PhC_6H_4)_3Sb(OH)_2$
Prepn.: by treating $(2-PhC_6H_4)_3SbBr_2$ with $EtOH-NH_3$ (1318).
Props.: m. 243-244° (1318).

TRIPHENYLSTIBINE BROMIDE HYDROXIDE $C_{18}H_{16}BrOSb$ $Ph_3Sb(OH)Br$
Prepn.: By treating Ph_5Sb with 40% HBr and extracting the prod. with hot H_2O; 28%
yield along with Ph_4SbBr (before extraction) (1311).
By heating Ph_3SbBr_2 with dild. alc. (1311).
Props.: m. 246-249° (1311).

PHENYLDI-2,4-XYLYLSTIBINE CHLORIDE HYDROXIDE $C_{22}H_{24}ClOSb$ $Ph(2,4-Me_2C_6H_3)_2Sb(OH)Cl$
Prepn.: by decompg. $PhSbCl_4.2,4-MeC_6H_3N_2Cl$ with Fe powder in Me_2CO (1069).
Rxn. with HCl → $Ph(2,4-Me_2C_6H_3)_2SbCl_2$ (1069).

TRI-p-TOLYLSTIBINE ACETATE HYDROXIDE $C_{23}H_{25}O_3Sb$ $(4\text{-MeC}_6H_4)_3Sb(OH)OAc$
Prepn.: by reacting $(4\text{-MeC}_6H_4)_2SbOAc$ with $4\text{-MeC}_6H_4N_2OAc$ in Me_2CO under reflux;
59% yield (910, 1063).
Props.: m. 166-167° (910).

OXYBIS(TRIPHENYLSTIBONIUM NITRATE) $C_{36}H_{30}N_2O_7Sb$ $(Ph_3SbNO_3)_2O$
Prepn.: by reacting $PhSbCl_2$ and $2\text{-}C_6H_4(CH_2Br)_2$ with Na in Et_2O in the presence
of EtOAc and oxidizing the prod. with concd. HNO_3 (757).
Props.: m. 226-227° (757).
Rxn. with HCl → Ph_3SbCl_2 (757)
    TRIPHENYLSTIBINE DIACETATE $C_{22}H_{21}O_4Sb$ $Ph_3Sb(OAc)_2$
Prepn.: by reacting PhN_2OAc with Ph_2SbOAc in Me_2CO under reflux; 40% yield, iso-
lated as $Ph_3Sb(OAc)_2.2AcOH$ (910). The rxn. in cold Me_2CO gave 26% yield (1063).
Rxn. with alc. HCl → Ph_3SbCl_2 (910).
Complex compd.: $Ph_3Sb(OAc)_2.2AcOH$, m. 204-205° (910).
Use: Corrosion inhibitor for dielectric compns. (599).

TRI-α-THIENYLSTIBINE DIBENZOATE $C_{26}H_{19}O_4S_3Sb$ $(\overline{SCH:CHCH:C})_3Sb(OBz)_2$
Use: Corrosion inhibitor for dielectric compns. (599).

TRIPHENYLSTIBINE DIBENZOATE $C_{32}H_{25}O_4Sb$ $Ph_3Sb(OBz)_2$
Use: Stabilizer for vinyl chloride polymers (382).

TRIORGANOSTIBINE OXIDES, SULFIDES, AND SELENIDES

Tertairy stibine oxides are formed by oxidation of tertiary stibines by atomospheric
oxygen in solutions, by mercury oxide in ether, or by selenium dioxide in benzene at
room temperature.

Tertiary stibine sulfides are formed by treating tertiary stibine dibromides with
hydrogen sulfide in alcohol saturated with ammonia, or by decomposing diaryliodonium sulfides with antimony powder.

Tertiary stibine selenides are formed along with tertiary stibine oxides by treating with selenium dioxide in benzene at room temperature.

    TRIPHENYLSTIBINE OXIDE $C_{18}H_{15}OSb$ Ph_3SbO
Prepn.: by treating Ph_3Sb with SeO_2 in C_6H_6 at r. t., a mixt. of Ph_3SbO and
Ph_3AsSe is formed (842).
Props.: m. 209° (842).
Rxn. with K in C_4H_8O → insol. black solid in a brown liquid (527).

    TRI-p-TOLYLSTIBINE OXIDE $C_{21}H_{21}OSb$ $(4\text{-MeC}_6H_4)_3SbO$
Props.: Dipole moment (604).

    TRIBENZYLSTIBINE OXIDE $C_{21}H_{21}OSb$ $(PhCH_2)_3SbO$
Prepn.: by oxidation of $(PhCH_2)_3Sb$ with atm. oxygen in soln. (1230).
Rxns. with: HCl → $(PhCH_2)_3SbCl_2$ (1230).
HBr → $(PhCH_2)_3SbBr_2$ (1230).

TRI-n-HEPTYLSTIBINE OXIDE $C_{21}H_{45}OSb$ $(n-C_7H_{15})_3SbO$
Prepn.: by oxidation of $(n-C_7H_{15})_3Sb$ with H_2O in Et_2O (1227).
Complex compd.: $(n-C_7H_{15})_3SbO.Sb_2O_3$ prepd. by evapg. an ether sol. of the stibine
oxide (1227).

2-BIPHENYLYL-(4-CARBOXYPHENYL)PHENYLSTIBINE OXIDE $C_{25}H_{19}O_3Sb$ $Ph(2-PhC_6H_4)-$
$(4-HO_2CC_6H_4)SbO$
Prepn.: by refluxing $Ph(2-PhC_6H_4)(4-HO_2CC_6H_4)Sb$ in decalin (178).
Props.: m. 290-294° (178).

TRIPHENYLSTIBINE SULFIDE $C_{18}H_{15}SSb$ Ph_3SbS
Prepn.: By passing H_2S into a soln. of Ph_3SbBr_2 in alc. satd. with NH_3 (736).
By decompg. diphenyliodonium sulfide with Sb powder (1099).
Props.: m. 112° (736). Dipole moment (604). Crystal structure (1338).

TRI-p-TOLYLSTIBINE SULFIDE $C_{21}H_{27}SSb$ $(4-MeC_6H_4)_3SbS$
Prepn.: by treating $(4-MeC_6H_4)_3SbBr_2$ with H_2S in EtOH with NH_3; 83% yield (491).
Props.: m. 111.5-112° (491). Crystal structure (1337).

TRIPHENYLSTIBINE SELENIDE $C_{18}H_{15}SbSe$ Ph_3SbSe
Prepn.: by treating Ph_3Sb with SeO_2 in C_6H_6 at r. t., a mixt. of Ph_3SbO and Ph_3-
SbSe is formed (842).
Props.: Crystal structure (1338).

TRI-p-TOLYLSTIBINE SELENIDE $C_{21}H_{21}SbSe$ $(4-MeC_6H_4)_3SbSe$
Props.: Crystal structure (1338).

QUATERNARY STIBONIUM COMPOUNDS

Tetraorganostibonium salts are prepared as follows:
  1. From tertiary stibines by:
      (a) Treating with alkyl or aralkyl halides in ether.
      (b) Heating triarylstibines with aluminum chloride in the presence or ab-
           sence of the corresponding aryl bromide.
      (c) Treating tertiary stibines with diaryliodinium fluoroborates, $R_2I.BF_4$,
           at an elevated temperature, or with trialkyloxonium fluoroborates,
           $(R_3O)BF_4$, in liquid sulfur dioxide to yield tetraorganostibonium
           fluoroborates, R_4SbBF_4.
  2. From pentaorganoantimony compounds by:
      (a) Treating with halogens in ether or benzene, or with hydrohalic acids,
           or with acetic acid at 60-70°, followed by alkali metal halides, to
           form tetraorganostibonium halides and perhalides when excess of halo-
           gens is used.
      (b) Treating with triphenylborine to form tetraorganostibonium organo-
           triphenylborates, $[R_4Sb][BRPh_3]$.
  3. From triorganostibine dihalides by treating with Grignard compounds in
      inert solvents at room temperature.

Quaternary stibonium halides are converted into the corresponding perhalides by treating with halogens in organic solvents. Metathetic reactions of stibonium bromides or fluoroborates with alkali metal iodides yield the corresponding stibonium iodides. Moreover, tetraphenylstibonium bromide was converted into tetraphenylstibonium chloride by passing the bromide through an anion-exchange column.

TETRAMETHYLSTIBONIUM IODIDE $C_4H_{12}ISb$ Me_4SbI
Prepn.: By reacting $SbCl_3$ with $MeMgI$ and treating Me_3Sb thus formed with MeI (297).
  By treating Me_3Sb with MeI in Et_2O (1314).
Props.: White large needles, sublime > 250° (1314); m. 288-302° (dec.) (297).
Rxns. with: D exchange catalyzed by D_2O (297).
$LiBH_4$ or $LiAlH_4$ → Me_3Sb (1288).
Ag_2O, followed by HCl → Me_4AsCl (1314).
MeLi in Et_2O, followed by MeI → Et_2Me_2AsI + CH_4 + C_2H_6 (1314).
PhLi in Et_2O, followed by MeI → Et_2Me_2AsI (1314).
PhLi in Et_2O, followed by H_2O and picric acid → Me_4As picrate (1314).
MeLi in Et_2O under N → Me_5Sb (1314).
I_2 in 96% EtOH and C_6H_6 → Me_4SbI_3 (1314).

DIBENZYLDIMETHYLSTIBONIUM BROMIDE $C_{16}H_{20}BrSb$ $Me_2(PhCH_2)_2SbBr$
Prepn.: by treating $Me_2(PhCH_2)Sb$ with $PhCH_2Br$ under N (1313).
Props.: m. 159-160° (1313).
Rxns. with: KI → the stibonium iodide (1313).
PhLi in Et_2O under N, followed by hydrolysis and treatment with MeI, after isolation of bibenzyl → $Me_3(PhCH_2CHPh)SbI$ (1313).

TRIMETHYL(1,2-DIPHENYLETHYL)STIBONIUM IODIDE $C_{17}H_{22}ISb$ $Me_3(PhCH_2CHPh)SbI$
Prepn.: by reacting $Me_2(PhCH_2)_2SbBr$ with PhLi in Et_2O under N, adding water, and treating the ether layer contg. $Me_2(PhCH_2CHPh)Sb$ with MeI; 77% yield (1313).
Props.: m. 209-214° (dec.) (1313).

TETRAPHENYLSTIBONIUM BROMIDE $C_{24}H_{20}BrSb$ Ph_4SbBr
Prepn.: By heating Ph_3Sb with $AlCl_3$ in the presence or absence of PhBr, at 230° and treating the rxn. prod. with KBr (201).
  By reacting Ph_3SbCl_2 with $PhMgBr$ in Et_2O-C_6H_6 at r. t. and decompg. with ice and 48% HBr (1302).
  By treating $Ph_5Sb.0.5C_6H_{12}$ with 1 mole Br in C_6H_6 at 30° (1311).
  By treating Ph_5Sb with 40% HBr (1311, 1314).
Props.: Colorless crystals, m. 210-218° (201); Soly. data (852); m. 210-215° (1302); m. 213-125° (1311). Inhibiting effect on corrosion of steel (30).
Rxns. with: KI → Ph_4SbI (201).
$LiAlH_4$ or $LiBH_4$ → Ph_3Sb (1290).
Cl^- in ion-exchange column → Ph_4AsCl (1302).
PhLi in C_6H_{12} → $Ph_5Sb.0.5C_6H_{12}$ (1311).
MeLi in Et_2O, followed by Ph_2CO and decompn. with H_2O → Ph_3COH (1314).

TABLE LXXII
QUATERNARY STIBONIUM COMPOUNDS

Stibonium salts R_4Sb^+	X	Prepd. from	Props.	Ref.
C4				
Me_4Sb	Br	$Me_5Sb + Br_2$ in Et_2O	m. 140° (dec.)	1314
	Br_3	Me_5Sb + excess Br in $CHCl_3$	Raman spectrum and force const.	1314
	Cl			1138
	I_3	Me_4SbI in EtOH + I in C_6H_6	m. 136° (dec.)	1314
	$BMePh_3$	$Me_5Sb + Ph_3B$	m.~240° (dec.)	1314
C8				
$(CH_2{:}CH)_4Sb$	Br	$(CH_2{:}CH)_5Sb + Br$	m. 53-54°	914
C12				
$(CH_3CH{:}CH)_4Sb$	Br	$(CH_3CH{:}CH)_5Sb + Br$	cis-isomer, m. 140-143°	913, 916
			trans-isomer, m. 45-48°, IR data	916
C16				
$Me_2(PhCH_2)_2Sb$	I	$Me_2(PhCH_2)_2SbBr + KI$	m. 122-124°	1313
$[{:}CSbMe(CH{-}Me_2)_2]_2$	I_2	$[(Me_2CH)_2SbC{:}]_2 + MeI$	Crystalline salt	515
C19				
$MePh_3Sb$	BF_4	$Ph_3Sb + (Me_3O)BF_4$	m. 133-134°	531
	I	$[MePh_3Sb]BF_4 + NaI$	m. 124-124.5°, rxn.[1]	531
C24				
Ph_4Sb	BF_4	$Ph_3Sb + Ph_2IBF_4$ at 213°	m. 265°	779, 905
	Br_3	$Ph_5Sb.0.5C_6H_{12}$ + Br	m. 130° (dec.), on heating at 135° → $Ph_3SbBr_2 + Ph_2$	1311, 1314
	Cl	$Ph_4SbBr + Cl^-$	m. 202-205°	1302
		$Ph_5Sb.0.5C_6H_{12}$ + Cl	m. 204-205°, use[2]	1311
	I	$Ph_4SbBr + KI$	Colorless needles, m.~200°	201
		$Ph_5Sb.0.5C_6H_{12}$ + I or AcOH, followed by NaI	Soly. data	852
			White needles, m. 225-226°	1311
	I_3	$Ph_5Sb.0.5C_6H_{12}$ + I (excess)	Tan leaflets, m. 175-176°	1311

BPh$_4$	Ph$_5$Sb·0.5C$_6$H$_{12}$ + Ph$_3$B	m. 256-260°	1311
	Ph$_4$SbBr$_3$ + LiBPh$_4$		1311
NO$_3$		Soly. data	852
F		Soly. data	852
ClO$_4$		Soly. data	852
(Ph$_4$Sb)$_2$ SO$_4$		Soly. data	852
(Ph$_4$Sb)$_2$ ON(SO$_3$)$_2$		Paramagnetic resonance absorption	230

1. Rxn. with PhLi in Et$_2$O under N, followed by BzPh under reflux and hydrolysis → Ph$_3$Sb, Ph$_2$CHCHO,
   Ph$_2$C:CH$_2$, and Ph$_3$SbO (531).

2. Use: in gravimetric detn. of MnO$_4^-$ and ClO$_4^-$ (1303).

523

PENTAORGANOANTIMONY COMPOUNDS

Pentaorganoantimony compounds are prepared by:
1. Reacting tertiary stibine- dihalides with organolithium compounds.
2. Reacting quaternary stibonium halides with organolithium compounds.
3. Reacting antimony pentachloride with organolithium compounds in ether under nitrogen at -70°.

PENTAMETHYLANTIMONY $C_5H_{15}Sb$ Me₅Sb
Prepn.: by reacting Me_4SbBr with MeLi in Et_2O under N (1314).
Props.: b. 126-127°, m. -18 to -16°; insol. in H_2O, decomp. to give a stronge alk. soln. (1314).
Rxns. with: Br in $CHCl_3 \rightarrow Me_4SbBr$ (1314).
Br(excess) in $CHCl_3 \rightarrow Me_4SbBr_3$ (1314).
Ph_3B under N $\rightarrow$ $(Me_4Sb)(BMePh_3)$ (1314).

PENTAVINYLANTIMONY $C_{10}H_{15}Sb$ $(CH_2:CH)_5Sb$
Prepn.: by reacting $(CH_2:CH)_3SbBr_2$ with $CH_2:CHMgBr$ (914).
Props.: n_D^{28} 1.5590; d_{20} 1.2986 (914, 195).
Rxn. with Br_2(L:1) $\rightarrow$ $(CH_2:CH)_4SbBr$ (914).

PENTAPROPENYLANTIMONY $C_{15}H_{25}Sb$ $(CH_3CH:CH)_5Sb$
Prepn.: by reacting cis- and transtriallylstibine dibromides with cis- and trans-$CH_3CH:CHLi$ respectively (913, 916).
Props.: The compound occurs in two stereoisomeric forms. Cis isomer: n_D^{20} 1.5620, d_{20} 1.1972, RSb 16.48 (915, 916). Trans isomer: n_D^{20} 1.5490, d_{20} 1.17025, R_{Sb} 16.48 (915, 916).
Rxns. with: $Br_2 \rightarrow$ cis- and trans-$(CH_3CH:CH)_4SbBr$ (913, 916).

PENTAPHENYLANTIMONY $C_{30}H_{25}$ Ph₅Sb
Prepn.: By treating Ph_4SbBr or Ph_3SbCl_2 with PhLi in Et_2O under N; 77% yield (1311).
By treating $SbCl_5$ with PhLi in Et_2O under N at -70° (1311).
By treating Ph_4SbBr with MeLi in Et_2O under N and decompg. the rxn. mixt. with H_2O (1314).
Props.: Crystallizes from C_6H_{12} with 0.5 mol. of the solvent ($Ph_5Sb.0.5C_6H_{12}$), in the form of large monoclinic prisms, m. 169-170°, loses the 05 C_6H_{12} when recrystallized from MeCN and heated at 100° in vacuo; decomp. at 175-200 into $Ph_3Sb + Ph_2$ (1311); m. 159-161° (from EtOH) (1314).
Rxns. with: PhLi in Et_2O under N $\rightarrow$ Li(SbPh₆) (1311).
I_2(1:1) in $C_6H_6 \rightarrow Ph_4SbI$ (1311).
I_2(1:1) in $C_6H_6 \rightarrow Ph_4SbI_3$ (1311).
Br_2(1:1) in $C_6H_6 \rightarrow Ph_4SbBr$ (1311, 1314).
Br_2(1:2) in $C_6H_6 \rightarrow Ph_4SbBr_3$ (1311, 1314).
Cl_2(1:1) in $CCl_4 \rightarrow Ph_4SbCl$ (1311).
Cl_2(1:2) in $CCl_4 \rightarrow Ph_3SbCl_2$ (1311).
40% HBr (heat) $\rightarrow$ $Ph_4SbBr + C_6H_6$ and $Ph_3Sb(CH)Br$ (extracted by hot H_2O) (1311).
45% HBr $\rightarrow Ph_3SbBr_2$ (1311).
AcOH, followed by NaI $\rightarrow Ph_4SbI$ (1311).
Ph_3B in Et_2O under N $\rightarrow$ $(Ph_4Sb)(BPh_4)$ (1311).

<u>Deriv.</u>: Li(SbPh$_6$), prep'd. by reacting Ph$_5$Sb.0.5C$_6$H$_{12}$ with PhLi in Et$_2$O under N,
forms microcrystals sintering at 165°, m. 185°; Crystallizes from tetrahydrofuran
with four molecules of the solvent; Li[SbPh$_6$].4C$_6$H$_8$O, m. 180-185°; fairly stable
in air but decomp. on contact with H$_2$O (1311).

MISCELLANEOUS DERIVATIVES OF PENTAVALENT ANTIMONY

DIBENZYLSTIBINIC ANHYDRIDE C$_{28}$H$_{28}$O$_3$Sb [(PhCH$_2$)$_2$SbO]$_2$O
<u>Prepn.</u>: by treating (PhCH$_2$)$_2$SbCl$_3$ with dild. Na$_2$CO$_3$ soln. (1230).
<u>Rxn.</u> with HBr → (PhCH$_2$)$_2$SbBr$_3$(1230).

DI-2-CHLOROPHENYLSTIBINIC ACID, ANHYDRIDE WITH ACETIC ACID C$_{14}$H$_{11}$Cl$_2$O$_3$Sb
(2-ClC$_6$H$_4$)$_2$SbOOAc
<u>Prepn.</u>: by reacting (2-ClC$_6$H$_4$)$_2$SbO$_2$H with aq. AcOH (904).

N-ACETYL-Sb,Sb,Sb-TRIPHENYLSTIBINIMIDE C$_{20}$H$_{18}$NOSb Ph$_3$SbNAc
<u>Prepn.</u>: by reacting Ph$_3$Sb with AcNNaBr in Me$_2$CO; 74.6% yield (996).
<u>Props.</u>: m. 157-159° (996).
<u>Rxn.</u> with 5% alc. HgCl$_2$ in CHCl$_3$ → C$_{20}$H$_{18}$NOSb.HgCl$_2$ (996).
<u>Complex compds.</u>: Ph$_3$SbNAc.HgCl$_2$, m. 162-128° (996).
Ph$_3$SbNAc.CuCl$_2$, m. 138-140° (996).

3. HETEROCYCLIC DERIVATIVES OF ANTIMONY

FIVE-MEMBERED RING SYSTEMS

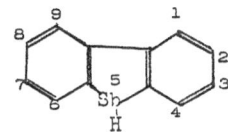

Dibenzostibiole

Dibenzostibiole derivatives are prepared by:
1. Reacting aryldihalostibines with 2,2'-biphenylenedilithium in benzene under
    reflux.
2. Cyclizing aryl-2-biphenylylstibinic acids in acetic anhydride, containing
    sulfuric acid, and reducing 5-aryldibenzostibiole 5,5-dihydroxides or the
    5-oxides thus formed by means of tin(II) chloride. Substituted aryl-2-
    biphenylylstibinic acids yield the corresponding substituted 5-aryldibenzo-
    stibiole derivatives. Aryl-2-biphenylylstibinic acids used in the prepara-
    tion of substituted dibenzostibioles are readily prepared by hydrolysis of
    aryl-2-biphenylylantimony trihalides.

Dibenzostibiole derivatives react with halogens to form the corresponding dibenzostibiole 5,5-dihalide compounds, which are readily hydrolyzed by alkali.

5-PHENYLDIBENZOSTIBIOLE $C_{18}H_{13}Sb$ *
Prepn.: by reacting $Ph.SbI_2$ with 2,2'-biphenylenedilithium in C_6H_6 under reflux (528).
Props.: m. 101° (528).

4-(2-BROMO-5-DIBENZOSTIBIOLYL)BENZOIC ACID $C_{19}H_{12}BrO_2Sb$ *
Prepn.: by hydrolyzing 4-(2-bromo-5-dibenzostibiolyl)benzoic acid ethyl ester with alc. NaOH (173).
Props.: m. 246-248° (173).

4-(3-NITRO-5-DIBENZOSTIBIOLYL)BENZOIC ACID $C_{19}H_{12}NO_4Sb$ *
Prepn.: by hydrolyzing ethyl 4-(3-nitro-5-dibenzostibiolyl)benzoate with alc. alkali.(175).
Props.: m. 212.5-213.5° (175).

2,5,5-TRIBROMO-5-p-TOLYLDIBENZOSTIBIOLE $C_{19}H_{13}Br_3Sb$ *
Prepn.: by treating 2-bromo-5-p-tolyldibenzostibiole with calcd. amt. Br_2 in CCl_4(182).
Props.: $C_{19}H_{13}Br_3Sb.CCl_4$ (from CCl_4), m. 190° (182).
Rxn. with 10% alc. NaOH → prod. m. 275-280° (182).

4-(5-DIBENZOSTIBIOLYL)BENZOIC ACID $C_{19}H_{13}O_2Sb$ *
Prepn.: by hydrolyzing ethyl 4-(5-dibenzostibiolyl)benzoate (175).
Props.: m. 231-233° (175).
Salts with optically active bases:
(+)-MeCHPhNH2, m. 208-209°; $[\alpha]_D$ + 29.0-29.5° (175).
(+)-Ephedrine, m. 184-185°; $[\alpha]_D$ + 5.47-5.67° (175).
(+)-Quinine, m. 203-205°; $[\alpha]_D$ - 94.7° (175).

2-BROMO-5-p-TOLYLDIBENZOSTIBIOLE $C_{19}H_{14}BrSb$ *
Prepn.: by hydrolyzing [4,2-Br(Ph)C_6H_4](4-MeC_6H_4)$SbCl_3$ in EtOH-Me_2CO by aq. NaOAc, cyclizing the stibinic acid thus obtained in Ac_2O in the presence of H_2SO_4, and reducing the prod. by $SnCl_2$ in Me_2CO-HCl (173).
Props.: m. 123-125° (173).
Rxn. with Br_2 in CCl_4 → the Sb,Sb-dibromide (182).

3-NITRO-5-p-TOLYLDIBENZOSTIBIOLE $C_{19}H_{14}NO_2Sb$ *
Prepn.: by hydrolyzing 4'-nitro-2-biphenylyl-p-tolylantimony trichloride, treating the resulting 4'-nitro-2-biphenylyl-p-tolylstibinic acid with Ac_2O-H_2SO_4, and reducing the 5-oxide deriv. by $SnCl_2$ in Me_2CO-HCl (175).
Props.: m. 157-159° (175).
Rxn. with Zn dust + $CaCl_2$ in aq. EtOH → 3-amino-5-p-tolyl-5-dibenzostibiole (175).

5-p-TOLYLBENZOSTIBIOLE $C_{19}H_{15}Sb$ *
Prepn.: by hydrolyzing 2-biphenylyl-p-tolylstibine trichloride in EtOH-Me_2CO by aq. NaOAc, cyclizing the resulting 2-biphenylyl-p-tolylstibinic acid in Ac_2O cont. H_2SO_4 to 9-tolyl-9-stibiafluorene 9-oxide, and reducing the oxide by $SnCl_2$ in Me_2CO-HCl (173).
Props.: m. 94° (173); UV spectrum (174, 178).

* For the structural formula see page 525.

3-AMINO-5-p-TOLYLDIBENZOSTIBIOLE $C_{19}H_{16}NSb$ *
Prepn.: by reducing 3-nitro-5-p-tolyl-5-dibenzostibiole by Zn dust and $CaCl_2$ in
aq. EtOH (175).
Props.: m. 132-133°; the compd. is optically active and was resolved via the H
tartrate into d- and l- antipodes, both m. 136-138° (175).

5,5-DIBROMO-5-p-TOLYL-3-DIBENZOSTIBIOLECARBOXYLIC ACID $C_{20}H_{15}Br_2O_2Sb$ *
Prepn.: by treating 5-p-tolyl-3-dibenzostibiolecarboxylic acid with calcd. amt.
of Br_2 in CCl_4 (182).
Props.: Crystallizes with 1 mol. CCl_4, m. 195° (182).
Rxn. with 10% alc. NaOH → gelatinous ppt. (dihydroxide monohydrate), m. > 300° (182).

4-(3-METHYL-5-DIBENZOSTIBIOLYL)BENZOIC ACID $C_{20}H_{15}O_2Sb$ *
Prepn.: by hydrolyzing ethyl 4-(3-methyl-5-dibenzostibiolyl)benzoate with 5% alc.
KOH (177).
Props.: m. 240°; optically active compd., was resolved into its antipodes by form-
ing a salt with ephedrine, m. 200-201° (177).

5-p-TOLYL-3-DIBENZOSTIBIOLECARBOXYLIC ACID $C_{20}H_{15}O_2Sb$ *
Prepn.: By hydrolyzing ethyl 5-p-tolyl-3-dibenzostibiolecarboxylate by 5% KOH
(173).
Props.: m. 208-209.5°, optically active compd., was resolved into its antipodes
by means of l-phenylethylamine (173).
Rxn. with Br_2 in CCl_4 → the Sb,Sb-dibromide (182).

4-(3-METHOXY-5-DIBENZOSTIBIOLYL)BENZOIC ACID $C_{20}H_{15}O_3Sb$ *
Prepn.: By reducing 4-(3-methoxy-5-dibenzostibiolyl)benzoic acid Sb,Sb-dihydroxide
with $SnCl_2$ in Me_2CO (182).
By hydrolyzing ethyl 4-(3-methoxy-5-dibenzostibiolyl)benzoate by 5% alc. KOH
(177).
Props.: m. 231°; optically active compd., resolved into its antipodes by forming
salts with ephedrine (177).

3-METHOXY-5-TOLYLDIBENZOSTIBIOLE $C_{20}H_{17}OSb$ *
Prepn.: by hydrolyzing $[2-(4-MeOC_6H_4)C_6H_4](p-MeC_6H_4)SbCl_3$ with aq. NaOAc in
Me_2CO, cyclizing the stibinic acid in $Ac_2O-H_2SO_4$, and reducing the stibiole 5-
oxide by $SnCl_2$ in Me_2CO (177).
Props.: m. 117-118° (177).

4-(3-METHOXY-5-DIBENZOSTIBIOLYL)BENZOIC ACID Sb,Sb-DIHYDROXIDE $C_{20}H_{17}Sb$ *
Prepn.: From 5-(4-carbethoxyphenyl)-3-methoxydibenzostibiole Sb,Sb-dihydroxide by
hydrolysis with alc. KOH (182).
By-prod. in the cyclization of $(4-EtO_2CC_6H_4)[2-(4-MeOC_6H_4)C_6H_4]SbO_2H$ in Ac_2O-
H_2SO_4 (182).
Props.: m. 285-290° (dec.) (182).
Rxn. with $SnCl_2$ in Me_2CO → 4-(3-methoxy-5-dibenzostibiolyl)benzoic acid (182).

---

* For the structural formula see page 525

4-(2-BROMO-5-DIBENZOSTIBIOLYL)BENZOIC ACID ETHYL ESTER $C_{21}H_{16}BrO_2Sb$ *
Prepn.: by hydrolyzing 4-$EtO_2CC_6H_4$[4,2-Br(Ph)C_6H_3]$SbCl_3$ in EtOH-Me_2CO with aq.
NaOAc, cyclizing the stibinic acid thus formed in Ac_2O contg. H_2SO_4, and reducing
the prod. by $SnCl_2$ in Me_2CO-HCl(173).
Props.: m. 157° (173).
Rxns. with: Alc. NaOH → hydrolysis of the CO_2Et group (182).
Br_2 in CCl_4 → the 5,5-dibromide (182).

4-(2-BROMO-5-DIBENZOSTIBIOLYL)BENZOIC ACID ETHYL ESTER 5,5-DIBROMIDE $C_{21}H_{16}Br_2O_2Sb$
Prepn.: by treating 4-(2-bromo-5-dibenzostibiolyl)benzoic acid ethyl ester with
Br_2 in CCl_4 (182).
Props.: Crystallizes from CCl_4 with one mol. of the solvent, m. 188-189°, from
C_6H_8 with 1/3 mol. of the solvent, m. 194-195° (182).
Rxn. with 10% alc. NaOH → prod. m. 302-304° (182).

4-(3-NITRO-5-DIBENZOSTIBIOLYL)BENZOIC ACID ETHYL ESTER $C_{21}H_{16}NO_4Sb$ *
Prepn.: by treating p-carbethoxyphenyl(4'-nitro-2-biphenylyl)stibinic acid with
Ac_2O-H_2SO_4 and reducing the prod. with $SnCl_2$ in Me_2CO-HCl (175).
Props.: m. 193.5-195° (175).
Rxn. with alc. alkali → hydrolysis of the -CO_2Et group (175).

4-(5-DIBENZOSTIBIOLYL)BENZOIC ACID ETHYL ESTER $C_{21}H_{17}O_2Sb$ *
Prepn.: by reducing 4-(5-dibenzostibiolyl)benzoic acid ethyl ester 5-oxide with
$SnCl_2$ in HCl (175).
Props.: m. 144-145° (175).

(5-p-TOLYL-3-DIBENZOSTIBIOLYLOXY)ACETIC ACID $C_{21}H_{17}O_3Sb$ *
Prepn.: by hydrolyzing ethyl (5-p-tolyl-3-dibenzostibiolyloxy)acetate with 5%
alc. KOH (177).
Props.: Optically active substance, m. 164-165°, resolved into its antipodes via
ephedrine salt (177).

4-(5-DIBENZOSTIBIOLYL)BENZOID ACID ETHYL ESTER Sb-OXIDE $C_{21}H_{17}O_3Sb$ *
Prepn.: by heating 2-biphenylyl-4-carbethoxyphenylstibinic acid at 90° with Ac_2O
contg. H_2SO_4 (175).
Props.: m. 172-173° (175).
Rxn. with $SnCl_2$ in Me_2CO → 4-(5-dibenzostibiolyl)benzoic acid ethyl ester (175).

(5-p-TOLYL-3-DIBENZOSTIBIOLYLOXY)ACETIC ACID 5,5-DIHYDROXIDE $C_{21}H_{19}O_5Sb$ *
Prepn.: from ethyl (5-p-tolyl-3-dibenzostibiolyloxy)acetate or from its 5,5-
dihydroxide derivative by hydrolysis with aq. KOH (182).
Props.: Monohydrate, m. 276° (dec.) (182).

5-p-TOLYL-3-DIBENZOSTIBIOLYLCARBOXYLIC ACID ETHYL ESTER $C_{22}H_{19}O_2Sb$ *
Prepn.: by treating 4-MeC_6H_4[2-(4-$EtO_2CC_6H_4$)C_6H_4]SbO_2H with Ac_2O-H_2SO_4 at 90°
and reducing the prod. with $SnCl_2$ in Me_2CO contg. HCl (173).
Props.: Rosettes of needles, m. 136-137° (173).
Rxn. with alc. KOH → hydrolysis of the -CO_2Et group (173).

---

* For the structural formula see page 525.

4-(3-METHYL-5-DIBENZOSTIBIOLYL)BENZOIC ACID ETHYL ESTER $C_{22}H_{19}O_2Sb$ *
Prepn.: by hydrolyzing $(4-EtO_2CC_6H_4)[2-(4-MeC_6H_4)C_6H_4]SbCl_3$ with aq. NaOAc
in Me_2CO, cyclizing the stibinic acid thus formed in $Ac_2O-H_2SO_4$, and reducing
the prod. with $SnCl_2$ in Me_2CO (177).
Props.: m. 155-156° (177).
Rxn. with 5% alc. KOH → hydrolysis of the $-CO_2Et$ group (177).

4-(3-METHOXY-5-DIBENZOSTIBIOLYL)BENZOIC ACID ETHYL ESTER $C_{22}H_{19}O_3Sb$ *
Prepn.: By hydrolyzing $(4-EtO_2CC_6H_4)[2-(4-MeOC_6H_4)C_6H_4]SbCl_3$ with aq. NaOAc in
Me_2CO, cyclizing the stibinic acid thus formed in $Ac_2O-H_2SO_4$, and reducing. the
prod. with $SnCl_2$ in Me_2CO (176, 177, 182).
Props.: m. 147-148° (177); m. 145-146° (176).
Rxn. with 5% alc. KOH → hydrolysis of the $-CO_2Et$ group (177).

4-(3-METHOXY-5-DIBENZOSTIBIOLYL)BENZOIC ACID ESTER 5,5-DIHYDROXIDE $C_{22}H_{21}O_5Sb$ *
Prepn.: by cyclization of $(4-EtO_2CC_6H_4)[2-(4-MeOC_6H_4)C_6H_4]AsO_2H$ in $Ac_2O-H_2SO_4$
(182).
Props.: m. 185-188° (182).
Rxns with: Alc. KOH → hydrolysis of the $-CO_2Et$ group (182).
$SnCl_2$ → redn. of the pentavalent Sb (Sb,Sb-dihydroxide) to ethyl 4-(3-methoxy-5-
dibenzostibiolyl)benzoate (182).

(5-p-TOLYL-3-DIBENZOSTIBIOLYLOXY)ACETIC ACID ETHYL ESTER $C_{23}H_{21}O_3Sb$ *
Prepn.: By hydrolyzing $[2-(4-EtO_2CCH_2OC_6H_4)C_6H_4](p-MeC_6H_4)SbCl_3$ with aq. NaOAc
in Me_2CO, cyclizing the stibinic acid in $Ac_2O-H_2SO_4$, and reducing the stibiole
5-oxide by $SnCl_2$ in Me_2CO (177, 182).
Props.: m. 132-133° (177, 182).
Rxn. with 5% alc. KOH or 10% aq. NaOH → hydrolysis of the ester group (177, 182).

5-p-TOLYL-3-DIBENZOSTIBIOLYLOXY)ACETIC ACID ETHYL ESTER 5,5-DIHYDROXIDE
$C_{23}H_{23}O_5Sb$ *
Prepn.: by cyclizing $(p-MeC_6H_4)[2-(4-EtO_2CCH_2OC_6H_4]SbO_2H$ in $Ac_2O-H_2SO_4$ (182).
Props.: m. 148-150° (dec.) (182).
Rxn. with $SnCl_2$ in Me_2CO → redn. of the pentavalent Sb (Sb,Sb-dihydroxide) to
ethyl (5-p-tolyl-3-dibenzostibiolyloxy)acetate (182).

DITHIASTIBIOLANE AND DITHIASTIBINDAN DERIVATIVES

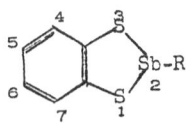

2-R-1,3,2-dithiastibiolane   2-R-1;3,2-dithiastibindan

* For the structural formula see page 525.

Substituted 1,3,2-dithiastibiolane derivatives are formed by reacting organodi-chlorostibines, $RSbCl_2$, or stiboso compounds, $RSbO$, with dithiols in aqueous alkali or in alcohol. Analogous reactions of organodichlorostibines or stiboso compounds with o-benzenedithiol yield 1,3,2-dithiastibindan derivatives. When stibonic acids, $RSbO_3H_2$, are used instead of organodihalostibines or stiboso compounds, they have to be reduced by thiols or mercaptocarboxylic acids in alkaline medium.

TABLE LXXIII

DITHIASTIBIOLANE AND DITHIASTIBINDAN DERIVATIVES

$RSbSCH_2CHRS$	Prepd. from	Ref.
C_9 4-$H_2NC_6H_4\overline{SbSCH_2CH(CH_2OH)S}$	$RSb(O)(CH)(ONa)$ + $HSCH_2CO_2K$ or $HSCH_2CO_2NH_4$ + $HSCH_2CHSHCH_2OH$ in EtOH	442, 444
	$RSbCl_2$ or $RSbO$ + $HSCH_2CHSHCH_2OH$ in EtOH or H_2O	443
4-$H_2NC_6H_4\overline{SbSCH_2CH(Me)S}$	$RSb(O)(OH)ONa$ + $HSCH_2CO_2NH_4$ + MeCHSH-CH_2SH in H_2O	442, 444
	$RSbCl_2$ or $RSbO$ + $MeCHSHCH_2SH$ in H_2O or EtOH	443
C_{11} 2,4-Cl(AcNH)$C_6H_3\overline{SbSCH_2CH(CH_2OH)S}$	$RSb(O)(OH)ONa$ + $HSCH_2CO_2NH_4$ + $HSCH_2-CHSHCH_2OH$ in H_2O	442, 444
	$RSbCl_2$ or $RSbO$ + $HSCH_2CHSHCH_2OH$ in H_2O or EtOH	443
4-$AcNHC_6H_4\overline{SbSCH_2CH(CH_2OH)S}$	$RSb(O)(OH)ONa$ + $HSCH_2CO_2NH_4$ + $HSCH_2-CHSHCH_2OH$ in H_2O	442, 444
	$RSbCl_2$ or $RSbO$ + $HSCH_2CHSHCH_2OH$ in H_2O or EtOH	443
C_{12} 4-$H_2NC_6H_4\overline{SbSC_6H_4S}$-o	$RSb(O)(OH)ONa$ + $HSCH_2CO_2K$ + o-C_6H_4-(SH)$_2$ in H_2O	442
4-[$N:C(NH_2)N:C(NH_2)N:CNH$]$C_6H_4\overline{SbSCH_2CH-}$ (CH_2OH)S	$RSb(O)(OH)ONa$ + $HSCH_2CO_2NH_4$ + $HSCH_2-CHSHCH_2OH$	444
4-[$N:C(NH_2)N:C(NH_2)N:CNH$]$C_6H_4\overline{SbSCH_2CH-}$ (CO_2H)S	$RSb(O)(OH)ONa$ + $HSCH_2CO_2NH_4$ + $HSCH_2-CHSHCO_2H$	442, 444
	$RSbCl_2$ or $RSbO$ + $HSCH_2CHSHCO_2H$	443
4-[$N:C(NH_2)N:C(NH_2)N:CNH$]$C_6H_4\overline{SbSCH_2CH-}$ (Me)S	$RSb(O)(OH)ONa$ + $HSCH_2CO_2NH_4$ + MeCH-$SHCH_2SH$	444
	$RSbCl_2$ or $RSbO$ + $MeCHSHCH_2SH$	443
C_{13} 4-[$N:C(NH_2)N:C(NH_2)N:CNH$]$C_6H_4\overline{SbS}$-(CO_2H)CH(CO_2H)S	$RSb(O)(OH)ONa$ + [-CH(SH)CO_2Na]$_2$ in H_2O at 40°	444e

TABLE LXXIII (cont'd.)
DITHIASTIBIOLANE AND DITHIASTIBINDAN DERIVATIVES

$RSbSCH_2CHRS$	Prepd. from	Ref.
C_{15} 4-[N:C(NH$_2$)N:C(NH$_2$)N:CNH]C$_6$H$_4$SbSC$_6$H$_4$S-o	RSb(O.)(OH)ONa + HSCH$_2$CO$_2$K + o-C$_6$H$_4$-(SH)$_2$ RSbCl$_2$ or RSbO + o-C$_6$H$_4$(SH)$_2$	442, 444 443
C_{19} CO[NH-p-C$_6$H$_4$SbSCH$_2$CH(CH$_2$OH)S]$_2$	CO[NH-p-C$_6$H$_4$Sb(O)(OH)ONa]$_2$ + HSCH$_2$-CO$_2$NH$_4$ + HSCH$_2$CHSHCH$_2$OH	442-4

SIX-MEMBERED RING SYSTEMS

PHENOXASTIBINE DERIVATIVES

10-R-phenoxastibine

Phenoxastibine derivatives are prepared by reacting dihaloörganostibines, $RSbX_2$, with 2,2'-oxydiphenylenebis(magnesium bromide) in suitable solvents, such as ether, benzene, dioxane, or their mixtures. Phenoxastibines react with halogens in carbon tetrachloride to yield the corresponding phenoxastibine dihalides.

10-(4-BROMOPHENYL-)-2-METHYLPHENOXASTIBINE 10,10-DICHLORIDE $C_{19}H_{14}BrCl_2OSb$
Prepn.: by treating 10-(4-bromophenyl)-2-methylphenoxastibine with chlorine in CCl$_4$ (172).
Props.: m. 202° (172).
Rxn. with H$_2$S → 10-[4-bromophenyl)-2-methylphenoxastibine (172).

10-(4-BROMOPHENYL-2-METHYLPHENOXASTIBINE $C_{19}H_{14}BrOSb$
Prepn.: by reacting 4-BrC$_6$H$_4$SbI$_2$ with 5-methyl(2,2'-oxydiphenylene)bis(magnesium bromide) in Et$_2$O-C$_6$H$_6$; 30-35% yield (172).
Props.: m. 116° (172).

2-METHYL-10-PHENYLPHENOXASTIBINE 10,10-DICHLORIDE $C_{19}H_{15}Cl_2OSb$
Prepn.: by treating 2-methyl-10-phenylphenoxastibine with Cl$_2$ in CCl$_4$ (172).
Props.: Colorless prisms from CCl$_4$, m. 112° (dec.); long flat plates from toluene m. 104-105°; small prisms from petroleum, m. 140-142°, free of solvent of crystallization (172).
Rxn. with H$_2$S → 2-methyl-10-phenylphenoxastibine (172).

2-METHYL-10-PHENYLPHENOXASTIBINE $C_{19}H_{15}OSb$ *

Prepn.: by reacting $PhSbI_2$ with 5-methyl(2,2'-oxydiphenylene)bis(magnesium bromide) in $Et_2O-C_6H_6$; 30-35% yield (172).
Props.: Feathery needles, m. 62-63° (172).
Rxn. with $Cl_2 \rightarrow$ 2-methyl-10-phenylphenoxastibine 10,10-dichloride (172).

4-(2-METHYL-10-PHENOXASTIBINYL)BENZONITRILE $C_{20}H_{14}NOSb$ *

Prepn.: by reacting $4-NCC_6H_4SbI_2$ with 5-methyl(2,2'-oxydiphenylene)bis(magnesium bromide) in $Et_2O-C_6H_6$; 34% yield (172).
Props.: m. 82° (172).
Rxn. with aq. acids $\rightarrow$ hydrolysis with elimination of Sb (172).

4-(2-METHYL-10-PHENOXASTIBINYL)BENZONITRILE Sb,Sb-DICHLORIDE $C_{20}H_{14}Cl_2NOSb$ *

Prepn.: by treating 4-(2-methyl-10-phenoxastibinyl)benzonitrile with Cl_2 in CCl_4 (172).
Props.: m. 180-181° (172).
Rxn. with $H_2S \rightarrow$ 4-(2-methyl-10-phenoxastibinyl)benzonitrile (172).

4-(2-METHYL)-10-PHENOXASTIBINYL)BENZOIC ACID $C_{20}H_{15}O_3Sb$ *

Prepn.: by hydrolyzing 4-(2-methyl-10-phenoxastibinyl)benzoic acid ethyl ester with alc. KOH under reflux (172).
Props.: Rosettes of small needles, m. 201°. The Sb atom is optically active, and the racemic product was resolved via the strychnine salt; pure d- acid m. 192°; a 50-50 mixt. of the d- and l- acid melts at 201° (172).

4-(2-METHYL-10-PHENOXASTIBINYL)BENZOIC ACID ETHYL ESTER $C_{22}H_{19}O_3Sb$ *

Prepn.: by reacting $4-EtO_2CC_6H_4SbCl_2$ with 5-methyl(2,2'-oxydiphenylene)bis-(magnesium bromide) in dioxan-Et_2O mixture, completing the rxn. under reflux, and hydrolyzing the rxn. prod. with HCl (172).
Props.: Flat needles, m. 136-137° (172).
Rxn. with alc. KOH under reflux $\rightarrow$ hydrolysis of the $-CO_2Et$ group (172).

MISCELLANEOUS HETEROCYCLIC DERIVATIVES OF ANTIMONY

2-[4-(4,6-DIAMINO-s-TRIAZIN-2-YLAMINO)PHENYL]-5-HYDROXY-1,3,2-DITHIASTIBINANE
    $C_{12}H_{15}N_6OS_2Sb$ $4-[\overset{\cdot}{N}:C(NH_2)N:C(NH_2)N:\overset{\cdot}{C}NH]C_6H_4SbSCH_2CHOHCH_2\overset{\cdot}{S}$
Prepn.: By reducing $4-[\overset{\cdot}{N}:C(NH_2)N:C(NH_2)N:\overset{\cdot}{C}NH]C_6H_4Sb(O)(OH)(ONa)$ with $HSCH_2CO_2$-NH_4 and reacting the reduced rxn. mixt. with $HOCH(CH_2SH)_2$ (442, 444).
    By reacting $4-[\overset{\cdot}{N}:C(NH_2)N:C(NH_2)N:\overset{\cdot}{C}NH]C_6H_4SbCl_2$ or the corresponding stiboso deriv. with $HOCH(CH_2SH)_2$ (443).

2-[4-ACETAMIDOPHENYL)-5-HYDROXYMETHYL-1,3,2-DITHIASTIBINANE $C_{13}H_{18}NO_3S_2Sb$
    $4-AcNHC_6H_4\overset{\cdot}{S}bSCH_2C(CH_2OH)_2CH_2\overset{\cdot}{S}$
Prepn.: By reducing $4-AcNHC_6H_4Sb(O)(OH)ONa$ with $HSCH_2CO_2NH_4$ and esterifying the prod. with $(HOCH_2)_2C(CH_2SH)_2$ in water (442, 444).
    By treating $4-AcNHC_6H_4SbCl_2$ or $4-AcNHC_6H_4SbO$ with $(HOCH_2)_2C(CH_2SH)_2$ in H_2O or EtOH (443).

---

* For the structural formula see page 531.

2-(4-SULFAMOYLPHENYL)-4,6-DIHYDROXY-4-HYDROXYMETHYL-1,3,2-DIOXASTIBEPANE-7-
  CARBOXYLIC ACID $C_{12}H_{16}NO_9SSb$ 4-$H_2NO_2SC_6H_4SbOCH(CH_2OH)CHOHCHOHCHCO_2H$
<u>Prepn.</u>: by reacting 4-$H_2NO_2SC_6H_4SbCl_2$ with Na gluconate in aq. soln. (20).

2-(4-CARBAMOYLPHENYL)-5,6-DIHYDROXY-4-HYDROXYMETHYL-1,3,2-DIOXASTIBEPANE-7-
  CARBOXYLIC ACID $C_{13}H_{16}NO_8Sb$ 4-$H_2NCOC_6H_4SbOCH(CH_2OH)CHOHCHOHCHCO_2H$
<u>Prepn.</u>: by reacting 4-$H_2NCOC_6H_4SbCl_2$ with Na gluconate in aq. soln. (20).

2-(4-ACETAMIDOPHENYL)-5,6-DIHYDROXY-4-HYDROXYMETHYL-1,3,2-DIOXASTIBEPANE-7-
  CARBOXYLIC ACID (?) $C_{14}H_{18}NO_8Sb$ 4-$AcNHC_6H_4SbOCH(CH_2OH)CHOHCHOHCHCO_2H$
<u>Prepn.</u>: by reacting 4-$AcNHC_6H_4SbCl_2$ with Na gluconate (20, 1166).

3-[4-(4,6-DIAMINO-s-TRIAZIN-2-YLAMINO]BENZO-2,4,3-DITHIASTIBEPIN $C_{17}H_{17}N_6S_2Sb$

<u>Prepn.</u>: By reducing 4-[$\overline{N:C(NH_2)N:C(NH_2)N:CNH}$]$C_6H_4Sb(O)(OH)ONa$ with $HSCH_2CO_2K$
and esterifying the prod. with o-$C_6H_4(CH_2SH)_2$ (442, 444).
  By esterifying 4-[$\overline{N:C(NH_2)N:C(NH_2)N:CNH}$]$C_6H_4SbX_2$ (X_2 = Cl_2 or O) with o-
$C_6H_4(CH_2SH)_2$ (443).

3-(p-AMINOPHENYL)-2,4,3-DITHIASTIBOBICYCLO[3.3.1]NONA-5,8,9-TRIENE $C_{12}H_{10}NS_2Sb$

<u>Prepn.</u>: by reducing 4-$H_2NC_6H_4Sb(O)(OH)ONa$ with $HSCH_2CO_2NH_4$ and esterifying the
prod. with m-$C_6H_4(SH)_2$ (444).
  By reacting 4-$H_2NC_6H_4SbX_2$ (X_2 = Cl_2 or O) with m-$C_6H_4(SH)_2$ in EtOH or H_2O (443).

BENZENESTIBONIC ACID CYCLIC ESTERS

Compounds of the general formula RSb(O)X, where X is gluconic, glucoheptanoic,
arabonic, ribonic, or tartaric acid bonded to the Sb through two alcoholic oxygen
atoms, and R may be 4-carbamoylphenyl, 4-sulfamoylphenyl, 4-(2-pyrimidylsulfamoyl)-
phenyl, 4-chlorophenyl , 4-anisyl, or 4-nitrophenyl group, were prepared by reacting the corresponding arylantimony tetrahalides or their pyridinium, quinolinium,
or ammonium halide double salts with the polyhydroxycarboxylic acid salts in the
presence of a weak base (20).

1. COMPOUNDS OF TRIVALENT BISMUTH

TERTIARY BISMUTHINES

SYMMETRICAL COMPOUNDS

Symmetrical tertiary bismuthines are parpared by:

1. Reacting bismuth trichloride with organometallic compounds, such as Grignard compounds, trialkylaluminum, tetraalkyllead, organosodium, and organolithium derivatives.
2. Decomposition of aryldiazonium chloride-bismuth trichloride double salts in ethyl acetate or in acetone with bismuth powder or in absolute ethanol with copper bronze, followed by reduction of triarylbismuthine dichloride which may be formed by hydrazine hydrate.
3. Decomposition of diaryl-chloronium, bromonium, or iodonium fluoroborates or chlorobismutates with bismuth powder in acetone.
4. Reacting bismuth trichloride with potassium aryldiazocarboxylates in acetone in the presence of bismuth powder.
5. Arylation of metallic bismuth with aryldiazonium fluoroborates in acetone.
6. Arylation of bismuth trichloride with arylhydrazines in hydrochloric acid in the presence of copper(II) and iron(III) chlorides.
7. Reduction of tertiary bismuthine dihalides, R_3BiX_2, with hydrazine hydrate in alcohol, or with mercury(II) chloride in ethanol, or with mercury oxide in alkaline solution, or with lithium aluminum hydride or lithium tetrahydroborate at very low temperatures.
8. Disproportionation diorganohalobismuthines in the presence of mercury(II) chloride in ethanol or with mercury oxide in alkaline solution.
9. Reaction of diarylbimuthine-alkali metal derivatives, R_2BiM, with aryl halides in liquid ammonia.
10. Metathetic reactions of triarylbismuthines with alkyllithium or alkylsodium compounds to form trialkylbismuthines.
11. Metathetic reaction of bismuth trichloride with cyanotrimethylsilane to form probably tricyanobismuthine.
12. Electrolysis of silver cyanide in pyridine, using a bismuth anode, to form tricyanobismuthine.

Radioactive trimethylbismuthine, Me_3Bi^{210}, is formed in the course of tetramethyllead decay in the gas phase.

TRICYANOBISMUTHINE C_3BiN_3 $Bi(CN)_3$

Prepn.: By electrolysis of AgCN in C_5H_5N with Bi anode (1108).
The compd. is probably formed in a reaction of bismuth trichloride with Me_3SiCN (87a).

TRIMETHYLBISMUTHINE C$_3$H$_9$Bi Me$_3$Bi

Radioactive Me$_3$Bi^{210} is formed during Me$_4$Pb decay in the gas phase (329).
Props.: b. 109.3° (extrapol.); m. -107.7° (extrapol.); d 2.313 (747).
b. 107.1°, f. p. -85.8°, log$_{10}$p = 7.659-1815/T (31).
Thermodynamic data (745, 747); Pyrolysis (1035a).
Intramolecular potential (1131). Force consts. (1138).
Relation between polarity and reactivity with O, CO$_2$, and H$_2$O (1098).

TRIVINYLBISMUTHINE C$_6$H$_9$Bi (CH$_2$:CH)$_3$Bi
Prepn.: by reacting CH$_2$:CHMgBr with BiCl$_3$ in C$_4$H$_8$O under reflux; 23% yield (773, 774).
Props.: b. 158.1° (calcd.); thermodynamic data; inflames spontaneously in air, but is stable in vacuo at 100° (773).

TRIETHYLBISMUTHINE C$_6$H$_{15}$Bi Et$_3$Bi
Prepn.: By reacting BiCl$_3$ with EtMgBr in Et$_2$O under N (464).
  By reacting BiCl$_3$ with Et$_4$Pb (465).
  By reacting BiCl$_3$ with Et$_3$Al in Et$_2$O; 86% yield (1331).
Props.: b. 123°/150 mm.(464); b. 104-105°/76 mm., d$_{20}$ 1.820; inflames in air (1331).
Heat of combustion (725b).
Rxns. with: Compds. contg. the -SH group → cleavage of the Et group (464, 895).
Compds. contg. the -SeH group → cleavage of the Et group (895).
O$_2$ at -25° → mixt. of [Et$_3$BiO$_2$], Et$_2$BiOEt, EtOBiO, Et$_2$O, Et$_2$O$_2$, EtOH, and C$_2$H$_4$ (171).
Br$_2$(1:1) → Et$_2$BiBr (171).

TRICYCLOPENTADIENYLBISMUTHINE C$_{15}$H$_{15}$Bi (C$_5$H$_5$)$_3$Bi
Prepn.: by reacting BiCl$_3$ with C$_5$H$_5$Na in C$_4$H$_8$O at r. t.; 60% yield (382a, 383).
Props.: The compd. occurs in two forms: orange needles obtd. by crystallization from petr. ether at -78°, which at 15-20° are transformed into a black modification. Both modifications are diamagnetic and sensitive to air (383). Sublimes at ~75° in high vacuum with decompn. (382a). IR spectrum (383, 444b). Magnetic data (383b). Dipole moment (383a). π-Bonded structure proposed (383).
Rxn. with: PCl$_3$ → (C$_5$H$_5$)$_2$BiCl (383).
O$_2$O$_2$ → violent rxn. with inflamation (383).
H$_2$O at r. t. → hydrolysis (C$_5$H$_6$) (382a).

TRI(4-BROMOPHENYL)BISMUTHINE C$_{18}$H$_{12}$BiBr$_3$ (4-BrC$_6$H$_4$)$_3$Bi
Prepn.: By treating (4-BrC$_6$H$_4$N$_2$)$_2$BiCl$_5$ with Cu-bronze in abs. EtOH, allowing the mixt. to reflux under autogenous heat and, after spontaneous cooling, adding N$_2$H$_4$.-H$_2$O; 7% yield (471).
  By decompg. (4-BrC$_6$H$_4$N$_2$)$_2$BiCl$_5$ with Bi powder in EtOAc at 5°; 39.6% yield (878).
  By decompn. of 4-BrC$_6$H$_4$N$_2$BF$_4$ with Bi powder in Me$_2$CO, with cooling, and neuralization of the rxn. mixt. with NH$_4$OH; 40% yield (909).
  By treating BiCl$_3$ with 4-BrC$_6$H$_4$N$_2$CO$_2$K in Me$_2$CO at -18° and completing the rxn. at <20° in the presence of Bi powder; 18% yield (1064).
  By treating 4-BrC$_6$H$_4$(Ph)I.I with BiCl$_3$ in Me$_2$CO in the presence of powdered Bi at r. t. and pouring the sol. prod. into NH$_4$OH; 16% yield (1073a).
Props.: m. 148-149° (878); m. 147-148° (909); m. 149° (1064).

TRI-4-CHLOROPHENYLBISMUTHINE C₁₈H₁₂BiCl₃ (4-ClC₆H₄)₃Bi

TRI-4-CHLOROPHENYLBISMUTHINE $C_{18}H_{12}BiCl_3$ $(4-ClC_6H_4)_3Bi$

Prepn.: By treating $(4-ClC_6H_4)_3BiCl_2$ with $N_2H_4.H_2O$ (468).

By treating $(4-ClC_6H_4N_2)_2.BiCl_5$ with Cu-bronze in abs. EtOH under reflux (exothermic rxn.) and, after spontaneous cooling, reducing the rxn. mixt. with $H_2H_4.H_2O$; 8.4% yield (471).

By decompg. $(4-ClC_6H_4N_2)_2BiCl_5$ with Bi powder in EtOAc at 5°; 15.5% yield (378).

By decompn. of $4-ClC_6H_4N_2BF_4$ with Bi powder in Me_2CO, with cooling, and neutralization of the rxn. mixt. with NH_4OH; 31% yield (909).

By treating $BiCl_3$ with $4-ClC_6H_4N_2CO_2K$ in Me_2CO at -18° and completing the rxn. at < 20° in the presence of Bi powder; 7% yield (1064).

By treating $Ph(4-ClC_6H_4)I.Cl$ or $(4-MeOC_3H_4)(4-ClC_6H_4)I.Cl$ with $BiCl_3$ in Me_2CO in the presence of powdered Bi at r. t. and pouring the sol. prod. into NH_4OH; 13 and 11% yield, resp. (1073a).

Props.: m. 115° (878, 1064); m. 115-116° (909).

Rxns. with: BuLi → Bu_3Bi (466).

RSH → $4-ClC_6H_4Bi(SR)_2$ (477).

Cl_2 → $(4-ClC_6H_4)_3BiCl_2$ (909).

TRIPHENYLBISMUTHINE $C_{18}H_{15}Bi$ Ph_3Bi

Prepn.: By treating Ph_2BiCl, Ph_2BiBr, Ph_3BiCl_2, Ph_3BiBr_2, or Ph_3BiF_2 with $N_2H_4.H_2O$ in EtOH (468), or with $HgCl_2$ in EtOH, or with HgO in alkali (777).

By treating Ph_2BiNa with PhBr, PhI, $o-MeOC_6H_4Br$, $1-C_{10}H_7Br$, or $o-Me_2NC_6H_4I$ in liq. NH_3 (470).

By reducing Ph_3BiCl_2 with liq. NH_3; 10% yield (470).

By decompg. $(PhN_2)_2BiCl_5$ or $(PhN_2)_3BiCl_6$ with Cu-bronze in abs. EtOH and reducing the rxn. prod. with $N_2H_4.H_2O$ (471).

By decompg. PhN_2BiCl_4 with Bi powder in Me_2CO and neutralizing the rxn. mixt. with NH_4OH; 50.3% yield (692).

By treating a mixt. of $BiCl_3$, $CuCl_2$, and $FeCl_3$ with $PhNHNH_2.HCl$ in aq. HCl in an open beaker and neutralizing the rxn. mixt. (137).

By treating $BiCl_3$ with PhMgBr in Et_2O and completing the rxn. at 150°; 82% yield (815).

By decompg. PhN_2BF_4 with Bi powder in Me_2CO, with cooling, and treating the prod. with NH_4OH; 69% yield (909).

By reacting $Ph_2Br.BF_4$ with Bi powder in Me_2CO (911).

By reacting $Ph_2Cl.BiCl_4$ or $Ph_2Br.BiCl_4$ with Bi powder; 74 and 68% yield, resp. (912, 1073).

By treating $BiCl_3$ in Me_2CO with PhN_2CO_2K at -18° and completing the rxn. at 20° in the presence of Bi powder; 23% yield (1063).

By reducing Ph_3BiCl_2 with $LiBH_4$ at -60° or with $LiAlH_4$ at -95° (1291).

By treating $BiBr_3$ with PhMgCl in toluene (isolated as dichloride) (1339).

By reacting $BiBr_3$ with PhLi in Et_2O; 44% yield (1339).

By treating $BiCl_3$ with $Ph(4-MeOC_6H_4)I.Cl$ in Me_2CO in the presence of powdered Bi at r. t. and pouring the sol. prod. into NH_4OH; 23% yield (1073a).

By-prod. in the rxn. of Ph HgCl with Na-Bi melt in xylene (1339).

Props.: m. 77-78° (137, 909); m. 77.5-78° (692); m. 75-76° (815); m. 77.5° (1063); m. 78-78.5° (1312). Vapor pressure (386). Relative reactivity (190). Crystal structure (593, 1176a, 1286). Dipole moment (1147b). UV spectrum (598). IR spectrum (711, 821, 1050). Near UV spectrum (1050). Magnetic susceptibility (967, 969, 975).

Rxns. with: Li in Et$_2$O or xylene → cleavage of the C-Bi bond (1192).
Na in liq. NH$_3$ → cleavage of the Ph groups (470).
RSH → PhBi(SR)$_2$ and Bi(SR)$_3$ (477, 690).
2-HSC$_6$H$_4$CO$_2$Me at 150° → PhBiSC$_6$H$_4$C(O)O (477).
RCO$_2$H → PhBi(O$_2$CR)$_2$ (477).
HCO$_2$H → Bi(O$_2$CH)$_3$, (HCO$_2$)$_2$BiOH, and C$_6$H$_6$ (685, 686).
AcOH → AcOBi(OH)$_2$ and AcOBiO (685, 686).
Phenols → cleavage of the phenyl groups (687).
Me$_3$NHCl at 130° → cleavage of the Ph groups (688, 689).
2-HSC$_6$H$_4$CO$_2$H → SC$_6$H$_4$COBiO$_2$H (691).
N$_2$O$_3$ + N$_2$O$_4$ → PhN$_2$NO$_3$ (778).
AlCl$_3$ in HCCl$_3$ → BiCl$_3$ and PhAlCl$_2$ (815).
FeCl$_3$ in HCCl$_3$ or in Et$_2$O → BiCl$_3$ (815, 1339).
BiCl$_3$ in HCCl$_3$ → Ph$_2$BiCl (817, 1339).
Cl in CHCl$_3$ at -5° → Ph$_3$BiCl$_2$ (1312).
Pb(OAc)$_4$ in CCl$_4$ → Ph$_3$Bi(OAc)$_2$ (1339).
SiCl$_4$ in CHCl$_3$ → cleavage of the Ph groups with quant. recovery of SiCl$_4$ (817a).
Use: Catalyst for tetramerization of acetylene to cycloöctatetraene (1323b).
Catalyst for polymerization of HCHO (306). Additive for monofuel (508). Additive
for curing polyepoxides (1132).

TRIS(4-SULFAMYLPHENYL)BISMUTHINE C$_{18}$H$_{18}$BiN$_3$O$_6$S$_3$ (4-H$_2$NSO$_2$C$_6$H$_4$)$_3$Bi
Prepn.: by decompg. (4-H$_2$NSO$_2$C$_6$H$_4$N$_2$Cl)$_2$BiCl$_3$ with 50% excess of Bi in Me$_2$CO at
30° or EtOH at 150°; 25 and 6% yield, resp. (900).
Props.: m. 233° (900).
Rxn. with Br in CCl$_4$ → 4-H$_2$NSO$_2$C$_6$H$_4$Br (900).

TRI(METHYLCYCLOPENTADIENYL)BISMUTHINE C$_{18}$H$_{21}$Bi (MeC$_5$H$_4$)$_3$Bi
Prepn.: from MeC$_5$H$_4$Na and BiCl$_3$ in C$_4$H$_8$O at r. t.; 68% yield (383).
Props.: The compd. occurs in two modifications; orange crystals, obtd. from petr.
ether at -78°; are transformed at 15-20° into a black prod.; both modifications
are stable in air and on contact with H$_2$O; IR spectra; π-bonded structure proposed (383).

TRI(2-BROMOBENZYL)BISMUTHINE C$_{21}$H$_{18}$BiBr$_3$ (2-BrC$_6$H$_4$CH$_2$)$_3$Bi
Prepn.: by reacting 2-BrC$_6$H$_4$CH$_2$MgBr with BiCl$_3$ in Et$_2$O under N in darkness (27a).
Props.: Orange-yellow crystals, m. 83-84°, decomp. 170-180° (27a).
Rxns. with: Atm. 0 → oxidation (27a).
I → BiI$_3$ (27a).
AgNO$_3$ in alc. → Ag (27a).
KMnO$_4$ in Me$_2$CO → MnO$_2$ (27a).
Na → Bi (27a).

TRI(2-CHLOROBENZYL)BISMUTHINE C$_{21}$H$_{18}$BiCl$_3$ (2-ClC$_6$H$_4$CH$_2$)$_3$Bi
Prepn.: by reacting 2-ClC$_6$H$_4$CH$_2$MgBr with BiCl$_3$ in ether under N in darkness and
extracting the prod. with C$_6$H$_6$ (27a).
Props.: Lemon-yellow needles, m. 54° (27a).
Rxns. with: Atm. 0 → oxidation (27a).
I → BiI$_3$ (27a).

AgNO$_3$ in alc. → Ag (27a).
MnO$_4$ in Me$_2$CO → MnO$_2$ (27a).
Na → Bi (27a).

TRIBENZYLBISMUTHINE C$_{21}$H$_{21}$Bi (PhCH$_2$)$_3$Bi
Prepn.: by reacting PhCH$_2$MgCl with BiCl$_3$ in Et$_2$O under N in darkness (27a).
Props.: Greenish-yellow scales, m. 64.5-65.5°, decomp. 150°, preserved under N
in darkness (27a).
Rxns. with: Atm. O → oxidation (27a).
I → BiI$_3$ (27a).
AgNO$_3$ in alc. → Ag (27a).
KMnO$_4$ in Me$_2$CO → MnO$_2$ (27a).
Na → Bi (27a).

TRI-o-TOLYLBISMUTHINE C$_{21}$H$_{21}$Bi (2-MeC$_6$H$_4$)$_3$Bi
Prepn.: By treating (2-MeC$_6$H$_4$)$_3$SbCl$_2$ or (2-MeC$_6$H$_4$)$_3$BiBr$_2$ with N$_2$H$_4$.H$_2$O (468).
By reacting (2-MeC$_6$H$_4$N$_2$)$_2$BiCl$_5$ with Cu-bronze in abs. EtOH under reflux (exc-
thermic rxn.) and, after spontaneous cooling, treating the rxn. mixt. with N$_2$H$_4$.-
H$_2$O; 6.4% yield (471).
By decompg. (2-MeC$_6$H$_4$N$_2$)$_2$BiCl$_5$ with Bi powder in Me$_2$CO and neutralizing the
rxn. prod. with NH$_4$OH; 38% yield (692).
By decompg. 2-MeC$_6$H$_4$N$_2$BF$_4$ with Bi powder in Me$_2$CO, with cooling, and neutraliz-
ing the prod. with NH$_4$OH; 54% yield (909).
Props.: m. 130-131° (692, 909).
Rxn. with BiBr$_3$ (1:2) → o-MeC$_6$H$_4$BiBr$_2$ (469).

TRI-p-TOLYLBISMUTHINE C$_{21}$H$_{21}$Bi (4-MeC$_6$H$_4$)$_3$Bi
Prepn.: By reacting 4-MeC$_6$H$_4$N$_2$Cl with BiCl$_3$ 'n concd. HCl and decompg. the
resulting (4-MeC$_6$H$_3$N$_2$)$_3$.BiCl$_6$ with Cu in Me$_2$CO, followed by NH$_4$OH (467); in abs.Et
instead of Me$_2$CO, the yield was 6.7% (471).
By reducing (4-MeC$_6$H$_4$)$_3$BiCl$_2$ with N$_2$H$_4$.H$_2$O (468).
By decompg. (4-MeC$_6$H$_4$N$_2$)$_3$BiCl$_6$ with Bi powder in Me$_2$CO and neutralizing the rxn
mixt. with NH$_4$OH; 31% yield (392).
By decompg. 4-MeC$_6$H$_4$N$_2$BF$_4$ with Bi powder in Me$_2$CO, with cooling, and neutraliz-
ing the rxn. mixt.; 50% yield (909).
By treating BiCl$_3$ in Me$_2$CO with 4-MeC$_6$H$_4$N$_2$CO$_2$K at -20° and completing the rxn.
at < 20° in the presence of Bi powder; 16.5% yield (1064).
Props.: m. 116-117° (467); m. 119-120° (692); m. 118-119° (909); m. 119° (1064).
Magnetic measurements (967, 969).
Rxns. with: Bu Li → Bu$_3$Bi (466, 1194).
BiCl$_3$ (2:1) → (p-MeC$_6$H$_4$)$_2$SbCl (469).
RSH → p-MeC$_6$H$_4$Bi(SR)$_2$ (477).

TRI-o-ANISYLBISMUTHINE C$_{21}$H$_{21}$BiO$_3$ (2-MeOC$_6$H$_4$)$_3$Bi
Prepn.: By treating (2-MeOC$_6$H$_4$N$_2$)$_2$BiCl$_5$ with Cu-bronze in abs. EtOH under reflux and, after spontaneous cooling, reducing the mixt. with N$_2$H$_4$.H$_2$O; 6.4%
yield (471).
By treating (2-MeOC$_6$H$_4$N$_2$)$_2$BiCl$_5$ with Cu or Bi powder in Me$_2$CO at 9°; 47.3%
yield (692).
Props.: m. 167-169° (692).

TRIMESITYLBISMUTHINE $C_{27}H_{33}Bi$ $(2,4,6-Me_3C_6H_2)_3Bi$

Prepn.: by reacting $2,4,6-Me_3C_6H_2MgBr$ with $BiCl_3$ (1339); see also (469).

Props.: m. 136-137° (1339); m. 134-135° (469).

Rxns. with: Dry Cl → $(2,4,6-Me_3C_6H_2)_3BiCl_2$ (1339).

Dry Br → $(2,4,6-Me_3C_6H_2)_3BiBr_2$ (1339).

$BiCl_3$ in Et_2O → white, completely inorg. ppt. (1339).

Br in $CHCl_3$ at -15 → yellow plates, m. 91-93° (469).

Cl_2 (excess) in C_6H_6 → pale green solid, m. 149-150°, $C_{27}H_{33}BiCl_5$ (469).

TRI-1-NAPHTHYLBISMUTHINE $C_{30}H_{21}Bi$ $(1-C_{10}H_7)_3Bi$

Prepn.: By treating $(1-C_{10}H_7N_2)_2BiCl_5$ with Cu-bronze in abs. EtOH under reflux (exothermic rxn.) and, after spontaneous cooling, reducing the prod. with $N_2H_4 \cdot H_2O$; only trace of the prod. is formed (471).

By decompg. $1-C_{10}H_7N_2BiCl_4$, with Bi powder in Me_2CO; 15.1% yield (692).

Props.: m. 235-236° (692).

Rxns. with: BuLi → Bu_3Bi (466).

$BiBr_3$ (1:2) → $1-C_{10}H_7BiBr_2$ (469).

RSH in xylene under reflux → no rxn. (477).

RCO_2H → cleavage of the C-Bi bond (686).

Phenols → cleavage of the C-Bi bond (687).

Me_3NHCl → $(1-C_{10}H_7)_2BiCl$, $C_{10}H_8$, $BiCl_3$, and Me_3N (688).

TRICARVACRYLBISMUTHINE $C_{30}H_{39}Bi$ $[2,5-Me(Me_2CH)C_6H_3]_3Bi$

Prepn.: method not given; 36.1% yield (469).

Props.: m. 87° (469).

Rxns. with: Cl_2 in $CHCl_3$ → the dichloride (469).

Br_2 in CCl_4 → the dibromide (469).

TRI-4-BIPHENYLYLBISMUTHINE $C_{36}H_{27}Bi$ $(4-PhC_6H_4)_3Bi$

Prepn.: by reacting $4-PhC_6H_4MgBr$ with $BiCl_3$ in Et_2O (1317).

Props.: Tiny flat needles, m. 182-183° (1317).

Rxns. with: Cl_2 in CCl_4 → $(4-PhC_6H_4)_3SbCl_2$ (1317).

Br_2 in CCl_4 → $(4-PhC_6H_4)_3SbBr_2$ (1317).

TABLE LXXIV

SYMMETRICAL TERTIARY BISMUTHINES

R_3Bi	Prepd. from	Yield %	Props.	Ref.
C$_9$				
Pr_3Bi	$BiCl_3 + Pr_3Al$	70.9	b. 89-91°/10 mm., d20 1.615	1331
C$_{12}$				
Bu_3Bi	$Ph_3Bi + BuLi$ or $BuNa$	90	Rxn. with O → explosive oxidation	465
$(Me_2CHCH_2)_3Bi$	$BiCl_3 + (Me_2CHCH_2)_3Al$	78.2	b. 115-116°/9 mm., d20 1.477	1331
$(Me_3SiCH_2)_3Bi$	$BiCl_3 + Me_3SiCH_2MgCl$	35	m. ~107-109°, decomp. 140°, ignites spontaneously in air	1121
C$_{18}$				
$(2-ClC_6H_4)_3Bi$		14.4	m. 143°	469
$(4-FC_6H_4)_3Bi$	$(RN_2)_2BiCl_5 + Bi$ in EtOAc	66	m. 93-94°	469
$(4-IC_6H_4)_3Bi$		21	m. 204-205°	878
C$_{21}$				
$(4-MeOC_6H_4)_3Bi$	$(RN_2)_2BiCl_5 + Cu$ in EtOH + $N_2H_4 \cdot H_2O$	1		471
C$_{24}$				
$(2,4-Me_2C_6H_3)_3Bi$		78	Magnetic data	957, 969
$(2-EtOC_6H_4)_3Bi$	$RMgCl + BiCl_3$		m. 121-122°, rxn. with BuLi → Bu_3Bi	465
$(4-EtOC_6H_4)_3Bi$	$RN_2BF_4 + Bi$	27	m. 87-88°, rxn. with Cl_2 → R_3BiCl_2	909
$(4-Me_2NC_6H_4)_3Bi$	$RLi + BiCl_3$	9.95	m. 230°, rxn. with HCl or AcOH → cleavage of $PhNMe_2$	469

ASYMMETRICAL BISMUTHINES

Asymmetrical tertiary bismuthines, $R_2R'Bi$, are prepared:
1. From organodihalo- or diorganohalobismuthines with Grignard compounds.
2. From diorganobismuth-sodium or -lithium derivatives by reacting with aryl halides. Diorganobismuth-sodium and -lithium compounds are formed by treating diorganohalobismuthines with sodium or lithium in liquid ammonia.
3. From asymmetrical triorganobismuthine dihalides by reduction with hydrazine hydrate.
4. From diorganohalobismuthines by reacting with sodium acetylide or ethynylenebis(magnesium bromide) to yield ethynylenebis(diorganobismuthines).

TABLE LXXV
ASYMMETRICAL TERTIARY BISMUTHINES.5

$R_2R'Bi$	Prepd. from	Yield %	Props.	Ref.
C_{13}				
Ph_2BiCN			Stermitatory effect	769a
C_{18}				
$Ph_2(4\text{-}ClC_6H_4)Bi$	$R_2BiCl + R'MgBr$	33	m. 83-83.5°	469
$Ph_2(4\text{-}HOC_6H_4)Bi$ (?)	$R_2BiI + R'Br$		m. 179-180°	470
C_{19}				
$(4\text{-}ClC_6H_4)_2(2\text{-}MeC_6H_4)Bi$	$R'BiBr_2 + RMgX$	45	m. 102-103°, rxns.[1-3]	469
$Ph_2(4\text{-}MeC_6H_4)Bi$	$R_2BiCl + RMgX$		Oil, rxn.[1]	469
C_{20}				
$(4\text{-}MeC_6H_4)_2(4\text{-}ClC_6H_4)Bi$	$R_2BiI + R'MgBr$	55	m. 96-97°	469
	$R_2BiNa + R'I$	19		470
C_{22}				
$(4\text{-}ClC_6H_4)_2(1\text{-}C_{10}H_7)Bi$	$R_2BiCl + R'MgX$	25	m. 138-139°, rxn.[1,2]	469
	$R'BiX_2 + RMgX$	46.8		469
	$R_2BiNa + R'I$	10.8		470
$Ph_2(1\text{-}C_{10}H_7)Bi$	$R_2R'BiCl_2 + H_2NNH_2$		Rxn.[4]	468
	$R_2BiNa + R'I$	30		470
C_{24}				
$(2\text{-}MeC_6H_4)_2(1\text{-}C_{10}H_7)Bi$	$R_2R'BiCl_2 + H_2NNH_2$	59.5	m. 112-114°, rxn.[2]	469
$(4\text{-}MeC_6H_4)_2(1\text{-}C_{10}H_7)Bi$	$R_2BiI + R'MgBr$	21	m. 129-130°, rxns.[1,2]	469
	$R'BiBr_2 + RMgBr$	42		469
$(4\text{-}MeOC_6H_4)_2(1\text{-}C_{10}H_7)Bi$	$R'BiBr_2 + RMgX$	10.9	m. 135-136°	469
C_{26}				
$Ph_2BiC:CBiPh_2$	$Ph_2BiCl + NaC:CH$		m. 145° (dec.), rxn.[5]	514
	$Ph_2BiCl + (:CMgBr)_2$			514
$(4\text{-}EtOC_6H_4)_2(1\text{-}C_{10}H_7)Bi$	$R'BiBr_2 + RMgX$	13.9	m. 131-132°	469
C_{28}				
$(2,4,6\text{-}Me_3C_6H_2)_2(1\text{-}C_{10}H_7)Bi$	$R'BiBr_2 + RMgX$	31.4	m. 151-151.5°	469

C_{31}	$R_2BiCl + R'MgCl$		m. 59–60° (by-prod.)	469
$(Ph_3C)Bi$				

1. Rxn. with Cl_2 in $HCCl_3 \rightarrow R_2R'BiCl_2$ (469).
2. Rxn. with Br_2 in $CCl_4 \rightarrow R_2R'BiBr_2$ (469).
3. Rxn. with BuLi in $Et_2O \rightarrow$ displacement of $4\text{-}ClC_6H_4$ groups (476).
4. Rxn. with thiols (RSH) $\rightarrow$ cleavage of the phenyl groups and formation of $1\text{-}C_{10}H_7Bi(SR)_2$ (477).
5. Rxn. with aq. KOH $\rightarrow$ evolution of acetylene (514).

DIORGANOHALOBISMUTHINES

Diaryl- and dicyclopentadienylhalobismuthines are prepared:
1. From tertiary bismuthines by reaction with bismuth trichloride or phosphorus trichloride in an inert solvent.
2. From tertiary bismuthine dihalides by pyrolysis.
3. From tetraarylbismuth perhalides, R_4BiX_3, by spontaneous decomposition.
4. From aryldiazonium chloride-bismuth trichloride double salts, $(RN_2)_2BiCl_5$, by decomposition with copper bronze in ethanol to form mixtures of diaryl-halo- and aryldihalobismuthines.

Diarylchlorobismuthines are readily converted into diaryliodobismuthines by treatment with sodium iodide in acetone.

CHLORODIPHENYLBISMUTHINE $C_{12}H_{10}BiCl$ Ph_2BiCl
Prepn.: by treating Ph_3Bi with $BiCl_3$ in $CHCl_3$ under reflux; 25.8% yield (817); in ether (1339).
Props.: m. 183-184° (817); m. 186° (1339). Sternutatory effect (769a).
Rxns. with: $N_2H_4.H_2O \rightarrow Ph_3Bi$ (468).
$RMgX \rightarrow Ph_2BiR$ (469).
Na in liq. $NH_3 \rightarrow$ dark red soln. which decomp. to Ph_3Bi (470).
Na in liq. NH_3, followed by $C_{10}H_7I \rightarrow Ph_2(C_{10}H_7)Bi$, Ph_3Bi, and $C_{10}H_8$ (470).
Na in liq. NH_3, followed by PhBr, PhI, o-$MeOC_6H_4Br$, 1-C_6H_7Br, o-$Me_2NC_6H_4I \rightarrow$ Ph_3Bi (470).
$RSH \rightarrow Ph_2BiSR$ (477).
NaC:CH $\rightarrow Ph_2BiC:CBiPh_2$ (514).
$(:CMgBr)_2 \rightarrow Ph_2BiC:CBiPh_2$ (514).
$HgCl_2$ in 96% $BiOCl$, C_6H_6, and $PhHgCl$ (777).
HgO in alk. soln. $\rightarrow Ph_3Bi$ (777).

IODODIPHENYLBISMUTHINE $C_{12}H_{10}BiI$ Ph_2BiI
Rxns. with: Li in liq. NH_3, followed by p-$BrC_6H_4OH \rightarrow Ph_3Bi$ and (p-$HOC_6H_4BiPh_2$) (?) (470).
Polarographic study (1094, 1223).
$LiBH_4$ or $LiAlH_4$ in Et_2O at 65° $\rightarrow Ph_3Bi$, $(PhBi)x$, and Ph_2 (1296).
Na in liq. $NH_3 \rightarrow$ green color (Ph_2Bi formation?) (1316).

TABLE LXXVIII
ARYLBIS(ARYLTHIO)BISMUTHINES

RBi(SR')$_2$	Prepd. from	Props.	Ref.
4-ClC$_6$H$_4$Bi(SPh)$_2$	(4-ClC$_6$H$_4$)$_3$Bi + PhSH in MePh under reflux	m. 170° (dec.)	477
PhBi(SPh)$_2$	Ph$_3$Bi + PhSH in CHCl$_3$ under reflux	Yellow plates, m. 170° (dec.)	477
4-MeC$_6$H$_4$Bi(SPh)$_2$	(4-MeC$_6$H$_4$)$_3$Bi + PhSH at 100°	m. 155° (dec.)	477
1-C$_{10}$H$_7$Bi(SPh)$_2$	Ph$_2$(1-C$_{10}$H$_7$)Bi + PhSH in CHCl$_3$ under reflux		477
PhBi(SC$_6$H$_4$CO$_2$Me-o)$_2$	Ph$_3$Bi + 2-HSC$_6$H$_4$CO$_2$Me on a steam bath	m. 107-108°	477

MISCELLANEOUS DERIVATIVES OF TRIVALENT BISMUTH

BISMUTHOBENZENE (C$_6$H$_5$Bi)n (PhBi)n
Prepn.: by reducing PhBiBr$_2$ or Ph$_2$BiCl with LiBH$_4$ or LiAlH$_4$ in Et$_2$O at -70° in high vacuum (1296).
Props.: Black, polymeric, oxidizable substance, insol. in Et$_2$O (1296).
Rxns. with: Br$_2$ → PhBiBr$_2$ (1296).
O$_2$ → PhBiO (1296).

CHLOROMETHYLBISMUTH OXIDE CH$_2$BiClO ClCH$_2$BiO
Prepn.: by reacting BiCl$_3$ with CH$_2$N$_2$ in Et$_2$O at 0-30° and hydrolyzing the prod. (1320).
Props.: Greenish powder (1321), readily decomp. with explosive violence (1320).
Rxn. with conc. HCl → ClCH$_2$BiCl$_2$ (1321).

PHENYLBISMUTH OXIDE C$_6$H$_5$BiO PhBiO
Prepn.: by treating PhBiCl$_2$ with H$_2$O (728).

PHENYLBISMUTH DIBENZOATE C$_{20}$H$_{15}$BiO$_4$ PhBi(OBz)$_2$
Prepn.: by heating Ph$_3$Bi with BzOH on a steam bath (477).
Props.: m. 215-216° (477).

PHENYLBISMUTH DICINNAMATE C$_{24}$H$_{19}$BiO$_4$ PhBi(O$_2$CCH:CHPh)$_2$
Prepn.: by heating Ph$_3$Bi with cinnamic acid at 150° (477).

4-OXO-2-PHENYL-3,1,2-BENZOXATHIABISMUTHINE C$_{13}$H$_9$BiO$_2$S

Prepn.: by heating Ph$_3$Bi with 2-HSC$_6$H$_4$CO$_2$H at 150° (477).
Props.: m. > 250° (477).

DIPHENYLBISMUTH BROMOACETATE $C_{14}H_{12}BiBrO_2$ $Ph_2BiO_2CCH_2Br$
Props.: Sternutatory effect (769a).

DIPHENYLBISMUTH IODOACETATE $C_{14}H_{12}BiIO_2$ $Ph_2BiO_2CCH_2I$
Props.: Sternutatory effect (769a).

DIPHENYLBISMUTH PROPIONATE $C_{15}H_{15}BiO_2$ Ph_2BiO_2CEt
Props.: Sternutatory effect (769a).

ETHOXYDIETHYLBISMUTHINE $C_6H_{15}BiO$ Et_2BiOEt
Prepn.: by treating Et_3Bi with $Br_2(1:1)$ in hexane-CCl_4 at -60° and treating
the resulting Et_2BiBr with NaOEt in EtOH; 49% yield (171).
Props.: Snow-white, sword-shaped crystals, sublime at 100°/ 1 mm. (171).
Rxn. with O_2 at 14-17° → $Et_{1.2}BiO_{1.8}$, Et_2O_2, Et_2O, and EtOH (171).

DIPHENYL(PHENYLTHIO)BISMUTHINE $C_{18}H_{15}BiS$ Ph_2BiSPh
Prepn.: by reacting Ph_2BiCl with PhSH in Et_2O; 80% yield (477).
Props.: m. 160° (dec.) (477).

2. COMPOUNDS OF PENTAVALENT BISMUTH

TRIORGANOBISMUTHINE DIHALIDES AND DINITRATES

Triorganobismuthine dihalides are prepared from:
1. Tertiary bismuthines by treating with halogens in chloroform or other organic solvents. Tertiary bismuthines can be prepared by arylation of bismuth powder with aryldiazonium fluoroborates in acetone, neutralization of
   the reaction mixture, and extraction of the product with chloroform or by
   arylation of organodihalobismuthines with Grignard compounds.
2. Tetraorganobismuth perhalides by spontaneous decomposition.
3. Aryldihalobismuthines by reacting with aryldiazonium chlorozincates,
   $(RN_2)_2ZnCl_4$, in acetone in the presence of iron powder.

The dichlorides are converted into the corresponding dinitrates by reacting with
silver nitrate in acetone.

TRIPHENYLBISMUTHINE DICHLORIDE $C_{18}H_{15}BiCl_2$ Ph_3BiCl_2
Prepn.: by treating Ph_3Bi with excess chlorine in $CHCl_3$ at -5° (1312).
Props.: m. 158-159° (1312). Dipole moment (603, 950); Magnetic data (967, 969).
Rxns. with: $N_2H_4.H_2O$ → Ph_3Bi (468).
Liq. NH_3 → Ph_3Bi (470).
Na in liq. NH_3, followed by $1-C_{10}H_7I$ → 71% $C_{10}H_8$ and 12.3% Ph_3Bi (470).
RCO_2M → $Ph_3Bi(O_2CR)_2$ (478).
$HgCl_2$ in 96% EtOH → BiOCl, PhHgCl, and C_6H_6 (777).
HgO in alk. soln. → Ph_3Bi (777).
Slow neutrons → radioactive Ph_3BiCl_2 (875).
$LiAlH_4$ or $LiBH_4$ → Ph_3Bi (1291).
PhLi in Et_2O at -75° → Ph_5Bi (1312).

TRIORGANOBISMUTHINE DIHALIDES AND DIHYDRATES

R_3BiX_2	Prepd. from	Yield %	Props.	Ref.
C_{18}				
$(3-O_2NC_6H_4)_3BiCl_2$	RN_2BF_4 + Bi in Me_2CO, followed by Cl_2 in $CHCl_3$	26	m. 131-132°	909
$(4-O_2NC_6H_4)_3BiCl_2$	RN_2BF_4 + Bi in Me_2CO, followed by Cl_2	5.5	Decomp. 160-161°	909
$(4-ClC_6H_4)_3BiCl_2$	R_3Bi + Cl_2 in $CHCl_3$-C_8H_{18}	17	m. 170°, rxn.[1]	909
Ph_3BiBr_2	R_4BiBr_3 by decompn.	79	m. 148-149°, rxn.[1]	1312, 1312
Ph_3BiF_2	R_3Bi + Br_2		Rxn.[1]	468
$Ph_3Bi(NO_3)_2$			Dipole moment	603
C_{19}				
$(4-ClC_6H_4)_2(2-MeC_6H_4)BiBr_2$	$R_2R'Bi$ + Br_2 in CCl_4	92	m. 109-110°	469
$(4-ClC_6H_4)_2(2-MeC_6H_4)BiCl_2$	$R_2R'Bi$ + Cl_2 in $CHCl_3$	86.5	m. 132-133°	469
$Ph_2(4-MeC_6H_4)BiCl_2$	$R_2R'Bi$ + Cl_2 in $CHCl_3$	46.8	m. 109-110°	469
C_{20}				
$(4-MeC_6H_4)_2PhBiCl_2$	$R'BiCl_2$ + $(RN_2)_2ZnCl_4$ in Me_2CO + Fe			1066
C_{21}				
$(2-MeC_6H_4)_3BiBr_2$			Rxn.[1]	468
$(2-MeC_6H_4)_3BiCl_2$			Rxn.[1]	468
$(3-MeC_6H_4)_3BiCl_2$	RN_2BF_4 + Bi in Me_2CO, followed by Cl_2 in $CHCl_3$	37	m. 151-153°	909
$(4-MeC_6H_4)_3BiCl_2$			Magnetic data, rxn.[1]	967, 969
C_{22}				
$(4-ClC_6H_4)_2(1-C_{10}H_7)BiBr_2$	$R_2R'Bi$ + Br_2 in CCl_4	88.6	m. 102-103°	469
$(4-ClC_6H_4)_2(1-C_{10}H_7)BiCl_2$	$R_2R'Bi$ + Cl_2 in $CHCl_3$	85	m. 132°	469
$Ph_2(1-C_{10}H_7)BiCl_2$			Rxn.[1]	468
C_{24}				
$(2-MeC_6H_4)_2(1-C_{10}H_7)BiBr_2$	$R_2R'Bi$ + Br_2 in CCl_4	76	m. 122°	469

1. Redn. by $H_2NNH_2.H_2O \rightarrow R_3Bi$ (468).

TABLE LXXIX (cont'd.)
TRIORGANOBISMUTHINE DIHALIDES AND DINITRATES

R_3BiX_2	Prepd. from	Yield %	Props.	Ref.
$(4\text{-}MeC_6H_4)_2(1\text{-}C_{10}H_7)BiBr_2$	$R_2R'Bi + Br_2$ in CCl_4	90	m. 126-127°	469
$(2\text{-}MeC_6H_4)_2(1\text{-}C_{10}H_7)BiCl_2$	$R'BiBr_2 + RMgBr$ (1:12), followed by Cl_2 in $CHCl_3$		m. 140°, rxn.[1]	469
$(4\text{-}MeC_6H_4)_2(1\text{-}C_{10}H_7)BiCl_2$	$R_2R'Bi + Cl_2$ in $CHCl_3$	95	m. 147°	967, 969
$(2,4\text{-}Me_2C_6H_3)_3BiCl_2$			Magnetic data	969
C_{27}				
$(4\text{-}EtO_2CC_6H_4)_3BiCl_2$	$RN_2BF_4 + Bi$ in Me_2CO, followed by Cl_2 in $n\text{-}C_8H_{18}EtOAc$	29	m. 138-139°	909
$(2,4,6\text{-}Me_3C_6H_2)_3BiBr_2$	$R_3Bi + Br_2$ in CCl_4		Bright yellow, m. > 250°	1339
$(2,4,6\text{-}Me_3C_6H_2)_3BiCl_2$	$R_3Bi + Cl_2$		Decomp. 150°, rxn.[2]	1339
C_{30}				
$[2,5\text{-}Me(Me_2CH)C_6H_3]_3BiBr_2$	$R_3Bi + Br_2$ in CCl_4	84	m. 101-103°	469
$[2,5\text{-}Me(Me_2CH)C_6H_3]_3BiCl_2$	$R_3Bi + Cl_2$ in $CHCl_3$	89	m. 163-164°	469
C_{36}				
$(4\text{-}PhC_6H_4)_3BiBr_2$	$R_3Bi + Br$ in CCl_4		Pale yellow plates, rxn.[3]	1317
$(4\text{-}PhC_6H_4)_3BiCl_2$	$R_3Bi + Cl_2$ in CCl_4		Small plates, m. 198-200°, rxns.[3,4]	1317
$(4\text{-}PhC_6H_4)_3Bi(NO_3)_2$	$R_3BiCl_2 + AgNO_3$ in Me_2CO		Small plates, decomp. ~162°	1317

2. Redn. by $Na_2S_2O_4$ in $Me_2CO\text{-}H_2O \rightarrow R_3Bi$ (468).
3. Rxn. with $HNO_3 \rightarrow$ nitro deriv. which deflagrates during a rapid heating (1317).
4. Rxn. with $AgNO_3 \rightarrow R_3Bi(NO_3)_2$ (1317).

TRIORGANOBISMUTHINE DICARBOXYLATES AND DIBENZENETHIOLATE

Triarylbismuthine dicarboxylates are prepared by:
1. Reacting triarylbismuthine dihalides with sodium or silver carboxylates in dioxan at room temperature.
2. Reacting triarylbismuthine carbonates with carboxylic acids in acetone under reflux.
3. Treating triarylbismuthines with lead tetraacetate in carbon tetrachloride.

Triphenylbismuthine dibenzenethiolate, $Ph_3Bi(SPh)_2$, was prepared by treating triphenylbismuthine dichloride with benzenethiol.

TRIPHENYLBISMUTHINE DIBENZENETHIOLATE $C_{30}H_{25}BiS_2$ $Ph_3Bi(SPh)_2$
Prepn.: by reacting Ph_3BiCl_2 with PhSH in dioxan at room temp.; 35% yield (478).
Props.: m. 44° (478).

TABLE LXXX

TRIARYLBISMUTHINE DICARBOXYLATES

$R_3Bi(O_2CR')_2$	Prepd. from	Yield %	Props.	Ref.
C22				
$Ph_3Bi(O_2CCH_2Cl)_2$	$R_3BiCO_3 + ClCH_2CO_2H$	96.5	m. 155-156°	478
$Ph_3Bi(OAc)_2$	$R_3Bi + Pb(OAc)_4$	23	m. 164°	1339
C26				
$Ph_3BiO_2CC_6H_4CO_2\text{-}o$	$R_3BiCl_2 + o\text{-}C_6H_4(CO_2Na)_2$	58.3	m. 155-165°	478
C32				
$(4\text{-}ClC_6H_4)_3Bi(O_2CC_6H_4OH\text{-}o)_2$	$R_3BiCl_2 + o\text{-}HOC_6H_4CO_2Na$	85.7	m. 187°	478
$Ph_3Bi(O_2CC_6H_4OH\text{-}o)_2$	$R_3BiCl_2 + o\text{-}HOC_6H_4CO_2Na$	52	m. 184-185°	478
$Ph_3Bi(O_2CC_6H_4OH\text{-}p)_2$	$R_3BiCO_3 + p\text{-}HOC_6H_4CO_2H$	81.6	m. 250°	478
$Ph_3Bi(O_2CC_6H_4NH_2\text{-}o)_2$	$R_3BiCO_3 + o\text{-}H_2NC_6H_4CO_2H$	90.2	with 1 mol. C_6H_6 m. 95-96°	478
$Ph_3Bi(O_2CC_6H_4NH_2\text{-}p)_2$	$R_3BiCO_3 + p\text{-}H_2NC_6H_4CO_2H$	69.8	with 2 mols. C_6H_6, m. 148°	478
C34				
$Ph_3Bi(O_2CC_6H_4CO_2H\text{-}o)_2$	$R_3BiCO_3 + o\text{-}C_6H_4(CO_2H)_2$	63.3	m. 168-169°	478
C35				
$(2\text{-}MeC_6H_4)_3Bi(O_2CC_6H_4OH\text{-}o)_2$	$R_3BiCl_2 + 2\text{-}HOC_6H_4CO_2Na$	51	with 1 mol. C_6H_6 m. 150-151°	478
$(4\text{-}MeC_6H_4)_3Bi(O_2CC_6H_4OH\text{-}o)_2$	$R_3BiCl_2 + 2\text{-}HOC_6H_4CO_2Na$	65.5	with 1 mol. C_6H_6 m. 164-165°	478
$(4\text{-}MeC_6H_4)_3Bi(OBz)_2$	$R_3BiCl_2 + BzONa$	69	m. 168-169°	478
C36				
$Ph_2(2\text{-}C_{10}H_7)Bi(OBz)_2$	$R_2R'BiCl_2 + BzOAg$		m. 138-140°	470
$Ph_3Bi(O_2CCH{:}CHPh)_2$	$R_3BiCl_2 + PhCH{:}CHCO_2Na$	50	m. 176-178°	478
C38				
$Ph_2(4\text{-}PhC_6H_4)Bi(OBz)_2$	$R_2R'BiCl_2 + BzOAg$		m. 145-147°	470

TETRAORGANOBISMUTH SALTS

Tetraphenylbismuth halides and tetraphenylborate were prepared from pentaphenylbismuth by treating with halogens in carbon tetrachloride or hydrogen halides
in ether at -80 to -70° and with triphenylborine in benzene, respectively. When
excess of bromine was used, tetraphenylbismuth perbromide was obtained. Tetraphenylbismuth chloride was converted into tetraphenylbismuth perchlorate and nitrate
by reacting with sodium perchlorate and aqueous nitric acid, respectively.

TETRAPHENYLBISMUTH BROMIDE $C_{24}H_{20}BiBr$ Ph_4BiBr
Prepn.: by treating Ph_5Bi in Et_2O at -70° under N with Br_2 in CCl_4 (1312).
Props.: Highly unstable, colorless crystals, m. -32 to -31°, decomp. at r.t.
into Ph_3Bi and PhBr (1312).

TETRAPHENYLBISMUTH PERBROMIDE $C_{24}H_{20}BiBr_3$ Ph_4BiBr_3
Prepn.: by treating Ph_5Bi in Et_2O with 2 moles Br_2 in CCl_4 at -80°; 92% yield (1312)
Props.: Orange-colored prod., decomp. on warming from -40° to +5° to give Ph_3BiBr_2;
on warming to r. t. gives Ph_2BiBr (1312).

TETRAPHENYLBISMUTH CHLORIDE $C_{24}H_{20}BiCl$ Ph_4BiCl
Prepn.: by treating Ph_5Bi in Et_2O with 2.6 N Et_2O-soln. of HCl under N at -70°
(1312).
Props.: Colorless, when dried in vacuo at -30°, decomp. at r. t. into Ph_3Bi +
PhCl (1312).
Rxns. with: PhLi in Et_2O → Ph_5Bi (1312).
KBr in H_2O → white ppt. decompg. to Ph_3Bi (1312).
Aq. KCN → white ppt. decompg. to Ph_3Bi and PhCN (1312).
Na(BPh_4) → (Ph_4Bi)(BPh_4) (1312).
$NaNO_2$ → $PhNO_2$ (1312).
$NaNO_3$ → Ph_4BiNO_3 (1312).
$NaClO_4$ → Ph_4BiClO_4 (1312).

TETRAPHENYLBISMUTH PERCHLORATE $C_{24}H_{20}BiClO_4$ Ph_4BiClO_4
Prepn.: by treating Ph_4BiCl with aq. $NaClO_4$; 94% yield (1312).
Props.: White ppt. (1312).

TETRAPHENYLBISMUTH NITRATE $C_{24}H_{20}BiNO_3$ Ph_4BiNO_3
Prepn.: by treating Ph_4BiCl with aq. HNO_3 (1312).
Props.: Very stable, white ppt. (1312).

TETRAPHENYLBISMUTH TETRAPHENYLBORATE $C_{48}H_{40}BBi$ (Ph_4Bi)(BPh_4)
Prepn.: By treating Ph_5Bi in Et_2O with BPh_3 in C_6H_6 under N; 49% yield (1312).
Props.: Colorless needles, m. 225-228° (1312).

PENTAORGANOBISMUTH

PENTAPHENYLBSIMUTH $C_{30}H_{25}$ Ph_5Bi

<u>Prepn.</u>: By reacting Ph_3BiCl_2 with PhLi at -75° in Et_2O under N; 81% yield (1312).
  By treating Ph_4BiCl with PhLi in Et_2O at -70°; 67% yield (1312).
<u>Props.</u>: Purple needles, m.~100-105° (dec.); on heating under N decomp. into Ph_3Bi, Ph_2, and C_6H_6 (1312).
<u>Rxns. with</u>: Br_2(1:1) in Et_2O-CCl_4 under N at -70° → Ph_4BiBr and PhBr (1312).
Br_2 (1:2) in Et_2O-CCl_4 at -80° → Ph_4BiBr_3 (1312).
Ph_3B in Et_2O-C_6H_6 → $(Ph_4Bi)(BPh_4)$ (1312).
HCl in Et_2O at -70° under N → Ph_4BiCl + C_6H_6 (1312).

III.   BIBLIOGRAPHY TO DATA SECTION

1. Abbott Laboratories, Brit. 762,776, Dec. 5, 1956; CA 51, 11383.
2. Abel, E. W., Bennett, M. A., and Wilkinson, G., J. Chem. Soc. 1959, 2323-7.
3. -----, and Wilkinson, G., ibid., 1501-5; CA 53, 15838.
4. Adler, S., Haskelberg, L., and Bergman, F., J. Chem. Soc. 1940, 576-8; CA 34, 5421.
5. Affsprung, H. E., Barnes, N. A., and Potretz, H. A., Anal. Chem. 23, 1680-3(1951); CA 46, 4421.
6. Ahmad. Y., Pakistan J. Sci. Research 1, No. 1, 36-40 (in Pakistan J. Sci. 1(1949); CA 46, 4545.
7. Ahrland, S., Chatt, J., Davies, N. R., and Williams, A. A., Nature 179, 1187-8(1957); CA 51, 15324.
8. Ainley, A. D., Brit. 736,816, Sept. 14, 1955; U. S. 2,803,626, Aug. 20, 1957; CA 50, 8724.
8a. -----, and Davey, D. G., Brit. J. Pharmacol. 13, 244-9(1958); CA 53, 10243.
9. Albers, H., Künzel, W., and Schuler, W., Chem. Ber. 85, 239-49(1952); CA 46, 11101.
10. Albert, H. E., U. S. 2,476,821, July 19, 1949; CA 43, 8732.
11. Albrand, L., Ann. Pharm. Franç. 5, 53-6(1947); CA 41, 7050.
12. Allison, J. A. C., and Mann, F. G., J. Chem. Soc. 1949, 2915-21; CA 45, 543.
13. Anderson, J. J., and Burrows, G. J., J. Proc. Roy. Soc. N. S. Wales, 70, 63-8(1936); CA 31, 333.
14. Andres, R. J., and Hamilton, C. S., J. Am. Chem. Soc. 67, 946-7(1945); CA 39, 3289.
15. Angier, R. B., Gazzola, A. L., Semb, J., Gadekar, S. M., and Williams, J. H., J. Am. Chem. Soc. 76, 902-4(1954); CA 49, 10307.
16. Anschütz, L., and Wirth, H., Chem. Ber. 89, 1530-2(1956); CA 51, 4982.
17. ----- and -----, Naturwissenschften 43, 59(1956); CA 51, 14671.
18. Aoyama, K., J. Soc. Org. Synthet. Chem. Japan 11, 347-52(1953); CA 48, 566.
19. -----, ibid., 391-3; CA 48, 1605.
20. Archer, S. and Bair, R. K., U. S. 2,647,910, Aug. 4, 1953; CA 48, 10765.
21. Arthur, P., Jr., and Pratt, B. C., U. S. 2,571,099, Oct. 16, 1951; CA 46, 3068.
22. Aylett, B. J., Eméleus, H. J., and H. J., and Maddock, A. G., J. Inorg. Nuclear Chem. 1, 187-93(1955); CA 50, 4699.
23. Ayscough, P. B., and Eméleus, H. J., J. Chem. Soc. 1954, 3381-8; CA 49, 14634.
23a. Azanovskaya, M. M., Zhur. Obshchei Khim. 27, 1363-4(1957); CA 52, 2738.
24. Backer, J. H., and Mulder, C. H. K., Rec. trav. chim. 55, 357-70(1936); CA 30, 7541.
25. -----, and -----, ibid., 594-601; CA 30, 7098.
26. -----, and Oosten, R. P. van, ibid., 59, 41-63(1940); CA 34, 2324.
27. -----, and Jonge, J. de, Rec. trav. chim. 60, 495-501(1941); CA 36, 5793.

27a. Bähr, G., and Zoche, G., Chem. Ber. 90, 1176-8(1957); CA 53, 8995.
28. Baker, M. C., Campbell, D. H., Epstein, S. I., and Singer, S. J., J. Am. Chem. Soc. 78, 312-16(1956); CA 50, 9592.
29. Balezin, S. A., and Ignat'eva, M. A., Zhur. Priklad. Khim. 29, 1647-56(1956); CA 51, 3421.
30. -----, and -----, Doklady Akad. Nauk. S. S. S. R. 109, 771-3(1956); CA 51, 3413.
31. Bamford, C. H., Livi, D. L., and Newitt, D. M., J. Chem. Soc. 1946, 468-71; CA 40, 6902.
32. -----, and Newitt, D. M., ibid., 695-701; CA 41, 641.
33. Banerjee, S., and Ghosh, T. N., J. Indian Chem. Soc. 23, 157-9(1946); CA 41, 3103.
34. -----, and -----, J. Indian Chem. Soc. 23, 229-31(1946); CA 41, 3103.
35. Banks, C. K., and Hamilton, C. S., J. Am. Chem. Soc. 60, 1370-1(1938); CA 32, 5800.
36. -----, and -----, ibid., 61, 357-60(1939); CA 33, 2115.
37. -----, and -- --, ibid., 62, 3142-4(1940); CA 35, 87.
38. -----, and Gruhzit, O. M., Tillitson, E. W., and Controulis, J., ibid., 66, 1771-5(1944); CA 39, 1170.
39. -----, and Controulis, J., ibid., 68, 944-5(1946); CA 40, 4728.
40. -----, ibid., 66, 1127-30(1944); CA 38, 4952.
41. -----, Controulis, J., and Holcomb, W. F., ibid., 68, 2102(1946); CA 41, 105.
42. -----, and Sultzaberger, J. A., ibid., 69, 1-4(1947); CA 41, 3250.
43. -----, Controulis, J., Walker, D. F., and Sultzaberger, J. A., ibid., 5-11; CA 41, 2401.
44. -----, et. al., ibid., 927-30(1947); CA 41, 4092.
45. -----, Kähler, F. H., and Hamilton, C. S., ibid., 933; CA 41, 4117.
46. -----, Brit. 584,917, Jan. 27, 1947; CA 41, 4176.
47. -----, Walker, D. F., Controulis, J., Tillitson, E. W., and Sweet, L. A., J. Am. Chem. Soc. 70, 738-9(1948); CA 42, 3349.
48. -----, Controulis, J., Walker, D. F., Tillitson, E. W., Sweet, L. A., and Gruhzit, O. M., ibid., 1762-5; CA 43, 598.
49. -----, J. Am. Pharm. Assoc. 38, 503-9(1943); CA 43, 9376.
50. -----, ibid., 509-15; CA 43, 9376.
51. -----, and Controulis, J., U. S. 2,559,061, July 3, 1951; CA 46, 3077.
51a. Banks, C. V., Thompson, J. A., and O'Laughlin, J. W., Anal. Chem. 30, 1792-5(1958); CA 53, 1996.
52. Barac, G., and Morreu, H., Bull. soc. chim. biol 35, 299-300(1953); CA 48, 8756.
53. Baranger, P. M., and Mercier, J. M., Biochem. J. 36, 703-5(1942); CA 37, 1733.
54. Barclay, G. A., and Nyholm, R. S., J. Proc. Roy. Soc. N. S. Wales 81, 77-9(1947); CA 42, 3349 and CA 44, 1437.
55. -----, and -----, Chemistry and Industry 1953, 378-9; CA 48, 5707.
56. Barker, R. L., Booth, E., Jones, W. E., and Woodward, F. N., J. Soc. Chem. Ind. (London) 68, 277-9(1949); CA 44, 3931.
57. -----, -----, -----, Millidge, A. F., and Woodward, F. N., J. Soc. Chem. Ind. (London) 68, 285-9(1949); CA 44, 3451.

58. Barker, R. L., Booth, E., Jones, W. E., Millidge, A. F., and Woodward, F. N., J. Soc. Chem. Ind. (London) 68, 289-95; CA 44, 3451.

59. -----, -----, -----, and Woodward, F. N., ibid., 295-8; CA 44, 3451.

60. Barnes, R. G., and Bray, P. J., J. Chem. Phys. 23, 1177-8(1955); CA 49, 12124.

61. -----, and -----, ibid., 407; CA 49, 5979.

62. Barr, P. O., and Hamilton, C. S., J. Am. Chem. Soc. 59, 2444-6(1937); CA 32, 517.

63. Bartholomew, E., and Cross, H. C., U. S. 2,546,421, Mar. 27, 1951; CA 45, 5920.

64. Bauer, H., J. Am. Chem. Soc. 67, 591-3(1945); CA 39, 2493.

65. Bartlett, P. D., Dauben, H. J., Jr., and Rosen, L. J., U. S. 2,465,834, Mar. 29, 1949; CA 43, 5412.

66. Baumhardt, G. C., Engenharia e quim. (Rio de Janeiro) 3, 164-5(1951); CA 46, 3216.

67. Beck, W. W., and Hamilton, C. S., J. Am. Chem. Soc. 60, 620-1(1938); CA 32, 3388.

68. Beeby, M. H., Cookson, G. N., and Mann, F. G., J. Chem. Soc. 1950, 1917-22; CA 45, 1119.

69. -----, Mann, F. G., and Turner, E. E., ibid., 1923-5; CA 45, 1120.

70. -----, and -----, ibid., 1951, 886-91; CA 46, 2069.

71. Beguin, A. E., and Hamilton, C. S., J. Am. Chem. Soc. 61, 355-7(1939); CA 33, 2115.

72. Belov, N. V., Ann. secteur platine, Inst. chim. gén. (U. S. S. R.) 18, 112-92(1945); CA 41, 4989.

73. Benda, L., and Sievers, O., U. S. 2,070,109, Feb. 9, 1937; CA 31, 2233; also Ger. 558,311.

74. Bennett, F. W., Brandt, G. R. A., Emeléus, H. J., and Haszeldine, R. N., Nature 166, 225(1950); CA 45, 5051.

75. Behrens, O. K., Corse, J., Huff, D. E., Jones, R. G., Soper, Q. F., and White-head, C. W., J. Biol. Chem. 175, 771-92; CA 43, 2271.

76. Belcher, R., and Nutten, A. J., J. Chem. Soc. 1951, 544-6; CA 45, 5560.

77. Berger, M., Buu-Hoi, Daudel, P, Daudel, R., Holovée, E., Lacassagne, M., and Royer, R., Bull. soc. chim. 1946, 51-2; CA 40, 6980.

78. Bergmann, E., and Haskelberg, L., J. Chem. Soc. 1939, 1-5; CA 33, 2495.

79. Berlin, A. Ya., J. Gen. Chem. (U. S. S. R.) 9, 1856-7(1939); CA 34, 4061.

80. Bert, L., Compt. rend. 222, 965(1946); CA 40, 4348.

81. Biilmann, E., and Jensen, K. A., Bull. soc. chim. [5], 3, 2306-9(1936); CA 31, 7846.

82. Binkley, S. B., and Hamilton, C. S., J. Am. Chem. Soc. 59, 1716-19(1937); CA 31, 7858.

83. -----, and -----, ibid., 60, 134-5(1938); CA 32, 1667.

84. Binz, A., and Maier-Bode, H., Angew. Chem. 49, 486-9(1936); CA 30, 6370.

85. -----, and Schichk, O. V., Ber. 69B, 1727-34(1936); CA 30, 5988.

86. -----, and -----, Ger. 633,867, Aug. 8, 1936; CA 31, 817.

87. -----, Ger. 671,563, Mar. 4, 1939; CA 33, 6531.

87a. Bither, T. A., Knoth, W. H., Lindsey, R. V., Jr., and Sharkey, W. H., J. Am. Chem. Soc. 80, 4151-3(1958); CA 53, 1207.

88. Blankenstein, W. E., and Capps, J. D., J. Am. Chem. Soc. $\underline{76}$, 3211-13(1954); CA $\underline{49}$, 8967.

89. Blas, L., Genie civil $\underline{115}$, 448-9(1939); CA $\underline{34}$, 2342; see also Anales soc. es-pan. fis. quim. $\underline{36}$, 107-14(1940); CA $\underline{34}$, 7286.

90. -----, ibid., $\underline{36}$, 127-31(1940); CA $\underline{37}$, 3069.

91. -----, ibid., $\underline{37}$, 116-18(1941); CA $\underline{36}$, 3160.

92. -----, and Arimany, L., Anales fis. y quim. (Madrid) $\underline{40}$, 832-6(1944); CA $\underline{42}$, 7723.

93. Blicke, F. F., and Webster, G. L., J. Am. Chem. Soc. $\underline{59}$, 534-7(1937); CA $\underline{31}$, 3019.

94. -----, and -----, ibid., 537-9; CA $\underline{31}$, 3020.

95. -----, Oneto, J. F., and Webster, G. L., ibid., 925-7: CA $\underline{31}$, 4289.

96. -----, and Cataline, E. L., ibid., $\underline{60}$, 419-22(1938); CA $\underline{32}$, 2913.

97. -----, and -----, ibid., 423-4; CA $\underline{32}$, 2519.

98. -----, Willard, H. H., and Taras, J. T., ibid., $\underline{61}$, 88-90(1939); CA $\underline{33}$, 1684.

99. -----, and Safir, S. R., ibid., $\underline{63}$, 575-6(1941); CA $\underline{35}$, 2123.

100. -----, and -----, ibid., 1493-6; CA $\underline{35}$, 5471.

101. -----, and -----, ibid., 1496-8; CA $\underline{35}$, 5472.

102. Bockmühl, M., and Ehrhardt, G., U. S. 2,185,972, Jan. 2, 1940; Can. 387,168, Feb. 27, 1940; CA $\underline{34}$, 2866.

103. Bode, H., Z. Anal. Chem. $\underline{133}$, 95-100(1951); CA $\underline{45}$, 6962.

104. -----, ibid., $\underline{134}$, 100-6(1951); CA $\underline{46}$, 1386.

105. Bogert, M. T., and Stickler, W. C., Science $\underline{100}$, 526-7(1944); CA $\underline{39}$, 698.

106. Borgström, L. H., Compt. rend. soc. géol. Finlande $\underline{24}$, Bull. comm. géol. Finlande $\underline{154}$, 238-9(1951); CA $\underline{46}$, 3909.

107. Borisov, A. E., Izvest. Akad. Nauk S. S. S. R., Otdel. Khim. Nauk $\underline{1951}$, 402-8; CA $\underline{46}$, 2995.

107a. Borodina, G. M., Khim. Nauka i Prom. $\underline{3}$, 681(1958); CA $\underline{53}$, 7183.

108. Bose, A. N., Bose, S., and Ghosh, T. N., Indian Med. Gaz. $\underline{85}$, 50-4(1950); CA $\underline{44}$, 8002.

109. Böttcher, B., Brit. 729,464, May 4, 1955; Ger. 865,903, Feb. 5, 1953; CA $\underline{50}$, 7142.

110. Bourgeois, L., and Bolle, J., Mém. services chim. état (Paris) $\underline{34}$, 411-13 (1948); CA $\underline{44}$, 5814.

110a. Bouvier, J. A. F., and Guest, R. J., Can., Dept. Mines and Tech. Surveys, Mines Branch, Research Dept. No. $\underline{34}$, 14pp. (1958); CA $\underline{53}$, 14837.

111. Bowden, K., and Braude, E. A., J. Chem. Soc. $\underline{1952}$, 1086-77; CA $\underline{46}$, 8960.

112. Bowden, H. J. M., Trans. Faraday Soc. $\underline{50}$, 463-70(1954); CA $\underline{49}$, 683.

113. Bowers, G. W., and Hamilton, C. S., J. Am. Chem. Soc. $\underline{58}$, 1573-5(1936); CA $\underline{30}$, 7567.

114. Bradlow, H. L., and Van der Werf, C. A., ibid., $\underline{70}$, 654-7(1948); CA $\underline{42}$, 3349.

115. Bralley, J. A., U. S. 2,436,710, Feb. 24, 1948; CA $\underline{42}$, 8814.

116. Brand, K., and Bosenkranz, E., Pharm. Zentralhalle $\underline{79}$, 489-93(1953); CA $\underline{32}$, 8692.

117. Brant, R. A., Emeléus, H. J., and Haszeldine, R. N., J. Chem. Soc. $\underline{1952}$, 2552-5; CA $\underline{47}$, 8007.

18. Brannon, J. L., U. S. 2,475,005, July 5, 1949; CA 43, 8208.
19. Eraunholtz, J. F., and Mann, F. G., J. Chem. Soc. 1957, 3285-3291; CA 52, 4534.
19a. -----, Hall, G. E., Mann, F. G., and Sheppard, N., ibid., 1959; 862-72; CA 53, 11994.
20. Braz, G. I., and Tuturin, N. V., J. Gen. Chem. (U. S. S. R.) 9, 992-5(1939); CA 33, 8583.
21. -----, and Yakubovich, A. Ya., ibid., 11, 41-4(1941); CA 35, 5459.
22. -----, Berlin, A. Ya., and Makarova, Yu. V., ibid., 18, 316-19(1948); CA 42, 6764.
23. Bressler, W. L., and Ciereczko, L. S., J. Am. Soc. 77, 2330-1(1955); CA 50, 3276.
24. Breyer, B., Ber. 71B, 163-71(1938); CA 32, 2812.
25. Brintzinger, H., and Scholz, A., Chem. Ber. 83, 141-5(1950); CA 44, 7222.
26. Brotherton, T. K., and Bunnett, J. F., Chem. and Ind. (London) 1957, 80; CA 51, 16465.
27. Bruker, A. B., J. Gen. Chem. (U. S. S. R.) 6, 1823-7(1936); CA 31, 4291.
28. -----, and Makhlis, E. S., ibid., 7, 1880-4(1937); CA 32, 72.
29. -----, and Maklyaev, F. L., Doklady Akad. Nauk S. S. S. R. 63, 271-4(1948); CA 43, 2592.
30. -----, and Nikiforova, N. M., Zhur. Obshchei Khim. 18, 1133-6(1948); CA 43, 1737.
31. -----, ibid., 1297-1311; CA 43, 4647.
32. -----, ibid., 27, 2223-6(1957); CA 52, 6736.
33. -----, ibid., 2593-8; CA 52, 7188.
34. -----, ibid., 2700-4; CA 52, 7188.
35. -----, Spiridonova, T. G., and Zaboroviskii, L. Z., Zhur. Obschchei Khim. 28, 350-5(1958); CA 52, 13615.
36. -----, and Nikiforova, N. M., ibid., 2407-12; CA 53, 2131.
37. -----, and Malkov, K. M., Doklady Akad. Nauk S. S. S. R. 128, 948-50(1959); CA 54, 7609.
38. Bryden, J. H., Acta Cryst. 12, 558(1959).
39. Burkin, A. R., J. Chem. Soc. 1954, 71-81; CA 48, 6206.
39a. Burg, A. B., and Grant, L. R., J. Am. Chem. Soc. 81, 1-5(1959); CA 53, 13990.
40. Burkin, A. R., J. Chem. Soc., 1956, 538-41; CA 50, 8361.
41. Burns, J. H., and Waser, J., J. Am. Chem. Soc. 79, 859-64(1957); CA 51, 7098.
42. Burrows, G. J., and Sanford, E. P., J. Proc. Roy. Soc. N. S. Wales, 69, 182-9(1936); CA 30, 4111.
43. -----, and Lench, A., ibid., 70, 218-21(1937); CA 31, 1318.
44. -----, and -----, ibid., 294-9; CA 31, 4965.
45. -----, and -----, ibid., 300-1; CA 31, 4965.
46. -----, and -----, J. Proc. Roy. Soc. N. S. Wales 70, 437-9(1937); CA 32, 6630.
47. Burschkies, K. and Rothermundt, M., Ber. 69B, 2721-4(1936); CA 31, 2180.
48. -----, Arch. Pharm. 275, 503-6(1937); CA 32, 722.

149. Burschkies, K., and Rothermundt, M., Arch. Pharm. 276, 226-34(1938); CA 32 5997.
150. -----, and -----, Ger. 728,324, Oct. 22, 1942; CA 37, 6676.
151. Burstall, F. H., and Nyholm, R. S., J. Chem. Soc. 1952, 3570-9; CA 47, 2622.
152. Busse, S. A., U. S. S. R. 47,299, June 30, 1936; CA 33, 3397.
153. Buu-Hoi, Ng. Pg., Hiong-Ki-Wei, and Royer, R., Bull. soc. chim. 12, 904-8(1954); CA 40, 3759.
154. -----, -----, and -----, Compt. rend. 220, 50-2(1945); CA 46, 2838.
155. -----, -----, and -----, ibid., 361-3; CA 41, 5534.
156. -----, and Royer, R., Bull. soc. chim. 1946, 379-82; CA 41, 2713.
157. -----, Caussé, R., Daudel, P., Flon, M., Hertzeg, C., Nguyen-Hoan, and Lacassagne, A., Compt. rend. 228, 868-70(1949); CA 44, 1436.
158. -----, Royer, R., J. Chem. Soc. 1951, 795-8; CA 45, 9065.
159. -----, Bihan, H. Le., and Binon, F., J. Org. Chem. 16, 185-91(1951); CA 45, 8489.
160. -----, Royer, R., Eckert, B., and Jacquignon, P., J. Chem. Soc. 1952, 4867-9; CA 48, 5192.
161. -----, Eckert, B., and Royer, R., J. Org. Chem. 17, 1000-4(1952); CA 47, 5914.
162. -----, Royer, R., and Eckert, B., Rec. trav. chim. 71, 1059-64(1952); CA 47, 8008.
163. -----, -----, Hubert-Habart, M., and Mabille, P., J. Chem. Soc. 1953, 3584-7; CA 49, 330.
164. -----, Xuong, Ng. D., and Nam, Hg. H., ibid., 1955, 1573-81; CA 50, 3407.
165. -----, Royer, R., and Hubert-Habart, M., ibid., 1956, 2048-51; CA 51, 411.
166. -----, Jacquignon, P., and Lavit, D., ibid., 2593-6; CA 51, 3599.
167. -----, -----, and Long, C. T., ibid., 1957, 505-9; CA 51, 9620.
168. -----, Binh, I. C., Loc, T. B., Xoung, Ng. D., and Jacquignon, P., J. Chem. Soc. 1957, 3126-9; CA 51, 17916.
169. -----, Bull. soc. chim. France 1959, 445-7; CA 53, 17144.
170. Bywater, S. E., and Pritchard, G. J, Brit. 584,196, Jan. 7, 1947; CA 41, 3253.
171. Calingaert, G., Soroos, H., and Hnizda, V., J. Am. Chem. Soc. 64, 392-7 (1942); CA 36, 1900.
172. Campbell, I. G. M., J. Chem. Soc. 1947, 4-10; CA 41, 3469.
173. Campbell, I. G. M., ibid., 1950, 3109-16; CA 45, 6595.
174. -----, and Poller, R. C., Chemistry and Industry 1953, 1126-7; CA 48, 4313.
175. -----, J. Chem. Soc. 1952, 4448-53; CA 48, 6433.
176. -----, and Morrill, D. J., Chemistry and Industry 1953, 1229; CA 48, 13664.
177. -----, and -----, J. Chem. Soc. 1955, 1662-70; CA 50, 3401.
178. -----, J. Chem. Soc. 1955, 3116-21; CA 50, 7073.
179. -----, and Poller, R. C., ibid., 1956, 1195-1203; CA 50, 15508.
180. -----, ibid., 1976-9; CA 51, 4364.
181. -----, and White, A. W., ibid., 1958, 1184-90; CA 52, 10924.
182. -----, and -----, ibid., 1959, 1491-4; CA 53, 21883.
183. Capps, J. D., and Hamilton, C. S., J. Am. Chem. Soc. 60, 2104-6(1938); CA 32, 8421.

184. Capps, J. D., J. Am. Chem. Soc. 69, 176-8(1947); CA 41, 2056.
185. -----, ibid., 179-81; CA 41, 2056.
186. CarnevaleBonino, R. C. d'Alessio de, Rev. asoc. bioquim. argentina 14, 117-21(1947); CA 42, 1384.
187. Cass, R. C., Coates, G. E., and Hayter, R. G., Chemistry and Industry 1954, 1485-6; CA 49, 5196.
188. -----, -----, and -----, J. Chem. Soc. 1955, 4007-16; CA 50, 8498.
189. Catch, J. R., Elliott, D. F., Hey, D. H., and Jones, E. R. H., J. Chem. Soc. 1949, 552-5; CA 43, 7452.
190. Catlin, W. E., Iowa State Coll. J. Sci. 10, 65-7(1935); CA 30, 935.
191. Chabrier, P., Compt. rend. 214, 362-5(1942); CA 37, 3737.
192. Chabrier de la Saulniére, P., Ann. chim. 17, 353-70(1942); CA 38, 3255.
193. Challenger, F., and Rawlings, A. A., J. Chem. Soc. 1936, 264-7; CA 30, 2919.
194. -----, J. Proc. Roy. Inst. Chem. (Gt. Britain and Ireland) 1945, 105-6 (1945); CA 39, 4350
195. -----, and Greenwood, D., J. Chem. Soc. 1950, 26-30; CA 44, 7755.
196. -----, Lisle, D. B., and Dransfield, P. B., ibid., 1954, 1760-71; CA 48, 10387.
196a. Challis, H. J. G., and Jones, J. T., Anal. Chim. Acta 21, 58-67(1959); CA 53, 19701.
197. Chang, C - C., and Sun, S-M., J. Chinese Chem. Soc. 16, 41-5(1949); CA 44, 1508.
198. Charlot, G., Anal. Chim. Acta 1, 218-48(1947); CA 42, 7192.
199. Chatt, J., and Mann, F. G., J. Chem. Soc. 1939, 610-15; CA 33, 5377.
200. -----, and -----, ibid., 1940, 1184-92; CA 34, 7914.
201. -----, and -----, ibid., 1192-6; CA 34, 7869.
202. -----, ibid., 1951, 652-8; CA 45, 6952.
203. -----, and Wilkins, R., ibid., 2532-3; CA 46, 6027.
204. -----, and -----, ibid., 1952, 4300-6; CA 47, 3740.
205. -----, and -----, ibid., 1953, 70-4; CA 47, 7361.
206. -----, Duncanson, A., and Venanzi, L., Atti accad. nazl. Lincei, Rend., classe sci. fis. mat. e nat. 17, 120-4(1954); CA 49, 11418.
207. -----, and Venanzi, L. M., J. Chem. Soc. 1955, 2787-93; CA 49, 15595.
208. -----, and -----, ibid., 3858-64; CA 50, 2349.
209. -----, Duncanson, L. A., and Venanzi, L. M., ibid., 4461-9; CA 50, 6991.
210. -----, and Wilkins, R. G., ibid., 1956, 525-9; CA 50, 8362.
211. -----, Duncanson, L. A., and Venanzi, L. M., ibid., 2712-25; CA 51, 84.
212. -----, and Venanzi, L. M., ibid., 1957, 2351-61; CA 51, 11906.
213. -----, and -----, ibid., 2445-7; CA 51, 16178.
213a. -----, and Hart, F. A., Chem. and Ind. (London) 1958, 1474-5; CA 53, 6865.
214. -----, Duncanson, L. A., and Shaw, B. L., Proc. Chem. Soc. 1957, 343, (CA 52, 7004); Chem. and Ind. (London) 1958, 859-60; CA 53, 859.
214a. -----, -----, and Venanzi, L. M., J. Inorg. and Nuclear Chem. 8, 67-74(1958); CA 53, 5867.
215. -----, Gamlen, G. A., and Orgel, L..E.; J. Chem. Soc. 1959, 1047-9; CA 53, 10947.
215a. -----, and Hayter, R. G., Proc. Chem. Soc. 1959, 153; CA 53, 19852.

216. Chatterjee, D. N., and Ghosh, T. N., J. Indian Chem. Soc. 23, 420-2(1946); CA 42, 141.

217. Chaudhury, S. C., Ghosh, T. N., and U. P. Basu, J. Indian Chem. Soc. 23, 211-13(1946); CA 41, 1629.

218. Chelintsev, G. V., and Kuskov, V. K., J. Gen. Chem. (U. S. S. R.) 16, 1481-4(1946); CA 41, 5441.

219. Chen, T-C., and Yeh, S-K., Hua Hsueh Hsueh Pao 23, 474-9(1957); CA 52, 16982.

220. Chernyaev, I. I., Izvest. Sektora Platiny i Drugikh Blagorodnykh Metal., Inst. Obshchei i Neorg. Khim., Akad. Nauk S. S. S. R. No. 21, 27-31 (1948); CA 45, 5481.

220a. Chi, Yu-Li, and Moh, Yo-Yung, K'e Hsueh Tung Pao No. 2, 50-1(1957); CA 53, 18896.

221. Childs, A. F., Plant, S. G. P., Tompsett, A. L. L., and Weeks, G. A., J. Chem. Soc. 1948, 2180-3; CA 43, 2927.

222. Ciocca, B., Ravazzoni, C. C., and Canonica, L., Chimica e industria (Milan) 28, No. 3/4, 42-4(1946); CA 40, 6440.

223. Chiswell, B., and Livingstone, S. E., J. Chem. Soc. 1959, 2931-6.

224. -----, and -----, ibid., 1960, 97-101.

225. -----, and -----, ibid., 1071-74.

226. Chi-Ting, C., and Zu-Yoong, K., Acta Chim. Sin. 25, 43-44(1959).

227. Christiansen, W. G., and Jurist, A. E., U. S. 2,137,237, Nov. 22, 1938; CA 33, 1883.

228. -----, U. S. 2,245,572, June 17, 1940; CA 35, 5912.

229. -----, and Green, L. W., U. S. 2,488,268, Nov. 15, 1949; CA 44, 1234.

230. Chu, T. L., Pake, G. E., Paul, D. E., Townsend, J., and Weissman, S. I., J. Phys. Chem. 57, 504-7(1953); CA 47, 8435.

231. Clark, R. L., and Hamilton, C. S., J. Am. Chem. Soc. 65, 635-7(1943); CA 37, 3069.

232. Clemence, L-R. W., Brit. 701,965, Jan.6, 1954; CA 49, 4029.

233. -----, and Krueger, W. D., U. S. 2,843,614, July 15, 1958; CA 52, 19023.

234. Clinch, J., Anal. Chim. Acta 14, 162-71(1956); CA 51, 2462.

235. Coates, G. E., J. Chem. Soc. 1951, 2003-13; CA 46, 51.

236. -----, and Whitcombe, R. A., ibid., 1956, 3351-4; CA 51, 1764.

237. -----, and Livingstone, J. G., Chem. and Ind. (London) 1958, 1366; CA 53, 8035.

238. Cochran, W., Hart, F. A., and Mann, F. G., J. Chem. Soc. 1957, 2816-28; CA 51, 14588.

239. Coles, R. F., and Hamilton, C. S., J. Am. Chem. Soc. 68, 1799-801(1946); CA 40, 7191.

240. -----, and -----, ibid., 2588-9; CA 41, 1667.

240a. Conti, L., and Leandri, G., Bull. sci. fac. chim. ind. Bologna 14, 60-2 (1956); CA 51, 5765.

241. Cook, C. D., U. S. 2,726,189, Dec. 6, 1955; CA 50, 6793.

242. Cookson, R. C., and Mann, F. G., J. Chem. Soc. 1947, 618-24; CA 41, 6216.

243. -----, and -----, ibid., 1949, 67-72; CA 43, 8351.

244. -----, and -----, ibid., 288-94; CA 44, 3991.

245. -----, and -----, ibid., 2895-8; CA 44, 3932.

246. Cooper, R. S., U. S. 2,664,411, Dec. 29, 1953; CA 48, 4251.

247. Coover, H. W., and Dickey, J. B., U. S. 2,675,372, April 13, 1954; CA 48, 8587.
248. Corse, J., and Rohrmann, E., J. Am. Chem. Soc. 70, 370-1(1948); CA 42, 2245.
249. Cox, H. E., Analyst 64, 807-13(1939); CA 34, 541.
250. Cragoe, E. J. Jr., and Hamilton, C. S., J. Am. Chem. Soc. 67, 536-9(1945); CA 39, 2505.
251. -----, Andres, R. J., Coles, R. F., Elpern, B., Morgan, J. F., and Hamilton, C. S., ibid., 69, 925-6(1947); CA 41, 4116.
252. Csabay, J., and Tanay, I., Ber. ungar. pharm. Ges. 15, 83-90(1939); CA 33, 5101.
253. Cullen, W. R., and Emeléus, H. J., J. Chem. Soc. 1959, 372-5; CA 53, 13989.
254. -----, Can. J. Chem. 38, 439-44(1960).
255. -----, ibid., 445-51(1960).
256. -----, and Walker, L. G., ibid., 472-5(1960).
256a. Curtis, N. F., Ferguson, J. E., and Nyholm, R. S., Chem. and Ind. (London) 1958, 625-6; CA 53, 2919.
257. Dacey, J. R., Smelko, J. F., and Young, D. M., J. Phys. Chem. 59, 1058-60(1955); CA 50, 645.
258. Dale, J. W., Emeléus, H. J., Haszeldine, R. N., and Moss, J. H., J. Chem. Soc. 1957, 3708-13; CA 52, 244.
259. Das, K. and Niyogy, S. C., J. Proc. Inst. Chemists (India) 27, 227-32(1955); CA 50, 11268.
260. Das-Gupta, H. N., J. Indian Chem. Soc. 12, 627-8(1935); CA 30, 465.
261. -----, ibid., 13, 305-8(1936); CA 30, 7098.
262. -----, ibid., 14, 231-6(1937); CA 31, 7423.
263. -----, ibid., 349-53; CA 31, 8532.
264. -----, ibid., 397-9; CA 32, 553.
265. -----, ibid., 400-5; CA 32, 553.
266. -----, ibid., 15, 495-7(1938); CA 33, 2135.
267. -----, ibid., 498-500; CA 33, 2135.
268. Datta, S., Ghosh, T. N., and Bose, A. N., Science and Culture 11, 142-5 (1945); CA 40, 1970.
269. -----, -----, -----, Science and Sulture 11, 385(1946); CA 41, 105.
270. -----, and -----, J. Indian Chem. Soc. 24, 403-6(1947); CA 42, 5870.
270a. Davis, R. E., Openshaw, H. T., Spring, F. S., Stanley, R. H., and Todd, A. R., J. Chem. Soc. 1948, 295-9; CA 42, 5026.
271. Davis, W. C., and Othen, C. W., J. Chem. Soc. 1936, 1236-9; CA 30, 7575.
272. -----, and Addis, H. W., ibid., 1937, 1622-7; CA 31, 8292.
273. -----, and Mann, F. G., ibid., 1944, 276-83; CA 38, 5803.
273a. Davies, R. V., Kennedy, J., Lane, E. S., and Willans, J. L., J. Appl. Chem. (London) 9, 368-71(1959); CA 53, 19447.
274. Delaby, R., Baronnet, R., and Villiger, W., Bull. soc. chim. France 1952, 315-18; CA 47, 3259.
275. -----, -----, and -----, ibid., 319-22; CA 47, 3260.
276. -----, -----, and -----, ibid., 322-3; CA 47, 3260.
277. Delmar, G. S., U. S. 2,675,389, Apr. 13, 1954; CA 49, 11026.
278. -----, and Macallum, E. N., U. S. 2,752,344, June 26, 1956; also Brit. 772,119, Apr. 10, 1957; CA 51, 2888.
279. Dennis, E. W., and Surrey, A. R., U. S. 2,714,106, July 26, 1955; CA 50, 7877.

280. Despois, R., U. S. 2,274,593, Feb. 24, 1942; CA 36, 4133.
281. Diaz de Arce, H., Greene, J. L., Jr., and Capps, J. D., J. Am. Chem. Soc. 72, 2971-4(1950); CA 44, 9965.
282. Doak, G. O., J. Am. Chem. Soc. 62, 167-8(1940); CA 34, 1632.
283. -----, Eagle, H., and Steinman, H. G., ibid., 168-70; CA 34, 1632.
284. -----, -----, and -----, ibid., 3010-11; CA 35, 85.
285. -----, Steinman, H. G., and Eagle, H., ibid., 3012-13; CA 35, 86.
286. -----, -----, and -----, ibid., 63, 99-101(1941); CA 35, 1389.
287. -----, Eagle, H., and Steinman, H. G., ibid., 64, 1064-6(1942); CA 36, 4109.
288. -----, Steinman, H. G., and Eagle, H., ibid., 66, 194-7(1944); CA 38, 1481.
289. -----, -----, and -----, ibid., 197-200; CA 38, 1481.
290. -----, -----, and -----, ibid., 67, 719-21(1945); CA 39, 2977.
291. -----, and Steinman, H. G., ibid., 68, 1989-91(1946); CA 41, 105.
292. -----, ibid., 1991-5; CA 41, 106.
293. -----, and Steinman, H. G., ibid., 1987-9; CA 41, 3068.
294. -----, and Jaffé, H. H., ibid., 72, 3025-7(1950); CA 44, 9931.
295. -----, and Freedman, L. D,, ibid., 73, 5656-7(1951); CA 47, 519.
296. -----, -----, and Efland, S. M., ibid., 74, 830-1(1952); CA 47, 9929.
297. Doering, W. V. E., and Hoffmann, A. K., ibid., 77, 521-6(1955); CA 50, 785.
298. Donohue, J., Humphrey, G., and Schomaker, V., ibid., 69, 1713-16(1947); CA 41, 6090.
299. Drefahl, G., Gerlach, E., and Degen, W., J. prakt. Chem. [4], 4, 119-23 (1956); CA 51, 11298.
300. Drevon, B., J. pharm. chim. [9], 2, 11-16(1942); CA 38, 2897.
301. -----, ibid., 54-9; CA 38, 2897.
302. Drozdov, N. S., J. Gen. Chem. (U. S. S. R.) 6, 1641-50(1936); CA 31, 2610.
303. -----, and Stavrovskaya, V. I., ibid., 9, 409-14(1939); CA 34, 103.
304. Duke, F. R., and Brown, L. M., J. Am. Chem. Soc. 76, 1443-6(1954); CA 48, 6786.
305. Du Pont de Nemours, E. I., Brit. 579,715, Aug. 13, 1946, CA 41, 1697.
306. -- --, Brit. 766,629, Jan. 23, 1957; CA 51, 10126.
307. Dwyer, F. P., and Nyholm, R. S., J. Proc. Roy. Soc. N. S. Wales 75, 127-9(1942); CA 36, 3446.
308. -----, and -----, ibid., 75, 140-3(1942); CA 36, 5103.
309. -----, and -----, ibid., 76, 129-132(1942); CA 37, 1666.
310. -----, and -----, ibid., 133-6; CA 37, 2283.
311. -----, and -- --, ibid., 77, 116-18(1944); CA 38, 4528.
312. -----, Gibson, N. A., and Nyholm, R. S., ibid., 78, 118-21(1945); CA 39, 4808.
313. -----, -----, and -----, ibid., 226-8(1946); CA 40, 3995.
314. -----, and Nyholm, R. S., ibid., 79, 121-5(1946); CA 40, 7032.
315. -----, Gibson, N. A., and Nyholm, R. S., ibid., 118-20(1946); CA 41, 1629.
316. -----, Humpoletz, J. E., and Nyholm, R. S., ibid., 80, 217-19(1947); CA 42, 2885.
317. -- --, and Gibson, N. A., Analyst 75, 201-3(1950); CA 44, 6335.
318. -----, and -----, ibid., 76, 548-51(1951); CA 45, 10126.
319. -----, and Stewart, D. M., J. Proc. Roy. Soc. N. S. Wales 83, 177-80(1949); CA 45, 7462.
320. -----, and Gibson, N. A., Chemistry and Industry 1953, 153-4; CA 48, 113.

321. Dwyer, F. P., Nyholm, R. S., and Tyson, B. T., J. Proc. Roy. Soc. N. S. Wales 81, Pt. 4, 272-5(1947); CA 43, 6535.
322. Dyke, W. J. C., and King, H., J. Chem. Soc. 1935, 1745-7; CA 30, 1772.
323. Dyson, G. M., and Renshaw, A., Brit. 458,487, Dec. 17, 1936; CA 31, 3639.
324. -----, Rec. trav. chim. 57, 1016-28(1938); CA 33, 1289.
325. -----, and Renshaw, A., U. S. 2,195,885, Apr. 2, 1940; CA 34, 5252.
326. -----, Brit. 517,682, Feb. 6, 1940; CA 35, 7116.
327. Eberly, K. C., and Smith, G. E. P., Jr., J. Org. Chem. 22, 1710(1957); CA 52, 9008.
328. Eckert, B., Royer, R., and Buu-Boi, Ng. Ph., Compt. rend. 233, 1461-3 (1951); CA 46, 7538.
329. Edwards, R. R., and Coryell, C. D., Brookhaven Natl. Lab. Conf. Chem. Effects of Nuclear Transformations (Hot Atom Chem.) AECU-50(BNL-C-7) 68-78(1948); CA 45, 4147.
330. Ellis, K. W., and Gibson, N. A., Anal. Chim. Acta 9, 275-80(1953); CA 48, 75.
331. -----, and -----, ibid., 368-73; CA 48, 496.
332. Elpern, B. and Hamilton, C. S., J. Am. Chem. Soc. 68, 1436-8(1946); CA 46, 7210.
333. Emeléus, H. J., Haszeldine, R. N., and Walaschewski, E. G., J. Chem. Soc. 1953, 1552-64; CA 48, 5790.
334. -----, ibid., 1954, 2979-86; CA 48, 12672.
335. -----, Hazeldine, R. N., and Paul, R. C., ibid., 881-6; CA 49, 1542.
336. -----, and Moss, J. H., Z. anorg. u. allgem. Chem. 282, 24-8(1955); CA 50, 14429.
337. -----, Haszeldine, R. N., and Paul, R. C., J. Chem. Soc. 1955, 563-74; CA 50, 790.
337a. Emi, K., Toei, K., and Furukawa, K., Nippon Kagaku Zasshi 79, 681-6 (1958); CA 53, 6876.
338. Ende, A., Brit. 501,229, Feb. 23, 1939; CA 33, 5872.
339. Englert, R. D., and Sweeting, O. J., J. Am. Chem. Soc. 70, 2977-9(1948); CA 43, 601.
340. Erdos, J., Ciencia, 4, 17(1943);CA 37, 6652.
341. -----, Quimica (Mex.) 3, No. 3, 58(1945); CA 39, 3632.
342. -----, Anales escuela nacl. cienc. biol. (Mex.) 4, 15-18(1945); CA 40, 172.
343. -----, ibid., 165-71; CA 41, 3441.
344. -----, ibid., 173-4; CA 41, 725.
345. Étienne, A., Compt. rend. 221, 562-4(1945); CA 40, 3753.
346. -----, ibid., 628-30; CA 40, 3754.
347. -----, Bull. soc. chim. France 1947, 47-50; CA 41, 4790.
348. -----, ibid., 50-1; CA 41, 4790.
349. -----, Mém. services chim. état (Paris) 33, 405-8(1947); CA 43, 6102.
350. -----, and Degent, C., Compt. rend. 238, 2093-5(1954); CA 49, 8896.
350a. -----, and -----, Fr. 1,085,860, Feb, 8, 1955; CA 53, 6191.
351. Euler, H. V., Hasselquist, H., and Euler, B. V., Arkiv Kemi 9, 583-90 (1956); CA 51, 2914.
352. Evans, R. C., Mann, F. G., Peiser, H. S., and Purdie, D., J. Chem. Soc. 1940, 1209-30; CA 35, 2433.

353. Evans, A. G., and Warhurst, E., Trans. Faraday Soc. 44, 189-95(1948); CA 43, 23.
354. Ewins, A. J., Brit. 468,363, June 30, 1937; CA 31, 8834.
355. -----, and Newbery, G., Brit. 475,042, Nov. 9, 1937; Fr. 829,219, June 16, 1938; CA 32, 3556.
356. -----, and -----, U. S. 2,209,876, July 30, 1940; CA 35, 281.
357. -----, and -----, Ger. 680,888, Aug. 24, 1939; CA 36, 1260.
358. Faith, H. E., J. Am. Chem. Soc. 72, 837-9(1950); CA 45, 3826; and U. S. 2,586,316, Feb. 19, 1952; CA 46, 8149.
359. Feher, F., and Kolb, W., Naturwissenschaften 27, 615-16(1939); CA 34, 28.
360. Fehrle, A., Streitwolf, K., and Fritzsche, P., Ger. 618,447, Sept. 7, 1935; U. S. 2,082,880, June 8, 1937; CA 30, 3595.
361. -----, -----, and -----, Ger. 623,450, Dec. 21, 1935; U. S. 2,085,305, June 29, 1937; and Fr. 787,025, Sept. 16, 1935; Brit. 452,153; CA 30, 2204.
362. -----, -----, and -----, Ger. 635,398, Sept. 23, 1936; Brit. 469,324, July 22, 1937; CA 31, 114.
363. -----, and Fritzsche, P., Ger. 638,265, Nov. 12, 1936; Brit. 469,323, July 22, 1937; CA 31, 3941.
364. -----, and -----, Ger. 646,409, June 15, 1937; Brit. 484,594, May 9, 1938; Fr. 47,968, Aug. 28, 1937; CA 31, 6258.
365. -----, Herrmann, W., Fritzsche, P., and Hilmer, H., Ger. 666,573, Oct. 22, 1938; Brit. 494,249, Oct. 24, 1938; CA 33, 2286.
366. -----, -----, -----, and -----, Ger. 667,844, Nov. 21, 1938; CA 33, 2909.
367. -----, and Fritzsche, P., U. S. 2,191,740; CA 34, 4392.
368. -----, Herrmann, W., Fritzsche, P., and Hilmer, H., U. S. 2,232,659, Feb. 18, 1941; CA 35, 3394.
369. -----, -----, and Hampe, F., Ger. 701,527, Dec. 19, 1940; CA 35, 7979.
370. -----, and Fritzsche, P., U. S. 2,265,424, Dec. 9, 1941; CA 36, 2088.
371. -----, Herrmann, W., and Hampe, F., U. S. 2,289,878, July 14, 1942; CA 37, 388.
372. -----, -----, and Hilmer, H., Ger. 727,403, Oct. 1, 1942; CA 37, 6278.
373. -----, Oesterlin, H., Herrmann, W., and Hampe, F., U. S. 2,336,853, Dec. 14, 1943; CA 38, 3422.
374. -----, Herrmann, W., and Hampe, F., Ger. 849,697, Sept. 18, 1952; CA 48, 10765.
375. -----, -----, and Hilmer, H., Ger. 852,089, Oct. 13, 1952; CA 52, 10174.
376. -----, -----, and -----, Ger. 853,167, Oct. 23, 1952; CA 52, 10204.
376a. Feinberg, G. J., and Grant, R. A., Biochem. J. 65, 40P(1957); CA 53, 5458.
376b. Ferguson, J. E., and Nyholm, R. S., Chem. and Ind. (London) 1958, 1555; CA 53, 7854.
376c. -----, and -----, Nature 183, 1039-40(1959); CA 53, 16791.
377. Feitelson, B. N., and Petrow, V., J. Chem. Soc. 1951, 2279-83; CA 46, 975.
378. Fierz-David, H. E., Jadassohn, W., and Stoll, W. G., Helv. Chim. Acta 20, 1059-77(1937); CA 32, 5897.
379. -----, -----, and Margot, A., Helv. Chim. Acta 21, 280-93(1938); CA 32, 4962.
380. -----, and Stoll, W. G., U. S. 2,166,681, July 18, 1939; CA 33, 8923.

81. Figgis, G., and Gibson, Anal. Chim. Acta 7, 313-18(1952); CA 47, 4705.
82. Fincke, J. K., U. S. 2,556,420, June 12, 1951; CA 45, 8813.
82a. Fisher, E. O., and Schreiner, S., Angew. Chem. 69, 205-6(1957); CA 52, 8978.
83. -----, and -----, Chem. Ber. 93, 1417-24(1960).
83a. -----, and -----, ibid., 92, 938-48(1959); CA 53, 11979.
83b. -----, Joos, G., and Meer, W., Z. Naturforsch. 13b, 456-7(1958); CA 53, 1875.
84. Fitzpatrick, H. D. N., Hughes, S. R., C., and Moelwyn-Hughes, E. A., J. Chem. Soc. 1950, 3542-7; CA 46, 5008.
85. Flés, D., and Muić, N., Archiv. Kemi 21, 168-73(1949); CA 45, 9526.
86. Forward, M. V., Bowden, S. T., and Jones, W. J., J. Chem. Soc. 1949, S121-6; CA 43, 8786.
87. Foss, M. E., and Gibson, C. S., ibid., 3075-9; CA 44, 4860.
88. Fox, H. H., J. Org. Chem. 12, 872-5(1947); CA 42, 1904.
89. -----, ibid., 13, 438-42(1948); CA 43, 175.
90. -----, and Wenner, W., U. S. 2,465,307. Mar. 22, 1949; CA 43, 5041.
91. -----, and -----, U. S. 2,465,308, Mar. 22, 1949; Brit. 633,459, Dec. 19, 1949; CA 43, 4694.
92. -----, and -----, U. S. 2,553,515, May 15, 1951; CA 45, 9561.
93. Freedman, L. D., and Doak, G. O., J. Am. Chem. Soc. 71, 779-80(1949); CA 43, 4237.
94. -----, and -----, ibid., 75, 4905-6(1953); CA 48, 12691.
95. -----, and -----, ibid., 77, 6223-4(1955); CA 50, 8550.
96. -----, and -----, ibid., 6374-6; CA 50, 5061.
97. -----, and -----, J. Org. Chem. 21, 1533-4(1956); CA 51, 17792.
98. -----, and -----, J. Org. Chem. 24, 1590-2(1959); CA 54, 6599.
99. Freund, W., J. Chem. Soc. 1951, 1943-4; CA 46, 946.
400. -----, Brit. 670,317, Apr. 16, 1952; U. S. 2,710,874, June 14, 1955; CA 46, 10201.
401. -----, J. Chem. Soc. 1952, 3072-3; CA 47, 1632.
402. -----, ibid., 3073-5; CA 47, 1632.
403. -----, and Komzak, A., ibid., 1953, 3707; CA 49, 209.
404. -----, Australian 147,045, July 10, 1952; CA 51, 15595.
405. Frevel, L. K., Rinn, H. W., and Anderson, H. C., Ind. Eng. Chem., Anal. Ed. 18, 83-93(1946); CA 40, 2049.
406. Friedheim, E. A. H., Swiss 209,035, June 1, 1940; CA 35, 4551.
407. -----, Swiss 214,110, July 1, 1941; CA 36, 4975.
408. -----, Swiss 214,345, July 16, 1941; CA 36, 4975.
409. -----, U. S. 2,235,478, Mar. 18, 1941; Brit. 538,450, Aug. 5, 1941; CA 35, 4161.
410. -----, Brit. 537,690, July 2, 1941; Ger. 726,430, Sept. 3, 1942; CA 36, 1737.
411. -----, Brit. 539,528, Sept. 15, 1941; CA 36, 4291.
412. -----, Swiss 213,145, May 1, 1941; Ger. 722,339, May 21, 1942; CA 36, 4979.
413. -----, Ger. 722,340, May 21, 1942; CA 37, 5080.
414. -----, U. S. 2,295,574, Sept. 15, 1942; CA 37, 1228.
415. -----, U. S. 2,334,321, Nov. 16, 1943; CA 38, 2666.

416. Friedheim, E. A. H., J. Am. Chem. Soc. 66, 1775-8(1944); CA 39, 1171.
417. -----, U. S. 2,386,204, Oct. 9, 1945; CA 40, 177.
418. -----, U. S. 2,390,529, Dec. 11, 1945; Brit. 576,361, Apr. 1, 1946; CA 40, 5070.
419. -----, U. S. 2,390,089, Dec. 4, 1945; Brit. 572,402, Oct. 5, 1945; CA 41, 160.
420. -----, U. S. 2,390,090, Dec. 4, 1945; CA 41, 161.
421. -----, U. S. 2,390,091, Dec. 4, 1945; CA 41, 161.
422. -----, U. S. 2,390,092, Dec. 4, 1945; CA 41, 161.
423. -----, U. S. 2,391,452, Dec. 25, 1945; CA 40, 4462.
424. -----, U. S. 2,400,547, May 21, 1946; CA 40, 5459.
425. -----, U. S. 2,415,555, Feb. 11, 1947; CA 41, 3133.
426. -----, U. S. 2,415,556, Feb. 11, 1947; CA 41, 3134.
427. -----, U. S. 2,418,115, Apr. 1, 1947; CA 41, 3816.
428. -----, U. S. 2,419,348, Apr. 22, 1947; CA 41, 6674 and 42, 608.
429. -----, U. S. 2,422,724, June 24, 1947; CA 41, 6900.
430. -----, U. S. 2,430,461, NOv. 11, 1947; Brit. 592,875, Oct. 1, 1947; CA 42, 1973.
431. -----, U. S. 2,430,462, Nov. 11, 1947; CA 42, 1973.
432. -----, Brit. 582,043, Nov. 4, 1946; CA 41, 6283.
433. -----, Brit. 585,678, Feb. 19, 1947; CA 42, 623.
434. -----, Brit. 587,015, Apr. 10, 1947; CA 42, 218.
435. -----, Brit. 598,975, Feb. 10, 1948; CA 42, 5056.
436. -----, Vogel, H. J., and Berman, R. L., J. Am. Chem. Soc. 69, 560-2; CA 41, 4157.
437. -----, Swiss 214,903, May 31, 1941; CA 43, 4296.
438. -----, Swiss 240,545, Dec. 31, 1945; CA 44, 7028.
439. -----, Swiss 240,546, Dec. 31, 1945; CA 44, 7029.
440. -----, U. S. 2,549,795, Apr. 24, 1951; CA 46, 529.
441. -----, U. S. 2,593,434, Apr. 22, 1952; CA 46, 7291.
442. -----, Brit. 655,435, July 18, 1951; U. S. 2,662,079, Dec. 12, 1953; CA 47, 144.
443. -----, U. S. 2,659,723, Nov. 17, 1953, and U. S. 2,664,432, Dec. 29, 1953; CA 49, 1816.
444. -----, U. S. 2,772,303, Nov. 7, 1956; CA 51, 5831.
444a. -----, U. S. 2,880,222, Mar. 31, 1959; CA 53, 16158.
444b. Fritz, H. P., Chem. Ber. 92, 780-91(1959); CA 53, 11994.
444c. Fritz, J. S., and Johnson-Richard, M., Anal. Chim. Acta 20, 164-71(1949); CA 53, 21424.
445. Fusco, R., and Cottignoli, T., Farm. ital. 11, 89-91(1943); CA 38, 6034.
445a. Furukawa, J., Tsurata, T., Funo, T., Sakata, R., and Ito, K., Makromol. Chem. 30, 190-22(1959);
446. Fuson, R. C., and Shive, W., J. Am. Chem. Soc. 69, 559-60(1947); CA 41, 409
447. -----, and -----, U. S. 2,756,245, July 24, 1956; CA 51, 2020.
448. Gailliot, P., and Baget, J., Fr. 1,015,570, Oct. 15, 1952; CA 51, 15072.
448a. -----, and -----, Fr. 1,096,712, June 23, 1955; CA 53, 4246.
449. Garascia, R. F., and Mattei, I. V., J. Am. Chem. Soc. 75, 4589-90(1953); CA 49, 7554.
450. -----, and Overberg, R. J., J. Org. Chem. 19, 27-30(1954); CA 49, 3095.

451. Garascia, R. J., Carr, A. A., and Hauser, T. R., J. Org. Chem. 21, 252-4 (1956); CA 50, 15508.
452. Gazzola, A. L., and Angier, R. B., U. S. 2,760,960, Aug. 28, 1956; CA 51, 3676.
453. Gelewitz, E. W., Riedeman, W. L., and Klotz, I. M., Arch. Biochem. and Biophys. 53, 411-24(1954); CA 49, 4749.
454. Gex, M., Arch. phys. biol. 17, 83-4(1944); CA 40, 5982.
455. Ghatt, J., and Mann, F. G., J. Chem. Soc. 1939, 1622-34; CA 34, 1270.
456. Ghose, T. N., and Bhattacharya, S., Science and Culture 13, 510(1948); CA 43, 5014.
457. Ghosh, T. N., and Mitra, S. N., ibid., 10, 452(1945); CA 39, 4061.
457a. -----, and Roy, S. C., Ann. Biochem. and Exptl. Med.(India) 15, 93-100 (1955); CA 50, 6691.
458. -----, and -----, J. Indian. Chem. Soc. 22, 39-40(1945); CA 40, 1512.
459. -----, and -----, ibid., 257-9(1945); CA 40, 4377.
460. -----, and Banerjee, S., J. Indian Chem. Soc. 24, 39-41(1947); CA 42, 1236.
461. Gibbs, J. H., J. Phys. Chem. 59, 644-9(1955); CA 49, 14401.
462. Gibson, C. S., and Kingam, R., Brit. 569,037, May 2, 1945; CA 42, 217.
463. Gibson, N. A., and White, R. A., Anal. Chim. Acta 12, 413-17(1955); CA 50, 2353.
464. Gilman, H., and Nelson, J. F., J. Am. Chem. Soc. 59, 935-7(1937); CA 31, 4265.
465. -----, and Apperson, L. D., J. Org. Chem. 4, 162-8(1939); CA 33, 5819.
466. -----, Yablunky, H. L., and Svigoon, A. C., J. Am. Chem. Soc. 61, 1170-2 (1939); CA 33, 5376.
467. -----, and Svigoon, A. C., J. Am. Chem. Soc. 61, 3586(1939); CA 34, 1000.
468. -----, and Yablunky, H. L., ibid., 62, 665-6(1940); CA 34, 2810.
469. -----, and -----, ibid., 63, 207-11(1941); CA 35, 1388.
470. -----, and -----, ibid., 212-16; CA 35, 1388.
471. -----, and -----, ibid., 949-54; CA 35, 3616.
472. -----, and Stuckwisch, C. G., J. Am. Chem. Soc. 63, 2844-5(1941); CA 36, 423.
473. -----, and Stuckwisch, C. G., ibid., 3532-3; CA 36, 1022.
474. -----, -----, and Nobis, J. F., ibid., 68, 326-8(1946); CA 40, 2150.
475. -----, Tolman, L., Yeoman, F., Woods, L. A., Shirley, D. A., and Avakiam, S., ibid., 426-8(1946); CA 40, 2807.
476. -----, and Yale, H. L., ibid., 72, 8-10(1950); CA 45, 1056.
477. -----, and -----, ibid., 73, 2880-1(1951); CA 46, 3974.
478. -----, and -----, ibid., 4470-1; CA 46 3974.
479. -----, Gregory, W. A., and Spatz, S. M., J. Org. Chem. 16, 1788-91 (1951); CA 46, 9100.
480. -----, and Avakian, S., J. Am. Chem. Soc. 76, 4031-2(1945); CA 49, 10212.
481. -----, et al., J. Org. Chem. 19, 1067-79(1954); CA 50, 224.
481a. Gill, N. S., Nyholm, R. S., and Pauling, P., Nature 182, 168-70(1958); CA 53, 961.
482. Gilta, G., Bull. soc. chim. Belg. 46, 263-74(1937); CA 32, 851.
483. -----, ibid., 275-83; CA 32, 851.
484. -----, ibid., 48, 444-6(1939); CA 34, 2342.
485. -----, ibid., 315-25; CA 34, 3151.
486. -----, ibid., 51, 175-85(1942); CA 38, 3887.

487. Giral, F., Ciencia (Mex.) 6, 253-7(1945); CA 40, 3563.
483. -----, ibid., 9, 137-8(1948); CA 43, 8099.
489. Glauert, R. H., and Mann, F. G., J. Chem. Soc. 1950, 682-4; CA 44, 6868.
490. Gleysteen, L. F., and Kraus, C. A., J. Am. Chem. Soc. 69, 451-4(1947);
       CA 41, 2630.
491. Glushkova, V. P., Talalaeva, T. V., Razmanova, Z. P., Zhdanov, G. S., and
       Kocheshkov, K. A., Sbornik Statei Obshchei Khim. 2, 992-6(1953);
       CA 49, 6859.
492. Goissedet, P. E. C., and Despois, R. L., Ger. 719,365, Mar. 12, 1942;
       CA 37, 1724 and Brit. 501,232, Feb. 13, 1939; CA 33, 6000.
493. Goldberg, A. A., J. Chem. Soc. 1942, 713-16; CA 37, 880.
494. Goldsworthy, L. J., Hook, W. H., John, J. A., Plant, S. G. P., Rushton,
       J., and Smith, L. M., ibid., 1948, 2208-12; CA 43, 3382.
494a. Goodwin, H. A., and Lions, F., J. Am. Chem. Soc. 81, 311-14(1959); CA 53,
       11364.
495. Gray, W. H., and Lamb, I. D., J. Chem. Soc. 1938, 401; CA 32, 4550.
495a. Grim, S. O., and Seyferth, D., Chem. and Ind. 1959, 849-850.
496. Gröber, J. C., Ger. 822,032, Nov. 22, 1951; CA 48, 2357.
497. Guha, J. R., Guha, S. S., Roy, A. C., and Guha, P. C., Current Sci. (India)
       21, 247-8(1952); CA 47, 9287.
498. Gupta, J., and Guha, M. P., Indian J. Phys. 22, 64-8(1948); CA 42, 8650.
499. Gutbier, H., and Plust, H. G., Chem. Ber. 88, 1777-86(1955); CA 50, 16662.
500. Hall, R. M. S., and Sutin, N., J. Inorg. Nuclear Chem. 2, 184-91(1956);
       CA 50, 9169.
501. Hamilton, C. S., U. S. 2,099,685, Nov. 23, 1937; CA 32, 1404.
502. -----, U. S. 2,099,686, Nov. 23, 1937; CA 32, 1404.
503. -----, U. S. 2,202,733, May 28, 1940; CA 34, 6772.
504. -----, U. S. 2,224,387, Dec. 10, 1941; CA 35, 2282.
505. -----, U. S. 2,435,392, Feb. 3, 1948; CA 42, 4614.
506. -----, U. S. 2,435,393, Feb. 3, 1948; CA 42, 2992.
507. Hanby, W. E., and Waters, W. A., J. Chem. Soc. 1946, 1029-31; CA 41, 1628.
508. Hanmim, J. A., U. S. 2,559,071, July 3, 1951; CA 45, 9836.
509. Harris, C. M., Nyholm, R. S., and Stephenson, N. A., Rec. trav. chim. 75,
       687-93(1956); CA 50, 16519.
510. -----, -----, and -----, Nature 177, 1127-8(1956); CA 50, 16522.
511. -----, and -----, J. Chem. Soc. 1956, 4375-83; CA 51, 2446.
512. -----, and -----, ibid., 1957, 63-70; CA 51, 7216.
512a. -----, Livingston, S. E., and Stephenson, N. C., ibid., 1958, 3697-3702;
       CA 53, 5001.
513. Harris, J. I., Bowden, S. T., and Jones, W. J., ibid., 1947, 1568-71; CA 42,
       2940.
513a. Hart, F. A., and Mann, F. G., ibid., 1957, 3939-44; CA 53, 402.
514. Hartmann, H., Beerman, C., and Czempik, H., Angew. Chem. 67, 233(1955)
515. -----, and Künl, G., ibid., 68, 619-20(1956); CA 52, 2743.
516. -----, and Nowak, G., Z. anorg. u allgem. Chem. 290, 348-51(1957);
       CA 52, 2787.
517. -----, Reiss, W., and Karbstein, B., Naturwissenschaften 46, 321(1959).
518. -----, Niemöller, H., Reis, W., and Karbstein, B., ibid., CA 54, 355.
519. Haskelberg, L., and Bergmann, F., J. Soc. Chem. Ind. 60, 166-8(1941); CA 36,
       423.

520. Haszeldine, R. N., and West., B. O., J. Chem. Soc. 1956, 3631-7; CA 51, 2537.
521. -----, and -----, ibid., 1957, 3880-4; CA 52, 2742.
522. Heaney, H., Mann, F. G., and Miller, I. T., ibid., 1956, 1-5; CA 50, 12962.
523. -----, -----, and -----, ibid., 1957, 3930-9; CA 52, 2002.
524. -----, Heineckey, D. M., Mann, F. G., and Millar, I. T., ibid., 1958, 3838-44; CA 53, 5279.
525. Heimbach, N., U. S. 2,432,419, Dec. 9, 1947; CA 42, 2193.
526. Hein, F., and Pobloth, H., Z. anorg. allgem. Chem. 248, 84-104(1941); CA 37, 2676.
527. -----, and Hecker, H., Z. Naturforsch. 11b, 677-8(1956); CA 51, 6418.
528. Heinekey, D. M., and Millar, I. T., J. Chem. Soc. 1959, 3101-2; CA 54, 6688.
529. Helin, A. F., and Van der Werf, C. A., J. Am. Chem. Soc. 73, 5884-5(1951); CA 47, 519.
530. Hennig, H., Gasschutz u. Luftschutz 7, 18-21(1937); CA 31, 2311.
531. Henry, M. C., and Wittig, G., J. Am. Chem. Soc. 82, 563-4(1960); CA 54, 9806.
532. Herrmann, W., Hampe, F., and Sievers, O., Ger. 675,959, May 22, 1939; CA 33, 7048.
533. Herrmann, W., and Oesterlin, H., Ger. 676,050, May 24, 1939; Brit. 510,167, July 27, 1939; CA 33, 6527.
534. -----, Hilmer, H., and Hampe, F., U. S. 2,232,232, Feb. 18, 1941; CA 35, 3390.
535. -----, and Hampe, F., U. S. 2,242,581, May 20, 1941; CA 35, 5648.
536. -----, Hilmer, H., and Hampe, F., Ger. 699,772, Nov. 7, 1940; Swiss 215,242, Sept. 1, 1941; CA 35, 7120.
537. Hewett, C. L., Lermit, L. J., Openshaw, H. T., Todd, A. R., Williams, A. H., and Woodward, F. N., J. Chem. Soc. 1948, 292-5; CA 42, 5026.
538. -----, ibid., 1203-5; CA 43, 561.
539. -----, Jones, W. E., Vallender, H. W., and Woodward, F. N., J. Soc. Chem. Ind. (London) 68, 363-8(1949); CA 44, 4414.
540. -----, and Holliday, A. K., J. Chem. Soc. 1953, 530-4; CA 47, 6295.
541. Hieber, W., and Heusinger, H., Angew. Chem. 68, 678-9(1956); CA 51, 12731.
542. -----, and Thalhofer, A., ibid., 679; CA 51, 12731.
543. -----, and Heusinger, H., J. Inorg. and Nuclear Chem. 4, 179-89(1957); CA 51, 12730.
544. -----, and Breu, R., Chem. Ber. 90, 1259-69(1957):
545. -----, Heusinger, H., and Vohler, O., Chem. Ber. 90, 2425-34(1957); CA 52, 13514.
546. -----, Ger. 822,993, Nov. 29, 1951; CA 49, 2518.
547. Hirans, N. et al., Japan. 23('51); Jan. 9; CA 47, 3343.
548. Hiratsuka, K., Inoue, C., Nagae, Y., and Umeda, K., Japan. 8566('56), Oct. 5.; CA 52, 11912.
549. Hiratuka, K., J. Chem. Soc. Japan 58, 935-53(1937); CA 32, 516.
550. -----, ibid., 1051-9; CA 33, 157.
551. -----, ibid., 1060-8; CA 33, 157.
552. --- -, ibid., 1063-76; CA 33, 158.

553. Hodgson, H. H., and Hathway, D. E., J. Chem. Soc. 1945, 123-6; CA 39, 2743.
554. Hoffmann-La Roche and Co., Ger. 640,697, Jan. 15, 1937; CA 31, 5518.
555. Hojo, N., Shinkai, T., and Kawai, A., Research Repts. Fac. Textile and
        Sericult., Shinschu Univ. 5, 138-41(1955); CA 51, 1508.
556. Holcomb, W. F., and Hamilton, C. S., J. Am. Chem. Soc. 61, 1236-7(1939);
        CA 33, 5377
557. Holliman, F. G., and Mann, F. G., J. Chem. Soc. 1943, 547-50; CA 38, 748.
558. -----, and -----, ibid., 550-4; CA 38, 748.
559. -----, and -----, ibid., 1945, 45-8; CA 39, 2757.
560. -----, -----, and Thornton, D. A., ibid., 1960, 9-16; CA 54, 8822.
561. Holmes, R. R., and Bertaut, E. F., J. Am. Chem. Soc. 80, 2983-5(1958);
        CA 52, 16968.
562. Horner, L., Hoffmann, H., and Wippel, H. G., Chem. Ber. 91, 64-7(1958).
563. -----, and Oediger, H., ibid., 437-42(1958); CA 52, 18285.
564. Howard, A. N., and Wild, F., Biochem. J. 65, 651-9(1957); CA 51, 8826.
565. Hunsberger, I. M., Shaw, E. R., Fugger, J., Ketcham, R., and Lednicer, D.,
        J. Org. Chem. 21, 394-9(1956); CA 51, 4297.
566. Ida, M., Japan. 130,481, June 13, 1939; CA 35, 5133.
567. -----, Toyoshima, Z., and Nakamura, T., J. Pharm. Soc. Japan 69, 178-80
        (1949); CA 44, 3929.
568. -----, ibid., 180-2; CA 44, 3929.
569. -----, ibid., 182-5; CA 44, 3929.
570. -----, ibid., 129-31; CA 44, 3930.
571. -----, ibid., 131-3; CA 44, 3930.
572. -----, Toyoshima, Z., and Nakamura, T., ibid., 437-9; CA 44, 3930.
573. -----, -----, and -----, ibid., 443-5; CA 44, 3930.
574. -----, -----, and -----, ibid., 445-6; CA 44, 3931.
575. I. G. Farbenindustrie, Brit. 452,153, Aug. 18, 1936; Fr. 47,245, Feb.20,
        1937 and Fr. 47,262, Mar. 5, 1937; CA 31, 708.
576. -----, Brit. 469,323, July 22, 1937(Addn. to 447,773); CA 32, 591.
577. -----, Brit. 484,101, May 2, 1938; CA 32, 7677.
578. -----, Fr. 826,342, Mar. 29, 1938; CA 32, 7677.
579. -----, Fr. 831,074, Aug. 22, 1938; Ger. 667,845, Nov. 21, 1938; Brit.
        507,186, June 12, 1939; CA 33, 1760.
580. -----, Brit. 504,199, Apr. 21, 1939; CA 33, 7319.
581. -----, Brit. 514,417, Nov. 7, 1939; CA 35, 4551.
582. Irving, R. I., and Magnussoo, E. A., J. Chem. Soc. 1956, 1860-3; CA 50,
        12723.
583. Irving, T. A., Greene, J. L., Jr., Peterson, J. G., and Capps, J. D.,
        J. Am. Chem. Soc. 72, 4069-71(1950); CA 45, 5688.
584. Isacescu, D. A., Bul. Soc. Chim. Rumania 18A, 131-4(1936); CA 31, 3019.
585. Isacescu, D., and Gruescu, C., Antigaz(Bucharest) 15, 85-99, 149-66(1941);
        CA 37, 3744.
586. Ishibashi, M. and Higashi, S., Japan Analyst 4, 14-16(1955); CA 50, 4708.
587. -----, Fujinaga, T., and Sato, M., Japan Analyst 8, 77-80(1959).
588. Ishihara, K., and Inami, K., Japan. 179,838, Aug. 9, 1949; CA 46, 1588.
589. Ishikawa, W., J. Chem. Soc. Japan 63, 801-3(1942); CA 41, 3463.
590. -----, ibid., 804-10(1942); CA 41, 3463.
591. -----, ibid., 811-13; CA 41, 3463.

592. Ishikawa, W., J. Chem. Soc. Japan 63, 1262-4; CA 41, 3463.
593. Iveronova, V. I., and Roitburd, F. M., Zhur. Fiz. Khim. 26, 810-12(1952); CA 46, 10767.
594. Izmail'skii, V. A., Russ. 35,831, Apr. 30, 1934; CA 30, 3343.
595. -----, and Simonov, A. M., J. Gen. Chem. (U. S. S. R.) 7, 499-507(1937); Bull. soc. chim. /5/, 3, 1739-53(1936); CA 31, 4290.
596. Jaffé, H. H., and Doak, G. O., J. Am. Chem. Soc. 71, 602-6(1949); CA 43, 3696.
597. -----, and -----, ibid., 72, 3027-9(1950); CA 44, 9931.
598. -----, J. Chem. Phys. 22, 1430-3(1954); CA 48, 13296.
599. Jenkins, R. L., U. S. 2,566,208, Aug. 28, 1951; CA 45, 10442.
600. Jensen, K. A., Z. anorg. allgem. Chem. 229, 225-51(1936); CA 31, 3753.
601. -----, ibid., 252-64; CA 31, 1721.
602. -----, and Frederiksen, E., Z. anorg. allgem. Chem. 230, 34-40(1936); CA 31, 2542.
603. -----, ibid., 250, 257-67(1943); CA 37, 5292.
604. -----, ibid., 268-76; CA 37, 5292.
605. Jerchel, D., Ber. 76B, 600-9(1943); CA 38, 60.
606. Johnson, A. W., J. Org. Chem. 25, 183-6(1960).
607. Johnson, W. C., and Pechukas, A., J. Am. Chem. Soc. 59, 2068-71(1937); CA 31, 8415.
608. Jones, C. T., Dissertation Abstr. 20, 888(1959).
609. Jones, E. R. H., and Mann, F. G., J. Chem. Soc. 1955, 401-5; CA 50, 2611.
610. -----, and -----, ibid., 405-10; CA 50, 2611.
611. -----, and -----, ibid., 411-22; CA 50, 2612.
612. -----, and -----, ibid., 4472-7; CA 50 10743.
613. -----, and -----, ibid., 1958, 294-9; CA 52, 10925.
614. -----, and -----, ibid., 1719-21; CA 52, 13662.
615. Jones, W. E., Rosser, R. J., and Woodward, F. N., J. Soc. Chem. Ind. (London) 68, 258-62(1949); CA 44, 4413.
616. Jones, W. J., Davies, W. C., Bowden, S. T., Edwards, C., Davis, V. E., and Thomas, L. H., J. Chem. Soc. 1947, 1446-50; CA 42, 4920.
617. Jurist, A. E., U. S. 2,112,244, Mar. 29, 1938; CA 32, 3913.
618. Kabesh, A., and Nyholm, R. S., J. Chem. Soc. 1951, 38-43; CA 45, 5052.
619. Kaesg, H. D., Stafford, S. L., and Stone, F. G. A., J. Am. Chem. Soc. 81, 6336(1959).
620. Kahn, M., ibid., 73, 479-80(1951); CA 46, 4393.
621. Kaiser, D. W., and Nagy, D. E., U. S. 2,481,758, Sept. 13, 1949; CA 44, 5925.
622. Kali-Chemie A. G., Brit. 768,765, Feb. 10, 1957; Fr. 1,120,344; CA 52, 421.
622a. Kamada, M., and Onishi, T., Nippon Kagaku Zasshi 80, 275-80(1959); CA 53, 11090.
623. Kamai, G. Kh., Ber. 68B, 1893-8(1935); CA 30, 4836.
624. -----, J. Gen. Chem. (U. S. S. R.) 10, 733-5(1940); CA 35, 2482.
625. -----, and Zoroastrova, V. M., ibid., 921-6; CA 35, 3241.
626. -----, and -----, ibid., 1568-72; CA 35, 2853.
627. -----, ibid., 12, 104-11(1942); CA 37, 1997.
628. -----, ibid., 14, 245-8(1944); CA 39, 2284.

629. Kamai, G., and Belorossova, O. N., Bull. acad. sci. U. R. S. S., Classe sci. chim. 1947, 191-6; CA 42, 4133.

630. -----, J. Gen. Chem. (U. S. S. R.) 17, 2178-81(1947); CA 42, 4521.

631. -----, and Belorossova, O. N., Izvest. Akad. Nauk S. S. S. R., Otdel. Khim. Nauk 1950, 198-202; CA 44, 888.

632. -----, and Chadaeva, N. A., Doklady Akad. Nauk S. S. S. R. 86, 71-3 (1952); CA 47, 6365.

633. -----, and -----, Zhur. Obshchei Khim. 23, 1431-2(1953); CA 48, 7890.

634. -----, and Bastanov, E. M. Sh., Doklady Akad. Nauk S. S. S. R. 89, 693-3 (1953); CA 48, 6574.

635. -----, Khisamova, Z. L., and Chadaeva, N. A., ibid., 1015-16; CA 48, 6391.

636. -----, and Starshov, I. M., ibid., 96, 995-997(1954); CA 48, 13318.

637. -----, and Khisamova, Z. L., ibid., 105, 489-91(1955); CA 50, 11234.

638. -----, and -----, Zhur. Obshchei Khim. 26, 126-30(1956); CA 50, 13730.

639. -----, and Chadaeva, N. A., Doklady Akad. Nauk S. S. S. R. 109, 309-11 (1956); CA 51, 1876.

640. -----, and Starshov, I. M., Zhur. Obshchei Khim. 26, 2209-11(1956); CA 51, 4980.

641. -----, and Chadaeva, N. A., ibid., 2468-74; CA 51, 4932.

642. -----, and Starshov, I. M., Trudy Kazan. Khim. Tekhnol. Inst., im. S. M. Kirova 1956, No. 21, 159-61; CA 51, 12018.

643. -----, and Chadaeva, N. A., Izvest. Akad. Nauk S. S. S. R., Otdel. Khim. Nauk 1957, 585-8; CA 51, 14585.

644. -----, and -----, Izvest. Kazan. Filiala Akad. Nauk S. S. S. R. Ser. Khim. Nauk 1955, No. 2, 19-24; CA 52, 292.

645. -----, and -----, Doklady Akad. Nauk S. S. S. R. 115, 305-7(1957); CA 52, 6161.

646. -----, and Chernokal'skii, B. D., Trudy Kazan. Khim. Tekhnol. Inst. im. S. M. Kirova 23, 143-7(1957); CA 52, 8938.

646a. -----, Üchenye Zapiski Kazan. Gosudarst. Univ. 115, No. 10, 43-5(1953); CA 53, 1205.

647. -----, and Chadaeva, N. A., Izvest. Kazan. Filiala Akad. Nauk S. S. S. R., Ser. Khim. Nauk 1957, No. 4, 69-77; CA 54, 6521.

648. -----, and Chernokal'skii, B. D., Zhur. Obshchei Khim. 29, 1596-9(1959); CA 54, 8606.

649. -----, and -----, Doklady Akad. Nauk S. S. S. R. 128, 299-301(1959); CA 54, 7538.

650. -----, and Chadaeva, N. A., Izvest. Vysshikh Ucheb. Zavedenii, Khim. i Khim. Tekhnol. 2, 601-7(1959); CA 54, 7606.

651. Kanehori, T., Ann. Report. Takamine Lab. 10, 39-41(1958).

652. Kano, S., and Toyoshima, S., Pharm. Bull. (Japan) 2, 301-5(1954); CA 50, 7326.

653. Kapeller-Adler, R., and Boxer, G., Biochem. Z. 285, 55-66(1936); CA 30, 4881.

654. Karsten, P., Kies, H. L., and Walraven, J. J., Anal. Chim. Acta 7, 355-9 (1952); CA 47, 4791.

655. Kary, R. M., U. S. 2,646,440, July 21, 1953; CA 48, 7049.

655a. Kary, R. M., and Frisch, K. C., U. S. 2,863,893, Dec. 9, 1958; CA 53, 9148.

656. Kary, R. M., and -----, J. Am. Chem. Soc. 79, 2140-2(1957); CA 51, 12862.

574

657. Kato, S. et al., Japan. 2472('56), Apr. 4.; CA 51, 12141.
658. -----, et al., ibid., 3126('56), Apr. 26; CA 51, 12973.
659. Kelen, G. P. van der, and Herman, M. A., Bull. soc. chim. Belges 65, 350-61(1956); CA 50, 16389.
660. -----, ibid., 343-9; CA 51, 13747.
661. Kharasch, M. S., Jensen, E. V., and Weinhouse, S., J. Org. Chem. 14, 429-32(1949); CA 43, 6971.
662. -----, and Weinhaus, S., U. S. 2,636,893-4; Apr. 28, 1950; CA 47, 7700.
663. -----, and -----, U. S. 2,615,043, Oct. 21, 1952; CA 47, 9346.
664. Kielbasinski, S., Ger. 653,643, Nov. 29, 1937; CA 32, 2957.
665. Kilpatrick, M. L., J. Am. Chem. Soc. 71, 2607-10(1949); CA 43, 7790.
666. Kimm, R.-H., Bull. Inst. Phys. Chem. Research (Tokyo) 21, 496-503(1942); CA 43, 8099.
667. King, H., and Ludford, R. J., J. Chem. Soc. 1950, 2086-8; CA 45, 1014.
668. Kinoshita, K., Yakugaku Zasshi 78, 41-3(1958); CA 52, 11079.
669. Kirkhgof, G. A., and Divinskii, A. F., Org. Chem. Ind. (U. S. S. R.) 1, 92(1936); CA 30, 5193.
670. Kirkhgof, G. A., and Terent'eva, I. V., Russ. 51,422, July 31, 1937; CA 33, 6882.
671. -----, and Korzina, O. I., Russ. 53,785, Aug. 31, 1938; CA 35, 1582.
672. Klages, F., and Rapp, W., Chem. Ber. 88, 384-8(1955); CA 50, 7077.
673. Kline, E. R., and Kraus, C. A., J. Am. Chem. Soc. 69, 814-16(1947); CA 41, 4361.
673a. Klygin, A. E., and Pavlova, V. K., Zhur. Anal. Khim. 14, 163-173(1959); CA 53, 12798.
674. Knunyants, I. L., and Pil'skaya, V. Ya., Izvest. Akad. Nauk S. S. S. R., Otdel. Khim. Nauk 1955, 472-9; CA 50, 6298.
675. Kolle, W., Hallensleben, J., Streitwolf, K., and Bauer, H., U. S. 2,050,574, Aug. 11, 1936; CA 30, 6893.
676. Kolthoff, I. M., and Johnson, R. A., J. Electrochem. Soc. 98, 231-3(1951); CA 46, 382.
677. Komissarov, Ya. F., Maleeva, A. Ya., and Sorokooumov, A. S., Compt. rend. acad. sci. U. R. S. S. 55, 719-22(1947); CA 42, 3721.
678. -----, Sorokooumov, A. S., and Maleeva, A. Ya., ibid., 56, 53-6(1947); CA 42, 520.
679. Kondo, K., Folia Pharmacol. Japon. 27, 235-49(Breviaria 27-9) (1939); CA 34, 819.
680. Koninklijke Industrielle Maatschappij voorkeen Noury and van der Lande N. V., Dutch 62,339, Jan. 15, 1949; CA 43, 4817.
681. Koshimura, S., Japan. J. Expt. Med. 21, 343-5(1951); CA 47, 8038.
682. Kosolapoff, G. M., and Priest, G. G., J. Am. Chem. Soc. 75, 4847-9(1953); CA 48, 1006.
683. Kotelko, A., Acta Polon. Pharm. 11, 199-204(1954); CA 49, 12342.
684. -----, ibid., 12, 57-60(1955); CA 51, 17792.
685. Koton, M. M., J. Gen. Chem. (U. S. S. R.) 9, 2283-6(1939); CA 34, 5049.
686. -----, ibid., 11, 379-81(1941); CA 35, 5870.
687. -----, ibid., 17, 1307-8(1947); CA 42, 1903.
688. -- ---, ibid., 18, 936-40(1948); CA 43, 559.
689. -----, Moskvina, E. P., and Florinskii, F. S., Zhur. Obshchei Khim. 19, 1675-8(1949); CA 44, 1436.

690. Koton, M. M., Mosvina, E. P., and Florinskii, F. S., Zhur. Obshchei Khim 20, 2093-5(1950); CA 45, 5644.
691. -----, ibid., 22, 643-7(1952); CA 47, 5376.
692. Kozminskaya, T. K., Nad, M. M., and Kocheshkov, K. A., J. Gen. Chem. (U. S. S. R.) 16, 891-6(1946); CA 41, 2014.
693. Kraft, M. Ya. and Bashchuk, I. A., Compt. rend. acad. sci. U. R. S. S. 55, 419-21(1947); CA 42, 577.
694. -----, and -----, ibid., 723-5; CA 42, 3742.
695. -----, and Rossina, S. A , ibid., 821-3; CA 42, 531.
696. -----, and Katyshkina, Doklady Akad. Nauk S. S. S. R. 66, 207-9(1949); CA 44, 127.
697. -----, and -----, ibid., 509; CA 44, 128.
698. -----, and Bashchuk, I. A., ibid., 65, 509-12(1949); CA 45, 2890.
699. -----, and Katyshkina, V. V., Doklady Akad. Nauk S. S. S. R. 99, 89-92 (1954); CA 49, 4233.
700. -----, Agracheva, E. B , and Sytina, E. N., ibid., 259-60(1954); CA 49, 15767.
701. -----, Korzina, O. I., and Morozova, A. S., Sbornik Statei Obshchei Khim. 2, 1356-9(1953); CA 49, 5347.
702. -----, Al'bitskaya, O. P., and Morozova, A. S., ibid., 1360-5; CA 49, 5347.
703. -----, and Agracheva, E. B., Doklady Akad. Nauk S. S. S. R. 100, 279-82 (1955); CA 50, 1644.
704. -----, and Sytina, E. N., ibid., 116, 89(1957); CA 52, 4930.
705. -----, ibid., 131, 1342-4(1960).
706. -----, Borodina, G. M., Streltsova, I. N., and Struchkov, I. T., ibid., 1074-6.
707. Krishnan, P. P., Iyer, B. H., and Guha, P. C., Science and Culture 11, 565-6(1946); CA 40, 5711; J. Indian Chem. Soc. 24, 433-6(1947); CA 42, 8174.
708. -----, -----, and -----, ibid., 566-7; J. Indian Chem. Soc. 24, 285-8(1947); CA 42, 5870.
709. -----, -----, and -----, ibid., 567(1946); J. Indian Chem. Soc. 24, 289-92 (1947); CA 42, 5871.
710. Kröhnke, F., Chem. Ber. 83, 291-6(1950); CA 44, 9963.
711. Kross, R. D., and Fassel, V. A., J. Am. Chem. Soc. 77, 5858-60(1955); CA 50, 3083.
712. Krueger, W. D., Clemence, L. R. W., and Leffler, M. T., ibid., 76, 4929-30; CA 49, 13142.
713. Kubli, H., and Préiswerk, E., Brit. 623,488, May 18, 1949; CA 44, 1537.
714. Kul'berg, L. M., and Barkovskii, V. F., Doklady Akad. Nauk S. S. S. R. 85, 335-6(1952); CA 47, 12280.
715. Kuskov, V. K. and Vasil'ev, V. N., Zhur. Obshchei Khim. 21, 90-2(1951); CA 45, 7518.
716. Kuz'min, K. I., and Kamai, G., Doklady Akad. Nauk S. S. S. R. 73, 709-10 (1950); CA 44, 10409.
717. -----, and -----, Sbornik Statei Obshchei Khim., Akad. Nauk S. S. S. R. 1, 223-8(1953); CA 49, 841.
718. -----, Zhur. Obshchei Khim. 26, 675-6(1956); CA 50, 14517.
719. -----, ibid., 3415-16; CA 51, 9474.
720. Kuznetsov, V. I., and Vasyunina, N. A., ibid., 10, 1203-9(1940); CA 35, 2869.
721. -----, Zhur. Anal. Khim. 7, 226-32(1952); CA 47, 1534.

722. Kuznetsov, V. I., and Golubtsova, R. B., Zavodskaya Lab. 21, 1422-6(1955); CA 50, 7004.

723. -----, Budanova, L. M., and Matrosova, T. V., ibid., 2, 406-12(1956); CA 51, 2464.

723a. -----, and Levin, I. S., Izvest. Sibir. Otdel. Akad. Nauk S. S. S. R. 1958, No. 7, 131-2; CA 53, 12943.

724. Landau, N. A., U. S. S. R. 101,731, Dec. 31, 1955; CA 51, 13658.

725. Landsteiner, K., and Van der Scheer, J., Exptl. Med. 67, 709-23(1938); CA 32, 5060.

725a. Lane, W. J., and Fritz, J. S., U. S. At. Energy Comm. ISC-945, 110 pp.(1957); CA 53, 11095.

725b. Lautsch, W. F., Chem. Tech. (Berlin) 10, 419; CA 53, 43.

726. Lawesson, S. O., Acta Chem. Scand. 9, 1017(1955); CA 50, 8810.

727. -----, Arkiv Kemi 10, 167-70(1956); CA 51, 8030.

728. Lecoq, H., J. pharm. Belg. 19, 133-269(1937); CA 31, 6729.

729. Le Fevre, R. J. W., and Parker, C. A., J. Chem. Soc. 1939, 677; CA 33, 5376.

730. Lesslie, M S., and Turner, E. E., ibid.; 1936, 730-1; CA 30, 5994.

731. -----, ibid., 1938, 1001-3; CA 32, 7467.

732. -----, ibid., 1939, 1050-4; CA 33, 6857.

733. -----, ibid., 1949, 1183-4; CA 44, 2991.

733a. Lewis, J., Irving, R. J., and Wilkinson, G., J. Inorg. and Nuclear Chem. 7, 32-7(1958); CA 53, 50.

734. Li, Shih-Tsin, J. Chinese Chem. Soc. 7, 117-20(1940); CA 35, 5094.

735. Lide, D. R., Jr., Spectrochim. Acta 1959, 473-9; CA 53, 21164.

736. Lile, W. J., and Menzies, R. J., J. Chem. Soc. 1950, 617-21; CA 44, 7178.

737. Linsker, F., and Bogert, M T., J. Am. Chem. Soc. 65, 932-4(1943); CA 37, 4374.

738. -----, and -----, ibid., 66, 191-192(1944); CA 38, 1479.

739. Livingstone, S. E., J. Chem. Soc. 1956, 437-40; CA 50, 6244.

740. -----, ibid., 1042-4; CA 50, 10592.

741. -----, ibid., 1989-94; CA 50, 13644.

742. -----, ibid., 1994-9; CA 50, 13645.

743. -----, ibid., 1958, 4222-6; Chem. and Ind. (London) 1957, 143-5; CA 51, 9397.

744. Long, L. H., Eméleus, H. J., and Broscoe, H. V. A., J. Chem. Soc. 1946, 1123-6; CA 41, 3045.

745. -----, and Sackman, J. F., Trans. Faraday Soc. 50, 1177-82(1954); CA 49, 7358.

746. -----, and -----, ibid., 51, 1062-4(1955); CA 50, 3871.

747. Long, L. H., and Sackmann, J F., Research Correspondence, Suppl. to Research (London) 8, No. 5, S23-4(1955); CA 50, 9081.

748. -----, and -----, Trans. Faraday Soc. 52, 1201-1207(1956); CA 51, 55280.

749. -----, J. Chem. Soc. 1956, 3410-16; CA 51, 8565.

750. Lott, W. A., and Van Winkle, R., U. S. 2,409,291, Oct. 15, 1946; CA 41, 993.

751. Loudon, J. D., J. Chem. Soc. 1937, 391-2; CA 31, 3901.

752. Løvens Kemiske Fabrik ved A. Kongsted, Dan. 67,974, Nov. 22, 1948; CA 43, 6231.

753. Lowe, W. G., and Hamilton, C. S., J. Am. Chem. Soc. 57, 2314-17(1935); CA 30, 99.

754. Lukin, A. M., and Petrova, G. S., Zhur. Obshchei Khim. <u>27</u>, 2171-4(1957); CA <u>52</u>, 6297.
754a. -----, -----, and Smirnova, K. A., U. S. S. R. 115,820, Aug. 20, 1958; CA <u>53</u>, 2944.
754b. -----, and Osetrova, E. D., Trudy Vsesoyuz. Nauch. Issledovatel. Inst. Khim. Reactivov <u>1956</u>, No. 21, 3-9; CA <u>53</u>, 6875.
755. Lyon, D. R., and Mann, F. G., J. Chem. Soc. <u>1942</u>, 666-71; CA <u>37</u>, 872.
756. -----, and -----, ibid., <u>1945</u>, 30-4; CA <u>39</u>, 2751.
757. -----, -----, and Cookson, G. H., ibid., <u>1947</u>, 662-70; CA <u>41</u>, 6239.
758. McCullough, J. D., Acta Cryst. <u>6</u>, 746(1953); CA <u>48</u>, 423.
759. McDowell, C. A., Emblem, H. G., and Moelwyn-Hughes, E. A., J. Chem. Soc. <u>1948</u>, 1206-8; CA <u>43</u>, 916.
760. McGeachin, R. L., J. Am. Chem. Soc. <u>71</u>, 3755-6(1949); CA <u>44</u>, 1050.
761. -----, ibid., 4133-4; CA 44, 2467.
762. -----, and Cox, R. E., ibid., <u>73</u>, 556(1951); CA <u>45</u>, 5645.
763. -----, and -----, ibid., 860; CA <u>45</u>, 5645.
764. -----, and -----, ibid., 1869-70; CA <u>46</u>, 450.
765. -----, ibid., 5476-7; CA <u>47</u>, 519.
766. -----, and Hunt, O. R., Jr., ibid., 5477; CA <u>47</u>, 519.
767. -----, ibid., <u>75</u>, 973-5(1953); CA <u>48</u>, 2631.
768. -----, Price, S. M., and Dajani, R. M., ibid., 6062-3; CA <u>48</u>, 13648.
769. -----, and Greenwald, M., ibid., 5430-1; CA <u>49</u>, 7508.
769a. McCombie, H., and Saunders, B. C., Nature <u>159</u>, 491-4(1941); CA <u>42</u>, 1190.
770. Mackie, A., and Raeburn, J., J. Chem. Soc. <u>1952</u>, 787-90; CA <u>47</u>, 1160.
771. Maddock, A. G., and Sutin, N., Research (London) <u>6</u>, 75-85(1953); CA <u>48</u>, 2508.
772. -----, and -----, Trans. Faraday Soc. <u>51</u>, 184-96(1955); CA <u>49</u>, 12151.
773. Maier, L., Seyferth, D., Stone, F. G. A., and Rochow, E. G., J. Am. Chem. Soc. <u>79</u>, 5884-9(1957); CA <u>52</u>, 6167.
774. -----, -----, -----, and -----, Z. Naturforschg. <u>12b</u>, 263-4(1957); CA <u>51</u>, 15396.
774a. -----, Rochow, E. G., and Fernelius, W. C., Abstracts of Papers, 134th Meeting of the A. C. S., Sept. 7-12, 1958 in Chicago, Ill., p. 35N.
775. Majumdar, A. K., and Sarma, R. N. S., J. Indian Chem. Soc. <u>28</u>, 654-6(1951); CA <u>46</u>, 5483.
776. Majumdar, A. K., and Mukherjee, A. K., Anal. Chim. Acta <u>21</u>, 330-3 (1959); CA <u>54</u>, 169
777. Makarova, L. G., J. Gen. Chem. (U. S. S. R.) <u>7</u>, 143-7(1957); CA <u>31</u>, 4290.
778. -----, and Nesmeyanov, A. N, ibid., <u>9</u>, 771-9(1939); CA <u>34</u>, 391.
779. -----, and------, Bull. acad. sci. U. R. S. S. Classe sci. chim. <u>1945</u>, 617-25; CA <u>40</u>, 4686.
780. Makin, F. B., and Waters, Wm. A. J. Chem. Soc. <u>1938</u>, 843-8; CA <u>32</u>, 6630.
781. Malatesta, L., and Araneo, A., Atti accad. nazl. Lincei, Rend., Classe sci. fis., mat. e nat. 20, 365-6(1956); CA <u>51</u>, 121.
782. -----, and Angoletta, M., J Chem. Soc. <u>1957</u>, 1186-8; CA <u>51</u>, 10408.
783. - --- , and Araneo, A., ibid., 3803-5; CA <u>51</u>, 17553.
784. -----, and Cariello, C., ibid., <u>1958</u>, 2323-8; CA <u>52</u>, 14407.
784a. -----, and Lonenzini, A., Ricerca sci. <u>28</u>, 1874-9(1958); CA <u>53</u>, 7864.
785. Malec, E., Med. Doswiadczalna Spoleczna <u>25</u>, 264-94(1947); CA <u>44</u>, 10260.

786. Malec, E., Med. Doswiadczalna i Microbiol. 1, 253-64(1949); CA 45, 4836.
787. Malinovskii, M. S , J. Gen. Chem. (U. S. S. R.) 10, 1918-22(1940); CA 35, 4736.
788. -----, Sci. Records Gorky State Univ. 7, 34-47(1939); CA 35, 443.
789. -----, Zhur. Obshchei Khim. 19, 130-3(1939); CA 43, 6178.
790. -----, and Olifirenko, S. P., ibid., 25, 122-5(1955); CA 50, 1646.
791. -----, and -----, ibid., 2437-40; CA 50, 9318.
792. -----, and -----, ibid., 26, 118-20(1956); CA 50, 13786.
793. -----, and -----, ibid., 1402-5; CA 50, 14670.
794. Mangini, A., Boll. sci. facolta chim. ind., Bologna 1940, 143-5; CA 37, 617.
795. -----, ibid., 146-7; CA 37, 617.
796. Mann, F. G., and Purdie, D., J. Chem. Soc. 1936, 873-90; CA 30, 6668.
797. -----, -----, and Wells, A. F , ibid., 1503-13; CA 31, 1720.
798. -----, and Wells, A. F., Nature 140, 503(1937); CA 31, 8413.
799. -----, and Chaplin, E. J., J. Chem. Soc. 1937, 527-35; CA 31, 4288.
800. -----, Wells, A F., and Purdie, D , ibid., 1828-36; CA 32, 2449.
801. -----, and -----, ibid., 1938, 702-710; CA 32, 7365.
802. -----, and Purdie, D., ibid., 1940, 1230-5; CA 35, 2434.
803. -----, and Holliman, F. G., Nature 151, 474-5(1943); CA 37, 4396.
804. -----, and Cookson, R. C., ibid., 157, 846(1946); CA 40, 6409.
805. -----, and Watson, J., J. Chem. So . 1947, 505-13; CA 41, 5473.
806. -----, and -----, J. Org. Chem. 13, 502-31(1948); CA 43, 1045.
807. -----, and Baker, F. C., J. Chem Soc. 1952, 4142-7; CA 48, 5194.
808. -----, and Smith, B. B , ibid., 4544-5; CA 48, 622.
809. -- --, Millar, I. T., and Smith, B. B., ibid., 1953, 1130-4; CA 48, 6446.
810. -----, and Stewart, F. H. C., ibid., 1955, 1269-73; CA 49, 12174.
811. -----, and Wilkinson, A. J., ibid., 1957, 3336-46; CA 52, 4648.
812. -----, and -----, ibid., 3346-52; CA 52, 4629.
813. --- -, and Watson, H. R., ibid., 3945-9; CA 53, 3115.
814. Manuel, T. A., and Stone, F. G., A., J Am. Chem. Soc. 82, 366-72(1960).
815. Manulkin, Z. M , and Tatarenko, A. N., Zhur. Obshchei Khim. 21, 93-8(1951); CA 45, 7038.
816. Manulkin, Z. M., Tatarenko, A. N., and Yusupov, F. Yu., Doklady Akad. Nauk S. S. S. R. 88, 687-90(1953); CA 48, 2631.
817. -----, -----, and -----, Sbornik Statei Obshchei Khim. 2, 1308-14(1953); CA 49, 5397.
817a. -----, Uzbek. Khim. Zhur., Akad. Nauk Uzbek. S. S. R. 1958, No. 2, 41-5; CA 53, 9112.
818. Marathe, S. S , Limaye, N. S., and Bhide, B. V , J. Univ. Bombay, Sect. A, 21, Pt. 3, 36-41(1952); CA 48, 5135.
819. Maren, T. H., J. Am. Chem. Soc. 68, 1864-5(1946); CA 40, 7175.
820. Margerum, D. W., Byrd, C. H., Reed, S. A., and Banks, C. V., Anal. Chem. 25, 1219-21(1953); CA 48, 10644.
821. Margoshes, M., and Fassel, V. A., Spectrochim. Acta 7, 14-24(1955); CA 49, 7979.
822. Markus, G., and Karush, F , J. Am. Chem. Soc. 80, 89-94(1958); CA 52, 7392.
823. Marquez, A. R., Anales asoc. quim. argentina, 27, 258-65(1939); CA 34, 4728.
824. -----, ibid., 28, 135-42(1940); CA 35, 2852.
825. -----, Rec. facultad cienc. quim. (Univ. nacl. La Plata) 16, 109-16(1941); CA 36, 6138.

826. Marquez, A. R., Rec. facultad cienc. quim. (Univ. nacl. La Plata) 17, 109-16(1942); CA 38, 3609.

827. -----, and Marquez, M. L. B. de, Anales asoc. quim. argentina 42, 147-51 (1954); CA 50, 160.

828. Marsh, D. F., and Woodbury, R A., J. Am. Chem. Soc. 71, 3748-9(1949); CA 44, 1098.

829. Maruyama, M., and Furuya, T., Bull. Chem. Soc. Japan 30, 650-7(1957); CA 52, 3566.

830. -----, and -----, ibid., 657-61(1957); CA 52, 5166.

831. Matson, E J., and Clemence, L.-R., U. S. 2,606,200, Aug. 5, 1952; CA 47, 6982.

832. Matsui, S., and Sato, M., Japan. 24('51), Jan. 9; CA 47, 3343.

833. Matsumura, K., J. Am. Chem. Soc. 60, 593-5(1938); CA 32, 3404.

833a. Matthews, C N , Magee, T. A., and Wotiz, J. H., ibid., 81, 2273-4(1959); CA 53, 19936.

834. Mayer, A., and Bradshaw, G., Analyst 77, 154-8(1952); CA 46, 4949.

835. Mazza, F. P., and Migliardi, C , Atti accad. Lincei, Classe sei. fis., mat. nat. 28, 152-7(1938); CA 33, 9300.

836. Mead, D. J., Ramsey, J. B., Rothrock, D. A., Jr., and Kraus, C. A., J. Am. Chem. Soc. 69, 528-30(1947); CA 41, 3347.

837. Meals, R. N., J. Org. Chem. 9, 211-18(1944); CA 38, 4563.

838. Medoks, G V., and Soshestvenskaya, E. M., Zhur. Obshchei Khim. 27, 271-2(1957); CA 51, 12845.

839. -----, Doklady Akad. Nauk S. S. S. R. 117, 993-5(1957); CA 52, 7923.

840. Mellor, D. P., Burrows, G. J., and Morris, B. S., Nature 141, 414-15(1938); CA 32, 3287.

841. -----, and Craig, D. P., J. Proc. Roy. Soc. N. S. Wales 75, 27-30(1941); CA 35, 7310.

842. Mel'nikov, N N., and Rokitskaya, M. S., J. Gen. Chem.(U. S. S. R.) 8, 834-8(1938); CA 33, 1267.

843. Menis, O., Ball, R. G., and Manning, L., Anal. Chem. 29, 245-8(1957); CA 51, 9405.

844. Meyers, D B., and Jones, J. W., J. Am. Chem. Soc. 39, 401-4(1950); CA 44, 9955.

845. Miller, G. E., and Seaton, G., U. S. 2,442,372, June 1, 1948; CA 42, 6842.

846. -----, and Reid, E. E., U. S 2,695,306, Nov. 23, 1954; CA 50, 4195.

847. Mills, G. S., U. S 2,803,620, Aug. 20, 1957; CA 51, 18684.

848. Mingoia, Q., and Perego, C., Arquiv. biol. (Sao Paulo) 28, 137-42(1944); CA 39, 4069.

849. Mingoia, Q., and Perego, C , ibid., 29, 12-14(1945); CA 39, 4597.

850. -----, and Poggi, S., ibid., 119-24(1945); CA 40, 3410.

851. Mödritzer, K., Ber. 92, 2637-40(1959); CA 54, 3180.

852. Moffett, K. D., Simmler, J. R , and Potratz, H. A., Anal. Chem. 28, 1356 (1956); CA 50, 16281.

853. Mohler, H., and Polya, J., Helv. Chim. Acta 19, 1222-39 and 1239-43(1936); CA 31, 1515.

854. -----, ibid., 21, 784-6(1938); CA 32, 6764.

855. -----, ibid., 789-92; CA 32, 6764.

856. -----, Protar 7, 78-85(1941); CA 35, 4868.

857. Mokhnach, V. O , and Bagniuk, V. S., Compt. rend. acad. sci. U. R. S S. 14, 553-8(1937); CA 31, 5322.
858. Mole, J. D. C., and Turner, E. E., J Chem. Soc. 1939, 1720-4; CA 34, 1001.
858a. Monnet, R., and Viala, A., Congr. soc. pharm. France, 9e, Clermont Ferrand 1957, 213-16; CA 53, 20689.
859. Mooney, R. C. L., J. Am. Chem. Soc. 62, 2955-9(1940).
860. -----, Phys. Rev. 61, 739(1942).
861. Moore, A. M , Elslager, E. F., and Short, F. W., Brit. 788,130, Dec. 23, 1957; CA 52, 11128.
862. Morgan, G. T., and Walton, E , J. Chem. Soc. 1936, 902-5, CA 30, 6717; and Chemistry and Industry 1937, 853-5; CA 32, 127.
863. -----, and -----, ibid., 1938, 442-4; CA 32, 4961.
864. -----, and Hamilton, C. S., J. Am. Chem. Soc. 66, 874-5(1944); CA 38, 3962.
865. -----, Cragoe, E. J., Jr., Andres, R. J., Elpern. B., Coles, R. F., Lawhead, J., Clark, R. L , Hatlelid, E. B., Kahler, F. H., Paxton, K. W., Banks, C. K., and Hamilton, C. S., ibid., 69, 930-2(1947); CA 41, 4464.
866. -----, -----, Elpern, B., and Hamilton, C. S., ibid., 932-3(1947); CA 41, 4142.
867. -----, U. S. 2,454,742, Nov. 23, 1948; CA 43, 2242.
868. Mortimer, C. T., and Skinner, H. A., J. Chem. Soc. 1952, 4331-4; CA 47, 3103.
869. -----, and -----, ibid., 1953, 3189-91; CA 48, 3782.
869a. -----, J. Chem. Educ. 35, 381-4(1958); CA 53, 12.
870. Muić, N. and Fleš, D., Arkiv. Kem. 20, 92-6(1948); CA 44, 6413.
871. -----, and -----, ibid., 22, 182-5(1950); CA 46, 4527.
872. -----, Liječnički Viesnik 71, 310-12(1949); CA 49, 210.
873. Mulay, L. N., Proc. Indian Acad. Sci. 34A, 245-9(1951); CA 46, 5908.
874. Murakami, M., Yukawa, Y., and Matsumura, E., Mem. Inst. Sci. Ind. Research Osaka Univ. 5, 147-50(1947); CA 47, 2759.
875. Murin, A N., and Nefedov, V. D., Primenenie Mechenykh Atomov v Anal. Khim., Akad. Nauk S. S S. R., Inst. Geokhim. i Anal. Khim. 1955, 75-8; CA 50, 3915.
876. Musil, A., and Pietsch, H., Z. anal. Chem. 140, 421-6(1953); CA 48, 4358.
877. -----, and Pietsch, R., ibid., 142, 81-5(1954); CA 48, 9261.
878. Nad, M. M., Kozminskaya, T. K., and Kocheshkov, K. A., J. Gen. Chem. (U. S. S. R.) 16, 897-900(1946); CA 41, 2014.
879. Naik, K. G., Trivedi, R. K , and Mehta, S. M., J. Indian Chem. Soc. 20, 365-8(1943); CA 38, 3620.
880. Nakai, R., and Yamakawa, Y., Repts. Inst. Chem. Research, Kyoto Univ. 15, 25-6(1946); CA 45, 7038.
881. -----, Toyoda, R., and Tomono, H., ibid., 18, 22-4(1949); CA 45, 7971.
882. -----, -----, and -----, ibid., 19, 71-3(1949); CA 45, 7972.
883. -----, and Yamakawa, Yu., Bull. Inst. Chem. Research, Kyoto Univ. 22, 91-2(1950); CA 45, 7040.
884. -----, Tomono, H., and Azuma, T., ibid., 92(1950); CA 46, 8033.
885. -----, -----, and -----, ibid., 25, 72(1951); CA 46, 8033.
886. -----, and Yamakawa, Yu., ibid., 24, 80(1951); CA 46, 3510.

887. Nakatsu, H., and Kawase, S., Takamine Kenkyujo Nempo 8, 44-7(1956); CA 52, 297.

887a. -----, -----, Aizawa, T., and Endo, S., Japan. 4418('58), June 5; CA 53, 12238.

888 Nakaya, I., and Uematsu, K., Japan. 1280('53), Mar. 26; CA 48, 12302.

889. -----, et al., Japan. 5086('54), Aug. 14; CA 50, 6508.

890. Neeb, R., Z. anal. Chem. 152, 158-62(1956); CA 51, 3359.

891. -----, ibid., 154, 17-22(1957); CA 51, 11920.

892. Neely, T. A., and Capps, J. D., J. Am. Chem. Soc. 77, 182-4(1955); CA 50, 1012.

893. Nefedov, V. D. and Evtikheev, L. N., Zhur. Fiz. Khim. 30, 2090-2(1956); CA 51, 7182.

894. Neiman, M. B., Plotnikov, A. Ya., Razuvaev, G. A., and Ryabov, A V., Doklady Akad. Nauk S. S. S. R. 64, 365-8(1949); CA 43, 4675.

895. Nelson, J. F., Iowa State Coll. J. Sci. 12, 145-7(1937); CA 32, 3756.

896 Nenitsescu, C. D., Isacescu, D. A., and Gruescu, C., Bull. Soc. Chim. Romania 20A, 135-8(1938); CA 34, 1979.

897. Nesmeyanov, A. N., and Kocheshkov, K. A., Bull. acad. sci. U. R. S. S., Classe sci. chim. 1944, 416-31; CA 39, 4320.

898. -----, ibid., 1945, 239-50; CA 40, 2122.

899. -----, and Borisov, A. E., ibid., 251-60; CA 40, 2123.

900. Nesmeyanov, A. N., and Kocheskov., K. A., ibid., 524-6; CA 42, 5870.

901. -----, and Borisov, A. E., Doklady Akad. Nauk S. S. S. R. 60, 67-72.

902. -----, Batuev, M. I., and Borisov, A. E., Izvest. Akad. Nauk S. S. S. R., Otdel. Khim. Nauk 1949, 567-9; CA 44, 2374.

903. -----, Gipp, N. K., Makarova, L. G., and Mozgova, K. K., ibid., 1953, 298-302; CA 48, 6391.

904. -----, Reutov, O. A., and Ptitsyna, O A., Doklady Akad. Nauk S. S. S. R. 91, 1341-4(1953); CA 48, 11375.

905 -----, and Makarova, L. G , Uchenye Zapiski Moskov. Gosudarst. Univ. im. M. V. Lomonosova No. 132, Org. Khim. 7, 109-16(1950); CA 49, 3903.

906. -----, Reutov, O. A , and Knol, P. G., Izvest. Akad. Nauk S. S. S. R., Otdel. Khim. Nauk 1954, 410-17; CA 49, 9651.

907. -----, and Ippolitov, E G., Vestnik Moskov. Univ. 10, No. 10, Ser. Fiz. Mat. i Estestven. Nauk No. 7, 87-90(1955); CA 50, 9906.

908. -----, Reutov, O. A., Undel, Yu. G., and Beletskaya, I. P., Izvest. Akad. Nauk S. S. S. R , Otdel. Khim. Nauk 1957, 929-41; CA 52, 4533.

909. -----, Tolstaya, T. P., and Isaeva, L. S., Doklady Akad. Nauk S. S. S. R. 122, 614-17(1958); CA 53, 4178.

910. -----, Reutov, O. A., Ptitsyna, O A., and Tsurkan, P. A., Izvest. Akad. Nauk S. S. S. R., Otdel. Khim. Nauk 1958, 1435-44; CA 53, 8037.

911. -----, Tolstaya, T. P., and Isaeva, L. S., Doklady Akad. Nauk S. S. S. R. 125, 330-3(1959).

912. -----, Reutov, O. A., Tolstaya, T. P., Ptitsyna, O. A., Isaeva, L. S., Turchinskii, M. F., and Bochkareva, G. P., ibid., 1265-8(1959); CA 53, 21757.

913. -----, Borisov, A. E., and Novikova, N. V., Izvest. Akad. Nauk S S. S. R., Otdel. Khim. Nauk 1960, 147; CA 54, 20853.

914. -----, -----, and -----, ibid., 952; CA 54, 24351.

915. Nesmeyanov, A. M., Borisov, M. I., and Novikova, N. V., Doklady Akad. Nauk S. S. S. R. 134, 100-1(1960); CA 54, 23558.
916. -----, -----, and -----, Tetrahedron Letters 1960, No. 8, 23-4; CA 54, 20848.
917. Neumann, L., Ger. 818,048, Oct. 22, 1951; CA 52, 20862.
918. Newell, M. P., Argus, M. F., and Ray, F. E., J. Am. Chem. Soc. 78, 6122-3(1956); CA 51, 2631.
919. Nigam, H. L., and Nyholm, R. S., Proc. Chem. Soc. 1957, 321-2; CA 52, 5191.
920. -----, -----, and Ramana Rao, D. V., J. Chem. Soc. 1959, 1397-402; CA 53, 13861.
921. -----, -----, and Stiddard, M. H. B., ibid., 1960, 1803-6.
922. -----, -----, and -----, ibid., 1806-12.
923. Niyogy, S., Proc. Indian Acad. Sci. 4A, 303-8(1936); CA 31, 1014.
924. -----, ibid., 309-13(1936); CA 31, 1014.
925. Northrup, J. M., Proc. Soc. Exptl. Biol. Med. 67, 15-17(1948); CA 42, 3287.
926. Nugent, L. J., Dissertation Abstr. 19, 3150(1959).
927. Nyburg, S. C., and Hilton, J., Acta Cryst. 8, 358-59(1955); CA 49, 11669.
928. Nyholm, R. S., J. Proc. Roy. Soc. N. S. Wales 78, 229-33(1946); CA 40, 3996.
929. -----, J. Chem. Soc. 1950, 843-8; CA 44, 7705.
930. -----, ibid., 848-51; CA 44, 7705.
931. -----, ibid., 851-6; CA 44, 7705.
932. -----, ibid., 857-9; CA 44, 7706.
933. -----, Nature 165, 154(1950); CA 44, 5255.
934. -----, J. Chem. Soc. 1950, 2061-71; CA 45, 969.
935. -----, ibid., 2071-8; CA 45, 970.
936. -----, ibid., 1951, 1767-74; CA 45, 9416.
937. -----, Nature 168, 705(1951); CA 46, 4411.
938. -----, J. Chem. Soc. 1951, 2602-7; CA 46, 2438.
939. -----, ibid., 1952, 1257-64; CA 46, 7455.
940. -----, ibid., 2906-10; CA 46, 11002.
941. -----, and Short, L. N., ibid., 1953, 2670-3; CA 48, 1193.
942. -----, and Parish, R. V., Chemistry and Industry 1956, 470; CA 50, 16520.
943. -----, and Sutton, G. J., J. Chem. Soc. 1958, 560-3; CA 52, 11642.
944. -----, and -----, ibid., 564-6; CA 52, 11642.
945. -----, and -----, ibid., 567-71; CA 52, 11643.
946. -----, and -----, ibid., 572-6; CA 52, 11643.
947. Nylen, P., Z. anorg. allgem. Chem. 246, 227-42(1941); CA 36, 1295.
948. Ochiai, E., and Suzuki, O., J. Pharm. Soc. Japan 60, 353-6(1940); CA 34, 7289.
949. O'Donnell, G. J., Iowa State Coll. J. Sci. 20, 34-6(1945); CA 40, 4689; Ph. D. Thesis No. 760, submitted Aug. 23, 1944 at Iowa State College.
950. Oesper, P. F., and Smyth, C. P., J. Am. Chem. Soc. 64, 173-5(1942); CA 36, 1222.
951. Oesterlin, H., Herrmann, W., and Hampe, F., Ger. 729,341, Nov. 19, 1942; CA 38, 376.
952. Ogilvie, J. D. B., Davis, S. G., Thompson, A. L., Grummitt, W. T., and Winkler, C. A., Can. J. Research 26F, 246-63(1948); CA 42, 7967.

953. Ockenden, D. W., U. K. At. Energy Authority, Ind. Group. Hdq. IGO-R/W-2 18pp.(1956); CA 53, 11090.
954. Omer, R. E., and Hamilton, C. S., J. Am. Chem. Soc. 59, 642-4(1937); CA 31, 3891.
955. Oneto, J. F., ibid., 60, 2058-9(1938); CA 32, 8381.
956. -----, and Way, E. L., ibid., 61, 2105-6(1939); CA 33, 7750.
957. -----, and -----, ibid., 62, 2157-8(1940); CA 34, 6586.
958. -----, and -----, ibid., 63, 762(1941); CA 35, 2869.
959. -----, and -----, ibid., 3068-70(1941); CA 36, 423.
960. Openshaw, H. T., and Todd, A. R., J. Chem. Soc. 1948, 374-6; CA 42, 4967.
961. Opfermann, A. C. J., Brit. 771,116, Mar. 27, 1957; CA 51, 15589.
962. -----, Brit. 782,887, Sept. 11, 1957; CA 52, 10192.
963. Orlando, A. M., Rec. fac. cienc. quim. (Univ. nacl. La Plata) 25, 97-102 (1950); CA 49, 916.
964. Orlić, S., Archiv. Hem. i Teknol. 12, 153-72(1938); CA 34, 6407.
965. Ozernova, E. G., U. S. S. R. 46,261, Mar. 31, 1936; CA 33, 3397.
965a. Palmer, A. R., Anal. Chim. Acta 19, 458-61(1958); CA 53, 21065.
966. Panchout, S., and Duval, C., Anal. Chim. Acta 5, 170-84(1951); CA 46, 4422.
967. Parab, N. K., and Desai, D. M., Science and Culture 23, 430(1958); CA 52, 15988.
968. -----, and -----, Current Sci. (India) 26, 389(1951); CA 52, 16856.
969. -----, and -----, J. Indian Chem. Soc. 35, 569-72, 573-5(1958); CA 53, 8749.
969a. Parish, Doctoral Thesis, London, 1958, cited by Nigam et al. in J. Chem. Soc. 1960, 1804.
970. Paris, F. L., Fr. 848,386, Oct. 27, 1939; CA 35, 6066.
970a. Park, H. F., U. S. 2,456,569, Dec. 14, 1948; CA 43, 2468.
971. Parke, Davis, and Co., Brit. 444,882, Mar. 25, 1936; CA 30, 6513.
972. -----, Brit. 496,733, Dec. 5, 1958; CA 33, 3533.
973. Parker, G. D., Clarke, R. J., and Thewlis, B. H., J. Chem. Soc. 1947, 429-31; CA 41, 5475.
974. Parrish, J. R., Chemistry and Industry 1956, 137; CA 50, 11561.
975. Pascal, P., Compt. rend. 218, 57-9(1944); CA 40, 1366.
976. Pascual, J. A., Brit. 607,490, Aug. 31, 1948; CA 43, 1804.
977. Pathak, B., and Ghosh, T. N., J. Indian Chem. Soc. 26, 53-4(1949); CA 43, 7449.
978. -----, and -----, ibid., 254-6(1949); CA 44, 4877.
979. -----, and -----, ibid., 293-5(1949); CA 44, 4877.
980. -----, and -----, ibid., 584-6(1949); CA 44, 6827.
981. -----, Science and Culture 16, 331-2(1951); CA 46, 1482.
982. -----, J. Indian. Chem. Soc. 28, 198-200(1951); CA 47, 7452.
983. -----, ibid., 561-2(1951); CA 48, 4457.
984. Pauling, L., Pressman, D., Campbell, D. H., Ikeda, C., and Ikawa, M., J. Am. Chem. Soc. 64, 2994-3003(1942); CA 37, 1181.
985. Pepe, F. A., and Singer, S. J., ibid., 78, 4583-6(1956); CA 51, 1332.
986. Pereira, J. Aurazo, Anales fac. farm. y bioquim., Univ. nacl. mayor San Marcos (Lima, Peru) 5, 538-41(1954); CA 51, 2453.
987. Peri, C. A., Ital. 569,527, Nov. 21, 1957; CA 53, 14936.
988. -----, Gaz. chim. ital. 89, 1315-23(1959).

989. Peronnet, M., and Remy, R. H., J. pharm. chim. 30, 353-64(1939); CA 34,
            5790.
990. Petit, G , Compt. rend. 209, 111-13(1939); CA 33, 8169.
991. -----, ibid., 211, 228-30(1940); CA 36, 1295.
992. -----, ibid., 205, 322-5(1937).
993. -----, Ann. chim. 16, 5-100(1941); CA 36, 2524.
994. -----, Compt. rend. 218, 414-15(1944); CA 39, 2930.
995. -----, Bull. soc. chim. France 1948, 140-1; CA 42, 4865.
996. Petreko, L. P., Zhur. Obshchei Khim. 24, 520-1(1954); CA 49, 6159.
997. Petrun'kin, V. E , Ukrain. Khim. Zhur. 22, 608-11(1956); CA 41, 5693.
998. Peyron, L., Depierre, F., and Jacob, J., Compt. rend. 283, 1626-8(1954);
            CA 48, 12303.
999. -----, Bull. soc. chim. France 1954, 727; CA 49, 10874.
1000. Pfeiffer, P., and Böttcher, H., Ber. 70B, 74-5(1937); CA 31, 2584.
1001. -----, and Schmidt, P., J. prakt. Chem. 152, 2744(1939); CA 33, 3347.
1002. Pharmaceutische Ges. "Vadag" A. G., Fr. 813,658, June 7, 1937; CA 32,
            958; Brit. 484,506, May 6, 1938; CA 32, 7478.
1003. Phillips, M. A., J. Chem. Soc. 1941, 192; CA 35, 6248.
1004. -----, Brit. 547,409, Aug. 26, 1942; U. S. 2,339,583, Jan. 18, 1944;
            CA 37, 6094.
1005. -----, Chemistry and Industry 1945, 247; CA 40, 4349.
1006. -----, ibid., 61; CA 43, 2436.
1007. -----, Mfg. Chemist 24, 471,474(1953); CA 48, 4182.
1008. Pietsch, R., Z. anal. Chem. 142, 85-8(1954); CA 48, 9261.
1009. -----, Microchim. Acta 1955, 954-61; CA 50, 3944.
1010. -----, ibid., 1019-25; CA 50, 3944.
1011. -----, Z. anal. Chem. 150, 190-2(1956); CA 50, 10597.
1012. -----, ibid., 152, 168-71(1956); CA 51, 3362.
1013. -----, Microchim. Acta 1957, 161; CA 51, 12736.
1014. -----, Z. anal. Chem. 155, 189-94(1957); CA 51, 14479.
1015. -----, Microchim. Acta 1957, 699-704.
1016. -----, ibid., 1957, 705-13.
1017. -----, ibid., 1958, 220-4; CA 53, 8927.
1018. -----, Z. anal. Chem. 159, 37-41(1957); CA 52, 5204.
1019. -----, ibid., 343-8(1958); CA 52, 7942.
1020. -----, Microchim. Acta 1959, 854-60.
1021. -----, Microchim. Acta 1959, 861-9.
1022. Plazek, E., and Tyka, R., Zeszyty Nauk Politech. Wroclaw., Chem. No. 4,
            79-81(1957); CA 52, 20156.
1023. Polya, J. F., J. Applied Chem. (London) 1, 473-4(1951); CA 46, 8033.
1023a. Popova, A. I., and Humphrey, R. E., J. Am. Chem. Soc. 81, 2043-7(1959);
            CA 53, 16689.
1024. Popp, F., Chem. Ber. 82, 152-6(1949); CA 44, 1014.
1025. Portnov, A. I., Zhur. Obshchei Khim. 18, 594-600(1948); CA 43, 57.
1026. -----, ibid., 601-4(1948); CA 43, 57.
1027. -----, ibid., 605-7(1948); CA 43, 57.
1028. Pozzi-Escot, E., Anales quim. labs. invest. cient. e ind. E. Pozzi-Escot
            (Peru) Oct., 1943, 11-12; CA 38, 1448.
1029. Prasad, M., and Mulay, L. N., J Chem. Phys. 19, 1051-6(1951); CA 46, 793.

1030. Prat, J., Chim. anal. 31, 111-12(1949); CA 43, 4975.

1031. -----, Mém. services chim. état (Paris) 33, 385-90(1947); CA 43, 6111.

1032. -----, and Vaganay, J., ibid., 391-4(1947); CA 43, 6114.

1033. -----, ibid., 395-404(1947); CA 44, 3435.

1034. Pressman, D., and Brown, H., J. Am. Chem. Soc. 65, 540-3(1943); CA 37, 3979.

1035. Price, A. H., J. Phys. Chem. 62, 773-7(1958); CA 52, 19305.

1035a. Price, S. J. W., and Trotman-Dickenson, A. F., Trans. Faraday Soc. 54, 1630-7(1958); CA 53, 19851.

1036. Pritchard, G. J., Brit. 603,463, June 12, 1948; CA 43, 690.

1037. -----, U. S. 2,546,274, Mar. 27, 1951; Brit. 547,564; Brit. 552,751; CA 45, 7143.

1038. Probey, T. F., and Harrison, W. T., U. S. Pub. Health Repts. 53, 939-49 (1938); CA 32, 6004.

1039. Produits Roche, Belg. 437,342, Dec. 12, 1939; CA 36, 2567.

1040. -----, Belg. 438,274, Mar. 11, 1940; CA 36, 3050.

1041. Proescher, F., and Sycheff, V. M., U. S. 2,348,417, May 9, 1944; CA 39, 1255.

1042. Protiva, M., and Pliml, J., Chem. Listy 46, 773-4(1952); CA 47, 11194.

1042a. Przheval'skii, E. G., Golovina, A. P., and Kuteinikov, A. F., Vestnik Moskov Univ., Ser. Mat., Mekh., Astron. Fiz. Khim. 13, No. 6, 99-104(1958).

1043. Pujari, H. K., and R'out, M. K., J. Sci. Ind. Research (India) 14B, 563-4 (1955); CA 50, 14713.

1044. Pyman, F. L., Garforth, B., and Anderson, L., Ger. 673,842, Mar. 30, 1939; Brit. 451,960, Aug. 14, 1936; Ind. 22,488, Aug. 1, 1936; CA 33, 6527.

1045. Rády, G., and Erdey, L., Z. anal. Chem. 152, 253-8(1956); CA 51, 3356.

1046. Raiziss, G. W., and Kremens, A. I., U. S. 2,074,757, Mar. 23, 1937; CA 31, 3214.

1047. -----, Clemence, L.-R. W., and Kremens, A. I., U. S. 2,088,608, Aug. 3, 1937; CA 31, 6825.

1048. -----, -----, and -----, U. S. 2,258,862, Oct. 14, 1941; CA 36, 869.

1048a. -----, Severac, M., and Clemence, L.-R. W., U. S. 2,476,508, July 19, 1949; CA 43, 8105.

1049. Ray, F. E., and Garascia, R. J., J. Org. Chem. 15, 1233-40(1950); CA 45, 42.

1050. Rao, C. N. R., Ramachandran, J., Iah, M. S. C., Somasekhara, S., and Rajakumar, T. V., Nature 183, 1475-6(1959); CA 53, 19563.

1051. Ravazzoni, C., Ann. chim. applicata 32, 285-9(1942); CA 38, 3610.

1052. Razumova, Z. A., Zhur. Neorg. Khim. 3, 1126-30(1958); CA 53, 3970.

1053. Razuvaev, G. E., Malinovskii, V. S., and Godina, D. A., J. Gen. Chem. (U. S. S. R.) 5, 721-7(1935); CA 30, 1057.

1054. -----, and Shubenko, M. A., Doklady Akad. Nauk S. S. S. R. 67, 1094-52(1959); CA 44, 1435.

1055. -----, and -----, Zhur. Obschchei Khim. 21, 1974-9(1951); CA 46, 3411.

1056. Reesor, J. W. B., and Wright, G. F., J. Org. Chem. 22, 382-5(1957); CA 52, 1087.

1057. Redemann, E., Chaikin, S. W., Fearing, R. B., and Benedict, D., J. Am. Chem. Soc. 70, 637-9(1948); CA 42, 4011.

1058. Reppe, W., and Sweckendiek, W. J., Ann. 560, 104-16(1948); CA 43, 6202.

1059. Reutov, O. A., and Ptitsyna, O. A., Doklady Akad. Nauk S. S. S. R. 79, 819-21(1951); CA 46, 6093.
1060. -----, and -----, Izvest. Akad. Nauk S. S. S. R., Otdel. Khim. Nauk 1952, 93-102; CA 47, 1631.
1061. -----, Doklady Akad. Nauk S. S. S. R. 87, 991-4(1952); CA 48, 143.
1062. -----, Bundel, Yu. G., Izvest. Akad. Nauk S.S.S. R., Otdel. Khim. Nauk 1952, 1041-8; CA 48, 623.
1063. -----, and Ptitsyna, O. A., Doklady Akad. Nauk S. S. S. R. 89, 877-80 (1953); CA 48, 5135.
1064. -----, Vestnik Moskov. Univ. 8, No. 3, Ser. Fiz. Mat. i Estestven. Nauk No. 2, 119-23(1953); CA 49, 3867.
1065. -----, and Kondrat'eva, V. V., Zhur. Obshchei Khim. 24, 1259-65(1954); CA 49, 12339.
1066. -----, and Bundel, Yu. G., ibid., 25, 2324-32(1955); CA 50, 9318. See also Vestnik Moskov. Univ. 10, No. 8, Ser. Fiz., Mat i Estestv. Nauk No. 5, 85-9(1955); CA 50, 11964.
1067. -----, and Markovskaya, A. G., Doklady Akad. Nauk S. S. S. R. 98, 979-82 (1954); CA 49, 2926.
1068. -----, -----, and Lovtsova, A. N., ibid., 99, 269-72(1954); CA 49, 15767.
1069. -----, and -----, ibid., 543-6(1954); CA 49, 15767.
1070. -----, and Ptitsyna, O. A., ibid., 102, 291-4(1955); CA 50, 6346.
1071. -----, Markovskaya, A. G., and Mardaleishvili, R. E., ibid., 104, 253-5(1955); CA 50, 6160. See also Zhur. Fiz. Khim. 30, 2533-8 (1956); CA 51, 9511; and Reutov, O. A., Tetrahedron 1, 67-74 (1957); CA 51, 13796.
1072. -----, Ptitsyna, O. A., and Ertel, G., Chem. Tech. (Berlin) 10, 201-202(1958).
1073. -----, Tolstaya, T. P., Ptitsyna, O. A., Isayeva, L. S., Turchinskii, M. F., and Bochkareva, G. P., Doklady Akad. Nauk S. S. S. R. 125, 1265-68(1959).
1073a. -----, Ptitsyna, O. A., and Styazhkina, N. B., ibid., 122, 1032-4(1958); CA 53, 4169.
1074. -----, Ptitsyna, O. A., Lovtsova, A. N., and Petrova, V. F., Zhur. Obshchei Khim. 29, 3888-94(1959).
1074a. -- --, and Lovtsova, A. N., Vestnik Moskov. Univ. Ser. Mat., Mekh., Astron., Fiz., Khim. 13, No. 3, 191-6(1958); CA 53, 11283.
1075. Ribas, I., and Seoane, E., Anales real soc. espan. fis. y quim. 45B, No. 4, 599-626(1949); CA 44, 5836.
1076. Riebsomer, J. L., J. Org. Chem. 13, 815-21(1948); CA 43, 2619.
1077. Rius, A., and Carrancio, H., Anales real soc. fis. y quim. (Madrid) 47B, 767-76(1951); CA 46, 10010.
1078. Rohrmann, E., U. S. 2,516,831, July 25, 1950; Brit. 661,284, Nov. 21, 1951; CA 45, 652.
1079. Rosenbaum, E. J., and Ashford, T. A., J. Chem. Phys. 7, 554(1939); CA 33, 6718.
1080. -----, Rubin, D. J., and Sandberg, C. R., J. Chem. Phys. 8, 366-8(1940); CA 34, 4664.
1081. -----, and Sandberg, C. R., J. Am. Chem. Soc. 62, 1622-3(1940); CA 34, 6218.

1082. Roy, A. C., Iyer, B. H., and Guha, P. C., Current Sci. (India) 17, 89 (1948); CA 42, 7267.
1083. -----, -----, and -----, ibid., 126-7(1948).
1084. -----, and Guha, P C., J. Sci. Ind. Research (India) 9B, 242-4(1950); CA 45, 7040.
1085. Royer, R., and Buu-Hoi, Compt. rend. 222, 558-60(1946); CA 40, 3755.
1086. -----, Ann. chim. 1, 395-445(1946); CA 41, 2716.
1087. -----, Bull. soc. chim. France 1953, 412-7; CA 48, 3925.
1088. Rozina, D. Sh., Zhur. Priklad. Khim. 23, 211-14(1950); CA 45, 1531.
1089. -----, ibid., 1110-12(1950); CA 46, 10121.
1090. Rubtsov, M. V., J. Gen. Chem. (U. S. S. R.) 16, 221-34(1946); CA 41, 431.
1091. Ruddy, A. W., Starkey, E. B., and Hartung, W. H., J. Am. Chem. Soc. 64, 828-9(1942); CA 36, 3160.
1092. -----, and -----, Org. Syntheses 26, 60-3(1946); CA 41, 725.
1093. Rueggeberg, W. H. C., Ginsburg, A., and Cook, W. A., J. Am. Chem. Soc. 68, 1860-2(1946); CA 40, 7189.
1094. Saikina, M. K., Uchenye Zapiski Kazan. Gosudarst. Univ. im. V. I. Ul'yanova-Lenina Khim. 116, No. 2, 129-86(1956); CA 51, 7191.
1095. Salellas, J. F., Rev. facultad cienc. quim. (Univ. nacl. La Plata) 11, 59-64(1936); CA 33, 556.
1096. -----, and Orlando, A. M., Anales asoc. quim. argentina 40, 74-9(1952); CA 47, 4316.
1097. Salmi, E. J., Merivuori, K., and Laaksonen, E., Suomen Kemistilehti 19B, 102-8(1946); CA 41, 5440.
1098. Sanderson, R. T., J. Am. Chem. Soc. 77, 4531-2(1955); CA 49, 1505.
1099. Sandin, R. B., McClure, F. T., and Irwin, F., J Am. Chem. Soc. 61, 2944-6(1939); CA 34, 389.
1100. Sartori, M., and Recchi, E., Ann. Chim. applicata 29, 128-30(1939); CA 33, 9175.
1101. Sato, M., J. Pharm. Soc. Japan. 69, 303-4(1949); CA 44, 3931.
1102. -----, Nagasawa, M., Kado, M., and Kubota, R., Japan. 5562('57), July 26; CA 52, 11119.
1103. Saunders, C. R., Smith, C. E., Jr., and Capps, J. D., J. Am. Chem. Soc. 73, 5910-11(1951); CA 46, 11196.
1104. Savolahti, A., Virtanen, P., and Penttila, E., Geologi (Helsinki) 9, 95-6 (1957); CA 52, 4277.
1105. Sawyers, J. L., Burrows, B., and Maren, T. H., Proc. Soc. Exptl. Biol. Med. 70, 194-7(1949); CA 43, 3931.
1106. Schmidt, H., U. S. 2,215,430, Sept. 17, 1940; CA 35, 857. Ind. 24,731, Dec. 13, 1939. Ger. 728,803, Nov. 5, 1942; CA 38, 377.
1107. -----, Chem. Ber. 81, 477-83(1948); CA 43, 5377.
1108. -----, and Meivert, H., Z. anorg. u. allgem. Chem. 295, 173-84(1958); CA 52, 18022.
1109. Schmitz, E., Ger. (East) 9024, Jan. 7, 1955; CA 42, 5465.
1110. Schwenk E., Papa, D., Whitman, B., and Ginsberg, H., J. Org. Chem. 9, 1-8 (1944); CA 38, 2018.
1111. Schweckendick, W., Ger. 834,991, Mar. 27, 1952; CA 52, 9192.
1111a. Schwerdle, A., U. S. 2,889,347, June 2, 1959; CA 53, 18864.
1112. Scott, A. B., Hummel, R. D., Tullar, B. F., and Wainwright, J., U. S. 2,221,817, Nov. 17, 1940; CA 35, 1808.

1113. Scott, A. B., and Sultzaberger, J. A., U. S. 2,280,132, Apr. 21, 1942; CA 36, 5321.
1114. -----, and Tullar, B. F., Can. 405,532, June 16, 1942; CA 36, 5618.
1115. Seifter, J., J. Am. Chem. Soc. 61, 530-1(1939); CA 33, 2105.
1116. -----, J. Pharmacol. 66, 366-77(1939); CA 33, 7384.
1117. Seoane, E., and Ribas, I., Anales real soc. espan. fis. y quim. 45B, 1235-58(1949); CA 46, 6094.
1118. Sergeev, P. G., and Kudryashev, D. G., J. Gen. Chem. (U. S. S. R) 7, 1488-94(1937); CA 31, 8517.
1119. -----, and Bruker, A. B., Zhur. Obshchei Khim. 27, 2220-3(1957); CA 52, 6236.
1120. Seyferth, D., and Rochow, E. G., J. Org. Chem. 20, 250-6(1955); CA 49, 13888.
1121. -----, J. Am. Chem. Soc. 80, 1336-7(1958); CA 52, 13616.
1121a. -----, and Weiner, M. A., Chem. and Ind. (London) 1959, 402; CA 53, 21756.
1122. Shamshurin, A. A., Trudy Usbekskogo Gosudarst. Univ. /N. S./, No. 25, Khim. No. 1, 19-22(1941); CA 35, 5869.
1123. Sharp, D. B., and Hamilton, C. S., J. Am. Chem. Soc. 68, 588-91(1946); CA 40, 3443.
1124. Sharp, T. M., and Gray, W. H., Brit. 619,658, Mar. 11, 1949; CA 44, 2023.
1125. Shemyakin, M. M., Bamdas, E. M., Vinogradova, E. I., Karapetyan, M. G., Kolosov, M. N., Khokhlov, A. S., Shvetsov, Yu. B., and Shchukina, L. A., Doklady Akad. Nauk S. S. S. R. 86, 565-8(1952); CA 48, 640.
1126. -----, -----, -----, -----, -----, -----, -----, and -----, Zhur. Obshchei Khim. 23, 1854-67(1953); Doklady Akad. Nauk S. S. S. R. 79, 601 (1951); CA 49, 946.
1127. Sherlin, S. M., and Berlin, A. Ya., J. Gen. Chem. (U. S. S. R.) 5, 938-42 (1935); CA 30, 1055.
1128. -----, Braz, G. I., Yakubovich, A. Ya., and Konoval'chik, A. I., J. Gen. Chem. (U. S. S. R.) 9, 985-91(1939); CA 33, 8584.
1128a. Shibata, S., Takeuchi, F., and Matsumae, T., Bull. Chem. Soc. Japan 31, 888-9(1958); CA 53, 14830.
1128b. -----, -----, and -----, Anal. Chim. Acta 21, 177-81(1959); CA 53, 19693.
1129. Shimada, A., Bull. Chem. Soc. 32, 309-10(1959); CA 54, 2873.
1130. -----, ibid., 33, 301-4(1960); CA 54, 18015.
1131. Shimuzu, K., J. Chem. Soc. Japan, Pure Chem. Sect. 77, 1103-5(1956); CA 51, 839.
1131a. Shinagawa, M., Matsuo, H., and Nakashina, F., Kogyo Kazaku Zasshi 60, 1409-12(1957); CA 53, 14781.
1132. Shokal, E. C., U. S. 2,768,153, Oct. 23, 1956; Brit. 781,416, Aug 21, 1957; CA 51, 4057.
1133. -----, U. S. 2,840,617, June 24, 1958; CA 52, 16745.
1134. Shpeir, L. F., and Bogunets, Zhur. Priklad. Khim. 21, 873-5(1948); CA 43, 6178.
1135. Shreve, R. N., and Bennet, R. B., J. Am. Chem. Soc. 65, 2243-5(1943); CA 38, 257.
1136. Shriner, R. L., and Wolf, C. N., Org. Syntheses 30, 95-6(1950); CA 45, 132.
1137. Shubenko, M. A., Sbornik Statei Obshchei Khim. 2, 1043-5(1953); CA 49, 6856.
1138. Siebert, H., Z. anorg. u. allgem. Chem. 273, 161-9(1953); CA 47, 11976.

1139. Siekierska, K. E., Sokolowska, A., and Campbell, I. G., J. Inorg. Nuclear Chem. 12, 18-29(1959).
1140. S. I. F. E. M., Fr. 965,206, Sept. 6, 1950; CA 46, 2471.
1141. Simons(Georg.) Chem. Fab. und Export Geschäft, Ger. 644,078, May 25, 1937; CA 31, 5951.
1142. Singh, G., and Singh, M., J. Indian Chem. Soc. 23, 224-8(1946); CA 41, 2419.
1143. -----, Singh, T. and Singh, Gh., ibid., 24, 51-6(1947); CA 42, 1278.
1144. -----, Singh, Gh., and Singh, M., ibid., 79-82(1947); CA 42, 1179.
1145. Skiles, B. F., and Hamilton, C. S., J. Am. Chem. Soc. 59, 1006-8(1937); CA 31, 5350.
1146. Skinner, H. A., and Sutton, L. E., Trans. Faraday Soc. 40, 164-84(1944); CA 38, 5701.
1147. Slater, R. C. L. Mooney, Acta Cryst. 12, 187-96(1959); CA 53, 9772.
1147a. Smith, W. C., Tullock, C. W., Muetterties, E. L., Hasek, W. R., Fawcett, F. S., Engelhardt, V. A., and Coffman, D. D., J. Am. Chem. Soc. 81, 3165-6(1959); CA 53, 21589.
1147b. Smyth, C. P., J. Org. Chem. 6, 421(1941); CA 35, 4647.
1148. Societe des usines chimiques Rhône-Poulenc, Brit. 642,457, Sept. 6, 1950; Fr. 948,870, Aug. 12, 1949; CA 45, 5184.
1149. Societe pour l'ind. chim. à Bâle, Brit. 451,081, July 29, 1936; CA 31, 217; Ger. 637,433, Oct. 28, 1936; CA 31, 818; and Swiss 183,800, July 16, 1936; CA 30, 8532.
1150. -----, Swiss 188,548-9, Apr. 1, 1937; CA 31, 6825.
1151. -----, Fr. 811,407, Apr. 14, 1937; CA 31, 8547.
1151a. Societe des usines chimique Rhône Poulenc, Brit. 796,486, June 11, 1958; CA 53, 11261.
1151b. Solvic Soc. Anon., Belg. 566,536, Oct. 6, 1958; CA 53, 11892.
1152. Sommer, L., Z. anal. Chem. 169, 342-6(1959); CA 54, 3079.
1153. Soper, Q. F., Whitehead, C. W., Behrens, O. K., Corse, J. J., and Jones, R. G., J. Am. Chem. Soc. 70, 2849-55; CA 43, 3364.
1154. Souchay, P., Bull. soc. chim. France 1948, 143-56; CA 42, 4428.
1155. Spada, A., Atti soc. nat. mat. Modena 71, 155-61(1940); CA 37, 2732.
1156. -----, ibid., 72, 34-42(1941); CA 38, 2331.
1157. Spasov, A., and Zhechev, M., Bull. inst. chim. acad. bulgar. sci. 2, 67-74(1953); CA 49, 5492.
1158. Springall, H. D., and Brockway, L. O., J. Am. Chem. Soc. 60, 996-1000 (1938); CA 32, 5269.
1158a. Stadtherr, R. J., and Widmer, R. E., Weeds 7, 82-7(1959); CA 53, 22696.
1159. Starik, I. E., Ratner, A. P., Pasvik, M. A., and Ginzburg, F. L., Radio-khimiya 1, 545-47(1959).
1160. Starshov, I. M., and Kamai, G., Zhur. Obshchei Khim. 24, 2044-9(1954); CA 49, 14663.
1161. Stauff, J., Koch, E., and Uhlein, E., Arzneimittel Forsch. 4, 142-6(1954); CA 48, 6652.
1162. Steiger, N., and Keller, O., U. S. 2,584,639, Feb. 5, 1952; Brit. 659,250, Oct. 17, 1951; CA 46, 11238.
1163. -----, U. S. 2,599,291, June 3, 1952; CA 46, 10525.
1164. Steinman, H. G., Doak, G. O., and Eagle, H., J. Am. Chem. Soc. 66, 192-4 (1944); CA 38, 1479.

1165. Sterlin, R. N., Pinkina, L. N., Yatsenko, R. D., and Knunyants, I. L., Khim. Nauka i Prom. 4, 800-1(1959); CA 54, 14103.
1166. Sterling Drug, Inc., Brit. 635,367, Apr. 5, 1950; CA 44, 7865.
1167. Stocken, L. A., J. Chem. Soc. 1947, 592-5; CA 41, 6197.
1168. -----, and Thompson, R. H. S., Biochem. J. 40, 529-35; CA 41, 215.
1169. Stone, F. G. A., and Burg, A. B., J. Am. Chem. Soc. 76, 386-5(1954); CA 49, 5187.
1170. Straube, R. L., Neal, W B., Jr., Kelly, T., and Ducoff, H. S., Proc. Soc. Exptl. Biol. Med. 69, 270-2(1948); CA 43, 1650.
1171. Streitwolf, K., Fehrle, A., and Herrmann, W., U. S. 2,161,538, June 6, 1939; CA 33, 7494.
1172. -----, and -----, U. S. 2,068,206, Jan. 19, 1937; Brit. 386,537, Feb. 9, 1933; CA 31, 1961.
1173. -----, -----, and Oesterlin, H., U. S. 2,070,145, Feb. 9, 1937; Ger. 515,207, Dec. 12, 1930; CA 31, 2360.
1174. -----, -----, and -----, U. S. 2,070,146, Feb. 9, 1937; Ger. 524,804, May 16, 1931; Brit. 347,083, May 14, 1931; CA 31, 2360.
1175. -----, -----, and Herrmann, W., U. S. 2,180,779, Nov. 21, 1941; Brit. 360,957, Dec. 10, 1931; Austrian 3150/1931; CA 34, 1819.
1176. -----, -----, and Fritzsche, P., Can. 386,659, Jan. 30, 1940; CA 34, 2140.
1176a. Stroganov, E. V., Vestnik Leningrad. Univ. 14, No. 4, Ser. Fiz. i Khim. No. 1, 103-6(1959); CA 53, 14631.
1177. Struchkov, Yu. T., and Khotsyanova, T. L., Doklady Akad. Nauk S. S. S. R. 91, 565-8(1953); CA 48, 422.
1178. -----, Kitaigorodskii, A. I., and Kotsyanova, T. L., Zhur. Fiz. Khim. 26, 530-7(1952); CA 49, 6686.
1179. Sutor, D. J., and Harper, F. R., Acta Cryst. 12, 585-9(1959); CA 53, 18588.
1179a. -----, Australian J. Chem. 11, 415-19(1958); CA 53, 2921.
1179b. -----, ibid., 420-5(1958); CA 53, 2922.
1179c. -----, ibid., 12, 122-6(1959); CA 53, 14811.
1180. Sweet, L. A., Gruhzit, O. M., and Hamilton, C. S., Brit. 510,683, Aug. 4, 1939; CA 34, 5093.
1181. -----, Controulis, J., Tillotson, E. W., and Banks, C. K., J. Am. Chem. Soc. 69, 2258-9(1957); CA 42, 1236.
1182. -----, Calkins, D. G., and Banks, C. D., ibid., 2260-1(1947); CA 42, 141.
1183. -----, and Tillitson, E. W., U. S. 2,566,382, Sept. 4, 1951; CA 46, 2576.
1184. Takahashi, T., J. Pharm. Soc. Japan 55, 875-9(in German 164-5(1935)); CA 30, 721.
1185. -----, and Morita, M., ibid., 64, No. 7A, 11(1944); CA 45, 8014.
1186. -----, ibid., 72, 523-6(1952); CA 46, 9790.
1187. -----, ibid., 529-32(1952); CA 46, 9791.
1188. -----, ibid., 533-7(1952); CA 46, 9791.
1189. -----, ibid., 1144-8(1952); CA 47, 6886.
1190. -----, ibid., 1148-52(1952); CA 47, 6887.
1191. -----, and Ueda, T., U. S. 2,701,812, Feb. 8, 1955; CA 50, 1907.
1192. Talalaeva, T. V., and Kocheshkov, K. A., J. Gen. Chem. (U. S. S. R.) 8, 1831-8(in French 1838) (1939); CA 33, 5819.
1193. -----, and -----, ibid., 16, 777-80(1946); CA 41, 1215.
1194. -----, and -----, Doklady Akad. Nauk S. S. S. S. 77, 621-4(1951); CA 45, 10191.

1195. Talalaeva, T. V., and Kocheshkov, K. A., Izvest. Akad. Nauk S. S. S. R., Otdel. Khim. Nauk 1953, 290-3; CA 48, 6389.
1195a. -----, and -----, ibid., 216-34; CA 48, 3285.
1196. Tani, C., Ohsaka, H., and Kondo, T., J. Pharm. Soc. Japan 70, 130-3(1950); CA 44, 5831.
1197. Tanner, C. C., U. S. 2,409,519, Oct. 15, 1946; CA 41, 573 and Brit. 573,-928, Dec. 13, 1945; CA 43, 5041.
1198. Tarbell, D. S., and Baughan, J. R., Jr., J. Am. Chem. Soc. 67, 41-3(1945); CA 39, 1398.
1199. -----, and Bunnett, J. F., ibid., 69, 263-5(1947); CA 41, 3752.
1200. Tatarenko, A. N., Doklady Akad. Nauk Uzbek S. S. R. 1955, No. 5, 35-9; Referat. Zhur., Khim. 1956, Abstr. No. 43232; CA 52, 20005.
1201. Taylorson, R. B., Rom, R. C., and Dana, M. N., Proc. N. Central Weed Control Conf., 13th Meeting 1956, 44; CA 51, 15866.
1202. Terent'ev, A. P., and Potapov, V. M., Zhur. Obshchei Khim. 28, 1161-6 (1958); CA 52, 19991.
1203. Thomas, S. L. S., and Jones, J. I., Brit. 762,085, Nov. 21, 1956; CA 51, 7616.
1204. Thurston, J. T., U. S. 2,482,076, Sept. 13, 1949; CA 44, 5926.
1205. Tillitson, E. W., U. S. 2,370,092, Feb. 20, 1945; CA 39, 3883.
1206. -----, U. S. 2,566,383, Sept. 4, 1951; CA 46, 2576.
1207. Tiollais, R., and Berthois, L., Bull. soc. chim. /5/, 5, 73-8(1938).
1208. -----, and Perdreau, H., ibid., 6, 631-8(1939); CA 33, 8475.
1209. -----, -----, and Berthois, L., ibid., 638-46(1939); CA 33, 8565.
1210. -----, ibid., /5/, 3, 70-87(1936); CA 30, 3305.
1211. -----, ibid., 87-95(1936); CA 30, 3305.
1212. -----, and Perdreau, H., ibid., /5/, 4, 1896-8(1937).
1213. Titov, A. I., and Levin, B. B., Sbornik Statei Obshchei Khim. 2; 1469-72(1953); CA 49, 4503.
1214. -----, and -----, ibid., 1473-7(1953); CA 49, 4504.
1215. -----, and -----, ibid., 1478-82(1953); CA 49, 4505.
1216. To, K., Japan. J. Med. Sci. IV Pharmacol. 12, No. 2/3, Proc. Japan. Pharmacol. Soc. 13, 110-11(1940); CA 34, 7429.
1217. Tom, T. B., U. S. 2,506,847, May 9, 1950; CA 44, 7525 and U. S. 2,543,-995, Mar. 6, 1951; CA 46, 2788.
1218. Tomita, M., and Nakao, S., Japan. 3722('50), Oct. 27; CA 47, 5962.
1219. Tomono, H., Bull. Inst. Chem. Research, Kyoto Univ. 21, 41-6(1950); CA 45, 7971.
1220. -----, ibid., 22, 49-52(1950); CA 45, 7971.
1221. -----, ibid., 23, 45-50(1950); CA 45, 7971.
1222. -----, Yamakawa, Y., and Nakai, R., ibid., 26, 99(1951); CA 46, 8033.
1223. Toropova, V. F., and Saikina, M, K., Sbornik Statei Obshchei Khim., Akad. Nauk S. S. S. R. 1, 210-214(1953); CA 48, 12579.
1224. Treffler, A., Chem. Industries (N. Y.) 54, 854-5(1944); CA 38, 6500.
1225. Trefouel, J., U. S. 2,616,913, Nov. 4, 1952; Brit. 689,447, Mar. 25, 1953; Ger. 883,156, July 16, 1953; CA 47, 10003.
1226. Tribalat, S., Ann. chim. 4, 289-351(1949); CA 45, 2352.
1227. Tseng, C.-L., and Shih, W.-Y., J. Chinese Chem. Soc. 4, 183-6(1936); CA 31, 669.

1228. Tsuji, K., Kageyama, M., Nakamura, S., and Ueda, T., J. Pharm. Soc. Japan 74, 1180-4(1954); CA 49, 14664.
1229. -----, and -----, ibid., 1184-6(1954); CA 49, 14664.
1230. Tsukervanik, I. P., and Smirnov, D., J. Gen. Chem. (U. S. S. R.) 7, 1527-31(1937); CA 31, 8518.
1231. Tullar, B. F., U. S. 2,516,276, July 25, 1950; CA 45, 652.
1232. Ueda, T., Mizuma, T., and Takahashi, K., J. Pharm. Soc. Japan 69, 301-2(1949); CA 44, 1921.
1233. -----, Mizuma, T., and Takahashi, K., ibid., 561-4(1949); CA 44, 3436.
1234. -----, and Kosuge, T., ibid., 71, 821-3(1959); CA 46, 1714.
1235. -----, and -----, ibid., 823-7; CA 46, 1714.
1236. -----, and -----, ibid., 974-7; CA 46, 1714.
1237. -----, and Kano, S., Japan. 1303('50), Apr. 19; CA 47, 2206.
1238. -----, and Mizuma, T., Japan. 4217('50), Nov. 30; CA 47, 3343.
1239. -----, and -----, Japan. 4376('50), Dec. 16; CA 47, 3343.
1240. -----, Japan. 3979('51), July 24; CA 47, 8092.
1241. -----, and Takahashi, K., Japan. 4227('53), Aug. 28; CA 48, 8822.
1242. -----, and -----, Japan. 5183('53), Oct. 9; CA 49, 7594.
1243. -----, and Kano, S., Japan. 7871('54), Nov. 29; CA 50, 13992.
1244. -----, Japan. 1073('55), Feb. 19; CA 51, 2859.
1245. -----, Toyoshima, S., and Nakata, I., Pharm. Bull. 1, 252-4(1953); CA 49, 571.
1246. Uematsu, K., and Nakaya, I., Japan. 1279('53), Mar. 26; CA 48, 12802.
1247. Ullmann, F., and Friedheim, E., U. S. 2,092,425, Sept. 7, 1937; CA 31, 8122.
1248. Urbschat, E., U. S. 2,644,005, June 30, 1953; CA 48, 5879.
1249. -----, and Brohberger, P. E., U. S., 2,767,114, Oct. 16, 1956; CA 51, 5354. Ger. 957,987.
1250. -----, and Schrader, G., Ger. 940,827, Mar. 29, 1956; CA 52, 14713.
1251. Vagenberg, I. D., J. Gen. Chem. (U. S. S. R.) 7, 808-14(1937); CA 31, 5777.
1252. Vallarino, L., J Chem. Soc. 1957, 2287-92; CA 51, 11908.
1253. -----, J. Inorg. and Nuclear Chem. 8, 288-90(1958); CA 53, 5944.
1254. -----, Rend. ist. lombardo sci. Pt. I Classe sci. mat. e. nat. 91, 399-400(1957); CA 52, 18064.
1255. Vanags, G., and Veinbergs, A., Ber. 75B, 1558-69(1942); CA 38, 1221.
1256. Varma, P. S., Raman, K. S. V., and Yashoda, K. M., J. Indian Chem. Soc. 16, 515-18(1939); CA 34, 2809.
1257. Vasil'ev, S V., J. Gen. Chem. (U. S. S. R.) 16, 451-4(1956); CA 41, 951.
1258. -----, Zhur. Obshchei Khim. 19, 350-5(1949); CA 43, 6593.
1259. -----, and Vovchenko, G. D., Vestnik Moskov. Univ. 5, No. 3, Ser. Fiz. Mat. i Estest. Nauk, No. 2, 73-80(1950); CA 45, 6594.
1260. Vaska, L., Z. Naturforsch. 15b, 56(1960).
1261. Vaughan, J. R., Jr., and Tarbell, D. S., J. Am. Chem. Soc. 67, 144-8(1945); CA 39, 397.
1261a. Viala, A., Congr. soc. pharm. France, 9e, Clermont-Ferrand 1957, 217-19; CA 53, 20690.
1262. Vijayarazhavan, K. V., J. Indian Chem. Soc. 22, 141-46(1945); CA 40, 2787.

1263. Voigt, A. F., Acta Chem. Scand. 1, 118-19(1947); CA 42, 1236.

1264. Wagner-Jauregg, T., and Helmert, E., Ber. 75B, 935-49(1942); CA 37, 4393.

1265. Ward, W. C., U. S. 2,808,414, Oct. 1, 1957; CA 52, 2922.

1266. Walaschewski,E. G., Chem. Ber. 86, 272-7(1953); CA 49, 7487.

1267. Walde, A. W., Van Essen, H. E., and Zbornik, T. W., U. S. 2,762,821, Sept. 11, 1956; CA 51, 4424.

1267a. Wallia, C. P., J. Electroanalyt. Chem. 1, 307-13(1960).

1268. Walton, E., J. Chem. Soc. 1938, 471-2; CA 32, 4962.

1269. -----, ibid., 1939, 156-8; CA 33, 2495.

1270. -----, Brit. 573,770, Dec. 5, 1945; CA 43, 4296.

1271. Waser, J., and Schomaker, V., J. Am. Chem. Soc. 67, 2014-18(1945); CA 40, 257.

1272. Waters, W. A., Nature 140, 466-7(1937); CA 31, 8517.

1273. -----, J. Chem. Soc. 1937, 2007-14; CA 32, 1667.

1274. -----, ibid., 1939, 864-70; CA 33, 6263. Nautre 142, 1077(1938); CA 33, 2115.

1275. -----, and Williams, J. H., J. Chem. Soc. 1950, 18-22; CA 44, 7758.

1276. Way, E. L., and Oneto, J F., J. Am. Chem. Soc. 64, 1287-8(1942); CA 36, 4485.

1277. Wegler, R., Regel, E., and Grewe, F., U. S. 2,784,138, Mar. 5, 1957; Ger. 956,547, July 26, 1956; CA 51, 9077.

1278. Weidenhagen, R., and Rienäcker, H., Ber. 72B, 57-67(1939); CA 33, 2137; Ger. 706,577, Apr. 30, 1941; CA 36, 2088.

1279. Weiss, U., J. Am. Chem. Soc. 69, 2682-4(1947); CA 42, 1220. U. S. 2,520,-293, Aug. 29, 1950; CA 46, 134.

1280. Weitkamp, A. W., and Hamilton, C. S., J. Am. Chem. Soc. 69, 2699-702(1937) CA 32, 938.

1281. Wells, A. F., Z. Krist. 94, 447-60(1936); CA 31, 590.

1282. -----, Proc. Roy. Soc. (London) A167, 169-89(1938); CA 33, 4488.

1283. Werbel, L. M., Dawson, T. P., Hooton, J. R., and Dalbey, T. E., J. Org. Chem. 22, 252-4(1957); CA 52, 1088.

1284. West, T. S., Metallurgia 47, 97-106(1953); CA 47, 4791.

1285. Weston, R. E., Jr., J. Am. Chem. Soc. 76, 2645-8(1954); CA 48, 11191.

1286. Wetzel, J., Z. Krist. 104, 305-47(1942); CA 37, 5298.

1287. Whiting, G. H., J. Chem. Soc. 1948, 1209-10; CA 43, 916.

1288. Wiberg, E., and Mödritzer, K., Z. Naturforsch. 11b, 750-1(1956); CA 51, 11902.

1288a. -----, and -----, ibid., 748-50(1956); CA 51, 11902.

1289. -----, and -----, ibid., 751-3(1953); CA 51, 11902.

1290. -----, and -----, ibid., 753-5(1953); CA 51, 11903.

1291. -----, and -----, ibid., 755-6(1953); CA 51, 11903.

1292. -----, and -----, ibid., 12b, 123-5(1957); CA 53, 149.

1293. -----, and -----, ibid., 127-8; CA 52, 290.

1294. -----, and -----, ibid., 128-30; CA 52, 290.

1295. -----, and -----, ibid., 131-2; CA 52, 290.

1296. -----, and -----, ibid., 132-4; CA 52, 290.

1297. -----, and -----, ibid., 135-6; CA 52, 290.

1298. Wilkins, C. J., J. Chem. Soc. 1951, 2726-8; CA 46, 2437.

1299. Willard, H. H., and Smith, G. M., Ind. Eng. Chem., Anal. Ed. 11, 186-8 (1939); CA 33, 4154.

1300. Willard, H. H., and Smith G. M.,Ind.Eng.Chem.,Anal.Ed. 11, 186-8(1939);
1301. -----, and -----, ibid., 305-6; CA 34, 5315.
1302. -----, Perkins, L. R., and Blicke, F. F., J. Am. Chem. Soc. 70, 737-8(1948); CA 42, 3349.
1303. -----, and --- -, Anal. Chem. 25, 1634-7(1953); CA 48, 2506.
1304. Williams, R. R., J. Phys. and Colloid Chem. 52, 603-11(1948); CA 42, 4459.
1305. Willstaedt, H.,Svensk Kem. Tid. 54, 223-35(1942); CA 38, 3627. Swed. 118,196, Feb. 25, 1947; CA 42, 606.
1306. Witt, I. H., and Hamilton, C. S., J. Am. Chem. Soc. 67, 1078-9(1945); CA 39, 4078.
1307. Witten, B., J. Am. Chem. Soc. 69, 1229-30(1947); CA 41, 4444.
1308. -----, U. S. 2,531,487, Nov. 28, 1950; CA 45, 2799.
1309. Wittenberg, D. and Gilman, H., J. Org. Chem. 23, 1063-5(1958); CA 53, 19937.
1310. Wittig, G., Jesaitis, M. A., and Glos, M., Ann. 577, 1-10(1952); CA 47, 3317.
1311. -----, and Clauss, K., ibid., 26-39; CA 47, 3260.
1312. -----, and -----, ibid., 578, 136-46(1952); CA 47, 12282.
1313. -----, and Laib, H., ibid., 580, 57-68(1953); CA 48, 7589. Abstracts of 135th ACS Meeting, April 1959, p. 67-0.
1314. -----, and Torssell, K., Acta Chem. Scand. 7, 1293-1301(1953); CA 49, 7511.
1315. Woods, L. A., and Gilman, H., Proc. Iowa Acad. Sci. 48, 251(1941); CA 36, 3492.
1316. -----, Iowa State Coll. J. Sci. 19, 61-3(1944); CA 39, 693.
1317. Worrall, D. E., J. Am. Chem. Soc. 58, 1820-1(1936); CA 30, 7564.
1318. -----, ibid., 62, 2514-15(1940); CA 34, 7885.
1319. Wu, Y.-H. and Hamilton, C. S., ibid., 74, 1863-4(1952); CA 48, 687.
1320. Yakubovich, A. Ya., Ginsburg, V. A., and Makarov, S. P., Doklady Akad. Nauk S. S. S. R. 71, 303-5(1950); CA 44, 8320.
1321. -----, and Makarov, S. P., Zhur. Obshchei Khim. 22, 1528-34(1952); CA 47, 8010.
1322. -----, and Motsarev, G. V., ibid., 23, 1414-17(1953); CA 47, 12281.
1323. Yamamoto, K., and Kunizaki, S., Japan. 5087('54), Aug. 14; CA 50, 6508.
1323a. Yamashita, Yu, and Shimamura, T., Kogyo Kazaku Zasshi 60, 423-6(1957); CA 53, 9025.
1323b. Yamazaki, H., and Haziwara, N., ibid., 61, 21-3(1958); CA 53, 18885.
1324. Yang, P. S., and Lo, C.-P., J. Chinese Chem. Soc. 4, 477-84(1936); CA 31, 3892.
1325. -----, and Wang, T. Y., ibid., 5, 89-95(1937); CA 31, 4657.
1325a. Yen, J.-Y. and Tao, T.-N., Hua Hsueh Hsueh Pao 24, 97-103(1958); CA 53, 971.
1326. Yoe, J. H., and Cogbill, E. C., Mikrochemie ver. Mikrochim. Acta 38, 492-7 (1951); CA 46, 2443.
1327. Yoshida, S., J. Pharm. Soc. Japan 67, 72-3(1947); CA 45, 9544.
1328. Yust, V. E., and Bame, J. L., U. S. 2,819,156, Jan. 7, 1958; CA 52, 8535.
1329. Yusupov, F. Yu. and Manulkin, Z. M., Doklady Akad. Nauk S. S. S. R. 97, 267-8(1954); CA 49, 8843.
1330. Zakharkin, L. I., and Okhlobystin, O. Yu., Doklady Akad. Nauk S. S. S. R. 116, 236-8(1957); CA 52, 6167.

1331. Zakharkin, L. I., and Okhlobystin, O. Yu., Izvest. Akad. Nauk S. S.S. R., Otdel. Khim. Nauk 1959, 1942-7; CA 54, 9738.
1332. Zappi, E. V., and Salellas, J. F., Anales asoc. quim. argentina 24, 65-72 (1936); CA 31, 3902.
1333. -----, and -----, ibid., 26, 21-9(1938); CA 32, 7026.
1334. -----, and Simonin, L. M., Ciencia (Mex.) 3, 160-1(1942); CA 37, 2009.
1335. Zaslow, B., and Rundle, R. E., J. Phys. Chem. 61, 490-4(1957); CA 51, 11805.
1336. Zeide, O. A., Sherlin, S. M., and Bruker, A. B., Zhur. Obshchei Khim. 28, 2404-7(1958); CA 53, 3114.
1337. Zhdanov, G. S., and Razmanova, Z. P., Doklady Akad. Nauk S. S. S. R. 72, 1055-7(1950); CA 45, 7844.
1338. -----, Pospelow, V. A., Umanskii, M. M , and Glushkova, V. P., ibid., 92, 983-5(1953); CA 49, 12075.
1339. Zhitkova, L. A., Sheverdina, N. I., and Kocheshkov, K. A., J. Gen. Chem. (U. S. S. R.) 8, 1839-43 (1938); CA 33, 5819.
1340. Ziegler, M., and Glemser, O., Angew. Chem. 68, 522(1956); CA 51, 23762.
1341. Ziegler, K., Brit. 814,609, June 10, 1959; CA 53, 17733.

Coats, G. E., "Antimony and Bismuth," in "Organometallic Compounds," John Wiley and Sons, Inc., New York, 1956, pp. 147-163.

Courtot, Ch., Doeuvre, J., Guillaumin, A., and Kipping, F. S., "Traite de Chimie Organique. Composes Azotés de l'Acide Carbonique. Composés Organoarseniés, Organophosphorés ou Organosiliciés," Paris, Masson et Cie., Vol. XIV.

Freidlina, R. Kh., "Synthetic Methods in the Field of Metalloörganic Compounds of Arsenic," Izdat. Akad. Nauk S.S.S.R., 1945, 180 pp.

Hamilton, C. S., and Morgan, J. F., "The Preparation of Aromatic Arsonic and Arsinic Acids by the Bart, Bechamp, and Rosenmund Reactions," in Organic Reactions, Roger Adams, ed., John Wiley and Sons, 1944, pp. 415-454.

Kocheshkov, K. A., and Skoldinov, A. P., "Synthetic Methods in the Field of Organometallic Compounds, No. 8. Antimony and Bismuth," Moscow: Publishing House Akad. Sci. U.S.S.R, 1947, 206 pp.

Mann, F. G., "The Heterocyclic Derivatives of Phosphorus, Arsenic, Antimony, Bismuth, and Silicon," Interscience Publishers, New York, 1950, 180 pp.

Rochow, E. G., Hurd, D. T., and Lewis, R. N., "Compounds of the Group V Elements," in "The Chemistry of Organometallic Compounds," John Wiley and Sons, New York, 1957, pp. 198-223.

Andrianov, K. A., "Novelties in the Area of Synthesis of Hetero-Chain Hetero-organic Polymers," Uspekhi Khim. 26, 895-922(1957).—A review with 108 references, including polymers of organic derivatives of Bi.

Berlin, A. A., and Parini, V. P., "Classification of High-Molecular Compounds and Basic Type of Polymers of Inorganic Compounds and of Organic Derivatives of Silicon Arsenic, Lead, etc.," Uspekhi Khim. 18, 546-56(1949).—A critical review.

Bernardo, P. S.," Acetylarsan. II. The Percentage Composition of Acetylarsan," Rev. asoc. bioquim. argentina 7, No, 22, 49-61(1941).—The technique of the elementary analysis is described and various reactions are tabulated.

Breyer, B., "Scientific Method in the Evolution of New Drugs. V. Phenylmercurials and Phenylarsonic Acids. Constitution and Action," Australian J. Sci. 6, 56-9 (1943).

Challenger, F., "Biological Methylation," Science Progress 35, 396-416(1947).—A review of the action of the mold Scopulariopsis brevicaulis in the methylation of As.

Delab, R., "Sulfamidoamidines   and Arsenicals," Bull. acad. natl. med. 114, 597-9(1950).—A lecture.

Diserens, L., "War Gases and Toxic Substances," Teintex 5, 45-54(1940).

Esposito, A., "Gosio Gas and Recent Investigations on Its Compositon," Chem. ind. agr. biol. 14, 6-8(1938).—A review including Me$_3$As.

Etienne, A., "Arsenic Compounds Derived from Thiophene and Furan. General Statement," Mem. services chim. etat (Paris) 34, 429-41(1948).

Gilman, H., and Yale, H. L., "Organobismuth Compounds," Chem. Rev. 30, 281-320 (1942).

Harwood, J. H., "Industrial Applications of the Metal Organic Compounds," Ind. Chemist 35, 348-350(1959).

Haszeldine, R. N.,"Recent Developments in Fluorine Chemistry. Organometallic and Metalloid Compounds of Fluorine," Angew. Chem. 66, 693-701(1954).—A review with 28 references, including As and Sb derivs.

Heal, H. G., "Analogs of the Ammonium Compounds in the Periodic Groups V, VI, and VII," J. Chem. Educ. 35, 192-7(1958).—A review.

Herrmann, W., and Hilmer, H., "Research in Arsenic Chemistry," Angew.Chem. 66, 349-59(1954).

Hoogeveen, A. P. J., "Chemical Weapons," Chem. Weekblad 34, 35-41(1937).—A review including As derivs.

Ida, M., Toyoshima, Z., and Nakamura, T., "Chemistry of Stibonic Acids," Japan. J. Pharm. and Chem. 19, 158-69(1947).—A review.

Jackson, K. E., "Smoke-Forming Chemicals," Chem Rev. 25, 67-119(1939).—A review including PhAsCl$_2$.

Kinoshita, J., "Organic Antimony Compounds," Nagoya Shiritsu Daigaku Yakugakubu Kiyô 5, 1-13(1957).—A review with 70 references.

Kirkhgof, G. A., and Korzina, O. I., "Preparation of the Principal Medicinal Aromatic Arsenic Compounds," Org. Chem. Ind. (U.S.S.R.) 5, 282-7(1938).—A review.

Koton, M. M., "Heteroörganic Unsaturated Compounds and Their Polymers," Uspekhi Him. 26, 1125-40(1957).

Leipert, T., "New War Gases," Wien. klin. Wochschr. 51, 549-51(1938).—A discussio

Mann, F. G., "Some Aspects of the Organic Chemistry of P and As," J. Chem. Soc. 1945, 65-73.—The Tilden lecture of 1944.

Mann, F. G., "Heterocyclic Derivatives of Phosphorus, Arsenic, and Antimony," Progr. Org. Chem. (Academic Press Inc., N. Y.) 4, 217-248(1958).

Matsui, M., "Recent Trend in the Organoarsenic Compounds," Yakugaku 2, 141-50(1959).

Myttenaere, F. de, "Composition and Toxicity of Arsenobenzenes," J. pharm. Belg. [N.S.] 2, 214-18(1947).

Nesmeyanov, A. N., "Organic Compounds and the Periodic System," Uspekhi Khim. 14, 261-81(1945).—A review with 81 references.

Nesmeyanov, A. N., and Reutov, O. A., "Synthesis of Metalloorganic Compounds of the Group V Elements of the Periodic System with Diazo Compounds," Uchenye Zapiski, Moskov. Gosudarst. Univ. im. M. V. Lamonosova, Org. Khim. 1956, No. 175, 55-69. A review.

Platunov, B. A.," Current Views on the Processes for the Precipitation of Tungsten with Organic Precipitants," Uchenye Zapiski Leningrad. Gosudarst. Univ. im. A. A. Zhdanova No. 150, Ser. Khim. Nauk No. 10, 3-8(1951).—A review; Ph4AsCl included.

Shiganova, M., and Matsuo, H., "Organic Onium Compounds; Their Application to Analytical Chemistry," Kagaku no Ryôiki 10, 111-19(1956).—A review; arsonium and stibonium compds. included.

Smith, R. G.,"Organic Arsenical Drugs," Australian J. Pharm. 27, 621-4(1946). A review.

Yakubovich, A. Ya., and Ginsburg, V. A.," The Diazo Method of Synthesis of Hetero-organic Compounds of the Aliphatic Series," Uspekhi Khim. 20, 734-58(1951).—A review with 57 references.

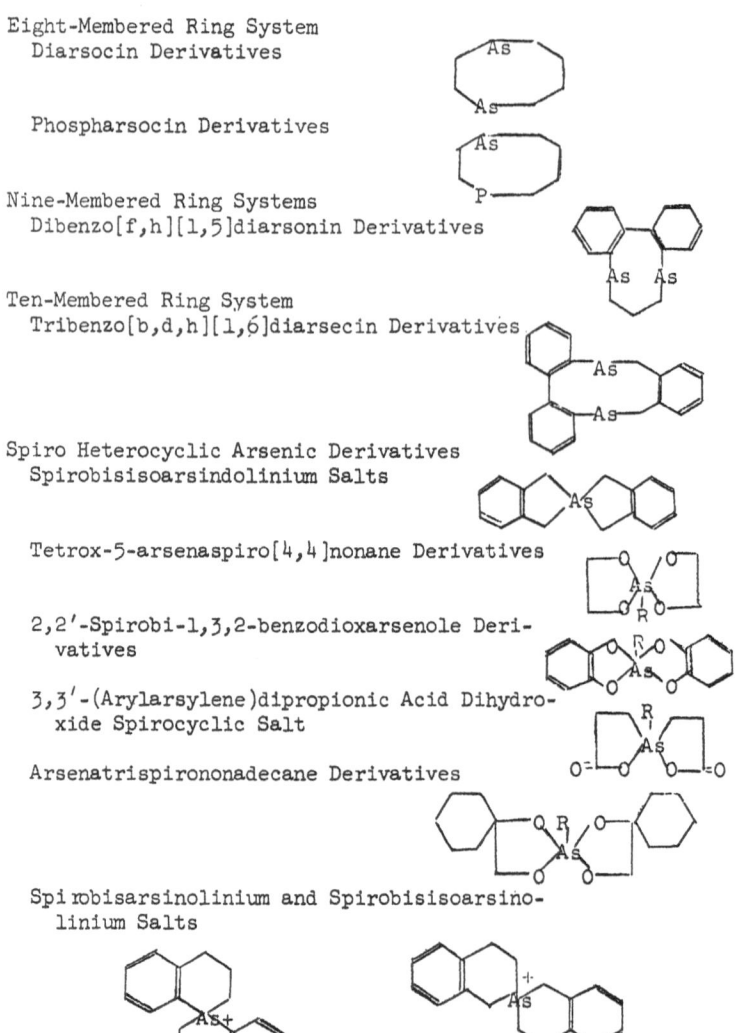